MATHEMATICS AS A SECOND LANGUAGE

MATHEMATICS AS A SECOND LANGUAGE

FOURTH EDITION

JOSEPH NEWMARK
The College of Staten Island of the
City University of New York

FRANCES LAKE
Late Professor
The College of Staten Island of the
City University of New York

ADDISON-WESLEY PUBLISHING COMPANY
Reading, Massachusetts · Menlo Park, California
Don Mills, Ontario · Wokingham, England · Amsterdam
Sydney · Singapore · Tokyo · Madrid · Bogotá
Santiago · San Juan

Sponsoring Editor: Jeffrey M. Pepper
Production Supervisor: Margaret Pinette
Production and Art Coordinator: Susan Vicenti
Copy Editor: Barbara Willette
Illustrator: Illustrated Arts
Manufacturing Supervisor: Ann DeLacey
Cover Designer: Marshall Henrichs
Text Designer: Marie McAdam

Cover Art: *Relief from Kumma: Temple of Amenhotep II*
Courtesy, Museum of Fine Arts, Boston

Library of Congress Cataloging-in-Publication Data

Newmark, Joseph.
  Mathematics as a second language.

  Includes index.
    1. Mathematics—1961–   . I. Lake, Frances.
QA39.2.N48 1987   510   85-30646
ISBN 0-201-19297-7

Reprinted with corrections, May 1988

ABCDEFGHIJ-HA-898

To Trudy, Sharon,
Rochelle, and Stephen

TO THE INSTRUCTOR

This revised edition of *Mathematics as a Second Language* reflects the many suggestions and recommendations of users of the earlier editions of the book. As the table of contents indicates, substantial material has been added to certain chapters while other chapters have been reorganized and streamlined. There is a new chapter on consumer mathematics as well as two chapters on computers and their uses, so that the reader may become computer literate. Thus the book conforms in scope, format, approach, and content to the mathematical requirements currently being established by numerous states.

As with the earlier editions, this text is designed to serve as an introduction to the basic concepts of math for students in all academic areas. Applications are drawn from such diverse areas as sociology, economics, business, ecology, education, medicine, psychology, and mathematics. Liberal arts students and prospective elementary school teachers must frequently demonstrate a knowledge of the language and methods of mathematics.

Since the concepts of mathematics are assuming an ever-increasing role in the social sciences (especially in business-related fields), a special effort has been made in this text to make these ideas available to the students who are not prepared for elaborate symbolism or complex arithmetic. Although the mathematical content is complete and correct, the language is rather elementary and understandable. Mathematical rigor has not been sacrificed. This makes the text comprehensible to students of varying backgrounds, especially those with little mathematical background. Words, phrases, and any modes of expression that students find difficult to comprehend have been avoided. The introduction of new terminology has been held to a minimum. Furthermore, each chapter starts off with chapter objectives and a newspaper clipping, to set the stage for the material to be covered, and concludes with a summary and a study guide to further reinforce the ideas covered in the chapter. Also, each idea introduced in a chapter is first explained with an example chosen from everyday situations with which the student can identify. Furthermore, each chapter has a Formulas-to-Remember section that serves as quick reference for students.

In selecting topics, we have allowed for a wide diversity of interest among students. We have found that many students can benefit from a solid survey in basic mathematical concepts, such as sets, logic, and real numbers, together with more meaningful applications. The business– or social-science–oriented student will probably find that topics such as game theory, linear programming, probability, and statistics have more immediate relevance. For the prospective elementary school teacher and the interested liberal arts student, we have included material on geometry, graphs, and the metric system. The chapter on consumer mathematics should prove useful to all students.

As with the earlier editions, the only prerequisite needed to use this text satisfactorily is a little high school algebra. Chapter 7 contains a review of the ideas of graphs and functions needed for those students who feel that their math background is a bit rusty. This chapter can be deleted by students who feel that they are adequately prepared. Moreover, algebra is reviewed throughout the text as needed.

FEATURES OF THE TEXT

You may be wondering how this fourth edition differs from our first three editions and from the competing texts on the market. To answer this question, consider the following items:

Chapter objectives A chapter objectives section opens each chapter, highlighting the ideas to be discussed.

Newspaper article or magazine clipping Each chapter starts with an article representing ideas discussed in the chapter. These motivate the student by showing how the concepts can be applied in real-life situations. Moreover, there are numerous newspaper clippings and pictures throughout the book to further enhance the material.

Introductions An introductory section sets the stage for the ideas contained in the chapter.

Readability The text is written on a level easily understood by students with a limited mathematical background.

Student-oriented comments In numerous places, "Comments" are included, covering ideas that students often miss or find confusing.

Historical notes Most chapters contain brief historical notes on mathematicians who contributed to the ideas discussed within the chapter.

Numerical illustrations Many charts, graphs, sketches, and cartoons illustrate the concepts.

Student aids Each chapter concludes with Summary, Study Guide, and Formulas to Remember sections. The Study Guide lists the significant new terms introduced in that chapter. Past experience indicates that students find these extremely useful in studying and preparing for exams.

Examples and exercises All the examples and exercises have been selected to maintain student interest. These cover many different areas, including the social sciences, psychology, education, medicine, ecology, and business, and relate directly to the student's own experience.

Suggested course outline A suggested course outline (following the Acknowledgments) is provided for one- and two-semester courses. This outline indicates how long each teacher might spend on each topic.

Calculators Some of the exercises are specifically designed to be worked out using a hand-held calculator. Such exercises are designated with the symbol 🖩.

Mastery tests Each chapter contains two mastery tests to measure a student's comprehension of the material covered in the chapter, as well as to present a few challenging problems for the more demanding student.

Test bank A test bank is available with several alternate forms of tests for each chapter.

TO THE STUDENT

This text was written for people who find mathematics a difficult subject. We frequently meet students who tell us "Why should I study math? I am never going to use it! It has no real practical applications for me."

In this text, we hope to show how mathematics affects almost every aspect of our lives. The possible applications of mathematics are increasing every day. Did you know that an insurance company can use mathematical logic in writing its policies, and that a dress manufacturer can use advanced mathematics to decide how many size 6's and how many size 16's to produce? Every gambler (from the casual card player to the professional) can profit from a knowledge of probability. The use of statistics is so widespread that we need not comment on its importance. The chapter on consumer mathematics should also prove extremely helpful to you. Regardless of your objectives in life, both career and recreational, a knowledge of mathematics is essential.

Perhaps you feel that your mathematical background is rather shaky. Don't worry about that! To read and understand this book, you do not need to have any extensive background in high-school algebra or geometry, but if you have this background, you will not find the material in this book repetitious. Instead, you will discover how mathematics can be put to work in interesting and important ways. Occasionally a section, explanation, or exercise does require a little algebra. Such material has a dagger (†) in front of it. If your background in this area is weak, just omit the material, since it will not affect your understanding of what follows. Occasionally a section or exercise is starred (**). This means that it is slightly more difficult, and may require more time and thought. Some of the exercises have a calculator 🖩 symbol in front of them. This means that it is advisable to solve these problems using a hand-held calculator.

We hope that you will find reading and using this book an enjoyable experience, and that it will prove helpful to you in much of what you do in your daily life.

ACKNOWLEDGMENTS

I am grateful to my many colleagues and former students who gave me valuable suggestions and constructive criticisms.

I also wish to thank the following people who reviewed the manuscript and made helpful suggestions for its improvement:

Donald C. Cathcart, Salisbury State College

Gladys C. Cummings, St. Petersburg Junior College

Richard Fritsche, Northeast Louisiana University

Lester C. Hartsell, East Tennessee State University

Nancy Johnson, Broward Community College

Hiram Johnston, Georgia State University

Curtis C. McKnight, University of Oklahoma

Dawn Ross, University of Missouri

Susan Luginbill Wirth, Indian River Community College

Finally, and most important, I wish to thank my family, Trudy, Sharon, Rochelle, and Stephen, for their understanding and continued encouragement as they patiently endured the enormous strain associated with such an undertaking.

Joseph Newmark

SUGGESTED COURSE OUTLINES

One-semester courses	Chapters
Liberal arts students	1, 2, 3, 4, 5
	1, 2, 5, 10, 11
	2, 4, 5, 10, 11
	1, 2, 4, 5, 6, 7
	1, 2, 3, 4, 10
Prospective elementary school teachers	1, 2, 3, 4, 7, 8, 9
	1, 2, 4, 7, 9, 10
	1, 3, 4, 7, 9, 10, 11
Business or social-science students	5, 7, 8, 10, 12, 13, 14
	5, 7, 10, 11, 12, 13, 14

Two-semester courses	Chapters
Liberal arts students	1, 2, 3, 4, 5, 7, 8, 10, 11, 13, 14
Prospective elementary school teachers	1, 2, 3, 4, 5, 6, 7, 8, 9, 10, 13, 14
Business or social-science students	1, 4, 5, 7, 8, 10, 11, 12, 13, 14

BACKGROUND NEEDED FOR EACH CHAPTER

Chapter	Prerequisite
1	None
2	None (except parts of Section 2.11 that depend on Chapter 1)
3	None
4	None
5	Chapter 4 (or equivalent)
6	Chapter 4 (or equivalent)
7	Chapter 4 or some high school algebra
8	Chapters 4 and 7 or some high school algebra
9	Chapter 4 or some high school algebra
10	None
11	Parts of Chapter 4 and parts of Chapter 7
12	Parts of Chapter 4 and parts of Chapter 7
13	None
14	Chapter 13

CONTENTS

3

DIFFERENT NUMBER SYSTEMS   117

4

THE REAL NUMBER SYSTEM   175

5
CONSUMER MATHEMATICS 243

6
MATHEMATICAL SYSTEMS 299

MATHEMATICS AS A SECOND LANGUAGE

1

SETS

CHAPTER OBJECTIVES

To learn what a set is and the notation used for it. (*Section 1.2*)

To distinguish between sets that are well-defined and those that are not. (*Section 1.2*)

To understand the different kinds of sets. (*Section 1.2*)

To discuss various set operations that can be performed as well as the concept of subsets. (*Section 1.3*)

To determine how to find the cross product between two sets and how this can be applied. (*Section 1.3*)

To describe sets through pictorial representations known as Venn diagrams. (*Section 1.4*)

To apply the concepts of sets to survey problems and voting coalitions. (*Section 1.5*)

To study some of the ideas and properties of rather unusual sets known as infinite sets. (*Section 1.6*)

To point out how we use sets in everyday life. (*Throughout chapter*)

Surplus Budget Presenting Problems

SAYERVILLE—Ever since the governor announced that there would be an unexpected 358.4-million-dollar budget surplus, many of the state's legislators have been engaged in a heated debate over how the state should spend its huge surplus. Many suggestions have been put forth. The latest survey of 2000 constituents conducted by Representative Ali Smith's staff indicates that 1354 believe the money should be spent on revamping the criminal justice system and hiring more police, 608 believe the money should be spent on building new roads and schools, 618 believe the money should be spent on combating the illicit drug business and establishing drug rehabilitation centers, 278 believe the money should be spent on both police and roads, 200 believe the money should be spent on building roads and combating illicit drugs, 178 believe the money should be spent on more police and combating illicit drugs, and 62 believe the money should be spent on all three proposals.

The governor's office refused to accept the report, claiming that the data were inconsistent. Representative Smith was unavailable for comment.

Middletown News, March 12, 1985

On the basis of the data collected by Representative Smith's staff it would seem that the money should be spent on revamping the criminal justice system and hiring more police. Nevertheless, the governor's office claimed that the data are inconsistent. Is there a way to determine whether the data are consistent or not?

1.1
INTRODUCTION

In recent years there has been much talk about the "new math." In reality there is nothing really new about the new math. Rather, it is a different way of looking at traditional math. The new math has its roots in the last half of the nineteenth century. At that time, mathematicians started to look very carefully at the basic ideas and methods of their subject, at the "foundations" of mathematics. They began to ask such questions as: What is a number? What is infinity? How do we know that our methods of reasoning are correct? Moreover, general philosophic questions were considered, questions such as: What is mathematics? Is mathematics part of science? What is the relationship between mathematics and the "real world"? Do mathematical concepts like "number," "point," and "circle" really exist?

One of the outstanding people involved in these investigations was a German mathematician, Georg Cantor (1845–1918). His theory of sets provided a basis for explaining many of the questions raised at that time. His theory also provided a new way of looking at old ideas such as numbers, addition, and multiplication. So important was this theory that it has been said that it has "largely transformed most branches of mathematics."

Basic to Cantor's work is the idea of a set. The word "set" in mathematics has exactly the same meaning as it does in everyday English usage. It is just any collection of things.

In our daily lives we come across sets all the time. Think of all your friends and relatives; they form a set. The pages in this book form a set. If you own a phonograph, your collection of records forms a set. Empty your pockets; the contents form a set.

In what follows, we will see how the ideas of sets are easy to understand and interesting to apply.

1.2

THE DIFFERENT KINDS OF SETS

Since the idea of a set is so important in mathematics, we will state again what we mean by it. A **set** is any collection of objects.

HISTORICAL NOTE

Georg Cantor was a German mathematician involved in the study of sets. Born in 1845 in St. Petersburg (now Leningrad), he spent most of his life in Germany. He decided at an early age that he wanted to be a mathematician, but his father was determined that Georg should go into engineering, which he believed to be a better-paying profession. Being an obedient son, Cantor did study engineering as his father wished; however, he was so miserable that when he was seventeen, his father finally allowed him to pursue a career in mathematics.

Cantor became a great mathematician. However, his work was so different and controversial that it was vigorously attacked by the mathematical community.

It is not hard to understand why Cantor's work met with such resistance, since many of his results were startling. For example, suppose that you could draw a line from your house to the moon. Now look at the horizontal line shown here:

Which line has more points on it? Did you say the line to the moon? Then you are wrong. Cantor showed that both lines contain the same number of points. Even Cantor himself found some of his results hard to accept. In a letter to another mathematician he wrote of one result, "I see it, but I don't believe it."

Cantor's theory of sets and its implications led to a prolonged and unpleasant dispute with one of his former teachers, Leopold Kronecker (1823–1891), himself an important mathematician. Possibly as a result of the hostility and criticism directed at his work, Cantor was never able to get a teaching position at the University of Berlin, which was his ambition, but taught all his life at a third-rate university. Unable to cope with the attacks and the disappointments, Cantor suffered several mental breakdowns. He died in 1918 in an insane asylum at the age of seventy-three.

The French mathematician Henri Poincaré stated that later generations would think of Cantor's work as a "disease from which one has recovered." It should be noted, however, that not all mathematicians were bitterly critical of Cantor's work. The German mathematician David Hilbert asserted that "No one shall expel us from the paradise which Cantor has created for us."

EXAMPLE 1   The set of all letters of the English alphabet.   ☐

EXAMPLE 2   The set of all American Indians.   ☐

EXAMPLE 3   The set of all people over 22 ft tall.   ☐

EXAMPLE 4   The set of all months with fewer than 30 days.   ☐

EXAMPLE 5   The set of all numbers larger than 1.   ☐

> **Definition 1.1**   *The objects in the set are called either **elements** or **members** of the set.*

In Example 1 above, the elements are $a, b, c, d, e, f, g, h, i, j, k, l, m, n, o, p, q, r, s, t, u, v, w, x, y$, and z. In Example 2, some of the elements are Pocahontas, Jim Thorpe, Geronimo, and Sitting Bull. In Example 4, there is only one element, February. How many elements are there in the set of Example 3? of Example 5?

Instead of writing out the words "the set of all letters of the English alphabet," we can write the same thing in shortened form as {all letters of the English alphabet}. Similarly, Example 2 can be written as {all American Indians}. The braces (curly brackets) stand for the set containing whatever is written inside them.

Very often we must refer to a set whose elements we do not know or do not wish to write down. We then denote the set by the capital letter A, or B, or C, etc. We denote the elements of such a set by small letters of the alphabet, a, b, c, etc.

In Example 1 above, we notice that "g" is an element of the set. If we denote this set by the letter A, we write $g \in A$. This means "g is an element of set A," since the symbol $\in$ stands for the words "is an element of." Similarly, $a \in A, b \in A, c \in A, \ldots, z \in A$. Observe that Geronimo is not an element of set A. We symbolize this by writing Geronimo $\notin A$; $\notin$ stands for "is not an element of." If B denotes the set of Example 2, then Geronimo $\in B$.

Notice that in Example 1 we have merely described the elements of the set without actually naming them. For someone not familiar with the English language, it would not be immediately obvious which elements are in the set. It would then be advisable to actually list the elements of the set, and write it as

$$A = \{a, b, c, d, e, f, g, h, i, j, k, l, m, n, o, p, q, r, s, t, u, v, w, x, y, z\}.$$

This set can also be written in an abbreviated form as $\{a, b, c, d, \ldots, z\}$, where the three dots stand for the letters between d and z. This notation assumes that the reader *is* familiar with the letters of the English alphabet. In the set $\{1, 2, 3, 4, \ldots\}$, the three dots stand for 5, 6, 7, 8, and so on.

When we list the elements of a set, we refer to the process as the **roster method,** as opposed to the **descriptive method** in which we *describe* the elements of a set rather than list them. Similarly, the set of Example 4 can be written in the roster method as {February}.

There is still another way of describing a set by using a method called **set-builder notation.** Suppose we are interested in the set of all whole numbers between 1 and 10. Using set-builder notation, we would write this as

$$\{X \mid X \text{ is a whole number between 1 and 10}\}.$$

This is read as the set of all elements X such that X is a whole number between 1 and 10. The vertical line stands for the words "such that." In this notation, X (or whatever letter is used) is called a *variable*, since it represents any element of a given set of numbers. (Variables will be discussed in greater detail in a later chapter.)

Some other examples of sets described by set-builder notation are

$$M = \{Y \mid Y \text{ is a state in the United States}\}.$$

$$N = \{a \mid a \text{ is a letter in the word mathematics}\}.$$

If we actually tabulate the elements in these sets, we find that set M has 50 elements and set N has 8 different elements. The elements of set N are

$$N = \{m, a, t, h, e, i, c, s\}.$$

In a West Coast town the local Clear-Air Committee and the Pure-Water activists decide to merge into one antipollution group. A list is drawn up of the membership of the new group. This list will, of course, contain the names of all those in either of the original groups. Mr. I. M. Covington belongs to both of the original groups. We write his name only *once* on the combined mailing list.

Consider the set $M = \{\text{all great baseball players}\}$. Is Reggie Jackson an element of that set or not? Some people may say yes; others may say no. There is no way of determining who is right. Such a set is said to be **not well-defined.**

Definition 1.2 *If there is a way of determining for sure whether an object belongs to a set or not, we say that the set is **well-defined**.*

EXAMPLE 6 The set of all cute children. ▢

EXAMPLE 7 The set of all tall men. ▢

EXAMPLE 8 The set of all ripe apples. ▢

EXAMPLE 9 The set of all two-door automobiles. ▢

EXAMPLE 10   The set of all planets on which water can be found.   □

Examples 6 and 7 are *not* well-defined because one may disagree as to whether Mary is cute or not, or as to whether John, who is 5 ft 8 in., is tall or not. Is Example 8 well-defined or not? Example 9 *is* well-defined. Example 10 *is* well-defined also, because there is a way of determining which planets are to be included in this set (even though, at present, technology has not advanced far enough for us to know). In mathematics we use only well-defined sets.

Consider the set {all months with less than 30 days}. This set has but one element, namely, February. Such a set is called a **unit** set.

> **Definition 1.3**   *Any set that has only one element in it is called a **unit** set.*

Now how many elements are there in the set {all people who weigh a ton} or the set {all people over 30 feet tall}? Your answer, of course, will be "none." Such a set is called a **null** set.

> **Definition 1.4**   *Any set that has no elements in it is called a **null** or **empty** set. We denote such a set by the symbol ∅ or {   }.*

We will show shortly that there is only one null set.

Let us look again at Examples 1–5. Example 1 has 26 elements in it. Example 2 has a specific number of elements in it, although we may not know the number offhand. (The latest U.S. census figures show that there are about 1,000,000 American Indians.) Example 3 has no elements and is thus a null set. Example 4 has only one element and is thus a unit set. All of these are examples of **finite sets.** How many elements are there in Example 5? Example 5 is called an **infinite set.** Another example of an infinite set is the set of all fractions. Do you think that the set of all grains of sand on the beaches of Florida is a finite or an infinite set?

John and Ann were asked to think of a set. John thought of the set of all the vowels in the English alphabet. Ann thought of the set {a, e, i, o, u}. Are they thinking of different sets? Of course not. Both sets obviously have the same elements.

> **Definition 1.5**   *Two sets that have exactly the same elements are called **equal sets.***

Mabel thought of the set {e, o, a, u, i}. Leon thought of the set {t, o, a, u, i}. Notice that Mabel's set is the same as Ann's and John's set, even though the elements are presented in a different order. *The order in which the elements of a set are written is not important.* What about Leon's set? Is it equal to the others? You probably answered that Leon's set was not equal to the others. However, it has the same number of elements as the others. Let us investigate this further.

Imagine that we have a set of three students: Maurice, Ben, and Arline. We also have a set of three chairs: a red chair, a blue chair, and a green chair. Each of the three students sits down on a chair. Obviously, each chair will be occupied, and each student will be seated. There will be no vacant chair and no standing student. We have matched each student with a chair and each chair with a student. Such a matching is called a one-to-one correspondence. More generally, the following holds.

Definition 1.6 *Set A and set B can be put in **one-to-one** (1–1) **correspondence** if each element of set A can be paired with exactly one element of set B and every element of set B can be paired with exactly one element of set A.*

EXAMPLE 11 Let

$$A = \{a, e, i, o, u\},$$
$$B = \{t, o, a, u, i\}.$$

Sets A and B can be put in 1–1 correspondence in many different ways. One way is

$$A = \{a, \quad e, \quad i, \quad o, \quad u\}$$
$$\updownarrow \quad \updownarrow \quad \updownarrow \quad \updownarrow \quad \updownarrow$$
$$B = \{t, \quad o, \quad a, \quad u, \quad i\}$$

Another possible way is the following:

$$A = \{a, \quad e, \quad i, \quad o, \quad u\}$$
$$B = \{t, \quad o, \quad a, \quad u, \quad i\}$$

How many possible ways can you find of putting set A and set B in 1–1 correspondence? □

EXAMPLE 12 Let

$$C = \{\triangle, \square, 0, \text{⚹} \},$$
$$D = \{a, b, c, d, e\}.$$

If we attempt to put these sets in 1–1 correspondence, we find that it cannot be done. One element of set D is always left out of any pairing. For example,

$$C = \{\triangle, \quad \square, \quad 0, \quad \text{⚹} \},$$
$$\updownarrow \quad \updownarrow \quad \updownarrow \quad \updownarrow$$
$$D = \{a, \quad b, \quad c, \quad d, \quad e\}$$

Observe that element *e* is not matched with any element of set *C*. If we try to match *e* with any element of set *C*, we will be forced to leave out another element of set *D*. Thus sets *C* and *D* cannot be put into 1–1 correspondence.

In Example 11 above, set *A* and set *B* are not equal, but they can be put in 1–1 correspondence. We say that they are **equivalent**. ☐

> **Definition 1.7**   *Sets A and B are said to be **equivalent** if they can be put into 1–1 correspondence.*

Comment   In everyday English, "equal" and "equivalent" mean the same thing. In our discussion of sets, equal sets and equivalent sets are not the same thing.

How many elements are there in set *A* and set *B* of Example 11? What about set *C* and set *D* of Example 12? In Example 11, both sets *A* and *B* have five elements. On the other hand, in Example 12, set *C* has four elements, whereas set *D* has five elements. This leads us to the following definition.

> **Definition 1.8**   *The **cardinal number** of any set A is defined as the number of elements in set A. This is denoted by the symbol n(A), read as "the number of elements in set A."*

EXAMPLE 13   The set $T = \{x, y, z\}$ has cardinal number 3, that is, $n(T) = 3$. ☐

EXAMPLE 14   The set $W = \{\square, \triangle, \star, \text{⚡}, m, 1, \text{Joe}\}$ has cardinal number 7; that is, $n(W) = 7$. ☐

EXAMPLE 15   The set $M = \{$all letters of the English alphabet$\}$ has 26 elements; $n(M) = 26$. ☐

EXAMPLE 16   If $D = \{1, 2, 3, \ldots, 10\}$, what is $n(D)$? ☐

In view of Definition 1.8 we could have defined "equivalent sets" as "sets that have the same cardinal number."

Comment   The following statements all say the same thing.

1. Sets *A* and *B* can be put into 1–1 correspondence.
2. Sets *A* and *B* are equivalent.
3. Sets *A* and *B* have the same cardinal number.

Comment   Any two null sets have the same elements, namely, *none*. Thus any two null sets are equal. This means that essentially there is only one null set. Therefore we speak of *the* null set.

Comment   The set $\{0\}$ is not an empty set, since this set contains the element 0.

EXERCISES

In Exercises 1–8, describe each of the indicated sets in words.

1. {January, June, July}
2. {4, 8, 12, 16, 20, 24, 28}
3. {2, 3, 5, 7, 11, 13, 17, 19}
4. {1, 8, 27, 64, 125}
5. {Tuesday, Thursday}
6. {3, 6, 9, . . . , 999}
7. {IBM, Atari, Apple, Compaq, Commodore}
8. {2, 4, 6, 8, 10, . . .}

In Exercises 9–18, list the elements of the set that is described.

9. The set of different letters in the word MISSISSIPPI.
10. The set of months with fewer than 31 days.
11. The set of U.S. Presidents since Franklin D. Roosevelt.
12. The set of all whole numbers less than 80 that are squares of whole numbers.
13. The set of all even integers greater than 12 and less than 19.
14. $\{X \mid X$ is an even number between 11 and 12$\}$.
15. The set of all states in the United States that have executed criminals since 1980.
16. The set of all countries in the world in which capital punishment has been outlawed.
17. The set of all odd numbers less than 50 that are divisible by 2.
18. The set of all females who have been or are currently president of the United States.

In Exercises 19–24, determine whether the set is well-defined or not.

19. The set of all expensive math books.
20. The set of all popular folk music singing groups.
21. The set of all college students with a high grade point average (GPA) on their college transcripts.
22. The set of all good comedians.
23. The set of all days with 30 hours.

24. The set of all large corporations that use Federal Express delivery service or Purolator delivery service when shipping urgent packages.

In Exercises 25–34, state whether the set is a finite nonempty set, an infinite set, a unit set, or a null set.

25. The set of all women who are 150 centimeters tall.
26. The set of all women who are 150 feet tall.
27. The set of U.S. astronauts who have died while participating in the space exploration program.
28. The set of polluted lakes (or rivers) to which fish have returned as a result of cleanup campaigns by environmentalists.
29. The set of all people who weigh a ton.
30. The set of all prime numbers.
31. The set of all animals in the world.
32. The set of all countries where women are assigned to combat units in the armed forces.
33. The set of all triangles having two and only two sides.
34. The set of all points that are on a straight line.

In Exercises 35–42, tell whether the statement is true or false.

35. The set {0} and the null set are equal sets.
36. $4 \in$ {all even numbers}.
37. $\{m\} = m$.
38. Tomato $\in$ {all fruits}.
39. $A \in \{a, b, c, d\}$.
40. $\varnothing = 0$.
41. {all male senators} and {all female senators} are equivalent sets.
42. $10 \in \{X \mid X$ is an odd number$\}$.
43. Are equal sets equivalent? Explain.
44. Are equivalent sets equal? Explain.

In Exercises 45–53, determine which of the pairs of sets are equivalent, equal, both, or neither.

45. {12, 10, 8, 6} and {6, 8, 10, 12}.

46. {a, b, c} and {1, 2, 3, 4}.
47. {Tom, Dick, Harry} and {Mary Ruth, Jane}.
48. {letters in the word STALE} and {letters in the word LEAST}.
49. {1, 4, 7} and {1, 4, 7, 8}.
50. {all fish that can program a computer} and {all elephants that can cook}.
51. {$X \mid X$ is an odd number between 18 and 24} and {$Y \mid Y$ is a multiple of 3 that is larger than 4 but less than 13}.
52. {Spring, Summer, Fall, Winter} and {four seasons of the year}.
53. {all male senators} and {all female senators}.
54. Determine several 1–1 correspondences between each of the following pairs of sets:

     a) A = {x, q, John, b},

       B = {$\triangle$, $\square$, α, c}.

     b) C = {Heather, Jason, Marlene},

       D = {Marlene, Joe, Bill}.

In Exercises 55–60, find the cardinal number of the indicated sets.

55. {Professional football leagues in the U.S.}.
56. { }.
57. {Females who have been executed since 1930 in the U.S.}.
58. {The states of the U.S. beginning with the letter Q}.
59. {All prime numbers less than 30}.
**60. {All dots on a line 7 centimeters long}.

61. If set A has 3 elements and set B has 4 elements, can set A and set B be put into 1–1 correspondence? Explain.
62. Is the null set a well-defined set? Explain your answer.
63. If set A = {a, b} and set B = {1, 9}, how many distinct 1–1 correspondences can be made between the elements of these sets?
**64. If two sets A and B are equivalent, then the number of distinct 1–1 correspondences that can be made between the elements of these sets is shown in the following table:

If the number of elements in sets A and B is	then the total number of distinct 1–1 correspondences that can be made is
1	1
2	2
3	6
4	24
5	120

Write a formula for determining the total number of distinct one-to-one correspondences that can be made if each of the equivalent sets has n elements.

**Exercises preceded by ** require careful thought and consideration.

1.3

SUBSETS AND SET OPERATIONS

Consider the set B = {all teachers at this school} and set A = {all math teachers at this school}. We notice that every member of set A is also a member of set B, that is, set A is part of set B. We then say that set A is a subset of set B.

    Definition 1.9   *Set A is said to be a **subset** of set B if every element of set A is an element of set B. We denote this by $A \subseteq B$. This symbol is read as "A is a subset of B" or "A is contained in B."*

EXAMPLE 1   {all basketball players in this school} is a subset of {all athletes in this school}.   ☐

EXAMPLE 2   {a, b, c} is a subset of {a, b, c, d}.   ☐

EXAMPLE 3   {Judas Priest, the Rolling Stones, the Go-Go's, the Talking Heads} ⊆ {all singing groups}   ☐

EXAMPLE 4   {Moe, Larry, Curly} ⊆ {Curly, Moe, Larry}   ☐

Dr. Chang, who teaches math at this school, is a member of the set A = {all math teachers at this school}. He is also a member of set B = {all teachers at this school}. Professor Rivera, who teaches history, is a member of set B, but not of set A. We say that A is a proper subset of B.

> **Definition 1.10**   *Set X is said to be a **proper subset** of set Y if every element of set X is an element of set Y, and also set Y has at least one other element that is not in X. We denote this by $X \subset Y$. This symbol is read as "X is a proper subset of Y."*

> **Definition 1.11**   *Set X is said to be an **improper subset** of set Y if every element of set X is an element of set Y, but Y does not have any other elements that are not in X. This really means X equals Y, and we denote this as $X = Y$.*

Comment   If we know definitely that A is a proper subset of B, then we write $A \subset B$. If we know only that A is a subset of B but we do not know if it is proper or improper, then we write $A \subseteq B$.

EXAMPLE 5   {Moe, Larry, Curly} is an improper subset of {Larry, Curly, Moe}, and thus we usually write this as {Moe, Larry, Curly} = {Larry, Curly, Moe}.   ☐

EXAMPLE 6   {a, e, u, w} is a proper subset of {a, e, u, w, t}. We write {a, e, u, w} ⊂ {a, e, u, w, t}.   ☐

EXAMPLE 7   {a, e, u, w} is also a proper subset of {all letters of the English alphabet}.   ☐

EXAMPLE 8   {a, e, u, w} is an improper subset of {a, e, u, w}. We write {a, e, u, w} = {a, e, u, w}.   ☐

Comment   The null set is assumed to be a subset of *every* set. Can you see why?

Beware   Do not confuse the symbol "⊂," which stands for "is a proper subset of," with "∈," which means "is an element of." For example, Sting ∈ {all singers} and Diana Ross ∈ {all singers}. However, {Sting, Diana Ross} is a proper subset of the set of all singers, i.e., {Sting, Diana Ross} ⊂ {all singers}.

Suppose in a journalism class we are discussing newspaper editorials. We might then consider the *New York Times*, the *Washington Post*, the *Atlanta Constitution*, the *San Francisco Examiner*, etc. It would not be appropriate for someone to introduce into the discussion an editorial heard on NBC News. The discussion is limited to newspapers only. We call the set of all newspapers the universal set for this discussion.

Definition 1.12   *If all sets in a given discussion are to be subsets of a fixed overall set U, then this set U is called the **universal set**.*

EXAMPLE 9   If we want to talk about the set of math teachers at this school, the set of English teachers at this school, etc., then a suitable universal set would be {all teachers at this school}.   □

EXAMPLE 10   If we are going to discuss positive numbers, negative numbers, fractions, etc., we might choose as our universal set {all numbers}.   □

EXAMPLE 11   Suppose we are considering Reggie Jackson, Tom Seaver, Willie Mays, etc. We could use {all baseball players} as our universal set. If we consider them as amateur golfers or tennis players as well as baseball players, then a more appropriate universal set would be {all athletes}.   □

Comment   The universal set may vary from discussion to discussion.

Comment   Every set is a subset of itself. Would you say that it is a proper or improper subset of itself?

Suppose we wanted to distribute a questionnaire to all students in Math 1 and all students in English 1 at our school. We would have to compile a mailing list of all students in both classes. Some students may be in both classes, but of course we would not list them twice. If we denote the set of students in Math 1 by *A* and the set of students in English 1 by *B*, our list would be an example of what is meant by the union of *A* and *B*.

Definition 1.13   *The **union** of two sets A and B, which is denoted as A ∪ B (read as "A union B"), means the set of all elements that are either in A or in B or both.*

EXAMPLE 12   **a)** If $A = \{1, x, \triangle\}$ and $B = \{y, \text{ŝ}\}$, then $A \cup B = \{1, x, \triangle, y, \text{ŝ}\}$.

b)

    c) If $A = \{m, \text{John, Bob, } y\}$ and $B = \{\text{Bob, Alice, } x\}$, then $A \cup B$ = $\{m, \text{John, Bob, } y, \text{Alice, } x\}$.

    Observe that even though Bob is in both sets, we list him *only once* in $A \cup B$. □

EXAMPLE 13   If $M = \{$all months of the year with only 6 days$\}$ and $N = \{$all months of the year with 30 days$\}$, then what is $M \cup N$? □

    Let us consider again our questionnaire project. If we want to send the questionnaire only to students in *both* classes (if there are any), then this set of students is called the intersection of sets A and B (remember $A = \{$Math 1 students$\}$ and $B = \{$English 1 students$\}$).

> **Definition 1.14**   *The **intersection** of two sets A and B, denoted as $A \cap B$ (read as "A intersect B"), means the set of all elements that are in both sets A and B at the same time (if there are any).*

> **Definition 1.15**   *If sets A and B have no members in common, then sets A and B are called **disjoint sets**. This means that if A and B are disjoint, then $A \cap B = \varnothing$.*

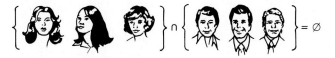

EXAMPLE 14   If $X = \{a, e, \text{ŝ}, \triangle\}$ and $Y = \{\text{ŝ}, \triangle, b, 1, \text{Joe}\}$, then $X \cap Y = \{\text{ŝ}, \triangle\}$. □

EXAMPLE 15   If $M = \{$all convertibles$\}$ and $N = \{$all green cars$\}$, then $M \cap N = \{$all green convertible cars$\}$. □

EXAMPLE 16   If $A = \{$all members of this school's baseball team$\}$ and $C = \{$all members of this school's basketball team$\}$, then $A \cap C = \{$all members of this school who are on both the basketball and baseball teams$\}$.

□

EXAMPLE 17   If M = {all men who have blond hair} and N = {all women who have blond hair}, then $M \cap N = \emptyset$. In this case, sets M and N are disjoint, since there are no elements common to both sets.   ☐

Comment   Note that the intersection is not {blond hair}. Having blond hair is a property of all the elements of each set. Blond hair itself is not an element of either set.

EXAMPLE 18   If A = {all guitar players} and B = {all musicians}, then $A \cap B$ = {all guitar players}.   ☐

EXAMPLE 19   If X = {all Volkswagens with air conditioning} and Y = {all Fords with air conditioning}, what is $X \cap Y$? Note that $X \cap Y$ is not {air-conditioned cars}. Why?   ☐

Imagine that we are sponsoring a school dance this Friday at 8:00 P.M. Let A be the set of students who have already bought tickets. Being on the dance committee, we would obviously be interested in the students who have not yet purchased tickets. If our universal set is the set of all students at this college, then we call the set of all students who have not yet purchased tickets the complement of set A.

> **Definition 1.16**   *The **complement** of set A, denoted by A' (read as "A prime" or "A complement"), is the set of all elements in the universal set that are not also in A.*

EXAMPLE 20   If the universal set = {1, 2, 3, 4, 5, 6, 7, 8, 9, 10} and A = {1, 4, 6, 8, 9}, then A' = {2, 3, 5, 7, 10}.   ☐

EXAMPLE 21   If the universal set = {all men} and X = {men who are bald}, then X' = {all men who have hair on their heads}.   ☐

EXAMPLE 22   If the universal set = {all American-made cars} and G = {all American-made cars with air conditioning}, then G' = {all American-made cars without air conditioning}.   ☐

EXAMPLE 23   If U = {all humans} and P = {all males}, then P' = {all females}.   ☐

EXAMPLE 24   If the universal set = {all vowels in the English language} and A = {a, e, i, o, u}, then what is A'?   ☐

Let A = {all students in this class} and B = {all students who read *Playboy*}. The set of all students in this class who do not read *Playboy* is called the difference between sets A and B.

Definition 1.17 *The **difference** between sets A and B, denoted as A − B (read as "A minus B"), means the set of all elements that belong to set A but not to set B.*

EXAMPLE 25 If M = {piano, guitar, drums, clarinet} and N = {piano, saxophone, organ, clarinet}, then $M - N$ = {guitar, drums}. Note that saxophone and organ are not part of the difference, since they are not in set M. ☐

EXAMPLE 26 If H = {3, 6, 9, 12} and J = {3, 6, 9}, then $H - J$ = {12}. What is $J - H$? If your answer is {12}, you are wrong. (Look at the definition again.) The answer is $\varnothing$. ☐

EXAMPLE 27 If V = {Milton, Bob, Carl} and F = {Bob, Carl, Milton}, then $V - F$ = $\varnothing$. What is $F - V$? The answer is $\varnothing$. ☐

EXAMPLE 28 If R = {all Mickey Mouse sweatshirts} and Q = {all triangles with two sides}, then $R - Q = R$, since Q is obviously the null set. What is $Q - R$? It is $\varnothing$. ☐

EXAMPLE 29 If S = {9} and T = {3}, then $S - T$ = {9} and $T - S$ = {3}. ☐

Bob and Rita are discussing their date for this Friday night. Their friends have recommended the following three restaurants: The China Palace, Tony's Grotto, and Esther's Soul Food. After eating, they plan to go to a discotheque or to a movie. Thus they can spend the evening in one of the following six ways.

Eat at	*After eating, go to*
1. The China Palace	a discotheque
2. The China Palace	a movie
3. Tony's Grotto	a discotheque
4. Tony's Grotto	a movie
5. Esther's Soul Food	a discotheque
6. Esther's Soul Food	a movie

This assumes that they plan to eat first. Let A = {The China Palace, Tony's Grotto, Esther's Soul Food} and B = {a discotheque, a movie}. Then each possibility can be thought of as an **ordered pair** of elements *where the first element comes from set A and the second element comes from set B*. For example, possibility (4) can be considered as an ordered pair where the first element is Tony's Grotto (from set A) and the second element is a movie (from set *B*). We write this as (Tony's Grotto, a movie).

> **Definition 1.18**   *An **ordered pair** of elements is any two elements written in a specific order. If the elements are a and b, we denote this by (a, b).*

Since the order is important, the ordered pair (*b*, *a*) is different from the ordered pair (*a*, *b*).

EXAMPLE 30   The ordered pair (The China Palace, a movie) is different from (a movie, The China Palace). In the ordered pair (The China Palace, a movie), Bob and Rita eat first and then go out. In the ordered pair (a movie, The China Palace), they go out first and then eat. (As we all know, if you are hungry, it makes a big difference which you do first!) ☐

> **Definition 1.19**   *The **cross product** of two sets A and B is the set of **all ordered pairs** that can be formed by taking the first element from set A and the second element from set B. We denote this by A × B. (We read this as "A cross B.")*

EXAMPLE 31   If A = {The China Palace, Tony's Grotto, Esther's Soul Food} and B = {a discotheque, a movie}, then $A \times B$ = {(The China Palace, a discotheque), (The China Palace, a movie), (Tony's Grotto, a discotheque), (Tony's Grotto, a movie), (Esther's Soul Food, a discotheque), (Esther's Soul Food, a movie)}. These correspond to the six possibilities listed above. ☐

EXAMPLE 32   If A and B are the same as in Example 31, then $B \times A$ = {(a movie, The China Palace), (a movie, Tony's Grotto), (a movie, Esther's Soul Food), (a discotheque, The China Palace), (a discotheque, Tony's Grotto), (a discotheque, Esther's Soul Food)}. This represents all the possibilities if they want to eat later in the evening. ☐

EXAMPLE 33   If C = {6, 8, 9} and D = {x, y, z}, then $C \times D$ = {(6, x), (6, y), (6, z), (8, x), (8, y), (8, z), (9, x), (9, y), (9, z)}. Also $D \times C$ = {(x, 6), (x, 8), (x, 9), (y, 6), (y, 8), (y, 9), (z, 6), (z, 8), (z, 9)}. ☐

EXAMPLE 34   If M = {□, △}, then $M \times M$ = {(□, △), (□, □), (△, □), (△, △)}. ☐

Sometimes the cross product is called the **Cartesian product.** Look at Examples 31 and 32. Observe that $A \times B$ is not the same as $B \times A$. This is usually the case.

Comment   We have already mentioned (in Section 1.2) that, when listing the elements of a set, the order in which they are written is not important. For example, the set {*a*, 2} is the same as the set

{2, *a*}. When writing an **ordered pair,** the order *is* important. For example, (*a*, 2) is not the same as (2, *a*). This idea of ordered pairs will be applied in Chapter 8, when we study graphs and functions.

The importance of ordered pairs and of cross products can be seen in everyday experiences. Imagine that Mr. Jones is planning a trip from New York to Paris in the spring. His travel agent has notified him that in the New York metropolitan area there are three airports from which he may depart: Kennedy, La Guardia, and Newark. In Paris he may land at either Le Bourget or Orly airport. Let the first element of an ordered pair represent the airport from which he leaves and the second element the airport at which he arrives. The ordered pair (Newark, Orly) means that he leaves from Newark airport and arrives at Orly. If *A* = {Kennedy, Newark, La Guardia} and *B* = {Orly, Le Bourget}, then *A* × *B* = {(Kennedy, Orly), (Kennedy, Le Bourget), (Newark, Orly) (Newark, Le Bourget), (La Guardia, Orly), (La Guardia, Le Bourget)}. This represents all the different ways that Mr. Jones can travel from New York to Paris.

In this example, *B* × *A* would represent all the different ways that Mr. Jones can go from Paris to New York. These are:

$$B \times A = \{(\text{Orly, Kennedy}), (\text{Orly, La Guardia}), (\text{Orly, Newark}),$$
$$(\text{Le Bourget, Kennedy}), (\text{Le Bourget, La Guardia}),$$
$$(\text{Le Bourget, Newark})\}.$$

Comment The definition of the cross product of two sets *A* and *B* *does not* require that the sets used in forming *A* × *B* be different. They *may* be the same sets, as the exercises below will show.

EXERCISES

In Exercises 1–8, give two different subsets of each of the indicated sets.

1. {Reagan}

2. {True, false}

3. {All U.S. senators}

4. {All vowels in the English language}

5. {1, 2, *α*, *β*}

6. {Environmental problems that are of concern to environmentalists}

7. {Hamburgers, beefburgers}

8. {Foreign import subcompact cars}

In Exercises 9–16, find an appropriate universal set for a discussion involving the indicated objects.

9. Bell Atlantic, NYNEX, Pacific Telesis Group, US West, Southwestern Bell, Ameritech.

10. Martina Navratilova, Chris Evert Lloyd, Andrea Jaeger, Tracy Austin, Sylvia Hanika.

11. Wordstar, Visicalc, Lotus, Symphony, dBase II.

12. Datsun, Honda, Volvo, Ferrari, Mecedes, Audi, Peugeot.

13. Federal Express, United Parcel Service, Purolator, Emery.

14. Catfish, trout, bass, salmon, whitefish, eel, whale, bluefish, snapper.

15. Apollo, Columbia.

16. Pepsi, Coca Cola, Seven-Up, Sprite.

In Exercises 17–24, state whether set A is a proper or an improper subset of set B or neither.

17. A = {all even numbers}, B = {all odd numbers}.

18. A = {all even numbers divisible by 3}, B = {all even numbers}.

19. A = {Bill, Mary, George}, B = {George, Mary, Bill}.

20. A = {sperm whales}, B = {mammals}.

21. A = {musical instruments}, B = {compact disc players}.

22. A = {Pluto, Mercury, Mars, Earth, Venus}, B = {planets}.

23. A = {people with three heads}, B = {people with six eyes}.

24. A = {#, &, @}, B = {#, &, ^, @}.

In Exercises 25–28, let A = {9, 11, 12}.

25. Find all the subsets of A that contain one element.

26. Find all the subsets of A that contain two elements.

27. Find all the subsets of A that contain three elements.

28. Find all the subsets of A that contain no elements.

In Exercises 29–34, list all the subsets of the indicated sets. (Be sure you do not leave any out.)

29. { }

30. {a}

31. {a, b}

32. {a, b, c}

33. {a, b, c, d}

34. {a, b, c, d, e}

How many different committees of people can be formed from each of the sets given in Exercises 35–38?

35. {Ronald}

36. {Joanna, Jill}

37. {Steve, Kim, Juan}

38. {Smitty, Margaret, Cathy, Beth}

39. On the basis of your answers to Exercises 29–38, complete the following chart:

Number of elements in set	Number of possible subsets
0	1
1	2
2	4
3	8
4	16
5	32
6	?
10	?
100	?
n	?

Classify each of the statements given in Exercises 40–48 as either true or false. Justify your answer.

40. {Sue, Joan} is a subset of {Maria, Joan, Sue}.

41. $\varnothing$ is a subset of {5, 8, 12}.

42. {D, R, U, G, S} $\subseteq$ {S, G, R, U, D}.

43. {7} $\subset$ {$n \mid n$ is an odd number}.

44. {5, 7, 9, 11} is a subset of {5, 9, 11}.

45. {7} $\subset$ {$y \mid y$ is an even number}.

46. $B \subseteq B$.

47. {fish} $\subset$ {salmon}.

48. {all airplanes over 200 years old} $\subset$ {all flying machines}.

For Exercises 49–62, let the universal set U = {a, b, c, d, e, f, g, h, i}, let A = {a, d, f, g}, B = {b, c, e, f, g, i}, and C = {b, f, i}. Find each of the following:

49. $A \cup B$

50. $A \cap B$

51. A'

52. $A' \cup B$

53. $A - B$

54. $B - A$

55. $C - B$

56. $A - C$

57. $A \times C$

58. $C \times A$

59. $(A' \cap B) \cup C$

60. $(A \cup B)' \cap C$

61. $A' \cap B' \cap C'$

62. $(A' \cap B') \cup C'$

For Exercises 63–74, let the universal set $U = \{a, b, c, d, e, f, g, h\}$ and let $A = \{a, b, e, g\}$. Find each of the following:

63. A' **64.** $A \cup A$ **65.** $A \cup A'$

66. $A \cup \varnothing$ **67.** $A \cap \varnothing$ **68.** $A - \varnothing$

69. $A \cup U'$ **70.** $A \cap A'$ **71.** $A \cap U$

72. $A - U$ **73.** $U - A$ **74.** $\varnothing - A$

75. If $A \cap B = A$, how are A and B related?

76. If $A \cup B = A$, how are A and B related?

77. If $A \cap B = B$, how are A and B related?

78. If $A \cup B = B$, how are A and B related?

For Exercises 79–84, let $A = \{x \mid x$ is a number between 1 and 20 inclusive$\}$, let $B = \{x \mid x$ is a number between 60 and 80 inclusive$\}$, and let $C = \{x \mid x$ is a number between 20 and 80 inclusive$\}$. Find each of the following:

79. $A \cup B$ **80.** $A \cap C$ **81.** $A \cap B$

82. $A \cup C$ **83.** $B \cup C$ **84.** $B \cap C$

85. If $A \cup A = \varnothing$, what can be said about A?

86. Is it ever possible for $A \times B$ to equal $B \times A$? Explain.

For Exercises 87–95, let the universal set $U = \{a, b, c, d\}$, $A = \{a, b\}$, $B = \{a, b, d\}$, $C = \{d\}$, and $D = \{c, d\}$. Indicate the answer to each set operation by writing the capital letter that names the resulting set. (For example, $A \cap B = A$.)

87. $A \cup C$ **88.** $A \cap C$ **89.** $B \cup C$

90. $B \cap C$ **91.** $B \cap D$ **92.** $A' \cap D$

93. $D' \cup B$ **94.** $B \cap U$ **95.** $B \cup U$

**96. Suppose that Mr. Smith is a small-town barber who shaves all the men and only those men in his town who do not shave themselves. (Everyone in the town is shaved or shaves daily.) Let $A = \{$all men in the town who shave themselves$\}$, and let $B = \{$all men in the town who do not shave themselves$\}$.

  a) Is Mr. Smith a member of set A?

  b) Is Mr. Smith a member of set B?

  c) What is $A \cup B$?

  d) Is Mr. Smith a member of $A \cup B$?

97. If $A = \{a, b\}$ and $B = \{c, d, e\}$, find

  a) $n(A)$ **b)** $n(B)$ **c)** $n(A \times B)$

98. If $n(A) = p$ and $n(B) = q$, how many elements are there in $n(A \times B)$?

99. A certain house can be purchased with or without a finished basement and with a choice of three possible sources of heat (solar, oil, or gas). List all the possible combinations of features that can be purchased with the house.

Buying a Used Car

For Exercises 100–109, use the following information. Recently, a nationwide survey by the Brown Associates of used car buyers in different age groups was conducted to determine which feature in a used car was of utmost concern to the buyer. The following results were obtained:

Age (in years)	Economy of operation (E)	Styling (S)	Size (Z)	Color (C)
Under 30 (U)	262	392	301	361
30–50 yrs (T)	427	309	296	203
Over 50 (O)	522	283	371	152

Using the letters indicated, find the number of people in each of the following sets.

100. T **101.** $O \cap E$

102. $U - S$ **103.** $S \cap T$

104. $U \cup C$ **105.** $E - T$

106. $E - S$ **107.** $(O \cap E) \cup (O \cap S)$

108. $Z \cup (E \cap O)$ **109.** $(C \cup U) \cap Z$

For Exercises 110–113, refer to the newspaper article given below. Using the letters given in the article, find the number of people in the following sets.

110. $(M \cap S) \cup N$ **112.** $D - (I \cup N)$

111. $F \cap (N \cup S)$ **113.** $O \cup (M \cap S)$

Tighter Security at Airports

WASHINGTON—As a direct result of the recent TWA hijacking, government officials have announced a series of security checks at airports of all international travelers and their luggage to thwart any possible terrorist attacks. Yesterday, the Tribune conducted a random survey of travelers at the nation's airports to determine whether they were satisfied with the new security arrangements and the resulting delays. The results are summarized below:

	Supports the checks enthusiastically despite delays (S)	Supports the checks moderately (D)	Is opposed to the checks because of delays (O)
Male			
Frequent traveler (M)	43	67	21
Infrequent traveler (I)	79	99	32
Female			
Frequent traveler (F)	48	58	17
Infrequent traveler (N)	69	95	27

Springfield Tribune, July 5, 1985

For Exercises 114–122, use the data obtained by the Acme Insurance Company in its recent inspection of 300 cars equipped with some anti-theft device. The survey revealed the information given at the top of the next column.

Type of anti-theft device

Type of car	Ignition shutoff (G)	Steering wheel lock (S)	Burglar alarm (B)
Compact (C)	48	27	53
Intermediate car (I)	32	19	46
Large size (L)	17	22	36

Determine the number of cars in each of the following categories:

114. $C \cap G \cap S'$ **115.** C **116.** $C \cap G' \cap S$
117. C' **118.** I **119.** $I \cap B$
120. $(I \cap S) \cup B$ **121.** $G \cup (S \cap L)$
122. $B \cup (I - G)$

For Exercises 123–126, let us analyze the medical records of the Prince Corporation, which reveal the following information about 316 employees who applied for coverage under the company's extended health coverage plans.

	Heavy smoker and drinker (S)	Heavy smoker and non-drinker (N)	Non-smoker but heavy drinker (H)	Non-smoker and non-drinker (D)
Male (M)	81	49	57	3
Female (F)	32	68	21	5

Find the number of people in the following categories:

123. $M \cap (N \cup H)$ **124.** $F - (D \cap M)$
125. $(M \cup F) - (H \cap D)$
126. $(H \cap D) - (M \cup N)$

1.4

USING VENN DIAGRAMS TO UNDERSTAND SETS

In your experience with mathematics you have probably found that a diagram is often a useful device. In working with sets, diagrams can also be very helpful. Such diagrams are called **Venn diagrams.**

The universal set is pictured as a rectangle as shown in Fig. 1.1.

The universal set U **Figure 1.1**

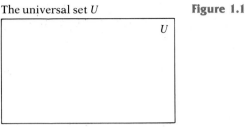

All other sets are drawn as circles within the rectangle. We shade the part that is in set A, as shown in Fig. 1.2.

Set A **Figure 1.2**

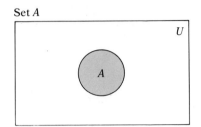

It is easy to see how to picture A'. This time we shade the part that is in set A'. This is shown in Fig. 1.3.

A complement A' **Figure 1.3**

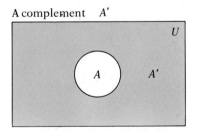

If A is a *proper subset* of B (see Definition 1.10), we draw the circle for A completely inside the circle for B, as shown in Fig. 1.4. We do

$A \subset B$ **Figure 1.4**
Set A is a proper subset of set B.

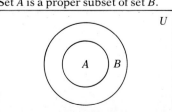

not shade anything here, since we are picturing a **relationship** between two sets. We shade the diagram only when we want to emphasize the set that we are picturing, rather than a relationship between two or more sets.

If A is an *improper subset* of B (remember, this means $A = B$), we draw only one circle and label it as both A and B, as shown in Fig. 1.5.

$A=B$
Set A is an improper subset of B.   **Figure 1.5**

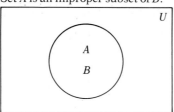

We picture $A \cap B$ as shown in Figs. 1.6 and 1.7, depending on whether $A \cap B \neq \varnothing$ or $A \cap B = \varnothing$.

$A \cap B$ is not the null set.

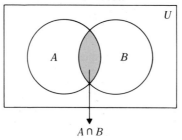

$A \cap B$

Figure 1.6

$A \cap B = \varnothing$. Nothing is shaded.

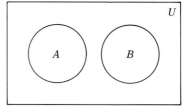

Figure 1.7

If A is a subset of B, we picture $A \cap B$ as shown in Fig. 1.8.

$A \cap B$ when A is a subset of B.   **Figure 1.8**

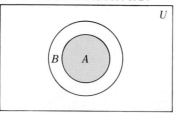

We picture $A \cup B$ as shown in Figs. 1.9, 1.10, and 1.11, depending on the relationship between A and B.

$A \cup B$ when A and B are not disjoint.

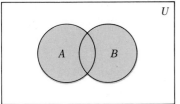

Figure 1.9

$A \cup B$ when A and B are disjoint.

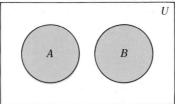

Figure 1.10

$A \cup B$ when A is a subset of B.

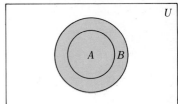

Figure 1.11

The Venn diagram for $A - B$ or $B - A$ is shown in Figs. 1.12, and 1.13 (if $A \cap B \neq \varnothing$).

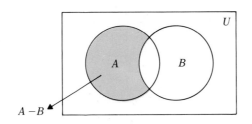

Figure 1.12

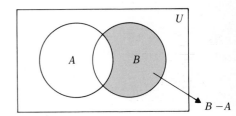

Figure 1.13

EXAMPLE 1   Let $U = \{$all people$\}$, $A = \{$all men with blonde hair$\}$, $B = \{$all women with blonde hair$\}$, and $C = \{$all women with blue eyes$\}$. The relationship between these sets is shown in Fig. 1.14.

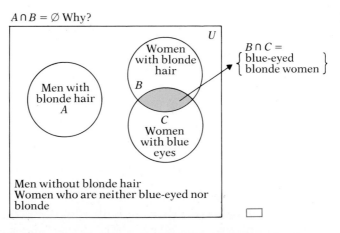

Figure 1.14

EXAMPLE 2   If U = {all college students}, A = {all female college students}, B = {all blonde college students}, and C = {all freshmen college students}, then we picture these sets as shown in Fig. 1.15.

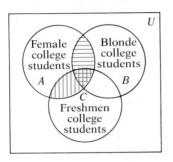

Figure 1.15

    The part that has been shaded horizontally represents $A \cap B$. This is the set of all female blonde college students.

    The part that has been shaded vertically represents $A \cap C$. This is the set of all female freshmen college students.

    Note that in the center part the shading overlaps. This represents $A \cap B \cap C$. How would you describe this set in words?   ☐

EXAMPLE 3   Using Venn diagrams, show that $(A \cap B)' = A' \cup B'$.

SOLUTION   We make two diagrams, one to represent $(A \cap B)'$ and one to represent $A' \cup B'$. To draw the diagram for $(A \cap B)'$, we first draw $A \cap B$ and then shade everything *outside* $A \cap B$. This is shown in Fig. 1.16.

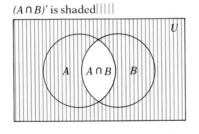

A' is shaded |||||||||||
B' is shaded ≡≡≡≡≡

Figure 1.16   **Figure 1.17**

    To draw the diagram for $A' \cup B'$, we first shade vertically everything outside A. This gives us A'. We then shade horizontally everything outside B. This gives us B'. The part that has *any* shading is $A' \cup B'$. This is shown in Fig. 1.17.

    Observe that in both diagrams the same regions have been shaded. This means that $(A \cap B)'$ and $A' \cup B'$ are equal.   ☐

Comment The previous results, along with the fact that $(A \cup B)' =$ $A' \cap B'$, are known as **De Morgan's laws** in honor of the famous mathematician Augustus De Morgan, who lived during the nineteenth century. Such formulas enable us to transform statements and formulas into alternative and often more useful forms. This will be especially helpful when we study logic.

EXERCISES

In Exercises 1–7, draw a Venn diagram similar to the diagram shown in Figure 1.18 and shade in the area indicated.

1. $A \cap B \cap C$ **2.** $A \cup B \cup C$

3. $(A \cap B \cap C)'$ **4.** $A \cup (B \cap C)$

5. $A \cap (B \cup C')$ **6.** $A' \cup (B \cap C)$

7. $[(A \cap B) \cup C]'$

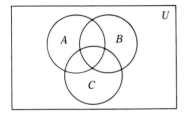

Figure 1.18

For Exercises 8–13, refer to Figure 1.19. Let $T = \{$tall people$\}$, $I = \{$intelligent people$\}$, and $P = \{$pleasant people$\}$. Describe in words the set represented by each of the indicated symbols.

8. $T \cap I \cap P'$ **9.** $(T \cup I) \cap P$

10. $T \cap (I' \cup P)$ **11.** $T' \cup (I \cap P)$

12. $(T' \cup I) \cap P$ **13.** $(T \cap I) - P$

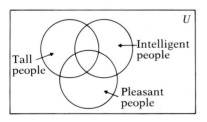

Figure 1.19

For Exercises 14–19, use Figure 1.20 to find the value of the indicated expression.

14. $n(A \cup B \cup C)$ **15.** $n(A \cap B \cap C)$

16. $n(B')$ **17.** $n[(A \cap B)']$

18. $n[(A \cup B') \cap C']$ **19.** $n[(A \cap B') \cup C']$

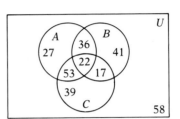

Figure 1.20

For Exercises 20–26, use Figure 1.21 to find the number of people in each category.

20. Overweight, nonsmoker, but drug user.

21. Overweight, smoker, but not drug user.

22. Not overweight but smoker and drug user.

23. Not overweight, not smoker, nor drug user.

24. Only drug user.

25. Overweight, smoker, and drug user.

26. Drug user, smoker, but not overweight.

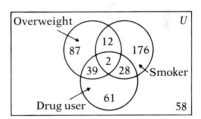

Figure 1.21

27. In Figure 1.22, the different parts of the circles are numbered from 1 to 8. For example, region 1 corresponds to the set $(A \cup B \cup C)'$. In a similar manner, describe symbolically the set that corresponds to each of the regions 2, 3, 4, 5, 6, 7, and 8.

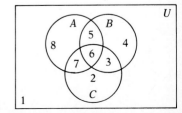

Figure 1.22

28. Using Venn diagrams, verify *De Morgan's law*, $(A \cup B)' = A' \cap B'$.

In Exercises 29–32, use set notation to describe the shaded portion of the diagram.

29.
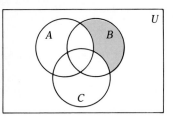

30.
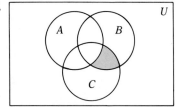

31.
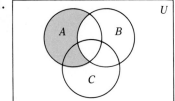

32.
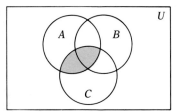

By using Venn diagrams, determine which of the statements given in Exercises 33–40 are true for all sets.

33. $A' \cup B' = A \cap B$

34. $(A \cup B)' = A' \cap B$

35. $A \cup (B \cap C) = (A \cup B) \cap C$

36. $A' \cup B' = A' \cap B$

37. $A' \cup (B \cap C) = A \cap (B \cup C)'$

38. $A \cap (B \cup C) = (A \cup B) \cap (A \cup C)$

39. $A \cap (B \cup C) = (A \cap B) \cup (A \cap C)$

40. $A \cap (B \cup C)' = A \cap (B' \cup C')$

For each of Exercises 41–49, draw a Venn diagram to represent the given statement.

41. $A \subset B$

42. $A \cup \varnothing$

43. $A \cap U$

44. $A \cup B$ when $A \cap B = \varnothing$

45. $(A' \cup B)'$ when A and B are disjoint

46. $A \cup B$ when $A = B$

47. $A \cap B = B$

48. $A \cup B = A$

49. $(A \cup B)'$ when $A \cap B = \varnothing$

In Exercises 50–55, use De Morgan's laws to simplify the given expression:

50. $A' \cup (A \cap B)'$ **51.** $A \cap (A \cup B)'$

52. $(A' \cup B)'$ **53.** $(A' \cap B')'$

54. $B \cup (A \cap B)'$ **55.** $(A' \cap B)' \cup A$

1.5

APPLICATIONS OF SETS   In this section we discuss two important applications of sets: survey problems and voting coalitions.

Survey Problems   The first application of sets is to analyzing **survey problems.** This is illustrated in the following examples.

EXAMPLE 1   Consider the newspaper article shown below.

Americans Love New Gadgets

LOS ANGELES—A recent survey of 600 families revealed that 348 of them had
a video cassette recorder (VCR), 198 had a microwave oven, and 75 had both a
VCR and a microwave oven. Americans seem to be fascinated with any new
technology.

Portland Star, June 3, 1985

Based upon the information contained in the article,

a) how many families had neither a VCR nor a microwave oven?
b) how many families had only a VCR?
c) how many families had only a microwave oven?

SOLUTION   Let

$$U = \{\text{families involved in the survey}\},$$
$$V = \{\text{families who own a VCR}\},$$
$$M = \{\text{families who own a microwave oven}\}.$$

The article provides us with the following information:

1. The number of families surveyed, that is, $n(U) = 600$.
2. The number of families who own a VCR, that is, $n(V) = 348$.
3. The number of families who own a microwave oven, that is, $n(M) = 198$.
4. The number of families who own both a VCR and a microwave oven, that is, $n(V \cap M) = 75$.

We can illustrate this information with a Venn diagram as shown in Figure 1.23. From the previous section we know that $V \cap M$ represents the region that is common to both circles. This represents region II in Figure 1.23. Since $n(V \cap M) = 75$, we write 75 in this region. Since we are told that 348 families own a VCR, this means that $n(V) = 348$ or that the total number of families in regions I and II together is 348. However, we already know that region II contains 75 families, so region I contains $348 - 75$ or 273. We indicate this on the diagram. Also, we are told that 198 families own a microwave oven. This means that the total number of families in regions II and III is 198. Again, we already know that region II contains 75, so region III contains $198 - 75 = 123$. Finally, we have accounted for $273 + 75 + 123$ in regions I, II, and III, respectively. This gives 471. Since

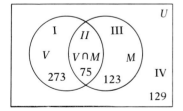

Figure 1.23

the survey involved 600 families, we must have 600 families in regions I, II, III, and IV together. Thus we have 600 − 471 or 129 for region IV alone.

a) How many families had neither a VCR nor a microwave oven? This is represented by region IV in the diagram. The number of families in region IV is 129.

b) How many families had only a VCR? This is represented by region I in the diagram. The number of families in region I is 273.

c) How many families had only a microwave oven? This is represented by region III in the diagram. The number of families in region III is 123. ☐

EXAMPLE 2   The student government at State University recently conducted a survey to gather more information on the high school backgrounds of entering freshmen. There were 100 students interviewed and the following data were collected:

> *28*   took physics,
>
> *31*   took biology,
>
> *42*   took geometry,
>
> *10*   took physics and geometry,
>
> *6*   took biology and geometry,
>
> *9*   took physics and biology, and
>
> *4*   took all three subjects.

On the basis of these figures,

1. How many students took none of the three subjects?

2. How many students took physics but not biology or geometry?

3. How many students took biology and physics but not geometry?

SOLUTION   This problem can be solved easily by means of Venn diagrams. Three circles are used to represent the students taking each of the subjects.

In Fig. 1.24 we first put in the number of students who took all 3 subjects. This was given to be 4. Since we know that 6 altogether took biology and geometry, we must have 6 − 4, or 2, who took biology and geometry *but not* physics. We enter this in Fig. 1.24. Similarly, we know that 10 took physics and geometry. Therefore we must have 10 − 4, or 6, who took physics and geometry *but not* biology. We fill this in on the diagram. Also we know that 9 took physics and biology. Therefore we must have 5 who took physics and biology *but not* geometry. We fill this in on the diagram.

According to the data, 42 took geometry. If we look at the geometry circle, we see that we have accounted for 6 + 4 + 2, or 12. This

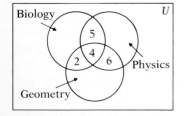

Figure 1.24

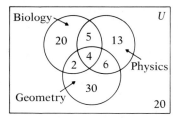

Figure 1.25

leaves 42 − 12, or 30 students to be put in the remainder of the geometry circle. We then have the diagram shown in Fig. 1.25.

Now 31 took biology. In Fig. 1.24, we have accounted for 5 + 4 + 2, or 11 students. This leaves 31 − 11, or 20 students to be put in the remaining portion of the biology circle. In a similar manner, there are 28 − 15, or 13 students for the remainder of the physics circle.

We now add together all the numbers appearing in the diagram: 20 + 5 + 4 + 2 + 6 + 13 + 30. This sum is 80. Since the survey involved 100 students, we are left with 100 − 80, or 20 students who did not take any of these subjects. This answers question 1. The answer to question 2 can be read from Fig. 1.25. It is obviously 13. The diagram also answers question 3. Five took biology and physics, but not geometry. ☐

Another interesting application of sets is to voting coalitions.

Voting Coalitions*

Recently, the student government at State College appointed a committee consisting of Mike, Richard, Helen, Bess, and David to consider a proposal to establish a drug clinic on campus. There is some disagreement on this committee as to whether they should accept or reject the proposal. Each member has one vote. Richard, Mike, and Helen agree with the proposal, whereas David and Bess oppose it. The committee operates under majority rule. Since Richard, Mike, and Helen agree, we call them a **winning coalition.** Since David and Bess have only two votes between them, they form a **losing coalition.**

Let the universal set U be some voting body such as the committee above, or the United States Senate, etc. Assume that there are no abstentions or absentees. We then have the following definitions.

Definition 1.20 *Any combination of people (that is, a subset of U) that can carry a proposal is called a **winning coalition.***

Definition 1.21 *A combination of people (that is, a subset of U) is called a **losing coalition** if a proposal will pass even though they vote against it.*

EXAMPLE 3 In the committee example above, the subset {Richard, Mike, Helen} is a winning coalition. The subset {David, Bess} is a losing coalition. ☐

EXAMPLE 4 Suppose a committee consists of three people whom we shall call p, q, r, so that $U = \{p, q, r\}$. If the committee operates under majority rule, then the winning coalitions are $\{p, q\}$, $\{p, r\}$, $\{q, r\}$, and $\{p, q, r\}$. The losing coalitions are $\{p\}$, $\{q\}$, and $\{r\}$. ☐

*The remainder of this section can be omitted without affecting the continuity of the text.

EXAMPLE 5   Suppose the committee in Example 4 elects p as its chairperson and gives her two votes. Everyone else has only one vote. Then the winning coalitions are $\{p, q, r\}$, $\{p, q\}$, and $\{p, r\}$. The coalition $\{q, r\}$ is no longer a winning coalition because p alone has two votes and can block q and r together. The losing coalitions are $\{q\}$ and $\{r\}$. Why is $\{p\}$ no longer a losing coalition?   ☐

In the last example, assume that p votes against a proposal and q and r each vote for it. What happens? The proposal will neither win nor lose. It is **blocked.** We call $\{p\}$ and $\{q, r\}$ **blocking coalitions.**

> **Definition 1.22**   *A coalition is called a **blocking coalition** if it can prevent any proposal from winning, but it cannot win by itself.*

EXAMPLE 6   In Examples 3 and 4 above, there are *no* blocking coalitions.   ☐

EXAMPLE 7   The U.C. Corporation of America has six members on its board of directors, each with one vote. Any coalition containing exactly three members is a blocking coalition.   ☐

In some voting bodies (for example, fascist states), there is one person who has the power to pass any measure on his or her vote alone (even if no one else votes for it). Such a person is called a **dictator.** A somewhat less powerful person is someone who is not a dictator, but who can **block** any proposal with his or her single vote alone. Such a person is said to have **veto power.** There may also be someone who has little power. Any winning coalition to which he or she belongs would be a winning coalition even without him or her. Such a person is called a **dummy.**

> **Definition 1.23**   *A **dictator** is someone who can pass any proposal with only his or her own vote. He or she needs no other votes.*

> **Definition 1.24**   *A member of a voting body has **veto power** if his or her vote alone is enough to **block** any proposal. (He or she cannot win alone.)*

> **Definition 1.25**   *A member of a voting body is called a **dummy** if any winning coalition of which he or she is a member will be a winning coalition even without him or her.*

EXAMPLE 8   The City Council of Atlantis consists of four members with the following voting strengths.

Mayor	5 votes
Treasurer	2 votes
County clerk	1 vote
Council president	1 vote

A simple majority, that is, more than half of the votes cast, is needed to pass any proposal. The mayor is obviously the dictator.

The treasurer, the county clerk, and the council president are each dummies. Why? ☐

EXAMPLE 9 In a neighboring town, the city council consists of four members with voting strength as follows:

Mayor	3 votes with veto power
Treasurer	3 votes
Sheriff	2 votes
Council president	1 vote

A simple majority is needed to pass any proposal. If M denotes mayor, T denotes treasurer, S denotes sheriff, and C denotes council president, then M has veto power. The winning coalitions are $\{M, T, S, C\}$, $\{M, T, S\}$, $\{M, T, C\}$, $\{M, S, C\}$, $\{M, S\}$, and $\{M, T\}$.

The losing coalitions are $\{C\}$, $\{S\}$, $\{T\}$, $\{S, C\}$, and $\{T, C\}$.

The council president C is the dummy. Are there any other dummies? Why is $\{T, S, C\}$ not a winning coalition? ☐

In ancient Greece the ostrakon was used in elections to ensure one vote for one person. (*Printed by permission of the American School of Classical Studies: Agora Excavations.*)

Look back at the last example. One of the winning coalitions was $\{M, T, S\}$. Another was $\{M, S\}$. Coalition $\{M, T, S\}$ would still have been winning, even if T or S (but not both) had voted against it. On the other hand, in $\{M, S\}$ all votes are essential. Similarly, in $\{M, T\}$ all votes are essential. Observe that $\{M, T\}$ and $\{M, S\}$ are proper subsets of $\{M, T, S\}$. (Refer back to Definition 1.10.) Thus $\{M, T, S\}$ has proper subsets that are also winning coalitions. On the other hand, $\{M, T\}$ and $\{M, S\}$ have no subsets that are winning coalitions. This leads us to the following definition.

> **Definition 1.26** *A **minimal winning coalition** is any winning coalition that has no proper subset that is a winning coalition (that is, each member's vote is essential or the proposal will not pass).*

EXAMPLE 10 In Example 9 above, the minimal winning coalitions are $\{M, S\}$ and $\{M, T\}$. ☐

EXAMPLE 11 A committee has six members $\{a, b, c, d, e, f\}$, each with one vote. A two-thirds vote is needed to carry any proposal. Any coalition with exactly four members in it is a minimal winning coalition. ☐

EXERCISES

For Exercises 1–3, use the following information. A total of 427 students were admitted to the business program at Technical University. Of these, 246 had taken intermediate algebra in high school, 159 had taken trigonometry, and 57 had taken both intermediate algebra and trigonometry.

1. How many students took trigonometry but not intermediate algebra?

2. How many students took intermediate algebra but not trigonometry?

3. How many students did not take either subject in high school?

For Exercises 4–6, use the following information. A survey of 600 farmers in the Midwest revealed that 382 of them raised hogs, 335 of them raised chickens, and 179 of them raised both hogs and chickens.

4. How many of the farmers raised only chickens?

5. How many of the farmers raised only hogs?

6. How many farmers raised neither hogs nor chickens?

The following results were reported by researchers at the Kingston Medical Center. Of the 100 people in a control group who were suffering from arthritis, 75 were given a new drug. The remaining patients were given a placebo. Sixty of the patients showed improvement; 54 of these people actually received the drug.

7. How many patients receiving the drug showed no improvement?

8. How many patients not receiving the drug showed no improvement?

Union Membership

A survey of the 200 workers at the Printex Corporation revealed that 115 of the workers were union members, 53 of the workers were satisfied with the working conditions, and 27 of the workers were union members who were satisfied with the working conditions.

9. How many union members were not satisfied with the working conditions?

10. How many workers were satisfied with the working conditions but were not union members?

11. How many workers were not satisfied with the working conditions and were not union members?

The 200 workers of the Printex Corporation have just negotiated a new contract with management whereby additional funds will be contributed to employee benefit programs. The employees were polled as to how the funds should be spent. The survey showed that

94 wanted better pension benefits,

99 wanted better health benefits,

113 wanted better dental insurance,

32 wanted better pension and health benefits,

41 wanted better dental insurance and health benefits,

47 wanted better pension and dental plans, and

14 wanted better pension, health, and dental plans.

12. How many employees wanted only a better pension plan?

13. How many employees wanted only a better dental plan?

14. How many employees wanted a better health and dental plan but were not interested in a pension plan?

15. How many employees wanted a better health and pension plan only?

16. How many employees were not interested in any of these programs?

Four hundred applicants for a stenographer's job at the United Nations in New York City indicated that, in addition to English, they were fluent in the following languages:

195 in Spanish,

143 in French,

184 in German,

56 in Spanish and French,

85 in Spanish and German,

69 in French and German, and

39 in all three languages.

17. How many were fluent only in Spanish?

18. How many were fluent only in French?

19. How many were fluent in Spanish and French but not German?

20. How many were fluent in French and German but not Spanish?

21. How many were not fluent in any of these three languages?

In a recent survey of 400 business executives, the following information was obtained:

190   read *Fortune*,
230   read *Time*,
110   read *Newsweek*,
60   read *Fortune* and *Time*,
50   read *Fortune* and *Newsweek*,
70   read *Time* and *Newsweek*, and
20   read all three.

22. How many did not read any of the magazines?
23. How many read *Fortune* but not *Time* or *Newsweek*?
24. How many read only *Newsweek* but not *Time* or *Fortune*?
25. How many read *Time* and *Newsweek* but not *Fortune*?
26. How many read *Newsweek* and *Fortune* but not *Time*?

As they entered Jason's Ice Cream Parlor, 400 customers were asked to indicate the ice cream flavor or flavors they liked. The following results were obtained:

180   liked vanilla,
170   liked chocolate,
110   liked strawberry,
70   liked vanilla and chocolate,
30   liked vanilla and strawberry,
40   liked chocolate and strawberry, and
10   liked all three flavors of ice cream.

27. How many customers liked only chocolate?
28. How many customers liked vanilla and chocolate but not strawberry?
29. How many customers liked strawberry and chocolate but not vanilla?
30. How many customers liked strawberry only?
31. How many customers liked none of these flavors?

A survey conducted by Bradley and Robbins of 700 Americans who exercise regularly to maintain their physical fitness disclosed the following:

350   jog,
279   swim,
321   cycle,
133   jog and swim,
86   swim and cycle,
105   jog and cycle,
47   jog, swim, and cycle.

32. How many of the people surveyed jog only?
33. How many of the people surveyed swim only?
34. How many of the people surveyed cycle only?
35. How many of the people surveyed jog and swim but do not cycle?
36. How many of the people surveyed jog and cycle but do not swim?
37. How many of the people surveyed do not participate in any of the activities mentioned?
38. All customers returning Christmas gifts at Rochelle's Department Store must complete the questionnaire shown below:

Why are you returning the gift?	
Wrong size	☐
Wrong color	☐
Already have it	☐

A clerk tallied the results and submitted the following report on 100 returned items. (It was known that each customer had checked at least one box.)

30   already have it,
23   wrong size,
50   wrong color,
20   wrong size and color,
8   already have it and wrong color,
5   already have it, wrong size, and wrong color.

The clerk was fired. Why?

39. Refer to the newspaper clipping at the beginning of this chapter. By analyzing the data, determine whether or not the governor's office is justified in refusing to accept the report.

40. Anita, Mary, and Heather are analyzing this Sunday's baseball schedule. The following teams are scheduled to play baseball games: Mets, Dodgers, Pirates, Cubs, Phillies, Giants, Astros, and Cardinals. Anita predicts that the winners will be the Mets, Dodgers, Pirates, and Phillies. Mary believes that the winners will be the Cubs, Pirates, Giants, and Mets. Heather believes that the winners will be the Phillies, Mets, Cardinals, and Giants. No one believes that the Astros will win. By means of Venn diagrams, determine which teams are scheduled to play each other.

A small town has a population of 3600 and only one movie house. During the week of December 1–7 the film *Star Trek* was shown, and 2800 people from the town saw it. The following week the film *E.T.* was shown and 1420 people saw it.

**41. Find the largest number of people who could have seen both movies.

**42. Find the smallest number of people who could have seen both movies.

A committee consists of six members Bob, Sue, Jo, Ann, Tom, and Ron, each with one vote. Four votes are needed to carry any proposal.

*43. Find all the winning coalitions.

*44. Find all the losing coalitions.

*45. Find all the minimal winning coalitions.

*46. Are there any blocking coalitions? If so, find them.

Refer to Exercises 43–46. Assume that, if there is a tie, then Ann has an additional vote to break the tie.

*47. Find all the winning coalitions.

*48. Find all the losing coalitions.

*49. Find all the minimal winning coalitions.

*50. Are there any blocking coalitions? If so, find them.

*51. Can a voting power ever have 2 dictators?

*52. If a voting power has a dictator, can it also have a dummy?

The Board of Directors of the Carle Corporation has eight members. The votes are distributed as follows:

Thomas Meringolo, Chairperson	4 votes
Lisa Palm, President	4 votes
Gary Yhap, Treasurer	4 votes
Judith Ramos	2 votes
Felix Sosa	2 votes
Kimberly Juliano	2 votes
Patrick Butler	2 votes
Robert Zimmer	2 votes

A simple majority carries an issue.

*53. Find six minimal winning coalitions.

*54. Find six blocking coalitions.

*55. The United Nations Security Council has fifteen member nations: the "Big Five" and ten smaller members. Each member of the Big Five has veto power. If nine votes are needed to pass any proposal, find the minimal winning, losing, and blocking coalitions.

*Exercises preceded by * correspond to material found in an optional section.

We have already seen many illustrations of infinite sets. For example,

$$A = \{1, 2, 3, 4, \ldots\}, \quad B = \{2, 4, 6, 8, \ldots\}, \quad C = \{1, 3, 5, 7, \ldots\}$$

are infinite sets. However, we have not yet said precisely what we mean by an infinite set. If you ask people what an infinite set is, you will get such answers as "it goes on forever," "it can't be counted," or "it has no end." In this section we will try to find a more precise and mathematical description of infinite sets. We will also see that infinite sets behave somewhat differently from finite sets. As we stated at the beginning of this chapter, many of their properties are very surprising.

We begin by going back to prehistoric days, where we imagine a caveman who keeps five sheep. Each morning he lets them out of the cave to graze. To keep track of them, he puts a stone in a pile, one for each sheep, as they leave. When they are all gone, he has a pile of stones like that shown in Fig. 1.26.

Figure 1.26

In the evening, when the sheep return, he takes one stone off the pile as each sheep comes back. Afterwards, if he has any stones left, he knows that some of his sheep are missing. On the other hand, if he has no more stones left and another sheep comes in, he knows that a strange sheep has wandered in. Thus he is **counting** his sheep by using stones.

Let us see what is involved in the caveman's counting system. It is really based on a **one-to-one correspondence** between the sheep and the stones. This idea is the basis of all counting.

To see how this works, suppose we have a football stadium with 54,327 seats. For a certain game the stadium is sold out, and there are no standees. (Only one person is allowed to sit in each seat.) If we want to know how many spectators are present, we do not have to count them. We know that there are 54,327 people present, because we have a one-to-one correspondence between the seats and the spectators. This one-to-one correspondence enables us to **count** the number of spectators.

Now count the bugs shown in Fig. 1.27.

Of course you will get seven by starting at the top left, and counting 1, 2, 3, 4, 5, 6, 7. You too are using a one-to-one correspondence, but you are not doing it with stones or seats. Instead, you are using an abstraction of these, namely, numbers.

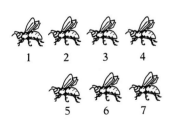

Figure 1.27

Now let us go back to the caveman and his five sheep. He has been having trouble. His children play tricks on him by taking stones off the pile when he is not looking or by adding stones to the pile. To prevent this, he decides to stop using stones; instead, he makes tally marks, one for each sheep, on the wall (up high, where the children can't reach them). Now he counts his five sheep by means of these tally marks:

The 1–1 correspondence is now between the sheep and the tally marks, but the idea is the same as before.

Gradually, the caveman realizes that he can replace the five tally marks by a single symbol such as V, or 5. (We will have more to say

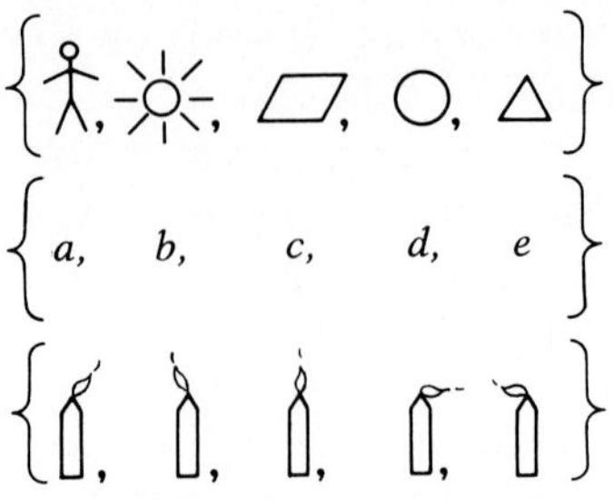

Figure 1.28

about this in Chapter 3.) These symbols stand for something that the five sheep, the five stones, and the five tally marks all have in common, that is, their "fiveness." Any other set that has five elements in it, such as the ones shown in Fig. 1.28, share this common property of "fiveness." What this means is that all sets with five elements can be put into 1–1 correspondence.

EXAMPLE 1   Let $A = \{1, 2, \ldots, 100\}$ and $B = \{101, 102, \ldots, 200\}$. Then A and B each have 100 elements, and we can put them in 1–1 correspondence as follows:

$$A = \{ \; 1 \;, \; 2 \;, \; 3 \;, \ldots, 100\}$$
$$\updownarrow \quad \updownarrow \quad \updownarrow \qquad \updownarrow$$
$$B = \{101, 102, 103, \ldots, 200\}. \quad \square$$

EXAMPLE 2   Let $C = \{1, 2, 3, \ldots, 100\}$ and $D = \{2, 4, 6, \ldots, 200\}$. Then C and D each have 100 elements and can be put in 1–1 correspondence as shown:

$$C = \{1, 2, 3, \ldots, 100\}$$
$$\updownarrow \updownarrow \updownarrow \qquad \updownarrow$$
$$D = \{2, 4, 6, \ldots, 200\}.$$

It may not be quite so obvious here as it was in Example 1 that each element of set C can be matched with a unique element of set D and vice versa. However, we can make it clear by giving a rule for the matching. The rule is to match each number in set C with twice that number in set D. Thus we have

      1 in set C matched with 2 times 1, or 2, in set D;

      2 in set C matched with 2 times 2, or 4, in set D;

      3 in set C matched with 2 times 3, or 6, in set D;

$$\cdot \qquad \cdot \qquad \cdot \qquad \cdot$$
$$\cdot \qquad \cdot \qquad \cdot \qquad \cdot$$
$$\cdot \qquad \cdot \qquad \cdot \qquad \cdot$$

      100 in set C matched with 2 times 100, or 200, in set D.

In other words, we can say that if n is an element of set C, we can match it with 2 times n (which we write as $2n$) in set D as shown:

$$C = \{1, 2, 3, \ldots, n, \ldots, 100\}$$
$$\updownarrow \updownarrow \updownarrow \qquad \searrow \qquad \searrow$$
$$D = \{2, 4, 6, \ldots, 2n, \ldots, 200\}. \quad \square$$

To further illustrate this, let us consider another example.

EXAMPLE 3 Find a 1–1 correspondence between the sets $O = \{1, 3, 5, 7, \ldots\}$, or all odd numbers, and $E = \{2, 4, 6, 8, \ldots\}$, or all even numbers.

SOLUTION We cannot show the correspondence by actually matching each element of O with an element of E; the sets are infinite, and we could never finish. Therefore to show that such a 1–1 correspondence exists, we must find a rule for matching all the elements of O with unique elements of E and vice versa. It makes sense to match as follows:

$$O = \{1, 3, 5, 7, \ldots\}$$
$$\updownarrow \updownarrow \updownarrow \updownarrow$$
$$E = \{2, 4, 6, 8, \ldots\}.$$

Looking at the first few matchings above, we see that a general rule would be "a number in O is matched to the same number plus 1 in E." To put it in symbols, each number n in O is matched to the number $n + 1$ in E. So the complete 1–1 correspondence is

$$O = \{1, 3, 5, 7, \ldots, n, \ldots\}$$
$$\updownarrow \updownarrow \updownarrow \updownarrow \qquad \searrow$$
$$E = \{2, 4, 6, 8, \ldots, n + 1 \ldots\}.$$

This means that even though O and E are infinite sets, they still have the same number of elements. Remember that sets that can be put in 1–1 correspondence have the same cardinal number (that is, they have the same number of elements). It may seem strange to speak of "the number of elements" in an infinite set, but there is no good reason why we cannot. We call the number of elements in the sets O and E by the name $\aleph_0$ (pronounced aleph-null). That is $\aleph_0$ is the cardinal number of these sets. $\square$

EXAMPLE 4 Put the sets $A = \{1, 2, 3, 4, \ldots\}$ and $B = \{2, 4, 6, 8, \ldots\}$ in 1–1 correspondence.

SOLUTION You may say immediately that this cannot be done. For if A and B can be put in 1–1 correspondence, they each have the same number of elements. But it seems clear that A has more elements than B. After all, A contains everything in B and, also, all the odd numbers 1, 3, 5, etc. We see that B is a proper subset of A.

But let us ignore common sense for a while and try to put these two sets in 1–1 correspondence anyway. To do so, we must find a "matching rule." Let us try this: Match every number, n, in A with twice n (written as $2n$) in B. This gives us

$$A = \{1,\quad 2,\quad 3,\ldots,\quad n,\ldots\}$$
$$\updownarrow \ \updownarrow \ \updownarrow \qquad\quad \updownarrow$$
$$B = \{2,\quad 4,\quad 6,\ldots,\quad 2n,\ldots\}.$$

This rule matches each element in A with a unique element in B and vice versa. No element of either set has been left out. So we do have a 1–1 correspondence. This is startling because, as we have said, it means that *A and B have the same number of elements, even though B is a proper subset of A.* How can part of A have the same number of elements as all of A? Perhaps you think that we have played a mathematical trick on you. Of course, we haven't. What is happening here is that we are dealing with *infinite sets*. It is not possible for a *finite* set to have the same number of elements as one of its proper subsets. However, for infinite sets, it is quite possible. In fact, this strange property is exactly the difference between finite and infinite sets. We can therefore use it to give a precise definition of infinite sets.   □

Definition 1.27   *A set is **infinite** if it can be put in 1–1 correspondence with a proper subset of itself.*

EXAMPLE 5   Prove that the set $A = \{10, 20, 30, \ldots\}$ is infinite.

SOLUTION   Let $B = \{20, 30, 40, 50, \ldots\}$. B is a proper subset of A. Note that A is

$$\left\{\begin{matrix}10 \text{ times } 1, & 10 \text{ times } 2, & 10 \text{ times } 3, \ldots \\ \text{or } 10 & \text{or } 20 & \text{or } 30\end{matrix}\right\}.$$

We can put A and B in 1–1 correspondence as follows:

$$A = \{10, 20, 30, \ldots, 10n, \ldots\}$$
$$\updownarrow \ \updownarrow \ \updownarrow \qquad\quad \searrow$$
$$B = \{20, 30, 40, \ldots, 10n + 10, \ldots\}.$$

So A can be put in 1–1 correspondence with a proper subset of itself. Therefore A is infinite.   □

Comment   Remember that in a 1–1 correspondence, each element of the first set must be matched with a *unique* element of the second set and vice versa. If any elements of either set are left out of the matching, it is *not* a 1–1 correspondence.

EXAMPLE 6   Let AB be part of a line as shown in Fig. 1.29. Several points on this line are shown: P, Q, R, S. Show that the set of all points on this line is infinite.

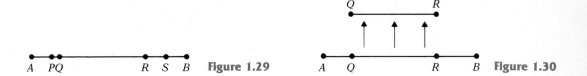

A PQ   R S B   **Figure 1.29**   **Figure 1.30**

SOLUTION   We take a part of this line from Q to R and redraw it above the line AB, as shown in Fig. 1.30.

Now erase the letters P, Q, R, S from the *original* line AB. Join A to Q and B to R as shown, and extend the joining lines so that they meet in point E, forming a triangle as shown in Fig. 1.31.

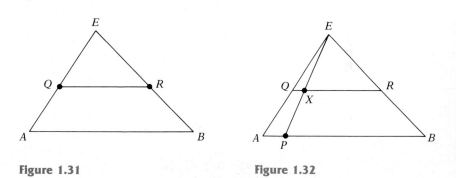

Figure 1.31   **Figure 1.32**

Let P be any point on AB. We match it with a point on QR by joining E and P as in Fig. 1.32. The line from E to P meets the line QR in X. So X is the point on QR that matches P.

Every point on line AB can be matched in a similar way to a *unique* point on QR and vice versa. Some of these matchings are shown in Fig. 1.33.

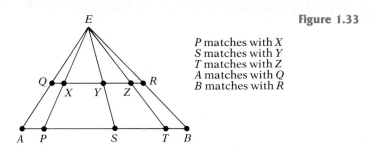

Figure 1.33

P matches with X
S matches with Y
T matches with Z
A matches with Q
B matches with R

Therefore the set of points on line AB is in 1–1 correspondence with the set of points on line QR (which is a proper subset of line AB). So the set of points on line AB is infinite. ☐

We have now seen many examples of infinite sets. The sets in Examples 3, 4, and 5, namely,

$$\{1, 3, 5, 7, \ldots\}, \quad \{1, 2, 3, 4, \ldots\},$$
$$\{2, 4, 6, 8, \ldots\}, \quad \{10, 20, 30, 40, \ldots\},$$

all had cardinal number $\aleph_0$. What about the set in Example 6? Does it also have cardinal number $\aleph_0$? Since it is also infinite, you might think that the answer must be yes. However, it is not. This fact is known as the *Second Diagonal Proof*, and it is quite remarkable. The cardinal number of the set in Example 6 (that is, the set of points on line AB) is called $\aleph_1$ (pronounced aleph-one).

You will remember that in Example 6 we showed that the set of points on line QR was in 1–1 correspondence with the set of points on line AB. So the set of points on line QR also has cardinal number $\aleph_1$. It can be shown that *any* line, even one from your house to the moon, also has exactly $\aleph_1$ points on it!

$\aleph_0$ and $\aleph_1$ are called *transfinite cardinal numbers*. We might now ask if there are any other transfinite cardinal numbers. The answer is yes. In fact, there are infinitely many, which we call $\aleph_2, \aleph_3, \aleph_4$, etc. It has been shown that $\aleph_0$ is the smallest of these "infinite" numbers.

Transfinite cardinal numbers can be added and multiplied just as finite numbers are, but they cannot be subtracted or divided. Moreover, the arithmetic of transfinite numbers has strange results—for example,

$$\aleph_0 + \aleph_0 = \aleph_0,$$
$$\aleph_1 + \aleph_0 = \aleph_1,$$
$$\aleph_1 + 7 = \aleph_1,$$
$$\aleph_0 \text{ times } 2 = \aleph_0,$$
$$\aleph_1 \text{ times } \aleph_1 = \aleph_1,$$
$$\aleph_0 \text{ times } \aleph_1 = \aleph_1.$$

EXERCISES

Put each of the pairs of sets given in Exercises 1–5 into 1–1 correspondence. Give a "matching rule" in each case.

*1. $\{3, 6, 9, 12, \ldots\}$ and $\{6, 12, 18, 24, \ldots\}$

*2. $\{7, 14, 21, 28, \ldots\}$ and $\{8, 16, 24, 32, \ldots\}$

*3. $\{1, 8, 27, 64, \ldots\}$ and $\{1, 1/8, 1/27, 1/64, \ldots\}$

*4. $\{1, 8, 27, 64, \ldots\}$ and $\{1, 16, 81, 256, \ldots\}$

*5. $\{4, 8, 12, 16, \ldots\}$ and $\{4, 16, 64, 256, \ldots\}$

*6. Figure 1.34 shows a circle within a square. Set up a 1–1 correspondence between the points on the circle and the points on the square.

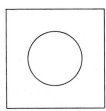

Figure 1.34

*7. For each of the diagrams in Figures 1.35 and 1.36, set up a 1–1 correspondence between the points on the inner figure and the points on the outer figure.

Figure 1.35   Figure 1.36

*8. For each of the diagrams shown in Figures 1.37 and 1.38 set up a 1–1 correspondence between the set of points on line *AB* and the set of points on line *CB*.

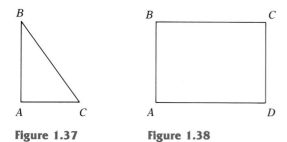

Figure 1.37   Figure 1.38

For Exercises 9–16, use the following information: In the "new math" approach to teaching arithmetic, children are taught to add by the following method. Suppose you wish to add 3 + 4. First you take a set *C* with three elements, say {*a*, *b*, *c*}. Then you take a set *D* with four elements, say {*m*, *n*, *o*, *p*}. It is essential that sets *C* and *D* be disjoint, that is, have no elements in common. Now we form $C \cup D$, which is {*a*, *b*, *c*, *m*, *n*, *o*, *p*}. Then 3 + 4 is called the cardinal number of $C \cup D$, which is 7.

*9. In this scheme, why is it important that $C \cap D = \varnothing$?

*10. Use this technique to compute 5 + 3.

*11. Use this technique to compute 8 + 7.

*12. Use this technique to compute 6 + 0.

*13. Use this technique to compute $\aleph_0 + 8$.

*14. Use this technique to compute $\aleph_0 + 3$.

*15. Use this technique to compute $\aleph_0 + \aleph_0$.

*16. Use this technique to compute $\aleph_1 + \aleph_1$.

*17. *Cantor's hotel.* A famous mathematician arrives at a hotel and asks for a room. The hotel is full, but the hotel manager, who would like to please the mathematician, is determined to find a room for her. The manager does it by moving the guest who is in room 1 to room 2, the guest who is in room 2 to room 3, and so on. In this manner, the manager is able to provide a room for the mathematician and for each of the original guests. Doubling up is not allowed. How many rooms does Cantor's hotel have?

1.7

SUMMARY   In this chapter we introduced the theory in mathematics dealing with sets. Sets play an important role in mathematics today as evidenced by the fact that many of the concepts of sets are being taught to elementary school children. Thus it is important that you have a thorough knowledge of sets, the notation used for them, the operations that can be performed with them, their pictorial representation, and finally their varied applications.

After thoroughly analyzing the different kinds of sets, the notations used for them, and the operations that can be performed with them, we applied the ideas to voting coalitions. By using Venn diagrams we were also able to apply sets to different kinds of survey problems.

Finally, we discussed infinite sets and how they can be analyzed and applied.

You should now be able to demonstrate your knowledge of the following ideas by giving definitions, descriptions, or specific examples. Page references are given in parentheses.

Basic Ideas

Set (p. 3)
Elements or members of a set (p. 4)
Set notation (p. 4)
Roster method (p. 5)
Descriptive method (p. 5)
Set-builder notation (p. 5)
Well-defined set (p. 5)
1–1 correspondence (p. 7)
Cardinal number (p. 8)

Ordered pair (p. 16)
Venn diagram (p. 20)
Voting coalition (p. 29)
Winning coalition (p. 29)
Losing coalition (p. 29)
Blocking coalition (p. 30)
Veto power (p. 30)
Dictator (p. 30)
Dummy (p. 30)

Kinds of Sets and Relations Between Sets

Unit (p. 6)
Null (p. 6)
Finite (p. 6)
Infinite (p. 6)
Equal (p. 6)
Equivalent (p. 8)

Subset (p. 10)
Proper subset (p. 11)
Improper subset (p. 11)
Universal (p. 12)
Disjoint (p. 13)

Set Operations

Union (p. 12)
Intersection (p. 13)
Complement (p. 14)

Difference (p. 15)
Cross product (p. 16)
Cartesian product (p. 16)

Applications of the Ideas of Sets Were Given for:

a) cardinal numbers (p. 8);

b) survey problems (p. 26);

c) voting coalitions (p. 29).

FORMULAS TO REMEMBER　When analyzing survey problems by means of Venn diagrams, be sure to first use the data about all three categories.

MASTERY TESTS

Form A

1. How many possible subsets does the set $\{a, b, c, d, e\}$ have?

   a) 5　　b) 10　　c) 16　　d) 32　　e) none of these?

2. The diagram at the right illustrates

   a) $A - B$　　b) $B - A$

   c) $A \cap B$　　d) $U \cap A$

   e) none of these

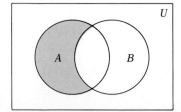

3. The diagram at the right illustrates

   a) $A \cup (B \cap C)$

   b) $(A \cap B) \cup (A \cap C)$

   c) $A \cap B' \cap C'$

   d) $(A \cup B) \cap (A \cup C)$

   e) none of these

4. The cardinal number of the set {1, 2, 4, 5, 6} is

   a) 4   b) 5   c) 6   d) 3   e) none of these

5. In a survey of 200 people the following information was obtained: 70 drank Coke, 55 drank Seven-Up, 60 drank Pepsi, 40 drank Coke and Pepsi, 35 drank Seven-Up and Pepsi, 50 drank Seven-Up and Coke, and 30 drank all three. How many drank only Coke?

   a) 10   b) 70   c) 90   d) 110   e) none of these

6. Refer to Exercise 5. How many drank none of the three drinks mentioned?

   a) 90   b) 110   c) 30   d) 80   e) none of these

For questions 7–11, let $U = \{1, 2, 3, 4, 5, 6, 7, 8, 9, 10\}$, $A = \{2, 4, 6, 8\}$, $B = \{2, 3, 6, 9\}$, $C = \{4, 2, 6, 8\}$, and $D = \{3, 9\}$.

7. $A \cup B =$

   a) {2, 6}   b) {2, 4, 6, 8}   c) $\varnothing$

   d) {2, 3, 4, 6, 8, 9}   e) none of these

8. $A \cap B =$

   a) $\varnothing$ b) U c) {2, 3, 4, 6, 8, 9}   d) {2, 6}   e) none of these

9. $A' =$

   a) B b) $\varnothing$ c) C d) $B \cap C'$ e) none of these

10. $(A \cup B)' =$

   a) $\varnothing$ b) U c) A d) $A' \cap B'$ e) none of these

11. $n(D) =$

   a) 10   b) 0   c) 4   d) 2   e) none of these

12. Which of the following sets is not well-defined?

   a) All females over 67 feet tall

   b) All disagreeable children

   c) All math books that cost under one dollar

   d) All curable cancers   e) none of these

13. Which of the following statements is false?

   a) Every set is a subset of itself

   b) Equal sets are equivalent sets

   c) All equivalent sets are equal

   d) If $A \cap B = \varnothing$, then A and B are disjoint

   e) none of these

14. Ellen, Jennifer, Jason, and Dominique form a committee. Ellen has 4 votes, Jennifer has 3 votes, Jason has 2 votes, and Dominique has 1 vote. Jennifer also has veto power. Then {Ellen, Jason, Dominique} forms

    a) a winning coalition   **b)** a blocking coalition

    c) a minimal winning coalition   **d)** a losing coalition

    e) none of these

15. In the committee given in Exercise 14, {Jennifer, Dominique} forms

    a) a winning coalition   **b)** a blocking coalition

    c) a minimal winning coalition   **d)** a losing coalition

    e) none of these

Form B

1. Set A is equivalent to set B if and only if

    a) they have exactly the same elements   **b)** they have the same number of elements   **c)** A is a subset of B **d)** B is a proper subset of A **e)** none of these

2. The set equal to $\varnothing$ is

    a) {0}   **b)** 0   **c)** { }   **d)** {$\varnothing$}   **e)** none of these

3. A is a subset of B if

    a) every element in A is an element in B

    b) every element in B is an element in A

    c) there are elements in A and B that are the same

    d) there are no elements in A and B that are the same

    e) none of these

4. If $A = \{1, 2, 3\}$ and $B = \{a, b\}$, then $A \times B$ is

    a) $\{(1, a), (1, b), (2, a), (2, b), (3, a), (3, b)\}$

    b) $\{(a, 1), (a, 2), (a, 3), (b, 1), (b, 2), (b, 3)\}$

    c) $B \times A$ **d)** $\{1, 2, 3, a, b\}$ **e)** none of these

5. Let B be any set. Then $B \cap \varnothing =$

    a) U **b)** B **c)** $\varnothing$ **d)** B' **e)** none of these

6. A survey of the residents living along Park Avenue revealed that 200 of them owned a home computer, 400 of them owned an exercise bike, and 70 of them owned both. How many residents own at least one of the items mentioned?

Use the following diagram to answer questions 7–11.

7. $n(A \cup B) =$

    a) 29   **b)** 14   **c)** 25

    d) 8   **e)** none of these

8. Find $n(A \cap B')$

    a) 6   **b)** 18   **c)** 10

    d) 8   **e)** none of these

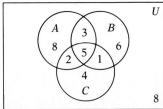

9. Find $n(A \cap C)$

    a) 7   **b)** 5   **c)** 21   **d)** 8   **e)** none of these

10. Find $n(A \cup B \cup C)$

    a) 24   **b)** 37   **c)** 29   **d)** 5   **e)** none of these

11. Find $n(A \cap B' \cap C')$

    a) 3   **b)** 8   **c)** 5   **d)** 10   **e)** none of these

12. If $n(C) = 5$, $n(D) = 4$, and $n(C \cap D) = 2$, find $n(C \cup D)$.

13. Using De Morgan's laws, simplify the expression $(A \cup B')'$.

14. Five roads lead from Springfield to Madison, and three roads lead from Madison to Endicott. In how many different ways can a person travel from Springfield to Endicott by way of Madison?

15. The business department of a small midwestern college is analyzing the status of its majors. It discovers that 54 of the majors are juniors, 162 are sophomores, and 162 are freshmen. No seniors are currently majoring in business. Furthermore, 171 females are majoring in business, and twice as many sophomore females as sophomore males are business majors. For the juniors the ratio is reversed. On the basis of this information, how many female freshmen are business majors?

2

LOGIC

CHAPTER OBJECTIVES

To analyze the difference between inductive and deductive reasoning. (*Section 2.2*)

To discuss the nature and patterns of inductive logic. (*Section 2.3*)

To study magic squares, which are arrangements of numbers in the shape of a square with special properties. (*Section 2.4*)

To see how arguments are tested using Venn diagrams. (*Section 2.5*)

To use the ideas of logic to test the validity of arguments by using a standard diagram. (*Section 2.6*)

To identify a compound statement as being either a negation, conditional, disjunction, or conjunction. (*Section 2.7*)

To distinguish between the various forms of the conditional. These include the converse, inverse, and contrapositive. (*Section 2.7*)

To determine when statements are tautologies (*always true*) or self-contradictions (*never true*) by using truth tables. (*Section 2.8*)

To apply the concepts of logic to arguments and switching circuits. (*Sections 2.9 and 2.10*)

To write a symbolic statement for a switching circuit; to construct a switching circuit from a given symbolic statement. (*Section 2.10*)

To indicate the relationship between sets and logic. (*Section 2.11*)

47

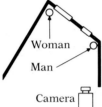

Is it logical to think that the room is rectangular? (The Exploratorium, San Francisco.)

Logic may be used to analyze this circuit board.

The metal figure shown here is made up of unrelated components (e.g., an auto for its head). Is the whole greater than the sum of its parts? Does the unrelatedness affect the logic of the composition? (Picasso, Pablo. *Baboon and Young.* 1951. Bronze [cast 1955], after found objects, 21″ high, base $13\frac{1}{4} \times 6\frac{7}{8}$″. Collection, The Museum of Modern Art, New York. Mrs. Simon Guggenheim Fund.)

If we were to analyze many of the arguments that exist in print or many of the speeches that are made by politicians, we would find that a good number of them are inconsistent. How do we determine whether something is logical or not? Would you consider the situations shown in the three photographs on page 48 to be logical?

In this chapter we will analyze some of the tools that are needed to determine whether something is logical or not.

2.1
INTRODUCTION

People's ability to reason distinguishes them from lower animals. Logic is not just a tool of the mathematician or logician; it is used by each of us every day in almost every aspect of our lives. As soon as we wake up in the morning, we must decide what to wear. In making our decision we consider the weather, the season, the activities planned for the day, what we wore yesterday, and so on. What we ultimately decide to wear is determined by a logical process. As the day progresses, we continually make decisions that involve logical thinking. Psychologists tell us that even though our dreams seem to be disconnected, there is actually a logical thought process connecting them. In mathematics, logic is an especially important tool, as you probably found in your high school studies.

In this chapter we will discuss inductive and deductive logic and how it is related to sets. We will also present many logic problems, some of which can be analyzed by the techniques to be discussed in this chapter and others that can be analyzed by using your reasoning ability.

HISTORICAL NOTE

Historically, the study of logic can be traced back to the ancient Greeks, specifically to Aristotle (384–322 B.C.). He is generally considered to be the "father" of logic. The logic of Aristotle was based on a formal kind of argument, called a *syllogism*. For example:

1. All men are mortal.
2. Socrates is a man.

   Therefore Socrates is mortal.

It can be shown that the statement "Socrates is mortal" follows logically from the first two sentences. Much of Aristotle's logic was devoted to a detailed study of such syllogisms. For many centuries, Aristotle was considered to be the ultimate authority on logic. In fact, it has been said that further developments in the study of logic were delayed for centuries because of the unquestioning faith that logicians had placed in Aristotle's work.

2.2

**INDUCTIVE AND
DEDUCTIVE
REASONING**

To begin, let us consider Dr. Smith, who recently announced that he has developed a serum for a certain disease. Over a period of ten years, he has administered his serum in varying dosages to 6341 patients. All those receiving this medicine recovered shortly thereafter. He therefore claims that his serum is a cure for this dreadful disease. Do you believe his claim?

You would probably say yes. Let us analyze the claim more carefully. Suppose on the 6342nd patient the serum fails. However, on the next 5000 patients it works. Does this one failure constitute a disproof of Dr. Smith's claim? Not really. We would not expect his serum to work in every case. It should work in *almost* every case. This is how we interpret his claim. In other words, we expect that anyone who gets the serum will *probably* recover.

Now consider a traffic light at the intersection of Main Street and Broadway. A taxi pulls up just as the light turns red. On the basis of past experience, the driver knows that the light changes every 30 seconds; consequently, at the end of 30 seconds she begins to move through the intersection. Since, in the past, the light has changed every 30 seconds, the driver assumes that the light will again change after 30 seconds and that she can proceed safely. In most cases this conclusion is correct. However, occasionally the light may be broken and will not change as anticipated. Although the driver's decision is probably justified, there still is a possibility that the light will not change.

Both of the above examples are illustrations of inductive reasoning.

> **Definition 2.1**   *In **inductive reasoning** we arrive at conclusions on the basis of a number of observations of specific instances. The conclusion is probably, but not necessarily, true.*

Inductive reasoning is widely used in science. It is also the kind of reasoning we all use daily in making decisions. In inductive reasoning we assume that the present or future will resemble the past and we act or reason accordingly. The following examples illustrate these ideas.

EXAMPLE 1   Before leaving the house in the morning, Alice looks out the window. The skies are overcast. She has heard the weather forecaster predict rain. She decides to take her umbrella. In making this decision, Alice is reasoning inductively. It *probably* will rain and Alice will need her umbrella. There is also a slim chance that it will clear up.   ☐

EXAMPLE 2   Johnny is crying. His mother has just told him that they are going to the dentist. His past visits to the dentist were quite painful, so he concludes that the present visit will be painful. Although his fears are *probably* justified, he may find that the visit will turn out otherwise. ☐

EXAMPLE 3 In its March 1985 issue, *Consumer Reports* published a report on four compact cars. It concluded that the Toyota Camry "gets our top billing among compact models." *Consumer Reports* arrives at such conclusions after extensively road testing numerous cars. Their conclusion about the Toyota Camry was made inductively from accumulated data. The report indicates that buyers will *probably* be satisfied with the car, since it had responsive performance and would provide good fuel economy in the future as it has in the past. ☐

EXAMPLE 4 In 1866 the Austrian monk Gregor Mendel published a major work on the theory of heredity. In experiments with garden peas, Mendel noticed that certain characteristics appeared in peas according to a recognizable pattern. For example, when he crossbred green and yellow peas, he found that out of every four peas produced, approximately three were green and one was yellow. On the basis of these experiments, Mendel was able to state general "laws" of heredity, not only for plants but also for humans. These enabled him to predict such things as eye color and hair color. ☐

EXAMPLE 5 In recent years, doctors in the United States have been experimenting with the drug lithium in the treatment of mentally depressed people. Approximately 80 percent of all such patients treated with lithium have reported feeling better. As a result, doctors are concluding that lithium may be a remedy for the symptoms of chronic depression. Their reasoning is inductive. There is no guarantee that a patient who takes lithium will feel better, but it is probable that he or she will. ☐

EXAMPLE 6 Suppose you are given the following sequence of numbers: 1, 4, 7, 10, 13, ... What is the next number in this sequence? Your answer is 16. How did you arrive at this answer? Are you 100 percent sure that you are right? Is there any possibility that there may be a different answer?

Now consider the following argument:

1. All musicians have beards.
2. Jane is a musician.
3. Therefore Jane has a beard. ☐

In this example we are using a different kind of reasoning called *deductive reasoning*. If you agree that statements (1) and (2) are true, then you *must* agree that statement (3) is also true. The three statements together make up an *argument*. Statements (1) and (2) are called the *hypotheses* or *premises* of the argument. Statement (3) is called the *conclusion* of the argument. It is important to note that this argument does not say that the statements (1) and (2) *are* true. It just says that *if* they are true, then so is (3).

> **Definition 2.2** *If we are given a series of statements with the claim that one must follow from the others, then this is called a **deductive argument**.*

The statement that follows from the others is called the **conclusion.** The other statements are called the **hypotheses,** or **premises.**

> **Definition 2.3** *If the conclusion of a deductive argument does follow logically from the hypotheses, then we say that the argument is **valid.** If the conclusion does not necessarily follow logically from the hypotheses, then the argument is **invalid.***

Comment In inductive reasoning, the conclusion is never more than probably true. In deductive reasoning, if the hypotheses are accepted as true, then the conclusion is *inescapable*.

Let us consider the following examples.

EXAMPLE 7   *Hypotheses:* 1. All Brooklynites live in New York State.
                    2. All people who live in New York State pay high taxes.

                 Conclusion: All Brooklynites pay high taxes.   ☐

EXAMPLE 8   *Hypotheses:* 1. All cats are dogs.
                    2. All dogs meow.

                 Conclusion: All cats meow.   ☐

EXAMPLE 9   *Hypotheses:* 1. All college students are nuts.
                    2. All nuts talk to themselves.

                 Conclusion: All college students talk to themselves.   ☐

EXAMPLE 10   *Hypotheses:* 1. All men are women.
                    2. All women are mammals.

                 Conclusion: All men are mammals.   ☐

EXAMPLE 11   *Hypotheses:* 1. All worms are Texans.
                    2. All Texans are U.S. residents.

                 Conclusion: All worms are U.S. residents.   ☐

In Example 7, both the hypotheses and the conclusion are true. In Example 8, both of the hypotheses are false, but the conclusion is true. In Example 9, both the hypotheses and conclusion are false. In Example 10 the first hypothesis is false, but the second hypothesis is true. The conclusion is true. What can you say about the truth or falsity of the hypotheses and conclusion of Example 11? All of the above arguments are *valid*.

Comment   The above examples show that an argument may be valid even if one or more of the statements in it (that is, hypotheses or conclusion) is false. In everyday English we often use the words "true" and "valid" interchangeably. In logic they have different meanings. A **statement** is either true or false. An **argument** is either valid or invalid.

Summary   In inductive reasoning, the conclusion is never more than *probably true*. In deductive reasoning, if the hypotheses are accepted as true, then the conclusion is *inescapable* (if the argument is valid).

EXERCISES

In Exercises 1–4, determine whether the reasoning used is inductive or deductive.

1. Since the Arctic Ocean has had icebergs for thousands of years, it will continue to have them in the foreseeable future.

2. All cats are afraid of water. All things that are afraid of water never take a shower. Therefore all cats never take a shower.

3. Alan is trying to get his television set to work. He plugs it into every outlet in the house, and the television set still does not work. He concludes that there is an electrical blackout.

4. All students receiving financial aid from the state while attending college are residents of the state. Melissa is a resident of the state attending college. Therefore she is receiving financial aid.

5. Assume that you are offered two equally attractive jobs, each with a starting salary of $40,000. However, one job offers annual raises of $4000, while the other job promises a semiannual raise of $2000. Which job would you take and why?

6. What is the next picture in the following sequence?

In Exercises 7–15, what is the next number or letter in the indicated sequence?

7. 1, 4, 9, 16, . . .

8. 2, 4, 6, 4, 6, 8, 6, . . .

9. 0, 1, 1, 2, 3, 5, 8, 13, . . .

10. 1, 3, 1, 6, 1, 9, . . .

11. 2, 3, 5, 7, 11, 13, . . .

12. O, T, T, F, F, S, S, E, N, . . .

13. 2, 5, 10, 17, 26, 37, . . .

14. 8, 5, 4, 9, 1, 7, 6, . . .

15. 1/4, 2/5, 3/6, 4/7, 5/8, . . .

** 16. *The missing two dollars:* The Smith's are analyzing how they spent the $100 that was budgeted for food during the month of February. They spent the money as follows:

Week of	Amount spent for food	Balance
Feb. 1–7	$ 40	$ 60
Feb. 8–14	30	30
Feb. 15–21	18	12
Feb. 22–28	12	0
	$100	$102

Adding, we find that the Smith's spent $100 for food. Yet the total of the balances is $102. What happened to the extra two dollars?

2.3

THE NATURE AND PATTERNS OF INDUCTIVE LOGIC

Look at the triangle shown in Fig. 2.1. Side *AC* is 4 in. and side *BC* is 3 in. Now measure side *AB*. How many inches did you get? You should get 5 in. If you didn't, try again. Notice that

$$(3 \times 3) + (4 \times 4) = 5 \times 5.$$

In mathematics, 3×3 is abbreviated as 3^2. Similarly, 4×4 is abbreviated as 4^2. The same is true for 5×5, which is written as 5^2. Using

Figure 2.1

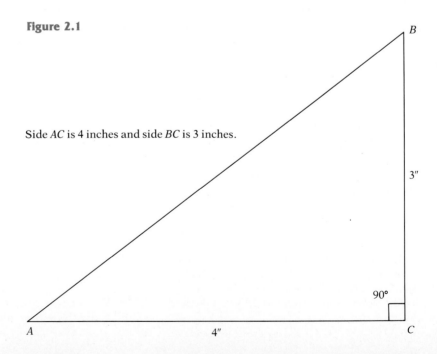

Side *AC* is 4 inches and side *BC* is 3 inches.

B

3"

90°

A   4"   C

this notation, we have

$$3^2 + 4^2 = 5^2.$$

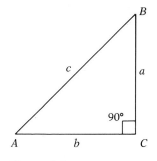

Figure 2.2

This is a special case of a well-known theorem in geometry known as the *Pythagorean theorem*. It states the following: *In a right triangle (that is, a triangle with a 90° angle) such as the one in Fig. 2.2, the sides are related by the formula $a^2 + b^2 = c^2$.*

The Pythagorean theorem is named for the Greek mathematician Pythagoras, who lived in the sixth century B.C. Actually, the theorem was known much earlier. According to some historians, the ancient Egyptians knew of at least one special case of the theorem. Tablets from the Babylonian era show that they also knew of at least one special case of this theorem. In fact, one historian claims that the Babylonians "did indeed make use of this theorem in its full generality."[1] In measuring land for tax assessment and in building, some of the basic ideas of geometry were developed. It seems reasonable to assume that in measuring areas that were in the form of right triangles, the Babylonians observed this important relationship between the sides.

The Pythagorean theorem was discovered inductively by observing special cases of right triangles. But inductive reasoning can tell us only that it is *probably* true. In order to *know* that it is *always* true, it must be *proved deductively*. Remember, in deductive reasoning, the conclusion *must* be true (if the hypotheses are true and the reasoning is valid). The proof of this theorem, which you may have studied in high school geometry, was given by Euclid about 200 years after Pythagoras. It is not known when it was *first* proved. The proof involves deductive reasoning.

We see, then, that both inductive and deductive reasoning have a place in mathematics, the former in *discovering* new truths and the latter in *proving* these truths.

EXAMPLE 1   *Goldbach's Conjecture.* A **prime number** is any whole number larger than 1 that can be evenly divided *only* by itself and 1 (assuming that we divide only by positive numbers). Some of the prime numbers are 2, 3, 5, 7, 11, 13, 17, 19, ... Note that

$$6 = 3 + 3 \qquad\qquad 16 = 11 + 5$$
$$8 = 5 + 3 \qquad\qquad 18 = 11 + 7 \text{ or } 13 + 5$$
$$10 = 7 + 3 \text{ or } 5 + 5 \qquad\qquad \cdot \qquad \cdot$$
$$12 = 5 + 7 \qquad\qquad\qquad \cdot \qquad \cdot$$
$$14 = 7 + 7 \text{ or } 11 + 3 \qquad\qquad \cdot \qquad \cdot$$

---

1. Asger Aaboe, *Episodes from the Early History of Mathematics*, p. 26. New York: Random House, 1964.

Observe that all the numbers on the right of the equals signs are prime numbers. All the numbers on the left are *even* numbers larger than four. The mathematician Goldbach, reasoning *inductively*, claimed that *any* even number larger than four can be written as the sum of two odd prime numbers. Although it seems reasonable, as yet no one has been able to prove this *deductively*. Do you agree with his claim? ☐

EXAMPLE 2   Consider the following arrangement of numbers.

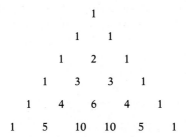

Can you find the next row of numbers? This arrangement of numbers is called **Pascal's triangle.** Although studied by Pascal (1623–1662), it was known to the Chinese much earlier. Among other things, this triangle has applications in the theory of probability, as we shall see later. ☐

EXAMPLE 3   Consider the following chart.

Numbers added	Sum	Another way of writing sum
$1 + 2$	3	$\dfrac{2 \times 3}{2}$
$1 + 2 + 3$	6	$\dfrac{3 \times 4}{2}$
$1 + 2 + 3 + 4$	10	$\dfrac{4 \times 5}{2}$
$1 + 2 + 3 + 4 + 5$	15	$\dfrac{5 \times 6}{2}$
$1 + 2 + 3 + 4 + 5 + 6$	21	$\dfrac{6 \times 7}{2}$
$1 + 2 + 3 + 4 + 5 + 6 + 7$	28	
$1 + 2 + 3 + 4 + 5 + 6 + 7 + 8$	36	
$1 + 2 + 3 + \cdots + 20$	210	
$1 + 2 + 3 + \cdots + n$		

Can you complete the third column? If you get the last entry in the last column right, you will have obtained a useful mathematical formula that gives the sum of the first n counting numbers, where n is any counting number. Your answer was obtained by observing the pattern in the last column. You reasoned inductively. This result can be proved deductively. □

EXAMPLE 4 *The Four-Color Problem.* For many years a famous problem of mathematics was concerned with map coloring. When we color maps drawn on flat surfaces (planes), such as sheets of paper, two countries having a common border must be colored with different colors. If two countries meet at only one point, they can have the same color. Some examples of maps are shown in Fig. 2.3. No one was ever able to draw a map, no matter how complicated, that required more than four colors. Therefore by inductive reasoning, it seemed probable that *every* map can be colored in *at most* four colors.

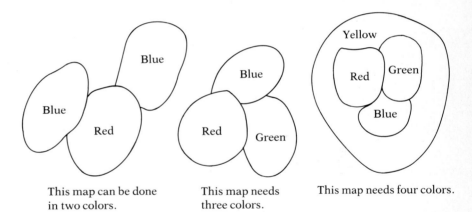

This map can be done in two colors.　This map needs three colors.　This map needs four colors.

Figure 2.3

In 1878, Arthur Cayley told the London Mathematical Society that he had tried to prove the theorem and could not. In 1889 an apparent proof was published. However, a hidden mistake was found the following year. In 1976, Kenneth Appel and Wolfgang Haken of the University of Illinois, using a combination of logic, ingenuity, and modern computer technology, succeeded in proving the theorem. The proof would have been impossible without a computer, since it would have been almost impossible to do a detailed graph-by-graph check using only human labor. As it was, Appel and Haken needed some 1200 hours of computer time to produce the proof.[2] □

---

2. For a detailed discussion of the proof, see *Science*, 13 August 1976, p. 564, and *Bulletin of the American Mathematical Society*, September 1976.

EXERCISES

1. Draw a map that has at least four countries on it and that can be colored in at most two colors.

2. Draw a map that has exactly five countries on it and that can be colored with at least four colors.

3. Draw a map that has at least six countries on it and that can be colored with at most two colors.

4. Draw a circle. If we mark off one point on the circle (as shown below), then there is only *one* region inside the circle. If we mark off two points on the circle and connect them with a straight line, then the line divides the circle into two regions (as shown below). If we mark off three points on the circle and connect them, then there are three lines and four regions. Complete the following table.

Number of points on the circle	Picture	Number of regions
1		1
2		2
3		4
4		8
5		?
6		?
7		?
8		?

For Exercises 5–8, use the following information: A circle separates a plane into two regions, the inside and outside of the circle. Two circles separate a plane into three regions if they do not intersect and four regions if they do.

5. What is the maximum number of regions into which three circles divide a plane?

6. What is the maximum number of regions into which four circles divide a plane?

7. What is the maximum number of regions into which five circles divide a plane?

8. Can you make a general statement relating the number of circles and the number of regions?

9. The Pythagorean theorem states that in a right triangle, labeled as shown in Fig. 2.2,

$$a^2 + b^2 = c^2.$$

One solution of this equation is $a = 3$, $b = 4$, and $c = 5$, since

$$3^2 + 4^2 = 5^2$$

or

$$9 + 16 = 25.$$

Another solution of this equation is $a = 5$, $b = 12$, and $c = 13$. (Verify this.) Now consider the equation

$$a^3 + b^3 = c^3.$$

Try to find values (natural numbers) for a, b, and c that are solutions to this equation. The seventeenth century mathematician Pierre Fermat claimed that no solution exists. However, no proof was ever found among his papers. More generally, he claimed that if the equation is of the form

$$x^n + y^n = z^n, \quad \text{(where } n \text{ is greater than 2),}$$

then no natural numbers can ever be found that would be solutions to this equation. Can you find any natural numbers that are solutions to this equation?

10. It has been suggested that the cube of every natural number larger than 1 can be written as the difference between the squares of two natural numbers. Thus

$$2^3 = 3^2 - 1^2,$$
$$3^3 = 6^2 - 3^2,$$
$$4^3 = 10^2 - 6^2.$$

Can you express 5^3 and 6^3 as the difference between the squares of two natural numbers?

11. Consider the following facts:

$$1^3 = 1,$$
$$2^3 = 3 + 5,$$
$$3^3 = 7 + 9 + 11,$$
$$4^3 = 13 + 15 + 17 + 19.$$

Adding the numbers on each side of the equation gives

$$1^3 + 2^3 + 3^3 + 4^3$$
$$= 1 + (3 + 5) + (7 + 9 + 11)$$
$$+ (13 + 15 + 17 + 19)$$
$$= 100$$
$$= 10^2.$$

Thus

$$1^3 + 2^3 + 3^3 + 4^3 = 10^2 = (1 + 2 + 3 + 4)^2.$$

Can you generalize the relationship demonstrated?

12. Explain how the diagram below "proves" the Pythagorean theorem for isosceles triangles. (An *isosceles triangle* is a triangle with two sides equal.) [*Hint*: It is as easy as 2 + 2.]

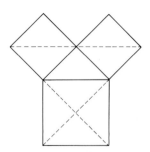

13. James Garfield, our twentieth president, discovered a new proof of the Pythagorean theorem while he was a congressman. His proof was based on the fact that the area of a whole object equals the sum of the areas of its parts and on the formula for the area of a trapezoid. Explain how his diagram (see Fig. 2.4) can be used to prove that $a^2 + b^2 = c^2$. [*Hint*: The area of a trapezoid (see Fig. 2.5) equals the average of the height times the width or Area equals $\frac{1}{2}(a + b) \cdot w$.]

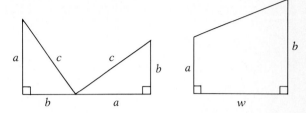

Figure 2.4   **Figure 2.5**

14. An interesting proof of the Pythagorean theorem was given by the Indian mathematician Bhaskara. Look at the triangle in Fig. 2.6 and draw squares on each of the three sides as shown in Fig. 2.7. The Pythagorean theorem says that (the area of square 1) = (the area of square 2) + (the area of square 3).

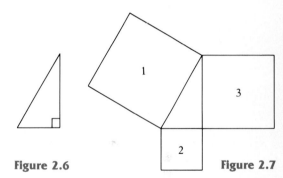

Figure 2.6   **Figure 2.7**

Unlike the proofs of the theorem that are usually given in high school geometry books, Bhaskara's proof consists only of cutting up square 1 and rearranging it as shown in Fig. 2.8. His only explanation was the word "Behold!"

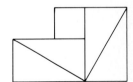

Figure 2.8

Can you explain how this diagram "proves" the Pythagorean theorem? (*Hint*: Copy the diagram on a piece of paper, cut it up, and rearrange the pieces.)

HISTORICAL NOTE

Bhaskara, a twelfth century Indian mathematician, was also an astrologer. There is a legend that he predicted that his daughter could marry only on a particular day at a given hour. On the wedding day the eager girl was bending over a water clock and a pearl from her headdress fell into the clock, stopping the flow of water. By the time the accident was discovered, the wedding hour had passed and the girl was doomed to spinsterhood. To console her, Bhaskara named his best-known mathematical work after her. It is called the *Lilavati*.

For Exercises 15–20 refer back to the Pascal triangle, which was given in Example 2 and is reproduced below. Notice that we have numbered the rows for easy reference. We first observe that each row begins with and ends with a 1. Also each row is found by adding the two previous numbers as shown by the arrows.

15. Write down the next two rows of the Pascal triangle shown on the left.

16. What is the sum of the numbers in row 1 of Pascal's triangle?

17. What is the sum of the numbers in row 2 of Pascal's triangle?

18. What is the sum of the numbers in row 3 of Pascal's triangle?

19. What is the sum of the numbers in row 4 of Pascal's triangle?

20. Can you generalize about the sum of the numbers in row n of Pascal's triangle?

Rows

```
0                      1
1                   1     1
2                1     2     1
3             1     3     3     1
4          1     4     6    4     1
5       1     5    10    10    5     1
```

We add the
6 and 4 to
obtain the 10

2.4

MAGIC SQUARES

A **magic square** is an arrangement of numbers in the shape of a square in which the sum of each vertical column, each horizontal row, and each diagonal is the same. The number of rows and the number of columns must be the same, and this number is known as the **order** of the magic square. Some examples of magic squares are shown below:

EXAMPLE 1

1

represents a magic square of order 1. ☐

EXAMPLE 2

4	9	2
3	5	7
8	1	6

represents a magic square of order 3. The sum of each row, each column, and along each diagonal is 15. ☐

EXAMPLE 3

4	14	15	1
9	7	6	12
5	11	10	8
16	2	3	13

represents a magic square of order 4. The sum of each row, each column, and along each diagonal is 34. ▢

Over the centuries, many mathematicians have been fascinated by the magic square problem. It is alleged that the first known example of a magic square was found on the back of a tortoise by the Chinese Emperor Yu around 2200 B.C. This form of Chinese magic is known as *lo-shu*. It appeared as an array of numerals indicated by knots in strings; black knots were used for even numbers, and white ones were used for odd numbers. In modern-day notation this represents a magic square whose order is 3. In this case the **magic sum** or the sum along any row, any column, or any diagonal is 15. See Figs. 2.9 and 2.10.

Figure 2.9

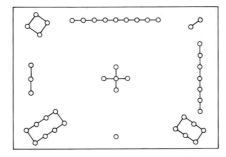

Figure 2.10 The *lo-shu*

One of the methods used by the ancient Chinese mathematicians to solve the magic square problem involved pairs of integers. The ancient Chinese mathematician Yang Hui (1275) solved the magic square problem by a balancing technique. He interchanged elements of different sets of numbers and reordered them. To best understand his procedure, we begin with a definition.

> **Definition 2.4**   *An interchange of a pair of numbers is called a **transposition**.*

To illustrate how the balancing technique was used by Yang Hui to solve the 3 × 3 magic square problem, let us arrange the integers $n = 1$ to $n = 9$ diagonally in their proper sequential order (from top to right) as shown in Fig. 2.11.

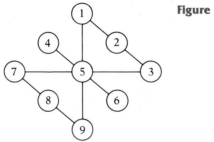

Figure 2.11

Now transpose the number in the top row and the number in the bottom row according to Definition 2.4; that is, interchange the 1 and the 9. This produces the scheme shown in Fig. 2.12.

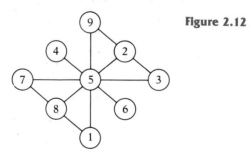

Figure 2.12

Then transpose the 7 on the left with the 3 on the right. The result of this transposition is given in Fig. 2.13.

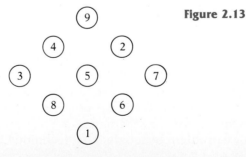

Figure 2.13

Finally, move the 2 and the 4 as well as the 8 and the 6 to corner positions. The result of all these moves, as given in Fig. 2.14, represents the solution to the 3 × 3 magic square problem.

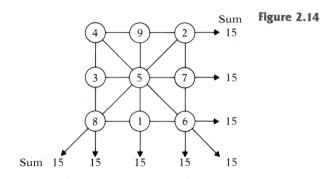

Figure 2.14

The sum of each row, each column, and along each of the main diagonals is 15. It should be noted that, although strictly speaking the movement of numbers to corner positions is not a transposition as defined in Definition 2.4, nevertheless this procedure was used by Yang Hui to solve the 3 × 3 magic square problem.

To solve the 4 × 4 magic square problem, Yang Hui arranged the integers $n = 1$ to $n = 16$ horizontally and vertically in reverse order if read from top to bottom and right to left as shown in Fig. 2.15.

This arrangement is obviously not a solution to the 4 × 4 magic square problem, since the horizontal, vertical, and diagonal sums are not the same. The solution to this magic square problem begins with the transposition of the diagonally opposite corner numbers; that is, interchange the 1 and the 16 as well as the 4 and the 13. The results are given in Fig. 2.16.

If we now interchange the 6 with the 11 and the 7 with the 10, we arrive at the solution to the 4 × 4 magic square problem as shown in Fig. 2.17.

Figure 2.15

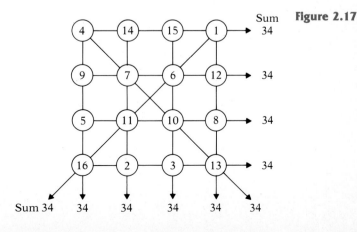

Figure 2.17

Figure 2.16

This process of rearranging a normal initial sequential ordering of numbers was also used by Yang Hui to solve the 5×5 and the 6×6 magic square problems.

Comment If we know one magic square, we can construct other magic squares by using rotations, reflections, and/or other techniques. However, our objective is merely to introduce you to magic squares. For further information the interested reader can consult the numerous articles that have been written about magic squares.

EXERCISES

In Exercises 1—7, find the missing numbers so that the resulting arrangement will form a magic square. Verify that you have a magic square by finding the sum of each row, each column, and along the diagonals.

1.

6	1	
7	5	3
2	9	4

2.

	3	8
9	5	1
2		6

3.

8	1	6
3		7
	9	2

4.

16	3	2	13
5	10		8
9			
	15	14	

5.

17		1	8	15
	5		14	16
4		13		22
10	12		21	
11	18			9

6.

39		23	30	37
	27		36	38
26		35		44
32	34		43	
33	40			31

7.

6	2	34	33	35	1
	11	27	10	8	
19		16	15	20	18
13					24
12		9	28	26	7
36		3	4	5	31

8. Find two magic squares of order 4. (Do not use any of those already given in the text.)

9. Make up a 7×7 magic square.

**10. In the Chinese magic squares problem the objective is to obtain a rearrangement of the n^2 integers from 1 to n in such a way that the sum of the numbers in all horizontal rows and vertical columns as well as both diagonals are the same. Find a formula for this sum.

11. Read the article "An Art-Full Application Using Magic Squares" by Margaret J. Kenney, which appeared in the January 1982 issue of *The Mathematics Teacher* (pp. 83–89) for an interesting discussion of the numerous designs and artistic patterns that can be made from a magic square.

12. Read the article "The Billiard Ball Problem Solved by the Ancient Chinese Concept of Pairing" by Joseph Newmark and I-Chen Chang, which appeared in the March–April 1986 issue of *International Journal of Mathematical Edu-cation in Science and Technology* (Vol. 17, No. 2, pp. 169–178) for an interesting discussion of how the ancient Chinese concept of pairing (and balancing), which was used by Yang Hui to solve the magic squares problem, can also be applied to the modern-day billiard ball problem.

2.5

USING VENN DIAGRAMS TO TEST VALIDITY

Many statements of the English language are of one of the following forms:

1. All *A* are *B*.
2. Some *A* are *B*.
3. Some *A* are not *B*.
4. No *A* are *B*.

For example:

"All songs are sad"	is of type (1).
"Some songs are sad"	is of type (2).
"Some songs are not sad"	is of type (3).
"No songs are sad"	is of type (4).

These statements can be represented by pictures known as *Venn diagrams*.

We draw two circles, one to represent songs and one to represent sad things. "All songs are sad" is pictured by one of the two possibilities shown in Figs. 2.18 and 2.19. In Fig. 2.18, since "all songs are sad," then anything that is a song must be sad; but there may be things that are sad that are not songs. In Fig. 2.19, anything that is a song is sad, and also anything that is sad is a song. *Both* of these are acceptable pictures for "all songs are sad." The given statement does not contain enough information to tell us which of these situations is the correct one. The two diagrams represent two different *interpretations* of the statement "all songs are sad." Both interpretations are in agreement with the given statement.

"Some songs are sad" is pictured by one of the possibilities shown in Figs. 2.20–2.23. Notice the "*x*" in each of the diagrams. This will be explained shortly.

Figure 2.18

Figure 2.19

Figure 2.20

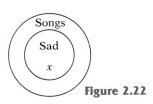

Figure 2.21

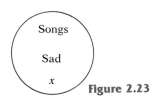

Figure 2.22

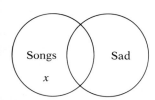

Figure 2.23

Figure 2.24

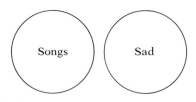

Figure 2.25   Figure 2.26   Figure 2.27

In Fig. 2.20 there are some things that are both songs and sad. There are also sad things that are not songs, and songs that are not sad things.

Although Fig. 2.21 implies that "*all* songs are sad," this is still a possible picture for "some songs are sad." In mathematics the word "some" does not have quite the same meaning that it does in everyday English. *In mathematics, "some" means "at least one and possibly all."* How does this differ from the everyday usage of the word?

In Fig. 2.22, some songs are sad. Moreover, there are no sad things that are not songs. This is another possible interpretation of "some songs are sad."

In Fig. 2.23, some songs are sad. As a matter of fact, *all* songs are sad, and all sad things are songs. This is yet another way of interpreting the statement "some songs are sad."

The statement "some songs are sad" means that there is at least one song that is sad. There may be many songs that are sad. Perhaps all songs are sad. All we can know from "some songs are sad" is that there is at least one song that is sad. Since we do not know which case is correct, we must allow for all the possibilities. The x in the diagrams represents the one song that we definitely know is sad.

"Some songs are not sad" is pictured by one of the possibilities shown in Figs. 2.24–2.26.

In Fig. 2.24 there is at least one song, as indicated by the x, that is not sad. How does this differ from Fig. 2.20, which represents "some songs are sad"?

In Fig. 2.25 there is also at least one song that is not sad. What else does this figure imply?

Figure 2.26 says that "no songs are sad." The statement "some songs are not sad" does allow for that possibility. The statement "some songs are not sad" does not tell us which of these three pictures is the right one. Therefore any one of the three is a possible interpretation of "some songs are not sad."

"No songs are sad" can be pictured only as shown in Fig. 2.27.

We will now show how to test the validity of arguments by means of these Venn diagrams. Consider the following example.

EXAMPLE 1   *Hypotheses:* 1. All cats are felines.
                            2. All felines are mammals.
          Conclusion: All cats are mammals.

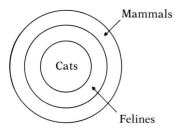

Figure 2.28

We draw a diagram (Fig. 2.28) to represent hypothesis (1). We place the circle of cats within the circle of felines. On the *same* diagram, we picture hypothesis (2). We place the circle of felines within a larger circle of mammals. Observe that in this diagram the circle of cats is within the circle of mammals. The conclusion "all cats are mammals" is then inescapable. We therefore say that this argument is valid.

Some other ways of drawing the diagrams for these hypotheses are given in Fig. 2.29. Can you find any others? In each of these cases the conclusion is inescapable, and therefore the argument is *valid*.

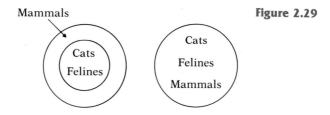

Figure 2.29

EXAMPLE 2   *Hypotheses:* 1. All college students are mermaids.
                            2. All mermaids are human beings.
          Conclusion: All college students are human beings.

We represent the hypotheses as shown in Fig. 2.30.

The conclusion "all college students are human beings" follows from the diagram, since the circle of college students is within the circle of human beings. Other diagrams are possible. (Draw as many as you can.) In each case the conclusion is inescapable. Hence the argument is *valid*.

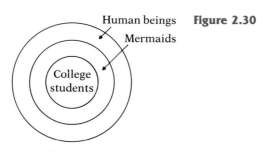

Figure 2.30

EXAMPLE 3

Hypotheses: 1. All gentle people are Republicans.
            2. Some dentists are gentle.

Conclusion: Some dentists are Republicans.

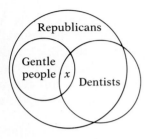

Figure 2.31

"All gentle people are Republicans" is pictured by placing the circle for "gentle people" within the circle for "Republicans." (See Fig. 2.31.)

Since "some dentists are gentle," the circle for dentists and the circle for gentle people must have some overlap. There must be at least one person who is in both circles. This person is designated by the *x* in the diagram. Therefore the conclusion "some dentists are Republicans" follows. There are many other possible diagrams. You should draw at least three of these. In each case the conclusion follows. Hence the argument is valid.   □

EXAMPLE 4

Hypotheses: 1. No children are nuisances.
            2. Some children are lovable.

Conclusion: No nuisances are lovable.

Two ways of picturing the hypotheses are shown (Figs. 2.32 and 2.33). In Fig. 2.32, since "some children are lovable," these two circles must overlap. We are also told that "no children are nuisances." Thus the circle of "children" and the circle of "nuisances" must be separate circles. From Fig. 2.32 the conclusion follows.

On the other hand, look at Fig. 2.33, which represents a different interpretation of the hypotheses. In this diagram it is possible for someone to be lovable and also a nuisance. The conclusion "no nuisances are lovable" does not follow.

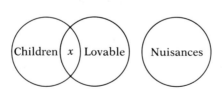

Figure 2.32

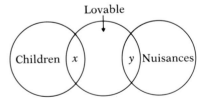

Figure 2.33

Other diagrams can be drawn. In some, the conclusion will follow. In others, it will not. In a *valid* deductive argument the conclusion *must* follow no matter how we interpret the hypotheses (that is, draw different diagrams). Therefore *in a valid argument the conclusion must follow from all possible pictures. If there is even one diagram that does not agree with the conclusion, then the argument is invalid.*

Getting back to our example, since we have at least one diagram (Fig. 2.33) that contradicts the conclusion, the argument is *invalid.*
□

EXAMPLE 5   *Hypotheses:* 1. Some women are angels.
                                 2. Some angels sing.

                 Conclusion: Some women sing.

One way of picturing the hypotheses is shown in Fig. 2.34. The diagram does not agree with the conclusion. We then know immediately that the argument is *invalid*.

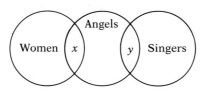

Figure 2.34

Note that there may be other diagrams for this argument. However, we do not need to draw them, since we have already found one diagram that does not agree with the conclusion. *We draw different diagrams until we find one that contradicts the conclusion. If we cannot find such a diagram, then the argument is valid.* ☐

EXAMPLE 6   *Hypotheses:* 1. Some men are clowns.

                                 2. No women are clowns.

                                 3. Some acrobats are women.

                 Conclusion: Some men are acrobats.

One way of picturing these hypotheses is shown in Fig. 2.35. From the diagram the conclusion "some men are acrobats" does not follow. Hence the argument is *invalid*. ☐

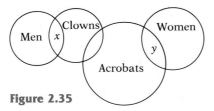

Figure 2.35

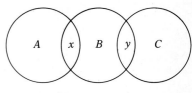

Figure 2.36

EXAMPLE 7   *Hypotheses:* 1. Some A's are B's.

                                 2. Some B's are C's.

                 Conclusion: Some A's are C's.

One way of picturing the hypotheses is shown in Fig. 2.36. The conclusion does not follow, so the argument is *invalid*. ☐

Compare Examples 5 and 7. They are really the same argument, except that in Example 5 we have used words, whereas in Example 7 we have used letters. Mathematicians and logicians often analyze arguments using letters instead of words. Perhaps this seems pointless to you. Much may be gained, however, by using this technique. In using words, your personal feelings may prejudice your decision as to the validity of the argument. No one can get emotionally involved (we hope!) with letters.

EXERCISES

In Exercises 1–15, use Venn diagrams to test the validity of the indicated arguments.

1. *Hypotheses:* (1) All drugs are hallucinatory.
    (2) Some hallucinatory things are dangerous to the user's health.

    Conclusion: All drugs are dangerous to the user's health.

2. *Hypotheses:* (1) All large cars are gas guzzlers.
    (2) Some gas guzzlers are expensive to maintain.

    Conclusion: All large cars are expensive to maintain.

3. *Hypotheses:* (1) Some judges are very lenient.
    (2) All lenient judges are often ridiculed.

    Conclusion: Some judges are often ridiculed.

4. *Hypotheses:* (1) All dogs are good swimmers.
    (2) All good swimmers like to go skin diving.

    Conclusion: All dogs like to go skin diving.

5. *Hypotheses:* (1) Some teenagers are athletic.
    (2) All teenagers like to drive a shiny new car.

    Conclusion: (a) Some athletic people like to drive a shiny new car.
    (b) Some people who drive a shiny new car are teenagers.

    (c) Some people who drive a shiny new car are athletic.
    (d) Some teenagers are not athletic.

6. *Hypotheses:* (1) Some doctors are reluctant to work at night.
    (2) No nurses are reluctant to work at night.

    Conclusion: (a) No nurses are doctors.
    (b) Some nurses are doctors.
    (c) Some doctors are not nurses.
    (d) Some doctors are not reluctant to work at night.

7. *Hypotheses:* (1) Some doctors pay high malpractice insurance premiums.
    (2) Some lawyers pay high malpractice insurance premiums.
    (3) Arlene is a doctor.

    Conclusion: (a) Arlene is a lawyer.
    (b) Arlene is not a lawyer.
    (c) Arlene pays high malpractice insurance premiums.

8. *Hypotheses:* (1) All 35-mm cameras sold in Bakersfield are imported from Japan.
    (2) Some 35-mm cameras sold in Bakersfield are automatic.
    (3) Nothing imported from Japan comes with a rechargeable battery.

Conclusion: (a) No 35-mm camera sold in Bakersfield comes with a rechargeable battery.

(b) Some 35-mm cameras sold in Bakersfield come with a rechargeable battery.

(c) Some things that come with a rechargeable battery are automatic.

(d) Some things that are automatic come with a rechargeable battery.

(e) Some things that are imported from Japan are automatic.

9. *Hypotheses:* (1) Some environmentalists are concerned about animal welfare.

(2) Some environmentalists are not very influential.

(3) Richard Kiley is an environmentalist.

Conclusion: (a) Richard Kiley is concerned about animal welfare.

(b) Richard Kiley is not concerned about animal welfare.

(c) Richard Kiley is very influential.

(d) Richard Kiley is not very influential.

(e) Some environmentalists who are concerned about animal welfare are not very influential.

10. *Hypotheses:* (1) All professional athletes are paid huge salaries.

(2) All professional athletes are strong.

(3) Some professional athletes are tall.

Conclusion: (a) Some tall people are strong.

(b) Some strong people are paid huge salaries.

(c) Some tall people are paid huge salaries.

(d) Some people who are paid huge salaries are strong.

(e) Some people who are not professional athletes are not tall.

(f) Some people who are not tall are not strong.

(g) Some people who are not tall are not paid huge salaries.

11. *Hypotheses:* (1) All latex paints are water-soluble.

(2) Some concrete floor paints are moisture-resistant.

(3) No water-soluble paints are concrete floor paints.

Conclusion: (a) Some moisture-resistant paints are latex paints.

(b) All water-soluble paints are latex paints.

(c) No concrete floor paints are water-soluble.

(d) No water-soluble paints are moisture-resistant.

(e) Some concrete floor paints are latex paints.

12. *Hypotheses:* (1) Some public beaches are overcrowded.

(2) No private swimming pools are overcrowded.

(3) Some lakes are used as private swimming pools.

Conclusion: (a) No public beaches are used as private swimming pools.

(b) Some public beaches are lakes.

(c) Some private swimming pools are not public beaches.

(d) Some lakes are overcrowded.

13. *Hypotheses:* (1) If Cathy buys Pete's used car, she will have to buy new tires.

(2) If Cathy has to buy new tires, she will also have to buy new wheels.

(3) If Cathy has to buy new wheels, she will have to get a part-time job at McDonald's to earn some money.

Conclusion: If Cathy buys Pete's used car, she will have to get a part-time job at McDonald's to earn some money.

14. *Hypotheses:* (a) All A's are B's.

(b) All C's are B's.

(c) All A's are D's.

Conclusion: (1) All C's are A's.

(2) Some B's are D's.

(3) Some C's are B's.

(4) No C's are B's.

(5) Some C's are not D's.

15. *Hypotheses:* (a) All A are B.

(b) All B are C.

(c) Some A are D.

Conclusion: (1) Some B are D.

(2) All A are C.

(3) Some C are D.

(4) No C are D.

(5) Some C are not D.

16. We mentioned that there are advantages to using letters rather than words in analyzing arguments. We gave one advantage. Can you give any others?

---

* 2.6

ANOTHER WAY OF TESTING VALIDITY— THE STANDARD DIAGRAM

There is another way of using Venn diagrams to test the validity of arguments. This is by using the **standard diagram.** The standard diagram shows all possibilities in one diagram. It has the advantage that when using it, you need only one diagram and no more. However, you may find the diagram a little more difficult to work with than other Venn diagrams.

We will illustrate the standard diagram by using it to test some of the examples of the previous section.

EXAMPLE 1

Hypotheses: 1. All cats are felines.

2. All felines are mammals.

Conclusion: All cats are mammals.

Let C stand for cats, F for felines, and M for mammals. We draw three intersecting circles to represent C, F, and M as in Fig. 2.37.

Hypothesis (1) tells us that all C's are inside the F circle. And there are *no C's outside the F* circle. We indicate this by putting a "$\varnothing$" (the null set) in *all* parts of the C that are *outside* the F circle. This is shown in Fig. 2.38.

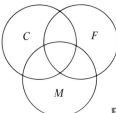

Figure 2.37

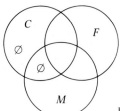

Figure 2.38

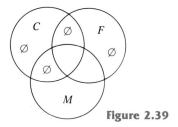

Figure 2.39

Hypothesis (2) tells us that there are no *F*'s outside the *M*'s. We represent this by putting ∅ in *all* parts of the *F*'s that are outside the *M* circle. We add this to the diagram, obtaining the result shown in Fig. 2.39.

The conclusion says that there are *no C*'s outside the *M* circle. This *does* follow from the final diagram (Fig. 2.39).   □

EXAMPLE 2   *Hypotheses:* 1. All college students are mermaids
                                          2. All mermaids are human beings.

Conclusion: All college students are human beings.

Let *C* stand for college students, *M* for mermaids, and *H* for human beings. Again we draw three intersecting circles to represent *C*, *M*, and *H*, as in Fig. 2.40.

Hypothesis (1) tells us that there are no *C*'s outside the *M* circle. Thus we put ∅ in every part of *C* that is outside the *M* circle, as shown in Fig. 2.40.

Figure 2.40

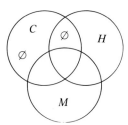

Figure 2.41

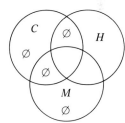

Hypothesis (2) tells us that there is nothing in the *M* circle that is outside the *H* circle. We indicate this by putting ∅ in each part of the *M* circle that is outside the *H* circle. The resulting diagram is shown in Fig. 2.41.

Looking at this final diagram (Fig. 2.41), we see that the conclusion does follow, since there are no *C*'s outside the *H* circle.   □

EXAMPLE 3   *Hypotheses:* 1. All gentle people are Republicans.
                                          2. Some dentists are gentle.

Conclusion: Some dentists are Republicans.

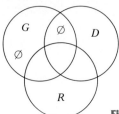

Figure 2.42

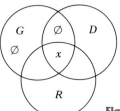

Figure 2.43

Let *G* stand for gentle people, *R* for Republicans, and *D* for dentists. Hypothesis (1) is pictured in Fig. 2.42.

To picture hypothesis (2), we must put an *x* where *D* and *G* overlap. There are two parts where *D* and *G* overlap. Since one of these already has $\emptyset$ in it, we know that the *x* can't go there. (The $\emptyset$ tells us that there is *nothing* there.) Thus the *x* must go in the other part. This is shown in Fig. 2.43.

The conclusion says that there is an *x* where *D* and *R* overlap. The diagram confirms this. Thus the argument is valid. □

EXAMPLE 4

Hypotheses: 1. No children are nuisances.

2. Some children are lovable.

Conclusion: No nuisances are lovable.

Let *C* stand for children, *N* for nuisances, and *L* for lovable.

Hypothesis (1) says that there is no one who is both a child and a nuisance. Thus we put $\emptyset$ in each part where *C* and *N* overlap, as in Fig. 2.44.

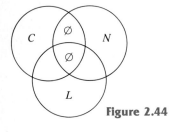

Figure 2.44

Since hypothesis (2) tells us that some children are lovable, we know that we must put an *x* in one of the two parts where *C* and *L* overlap. One of these parts already has a $\emptyset$ in it. Thus we must put the *x* in the other part. This is shown in Fig. 2.45.

The conclusion says there is nothing in the overlap of *N* and *L*. In our diagram the overlap of *N* and *L* has two parts. One of these definitely has a $\emptyset$ in it. The other part we know nothing about. In Fig. 2.45 we indicate this by a "?" The ? tells us that we do not know whether anything is there or not (from the given hypotheses). Thus the conclusion does not follow, and the argument is *not* valid. □

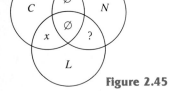

Figure 2.45

EXAMPLE 5

Hypotheses: 1. Some women are angels.

2. Some angels sing.

Conclusion: Some women sing.

Let *W* stand for women, *A* for angels, and *S* for singers.

Hypothesis (1) tells us that there is an *x* in one of the two parts where *W* and *A* overlap, but it does not tell us which. Thus we put it on the border between the two parts, as shown in Fig. 2.46.

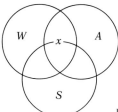

Figure 2.46

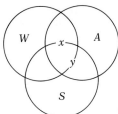

Figure 2.47

Hypothesis (2) tells us that there is a *y* in one of the two parts where *A* and *S* overlap, but again it does not tell us in which of the two parts to put it. Therefore we again put it on the borderline between *A* and *S* as shown in Fig. 2.47.

The conclusion says that there is *one x* or *y* that is in *both W* and *S* at the same time. In the diagram there is no such *x* or *y*, since the *y* is on the border between *W* and *S*. It may be in the part where *W* and *S* overlap, but, on the other hand, it may be in the part where they do not overlap. We do not know which situation is correct. The same is true for the *x*. Thus the conclusion is not necessarily true, and therefore the argument is invalid. ☐

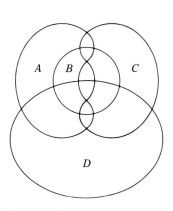

Figure 2.48

The standard diagram can also be applied when there are more than three letters. If there are four letters, then the diagram looks like the one in Fig. 2.48. This diagram cannot be made by drawing four circles for *A, B, C,* and *D*. In order to get all the possibilities, we have to "wiggle" the *A* and *C*, as shown.

EXERCISES

In Exercises 1–12, use the standard diagrams to test the validity of the indicated arguments.

*1. *Hypotheses:* (1) Some senior citizens over 65 years of age need a hearing aid.

(2) Ernest Smith is a senior citizen over 65 years of age.

Conclusion: Ernest Smith needs a hearing aid.

*2. *Hypotheses:* (a) Some senior citizen centers distribute U.S.D.A. surplus food.

(b) Some senior centers distributing U.S.D.A. surplus food are privately owned.

Conclusion: Some senior citizen centers are privately owned.

*3. *Hypotheses:* (a) All low-calorie soft drinks are artificially sweetened.

(b) Some artificial sweeteners are dangerous to your health.

Conclusion: All low-calorie soft drinks are dangerous to your health.

*4. *Hypotheses:* (a) All eyeglasses sold in the United States must have shatterproof lenses.

(b) All plastic lenses are shatterproof.

Conclusion: All eyeglasses sold in the United States must have plastic lenses.

*5. *Hypotheses:* (a) All cattle in the United States are fed with food that contains antibiotics.

        (b) Some antibiotics are harmful to humans when eaten.

Conclusion: Some cattle in the United States are harmful to humans when eaten.

*6. *Hypotheses:* (a) Some cans of tuna fish contain large amounts of sodium.

        (b) Some cans of sardines contain large amounts of sodium.

Conclusion: Some cans of tuna fish contain sardines.

*7. *Hypotheses:* (a) Some professional athletes are married.

        (b) Some professional athletes drive sports cars.

Conclusion: Some professional athletes who are married drive sports cars.

*8. *Hypotheses:* (a) All houses built during the 1970s by the Ace Construction Company are extremely well insulated.

        (b) No houses that are extremely well insulated allow any naturally occurring accumulated radon gas to escape.

Conclusion: Some houses built during the 1970s by the Ace Construction Company allow naturally occurring accumulated radon gas to escape.

*9. *Hypotheses:* (a) All people over 60 years of age with a family history of colon cancer should have an annual medical examination.

        (b) Some people who have an annual medical examination often discover that they have high blood pressure.

Conclusion: Some people over 60 years of age with a family history of colon cancer often discover that they have high blood pressure.

*10. *Hypotheses:* (a) No nuclear electricity generating stations operating in the United States are older than 50 years.

        (b) Some nuclear electricity generating stations operating in the United States are searching for new dumping sites for their nuclear waste material.

Conclusion: Some nuclear electricity generating stations operating in the United States that are searching for new dumping sites for their nuclear waste material are older than 50 years.

*11. *Hypotheses:* (a) No cats can swim.

        (b) No birds can swim.

Conclusion: No cats are birds.

*12. *Hypotheses:* (a) No obstetricians are professional athletes.

        (b) All major league baseball players are professional athletes.

Conclusion: No obstetricians are major league baseball players.

*13. Use the standard diagram to test the validity of Exercises 1–15 of Section 2.5.

2.7

PROPOSITIONS AND TRUTH TABLES

When doing the exercises of Section 2.5 you probably found that in some (especially those with three hypotheses) the number of possible diagrams became very large. Even if you used the standard diagram discussed in Section 2.6, you may have had difficulty. Perhaps in some cases you failed to obtain the correct answer simply because you did not think of the correct diagram. Fortunately, we have available another method of testing validity called **truth tables.** This method is purely mechanical, as we shall see shortly. Nothing is left to chance.

This new technique also has the advantage of working for statements for which Venn diagrams do not apply. For example, we are unable to draw a Venn diagram for the statement "Sherry will have either fish or meat for dinner."

We will start our discussion with simple sentences. Examples of these are:

1. All humans breathe oxygen.

2. $1 + 1 = 2$.

3. *Hamlet* was written by John Davidson.

4. Ronald Reagan is the king of Iran.

5. How are you?

6. Marijuana is a popular type of hair spray.

7. Shut up!

8. The word "drug" is a four-letter word.

9. $x + 3 = 5$.

10. Right on!

In the above examples, sentences (1), (2), and (8), are definitely true. Sentences (3), (4), and (6) are false. Sentences (5), (7), (9), and (10) are neither true nor false. In sentence (9) if we replace x by 2, then it becomes a true sentence. If we replace x by any other value, then it becomes a false sentence. We will be concerned only with true or false sentences.

Definition 2.5 *Any sentence that is either true or false is called a **proposition** or **statement.** Propositions are denoted by the lower case letters $p, q, r, \ldots$.*

Definition 2.6 *If a proposition is a true statement, then we say its **truth value** is true, denoted by* T. *If a proposition is a false statement, then we say its **truth value** is false, denoted by* F.

The examples given above are simple statements. Each expresses but one idea. If we were to use only simple sentences in everyday speech and thought, then what we could express would be extremely limited. (Try for just one hour to express your ideas using only simple sentences.) For this reason we will also deal with compound statements.

> **Definition 2.7** *Simple statements joined together in different ways are called **compound statements.***

One way of combining statements is by means of the word "and." For example, if we have the simple statements

1. Bruce failed math,

and

2. Bruce passed folk dancing,

then we can form the compound statement "Bruce failed math and passed folk dancing." Such a statement is called a *conjunction*.

> **Definition 2.8** *A **conjunction** consists of two or more statements joined together by means of the word "and." We use the symbol "∧" to stand for the word "and."*

If p stands for "Bruce failed math" and q stands for "Bruce passed folk dancing," then the statement "Bruce failed math and passed folk dancing" can be symbolized by $p \wedge q$.

As another example, if we let p = "all dogs wear glasses" and q = "flowers are intelligent," then the symbol $p \wedge q$ stands for "all dogs wear glasses and flowers are intelligent." Notice that the two simple statements forming the conjunction are not related. There is no rule that the statements must be related in any way.

It is necessary for us to have a method of determining the truth value for a conjunction, depending on the truth values of its components (that is, the simple statements out of which it is formed). We will make up a table (similar to the multiplication and addition tables learned in elementary school). This table will show all the different combinations of truth values for p and q and the resulting value for $p \wedge q$. Such a table is called a **truth table.**

Conjunction table

p	q	$p \wedge q$
T	T	T
T	F	F
F	T	F
F	F	F

Look at the first line of this table. It says that if p is true and q is true, then $p \wedge q$ is true. This seems reasonable, and it agrees with everyday usage. The second line says that if p is true and q is false, then $p \wedge q$ is false. The third line says that if p is false and q is true, then $p \wedge q$ is false. Finally, the last line says that when p is false and q is false, we conclude that $p \wedge q$ is false. These also agree with everyday usage. Thus the conjunction of two statements is true whenever each of the statements individually is true. In any other case it is not true.

Now, if we are given any conjunction, we do not have to consider its meaning to decide if it is true or false. We can just look at the table and determine *mechanically* its truth value, depending on the truth or falsity of the individual parts. We will see in Section 2.8 how this is done.

If we join together simple statements by the word "or," we have what is known as a **disjunction.**

> **Definition 2.9** A *disjunction* consists of two or more statements joined together by the word "or." We use the symbol "$\vee$" to stand for the word "or."

If p stands for "Joe will get an A in his math course" and q stands for "Joe will get an A in his English course," then the statement "Joe will get an A in his math course or an A in his English course" can be symbolized by $p \vee q$. In this example, it is quite possible that Joe will get an A in both courses, so that both components of the disjunction may be true. This is referred to as the **inclusive disjunction.**

As another example, let p = "Bill Murray will be elected president of the United States in 1992" and q = "Jane Smith will be elected president of the United States in 1992." Then the symbol $p \vee q$ stands for "Bill Murray will be elected president of the United States in 1992 or Jane Smith will be elected president of the United States in 1992." In this case it is obvious that both cannot be true. At most, one of the statements is true. It may turn out that neither is true. If both components of a disjunction cannot be true at the same time, then this is referred to as the **exclusive disjunction.**

In our discussion we will use only the inclusive disjunction. Again, this means the disjunction $p \vee q$ is true if p is true, or if q is true, or if both p and q are true.

It is easy to see how to construct the truth table for disjunction:

**Disjunction table
(inclusive case)**

p	q	$p \vee q$
T	T	T
T	F	T
F	T	T
F	F	F

Look at the first line of this table. It says that if p is true and q is true, then $p \lor q$ is true. This conforms to the inclusive use of the disjunction. The second and third lines say that $p \lor q$ is true if either of the components (p or q) is true. The fourth line says that a disjunction is false if both components are false.

Suppose we are given the statement "it is raining." From this we can form another statement, "it is not raining," by inserting the word "not." This is referred to as the **negation** of the original statement. Alternative ways of writing this are "it is false that it is raining" and "it is not true that it is raining."

> **Definition 2.10**   *The **negation** of statement p is the statement "p is not true" denoted as ~p (read as "not p"). Often, to avoid awkward English, we form the negation by simply inserting "not" in the appropriate place.*

EXAMPLE 1   If p is "Sharon is a genius," then ~p is "Sharon is not a genius."   ☐

EXAMPLE 2   If p is "2 + 2 = 4," then ~p is "2 + 2 is not equal to 4."   ☐

EXAMPLE 3   If p is "I was a fool to take this course," then ~p is "I was not a fool to take this course."   ☐

EXAMPLE 4   If p is "I take a bath once a month," then ~p is "I do not take a bath once a month."   ☐

The truth table for the negation is rather obvious. It is

Negation table

p	~p
T	F
F	T

In everyday language, conjunctions, disjunctions, and negations are often combined in the same sentence. Sentences of this kind can be symbolized as illustrated in the following examples.

EXAMPLE 5   Symbolize the following.

   a) My cat is stuck in the tree, and the firemen are not coming to rescue it.

   b) My cat is not stuck in the tree, and the firemen are not coming to rescue it.

SOLUTION   Let p = "my cat is stuck in the tree," and let q = "the firemen are coming to rescue it." Then we can symbolize statement (a) as

$p \wedge (\sim q)$. Using the same notation, we can symbolize statement (b) as $(\sim p) \wedge (\sim q)$. ▢

EXAMPLE 6 If p stands for "my car rattles" and q stands for "my mouse squeaks," then express in words each of the following symbolic expressions.

   a) $(\sim p) \wedge q$

   b) $(\sim p) \wedge (\sim q)$

   c) $(\sim p) \vee q$

SOLUTION   **a)** My car does not rattle and my mouse squeaks.

   b) My car does not rattle and my mouse does not squeak.

   c) Either my car does not rattle or my mouse squeaks. ▢

What would you say is the negation of "all violets are blue"? Here are some answers suggested by students recently.

1. All violets are not blue.

2. No violets are blue.

3. All violets are green.

4. Some violets are not blue.

Let us analyze each of these answers: Answer (1) says "all violets are not blue." This has more than one meaning. Two possible interpretations are:

a) "Anything that is a violet is not blue," or to put it another way, "no violets are blue."

b) "It is not true that each and every violet is blue," or to put it another way, "some violets are not blue."

We know that both white violets and blue violets exist. Thus the statements "all violets are blue" and "no violets are blue" are both false. But the negation table tells us that the negation of a false statement must be true. Thus "no violets are blue" cannot be the negation of "all violets are blue."

On the other hand, *"some violets are not blue" is a correct negation of "all violets are blue."*

We have seen that answer (2) is incorrect. Answer (4) is a correct negation.

Finally, look at answer (3). This statement "all violets are green" is false. So is the given statement "all violets are blue." Therefore the negation table shows us that this answer cannot be the negation we are seeking. Answer (4) is the only correct negation.

Similarly, the negation of "all teachers are sexy" is "some teachers are not sexy."

Now consider the statement "some violets are blue." What is its negation? The first answer that comes to mind is "some violets are not blue." Is this correct? Both of the statements "some violets are blue" and "some violets are not blue" are true statements. Hence one cannot be the negation of the other; the negation table requires that the negation of a true statement must be false.

"Some violets are blue" means that there is at least one violet that is blue. If this is false, then there can be no violets that are blue. Thus *the negation of "some violets are blue" is "no violets are blue."*

Similarly, the negation of "some women beat their husbands" is "no women beat their husbands."

Now consider the statement "some violets are not blue." This means that there is at least one violet that is not blue. To put it another way, it is possible to find a violet that is not blue. The negation would then say that it is *not* possible to find a violet that is not blue. Another way of saying this is "all violets are blue." Thus *the negation of "some violets are not blue" is "all violets are blue."*

Similarly, the negation of "some cigarettes are not menthol flavored" is "all cigarettes are menthol flavored."

Now consider the statement "no violets are blue." This says that it is *not* possible to find a violet that is blue. Then the negation says it *is* possible to find a violet that is blue, or that "some violets are blue." Thus *the negation of "no violets are blue" is "some violets are blue."*

Similarly, the negation of "no moon people are green" is "some moon people are green."

The results of the above discussions are summarized for the reader's convenience in the following chart:

Original statement	Negation
All violets are blue.	Some violets are not blue.
Some violets are blue.	No violets are blue.
Some violets are not blue.	All violets are blue.
No violets are blue.	Some violets are blue.

An important statement that occurs frequently is the **implication** or **conditional.** These are statements such as the following.

1. If Mary likes Pete, then she is insane.

2. If I eat my spinach, then I will be strong.

3. If I were a cat, then I would bark.

4. If I were a bell, then I would ring.

5. If I were you, I'd run.

The implication can be defined in the following way.

Definition 2.11 *An **implication** or **conditional** is any statement of the form "if p, then q." We symbolize this as $p \rightarrow q$. This is read as "p implies q" and means that if p is true, then q must be true.*

Definition 2.12 *The p statement of the conditional "$p \rightarrow q$" is called the **hypothesis**, and the q statement is called the **conclusion**.*

EXAMPLE 7 If p stands for "Leon passes the Math 5 exam" and q stands for "I will eat my hat," then the implication "if Leon passes the Math 5 exam, then I will eat my hat" can be symbolized as $p \rightarrow q$. The hypothesis is "Leon passes the Math 5 exam." The conclusion is "I will eat my hat." ☐

EXAMPLE 8 Let q stand for "you are out of Schlitz" and let r stand for "you are out of beer." The implication "if you are out of Schlitz, you are out of beer" can be symbolized as $q \rightarrow r$. The hypothesis is "you are out of Schlitz," and the conclusion is "you are out of beer." What is $r \rightarrow q$? ☐

EXAMPLE 9 Let p stand for "it is raining" and q stand for "the streets are wet." The implication "if it is raining, then the streets are wet" can be symbolized as $p \rightarrow q$. What does the symbol $q \rightarrow p$ represent? ☐

Christina's father has made the following promise to her: If she gets A's in all her courses this semester, then he will buy her a new car. Four different things can happen.

1. Hypothesis true: She gets A's in all her courses.
   Conclusion true: Her father buys her a new car.
2. Hypothesis true: She gets A's in all her courses.
   Conclusion false: Her father does not buy her a new car.
3. Hypothesis false: She does not get A's in all her courses.
   Conclusion true: Her father buys her a new car.
4. Hypothesis false: She does not get A's in all her courses.
   Conclusion false: Her father does not buy her a new car.

In case (1) her father certainly kept his promise. Thus the implication "if she gets A's in all her courses this semester, then he will buy her a new car" is true. *The hypothesis and conclusion are true, and this makes the implication true.*

In case (2) her father definitely did not keep his promise. Thus the implication "if she gets A's in all her courses this semester, then he will buy her a new car" is false. *The hypothesis is true, but since the conclusion is false, this makes the implication false.*

In cases (3) and (4), Christina does not live up to her end of the deal. Therefore her father's promise is never put to the test. Whether he buys her a car or not, he cannot be said to have broken his promise. Thus we cannot call his implication false. Since all statements are either true or false, we then classify his implication as true. This leads us to the following truth table.

Implication table

p	q	$p \rightarrow q$
T	T	T
T	F	F
F	T	T
F	F	T

The first two lines of the implication table seem to agree with everyday usage. You may not yet be convinced of the third and fourth lines. This is because in everyday speech we do not make implications that have false hypotheses. To prove to yourself that they are reasonable, consider the following discussion overheard in the student cafeteria.

Eric: My uncle is a famous movie producer.

José: If your uncle is a famous movie producer, then I am the king of France.

José's comment indicates his strong disbelief in Eric's claim that his uncle is a famous movie producer. José is sure that Eric's statement is false. José knows that his own statement "I am the king of France" is false. He definitely intends his own implication to be true. Thus José is really saying that "false implies false" results in a true statement. This corresponds to line 4 of the implication table. This is an example of an implication with a false hypothesis being used in everyday conversation. Can you find an example from everyday usage that illustrates line 3 of the implication table?

Variations of the Conditional

Let us look at the implication (or conditional) "if it is raining, then the streets are wet." Now consider the following variations (changes).

Original statement: If it is raining, then the streets are wet.

1. *Converse:* If the streets are wet, then it is raining.

2. *Inverse:* If it is not raining, then the streets are not wet.

3. *Contrapositive:* If the streets are not wet, then it is not raining.

If we let p stand for "it is raining" and q stand for "the streets are wet," then the original statement can be symbolized as $p \rightarrow q$.

The **converse** (statement 1) is formed by interchanging the hypothesis ("it is raining") and the conclusion ("the streets are wet"). Thus the converse is symbolized as $q \rightarrow p$.

In Example 8 we asked you to find $r \rightarrow q$. This was just the converse of the given implication $q \rightarrow r$. Similarly, in Example 9, $q \rightarrow p$ is the converse of the given implication $p \rightarrow q$.

The **inverse** (statement 2) is formed by inserting the word "not" in the hypothesis ("it is raining") and the word "not" in the conclusion ("the streets are wet"). Thus we have "if it is not raining, then the streets are not wet." So the inverse is $(\sim p) \rightarrow (\sim q)$.

The **contrapositive** (statement 3) is formed by switching around the hypothesis and the conclusion *and* then inserting the word "not" in both the hypothesis and the conclusion. Thus we symbolize the contrapositive of $p \rightarrow q$ as $(\sim q) \rightarrow (\sim p)$.

Comment The contrapositive of an implication is really obtained by first taking the converse of the given statement and then taking the inverse of the result. Try it for our original statement.

Now it is obvious that the original statement "if it is raining, then the streets are wet" is true. Must the converse also be true? Obviously not. It is possible that the street cleaners have just washed the streets. Thus if the streets are wet, it does not necessarily mean that it is raining. In other words, the converse is not necessarily true. (It may be either true or false.)

What can we say about the inverse? Must it be true? Again your answer should be "no." If it is not raining, it does not automatically follow that the streets are not wet. The street cleaners may have just washed the street. Obviously, the inverse may be false.

How about the contrapositive? Everyone would agree that if the streets are not wet, then it can't be raining. Thus the contrapositive must be true.

Summarizing If the original implication is true, then *only* the contrapositive *must* be true. The converse and inverse may or may not be true. In the next section, when we discuss truth tables in more detail, we will *prove* this.

As another example, consider the implication "if John studies hard, then he will pass this course." We have the following.

Original statement: If John studies hard, then he will pass this course.

Converse: If John passes this course, then he will have studied hard.

Inverse:	If John does not study hard, then he will not pass this course.
Contrapositive:	If John does not pass this course, then he will not have studied hard.

Look at the original statement. The hypothesis is "John studies hard." In forming the converse we rephrased this as "he will have studied hard." This change in the English was done merely to avoid an awkward sentence. (Had we not made this change, the converse would have read, "if he will pass this course, then John studies hard." Sounds peculiar, doesn't it?) This change does not affect the statement itself. Now look at the inverse and contrapositive. We have made similar changes. What would the inverse and contrapositive be without these adjustments?

Comment   When asked for the converse of "if John studies hard, then he will pass this course," some students answer, "John will pass this course if he studies hard." This is wrong because it says the same thing as the original statement.

It may be helpful to put these results into a table.

		Symbolic form	Truth value of this statement, assuming original statement is true
Original statement	If p, then q.	$p \rightarrow q$	True
Converse	If q, then p.	$q \rightarrow p$	May be true or false
Inverse	If $\sim p$, then $\sim q$.	$(\sim p) \rightarrow (\sim q)$	May be true or false
Contrapositive	If $\sim q$, then $\sim p$.	$(\sim q) \rightarrow (\sim p)$	*Must* be true

In the rest of this chapter we will make frequent use of the truth tables for conjunction, disjunction, implication, and negation. Therefore for handy reference, we summarize them here.

Conjunction table

p	q	$p \wedge q$
T	T	T
T	F	F
F	T	F
F	F	F

Disjunction table

p	q	$p \vee q$
T	T	T
T	F	T
F	T	T
F	F	F

Implication table

p	q	$p \rightarrow q$
T	T	T
T	F	F
F	T	T
F	F	T

Negation table

p	$\sim p$
T	F
F	T

EXERCISES

In Exercises 1–8, determine which of the expressions are propositions and which are not.

1. Gosh!

2. You have to be a genius to operate those new compact disc players.

3. Do you have a buck?

4. People who are nearsighted cannot use a binoculars.

5. Excuse me!

6. Credit card users are always being ripped off.

7. Can you lend me your wheels?

8. Kerosene heaters are so safe that everybody should own one.

In Exercises 9–18, let p stand for "I love music" and let q stand for "I attend rock concerts." Express in words each of the indicated symbolic expressions:

9. $p \wedge q$

10. $\sim q$

11. $p \vee q$

12. $(\sim p) \wedge q$

13. $(\sim p) \wedge (\sim q)$

14. $p \wedge (\sim q)$

15. $\sim(p \vee q)$

16. $(\sim p) \vee (q)$

17. $(\sim p) \vee (\sim q)$

18. $\sim(p \wedge q)$

Let p stand for "I am a salesman" and let q stand for "I fly frequently between New York and California." Express in symbolic form each of the sentences given in Exercises 19–24.

19. If I am a salesman, then I fly frequently between New York and California.

20. I am a salesman, but I do not fly frequently between New York and California.

21. I am not a salesman, and I do not fly frequently between New York and California.

22. It is not true that I am a salesman and that I fly frequently between New York and California.

23. Although I fly frequently between New York and California, I am not a salesman.

24. If I do not fly frequently between New York and California, then I am not a salesman.

Write the negation of each of the statements given in Exercises 25–28.

25. Some airlines have a better safety record than others.

26. All instant coffees raise the blood cholesterol level of humans.

27. Some cardiologists do not carry extensive medical malpractice insurance.

28. No president of the United States has ever died of cancer while in office.

If p stands for "I am afraid of heights" and q stands for "I have flown in an airplane," then express in words each of the expressions given in Exercises 29–34.

29. $p \rightarrow q$

30. $q \rightarrow p$

31. $p \rightarrow (\sim q)$

32. $(\sim p) \rightarrow q$

33. $(\sim p) \rightarrow (\sim q)$

34. $(\sim q) \rightarrow (\sim p)$

Write the inverse, converse, and contrapositive of each of the implications given in Exercises 35–37.

35. If an airplane crashes into the ocean, then the seats will float.

36. If you suffer from migraine headaches, then you use a lot of aspirin.

37. If the consumer price index rises by more than 4 percent, then college tuition costs will increase moderately.

38. Assume that the following implication is true: "If there is a water shortage, then people will conserve water." Which of the following statements must be true? Explain.

   a) If there is no water shortage, then people will not conserve water.

   b) If people are not conserving water, then there is no water shortage.

   c) If people are conserving water, then there is a water shortage.

   d) People will conserve water if there is a water shortage.

39. Mack was told by the parole board of his prison that he would be released on probation after a

year *only if* he would be a model prisoner. By the end of the year, Mack was rated as a model prisoner but still was not released on probation. He appealed to the governor, claiming that the parole board lied. Is Mack's appeal justified? Why or why not?

40. The New Dorp Insurance Company will promote its sales representatives to vice-presidents *only if* they generate at least $500,000 of sales during a year. For the calendar year of 1985, Maureen O'Hara generated more than $500,000 of sales and was not promoted. She has accused the company of sex discrimination in not promoting her. Is her claim valid? Explain.

41. Consider the following statistical statement: "If your luggage is fastened securely on all over-

seas flights, then it will usually arrive safely at its destination."

   a) Write another English statement that you know is true based upon this sentence.

   b) Write another English statement that you know is false based upon this sentence.

Let *p* stand for "Joan has good recommendations," let *q* stand for "Joan has a high college grade point average," and let *r* stand for "Joan was accepted to medical school." Express in words each of the symbolic expressions given in Exercises 42–47.

42. $p \land (q \rightarrow r)$ **43.** $p \lor (q \land r)$

44. $(\sim p) \lor [(\sim q) \land (\sim r)]$ **45.** $p \lor [q \rightarrow (\sim r)]$

46. $(\sim p) \rightarrow (r \land q)$ **47.** $[(\sim q) \lor (\sim r)] \land p$

2.8

TAUTOLOGIES AND SELF-CONTRADICTIONS

In everyday language, as well as in mathematics, we usually make statements that combine conjunctions, negations, disjunctions, and implications in different forms. Take, for example, the following statement: "If it rains this evening, I will either go to the movies or go bowling." By means of truth tables we can determine when such statements are true or false. These truth tables will involve combinations of $\land$, $\lor$, $\sim$, and $\rightarrow$. Let us now look at some truth tables and analyze them.

EXAMPLE 1   Determine the truth table for $(\sim p) \lor q$.

SOLUTION   Two letters are involved, *p* and *q*. So we have a column for each of these.

Again looking at the given expression, we notice that the negation of *p* occurs as one of the parts. So we add another column, $\sim p$. We then have the following:

p	q	$\sim p$

Now we need a column for $(\sim p) \lor q$. This gives us the following:

p	q	$\sim p$	$(\sim p) \lor q$

Comment We arrive at the last column by working our way across. We start *first* with the simple letters and *then* add the necessary $\sim$, $\land$, $\lor$, and $\rightarrow$ where needed, in the order in which they occur. The parentheses in the above example tell us that we must *first* take the negation of p and *then* the disjunction of this with q.

Comment Notice that in this example we do not need a column for $p \lor q$, since this is not part of the original expression. We need a column *only* for $(\sim p) \lor q$.

Getting back to this example, we now complete the table, using the basic truth tables given on p. 86. We work from left to right, filling in the blanks.

TABLE 2.1

p	q	$(\sim p)$	$(\sim p) \lor q$
T	T	F	
T	F	F	
F	T	T	
F	F	T	

In the first two columns we have written all possible combinations of T and F (there are exactly four). The third column is obtained from the negation table. To get the final column, we go to the disjunction table and apply it to the disjunction of $(\sim p)$ and q. Notice that to do this, we cannot use the *headings* of the disjunction table. We just work from the *entries* as follows. We look at the first row in Table 2.1. The entries for $(\sim p)$ and q are F and T *in that order.* In the disjunction table we find the entries F and T *in that order.* They appear on the third row. The result for this row is T. So we enter T in the first row of Table 2.1.

Now we look at the second row of Table 2.1. The entries for $(\sim p)$ and q are F and F in that order. In the disjunction table, we find the entries F and F in that order. These are on the fourth row and their result is F. So we enter F on the second row of Table 2.1. We now look at the third row of Table 2.1. The entries for $(\sim p)$ and q are T and T, in that order. In the disjunction table these entries are on the first row and their result is T. So we enter T in the third row of Table 2.1. Finally, on the fourth row of Table 2.1 the entries for $(\sim p)$ and q are T and F in that order. We look for T and F in that order in the disjunction table, and find them on the second row. The result is T, so we enter T on the fourth row of Table 2.1.

The final table then looks like this:

p	q	$(\sim p)$	$(\sim p) \vee q$
T	T	F	T
T	F	F	F
F	T	T	T
F	F	T	T

The final column of the table is circled. This means:

1. If p and q are both true, then the whole statement is true.

2. If p is true and q is false, then the whole statement is false.

3. If p is false and q is true, then the whole statement is true.

4. If p is false and q is false, then the whole statement is true.

The truth value of the entire statement may be true or false depending on the truth values of p and q. $\square$

EXAMPLE 2 Determine the truth table for $(p \wedge q) \rightarrow (p \vee q)$.

SOLUTION The parentheses tell us that we first take $p \wedge q$. Then we take $p \vee q$. Finally we connect $(p \wedge q)$ with $(p \vee q)$ by means of $\rightarrow$. Again two letters, p, and q, are involved. So we need a column for each of these. Looking at the expression, we see that we also need columns for $p \wedge q$ and $p \vee q$. Finally, we need a column for the whole expression [which connects $(p \wedge q)$ with $(p \vee q)$ by means of $\rightarrow$]. The truth table is as follows:

p	q	$p \wedge q$	$p \vee q$	$(p \wedge q) \rightarrow (p \vee q)$
T	T	T	T	T
T	F	F	T	T
F	T	F	T	T
F	F	F	F	T

As in the previous example, we must consider all possible combinations of truth values for the letters p and q. This gives us the first two columns. The entries in the third and fourth columns are obtained from the conjunction and disjunction tables, respectively.

The final column is obtained by applying the implication table to the entries in the $(p \wedge q)$ and $(p \vee q)$ columns. This is because the last column states that the third and fourth columns are connected by means of implication. For example, the entries in the second row of the $(p \wedge q)$ and $(p \vee q)$ columns are F and T *in that order*. So we go to the implication table and look for the row containing F and T in that order. It is in the third row. The result is T, so we enter T in the final column of our chart. In a similar manner we complete the final column.

Again we circle the final column. Notice that in this example all the entries in the final column are T. This means that $(p \wedge q) \rightarrow (p \vee q)$ is true regardless of the truth values of p and q. In other words, the statement $(p \wedge q) \rightarrow (p \vee q)$ is *always* true. Such a statement is called a **tautology.** ☐

Definition 2.13 *Any statement that is always true is called a **tautology.** This means that in the final column of its truth table there are only T's.*

Example 1 is not a tautology. Why not?

EXAMPLE 3   Write the truth table for $(p \rightarrow q) \wedge (q \rightarrow p)$.

SOLUTION   We need columns for p, q, $(p \rightarrow q)$, $(q \rightarrow p)$, and finally for $(p \rightarrow q) \wedge (q \rightarrow p)$. The truth table is as follows:

p	q	$p \rightarrow q$	$q \rightarrow p$	$(p \rightarrow q) \wedge (q \rightarrow p)$
T	T	T	T	T
T	F	F	T	F
F	T	T	F	F
F	F	T	T	T

☐

EXAMPLE 4   Construct the truth table for $(p \rightarrow q) \wedge [p \wedge (\sim q)]$.

SOLUTION   Proceeding in the same manner as we did in Examples 2 and 3, we make columns for p, q, $\sim q$, $(p \rightarrow q)$, $p \wedge (\sim q)$, and finally for $(p \rightarrow q) \wedge [p \wedge (\sim q)]$. We then have the following truth table:

p	q	$\sim q$	$p \rightarrow q$	$p \wedge (\sim q)$	$(p \rightarrow q) \wedge [p \wedge (\sim q)]$
T	T	F	T	F	F
T	F	T	F	T	F
F	T	F	T	F	F
F	F	T	T	F	F

The final column of this table consists only of F's. This statement is *never* true. Such a statement is called a ***self-contradiction.***

> **Definition 2.14**   *Any statement that is always false is called a **self-contradiction.** This means that in the final column of its truth table there are only F's.*

To better understand the idea of a self-contradiction, consider the following: "I have read *Hamlet*, and I have not read *Hamlet*." Clearly, this statement cannot be true under any circumstances. We can examine it symbolically by letting p stand for "I have read *Hamlet*." Symbolized, this statement is $p \wedge (\sim p)$. Its truth table is:

p	$\sim p$	$p \wedge (\sim p)$
T	F	F
F	T	F

Since we have only F's in the final column, it is a self-contradiction.

EXAMPLE 5   Determine the truth table for $[(p \rightarrow q) \wedge (q \rightarrow r)] \rightarrow (p \rightarrow r)$.

SOLUTION   In this case we need columns for p, q, r, $(p \rightarrow q)$, $(q \rightarrow r)$, $(p \rightarrow q) \wedge (q \rightarrow r)$, $(p \rightarrow r)$, and finally for the whole statement $[(p \rightarrow q) \wedge (q \rightarrow r)] \rightarrow (p \rightarrow r)$. The truth table for this is as follows:

p	q	r	$(p \rightarrow q)$	$(q \rightarrow r)$	$(p \rightarrow q) \wedge (q \rightarrow r)$	$p \rightarrow r$	$[(p \rightarrow q) \wedge (q \rightarrow r)] \rightarrow (p \rightarrow r)$
T	T	T	T	T	T	T	T
T	T	F	T	F	F	F	T
T	F	T	F	T	F	T	T
T	F	F	F	T	F	F	T
F	T	T	T	T	T	T	T
F	T	F	T	F	F	T	T
F	F	T	T	T	T	T	T
F	F	F	T	T	T	T	T

The table has eight lines, since there are eight possible combinations of T and F for the three letters. If there were four letters, how many lines would we need for the truth table? Notice that this expression is a tautology, since we have only T's in the final column. ☐

EXAMPLE 6   Show by means of truth tables that if an implication is true, then its converse does not necessarily have to be true.

SOLUTION   Let the implication be $(p \rightarrow q)$. Then its converse is $(q \rightarrow p)$. We compare their truth tables.

p	q	$p \rightarrow q$		p	q	$q \rightarrow p$
T	T	T		T	T	T
T	F	F		T	F	T
F	T	T		F	T	F
F	F	T		F	F	T

Looking at line 3 of each table, we see that if p is false and q is true, then $(p \rightarrow q)$ is true, whereas $(q \rightarrow p)$ is false. Compare the other lines of the truth tables; $(p \rightarrow q)$ and $(q \rightarrow p)$ may have different or the same truth values depending on the truth values of p and q. ☐

EXAMPLE 7   Show by means of truth tables that if the implication $p \rightarrow q$ is true, then its contrapositive $(\sim q) \rightarrow (\sim p)$ must be true.

SOLUTION   In this case we need columns for p, q, $(\sim p)$, $(\sim q)$, $p \rightarrow q$, and finally a column for $(\sim q) \rightarrow (\sim p)$. The truth table for this is the following:

p	q	$\sim p$	$\sim q$	$p \rightarrow q$	$(\sim q) \rightarrow (\sim p)$
T	T	F	F	T	T
T	F	F	T	F	F
F	T	T	F	T	T
F	F	T	T	T	T

Notice that the entries in the column for $(\sim q) \rightarrow (\sim p)$ are identical to the entries for $p \rightarrow q$. This means that the two statements $p \rightarrow q$ and $(\sim q) \rightarrow (\sim p)$ are equivalent and that, when one is true, the other is true. ☐

More generally, we say that

Definition 2.15   *Two statements are said to be **equivalent** if the final columns of their truth tables are identical (assuming, of course, that in both tables, the truth values of p and q are written in the same order.)*

Thus to determine whether two statements are equivalent, we first write the truth tables for each statement. Then we compare the final columns of each table. If the entries are identical, then the statements are equivalent. Otherwise, they are not equivalent.

Comment The idea of equivalent statements will be used later in this chapter when we study switching circuits.

EXERCISES

Construct a truth table for each of the statements in Exercises 1–20, and then determine whether the statements are tautologies, self-contradictions, or neither.

1. $\sim(p \rightarrow q)$

2. $\sim(p \vee q)$

3. $(p \wedge q) \rightarrow (p \vee q)$

4. $(p \wedge q) \rightarrow p$

5. $\sim[(p \vee q) \rightarrow (q \vee p)]$

6. $(p \wedge q) \rightarrow (\sim r)$

7. $(\sim p) \rightarrow (q \vee r)$

8. $[\sim(p \wedge q)] \vee [(\sim p) \wedge (\sim q)]$

9. $(p \rightarrow q) \wedge (\sim s)$

10. $(p \wedge q) \rightarrow (\sim r)$

11. $[(p \rightarrow q) \wedge (\sim r)] \rightarrow r$

12. $(p \vee q) \rightarrow [(\sim q) \wedge r]$

13. $\sim[(\sim p) \rightarrow ((\sim q) \vee r)]$

14. $[(p \vee q) \wedge (\sim r)] \rightarrow r$

15. $[(p \wedge (\sim q)) \rightarrow r] \rightarrow (\sim r)$

16. $[(\sim q) \vee p)] \wedge [(\sim s) \rightarrow (\sim r)]$

17. $(p \rightarrow q) \vee [s \rightarrow (\sim r)]$

18. $[(\sim p) \wedge s] \rightarrow (r \vee q)$

19. $[(p \wedge q) \rightarrow (\sim r)] \rightarrow (\sim p)$

20. $[(q \wedge (\sim r)) \vee p] \rightarrow (\sim p)$

21. Show by means of truth tables that if the implication $p \rightarrow q$ is true, then its inverse $(\sim p) \rightarrow (\sim q)$ may be false. (You will appreciate this exercise more if you reread the discussion of this on p. 86 before attempting this exercise.)

22. If an expression involves four different lines, how many lines are needed for the truth table?

23. If an expression involves five different lines, how many lines are needed for the truth table?

24. If an expression involves n different lines, how many lines are needed for the truth table?

Assuming that p is true and q is false, find the truth value of each of the statements given in Exercises 25–33.

25. $\sim(p \rightarrow q)$

26. $\sim[p \rightarrow (\sim q)]$

27. $\sim[p \wedge (\sim q)]$

28. $(p \wedge q) \rightarrow (\sim q)$

29. $[(\sim p) \wedge q] \vee (\sim q)$

30. $(p \wedge q) \vee (\sim q)$

31. $(p \rightarrow q) \rightarrow (\sim p)$

32. $\sim[(p \vee q) \rightarrow (q \rightarrow p)]$

33. $(\sim p) \wedge [(\sim q) \vee p]$

34. Consider the truth table shown below. Find compound statements to replace the question marks.

p	q	?	?	?
T	T	F	F	F
T	F	F	F	F
F	T	F	T	T
F	F	T	F	T

35. Consider the truth table shown below. Find compound statements to replace the question marks.

p	q	?	?	?
T	T	T	F	F
T	F	F	T	T
F	T	T	T	T
F	F	T	F	T

Find the truth values for each of the statements given in Exercises 36–39.

36. If $2 + 2 = 4$, then $5 + 2 = 8$.

37. Either the nation's health food craze is a fraud or America's 1985 national budget had a huge surplus.

38. If Abraham Lincoln was shot while president of the United States and John Kennedy's brother was assassinated, then the name of this chapter is "Logic."

39. Cigarette smoking is dangerous to your health, but the American Cancer Society has found a cure for cancer.

Determine which of the pairs of statements in Exercises 40–45 are equivalent.

40. $\sim(p \vee q)$ and $(\sim p) \vee (\sim q)$

41. $p \wedge (q \wedge r)$ and $(p \wedge q) \vee r$

42. $\sim(p \vee q)$ and $p \rightarrow q$

43. $p \wedge (q \vee r)$ and $(p \wedge q) \vee (p \wedge r)$

44. $(p \rightarrow q) \wedge p$ and q

45. $p \rightarrow (\sim q)$ and $(\sim p) \vee (\sim q)$

46. The symbol $p \veebar q$ means that *either p or q is true but not both*. This is the **exclusive disjunction** discussed in Section 2.7. Write the truth table for $p \veebar q$.

47. The symbol $p \downarrow q$ means that *p and q must both be false*. This is called the **joint denial** of p and q. Write the truth table for $p \downarrow q$.

48. Construct the truth table for $\sim(p \downarrow q) \wedge [(\sim p) \veebar q]$.

2.9

APPLICATION TO ARGUMENTS

In Section 2.5 we learned how to test the validity of arguments by using Venn diagrams. We pointed out then that this method does not work in every case and that even when it does work, there may be too many diagrams to consider.

Truth tables provide a method of testing the validity of arguments that overcomes these difficulties. The best way to see how this method works is to actually try it on some examples.

EXAMPLE 1 Test the validity of the following argument: "If Bigmouth wins the election, then I will leave this state. I am not leaving this state. Therefore Bigmouth will not win the election."

SOLUTION Let p stand for "Bigmouth wins the election" and let q stand for "I will leave this state." We can rewrite each sentence as follows:

	Verbally	*Symbolically*
Hypotheses:	If Bigmouth wins the election, then I will leave this state.	$p \rightarrow q$
	I am not leaving this state.	$\sim q$
Conclusion:	Bigmouth will not win the election.	$\sim p$

Remember that an argument is valid if the conclusion follows logically from the hypotheses. In other words, it is valid if the hypotheses imply the conclusion. In our example we must determine whether *the hypotheses* $(p \rightarrow q)$ *and* $(\sim q)$ *together imply the conclusion* $(\sim p)$. That is, we are asked to determine whether $[(p \rightarrow q) \wedge (\sim q)] \rightarrow (\sim p)$ is always true. (Remember, $\wedge$ stands for "and." We are asking if the conclusion follows from *both* of the hypotheses and not from one or the other alone. Therefore we connect them by "and.") This gives us the truth table shown below. Since the last column has only T's, the expression is always true. Hence whether the hypotheses are true or false, the conclusion *always* follows from them. Therefore the argument is valid.

p	q	$\sim p$	$\sim q$	$p \rightarrow q$	$(p \rightarrow q) \wedge (\sim q)$	$[(p \rightarrow q) \wedge (\sim q)] \rightarrow (\sim p)$
T	T	F	F	T	F	T
T	F	F	T	F	F	T
F	T	T	F	T	F	T
F	F	T	T	T	T	T

EXAMPLE 2   Test the validity of the following argument: "If the teacher talks too long, Arthur gets a headache. Arthur has a headache. Therefore the teacher is talking too long."

SOLUTION   Let r stand for "the teacher talks too long" and let s stand for "Arthur gets a headache." We can symbolize each sentence of the argument as shown below.

	Verbally	*Symbolically*
Hypotheses:	If the teacher talks too long, Arthur gets a headache.	$r \rightarrow s$
	Arthur has a headache.	s
Conclusion:	The teacher is talking too long.	r

We must now determine whether $(r \rightarrow s)$ and s together imply r. We want to know if $[(r \rightarrow s) \wedge s] \rightarrow r$ is always true. The truth table is as follows:

r	s	$r \rightarrow s$	$(r \rightarrow s) \wedge s$	$[(r \rightarrow s) \wedge s] \rightarrow r$
T	T	T	T	T
T	F	F	F	T
F	T	T	T	F
F	F	T	F	T

The final F of line 3 shows us that if r is false and s is true, then the conclusion does *not* follow from the hypotheses. A deductive argument is valid only if the conclusion *always* follows from the hypotheses. Hence this argument is *invalid.* ☐

Comment In Example 1 we had all T's in the truth table, so the expression $[(p \rightarrow q) \wedge (\sim q)] \rightarrow (\sim p)$ was a tautology, and the argument was valid. In Example 2, since we had one F, the statement was not a tautology, and the argument was invalid. *If an argument is valid, its truth table must be a tautology.*

EXAMPLE 3 Test the validity of the following argument: "If Dave sings, the cat howls. Either the baby cries or the cat howls. The baby is not crying. Therefore Dave is not singing."

SOLUTION Let p stand for "Dave sings," q for "the cat howls," and r for "the baby cries." We have:

	Verbally	*Symbolically*
Hypotheses:	If Dave sings, the cat howls.	$p \rightarrow q$
	Either the baby cries or the cat howls.	$r \vee q$
	The baby is not crying.	$\sim r$
Conclusion:	Dave is not singing.	$\sim p$

The argument is symbolized as $[(p \rightarrow q) \wedge (r \vee q) \wedge (\sim r)] \rightarrow (\sim p)$. Its truth table is as follows:

p	q	r	$\sim p$	$\sim r$	$p \rightarrow q$	$r \vee q$	$(p \rightarrow q)$ $\wedge (r \vee q)$	$(p \rightarrow q) \wedge (r \vee q)$ $\wedge (\sim r)$	$[(p \rightarrow q) \wedge (r \vee q)$ $\wedge (\sim r)] \rightarrow (\sim p)$
T	T	T	F	F	T	T	T	F	T
T	T	F	F	T	T	T	T	T	F
T	F	T	F	F	F	T	F	F	T
T	F	F	F	T	F	F	F	F	T
F	T	T	T	F	T	T	T	F	T
F	T	F	T	T	T	T	T	T	T
F	F	T	T	F	T	T	T	F	T
F	F	F	T	T	T	F	F	F	T

Since the final column contains one F, this statement is not a tautology. Hence the argument is invalid. ☐

EXAMPLE 4 Test the validity of the following argument: "If I kiss that frog, it will turn into a handsome prince. If the frog turns into a handsome prince, then I will marry him. Therefore if I kiss that frog, then I will marry him."

SOLUTION   Let p stand for "I kiss that frog," q for "the frog will turn into a handsome prince," and r for "I will marry him." Then the argument can be symbolized as:

Hypotheses: $p \rightarrow q$

$q \rightarrow r$

Conclusion: $p \rightarrow r$

The expression we must test is $[(p \rightarrow q) \land (q \rightarrow r)] \rightarrow (p \rightarrow r)$. The truth table is as follows:

p	q	r	$p \rightarrow q$	$q \rightarrow r$	$p \rightarrow r$	$(p \rightarrow q) \land (q \rightarrow r)$	$[(p \rightarrow q) \land (q \rightarrow r)] \rightarrow (p \rightarrow r)$
T	T	T	T	T	T	T	T
T	T	F	T	F	F	F	T
T	F	T	F	T	T	F	T
T	F	F	F	T	F	F	T
F	T	T	T	T	T	T	T
F	T	F	T	F	T	F	T
F	F	T	T	T	T	T	T
F	F	F	T	T	T	T	T

Since this statement is a tautology, the argument is valid. ☐

EXERCISES

Determine whether the argument forms given in Exercises 1–6 are valid or invalid.

1. Hypotheses: $p \rightarrow q$

   $\sim q$

   Conclusion: $\sim p$

2. Hypotheses: $p \lor q$

   $\sim p$

   Conclusion: $\sim q$

3. Hypotheses: $p \land q$

   $q \rightarrow r$

   r

   Conclusion: $\sim p$

4. Hypotheses: $p \rightarrow q$

   $(\sim q) \land r$

   $\sim p$

   Conclusion: $\sim q$

5. Hypotheses: $p \rightarrow r$

   $q \lor r$

   $(\sim q) \rightarrow (\sim p)$

   Conclusion: $(\sim p) \rightarrow (\sim r)$

6. Hypotheses: $\sim (p \land q)$

   $r \rightarrow (\sim p)$

   $q \lor r$

   Conclusion: $p \rightarrow r$

Translate each of the arguments given in Exercises 7–20 into symbolic form and then test its validity.

7. If Felicia exercises too much, she gets leg cramps. Felicia has leg cramps. Therefore Felicia has exercised too much.

8. Either Tracy will go on to medical school or she will become a lawyer. Tracy is not a lawyer. Therefore Tracy has gone on to medical school.

9. If Roger wins the lottery, then he will be rich. Roger is not rich. Therefore Roger has not won the lottery.

10. If Richard goes to England, then his girlfriend will date someone else. If his girlfriend dates someone else, then Richard's parents will be very upset. Therefore if Richard goes to England, then his parents will be very upset.

11. If Jennifer eats too much red meat, then her cholesterol level will be elevated. If her cholesterol level is elevated, then she will suffer from clogged arteries. Jennifer's cholesterol level is not elevated. Therefore if Jennifer eats too much red meat, then she will not suffer from clogged arteries.

12. If college tuition costs are raised, then students will either take a loan or be forced to drop out of college. If students take a loan, then they will not be forced to drop out of college. Therefore college tuition costs will not be raised.

13. If you're out of Schlitz, then you're out of beer. If you're out of beer, then you won't enjoy the game. Either you're out of beer or you will enjoy the game. Therefore if you're out of Schlitz, then you will enjoy the game.

14. If Cleon is hungry, then he will have a snack at McDonald's or at Burger King. If Cleon has a snack at Burger King, then he is not hungry. Cleon is hungry. Therefore Cleon will have a snack at McDonald's.

15. Paul has been told by the dean to improve his grade performance or be dropped from the college team. Paul does not improve his grade performance. Therefore Paul will be dropped from the college team.

16. If nuclear disarmament is not started quickly, then the world will become unsafe to live in. If nuclear disarmament is started quickly, then our children will enjoy healthy lives. The world will not become unsafe to live in. Therefore our children will not enjoy healthy lives.

17. For the museum to remain open, admission fees must be raised or new benefactors must be found. Since new benefactors have not been found, then the admission fees will not be raised. Therefore the museum will not remain open.

**18. If you use laundry detergent, make sure it is biodegradable. If the laundry detergent is biodegradable, then chemical action will break it down into simpler compounds and our lakes will not be harmed. Therefore if our lakes are harmed, then you are using laundry detergent that is not biodegradable.

**19. If you do not report all of your income, then you will be audited by the I.R.S. If you are audited by the I.R.S., then you will need an accountant and be fined a steep penalty. Therefore if you were not fined a steep penalty, then you did report all of your income.

20. If you sit in front of a computer monitor for too many hours, then your back may hurt or your eyesight will be temporarily blurred. Therefore if your back does not hurt, then you do not sit in front of a computer monitor for too many hours.

2.10

APPLICATION TO SWITCHING CIRCUITS

An interesting application of truth tables is to switching circuits.

A **switching circuit** consists of wires and switches through which electricity flows from one point to another. These points are called **terminals.** The purpose of a switch is either to allow electricity to flow or to stop it from flowing. The electrical system in your home is an example of a switching circuit. When you turn a light switch on, the electrical current will flow, causing the light to burn. When you turn the switch off, the electrical current is interrupted, and the light goes off.

Open switch

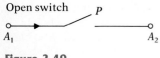

Figure 2.49

Closed switch

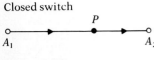

Figure 2.50

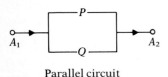

Parallel circuit

Figure 2.51

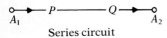

Series circuit

Figure 2.52

In a car, when you put the key into the ignition and turn it on, current will flow (the circuit is completed), and the engine starts. When you turn the ignition off, current will no longer flow, and the engine stops. (The circuit is broken.) Diagrams of some simple circuits are shown in Figs. 2.49 and 2.50.

In both Fig. 2.49 and Fig. 2.50, current can flow from terminal A_1 to terminal A_2. The arrows indicate the direction of the flow. Both circuits have a switch at P. In Fig. 2.49 the switch is open. Current *will not* flow from A_1 to A_2. The circuit is broken. In Fig. 2.50 the switch is closed. Current *will* flow from A_1 to A_2. The circuit is complete.

In Figs. 2.51 and 2.52, more complicated circuits are shown. Each has two switches that control the flow of current.

In Fig. 2.51, current starting at A_1 can go through either switch P or switch Q to get to A_2. Thus current will flow if either switch is closed. It is not necessary for both switches to be closed. If both switches are closed, then current can flow through either P or Q or both. If both switches are open, then current cannot flow at all from A_1 to A_2. Switches P and Q are said to be connected in **parallel.**

> **Definition 2.16** *Two switches P and Q are said to be **connected in parallel** if current will flow when either or both switches are closed. Current cannot flow if both switches are open.*

Now look at Fig. 2.52. The only way for current to flow from terminal A_1 to terminal A_2 is for both switches to be closed. If either switch P or switch Q is open, then current will not flow. What happens if both switches are open? In this circuit, switches P and Q are said to be connected in **series.**

> **Definition 2.17** *Two switches P and Q are said to be **connected in series** if current will flow only when both switches are closed. If either or both switches are open, current will not flow.*

To see the connection between switching circuits and truth tables, we use the following notation.

1. If a switch is closed, then we will assign to it the letter T. If a switch is open, we will assign to it the letter F.

2. If current will flow in a circuit, we will assign the letter T to the circuit. Otherwise, we will assign the letter F to the circuit.

In Fig. 2.49, switch P is open and is assigned F. Current will not flow in the circuit. So we assign F to the circuit.

In Fig. 2.50, switch P is closed. We assign T to it. Current will flow. We assign T to the circuit.

Look at Figs. 2.51 and 2.52. Using this notation, we can describe what will happen in each circuit by means of truth tables.

Table for parallel circuit (Fig. 2.51)				Table for series circuit (Fig. 2.52)		
P	*Q*	*Circuit*		*P*	*Q*	*Circuit*
T	T	T		T	T	T
T	F	T		T	F	F
F	T	T		F	T	F
F	F	F		F	F	F

Look at the table for parallel circuits. Line 1 says that if both switches are closed, then current will flow. Lines 2 and 3 say that current will flow if either switch is open and the other closed. Line 4 says current will not flow if both switches are open. Does this remind you of another truth table? Refer back to p. 86. You will see that this is exactly the disjunction table. If switches P and Q are connected in parallel, then another way of describing this is to say $P \vee Q$.

Now look at the table for series circuits. Line 1 says current will flow if both switches are closed. Lines 2, 3, and 4 show that current cannot flow if either or both of the switches are open. Does this table look familiar? It should (see p. 86). This is exactly the conjunction table. Thus another way of saying that P and Q are connected in series is to say $P \wedge Q$.

More complicated circuits can be constructed by using different combinations of parallel and series circuits. One example is given in Fig. 2.53.

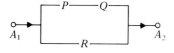

Figure 2.53

Switching circuits.

In this circuit, switches P and Q are connected in series. This combination of two switches is itself connected in parallel with R. This circuit can be symbolized as $(P \wedge Q) \vee R$. The truth table is as follows:

P	Q	R	$P \wedge Q$	$(P \wedge Q) \vee R$
T	T	T	T	T
T	T	F	T	T
T	F	T	F	T
T	F	F	F	F
F	T	T	F	T
F	T	F	F	F
F	F	T	F	T
F	F	F	F	F

Current will flow in the circuit if the switches are set up under the conditions of lines 1, 2, 3, 5, or 7 of the table.

As another example, consider the circuit shown in Fig. 2.54. Notice that one switch has the peculiar label $\sim Q$. This means that this switch is open if switch Q is closed and is closed if switch Q is open. P and $\sim Q$ are connected in parallel. The combination is connected in parallel with Q. This circuit can be symbolized as $[P \vee (\sim Q)] \vee Q$. The truth table is as follows:

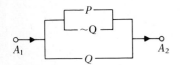

Figure 2.54

P	Q	$\sim Q$	$P \vee (\sim Q)$	$[P \vee (\sim Q)] \vee Q$
T	T	F	T	T
T	F	T	T	T
F	T	F	F	T
F	F	T	T	T

Notice that current will *always* flow in this circuit no matter which switches are open or closed. This is because we have only T's in the final column of the truth table. Therefore the switches may be removed from the circuit. They perform no useful function in allowing current to flow. They may be needed for other purposes.

Another example of a circuit is shown in Fig. 2.55. We can symbolize this as $[P \wedge (\sim Q)] \vee (P \wedge Q)$. The truth table is as follows:

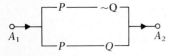

Figure 2.55

P	Q	$\sim Q$	$P \wedge (\sim Q)$	$P \wedge Q$	$[P \wedge (\sim Q)] \vee (P \wedge Q)$
T	T	F	F	T	T
T	F	T	T	F	T
F	T	F	F	F	F
F	F	T	F	F	F

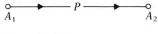

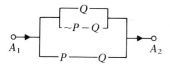

Figure 2.56

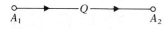

Figure 2.57

There is something interesting about this table. The first and last columns are identical. This means that current will flow if P is closed (lines 1 and 2) and will not flow if P is open (lines 3 and 4). Thus P alone will determine whether current flows. The other switches are unnecessary. Therefore we can **simplify** this circuit by eliminating all switches but switch P. The simplified circuit is shown in Fig. 2.56.

Why would we want to simplify the circuit?

As another example, consider the circuit shown in Fig. 2.57. In this circuit there are three Q switches. This means that when one of them is on, they are all on. When one of them is off, they are all off.

We can symbolize this circuit as $\{Q \lor [(\sim P) \land Q]\} \lor (P \land Q)$. The truth table is as follows:

P	Q	$\sim P$	$P \land Q$	$(\sim P) \land Q$	$Q \lor [(\sim P) \land Q]$	$\{Q \lor [(\sim P) \land Q]\} \lor (P \land Q)$
T	T	F	T	F	T	T
T	F	F	F	F	F	F
F	T	T	F	T	T	T
F	F	T	F	F	F	F

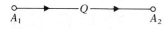

Figure 2.58

Look at the last column. It should look familiar. It is exactly the same as the column for Q. Thus the above circuit is exactly the same as the circuit that would be symbolized as Q. This would be a simple circuit as shown in Fig. 2.58.

Since this circuit is much simpler than the original one and does exactly the same job, we would probably prefer to use it rather than the original. Again we may ask, "Why would a simpler circuit be better?" There are several answers. One obvious reason is that a simpler circuit has fewer switches and is therefore cheaper and easier to construct. Also, if something goes wrong in the circuit, it is easier to locate the trouble and make repairs when there are fewer switches.

Our final example illustrates an interesting application of switching circuits to a familiar game. Everyone has at one time or another played the game "matching pennies." The game works as follows: Two players, Moe and Larry, flip a coin at the same time. If both coins come up heads or both coins come up tails, then Moe wins. Otherwise, Larry wins. A toy manufacturer is interested in making an electrical version of this game. A bulb is to light up if Moe wins. No bulb will light up if Larry wins. Instead of both players flipping coins, they will both push a button that will open or close a switch. We want to design the circuit for this game.

Let P be Larry's switch and Q be Moe's switch. At a given signal, each player pushes a button. This corresponds to flipping a coin. Just as a coin may land heads or tails, pushing a button may open or close a switch. If both switches are in the same position (open or closed), the bulb will light up. Otherwise, it will not light up. The simplest circuit for this is given in Fig. 2.59.

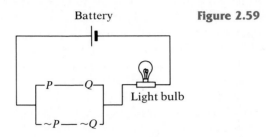

Battery   **Figure 2.59**

Light bulb

The formula for this circuit is $(P \wedge Q) \vee [(\sim P) \wedge (\sim Q)]$. Its truth table is shown below. This shows that current will flow and the bulb will light up if both switches are open (line 4) or if both switches are closed (line 1). If one switch is closed and one is open, then current will not flow and the bulb will not light up (lines 2 and 3).

P	Q	$\sim P$	$\sim Q$	$P \wedge Q$	$(\sim P) \wedge (\sim Q)$	$(P \wedge Q) \vee [(\sim P) \wedge (\sim Q)]$
T	T	F	F	T	F	T
T	F	F	T	F	F	F
F	T	T	F	F	F	F
F	F	T	T	F	T	T

EXERCISES

Symbolize and construct a truth table for each of the circuits shown in Exercises 1–7.

1.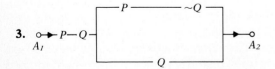

2.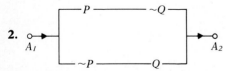

3.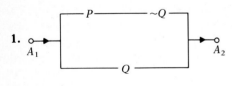

4.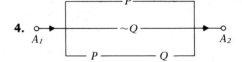

5.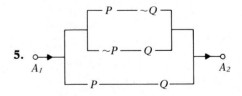

6.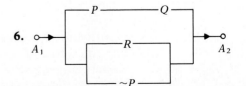

7.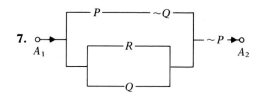

In Exercises 8–9, determine whether the pairs of circuits are electrically equivalent. (Hint: Use truth tables.)

8. 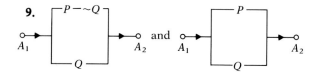 and

9. 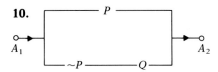 and

Simplify each of the circuits shown in Exercises 10–13.

10.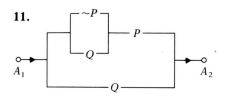

11.

12.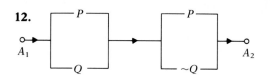

13.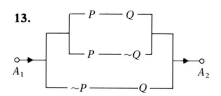

14. Can you simplify any of the circuits shown in Exercises 1–7?

Draw a diagram for the circuits corresponding to the formulas given in Exercises 15–18.

15. $(\sim P) \wedge (Q \vee R)$

16. $[P \wedge (\sim Q)] \vee [(\sim P) \wedge Q]$

17. $[(P \vee Q) \wedge R] \vee [(\sim P) \wedge Q]$

18. $[(P \wedge R) \vee Q] \vee [((\sim R) \vee (\sim P)) \wedge ((\sim Q) \vee P)]$

**19. Three judges must decide whether a convicted murderer should be executed or not. This will be decided by a secret vote. Design a switching circuit for determining a majority vote.

2.11

SOME FURTHER THOUGHTS

In the previous material on logic, a number of interesting points were purposely omitted because they were not in the mainstream of our discussion. However, they are too important to be ignored entirely. Moreover, you may have thought about them yourselves and have unanswered questions. Therefore we conclude this chapter with these "logical afterthoughts."

Symbolizing Other Kinds of Statements

Consider the statement "there is a man with seven wives and there are men who do not have seven wives." This statement cannot be usefully symbolized by the techniques we have learned so far. If you are wondering why not, then try it.

What about the statements "all teachers except mathematics teachers are human," and "not all dogs like children"? If you try to symbolize these statements with what we have learned so far, you will run into trouble. Try it!

You may then wonder if there are statements that cannot be symbolized at all. By more advanced techniques of mathematical logic, it is possible to symbolize the above propositions and others like them. Such statements require the use of other symbols. For the most part, truth tables cannot be used to determine the truth value of these more complicated statements.

Many-Valued Logic

Up to now we have worked with propositions that were either true or false. It has probably occurred to you that one may not always know for sure whether a statement is definitely true or false. There are statements that we may not want to classify as definitely true or false, but rather as maybe. For example, consider the statement "Richard likes spinach." This statement may or may not be true, depending on the condition in which it is served. For statements like this, we may want to consider three possible truth values: True, False, and Maybe. A logic that allows any statement to have three possible truth values is called a **three-valued logic.** Truth tables can be constructed accordingly. A simple example of such a table is the negation. If M stands for the truth value of "Maybe," then we have:

Negation table for three-valued logic

p	$\sim p$
T	F
M	M
F	T

A logic system may also be constructed with more than three possible truth values for any statement. Such systems are called **many-valued logics.** There are some situations to which such systems can be applied, but they are beyond the scope of this book.

The Connection between Sets and Logic

The subjects of sets and logic may not appear to have any significant connection. However, the two are closely related. To see the connection between them, consider the logical expression $p \wedge q$ and the set operation $P \cap Q$. We know that $p \wedge q$ is true *only* when both p and q are true.

We know that if an element is in $P \cap Q$, then it must be in both set P and set Q.

Do you see any connection? If not, replace p by P, q by Q, and the symbol $\wedge$ by the symbol $\cap$.

Again consider the logical expression $\sim p$ and the set operation P'. If p is true, then $\sim p$ is false. If an element belongs to set P, then it cannot belong to set P'. Now do you see the connection?

Comment Notice that a similar pattern occurs in both sets and logic. In mathematics it often happens that patterns appearing in one area reoccur in others. This is one of the beauties of mathematics. Much of the usefulness of mathematics can be attributed to this property.

Comment In view of the connection between sets and logic and the possible interchange of symbols, many problems in logic can be solved by the technique of sets, and vice versa.

Other Applications of Logic

We have discussed only a few of the possible applications of logic. It can be applied to many different fields, some outside mathematics. For example, if you have ever examined an insurance policy, you will find that it has a large number of confusing clauses. Some of the clauses may even appear to be contradictory or repetitious. Insurance companies often hire experts in logic to analyze and simplify policies by the methods of symbolic logic.

Similarly, logic can be used to examine complicated business contracts and legal situations for possible contradictions and repetitions.

Logic can also be used to test the accuracy of surveys and censuses.

EXERCISES

1. Make up a truth table for conjunction in three-valued logic.

2. Make up a truth table for disjunction in three-valued logic.

3. Make up a truth table for $[(\sim p) \wedge q] \vee p$ in three-valued logic.

4. We have pointed out that the operation of $\cap$ for sets corresponds to the logic symbol $\wedge$. Using this idea, complete the following chart, which connects the notation for logic with the notation for sets.

Logic expression	Set expression
$p \wedge q$	
$p \vee q$	
$\sim p$	
$p \rightarrow q$	

Some Mathematical Recreations

Some of the following problems can be solved by using the techniques we have discussed throughout this chapter. Others can be solved by using informal but correct reasoning. Try to solve as many as you can. Have fun!

** **5.** Suppose that Mr. Smith is a small-town barber who shaves those men and only those men of the town who do not shave themselves. Which of the following statements are true? (Everyone in the town is shaved or shaves himself.)

> Mr. Smith shaves himself.

> Mr. Smith does not shave himself.

** **6.** A travel agent has just booked flights for three of her clients, Bill Holland, Pat Canada, and Debbie England. One of them is going to Holland, one is going to Canada, and one is going to England. Bill is not going to Holland, Pat is not going to Canada, and Debbie is not going to England. If Pat is not going to England either, to which country is each going?

** **7.** There were three prisoners, one of whom was blind. Their jailor offered to free them all if any one could succeed in the following game. The jailor produced three white hats and two red hats and, in a dark room, placed a hat on each prisoner. Then the prisoners were taken into the light where, except for the blind one, they could see one another. (None could see the hat on his own head.) The game was for any prisoner to state correctly what color hat he himself was wearing. The jailor asked one of those who could see if he knew, and the man answered no. Then the jailor asked the other prisoner who could see, and he answered no. The blind prisoner at this point correctly stated the color of his own hat, winning the game for all three. How did he know?[3]

** **8.** The six employees of the Southshore Loan Company are Bill Black, Jane Coffee, Arlene McCarthy, George Kelly, Phil White, and Phyllis Pagano. The positions they occupy (not necessarily in order) are manager, assistant manager, cashier, stenographer, teller, and computer programmer. The assistant manager is the manager's grandson; the cashier is the stenographer's son-in-law; Bill Black is a bach-

---

3. Irving M. Copi, *Introduction to Logic*, pp. 16–17. New York: MacMillan, 1961. Reprinted by permission.

elor; Phil White is 22 years old; Arlene McCarthy is the teller's stepsister; and George Kelly is the manager's neighbor. Which position does each person occupy?

** **9.** *Should I do my homework assignment?* Professor Gertrude Hoffman teaches mathematics at a midwestern university. The class meets five days a week (Monday through Friday). Homework assignments, which are very lengthy and require at least three hours of intensive work, are assigned daily. Professor Hoffman has found that many students have not been doing their homework assignments. She announces to the class that she will send students to the blackboard to do the homework assignments on one of the days of the following week. (Students will be graded on this work.) She does not specify on which day of the week this will occur, since that would encourage last minute cramming. However, she does promise that she will cancel the assignment completely if any student figures out, in advance (not necessarily before the week begins), the actual day on which students will be sent to the board (the "event").

Hilda Lichtenfeld is a student in the class. She boldly claims that under the conditions mentioned, the event can never occur. She reasons as follows: First, Friday is excluded as a possible day, since if no one is called to the board by Thursday, then everyone would know that Friday is the designated day. Professor Hoffman would thus be forced to cancel the assignment. Second, Thursday is excluded. Since Friday has already been eliminated, students will be able to recognize Thursday plans as soon as Wednesday ends without students being sent to the board. Professor Hoffman would thus be forced to cancel the assignment. In a similar manner, Wednesday, Tuesday, and Monday are successively eliminated as days on which Professor Hoffman will send students to the board. Do you agree with Hilda's reasoning? Explain your answer.

** **10.** A father wished to leave his fortune to the most intelligent of his three sons. He said to them, "I shall presently take each one of you away separately and paint either a white or a blue mark on each of your foreheads, and none of you will have any chance to know the color of the mark on his own head. Then I shall bring you together again, and anybody who is able to see two blue marks on the heads of his companions is to laugh. The first of you to figure out his own color is to raise his hand, and on convincing me that his solution is correct, will become my heir." After all three had agreed to the conditions, the father took them apart and painted a white mark on each forehead. When they met again, there was silence for some time, at the end of which the youngest brother raised his hand, saying: "I'm white." How was he able to deduce the color of the mark on his forehead?[4]

**11. Three golfers, Tom, Dick, and Harry, are walking to the clubhouse. Tom always tells the truth, Dick sometimes tells the truth, and Harry never tells the truth. The three golfers are lined up from left to right. The golfer on the left says, "That guy in the middle is Tom." However, the golfer in the middle claims that he's Dick. Furthermore, the golfer on the right says, "No, the guy in the middle is Harry." Find out who is on the left, who is in the middle, and who is on the right. (*Hint:* First try to determine which one is Tom.)

**12. Four bank employees have been arrested for embezzling funds from the bank. It is known that one of these employees definitely embezzled the funds. In interviews with the bank investigators, Arthur claimed that Mike was the embezzler, Mike said that it was Georgina, Heather denied embezzling the funds, and Georgina declared that Mike had lied when he said that she was the embezzler. Each of these employees is then given a polygraph (lie detector) test. Assume that the results of the polygraph test are accurate. If the results of the polygraph tests show that only one of these employees is

a) lying, who is the guilty one?

b) telling the truth, who is the guilty one?

---

4. Max Black, *Critical Thinking*, 2nd ed., p. 12. New York: Prentice-Hall, 1952. Reprinted by permission.

**13. Fingers Nelson was shot dead in a recent gang war, and his body was dumped into an unused mine shaft. After a lengthy investigation the sheriff arrested five men and charged them with the crime. Each man claimed that he was innocent. Nevertheless, each made three statements to the sheriff, two of which were true and one of which was false.

> Rocky said: "I did not kill Fingers. I never owned a gun. Lucky did it."
>
> Ricky said: "I am innocent. I never saw Rico before. Lucky did it."
>
> Lucky said: "I am innocent. Rico is the guilty man. Rocky lied when he said I did it."
>
> Rico said: "I did not kill Fingers. Fats is the guilty man. Ricky and I are old buddies."
>
> Fats said: "I did not kill Fingers. I never owned a gun. The other guys are all passing the buck."

After analyzing their statements the sheriff was able to determine who was guilty. Who murdered Fingers Nelson?

**14. Mrs. Ada Gusher, wife of the oil billionaire Tex Gusher, was found murdered. Rock Head, the dashing private eye, was called in to solve the mystery. He discovered the following clues.

> The Gushers' maid, Sarah, was not home when the crime was committed.
>
> Either Sarah was home or the Gushers' son Rodney was out.
>
> If the stereo was on, Rodney was home.
>
> If the stereo was not on, Mr. Gusher did it.

Rock Head solved the crime the same day. How? And who did it?

**15. Of two tribes inhabiting a tropical island, members of one tribe always tell the truth and members of the other always lie. A math teacher vacationing on the island comes to a fork in the road and has to ask a native bystander which branch she should take to reach the nearest village. She doesn't know whether the native is a truth-teller or a liar, but she nevertheless manages to ask only one true–false question so cleverly phrased that she will know from the reply which road to take. What question could she ask?

**16. In a certain mythical community, politicians always lie, and nonpoliticians always tell the truth. A stranger meets three natives and asks the first of them if he is a politician. The first native answers the question. The second native then reports that the first native denied being a politician. Then the third native asserts that the first native is really a politician. How many of these three natives are politicians?[5]

**17. Mabel tells her friends, "I always lie; I never tell the truth." Is Mabel lying or telling the truth? Explain your answer.

**18. Frank has a cube that is three meters on each side. He wishes to cut it up into 27 one-meter cubes. There are several ways in which this can be done. One way is to make a series of six cuts through the cube while keeping it together in one block. Can you find another way to do

5. Copi, op. cit., p. 16.

this in which *fewer than* six cuts are needed? (*Hint:* The pieces may be rearranged between each cut.)

**19. There are three boxes of marbles on a table. It is known that one box has two red marbles, one box has one red and one white marble, and the third box has two white marbles. The boxes have been labeled red–red, red–white, and white–white. However, the wrong labels were attached to each box. Thus a box that has the label red–white does not really have a red and a white marble in it. A volunteer is removing one marble at a time from a box. Find the *minimum* number of marbles that must be removed before she knows what is in a particular box.

**20. Imagine that you have a checkerboard that has 64 squares. Since each domino can cover 2 squares, you would need 32 dominoes to cover the whole board. Imagine that we now cut off the 2 squares from opposite ends of the checkerboard. Can the checkerboard now be covered with only 31 dominoes? Explain your answer by using logic.

2.12

SUMMARY   In this chapter we discussed logic in general and the difference between inductive and deductive logic. Many arguments that are presented by people or that appear in print turn out to be invalid when analyzed by one of the several methods discussed in this chapter; Venn diagrams, the standard diagram, or truth tables. We applied truth tables to switching circuits.

We also discussed the connection between sets and logic and presented some challenging exercises for your enjoyment.

STUDY GUIDE   You should now be able to demonstrate your knowledge of the following ideas by giving definitions, descriptions, or specific examples. Page references are given in parentheses.

Basic Ideas
Inductive logic (p. 50)
Deductive logic (p. 52)
Hypothesis (premise) (p. 52)
Conclusion (p. 52)
Valid argument (p. 52)

Invalid argument (p. 52)
Pythagorean theorem (p. 55)
Prime number (p. 55)
Pascal's triangle (p. 56)
Four-color problem (p. 57)

Testing the Validity of Arguments by:

a) Venn diagrams (p. 65)

b) the standard diagram (p. 72)

c) truth tables (p. 95)

Applications of Logic Were Given for:

a) mathematical patterns (p. 54)

b) arguments (pp. 65, 72, 95)

c) switching circuits (p. 99)

d) mathematical recreations (p. 107)

FORMULAS TO REMEMBER

The following list summarizes all the formulas given in this chapter.

Negation table

p	$\sim p$
T	F
F	T

Conjunction table

p	q	$p \wedge q$
T	T	T
T	F	F
F	T	F
F	F	F

Disjunction table

p	q	$p \vee q$
T	T	T
T	F	T
F	T	T
F	F	F

Implication table

p	q	$p \rightarrow q$
T	T	T
T	F	F
F	T	T
F	F	T

Original statement:	If p, then q.
Converse:	If q, then p.
Inverse:	If $\sim p$, then $\sim q$.
Contrapositive:	If $\sim q$, then $\sim p$.

MASTERY TESTS

Form A   **1.** Let p = "Some blondes have more fun." Then $\sim p$ is

    a) All blondes have more fun.

    b) Some blondes do not have more fun.

    c) No blondes have more fun.

    d) Some blondes do have more fun.

    e) none of these

2. A man drinks 6 scotches and 6 sodas on Monday and gets drunk. On Tuesday he drinks 6 vodkas and 6 sodas and gets drunk. On Wednesday, he drinks 6 bourbons and 6 sodas and gets drunk. On Thursday, he drinks 6 brandies and 6 sodas and gets drunk. He concludes that the soda is making him drunk. This is an example of   **a)** inductive reasoning **b)** deductive   reasoning   **c)** illogical   reasoning   **d)** conductive reasoning   **e)** none of these

3. The statement $[(\sim p) \rightarrow q] \wedge [(\sim q) \rightarrow p]$ is   **a)** a tautology   **b)** a self-contradiction   **c)** neither of these

4. The converse of "If Ronald is handsome, then I will date him" is

    a) If I date Ronald, then he is handsome.

    b) If Ronald is not handsome, then I will not date him.

    c) If I do not date Ronald, then he is not handsome.

    d) If Ronald is not handsome, then I will date him.

    e) none of these

For questions 5–10, use the following information:

Hypotheses: (1) All drugs are expensive.

          (2) Some drugs are beneficial.

          (3) All drugs are habit-forming.

5. The conclusion "some drugs are not habit-forming" is

    a) valid   **b)** invalid   **c)** not enough information

6. The conclusion "some habit-forming things are beneficial" is

    a) valid   **b)** invalid   **c)** not enough information

7. The conclusion "some habit-forming things are not beneficial" is

    a) valid   **b)** invalid   **c)** not enough information

8. The conclusion "some expensive things are not habit-forming" is

    a) valid   **b)** invalid   **c)** not enough information

9. The conclusion "some expensive things are beneficial" is

    a) valid   **b)** invalid   **c)** not enough information

10. The conclusion "some expensive things are habit-forming" is

    a) valid   **b)** invalid   **c)** not enough information

11. Five hundred doctors in the metropolitan area were asked to recommend a sugarless gum. Since 50 percent of them recommended Tishman's gum, it was concluded that 50 percent of *all* doctors advise their patients to chew Tishman's gum. This is an example of

   a) deductive reasoning   **b)** inductive reasoning   **c)** reductive reasoning   **d)** an indirect proof   **e)** none of these

12. Which of the following is *not* a proposition?

   a) That's so obvious.   **b)** Not now.   **c)** $3 + 2 = 7$

   d) $3 + 2 = 5$ **e)** none of these

13. A formula representing the circuit shown in Fig. 2.60 is given by which of the following?

   a) $(Q \wedge P) \wedge (\sim Q \vee P)$

   b) $(Q \vee P) \vee (\sim Q \wedge P)$

   c) $(Q \vee \sim P) \wedge (\sim Q \wedge P)$

   d) $(Q \vee \sim P) \vee (\sim Q \wedge P)$

   e) none of these

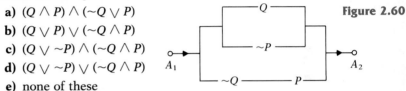

Figure 2.60

14. If p is true and q is false, then $[(\sim p) \wedge (\sim q)] \rightarrow p$ is

   a) true   **b)** false   **c)** not enough information given

15. Find the truth value of the statement "If $2 + 2 = 7$, then I am a genius."

   a) true   **b)** false   **c)** not enough information given

Form B

1. Write the inverse of the statement "If I win the lottery, then I will be a millionaire."

   a) If I will not win the lottery, then I will be a millionaire.

   b) If I am a millionaire, then I won the lottery.

   c) If I have not won the lottery, then I am not a millionaire.

   d) If I am not a millionaire, then I have not won the lottery.

   e) none of these

2. Which diagram is a possibility for the statement "Some senior citizens (*S*) are in excellent health (*E*)"?

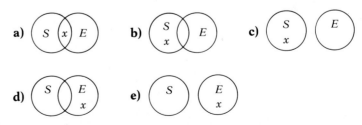

3. Draw a circuit that corresponds to the formula $[(\sim P) \vee (Q \wedge P)] \wedge P$.

4. Test the following conclusion for validity by using truth tables:

Hypotheses: (a) If you are planning to travel overseas, then you will need a visa.

(b) If you are planning to travel overseas, then you will need to be vaccinated against smallpox.

Conclusion: If you need a visa, then you will need to be vaccinated against smallpox.

a) valid   **b)** invalid   **c)** not enough information given

5. Test the following conclusion for validity by using the standard diagram:

Hypotheses: (a) All things injected with growth hormones are unhealthy.

(b) Some chickens are not injected with growth hormones.

Conclusion: Some chickens are unhealthy.

a) valid   **b)** invalid   **c)** not enough information given

6. The statement $p \rightarrow [(\sim p) \vee q)]$ is an example of

a) a tautology   **b)** a self-contradiction   **c)** neither of these

7. If p is false and q is true, then the statement $[(\sim p) \wedge q)] \rightarrow [p \vee (\sim q)]$ is

a) true   **b)** false   **c)** not enough information given

8. Find the truth value of the statement "Either Agent Orange is a chemical that is dangerous to your health or lowering your blood pressure can reduce your chances of having a heart attack."

a) true   **b)** false   **c)** not enough information given

9. Test the following for validity:

Hypotheses: (a) All college graduates get good-paying jobs.

(b) Some women get good-paying jobs.

Conclusion: Some women are college graduates.

a) valid   **b)** invalid   **c)** not enough information given

10. The following chart is the truth table for

p	q	
T	T	F
T	F	T
F	T	T
F	F	F

a) exclusive disjunction

b) inclusive disjunction

c) conjunction

d) joint denial

e) none of these

11. Write a symbolic statement for the circuit shown in Fig. 2.61.

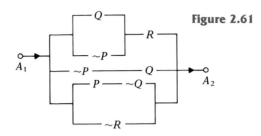

Figure 2.61

12. In inductive reasoning, the conclusion of an argument is
    a) probably true   **b)** probably false   **c)** definitely false
    d) none of these

13. Write the contrapositive of the statement "If you donate money to the United Way Fund, then you will not be audited by the I.R.S."

14. Consider the magic square shown on the right. Find the value of x.

6	1	8
7		3
		x

15. One solution to the 6×6 magic square problem is shown below. What is the magic sum?

6	2	34	33	35	1
25	11	27	10	8	30
19	23	16	15	20	18
13	14	22	21	17	24
12	29	9	28	26	7
36	32	3	4	5	31

3

DIFFERENT NUMBER SYSTEMS

CHAPTER OBJECTIVES

To study many different number systems including the Babylonian, Egyptian, Tamil, Mayan, and Roman systems. (*Sections 3.2 and 3.3*)

To analyze how computations are performed in these systems. (*Sections 3.2 and 3.3*)

To compare calculations performed in these systems with the calculations performed in the Hindu-Arabic decimal system. (*Section 3.3*)

To describe several different ways of performing multiplication. (*Section 3.3*)

To discuss different number bases and how the four basic operations are performed in these other bases. (*Section 3.4*)

To indicate how to convert numbers from base 10 into some other base and how to convert numbers from some other base back into base 10. (*Sections 3.5 and 3.6*)

To apply the idea of different number bases to computers and to the game of *Nim*. (*Sections 3.7 and 3.8*)

To introduce you to some interesting numbers. (*Section 3.9*)

To learn about prime numbers and how to express any number as a product of primes. (*Section 3.10*)

To use scientific notation when working with very large numbers or very small numbers. (*Section 3.11*)

117

Chisanbop to Replace the Pocket Calculator

NEW YORK—In a day and age when schoolchildren are performing various calculations using pocket calculators instead of their brains, it is a pleasure to report on a new method of performing the four basic mathematical functions rather proficiently. The method is known as **Chisanbop** and was invented 20 years ago by Sung Jin Pai, an expert Korean mathematician. In this "finger calculation" method, which is based on our familiar decimal system, children can do any form of addition, multiplication, division, or subtraction. The fingers on the hand replace the calculator. The functions of accumulating data and computing answers are done by the fingers and not the mind. To accomplish this, the fingers are assigned different values as shown in the figure at right. The method is currently being tested in several schools in New York City.

TRENDS IN EDUCATION, March 27, 1980

The ten fingers are marked with the values that are assigned to them in the Chisanbop system.

The magazine clipping above indicates that there is more than one way of performing the four basic operations (some of these are either already in use or are constantly being tried). How do these approaches differ from our familiar way of performing calculations in the decimal system? Are these other approaches more efficient? Do these alternative approaches instill in youngsters a real sense of the *meaning* of numbers?

3.1

INTRODUCTION

Humans learned how to count early in their development. Tally marks have been found on cave walls that indicate that even while still living in caves, people knew how to count and had progressed to the point at which they could record their results. Over the years, various bones have been dug up that have notches on them to represent different numbers. Some of these date back as far as 30,000 years.

However, it seems likely that the human ability to count dates back even farther than is indicated by any records that have been found. Counting appears to be a rather basic activity that does not require advanced mental development. It has been demonstrated that even some animals and birds (for example, crows) have the ability

HISTORICAL NOTE

Originally, early humans distinguished only between the numbers one and two. Everything else was just many. Thus if Ug had three children and his brother Og had ten children, each would say that he had "many" children. Even today there are some primitive tribes that still count this way.

As soon as people learned how to distinguish between larger numbers, they found it more convenient to count by groups than by ones. Today we count in groups of 10. Historically, counting has been done using groups of 2's, 3's, 5's, 10's, and other numbers. Groupings by 2's and 3's were widely used earlier in time. However, these were almost always replaced by groupings of 5's and 10's. From this came our present decimal system (from the Latin *decem*, meaning "ten").

As we pointed out earlier, piles of stones were often used in counting. For example, six stones represented the number "six." Because piles of

stones can easily be disturbed, numbers were sometimes recorded by carving notches in rocks, bones, and sticks. In the nineteenth century a wolf bone was found in Czechoslovakia with 55 notches in it, arranged in groups of 5.

In some cultures, groups of 20 were used. These systems, based on 20, are still to be found in various parts of the world. The French word for 80 is *quatre-vingts*, which means "four 20's." This suggests that at some time in the past, counting in some parts of France was done in groups of 20.

It is easy to see how systems developed based on 2, 5, 10, and 20 (remember, cave people did not wear shoes). However, it is surprising to find evidence that some cultures may have counted in groups of four or eight. The Latin word for "nine" is *novem*, which may be connected with the Latin word *novus* meaning "new." This suggests that nine was the start of a new group.

to distinguish between groups of up to four objects.[1] You may have seen, on television or in the circus, dogs and horses that can count small quantities.

In Section 3.2 we will examine one early number system. Then we will examine the decimal system, with which we all count today. We will also discuss other number systems and their applications.

3.2

THE BABYLONIAN NUMBER SYSTEM

The ancient Babylonians counted in groups of 60. Although it might seem strange that such a large number was used, we will see shortly that it has many advantages. In the late nineteenth century, tablets were found, most of which date back to around 1700 B.C. From these tablets we have been able to learn much about Babylonian mathematics.

In the Babylonian system, two basic symbols were used: a vertical wedge Y and a corner wedge ◄ . The vertical wedge represented

---

1. See Levi Conant, "The Number Concept: Its Origin and Development," 1923. In James R. Newman, ed., *The World of Mathematics*, pp. 432–441. New York: Simon and Schuster, 1956. Cf. H. Kalmus, "Animals as Mathematicians," *Nature* 202:1156–1160 (1964).

The University Museum, University of Pennsylvania

An early system of writing, in Mesopotamia, was made up of small, simple drawings called pictograms. Each pictogram stood for an object or idea, or sometimes for several words. For numerals, astronomers used a base of 60, repeating number signs up to nine times.

the number 1, and the corner wedge represented the number 10. For example, the symbol YYY meant the number 3, the symbol ◁YY meant the number 12, and the symbol ◁YYY meant the number 14. The Babylonians wrote all the numbers 1, 2, 3, . . . , 59 in this manner. When they got to 60, they moved over one place to the left, as we do in our decimal system when we get to 10. Thus the number 72 would be written as Y◁YY. The first Y symbol represents not "1," but one "60," because of its position. The last two Y 's (on the right) represent two "1's" because of their position.

The number 146 would be written as

YY   ◁◁   YYY
Two 60's Two 10's Six 1's

The first two YY symbols represent two 60's, or 120. The two ◁◁ symbols mean two 10's, and the YYY symbols mean 6. Thus we have 120 + 20 + 6 or 146.

When the Babylonians got to 60 × 60, or 3600, they moved over two places to the left as we do in our decimal system when we get to 10 × 10, or 100. Thus the number 4331 would be written as

Y   ◁YY   ◁Y
One 3600 Twelve 60's Eleven 1's

The first Y means one 60 × 60, or one 3600. The symbols ◁YY mean twelve 60's, or 720. Finally the symbol ◁Y on the right of the number means eleven 1's. This gives 3600 + 720 + 11 = 4331.

You will notice that in this system the value of a symbol is determined by its position. The same is true in our decimal system, as we will see later in this chapter.

One important difference between the Babylonian system and our decimal system is that the Babylonians never really had a symbol for zero as we use it today. Thus the symbol Y might stand for 1, 60, 60 × 60, etc. In later texts they did use a sign that looked like this ⌃ to indicate the empty spaces if they occurred *inside* numbers. Thus this symbol acted as a placeholder, much as our number 0 does. However, the Babylonians never used this symbol at the end of a number. So, as we have said, the symbol Y could stand for 1, 60, 60 × 60, etc.

To reemphasize the point, the importance of the Babylonian system was that the value of a number was determined by its position.

This idea of **positional notation** is fundamental in our own decimal system, as we shall see in Section 3.3.

The Babylonians used 60 as a grouping number for several reasons. One reason was that 60 can be evenly divided by many numbers. For example, 60 can be divided by 2, 3, 4, 5, ... This made division and work with fractions much easier. The choice of 60 may also have been due to the Babylonian interest in astronomy. As a matter of fact, the Babylonian year was divided into 12 months of 30 days each, with an additional 5 feast days (12 times 30 equals 360, which can be evenly divided by 60). It has also been suggested by some historians that 60 was used as a natural combination of two earlier systems, one using 10 and the other using 6.

The Babylonian system was taken over by the Greek astronomers. In fact, it was used for many mathematical and practically all astronomical calculations as late as the seventeenth century. Many traces are to be found even today. For example, hours are divided into 60 minutes, and minutes are divided into 60 seconds. In geometry, angles are divided into degrees. Each degree is divided into 60 minutes.

EXERCISES

The Egyptian Numeration System

The early Egyptians used the following symbols to represent numbers (the order or position of symbols was not important).

Number	1	2	3	4	5	6	7
Symbol	I	II	III	IIII	III II	III III	IIII III

Number	8	9	10	11	12	13
Symbol	IIII IIII	III III III	∩	'I∩	II∩	III∩

Number	14	15	16	17	18
Symbol	II II∩	III II∩	III III∩	IIII III∩	IIII IIII∩

Number	19	20	100	1000	10,000
Symbol	III III III∩	∩∩	9	⚮	⟍

For example, the symbol 999∩∩II would represent the number 322. The symbol 9 III∩ would represent the number 119. The number 21,318 could have been written by the Egyptians as

⟍⟍ ⚮ IIII∩
999

What does each of the numbers in Exercises 1–6, written in the **Egyptian system**, represent in our system?

1. 999 III ∩∩

2. ⟍⟍ ⚮ ⚮ ∩∩ III 9

3. ⚮ ∩∩∩∩ 99 IIII

4. ⟍ ⚮ ⚮ 9999 ∩∩ IIII

5. ⚮ III III III ∩ 9

6. ⟍ III 99 ⚮ ∩∩ III

In Exercises 7–14, translate each of the indicated numbers into the Egyptian system.

7. 376	**8.** 4237
9. 98	**10.** 27,695
11. 5432	**12.** 6002
13. 10019	**14.** 11111

Addition and Subtraction in the Egyptian System

Addition and subtraction of numbers was not difficult in the Egyptian system. For example, 1231 and 3412 would be added as follows:

Sometimes regrouping (or "carrying") is needed when we obtain more than nine strokes ||||||||||. We just replace ten of these strokes (or heelbones) by the symbol ∩. Thus we have the following:

More simply, the sum is

In Exercises 15–18, perform the indicated operations in the Egyptian system and check your answer by converting the numbers to our decimal system.

15. Plus

16. Plus

17. Minus

18. Minus

19. Explain why the Egyptian system had no need for a symbol for zero.

Tamil Numerals

In the Tamil language (south India) the following symbols were used to represent numbers. (In this case the order or position of symbols was important.)

Number	1	2	3	4	5	6
Symbol	𝆑	2	𝆑𝆑	𝆑	℺	𝆑𝆑
Number	7	8	9	10	100	1000
Symbol	6	2/	𝆑	ω	η	𝆑

For example, the number 3456 would be written as

𝆑𝆑 𝆑 𝆑 η ℺ ω 𝆑𝆑

Translate each of the numbers in Exercises 20–27 into the Tamil system.

20. 47   **21.** 83   **22.** 398   **23.** 476
24. 987   **25.** 654   **26.** 8765   **27.** 4321

The numbers in Exercises 28–35 are written in the Egyptian system. Translate each of them into the Tamil system and then into the Babylonian system.

28. ∩∩ |||
|||

29. ⚹ ⚹ 99 ∩ ||

30. ∩ 99 9 |||
||

31. ∩∩ 9999 ⚹ ||||

32. 𝒞 ⚹ ⚹ 99 ∩∩ ||||
|||

33. 𝒞 ⚹ ⚹ ⚹ 999 ∩∩ ||||

34. 999 ∩∩∩∩⚹ |||

35. 𝒞 ∩ |||
||| ∩ 99 ⚹

The Mayan Numeration System

In the Mayan number system, dots and dashes were used to represent numbers. The dots were grouped horizontally above the dashes, and the dashes were stacked vertically. Thus the Mayan number system was a vertical system, quite different from the Egyptian system and our Hindu-Arabic system. In the Mayan system the following symbols were used to represent the numbers 1 through 19:

Number	1	2	3	4	5	6	7	8	9	10
Symbol	•	••	•••	••••	—	•̲	••̲	•••̲	••••̲	̳
Number	11	12	13	14	15	16	17	18	19	
Symbol	•̳	••̳	•••̳	••••̳	̿	•̿	••̿	•••̿	••••̿	

For any number greater than 20 they used vertical groupings. Thus the Mayan number ̈•••̲ ̲ ̲ consists

of two groups of dots and dashes. The top group represents 8, and the bottom group represents 12 as shown below.

•••̲ } This represents 8
••̲̲ } This represents 12

The numeral in the bottom group represents the number of ones or units in the number, and the

---

HISTORICAL NOTE

The Mayas, a group of related Indian tribes of the Mayan linguistic stock, lived in what are now the Mexican states of Veracruz, Yucatán, the whole Yucatán peninsula, Guatemala, and parts of British Honduras. Historical records dating back to about 200 A.D. indicate that their civilization was quite advanced. Their highly complex calendar was the most accurate until the introduction of the Gregorian calendar. The year began on July 16 and consisted of 365 days; 364 of the days were divided into 28 weeks of 13 days each. The new year began on the 365th day. Additionally, 360 days of the year were divided into 18 months of 20 days each. The series of weeks and months ran both consecutively and independently of each other.

numeral in the top group represents the number of 20's in the number. So we have

••• } This represents $8 \times 20 = 160$

•• } This represents $12 \times 1 = \underline{12}$

172

Thus the Mayan number ••• represents 172. When

••

three vertical groupings of dots and dashes appear, then the bottom group denotes 1's, the middle group denotes 20's, and the top group denotes 360's.

Thus the Mayan number • would be 2392 in our
•
••
system.

• } This represents $6 \times 360 = 2160$

• } This represents $11 \times 20 = 220$

•• } This represents $12 \times 1 = \underline{12}$

2392

In Exercises 36–41, translate the numbers into the Mayan system.

36. 31   **37.** 57   **38.** 317

39. 1207   **40.** 1234   **41.** 2576

****42.** The Mayan numeral for 20 is not ☰ . What is the numeral for 20? (*Hint:* In the Mayan system a string of dots cannot have more than four, and a pile of dashes cannot have more than three.)

43. What is the Mayan numeral for 360?

44. What does the following Mayan numeral represent in our system?

••••

(*Hint:* You might be tempted to say that this represents 9, since there are 4 dots and 1 dash. However, this is wrong, since the four dots are not arranged horizontally.)

3.3

THE BASE 10 NUMBER SYSTEM

In this section we will discuss the decimal system that is commonly used throughout the world today. We will start by looking at the Roman numeral system. In this system the following symbols are used:

I stands for 1,

V stands for 5,

X stands for 10,

L stands for 50,

C stands for 100,

D stands for 500,

M stands for 1000,

and so forth.

The number 43 is written as XLIII.

The number 63 is written as LXIII.

The position of the numerals is important. In 43, the X goes before the L to indicate that it is 10 less than 50. In 63, X goes after the L. This means that 10 is to be added to 50. As we will see shortly, in

Bruce Anderson

our number system the position of the digits 0, 1, 2, 3, 4, 5, 6, 7, 8, 9 is even more important than in the Roman system. For example, the number 31 is different from the number 13 even though both numbers contain the same digits, 1 and 3. It is the use of position or *place* that makes our system so convenient to work with. Note that the use of position in the Roman system is different from our use of position.

To better appreciate *our* system (which is known as the **Hindu-Arabic system**), try to add the numbers 43 and 63 *using Roman numerals*.

$$\begin{array}{r} \text{XLIII} \\ +\underline{\text{LXIII}} \end{array}$$

What is your answer? It should be CVI. If you did get this answer, was it by adding the Roman numerals? Or did you add 43 and 63 in our system and convert the answer to Roman numerals? If you did, it is understandable, since doing arithmetic in the Roman system is quite complicated. As we look at our system in detail, we will see exactly why it makes arithmetic so much easier.

To get started, consider the number 4683. This means

$$4(\text{thousands}) + 6(\text{hundreds}) + 8(\text{tens}) + 3(\text{ones}).$$

This can be restated as

$$4(1000) \quad + 6(100) \quad + 8(10) \quad + 3(1).$$

Notice something interesting about this:

$$1000 = 10 \times 10 \times 10.$$

This is usually abbreviated as 10^3, which means 10 multiplied by itself 3 times. Similarly, $100 = 10 \times 10$. This is abbreviated as 10^2, which means 10 multiplied by itself. Moreover, 10 can be written as 10^1.

We can now write the number 4683 as $4(10^3) + 6(10^2) + 8(10^1) + 3$. This can be neatly summarized in the following chart.

10^3	10^2	10^1	1's
4	6	8	3

In each of these columns, one of ten possible **digits** can appear. These digits are 0, 1, 2, 3, 4, 5, 6, 7, 8, 9. Every number can be expressed as some combination of these digits. Why are these ten sufficient? Why don't we need more?

Comment   In this system of writing numbers, the *place* of the number determines its value. In the example above, the *place* of the 4 tells us it stands not for 4 ones, 4 tens, or 4 hundreds, but for 4 thousands. Similarly, the place of the 8 tells us it stands for 8 tens, and so on.

As another example, consider the number 20,694. This stands for 2(ten-thousands) + 0(thousands) + 6(hundreds) + 9(tens) + 4(ones), or $2(10^4) + 0(10^3) + 6(10^2) + 9(10^1) + 4$. This can be written as

10^4	10^3	10^2	10^1	1's
2	0	6	9	4

The zero that appears in the 10^3 column is very important. If we leave it out, the number would read 2694, and that is not the number we want. The zero is called a **placeholder.**

Now let us see how we add these two numbers. For convenience we put them in one chart.

	10^4	10^3	10^2	10^1	1's
		4	6	8	3
+	2	0	6	9	4
Sum	2	5	3	7	7

In the 1's column we add 3 and 4 and get 7 1's.

In the 10^1 column we add 8 and 9 and get 17. This means that we have 17 tens or 10(tens) + 7(tens). We know that 10(tens) is 10×10, or 10^2. Thus we have $1(10^2)$ and $7(10^1)$. So the 1 must be **carried over** to the 10^2 column where it belongs. The 7 remains in the 10^1 column.

In the 10^2 column we add 6 and 6 and the 1 that was carried. This gives $13(10^2)$, which equals $10(10^2) + 3(10^2)$. Now $10(10^2)$ is $10 \times 10 \times 10$, which equals 10^3. Thus $13(10^2)$ equals $1(10^3) + 3(10^2)$. The 3 is left in the 10^2 column, and the 1 is *carried over* to the 10^3 column.

This procedure is repeated for each column until we are finished. We can summarize this procedure as follows:

1. In each column, the only digits allowed are 0, 1, 2, 3, 4, 5, 6, 7, 8, 9.

2. In any column, when the sum is more than 9, we carry everything over 9 to the next column.

This convenient method of addition works because of the way in which numbers are written in the Hindu-Arabic system. It will not work for the Roman system. In the Hindu-Arabic system the value of each digit depends on its *place*. For example, 53 means 5 tens plus 3, whereas 35 means 3 tens plus 5.

In doing multiplication we use a similar technique. For example,

$$\begin{array}{r} 23 \\ \times 45 \\ \hline 115 \\ 92 \\ \hline 1035 \end{array}$$

In doing this multiplication we first multiply the 23 by 5, giving 115. Then we multiply the 23 by 4. This gives 92. We write the 92 under the 115 as shown. We then add and get 1035 as our final answer. Why did we move the 92 one place to the left?

Division and subtraction are done similarly.

The system we have just described is also called the **decimal system.** The decimal system was not widely accepted until the late Middle Ages. It took a long time for this system to be put into everyday use. In fact, in the late thirteenth century the city of Florence in Italy passed laws against the use of the Hindu-Arabic numerals. This was done to protect its citizens from the dishonest persons who did such things as interchanging the numbers 0, 6, and 9.

Because the decimal system uses groupings of 10's, we refer to it as a **base 10 number system.**

Comment   Although all of us are familiar with how the 10 digits are written in the decimal system, it should be noted that the digits that appear on the bottom of bank checks are often written in a slightly different form, as shown below:

0 1 2 3 4 5 6 7 8 9

EXERCISES

Translate each of the numbers given in Exercises 1–4 into the Roman numeral system.

1. 59   **2.** 94

3. 69   **4.** 1234

Translate each of the numbers given in Exercises 5–8 from the Roman numeral system into the Hindu-Arabic system.

5. LXXXIV   **6.** CCIX

7. MCXLIX   **8.** MDCCCXXII

In Exercises 9–10, try to perform the indicated operations in the Roman numeral system.

9.   CXIX
      +MDXLI

10.   MDCCCLXXII
       −   DCVI

11. Our number system is known as the Hindu-Arabic system because it originated in India and was spread by the Arabs. Look up the Hindu-Arabic numeration system in encyclopedias and books on the history of mathematics to determine exactly how the number system originated in India and how it was spread by the Arabs to Egypt.

Galley or Gelosia Multiplication

Another way of doing multiplication is by the so-called **Gelosia method.** It works as follows: Suppose we want to multiply the two numbers, 257 and 49. We set it up as shown:

First we multiply 7 by 4. This gives 28, which is placed as shown below:

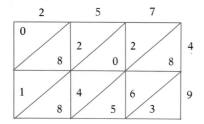

Then we multiply 9 by 7. This gives 63. We put this in as shown (second line, right column). Now we multiply 5 by 4. This gives 20, which we write in (first line, middle column). We complete the diagram in this manner. Note that 2 times 4 is entered as 08 (first line, left column). Now we add along the diagonals starting from the lower right corner, carrying where necessary. This gives the following:

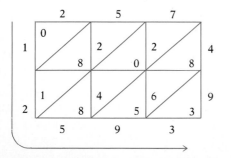

The first diagonal contains only 3. So 3 is its sum. The next diagonal contains 5, 6, and 8, which add to 19. We put the 9 down and carry the 1. The next diagonal contains 8, 4, 0, 2, and the 1 we carried. These add to 15, so we put down 5 and carry the 1. We continue, in a similar manner, until finished. Our answer is then read off as 12,593.

You can check the answer to the above problem by using regular multiplication.

Multiply each of the numbers given in Exercises 12–17 by using the Gelosia method. Check your answer by using regular multiplication.

12.	28	13.	368	14.	459
	×69		× 54		× 86

15.	1978	16.	6962	17.	7692
	× 532		× 439		× 635

18. Can you explain why Gelosia multiplication works?

Napier's Bones

The English mathematician John Napier (1550–1617) used the Gelosia system of multiplication to construct what we would today call a computing machine. His gadget is referred to as Napier's rods or Napier's bones (see Fig. 3.1a). We can construct a variation of Napier's bones using 10 popsicle

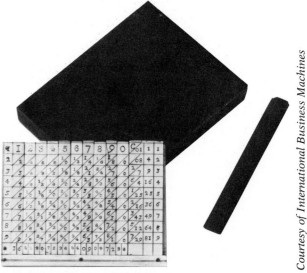

Figure 3.1(a) Napier's bones

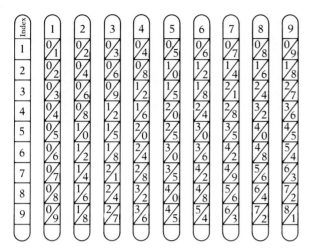

Figure 3.1(b)

sticks and numbering them as shown in Fig. 3.1(b). The first stick is called the index and lists all the digits from 1 through 9. On top of each of the other sticks we write one of the digits 1 through 9. Each stick gives the product of an index number with the number on the stick. The ones and tens of each product are separated by a diagonal line as in Gelosia multiplication. The different products are separated by horizontal lines.

To see how we can use these sticks to multiply two numbers, let us multiply 435 × 8. We place the index alongside the sticks headed by the digits 4, 3, and 5 as shown in Fig. 3.2. Next we locate the

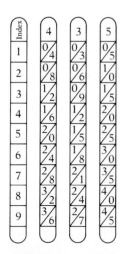

Figure 3.2

row of numbers that are on the same line as 8 on the index:

Finally, we add along the diagonals as we do in Gelosia multiplication:

Our answer is 3480. Thus 435 × 8 = 3480.

Using a similar procedure, we find that the product of 435 and 9 is 3915 as shown:

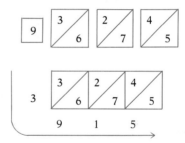

Thus 435 × 9 = 3915.

We can combine the two previous results to obtain the product 435 × 89. We have

$$
\begin{array}{r}
435 \\
\times 89
\end{array}
\qquad
\begin{array}{rl}
9 \times 435 = & 3{,}915 \\
80 \times 435 = & 34{,}800. \\
\hline
89 \times 435 = & 38{,}715
\end{array}
$$

Thus 435 × 89 = 38,715.

Using popsicle sticks, make up a set of Napier rods as described above and use them to find the product in each of Exercises 19–28.

19. 53 × 8

20. 65 × 9

21. 374 × 64

22. 586 × 76

23. 639 × 57

24. 768 × 54

25. 395 × 563

26. 641 × 293

27. 507 × 6143

28. 5324 × 8276

29. An everyday application of the base 10 number system occurs in our electric meter readings or in our natural gas meter readings.

   a) Determine the number of cubic feet of gas if the dials of the meter are as shown below.

   b) What role does 0 play in your answer to part (a)?

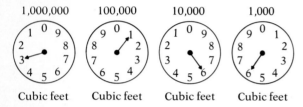

1,000,000   100,000   10,000   1,000

Cubic feet   Cubic feet   Cubic feet   Cubic feet

Egyptian Duplation

The Egyptians multiplied by a process we call **duplation.** This method of multiplication is based on the fact that any number can be expressed as the sum of powers of 2. Thus 19 can be expressed as 19 = 1 + 2 + 16. The product of two numbers is obtained by multiplying any one number by the different powers of 2 needed to get the other number. The partial products are added to obtain our answer. For example, let us multiply 19 by 58. We know that 19 = 1 + 2 + 16, so we set up the following chart:

Powers of 2	×	*Original Number*	=	*Product*
①	×	58	=	㊽ 58
②	×	58	=	⑯116
4	×	58	=	232
8	×	58	=	464
⑯	×	58	=	⑨928
19				1102

Adding the circled numbers gives 19 on the left side and 1102 on the right. Using the Egyptian numerals given in Section 3.2, we can multiply 19 × 7 as

follows:

Powers of 2 × Original Number = Product

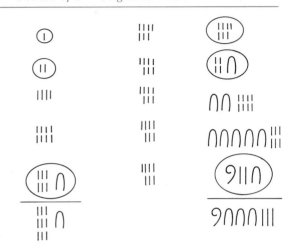

Using the Egyptian system of duplation, multiply each of the numbers given in Exercises 30–34.

30. 42 × 57

31. 17 × 78

32. 27 × 55

33. ∩|||| × ⫶∩∩∩∩

34. ∩∩∩⫶⫶ × ∩∩∩⫶⫶ ... ∩∩∩ ||| × ∩∩

Russian Peasant Method

An unusual way of doing multiplication, which was used by the ancient Egyptians and until recently by the Russian peasants, is known as the **Russian peasant method.** To use it, you need only know how to multiply and divide by 2. We illustrate the technique by multiplying 67 × 18. This is shown in Fig. 3.3.

Figure 3.3

67	18
33	36
16	~~72~~
8	~~144~~
4	~~288~~
2	~~576~~
1	1152
	1206

What we did is the following. In the left column of the figure we divide 67 by 2, *disregarding the remainder*. This gives 33. Then we divide 33 by 2, disregarding the remainder. This gives 16. We repeat the same procedure until we get 1 on the left. On the right, we double 18 and get 36. Then we double 36 and get 72, etc.

Finally, we cross off the numbers on the right that are opposite *even* numbers on the left. We add what remains in the right column. The sum, 1206, is our answer.

You can check the answer by using ordinary multiplication.

Using the Russian peasant multiplication technique, multiply the numbers in Exercises 35–45.

35. 52
× 33

36. 423
× 39

37. 227
× 58

38. 547
× 68

39. 234
× 49

40. 831
× 63

41. 783
× 68

42. 219
× 56

43. 628
× 49

44. 457
× 79

45. 299
× 68

46. How does the Russian peasant multiplication technique compare with the Egyptian method of duplation?

Finger Multiplication

Pick a number from 1 to 9. You can multiply it by 9 using **finger multiplication** (see Fig. 3.4). Suppose you are multiplying 4 by 9. Hold up your hands.

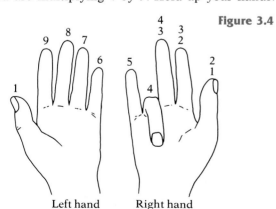

Figure 3.4

Left hand   Right hand

Count off, in succession, 9 fingers and then 4 fingers (after you get to the tenth finger start again) going from right to left. When you finish counting, bend down the next finger.

You can now read off your answer. The first digit is the number of fingers to the right of the bent finger. This is 3. The second digit is the number of fingers to the left of the bent finger. This is 6. Our answer then is 36.

47. Using finger multiplication, multiply 7 × 9.

48. Using finger multiplication, multiply 5 × 9.

49. Using finger multiplication, multiply 8 × 9.

50. Using finger multiplication, multiply 6 × 9.

51. Can you explain why the method of finger multiplication works?

Subtraction by Complements

Another method of performing subtraction involves using complements. The **complement** of a digit x is $9 - x$. Thus the complement of 5 is 4. The complement of 7 is 2. The complement of 0 is 9. The method of subtraction by complements involves finding the complement of each digit in the subtrahend. The resulting numerals are then added to the minuend. The answer to the original subtraction problem is now found by deducting 1 from the left digit and adding 1 to the right digit. To illustrate, let us subtract 4356 from 8967. We have

$$8967 \quad \text{Minuend}$$
$$-4356 \quad \text{Subtrahend}$$

The complement of the digits in the subtrahend is

$$\begin{array}{cccc} 9 & 9 & 9 & 9 \\ -4 & -3 & -5 & -6 \\ \hline 5 & 6 & 4 & 3 \end{array}$$

or 5643. This number is now added to 8967, yielding

$$8967$$
$$+5643$$
$$\overline{14610}$$

Subtracting 1 from the extreme left digit and adding 1 to the extreme right digit gives

$$14610$$
$$-1 \quad +1$$
$$\overline{4611}$$

Thus our answer to the original subtraction problem is

$$8967$$
$$-4356 \quad \text{or} \quad 4611.$$
$$\overline{4611}$$

(You should verify that 4611 is indeed the answer.)

Using the method of subtraction by complements, find each of the differences in Exercises 52–56.

52. 7961
      −4358

53. 6901
      −4653

54. 7654
      −2976

55. 6900
      −4276

56. 623
      −599

3.4

OTHER BASES

In Section 3.1 we mentioned that some primitive people counted in groupings of five. Suppose we had continued this method today and still counted by groupings of five. What would numbers look like in such a system? The setup is the same as that for the base 10 system. The only difference is that now we work with groupings of five. We will have only five digits. These are 0, 1, 2, 3, 4.

Let us consider the number written as 324 in the **base 5 system.**

5^2	5^1	1's
3	2	4

This means that we have $3(5^2) + 2(5^1) + 4(1\text{'s})$.

To indicate that we are working in base 5, we write this as $324_{(5)}$. This is read as "three-two-four" and not as "three hundred and twenty-four." The words "thousand," "hundred," "twenty," "tens," etc., are words that are used only in the base 10 system.

In base 10, when we write 324, we mean

10^2	10^1	1's
3	2	4

This would mean $3(10^2) + 2(10^1) + 4(1\text{'s})$. This, of course, we read as "three hundred and twenty-four."

Comment   Although we use the same symbol 324 in both bases, they have different meanings; 324 in base 5 represents a different quantity than 324 in base 10. The symbol 324 is a *numeral* that represents different *numbers* in different bases. A **number** is a quantity, whereas a **numeral** is a symbol used to represent it.

To better understand the difference between number and numeral, consider the *number* seven. We can represent it by any of the following numerals:

$$\text{VII}, \quad 7, \quad \bar{7}, \quad |||||| \text{ (Egyptian)}.$$

As another example of a number written in a different base, consider 3203 in base 5, written as $3203_{(5)}$. This means

5^3	5^2	5^1	1's
3	2	0	3

$$3(5^3) + 2(5^2) + 0(5^1) + 3(1\text{'s}).$$

Let us now add $324_{(5)}$ and $3203_{(5)}$ in base 5.

	5^3	5^2	5^1	1's
	3	2	0	3
+		3	2	4
Numbers carried	1		1	
Answer	4	0	3	2

The procedure is exactly the same as in base 10, except that we are now working with groupings of five.

First we add 3 and 4. This gives 7, which is the same as 5 plus 2 or $1(5^1) + 2(1\text{'s})$. The $1(5^1)$ must be "carried" to the next column, which is the 5^1 column. The 2 remains in the 1's column.

Next we add 2 and 0 and the 1 we carried in the 5^1 column. This gives 3. No carrying is needed. We then add 2 and 3 in the 5^2 column. This gives $5(5^2)$ or $1(5^3)$. We put down 0 in the 5^2 column and carry a 1 into the 5^3 column.

Finally, we have in the 5^3 column a 3 and the 1 carried, which gives 4. Our final answer is $4032_{(5)}$.

We can summarize the procedure as follows.

1. In each column the only digits allowed are 0, 1, 2, 3, and 4.

2. In any column, when the sum is more than 4, we carry everything over 4 to the next column.

Compare this to the technique used in base 10, summarized earlier.

To further illustrate base 5 arithmetic, we give several other examples.

EXAMPLE 1   Add $1204_{(5)}$
$\quad\quad\quad\quad\quad \underline{+332_{(5)}}$
$\quad\quad\quad\quad\quad 2041_{(5)}$

SOLUTION   In the 1's column, 2 + 4 gives 6, which is $1(5^1) + 1$. We carry a 1 and leave 1. In the 5^1 column, 0 + 3 + the 1 carried gives 4. Since there is no carrying needed, we just leave 4. In the 5^2 column, 3 + 2 gives 5, which is $5(5^2)$ or $1(5^3)$. We carry a 1 and leave 0. In the 5^3 column we have a 1 and the 1 carried, which gives 2. There is nothing to carry to the next column. Our answer is $2041_{(5)}$. ▭

EXAMPLE 2   Multiply $23_{(5)}$

$$\begin{array}{r} 23_{(5)} \\ \times\,34_{(5)} \\ \hline 202 \\ 124 \\ \hline 1442_{(5)} \end{array}$$

SOLUTION   First we multiply 3 by 4, which gives 12. We have $2(5^1)$ to carry and 2(1's) left over. Next we multiply 4 by 2, which is 8. Adding the carried 2, we have 10 altogether. This gives $2(5^1)$ to carry and 0 left over. So on the first line we have 202. Now we multiply the 23 by the 3. This gives 124. Notice that we move the 124 one space to the left. Compare this to the procedure we use for base 10. Why do we move one space to the left?

Finally, we add and get $1442_{(5)}$. ▭

EXAMPLE 3   Subtract $123_{(5)}$

$$\begin{array}{r} 123_{(5)} \\ -104_{(5)} \\ \hline 14_{(5)} \end{array}$$

SOLUTION   We cannot subtract 4 from 3, so we have to borrow. Since we are working in base 5, we borrow a 5. (Compare this to what we do in base 10.) Now we have 4 from 3 + the borrowed 5, or 8. This leaves 4. The 2 in the second column is now a 1. (Why?) 1 minus 0 is 1. Finally, we have 1 minus 1 in the last column, which gives 0. Our final answer is $14_{(5)}$. (As in base 10, a zero at the beginning of a number is not written down.) ▭

Many primitive peoples have only two numbers. Such a system is called a **base 2** or **binary system.** Let the two digits be 0 and 1. How would we do arithmetic in the binary system? Remember that we will use groupings of two.

EXAMPLE 4   Let us consider the number 110110 in the base 2 system. We can represent it in a chart as

2^5	2^4	2^3	2^2	2^1	1's
1	1	0	1	1	0

This means that we have $1(2^5) + 1(2^4) + 0(2^3) + 1(2^2) + 1(2^1) + 0(1's)$.

▭

EXAMPLE 5 Add $101_{(2)}$ and $110_{(2)}$.

SOLUTION
$$101_{(2)}$$
$$\underline{+110_{(2)}}$$
$$1011_{(2)}$$

First we add 1 and 0, which gives 1. Then we add 0 and 1, which gives 1. Finally, we add 1 and 1, which gives 2. We carry this "2" (as a 1) to the next column and leave behind 0. Our final answer is $1011_{(2)}$. (Since the only digits are 0 and 1, we cannot have 2 in any column.) ☐

EXAMPLE 6 Multiply
$$110_{(2)}$$
$$\underline{\times 11_{(2)}}$$
$$110$$
$$\underline{110}$$
$$10010_{(2)} \quad ☐$$

EXAMPLE 7 Subtract
$$1101_{(2)}$$
$$\underline{-111_{(2)}}$$
$$110_{(2)} \quad ☐$$

It is possible to use any number larger than 1 as a base. The following examples illustrate some of these other bases.

EXAMPLE 8 Add
$$456_{(7)}$$
$$\underline{+324_{(7)}}$$
$$1113_{(7)} \quad ☐$$

EXAMPLE 9 Multiply
$$357_{(8)}$$
$$\underline{\times 65_{(8)}}$$
$$2253$$
$$\underline{2632}$$
$$30573_{(8)} \quad ☐$$

EXAMPLE 10 Subtract
$$3101_{(4)}$$
$$\underline{-233_{(4)}}$$
$$2202_{(4)} \quad ☐$$

EXAMPLE 11 Multiply
$$468_{(9)}$$
$$\underline{\times 57_{(9)}}$$
$$3632$$
$$\underline{2574}$$
$$30472_{(9)} \quad ☐$$

In a base 12 system we need 12 digits. (Why?)

EXAMPLE 12   Let the digits of a base 12 system be 0, 1, 2, 3, 4, 5, 6, 7, 8, 9, t, and e. We have created two new symbols: t, which equals $9 + 1$, and e, which equals $9 + 2$. Let us add $12e_{(12)}$ and $t1_{(12)}$.

SOLUTION
$$12e_{(12)}$$
$$+ \ t1_{(12)}$$
$$\overline{210_{(12)}} \ \square$$

EXAMPLE 13   Multiply
$$12e_{(12)}$$
$$\times \ t1_{(12)}$$
$$\overline{12e}$$
$$\underline{1052}$$
$$1064e_{(12)} \ \square$$

EXAMPLE 14   In some base b, $57_{(b)}$ equals 52 in base 10. Find b.

SOLUTION   We know that
$$57_{(b)} \quad \text{means} \quad 5(b^1) + 7(1\text{'s}).$$
Therefore if $57_{(b)} = 52_{(10)}$, we have
$$57_{(b)} = 5 \cdot b + 7 = 52.$$
By trial and error you find that $b = 9$. (If you are familiar with algebra, you can solve it directly.) $\square$

Comment   Why should we study numbers written in bases other than 10? After all, the decimal (base 10) system is used throughout the world today. There are several reasons. As we shall see in Section 3.7, all modern computers work in base 2, 8, or 16. So if you ever work much with computers, you will need this background. In Section 3.8 we will see how some popular games, many of them thousands of years old, can be analyzed mathematically by using base 2. Furthermore, as we have seen, the difficulties we have in doing arithmetic in other bases are similar to those a beginner has in base 10. So if you ever teach a child arithmetic, either as a parent or a teacher, you will understand his or her struggles and be better able to help. We are so familiar with base 10 that we really have to study other bases to get a better understanding of base 10.

Division in other bases is performed in exactly the same way that we do division in base 10. This is because division is thought of as the opposite of multiplication. For example, in the decimal system, $18 \div 6 = 3$, since $6 \cdot 3 = 18$. To illustrate the procedure for dividing in base 5, consider the following.

EXAMPLE 15   Divide $1434_{(5)}$ by $4_{(5)}$.

SOLUTION   We set up the division problem as in base 10. We have

$$
\begin{array}{r}
221 \\
4\overline{)1434} \\
13 \\
\hline
13 \\
13 \\
\hline
4 \\
4 \\
\hline
0
\end{array}
\qquad \text{(In base 5, we know that } 4 \times 2 = 13.\text{)}
$$

Thus our answer is $221_{(5)}$. □

EXAMPLE 16   Divide $21104_{(5)}$ by $413_{(5)}$.

SOLUTION   Again we set up the problem as in base 10. We have

$$
\begin{array}{r}
23 \\
413\overline{)21104} \\
1331 \\
\hline
2244 \\
2244 \\
\hline
0
\end{array}
$$

Thus our answer is $23_{(5)}$. □

Comment   We can verify that our answer is correct by multiplying $23_{(5)}$ with $413_{(5)}$. Our answer is indeed $21104_{(5)}$.

EXERCISES

In Exercises 1–15, perform the indicated operations in the given base.

1. $11101_{(2)}$
    $+11111_{(2)}$

2. $21022_{(3)}$
    $+\ 1102_{(3)}$

3. $6834_{(9)}$
    $-4565_{(9)}$

4. $14te_{(12)}$
    $+\ te1_{(12)}$

5. $1232_{(4)}$
    $+\ 232_{(4)}$

6. $2012_{(7)}$
    $-1236_{(7)}$

7. $4764_{(8)}$
    $-2175_{(8)}$

8. $23232_{(4)}$
    $-\ 2323_{(4)}$

9. $354_{(6)}$
    $-245_{(6)}$

10. $425_{(6)}$
     $\times\ 34_{(6)}$

11. $543_{(7)}$
     $\times\ 65_{(7)}$

12. $et5_{(12)}$
     $\times 2te_{(12)}$

13. $22122_{(3)}$
     $\times\ 2221_{(3)}$

14. $878_{(9)}$
     $\times\ 57_{(9)}$

15. $534_{(9)}$
      $678_{(9)}$
     $+837_{(9)}$

In Exercises 16–20, divide the first number by the second number in the indicated base and check your answer.

16. Divide $13332_{(5)}$ by $23_{(5)}$.

17. Divide $25454_{(6)}$ by $34_{(6)}$.

18. Divide $101010101_{(2)}$ by $1011_{(2)}$.

19. Divide $133t6t_{(12)}$ by $te1_{(12)}$.

20. Divide $30230_{(4)}$ by $32_{(4)}$.

In Exercises 21–24, find the base b in which the numbers are written.

21. $56_{(b)} = 51$ **22.** $35_{(b)} = 32$

23. $77_{(b)} = 70$ ****24.** $345_{(b)} = 137$

25. Why must the base of a number system be greater than 1?

26. What is wrong with the following calculation?

$$7563_{(9)}$$
$$-2315_{(9)}$$
$$\overline{5248_{(9)}}$$

**27. In base 10 the number 32 is even. Are there any bases in which this number is odd? Explain your answer.

In Exercises 28–31, a computation was performed in a base other than 10. Can you find the base?

28. $\begin{array}{r} 324 \\ +\ 513 \\ \hline 1140 \end{array}$ **29.** $\begin{array}{r} 543 \\ -256 \\ \hline 265 \end{array}$ **30.** $\begin{array}{r} 245 \\ \times\ 32 \\ \hline 10302 \end{array}$ **31.** $\begin{array}{r} 527 \\ \times\ 32 \\ \hline 14622 \end{array}$

32. The following is an addition problem in base 3. However, instead of using the digits 0, 1, and 2, we have used the letters A, D, and M. Each letter stands for the same digit every time it is

used. Determine which of the digits 0, 1, and 2 is represented by each letter.

$$\begin{array}{r} M\,A\,M\,A \\ +\ \ D\,A\,D\,A \\ \hline M\,A\,M\,A\,A \end{array}$$

33. The following is an addition problem in base 4. However, instead of using the digits 0, 1, 2, and 3, we have used the letters A, D, M, and P. Each letter used represents the same digit every time it is used. Determine which of the digits 0, 1, 2, and 3 is represented by each letter.

$$\begin{array}{r} M\,A\,M\,A \\ +D\,A\,D\,A \\ \hline P\,A\,P\,A \end{array}$$

**34. The following is an addition problem in base 10. However, instead of using the digits 0, 1, 2, 3, 4, 5, 6, 7, 8, and 9, we have used letters. Each letter used represents the same digit every time it is used. Determine which of the digits 0, 1, 2, . . . , 9 is represented by each letter.

$$\begin{array}{r} T\,W\,O \\ T\,H\,R\,E\,E \\ +\ \ S\,E\,V\,E\,N \\ \hline T\,W\,E\,L\,V\,E \end{array}$$

3.5

CONVERTING FROM BASE 10 TO OTHER BASES

Imagine one person working in base 10 and another person working in base 5. To communicate with each other, they would need a method for changing a base 10 number to a base 5 number and vice versa. In this section we will discuss how we convert a base 10 number to any other base.

We will illustrate the technique by an example. Let us convert 258 written in base 10, to base 5. We write down 258 and divide it by 5. We get 51 with a remainder of 3. This is written as shown.

$$\begin{array}{rl} & \quad\ \textit{Remainder} \\ 5\,\lfloor\underline{258} & \quad\quad 3 \\ 5\,\lfloor\underline{\ 51} & \quad\quad 1 \\ 5\,\lfloor\underline{\ 10} & \quad\quad 0 \\ 5\,\lfloor\underline{\ \ 2} & \quad\quad 2 \\ \quad\ 0 & \end{array}$$

Then we divide the 51 by 5. This gives 10 with a remainder of 1. We write the 1 under the 3 (as shown) in the remainder column. Now we divide 10 by 5. This gives 2 with a remainder of 0. The remainder of 0 is written under the 1 in the remainder column. Finally, we divide the 2 by 5. Since 5 does not "go" into 2, we write down 0 with 2 in the remainder column. We now read off the answer *from bottom to top in the remainder column*. Our answer is 2013. Therefore

$$258_{(10)} = 2013_{(5)}.$$

This technique will work for conversion from base 10 to *any* other base.

EXAMPLE 1   Convert $152_{(10)}$ to base 5.

SOLUTION

	Remainder
5 | 152	2
5 | 30	0
5 | 6	1
5 | 1	1
0	

Our answer: $152_{(10)} = 1102_{(5)}$.   ☐

EXAMPLE 2   Convert $78_{(10)}$ to base 4.

SOLUTION

	Remainder
4 | 78	2
4 | 19	3
4 | 4	0
4 | 1	1
0	

Our answer: $78_{(10)} = 1032_{(4)}$.   ☐

EXAMPLE 3   Convert $167_{(10)}$ to base 2.

SOLUTION

	Remainder
2 $\lfloor 167$	1
2 $\lfloor\ 83$	1
2 $\lfloor\ 41$	1
2 $\lfloor\ 20$	0
2 $\lfloor\ 10$	0
2 $\lfloor\ \ 5$	1
2 $\lfloor\ \ 2$	0
2 $\lfloor\ \ 1$	1
0	

Our answer: $167_{(10)} = 10100111_{(2)}$.   ☐

EXAMPLE 4   Convert $138_{(10)}$ to base 3.

SOLUTION

	Remainder
3 $\lfloor 138$	0
3 $\lfloor\ 46$	1
3 $\lfloor\ 15$	0
3 $\lfloor\ \ 5$	2
3 $\lfloor\ \ 1$	1
0	

Our answer: $138_{(10)} = 12010_{(3)}$.   ☐

EXAMPLE 5   Convert $428_{(10)}$ to base 9.

SOLUTION

	Remainder
9 $\lfloor 428$	5
9 $\lfloor\ 47$	2
9 $\lfloor\ \ 5$	5
0	

Our answer: $428_{(10)} = 525_{(9)}$.   ☐

EXAMPLE 6   Convert $134_{(10)}$ to base 12.

SOLUTION

Remainder

12 $\lfloor$ 134   2

12 $\lfloor$ 11   e

0

Our answer: $134_{(10)} = e2_{(12)}$. $\square$

Beware For this method to work you must read *up* the remainder column. If you read down the remainder column, your answer will be wrong.

EXERCISES

In Exercises 1–16, convert each of the base 10 numbers to the indicated base

1. 423 to base 4
2. 178 to base 5
3. 129 to base 3
4. 426 to base 8
5. 546 to base 9
6. 478 to base 7
7. 386 to base 6
8. 329 to base 12
9. 278 to base 16
10. 288 to base 5
11. 84 to base 2
12. 193 to base 7
13. 129 to base 2
14. 278 to base 12
15. 627 to base 8
16. 4600 to base 12

17. Why does the technique discussed in this section work?

18. Write the number $217_{(10)}$ first in base 2 and then in base 8. Now do the same thing for the number 301. By examining your answers, can you find any relationship between base 2 and base 8?

19. What are some advantages to having a small-number base instead of a large-number base? What are some of the disadvantages?

**20. Mysterio the magician holds up four cards marked as shown below. He asks someone in the audience to think of a number from 1 through 15 and to tell him on which of the cards it appears. The first person tells him that the number he is thinking of is on cards 1, 3, and 4. Mysterio then correctly tells him that the number is 13. The second person then tells him that he is thinking of a number that appears only on cards 1 and 2. Mysterio tells him that his number is 3.

a) How does Mysterio do it?

b) Why does the trick work? (*Hint:* Write all the numbers 1 through 15 in the binary system.)

Card 1		Card 2		Card 3		Card 4	
1	9	2	10	4	12	8	12
3	11	3	11	5	13	9	13
5	13	6	14	6	14	10	14
7	15	7	15	7	15	11	15

21. Since base 2 involves only the digits 0 and 1, we can count in base 2 on our fingers. This can be done as follows. Keep all fingers down to represent 0 and a finger up to represent a 1. The position of the "up" finger denotes the position of the "1" in the base 2 representation. Thus

we can represent the first six counting numbers as shown below.

$00001_{(2)}$ $00010_{(2)}$ $00011_{(2)}$ $00100_{(2)}$ $00101_{(2)}$ $00110_{(2)}$
The number 1   The number 2   The number 3   The number 4   The number 5   The number 6

Using a similar procedure, show how we can represent the numbers 17 through 31 on one hand.

3.6

CONVERTING FROM OTHER BASES TO BASE 10

In the preceding section we learned how to convert a number from base 10 to any other base. In this section we will show how to convert a number written in any other base back into base 10.

Remember what 123 in base 5 really means. To refresh your memory, we will write it in the following form:

5^2	5^1	1's
1	2	3

This means that we have $1(5^2) + 2(5^1) + 3(1\text{'s})$.
Since $5^2 = 5 \times 5 = 25$, we then have

$$1(25) + 2(5^1) + 3(1\text{'s})$$
$$25 + 10 + 3$$
$$38.$$

Therefore $123_{(5)} = 38_{(10)}$.
Some further examples will be helpful.

EXAMPLE 1   Convert $314_{(6)}$ to base 10.

SOLUTION   We rewrite 314 in base 6 in the following form:

6^2	6^1	1's
3	1	4

This means that we have $3(6^2) + 1(6^1) + 4(1\text{'s})$.
Since $6^2 = 6 \times 6 = 36$, we have

$$3(36) + 1(6) + 4(1\text{'s})$$
$$108 + 6 + 4$$
$$118.$$

Our answer: $314_{(6)} = 118_{(10)}$. □

EXAMPLE 2   Convert $111011_{(2)}$ to base 10.

SOLUTION   We rewrite $111011_{(2)}$ as follows:

2^5	2^4	2^3	2^2	2^1	1's
1	1	1	0	1	1

Since

$$2^5 = 2 \times 2 \times 2 \times 2 \times 2 = 32,$$
$$2^4 = 2 \times 2 \times 2 \times 2 \quad\; = 16,$$
$$2^3 = 2 \times 2 \times 2 \qquad\quad\;\; = 8, \quad \text{and}$$
$$2^2 = 2 \times 2 \qquad\qquad\quad\; = 4,$$

we then have

$$1(32) + 1(16) + 1(8) + 0(4) + 1(2) + 1(1\text{'s})$$
$$32 + 16 + 8 + 0 + 2 + 1$$
$$59.$$

Our answer: $111011_{(2)} = 59_{(10)}$. □

EXAMPLE 3   Convert $673_{(8)}$ to base 10.

SOLUTION   Rewrite $673_{(8)}$ as follows:

8^2	8^1	1's
6	7	3

Again $8^2 = 8 \times 8 = 64$, so we have

$$6(64) + 7(8^1) + 3(1\text{'s})$$
$$384 + 56 + 3$$
$$443.$$

Our answer: $673_{(8)} = 443_{(10)}$. □

EXAMPLE 4   Convert $158_{(12)}$ to base 10.

SOLUTION   Rewrite $158_{(12)}$ as follows:

12^2	12^1	1's
1	5	8

Since $12^2 = 12 \times 12 = 144$, we get

$$1(144) + 5(12^1) + 8(1\text{'s})$$
$$144 + 60 + 8$$
$$212.$$

Our answer: $158_{(12)} = 212_{(10)}$. □

EXERCISES

In Exercises 1–16, convert each of the numbers from the indicated base to a base 10 number.

1. $12132_{(4)}$

2. $1111011_{(2)}$

3. $54321_{(6)}$

4. $tee_{(12)}$

5. $3231_{(7)}$

6. $4144_{(5)}$

7. $86t2_{(12)}$

8. $5486_{(9)}$

9. $1427_{(8)}$

10. $3114_{(5)}$

11. $7481_{(9)}$

12. $12102_{(3)}$

13. $210111_{(3)}$

14. $1231_{(4)}$

15. $12321_{(5)}$

16. $21te1_{(12)}$

17. Perform the following division: $13585_{(9)} \div et_{(12)}$. Express your answer in base 5.

18. *Social Security.* Joanne Pucci is at a Social Security office applying for retirement benefits. She says that her age is $11_{(64)}$ years old. The clerk believes that she is much too young. Is Joanne really as young as the clerk believes? Explain.

19. Three friends, who have not seen each other since they graduated from college, meet at a class reunion. Each friend is trying to impress the other by claiming that his or her salary is quite high. The actual salaries are as follows:

Person	Salary (in dollars)
Arlene Anderson	$264563_{(7)}$
Tom McDermott	$42782_{(9)}$
Bill Sommers	$30213_{(12)}$

Which of the three friends has the highest salary?

3.7

APPLICATIONS TO THE COMPUTER

It has been said that we live in the computer age. Computers are involved in almost every aspect of our lives. Let us look at a day in the life of John Doe.

He wakes up in the morning and turns on the light. Nothing happens. Because of a computer error in processing his electric bill, his electricity has been turned off. On the way to work he uses his credit card to buy gas, for which the computer will bill him at the end of the month. Unfortunately, stopping for gas has made him late. Speeding to work, he is stopped by a policeman, who asks to see his computerized driver's license. He is then issued a computerized summons. At work he "punches in" on a computerized card. At the end of the day he receives his paycheck, but the computer made a mistake and underpaid him five dollars.

John decides to register with a computer dating service, and at lunchtime he mails them a check. (All checks are processed by computer.) After work, John registers for an evening course at City College. Of course, the entire registration process is computerized. When John gets home, he finds a letter from the Internal Revenue Service stating that the computer has found an error in his income tax return.

Although we are all affected by computers, many of us are unaware of what a computer really is, how it works, and what it can and cannot do.

A computer can do *nothing* that a human could not do if given enough time. The usefulness of computers is in their tremendous speed and accuracy. Computer errors occur only where there is a mechanical failure or when the human operator makes a mistake.

How does a computer do things so quickly? It works on electrical impulse, and electricity travels at the speed of light (which is about as fast as you can get).

Most computers work in base 2 because of its simplicity. (Remember, in base 2 there are only two digits, 0 and 1.) Furthermore, the fact that there are only two digits makes it easy to represent numbers on the computer in many different ways.

1. If we are using light bulbs on the console (the front of the computer), then a bulb off means 0 and a bulb on means 1. If the console displays the bulbs lit up as in Fig. 3.5, this means 101100 in base 2, or 44 in base 10.

Figure 3.5

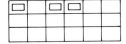

Figure 3.6

2. If we use a punched card, then we can represent 1 with a hole and 0 with no hole. Thus 101100 in base 2 would appear as shown in Fig. 3.6.

3. If we use magnetic tape or discs, then we can represent 1 by a magnetized spot and 0 by no magnetic spot.

4. If we use switches, then switch on means 1 and switch off means 0.

5. There are still other ways of representing base 2 numbers using magnetic cores and the direction of the current in a circuit.

Bruce Read

In 1804, Joseph Marie Jacquard made a loom for weaving intricate patterns. It used punched cards, pressed against the bank of needles, for each pass of the shuttle. Certain needles were pushed back by the card; others stayed in place because of holes in the card. This use of punched cards foreshadowed our present-day computers.

In the modern-day computer, information is fed in by using base 10 (this is usually done by means of punch cards, tape, or typewriter). Words are also converted into base 10 numerals. The information is then converted into base 2 and stored for future use in the computer's memory. Some computers store information in base 8 or 16 to save space. A typical computer with 100,000 "spaces" uses about 80,000 of these spaces for memory, that is, storage of information. Only about 20,000 spaces are actually used for computations.

To use the computer, we must feed information into the memory and then give precise instructions as to the computations to be performed with this information. We must also tell the machine how to write down the result. This is the job of the computer programmer.

All information and instructions must be given to the computer in a specific form. If even a comma is left out, the computer will not be able to understand what to do. The computer *cannot think* on its own. It merely follows human-given directions.

Suppose on an exam you were given the problem: "Dvide 6 by 2." You of course would know that the word "divide" is misspelled. This would not stop you from doing the problem. Since a computer cannot reason, it would not understand what "dvide" meant. It would stop and await further instructions.

Computers cannot do our thinking for us. However, their ability to do computations accurately at lightning speed does relieve us of many of the time-consuming and boring calculations associated with mathematical and scientific projects. Many mathematical problems have been solved with the help of computers doing the "dirty work." Some problems involving long truth tables can be analyzed in a fraction of a second using the computer.

In Chapter 13 we will discuss computers in more detail.

EXERCISES

1. Suppose you have just dropped on the floor 1000 cards numbered 0 to 999. You must pick them up and put them back into numerical order. It will be a long, tedious job, although a computer could do it in a fraction of a second. Surprisingly, by imitating the computer's technique you too could do the job, not as fast as the machine, but in a fairly short time. We will illustrate the method, not on 1000 cards, but on 32 cards numbered 0 to 31. First write the number of each card in base 2. Next cut holes in each card as shown in Fig. 3.7. A "closed" hole does not go through to the top of the card as does an "open" hole. A closed hole represents 1; an open hole represents 0. (See Fig. 3.8.)

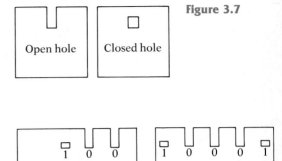

Figure 3.7

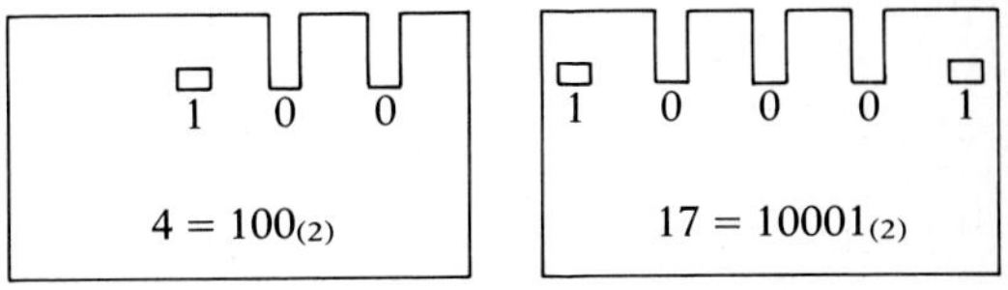

Figure 3.8

Each card will have 6 holes. Now take a long, thin object like a knitting needle, pass it though the first hole on the right, and lift it up. The cards with closed holes in that place will be lifted with it. Place these behind the other cards. Repeat this on the second hole, then the third, fourth, fifth, and sixth. When you are finished, the cards will be sorted in numerical order from 0 to 31. The sorting takes only 6 "passes" through the pile.

a) Make a set of cards and perform this experiment.

b) Why does each card have 6 holes?

c) Explain why this procedure works.

d) How many passes would be needed if there were only 16 cards? (*Hint:* How many holes would each card have?)

e) How many passes are needed for 100 cards?

2. Read one of the articles 44–50 in *Mathematics in the Modern World* (Readings from *Scientific American*). San Francisco: Freeman, 1968.

---

* 3.8

APPLICATION TO THE GAME OF *NIM*

Nim is a game that has been played in different forms for many centuries. Some time ago, a mechanical version for children was marketed in which the child played against a plastic mechanical computer. The word "Nim" comes from the German word "nehmen" meaning "to take." We will learn how to play the simplest version of the game.

The Game of *Nim*

We start with several piles of matches. Each pile can have any number of matches, and each pile can have a different number of matches. For example, we may have five piles with 4, 7, 3, 2, and 2 matches. The rules of the game are as follows.

1. There are two players who take alternate turns.

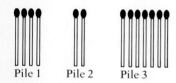

Figure 3.9

2. Each player in turn takes as many matches as he or she wants from *one and only one* pile. The player must take at least one match and may take them all. On the player's next turn he or she may select the same pile or a different pile.

3. The person who takes the last match wins.

To illustrate how *Nim* works, consider the following sample game between Joe and Frances. There are three piles with 4, 2, and 7 matches, as shown in Fig. 3.9. We now illustrate the situation that results after each move.

Move	*Situation after move*		
	Pile 1	*Pile 2*	*Pile 3*
Start of game			
Frances takes 2 matches from pile 3.			
Joe takes all the matches in pile 1.			
Frances takes 1 match from pile 2.			
Joe takes 1 match from pile 3.			
Frances takes 1 match from pile 2.			
Joe takes the 4 matches from pile 3 and wins.			

It may seem to you that winning the game is a matter of skill. However, this is not entirely so. Someone who knows the secret of the game can win, no matter how clever his or her opponent is, provided he or she is allowed to choose who goes first. Gamblers who know this secret have won a lot of money from innocent "suckers."

How to Win at *Nim*   The explanation is lengthy but not difficult. It is given in three parts. Read each of parts I, II, and III carefully before going on to the next part.

Part I   Write the number of matches in each pile in the binary system. In our sample game this would be 100, 10, and 111. We write these numbers in a column and then add them *as if they were in base 10:*

$$
\begin{array}{r}
100 \\
10 \\
\underline{111} \\
221
\end{array}
$$

If the result has all even digits, then we call it a **bad combination**. Otherwise, we call it a **good combination**.

EXAMPLE 1   If there are 3, 5, 8, and 4 matches, these numbers converted to base 2 and added are

$$
\begin{array}{r}
11 \\
101 \\
1000 \\
\underline{100} \\
1212
\end{array}
$$

This is a good combination.   ☐

EXAMPLE 2   If there are 10, 13, 2, 9, and 6 matches, these numbers converted to base 2 and added are

$$
\begin{array}{r}
1010 \\
1101 \\
10 \\
1001 \\
\underline{110} \\
3232
\end{array}
$$

Again this is a good combination.   ☐

EXAMPLE 3 If there are 11, 13, and 6 matches, these numbers converted to base 2 and added are

$$
\begin{array}{r}
1011 \\
1101 \\
\underline{110} \\
2222
\end{array}
$$

This is a bad combination. ☐

Part II Notice that just before the final winning play is made, there is only one pile. This means that there is only one number (written in base 2) in the sum of part I. Thus the total will consist just of this one number. Since numbers in base 2 contain only the digits 0 and 1, this total will also contain only the digits 0 and 1. Thus we must have a *good combination* (since it contains an odd number, 1). In our sample game, before Joe makes the last move, there are four matches in pile 3. Remember that 4 in base 10 is 100 in base 2 so we have the following:

$$
\begin{array}{ll}
\text{Pile 1} & 0 \\
\text{Pile 2} & 0 \\
\text{Pile 3} & \underline{100} \\
\text{Sum} & 100
\end{array}
$$

This is a good combination.

Summarizing   *The final winning move can only be made from a good combination.*

Part III Therefore our strategy is to keep our opponent from ever getting a good combination and to make sure at the end we ourselves get one. This can be accomplished in the following way.

1. *A play from a bad combination will always leave a good combination.* The reason for this is that in any given move you can change only one number in the sum (since you can take only from one pile). Thus in each column of the sum you can change only one digit (and you must change at least one digit). The only digit change can be from 0 to 1 or from 1 to 0 (because we are in base 2, these are the only digits we have). Since the sum contained all even digits to begin with, this change will make some of them odd. This will result in a good combination.

   To see how this works, consider the bad combination of 11, 13, and 6 matches given in Example 3 above. This combination was

$$
\begin{array}{r}
1011 \\
1101 \\
\underline{110} \\
2222
\end{array}
$$

Suppose we take 7 matches from pile 1. This changes the first number to 100, leaving

$$
\begin{array}{r}
100 \\
1101 \\
\underline{110} \\
1311
\end{array}
$$

This is a good combination. Try making the following plays on the original combination and verify that you always get a good combination.

   a) Take 2 from pile 3.

   b) Take 8 from pile 2.

   c) Take all of pile 1.

2. *A play from a good combination can always be made into a bad combination.* We do this as follows: Go to the first odd column from the left of the sum in the good combination. Pick any number that has a 1 in this column. Circle this 1. Now go to the next odd column. Circle the digit on the same line in this column. Repeat this for all odd columns and only odd columns. Now change all the circled digits. Since we are working in base 2, we can only change 0's to 1's and 1's to 0's. Change *only* circled digits. After we change this number, we play to *leave* this amount of matches in this pile.

Comment   We will not explain here why this strategy works. It is given as an exercise.

EXAMPLE 4   Suppose we are playing the good combination of Example 1.

$$
\begin{array}{cccc}
 & & 1 & 1 \\
 & 1 & 0 & 1 \\
\textcircled{1}0 & & \textcircled{0} & 0 \\
 & 1 & 0 & 0 \\
\hline
1 & 2 & 1 & 2
\end{array}
$$

The first and third columns from the left (in the sum) are odd. There is only one number with a 1 in the first column. We must use it. We circle it. In column 3 we circle the digit on the same line. We change these and get 0010 which is 2 in base 10. Thus we take all but 2 in pile 3. Notice that after we play, we leave

$$
\begin{array}{r}
11 \\
101 \\
10 \\
\underline{100} \\
222
\end{array}
$$

This is a bad combination.   ☐

Summary

1. Determine whether the original combination is good or bad.

2. If the original is good, choose to go first. Otherwise, let your opponent go first.

3. Whenever you move, use the strategy discussed to leave your opponent with a bad combination.

4. No matter what your opponent does, he or she must leave you a good combination.

EXERCISES

In Exercises 1–5, determine which are good combinations and which are bad combinations.

1. 8, 7, 2

2. 3, 3, 4, 5

3. 9, 13, 4

4. 15, 9, 6, 4

5. 8, 7, 9, 11, 5

Using the strategy discussed in this section, change each of the good combinations given in Exercises 6–8 to bad combinations.

6. 17, 9, 5   7. 8, 13, 11, 7   8. 5, 10, 1

9. Why does the strategy for changing a good combination to a bad combination discussed in this section work?

Play the game of *Nim* with an opponent, using the combinations given in Exercises 10–14.

10. 5, 8, 11   11. 4, 8, 6, 10

12. 5, 11, 3, 7, 6

13. 4, 2, 7, 5, 3, 8

14. 5, 8, 6, 1, 3, 9

15. *Nim: Alternative version:* Suppose we change our rules slightly by allowing a player whose turn it is to pick up at least one stick, but that this can be done by selecting as many sticks as the player wants from at most x piles. How does this affect the game?

16. Rework the game discussed between Joe and Frances but use the rules given in Exercise 15. Who wins now?

17. Another interesting game involving base 2 is the *Tower of Brahma* or *Tower of Hanoi*. This game was invented by Edouard Lucas in 1883 and involves transferring discs from one needle to another in a minimum number of moves. By going to your library, try to discover how to play the *Tower of Hanoi* game.

3.9
SOME INTERESTING NUMBERS

In this section we will discuss some special types of numbers that have fascinated mathematicians over the years because of their unusual properties. We will start with the Fibonacci numbers.

Fibonacci Numbers

Suppose we have decided to breed rabbits. Obviously, we must start with a pair, whom we shall call Jack Rabbit and Bunny Rabbit. We will assume the following.

1. Jack and Bunny are newborn when we start.
2. Rabbits begin to reproduce exactly two months after their own birth.
3. Thereafter, every month a pair of rabbits will produce exactly *one* other pair (a male and a female).
4. None of the rabbits dies.

Let us see how fast the number of pairs of rabbits increases.

The first pair, Jack and Bunny Rabbit, will have their first pair of children after two months. This gives us two pairs of rabbits.

After three months, Jack and Bunny will have another pair. We now have three pairs.

After four months, Jack and Bunny will again have another pair. Also, their first pair of children will have *their* first pair. This brings the total to five pairs.

Figure 3.10

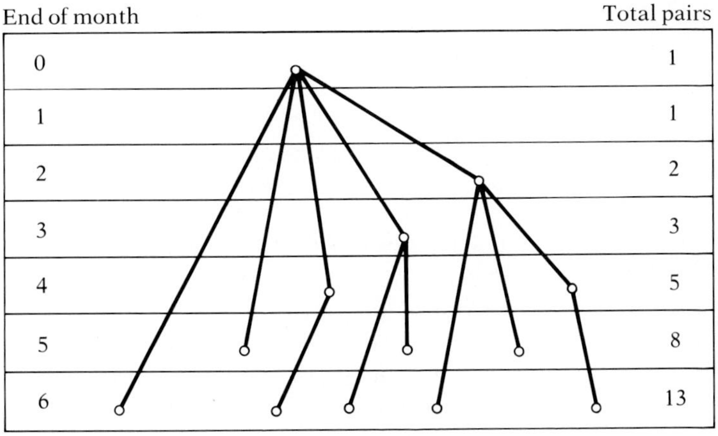

This process will continue. Figure 3.10 shows the number of rabbits at the end of each month through the first six months. The total number of pairs of rabbits for the first twelve months is shown in Table 3.1.

TABLE 3.1

Number of Months Passed	Total Number of Pairs of Rabbits
0	1
1	1
2	2
3	3
4	5
5	8
6	13
7	21
8	34
9	55
10	89
11	144
12	233

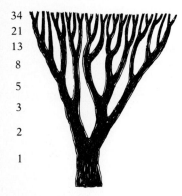

34
21
13
8
5
3
2
1

What connection does this picture have with Fibonacci numbers?

We see that at the end of the first year there will be 233 pairs of rabbits! (Maybe we should have started with mink!) If we were to continue this process beyond 12 months, we would get more and more pairs of rabbits.

Now let us examine the numbers in the second column of the table. They are 1, 1, 2, 3, 5, 8, 13, 21, 34, 55, 89, 144, 233. There is something especially interesting about these numbers (even to mathematicians who don't like rabbits). If we look at them carefully, we see that when we add the first two numbers, we get 1 + 1 = 2. This is the next number. Then 1 + 2 = 3. This is the next number. And 2 + 3 = 5. This again is the next number.

In general, we see that if we add any two consecutive numbers, we always get the next number in the list. The list of numbers 1, 1, 2, 3, 5, 8, 13, 21, 34, 55, 89, 144, 233, . . . is called a **Fibonacci sequence** in honor of its discoverer. He was Leonardo of Pisa, a remarkable Italian mathematician who lived in the last part of the twelfth century and the early thirteenth century. His father's name was Bonaccus, and Leonardo was nicknamed Filius Bonacci (which is Latin for "son of Bonaccus"). This nickname was shortened to Fibonacci.

Another entertaining application of Fibonacci numbers is to the game of *Fibonacci Nim*. This game is played by two players. There is one pile of matches. The first player takes one or more matches, but cannot take the whole pile. The next player can take up to twice the number of matches his or her opponent took, but no more. The first player can now also take up to twice the number of matches his or her opponent took on the last play, but no more. The play continues in this way. For example, if one player takes 4 matches on any play, the other may take as many as 8 matches on the next play, but no more. The game continues until all the matches are gone. The player who takes the last match wins.

It can be shown that if the number of matches in the original pile is a Fibonacci number, then the second player can *always* win if he or she plays correctly. If the number of matches in the original pile is not a Fibonacci number, then the first player can *always* win if he or she plays correctly. Play this game with a friend and figure out the winning strategy.

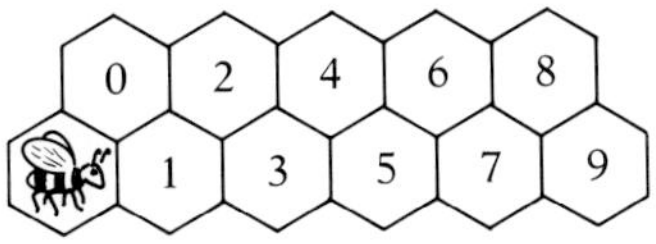

Fibonacci numbers can also be applied to the situation shown in Fig. 3.11. Suppose the bee wants to go to cell 3. The bee must *always move to the right* and always to a cell right next to it. In how many different ways can the bee get to cell 3?

One possible way is for the bee first to go to cell 1 and then to cell 3. We write this as

$$1 \rightarrow 3.$$

Another possible path is

$$0 \rightarrow 2 \rightarrow 3.$$

The other possible paths are

$$0 \rightarrow 1 \rightarrow 2 \rightarrow 3,$$
$$1 \rightarrow 2 \rightarrow 3,$$
$$0 \rightarrow 1 \rightarrow 3.$$

Thus the bee has 5 possible paths by which to get to cell 3. If you also calculate the number of possible paths by which the bee can get to cells 0, 1, 2, and 4, you get the results shown in Table 3.2. Notice that the number of possible paths form a Fibonacci sequence (except for the first term).

The Fibonacci sequence has many other interesting properties and applications. If you are interested, you can find some of these in the suggested further readings for this chapter.

Figure 3.11

TABLE 3.2

Cell	Number of Possible Paths
0	1
1	2
2	3
3	5
4	8

Perfect Numbers   The numbers that divide 6 evenly, excluding 6 itself, are 1, 2, and 3. Notice that $1 + 2 + 3 = 6$. We call a number **perfect** if it equals the sum of all the numbers that divide it and if these numbers are smaller than the number itself.

EXAMPLE 1   The number 12 is *not* a perfect number, since the numbers that divide

The diagram at right shows two opposite sets of spirals in a daisy—21 spirals in a clockwise direction and 34 counterclockwise. This ratio corresponds to the 21:34 sequence in the Fibonacci series. In pine cones, opposing spirals go 5 one way and 8 the other; in pineapples, they go 8 one way and 13 the other.

Bruce Anderson

it (that are less than 12) are 1, 2, 3, 4, and 6. We find that

$$1 + 2 + 3 + 4 + 6 = 16.$$

So 12 is *not* equal to the sum of these numbers. ☐

The number 15 is *not* perfect, since the numbers that divide it (that are less than 15) are 1, 3, and 5. If we add these numbers, we get $1 + 3 + 5$, which equals 9. Thus 15 is *not* equal to the sum of these numbers.

The number 28 *is* perfect, since the numbers less than 28 that divide it are 1, 2, 4, 7, and 14. Adding these, we get $1 + 2 + 4 + 7 + 14 = 28$.

It was once believed that perfect numbers had magical properties. In fact, it was believed that God created the world in 6 days because 6 is the smallest perfect number.

The first four perfect numbers are 6, 28, 496, and 8128. No one knows how many perfect numbers there are, although it is likely that there are infinitely many. All the known perfect numbers are even, for no one has yet discovered an odd one. However, there may be odd perfect numbers. It *is* known that every even perfect number ends in either 6 or 8.

There is a formula that will give all the even perfect numbers. It is

$$2^{P-1}(2^P - 1),$$

where P is a prime number and where $2^P - 1$ is also a prime number.

EXERCISES

1. We have given the first twelve numbers in the Fibonacci sequence. Write the next eight.

2. **a)** Add the first three numbers of the Fibonacci sequence. You get $1 + 1 + 2 = 4$. This is one less than the fifth number, which is 5. Now add the first four numbers. How does this sum compare to the sixth number?

   b) Add the first five numbers of the Fibonacci sequence. How does this sum compare to the seventh number?

   c) Without actually adding them, can you say what would be the sum of the first six numbers in the Fibonacci sequence?

3. Divide each number of the Fibonacci sequence by 4 and write down the remainders. Do this for the first thirty numbers. Do you notice any pattern?

4. Consider the ratios of successive terms of the Fibonacci sequence, that is, divide any term of the Fibonacci sequence by the next term. We get

Ratio	Value	Ratio	Value
$\frac{1}{1}$	$= 1.000$	$\frac{13}{8}$	$= 1.625$
$\frac{2}{1}$	$= 2.000$	$\frac{21}{13}$	≈ 1.615
$\frac{3}{2}$	$= 1.500$	$\frac{34}{21}$	≈ 1.618
$\frac{5}{3}$	≈ 1.667	$\frac{55}{34}$	≈ 1.619
$\frac{8}{5}$	$= 1.600$	$\frac{89}{55}$	≈ 1.618

Notice that this sequence oscillates and is approximately equal to 1.618. This number is called the **golden ratio.** It occurs frequently in mathematics. Select any two nonzero numbers. Add them together to obtain a third number. Form a sequence of numbers by adding the two previous terms to obtain the next term, much the same way that we obtained the numbers in the Fibonacci sequence. Verify that if you form the ratios of successive terms, then after a while the values oscillate around the golden ratio.

5. In the application of Fibonacci numbers to the bee's path, find all the possible paths by which it can get to cells 5, 6, and 7.

6. Closely related to the Fibonacci numbers are the **Lucas numbers,** named after the nineteenth century French mathematician Edouard Lucas (1842–1891). The Lucas numbers are generated in the same way as the Fibonacci sequence but start differently. The first six Lucas numbers are 1, 3, 4, 7, 11, 18. Find the next six Lucas numbers.

7. Can you locate the Fibonacci numbers embedded in Pascal's triangle?

8. Show that 10 is not a perfect number.

9. Write each of the four perfect numbers 6, 28, 496, and 8128 in binary notation. Do you notice any pattern?

Consider the sequence of numbers 1, 4, 7, 10, 13, ... Each number is 3 more than the one before it. Such a sequence is called an **arithmetic progression.** We can find the 100th term in this sequence without actually listing all the numbers by rewriting them as follows.

Number of Term in Sequence	Number	Rewritten Form
1	1	1
2	4	$1 + (1 \cdot 3)$
3	7	$1 + (2 \cdot 3)$
4	10	$1 + (3 \cdot 3)$
5	13	$1 + (4 \cdot 3)$
6	16	$1 + (5 \cdot 3)$
7	19	$1 + (6 \cdot 3)$
⋮	⋮	⋮
In general, we have n	?	$1 + (n - 1)3$

Number of Term in Sequence	Number	Rewritten Form
1	1	1 (also written as 2^0)
2	2	2^1
3	4	2^2
4	8	2^3
5	16	2^4
⋮	⋮	⋮
In general, we have n	?	2^{n-1}

Thus to find the 100th term, we take $n = 100$ and get

$$1 + (n - 1)3$$
$$= 1 + (100 - 1)3$$
$$= 1 + (99)3$$
$$= 298.$$

Therefore the 100th term in the above sequence is 298. To convince yourself that this is true, write out the first 100 numbers in the sequence.

Using a similar method, find the indicated term of each of the arithmetic progressions in Exercises 10–14.

10. 3, 6, 9, . . .   the 40th term

11. 1, 9, 17, . . .   the 17th term

12. 4, 15, 26, . . .   the 23rd term

13. 68, 62, 56, . . .   the 11th term

14. 103, 100, 97, . . .   the 17th term

15. Consider the sequence 1, 2, 4, 8, 16, 32, ... In this sequence, each number is twice the one before it. Such a sequence is called a **geometric progression**. We can find the tenth term in this progression without actually listing all the numbers by rewriting them as follows.

Thus to find the 10th term, we take $n = 10$ and get

$$2^{n-1}$$
$$= 2^{10-1} = 2^9$$
$$= 512.$$

Therefore the tenth term in the sequence is 512.

Using a similar procedure, find the ninth term in the sequence 1, 5, 25, 125, ...

Classify the sequences given in Exercises 16–20 as arithmetic, geometric, both, or neither.

16. 1, 5, 9, 13, . . .   **17.** 1, 4, 16, 64, . . .

18. 1, 4, 7, 10, . . .   **19.** 96, 90, 84, 78, . . .

20. 100, 50, 25, 12.5, . . .

21. The president of Geometric Progressions of America, Inc. is paid a monthly salary (31 days) of $25,000. The stockholders believe that she is overpaid. Being familiar with geometric progressions, she agrees to take an immediate drastic salary cut. She offers to be paid according to the following schedule:

1¢ first day of month,

2¢ second day of month,

4¢ third day of month,

8¢ fourth day of month,

that is, each day's salary is double the previous day's salary. The stockholders eagerly agree to this proposal. Is the stockholder's decision a wise one? Explain your answer.

3.10

**PRIME NUMBERS
AND PRIME
FACTORIZATION**

You will recall that a prime number is any number larger than 1 that is divisible by only itself and 1, assuming we divide only by positive numbers. The first few prime numbers are 2, 3, 5, 7, 11, 13, ... A number that is *not* prime is called **composite.**

EXAMPLE 1 The number 4 is not a prime number, since it can be divided by 2 as well as by itself and 1. It is a composite number. We know that 17 is a prime number, since the only numbers that divide it are 17 and 1. But 12 is not a prime number. It can be divided by 2, 3, 4, and 6. It is a composite number. ☐

How can we determine whether a number is prime or not? One method was developed by the Greek mathematician Eratosthenes (approximately 276–194 B.C.). His procedure, called the **sieve of Eratosthenes,** enables us to find all primes less than a given number.

We will illustrate the technique by finding all the primes up to 50. We write down all the numbers from 1 to 50 as shown below.

First cross out 1, which is not a prime. Circle 2, which is a prime, and cross out every *second* number after it. Now circle the next uncrossed number, 3, which is a prime, and cross out every *third* number after it. (Some of these will already have been crossed out.) Circle the next uncrossed number, 5, which is a prime, and cross off every fifth number after it.

$$1 \quad ② \quad ③ \quad 4 \quad ⑤ \quad 6 \quad ⑦ \quad 8 \quad 9 \quad 10$$
$$⑪ \quad 12 \quad ⑬ \quad 14 \quad 15 \quad 16 \quad ⑰ \quad 18 \quad ⑲ \quad 20$$
$$21 \quad 22 \quad ㉓ \quad 24 \quad 25 \quad 26 \quad 27 \quad 28 \quad ㉙ \quad 30$$
$$㉛ \quad 32 \quad 33 \quad 34 \quad 35 \quad 36 \quad ㊲ \quad 38 \quad 39 \quad 40$$
$$㊶ \quad 42 \quad ㊸ \quad 44 \quad 45 \quad 46 \quad ㊼ \quad 48 \quad 49 \quad 50$$

Continuing in this manner, we find all the prime numbers up to 50. We find that all the prime numbers have been circled and all the nonprime numbers have been crossed off. Using the same technique, we can find all the prime numbers less than any given number. Of course, if the given number is fairly large, this process may become very long and tiresome.

Now that we know how to find prime numbers, let us look at a nonprime number, such as 12. It can be written as 6×2. Notice that 2 is a prime, but 6 is not. However, 6 can be written as 2×3, and 2 and 3 are both primes. Thus we can write 12 as

$$12 = 2 \times 3 \times 2,$$

which is a product of prime numbers.

Can 12 be written as a product of primes in a different way from $2 \times 3 \times 2$? The answer is obviously yes. It can also be written as

$$12 = 2 \times 2 \times 3$$

or

$$12 = 3 \times 2 \times 2.$$

However, these are really the same as the original one, $2 \times 3 \times 2$, except for the order. We see, then, that 12 can be written as a product of primes in exactly *one* way if order is not considered.

Similarly, 20 can be written as a product of primes in only one way (except for order) as $2 \times 2 \times 5$. And 15 can be written as a product of primes in only one way, 3×5.

Actually, every integer greater than 1 (that is, 2, 3, 4, etc.) is either a prime number or can be written as a product of primes in exactly one way if order does not count. This fact is called the **fundamental theorem of arithmetic,** and we restate it formally in the following way.

> **Fundamental Theorem of Arithmetic**   *Any integer greater than 1 is either a prime number or can be written as a product of primes in only one way, except for order.*

EXAMPLE 2   The number 21 can be written as 3×7, which is a product of primes.

The number 18 can be written as $2 \times 3 \times 3$, which is also a product of primes.

The number 100 can be written as $2 \times 2 \times 5 \times 5$, which is also a product of primes.

The number 385 can be written as $5 \times 7 \times 11$, which is also a product of primes. ☐

Many interesting questions can be asked about prime numbers. For example, we may ask: How many prime numbers are there? Is there a formula that tells us which numbers are primes and which are not? In the remainder of this section we will attempt to answer some of these questions.

Let us begin with the first question: How many prime numbers are there? This was answered over 2000 years ago by Euclid, who *proved* that there are infinitely many prime numbers. He used the method of proof called **proof by contradiction.**

*   Euclid's Proof That the Number of Primes Is Infinite

There are two possibilities:

Possibility A.   There are only a finite (specific) number of primes.

Possibility B.   There are infinitely many primes.

Suppose possibility A is true. Then (if we have enough time), we can list *all* the primes. We let P_1 stand for 2, which is the first prime; P_2 will stand for 3, which is the second prime; P_3 will stand for 5, which is the third prime. Similarly, P_4 will denote the next prime, P_5 the prime after that, and so on. Eventually, we will come to the last prime, which we call P_n. (We must come to a last prime if Possibility A is true, since there is only a finite number of them.)

Let us now multiply all these primes together. We get

$$P_1 \times P_2 \times P_3 \times P_4 \times \cdots \times P_n.$$

(The dots stand for the primes between P_4 and P_n.)

Let Q be the number we get by adding 1 to this product. Thus

$$Q = P_1 \cdot P_2 \cdot P_3 \cdot P_4 \cdots P_n + 1.$$

What type of number is Q, prime or composite? If Q is *not* a prime number, then it must have divisors, and some of these are primes. (Why?) Let P denote one of these divisors that is a prime. Since P is prime, it is one of the primes $P_1, P_2, P_3, \ldots, P_n$ (since these are all the primes that there are). Therefore P *must* divide the product $P_1 \times P_2 \times \cdots \times P_n$. However, we know that P cannot divide 1. (Why?)

Thus we have that P *divides* $P_1 \times P_2 \times \cdots \times P_n$ and P does not divide 1. As we will see in Statement 2 of Section 4.5, P *cannot* divide the sum

$$P_1 \times P_2 \times P_3 \times \cdots \times P_n + 1.$$

Since this sum is just Q, we conclude that P *does not divide Q*. But this contradicts our earlier statement that Q was divisible by the prime P.

The contradiction means that possibility A is wrong. This leaves us with possibility B, which must be correct. Thus we see that the number of primes is infinite.

Euclid's proof also answers the next question: Is there a *largest* prime number? Clearly, the answer must be no. Why? The largest number that is definitely known to be prime (at least, up to the time that this book was written) is the number $2^{44,497} - 1$. This means it is the number we get when multiplying 2 by itself 44,497 times and then subtracting 1 from the result. If you do not believe that this is a prime number, why don't you try to multiply 2 by itself 44,497 times and subtract 1 from the result. If you survive this, you may want to convince yourself that this number is prime by trying to find a number that divides it other than itself and 1. (Don't spend too much time on this!) This was actually shown to be a prime number by two 18-year-old students, Laura Nickel and Curt Noll, at California State University, Hayward, using a computer.

Obviously, Euclid's proof shows us that this is definitely *not* the largest prime that exists. There is no such number. The number $2^{44,497} - 1$ is just the largest number that is known for sure to be a prime.

There is another question that has puzzled mathematicians for a long time and that has not yet been answered: Is there a formula that tells us which numbers are prime and which are not; a formula that, when numbers are substituted into it, will produce only prime numbers? Many attempts have been made to find such a formula, but none have been successful.

The mathematician Pierre Fermat (1601–1665) believed that he had succeeded in finding such a formula. He thought that the formula $2^{2^n} + 1$ would give only prime numbers, no matter what was substituted for n. Let us try it and see what happens!

If $n = 0$, the formula becomes

$$2^{2^0} + 1 = 2^1 + 1 \quad \text{(remember } 2^0 = 1)$$
$$= 2 + 1 \quad \text{(since } 2^1 = 2)$$
$$= 3, \quad \text{which is prime.}$$

Pierre Fermat (1601–1665).

If $n = 1$, the formula is

$$2^{2^1} + 1 = 2^2 + 1 \quad \text{(since } 2^1 = 2)$$
$$= 4 + 1 \quad \text{(since } 2^2 = 4)$$
$$= 5, \quad \text{which is prime.}$$

If $n = 2$, the formula gives

$$2^{2^2} + 1 = 2^4 + 1 \quad \text{(since } 2^2 = 4)$$
$$= 16 + 1 \quad \text{(since } 2^4 = 16)$$
$$= 17, \quad \text{which is prime.}$$

If $n = 3$, we get from the formula

$$2^{2^3} + 1 = 2^8 + 1 \quad \text{(since } 2^3 = 8)$$
$$= 256 + 1 \quad \text{(since } 2^8 = 256)$$
$$= 257, \quad \text{and this is prime.}$$

If $n = 4$, the formula yields

$$2^{2^4} + 1 = 2^{16} + 1 \quad \text{(since } 2^4 = 16)$$
$$= 65536 + 1 \quad \text{(since } 2^{16} = 65,536)$$
$$= 65537, \quad \text{and this too can be shown to be prime.}$$

So far, things look good. However, the next "Fermat number" is

$$2^{2^5} + 1 = 2^{32} + 1 \quad \text{(since } 2^5 = 32)$$
$$= 4,294,967,297,$$

and it has been shown that this number is *not* prime. In fact, it is divisible by 641. Try it!

It has also been shown that Fermat was wrong when $n = 6$. If $n = 6$, then $2^{2^6} + 1$ is a composite number.

Another attempt to produce a "prime-number–generating" formula was made by the Swiss mathematician Leonhard Euler (1707–1783). In addition to being a great mathematician, Euler had an amazing ability to perform calculations in his head. For the last 17 years of his life he was blind. Yet he was able to do accurate calculations to 50 decimal places. His formula was

$$n^2 - n + 41.$$

Leonhard Euler. (Although unable to produce a prime generating formula, Euler did produce a prime number of children, thirteen.)

If we let $n = 1$ in this formula, we get

$$1^2 - 1 + 41 = 1 - 1 + 41 = 41,$$

which is prime. If $n = 2$, we get

$$2^2 - 2 + 41 = 4 - 2 + 41 = 43,$$

which is prime. If $n = 3$, we get

$$3^2 - 3 + 41 = 9 - 3 + 41 = 47,$$

which is prime.

This formula also gives primes if $n = 3, 4, 5$, and up to $n = 40$. However, if $n = 41$, we get

$$41^2 - 41 + 41 = 41^2,$$

which is *not* prime, since it equals $41 \cdot 41$, and thus has 41 as a divisor, in addition to itself and 1.

Euler knew that his formula would not give primes *all* the time. However, he found that it seemed to give primes for about half of the possible values of n. No one has been able to prove for certain whether this percentage of primes remains the same as more and more values of n are tried.

Up to the present, no formula has been found that will give *only* prime numbers, and no one has come up with a formula that will give *all* the prime numbers.

EXERCISES

1. Write down the first nineteen prime numbers.
2. Find two consecutive numbers, both of which are prime.
3. Find three consecutive numbers, none of which is prime.
4. Find four consecutive numbers, none of which is prime.

5. Using the sieve of Eratosthenes, find all the prime numbers up to 100.

Write each of the numbers in Exercises 6–15 as a product of primes.

6. 72	**7.** 49	**8.** 1000
9. 525	**10.** 602	**11.** 327

12. 400   **13.** 780   **14.** 925

15. 1100

16. Can a prime number ever end in the digit 6? In 5?

17. The prime number 13 can be written as $4 \cdot 3 + 1$. Can you find three other primes that can be written in the form $4n + 1$, where n is any whole number?

18. The prime number 7 can be written as $4 \cdot 1 + 3$. Can you find three other primes that can be written as $4n + 3$, where n is any whole number?

19. Find three prime numbers that can be written in the form $2^P - 1$, where P is itself a prime number.

20. Find two values of n for which the formula $n! + 1$ gives primes. (See p. 528 for the meaning of $n!$.)

21. Find two values of n for which the formula $n! + 1$ gives composite numbers.

22. Two prime numbers that differ by 2 are called **twin primes.** For example, 3 and 5 are twin primes. Find three other pairs of twin primes.

23. Find a set of three prime numbers that differ from each other by 2. Can you find another such set?

24. Can a prime number ever be perfect?

25. **a)** List all the prime numbers that are less than 50. How many are there?

    b) List all the prime numbers that are less than 100. How many are there?

**26. There are 168 prime numbers that are less than 1000. There are 303 prime numbers that are less than 2000. There are 430 prime numbers that are less than 3000. How many prime numbers are there that are less than 1,000,000? As of now, the only way to answer this question is to write down the primes and count them. (Good luck!) No one has yet discovered a formula that will tell you how many primes there are that are less than a given number.

27. Pick a number—say, 100. Double it to get 200. There is at least one prime number between 100 and 200. One such prime number is 101. It has been proved that given any number n and its double $2n$, there is always at least one prime number between these two numbers.

    a) Find a prime number between 18 and 39.

    b) Find a prime number between 250 and 290.

    c) Find a prime number between 2000 and 3000.

28. **a)** Complete the following chart.

Prime number	Prime number + 1	Prime number − 1
5	6	4
7	8	6
11		
13		
17		
19		

    b) Divide each of the numbers in the second and third columns by 6 and consider their remainders (if any).

    **c) Examine the results of part (b). Make a general statement about prime numbers larger than 3. (*Hint:* It involves division by 6.)

29. There are seven prime number years in the period 1950–2000. Can you find three of them?

30. Using numbers that are part of twin primes, write 18 as a sum in two distinct ways.

31. *Goldbach's conjecture.* As we mentioned in Chapter 2, the famous mathematician Christian Goldbach stated that "Every even number greater than 2 can be expressed as the sum of *two* prime numbers." Show that Goldbach's conjecture is true for all even numbers (except 2), up to and including 36. The first few numbers are

$$4 = 2 + 2,$$
$$6 = 3 + 3,$$
$$8 = 3 + 5.$$

32. *Another Goldbach conjecture.* One of Goldbach's conjectures is that every odd number larger than 5 can be expressed as the sum of three primes. For example, 7 can be written as 2 + 2 + 3; also 9 = 2 + 2 + 5; and 11 = 7 + 2 + 2. Verify that this conjecture is true by writing each of the odd numbers between 19 and 39 as the sum of three primes.

Several prime numbers can be expressed as 1 more than the square of a natural number, whereas others can be written as 1 less than the square of a natural number. For example, the prime number 5 can be written as $2^2 + 1$, whereas the prime number 3 can be written as $2^2 - 1$.

33. Find three prime numbers that can be written as 1 more than the square of a number.

34. Find two prime numbers that can be written as 1 less than the square of a number.

A formula that often yields prime numbers is $n^2 - n + 41$. In Exercises 35–39, determine whether this formula yields primes, using the indicated value of n.

35. 60   **36.** 65   **37.** 70

38. 73   **39.** 81

3.11

LARGE AND SMALL NUMBERS— SCIENTIFIC NOTATION

Very few of us ever need to use extremely large or extremely small numbers in our daily lives, but we often encounter them in news reports. It can be mind boggling just to think of such numbers. For example, the deficit in the U.S. budget is measured in billions of dollars. Just how large is a billion? How long would it take for us to count (starting at 1) to a billion? What is the difference between a billion and a trillion?

Now consider a medical researcher who finds that the size of a certain virus is 0.000000008 millimeters. Just how small is this?

Recall that in the Hindu-Arabic system all numbers are written as powers of 10. Hence the name decimal system. Thus the number 365 really means

$$3 \text{ (hundreds)} + 6 \text{ (tens)} + 5 \text{ (ones)} \quad \text{or}$$
$$= 3 \times 100 \quad\quad + 6 \times 10 + 5 \times 1$$
$$= 3 \times 10^2 \quad\quad + 6 \times 10^1 + 5 \times 10^0.$$

Here we have used exponents according to the following.

Definition 3.1   *Exponents*   *For any nonzero number b and any natural number n,*

$$b^n = \underbrace{b \times b \times b \times \cdots \times b}_{n \text{ of them}}$$

$$b^0 = 1$$

and

$$b^{-n} = \frac{1}{b^n}.$$

*The number b is called the **base**, and n is called the **exponent**.*

Notation In base 10 the following alternate forms of various numbers with exponents given in Table 3.3 are true.

TABLE 3.3

Positive Powers of Ten	*Negative Powers of Ten*
$10^0 = 1$	$10^{-1} = \dfrac{1}{10^1} = 0.1$
$10^1 = 10$	$10^{-2} = \dfrac{1}{10^2} = 0.01$
$10^2 = 10 \times 10 = 100$	$10^{-3} = \dfrac{1}{10^3} = 0.001$
$10^3 = 10 \times 10 \times 10 = 1000$	$10^{-4} = \dfrac{1}{10^4} = 0.0001$
$10^4 = 10,000$	$10^{-5} = \dfrac{1}{10^5} = 0.00001$
$10^5 = 100,000$	$10^{-6} = \dfrac{1}{10^6} = 0.000001$
$10^6 = 1,000,000$	$10^{-7} = \dfrac{1}{10^7} = 0.0000001$
$10^7 = 10,000,000$	$10^{-8} = \dfrac{1}{10^8} = 0.00000001$
$10^8 = 100,000,000$	

Comment When working with positive exponents (powers of 10), the number of zeros following the 1 in the value of a power of 10 is the same as the exponent of 10. Thus in $10^4 = 10,000$ the exponent is 4, and there are 4 zeros after the 1 in 10,000.

Comment When working with negative exponents (powers of 10), the number of places to the right of the decimal point is equal to the exponent itself when we disregard the minus sign. Thus in $10^{-6} = 0.000001$ there are 6 places to the right of the decimal point. This is the same as the exponent when we disregard the minus sign.

Comment Often we use a **decimal point** in the decimal system to separate the whole part of a number from the fraction.

EXAMPLE 1 Write 456.789 in expanded notation using exponents.

SOLUTION The number 456.789 can be written with exponents as

4 (hundreds) + 5 (tens) + 6 (ones) + 7 (tenths) + 8 (hundredths)
+ 9 (thousandths)
$$= (4 \times 10^2) + (5 \times 10^1) + (6 \times 10^0) + (7 \times 10^{-1})$$
$$+ (8 \times 10^{-2}) + (9 \times 10^{-3}). \quad \square$$

When working with numbers that are very large or very small, we can express them in a more convenient form called scientific notation.

> **Definition 3.2** *Scientific Notation of a Number* A number written in **scientific notation** is expressed as a product of two quantities: the first is a number equal to or greater than 1 but less than 10, and the second is a power of 10.

Let us see what this definition really says.

EXAMPLE 2 The population of the United States in 1983 was approximately 225 million. Write this number in scientific notation.

SOLUTION We first write 225 million as 225,000,000. Now we apply Definition 3.2. We first express 225,000,000 as a product of two numbers where one of the numbers must be between 1 and 10. We have

$$225{,}000{,}000 = 2.25 \times 100{,}000{,}000.$$

Now we express the second number as a power of 10 using Table 3.3 so that

$$225{,}000{,}000 = 2.25 \times 10^8. \quad \square$$

EXAMPLE 3 The distance from the earth to the moon is approximately 2.4×10^5 miles. Write this number in ordinary decimal notation.

SOLUTION Using Table 3.3, we know that $10^5 = 100{,}000$. Thus we first determine the value of the power of 10. We then multiply the results with the first number. This gives

$$2.4 \times 10^5 = 2.4 \times 100{,}000$$
$$= 240{,}000 \text{ miles.} \quad \square$$

Comment In Example 3 we could have obtained our answer quickly by moving the decimal point in 2.4 five places to the right.

EXAMPLE 4 A scientist measured the diameter of a certain cell and found it to be 0.00003 inches. Express this number in scientific notation.

SOLUTION   We first write 0.00003 as a product of two numbers where one of the numbers must be between 1 and 10. We have

$$0.00003 = 3 \times 0.00001.$$

Now we express the second number as a power of 10 using Table 3.3 so that

$$0.00003 = 3 \times 10^{-5}. \quad \square$$

EXAMPLE 5   In the metric system, 1 millimeter is equal to 1×10^{-5} km. Rewrite this in ordinary decimal notation.

SOLUTION   We first determine the value of 10^{-5} (using Table 3.3). We have $10^{-5} = 1 \times 0.00001$. We then multiply this result with the first number. This gives

$$1 \times 10^{-5} = 1 \times 0.00001$$
$$= 0.00001. \quad \square$$

Comment   In Example 5 we could have obtained our answer quickly by moving the decimal point in 1 (understood to be at the end of the number) five places to the left.

Comment   In a later chapter, when we discuss hand-held calculators, we will indicate how the various operations are performed when numbers are expressed in scientific notation.

EXERCISES

In Exercises 1–9, rewrite the given numbers in expanded notation using exponents.

1. 78.1   **2.** 316.01   **3.** 412.37

4. 51.001   **5.** 619.823   **6.** 3472.13

7. 4609.182   **8.** 3247.193   **9.** 453.0258

In Exercises 10–20, write the number in scientific notation.

10. 400   **11.** 7000   **12.** 0.003

13. 0.0005   **14.** 17,000,000   **15.** 8,000,000

16. 0.00000078   **17.** 0.0000000879

18. 7654.32   **19.** 0.44

20. 0.0006123

In Exercises 21–30, find the number that can replace the question mark so that the resulting statement is true.

21. $0.029 = 2.9 \times 10^{?}$

22. $125 = 1.25 \times 10^{?}$

23. $691,000 = 6.91 \times 10^{?}$

24. $3,810,000 = 3.81 \times 10^{?}$

25. $4290 = 4.29 \times 10^{?}$

26. $0.000000000012 = 1.2 \times 10^{?}$

27. $0.00001111 = 1.111 \times 10^{?}$

28. $423.60 = 4.236 \times 10^{?}$

29. $0.0000000001 = 1.0 \times 10^{?}$

30. $40,000,000,000 = 4 \times 10^{?}$

In Exercises 31–40, write the number in ordinary decimal notation.

31. A ton is approximately 9.07×10^2 kilograms.

32. The sun weighs approximately 1.8×10^{27} tons.

33. The velocity of light in a vacuum is approximately 3×10^{10} cm/sec.

34. The distance between the earth and its nearest star (other than the sun), Alpha Centauri, is 2.6×10^{13} miles.

35. The age of the earth's crust is approximately 5×10^9 years.

36. The mass of the earth is 1.32×10^{25} pounds or 5.99×10^{24} kilograms.

37. The diameter of the orbit of an electron of a certain atom is 5.4×10^{-6} millimeters.

38. The distance from the earth to the sun is 9.3×10^7 miles.

39. The total area of the continental United States is 3.022387×10^6 square miles.

40. The volume of a certain neuron is about 4.37×10^{-9} cubic centimeters.

In Exercises 41–50, express the number in each statement in scientific notation.

41. A light year is the distance light travels in a year. A light year is approximately 9,500,000,000,000 kilometers.

42. The diameter of the universe is about 2,000,000,000 light years.

43. The diameter of a red blood corpuscle is 0.00075 centimeters.

44. The diameter of the smallest particle visible to the naked eye is about 0.01016 centimeters.

45. The number of atoms in a gram of hydrogen is about 600,000,000,000,000,000,000,000.

46. The population of the world was approximately 3,600,000,000 in 1984.

47. In one year there are about 31,557,600 seconds.

48. A spaceship travels at the rate of 18,000 miles per hour in space.

49. The planet Uranus is about 2,000,000,000 miles from the sun.

50. The first stage thrust of a rocket's engine launched into space is about 9,000,000 lb just prior to burnout.

****51.** A computer has been programmed to print all the counting numbers between 1 and 100,000 inclusive. The numbers will be separated by commas as they are printed by the printer. In executing this task, how many times will the digit zero be printed?

3.12

SUMMARY In this chapter we analyzed several number systems and different number bases. After studying these number systems you should appreciate our decimal system and the simplicity with which we are able to perform the various calculations. As was indicated, our familiar Hindu-Arabic system was not firmly established in Europe until about the thirteenth century. Yet computations had to be performed before the Hindu-Arabic system was adopted. Thus an understanding of how these various systems perform calculations is important to help us understand how our decimal system works.

Since most computers perform their calculations in base 2, it is important to understand how these computations are performed. In addition, *Nim* and *Tower of Hanoi* are two games whose solution depends on a knowledge of base 2 computations.

We also presented some interesting numbers and some of their properties. This led to a discussion of prime numbers and their uses. Of course, we will apply these ideas in the next chapter. Finally, we introduced scientific numbers, which are used when we have very large or very small numbers.

Although many other applications and/or interesting ideas are associated with different number bases, our purpose in this chapter was merely to introduce you to the basic ideas involved.

STUDY GUIDE You should now be able to demonstrate your knowledge of the following ideas by giving definitions, descriptions, or specific examples. Page references are given in parentheses.

Basic Ideas

The Babylonian number system (p. 119)
Positional notation (p. 121)
Egyptian numeration system (p. 121)
Tamil numerals (p. 122)
The Mayan numeration system (p. 123)
Vertical grouping system (p. 124)
Base 10 number system (p. 124)
Hindu-Arabic system (p. 125)
Digits (p. 125)
Roman numeral system (p. 125)
Base 10 arithmetic (p. 125)
Zero as a placeholder (p. 126)
Carrying (p. 126)
Decimal system (p. 127)
Galley or Gelosia multiplication (p. 128)
Napier's bones (p. 128)
Egyptian duplation (p. 130)
Russian peasant multiplication (p. 130)
Finger multiplication (p. 131)
Subtraction by complements (p. 131)

Other bases (p. 132)
Numerals (p. 132)
Binary system (p. 134)
Arithmetic in other bases (p. 134)
Converting from base 10 to other bases (p. 138)
Nim (p. 147)
Good combination (p. 149)
Bad combination (p. 149)
Fibonacci numbers (p. 153)
Fibonacci sequence (p. 154)
Perfect numbers (p. 155)
Golden ratio (p. 157)
Lucas numbers (p. 157)
Arithmetic progressions (p. 157)
Geometric progressions (p. 158)
Prime numbers (p. 159)
Composite numbers (p. 159)
Sieve of Eratosthenes (p. 159)
Fundamental theorem of arithmetic (p. 160)
Proof by contradiction (p. 160)
Twin primes (p. 164)
Exponents (p. 165)
Decimal point (p. 166)
Scientific notation (p. 167)

The Ideas of Numbers and Bases Were Applied To

a) computers (p. 144)
b) the game of Nim (p. 147)
c) prime factorization (p. 160)
d) scientific notation (p. 167)

FORMULAS TO REMEMBER

In this chapter, different number systems were discussed. You should review the symbols used in each of these systems. In addition, we discussed the following different ways of performing the various operations:

Technique	Used for
a) Gelosia multiplication (p. 128)	multiplication
b) Napier's bones (p. 128)	multiplication
c) Egyptian duplation (p. 130)	multiplication
d) Russian peasant multiplication (p. 130)	multiplication
e) Finger multiplication (p. 131)	multiplication
f) Scientific notation (p. 167)	simplifying large or small numbers

MASTERY TESTS

Form A

1. If $42_{(B)} = 26_{(10)}$ then $B = $?

   a) 4   **b)** 6   **c)** 7   **d)** 2   **e)** none of these

2. $1tee_{(12)}$
   $+ \quad te_{(12)}$

   a) $11071_{(12)}$ **b)** $1ett_{(12)}$ **c)** $1teete_{(12)}$
   d) $eeete_{(12)}$ **e)** none of these

3. $301_{(10)} = $?

   a) $100101101_{(2)}$ **b)** $101101001_{(2)}$ **c)** $10101101_{(2)}$ **d)** $10010101_{(2)}$
   e) none of these

4. The following is part of a Russian peasant multiplication problem. What is the value of the question mark?

21	17
10	34
5	?
2	136
1	272

   a) 61   **b)** 68   **c)** 21   **d)** 17   **e)** none of these

5. Refer to question 4. What is the answer to the multiplication problem?

   a) 357   **b)** 170   **c)** 340   **d)** 289   **e)** none of these

6. Which of the following is true?

   a) $212_{(6)}$ is greater than $66_{(12)}$ **b)** $212_{(6)}$ is equal to $66_{(12)}$
   c) $212_{(6)}$ is less than $66_{(12)}$ **d)** none of these

7. Mary is $25_{(8)}$ years old, and Christopher is $1012_{(3)}$ years old. By how many years is Christopher older than Mary?

   a) 27   **b)** 32   **c)** 21   **d)** 11   **e)** none of these

8. Add: $11011_{(2)}$
   $\quad \quad 1011_{(2)}$
   $\quad \quad 1101_{(2)}$

   a) $110011_{(2)}$ **b)** $100011_{(2)}$ **c)** $11011_{(2)}$ **d)** $101111_{(2)}$
   e) none of these

9. If the number 0.0000173 is written in the form 1.73×10^n, what is the value of n?

   a) -4

   b) -5

   c) -6

   d) -3

   e) none of these

10. The following chart illustrates Gelosia multiplication. Which two numbers are being multiplied?

   a) 368 and 16   **b)** 368 and 45   **c)** 16 and 560   **d)** 560 and 45

   e) none of these

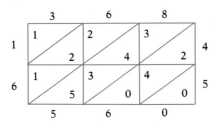

11. Refer to question 10. What is the answer to the multiplication problem?

   a) 36845   **b)** 56045   **c)** 16560   **d)** 61368   **e)** none of these

12. In the Egyptian numeration system, what number is represented

   by 𝟿∩ ⫴⫴ ?

   a) 116   **b)** 916   **c)** 812   **d)** 423   **e)** none of these

13. Multiply: $tel_{(12)}$
                $\times \; e6_{(12)}$

14. Subtract: $156_{(9)}$
               $- \; 78_{(9)}$

15. Perform the following division: $13232_{(4)} \div 212_{(4)}$.

Form B   1. Convert $624_{(7)}$ to a base 8 number.

2. In what base is the following computation being performed?

$$123$$
$$+231$$
$$\overline{404}$$

   a) 2   **b)** 3   **c)** 4   **d)** 5   **e)** none of these

3. When playing the game of *Nim*, the combination of 5, 7, 11, and 14 is

   a) a good combination   **b)** a bad combination

   c) a neutral combination   **d)** a reverse combination

   e) none of these

4. Write the number 4.23×10^6 in ordinary decimal notation.

5. $518_{(10)}$ equals what number in base 4?

   a) 20012   **b)** 21002   **c)** 1221   **d)** 10221   **e)** none of these

6. Express 48 as a product of primes.

7. How many prime numbers are there that are less than 55?

8. Write 451.623 in expanded notation using exponents.

9. Write the number 143,000,000,000 in scientific notation.

10. Perform the division $11270_{(12)} \div 1t_{(12)}$.

11. In what base is the following computation being performed?

$$25_{(B)} = 37$$

12. Write the eleventh term in the sequence 1, 4, 7, 10, ...

13. What is the value of 10^0?

    a) 0   **b)** 10   **c)** 1   **d)** undefined   **e)** none of these

14. Perform the following subtraction:

$$\begin{array}{r} 241_{(6)} \\ -\ 42_{(6)} \\ \hline \end{array}$$

15. Convert $56_{(8)}$ to a base 2 number.

4

THE REAL NUMBER SYSTEM

CHAPTER OBJECTIVES

To discuss the different types of number systems such as the natural numbers, the whole numbers, and the integers. (*Sections 4.2, 4.3, and 4.4*)

To analyze some properties that each of these systems must satisfy. (*Sections 4.2, 4.3, and 4.4*)

To describe various divisibility tests so that we can determine when a number can be divided evenly by 2, 3, 5, ... (*Section 4.5*)

To understand why rational numbers are needed as well as how to add, multiply, divide, and subtract two rational numbers. (*Section 4.6*)

To determine which decimals are rational numbers, that is, which can be expressed in the form $\frac{a}{b}$, where b ≠ 0, and which decimals are not rational. (*Sections 4.6 and 4.7*)

To apply some of the ideas of rational numbers. (*Section 4.6*)

To learn how to round off numbers. (*Section 4.6*)

To specify an order of operations when performing calculations. (*Section 4.8*)

Here is a listing of New York Stock Exchange transactions on the day of the big crash in 1929. Negative numbers indicate the extent of the disaster.

An Interesting Number

by Sam Hoffman

The following interesting number was brought to my attention by one of my students. It is the number 142857. The number has six digits, and no digit is repeated. Furthermore, it has no zeroes, threes, sixes, or nines. When this number is multiplied by two, the same six digits appear in the product. When this number is multiplied by three, four, five, or six, the product always maintains the same order of the digits as in the original number. The only difference is in the first and last digit.

However, a strange thing happens when the original number is multiplied by seven. Try it and see what happens.

The Island Teacher

Ever since the various number systems were invented to count and do computations, people have been fascinated by the properties that some numbers have and others do not have. In this chapter we will study the various number systems and then some very interesting numbers.

4.1
INTRODUCTION

The elementary **theory of numbers** usually deals with only the numbers 0, 1, 2, 3, 4, ... These numbers are called **whole numbers.** Since these numbers are so simple, you may think that this area of mathematics is especially easy in comparison to other areas. However, this is not the case. There are problems in number theory that have been handed down to us by the ancient Greeks and, after 2500 years, still remain unsolved.

Every day we use many different types of numbers. If you read the stock market listings in the daily newspapers, you will find, next to each stock, numbers such as $+\frac{1}{2}$, $+5$, -3, $-\frac{1}{4}$. These indicate how much the price of stock has gone up or down. You use numbers if you close out your bank account, and your bankbook then shows a balance of $0.00!

We are so accustomed to using different kinds of numbers—positive, negative, zero, fractions, decimals—that we take them for granted. They are so much a part of our lives that some people think of them as being God-given or as being part of nature. Actually, this is not true. Numbers are human-made. Different kinds of numbers were created by humans at various stages of their mathematical development to meet their changing needs. First came the counting numbers 1, 2, 3, ... Negative numbers were invented much later and were not generally accepted until the sixteenth century. The concept of a fraction arose during the Bronze age, but it was not fully developed for some time. As late as the year 1650 B.C.,[1] the Egyptians, who were rather good mathematicians, did not use fractions as we use them today. They used mostly unit fractions, that is, fractions whose numerators are 1 such as $\frac{1}{2}$, $\frac{1}{3}$, $\frac{1}{4}$, etc., and the fraction $\frac{2}{3}$. Other fractions were written as combinations of unit fractions. Although the Egyptians had an idea of the number zero, they had no symbol for it. Symbols for zero were invented at various times by different civilizations.

Whatever the reason, number theory has attracted, and still attracts, the attention of many first-rate mathematicians. The great

---

1. Our knowledge of Egyptian mathematics is derived from the Rhind Papyrus, which dates back to approximately 1650 B.C. See the historical note on page 122.

German mathematician Gauss once said, "Mathematics is the queen of the sciences, and number theory is the queen of mathematics."

One of the reasons that the numbers are so fascinating is that in working with them we discover many surprising and interesting patterns.

EXAMPLE 1   Let us examine the following number facts carefully to discover a method for finding the square of a number.

$$1^2 = 1$$
$$2^2 = 1 + 2 + 1 = 4$$
$$3^2 = 1 + 2 + 3 + 2 + 1 = 9$$
$$4^2 = 1 + 2 + 3 + 4 + 3 + 2 + 1 = 16$$
$$5^2 = 1 + 2 + 3 + 4 + 5 + 4 + 3 + 2 + 1 = 25$$
$$6^2 = 1 + 2 + 3 + 4 + 5 + 6 + 5 + 4 + 3 + 2 + 1 = 36$$
$$7^2 = 1 + 2 + 3 + 4 + 5 + 6 + 7 + 6 + 5 + 4 + 3 + 2 + 1 = 49$$

From the above patterns it is obvious that to find the square of a number, we start with 1 and keep adding 1 until we get to our number. Then we go down by 1's until we get back to 1 again. Thus to find 8^2, we have

$$1 + 2 + 3 + 4 + 5 + 6 + 7 + 8 + 7 + 6 + 5 + 4 + 3 + 2 + 1,$$

which equals 64. Therefore $8^2 = 64$. We can generalize this to obtain the square of any number n. We have

$$1 + 2 + 3 + \cdots + (n - 1) + n + (n - 1) + \cdots + 3 + 2 + 1$$

When we add these, we get n^2.   □

EXAMPLE 2   The mathematician Nichomachus of Gerasa (about 100 A.D.) noticed that if you write the odd numbers in the pattern

1   3, 5   7, 9, 11   13, 15, 17, 19   21, 23, 25, 27, 29, . . .

then the sums of the above groupings in order are

$$1 = 1 = 1^3$$
$$3 + 5 = 8 = 2^3$$
$$7 + 9 + 11 = 27 = 3^3$$
$$13 + 15 + 17 + 19 = 64 = 4^3$$

$$\vdots \qquad \vdots \qquad \vdots$$

Thus we see that these successive sums are the cubes of the integers.   □

HISTORICAL NOTE

We know that the Greeks, especially the Pythagoreans, were fascinated by number theory. It is likely that even before them the Babylonians studied it. Why should this subject arouse so much interest? There are several answers.

In the first place, as we have already pointed out, this branch of mathematics deals with rather simple numbers. This makes it easy, even for the nonmathematician, to state and understand many number theory problems. (Solving them is quite a different story.)

Another reason for the Greeks' interest in number theory was their belief that numbers had certain magical powers. Some of these superstitions survive even today. We consider 7 lucky, except if you break a mirror, in which case you get seven years of bad luck. We also think of 11 as being a lucky number. The number 13 is unlucky to many people, and some tall buildings have no thirteenth floor because it cannot be rented easily. How do you feel about taking a exam on Friday the thirteenth? The number 3 is usually considered lucky, but there are some smokers who will not light three cigarettes with one match.

The Pythagoreans went even further. They believed that the entire universe could be explained in terms of the numbers 0, 1, 2, 3, ... According to their beliefs, the number 1 gave rise to all other numbers, and they considered it the "number of reason." They called 2 the first female number (all other even numbers were also female). On the other hand, 3 was the first male number (all other odd numbers were also male). We do not care to comment on their belief that 2 was the "number of opinion" and 3 the "number of harmony." They believed that 5 was the "number of marriage" (can you see why?) and that 6 was the "number of creation" (remember, the Bible tells us that God created the world in six days).

The Pythagoreans believed that 10 was the holiest number. It was the number of the universe and the symbol of health and harmony. Because they believed that 10 was the perfect number, they believed that there were 10 heavenly bodies. At the center of the universe there was a fire. Around this fire the sun, earth, moon, and the five known planets revolved. Since this totaled only 9 heavenly bodies, they invented a tenth one called a "counterearth."

Finally, some of the ideas of number theory are closely related to geometry. This may not seem like much of an attraction to you, but the study of geometry was itself of major concern to both the Babylonians and the Greeks.

EXAMPLE 3 The odd numbers are the numbers 1, 3, 5, 7, 9, 11, ... We see that

$$1 = 1 \quad \text{which is the same as } 1^2$$
$$1 + 3 = 4 \quad \text{which is the same as } 2^2$$
$$1 + 3 + 5 = 9 \quad \text{which is the same as } 3^2$$
$$1 + 3 + 5 + 7 = 16 \quad \text{which is the same as } 4^2$$
$$1 + 3 + 5 + 7 + 9 = 25 \quad \text{which is the same as } 5^2$$
$$1 + 3 + 5 + 7 + 9 + 11 = 36 \quad \text{which is the same as } 6^2$$

There is a relationship between the number of odd numbers we are adding and the sum. The relationship can be seen when we rewrite the above list as follows:

Sum of first odd number equals 1, which is 1^2;

Sum of first two odd numbers equals 4, which is 2^2;

Sum of first three odd numbers equals 9, which is 3^2;

In general, the sum of first n odd numbers equals n^2.

Thus, for example, the sum of the first ten odd numbers ($1 + 3 + 5 + 7 + 9 + 11 + 13 + 15 + 17 + 19$) is 100 or 10^2. Compare this result with Example 1. What do you notice? ☐

It is with questions of this nature that mathematicians concern themselves. In this chapter we will discuss how and why our present-day number system developed. We will also analyze some interesting **number patterns.**

4.2

THE NATURAL NUMBERS

Humans must have invented natural numbers shortly after they learned how to count. At first they had only the numbers 1 and 2. A few more talented societies also had the number 3. Everything else was "many." Later, as society became more complex, it became necessary to count accurately larger quantities. At this point, numbers larger than 2 and 3 were introduced. Thus the natural numbers were born. Because they are used in counting, the natural numbers are also called the counting numbers.

> **Definition 4.1**   *The **natural numbers,** or **counting numbers,** are the numbers 1, 2, 3, 4, 5, 6, 7, 8, 9, 10, 11, ...*

Addition   Let us go back 25,000 years in history. Zaftig, the caveman, has 3 sheep. His brother has just given him 2 sheep. Zaftig now counts his sheep and discovers that he has 5 altogether. Zaftig, who is the "brain" of the tribe, sees that if you have 3 things and you then get 2 more, your total will *always* be 5 things. This is an example of what we call addition. Roughly speaking, by **addition** we mean that we combine *two numbers* to obtain a third number called the **sum.** This sum must be *unique.* (This means that there is *only one* answer. There can never be more than one answer.) For example,

$$4 + 7 = 11 \text{ (11 is the only answer),} \quad \text{and}$$
$$5 + 8 = 13 \text{ (13 is the only answer).}$$

Do the following simple addition problem:

$$\begin{array}{r} 7 \\ 8 \\ 9 \\ \hline 24 \end{array}$$

Your answer is 24, of course. How did you get it? You could add from bottom to top as follows: $9 + 8$ equals 17, and then $17 + 7$ equals 24.

Herds of cattle in this Egyptian tomb painting are being led past for census taking. Scribes are writing down the count. *(Reproduced by Courtesy of the Trustees of the British Museum.)*

You could also add from top to bottom as follows: 7 + 8 equals 15, and then 15 + 9 equals 24.

Notice that no matter which way we add, *we can only add two numbers at a time.* To add three numbers, we first add any two, and then add the third to the sum. Addition is an example of a **binary operation.**

> **Definition 4.2**   *A **binary operation** on numbers is a process by which **two** numbers are combined to obtain a **unique** third number.*

Suppose you were to ask a small child who is just learning arithmetic how much 2 + 3 is. The child would first put up 2 fingers and then 3 fingers and would count the total, getting 5. If you now ask the child to add 3 + 2, he or she would put up 3 fingers and then 2 more, again getting 5. The child soon discovers that

$$3 + 2 = 2 + 3$$

This is known as the **commutative law of addition.** What this tells you is that the *order* in which you add two numbers does not matter. Your answer is always the same. We state this formally as the following law.

> **Commutative law of addition**   *If a and b stand for any natural numbers, then a + b = b + a.*

This law may seem rather obvious to you. You may even feel that it does not say much. However, the commutative law is not always true for operations different from addition. For example, subtraction is not commutative:

$$5 - 3 \quad \text{is not equal to} \quad 3 - 5$$

When you add a column of numbers, you may add from the top down (or vice versa). The usual way of checking is to add again in reverse order. You know that if your addition is correct, the answer should be the same in both cases. You are using the commutative law of addition, which says that the order in which numbers are added is not important.

In everyday life situations the order in which we do things may or may not make a difference. For example, combing your hair and brushing your teeth in the morning is commutative. On the other hand, putting on your shoes and putting on your socks is not commutative.

Ask a chemist whether mixing water and sulfuric acid is commutative. That is, does it make a difference if we pour water into sulfuric acid or sulfuric acid into water?

In a certain town there are two schools, a two-story high school and a six-story elementary school. Recently the following headline appeared in the local newspaper: "High School Building Burns Down." Which one was it? The meaning is unclear, since we do not know whether the word "school" goes with "high" or with "building." If "school" is grouped with "high," then obviously the high school (two-story building) burned down. In other words, the grouping will make a big difference in the meaning. In English, to avoid this confusion, we can put a hyphen between the words "high" and "school" when we mean the two-story building.

Now consider Mr. Jones who has decided to buy a Volkswagen. Since the local bank is advertising "small car loans," he applies to the bank for a $2000 loan. The bank officer tells him that their maximum car loan is $300. Mr. Jones protests that they are advertising loans to buy small cars and that you cannot buy any small car for $300. The bank official responds that Mr. Jones is misinterpreting the advertisement, which offers small loans to buy cars. Who is right? In this case the disagreement is also about grouping. Mr. Jones is grouping "car" with "small." The bank official is grouping "car" with "loan."

In these two examples grouping made a difference in the meaning. This is not always the case. In the phrase "my friend Klutz the mechanic" it does not make any difference whether we group the word "Klutz" with "friend" or with "the mechanic."

In mathematics, grouping may or may not be important. Suppose we are asked to add

$$2 + 3 + 4$$

without changing the order of the numbers. Since we can add only two numbers at a time, this can be done in two ways. We have either

(2 + 3) + 4	(We always do what is in the parentheses
5 + 4	first. In this case we first add the 2 + 3.)
9	

or

$$2 + (3 + 4) \qquad \text{(This time we first add the 3 + 4.)}$$
$$2 + 7$$
$$9$$

Note that in both cases our answer is the same, that is,

$$(2 + 3) + 4 = 2 + (3 + 4)$$

In other words, it makes no difference whether we group the 3 with the 2 or with the 4. This is called the **associative law of addition.** We state it here.

> **Associative Law of Addition** *If a, b, and c are any natural numbers, then $(a + b) + c = a + (b + c)$.*

Comment Since the position of the parentheses does not matter in doing addition, we usually leave them out and write, for example,

$$2 + 3 + 4$$

instead of $(2 + 3) + 4$ or $2 + (3 + 4)$.

Beware Do not confuse the associate law for addition with the commutative law for addition. The associative law keeps the numbers in the same order but merely groups them differently. The commutative law involves changing the order of the numbers.

Although the associative law holds for addition, it does not work for all the binary operations. In particular, the associative law does not hold for subtraction. For example, $(15 - 4) - 3$ is not equal to $15 - (4 - 3)$. This is so because

$$(15 - 4) - 3 = 11 - 3 = 8$$

Remember, we do what is in the parentheses first. On the other hand,

$$15 - (4 - 3) = 15 - 1 = 14.$$

Thus $(15 - 4) - 3$ is not equal to $15 - (4 - 3)$.

Suppose you were restricted for some reason to using *only* the odd numbers 1, 3, 5, 7, 9, 11, etc. Then it would be impossible to add. If you add any two odd numbers, you will always get an *even* number. Since we are limited to *only* the odd numbers, this answer is not acceptable.

Consider now *all* the natural numbers. If we add any two natural numbers, our result will also be a natural number. This is called the **closure property for addition of natural numbers.** We say that the natural numbers are closed under addition. The odd numbers are not closed under addition.

> **Law of closure for addition**   *If a and b are any natural numbers, then the sum, a + b, is also a natural number.*

EXAMPLE 1   If we are restricted to the set of numbers {1, 2, 3}, then this set is *not* closed under the operation of addition. The numbers 2 and 3 are in the set, but their sum, which is 5, is not in the set. □

EXAMPLE 2   Let S = {all even numbers}. Then S is closed under the operation of addition, since when we add two even numbers, the sum is *also* an even number. □

EXAMPLE 3   Let A = {1}. S is not closed under addition since 1 + 1 equals 2, and 2 is not in S. □

Comment   Are the natural numbers closed under the operation of subtraction? That is, if we subtract one natural number from another natural number, is our answer also a natural number?

Multiplication   Let us return to our friend Zaftig, the caveman. He now has a herd of cattle. Each day he takes them out to pasture, walking them 5 abreast.

As they return from pasture, Zaftig counts them to make sure that they are all there. Instead of counting them one by one, he discovers that it is easier to count them in groups of 5's. He has three groups of 5 each, that is, he has

$$\underline{5 + 5 + 5}$$

3 groups of 5

This means that Zaftig has 15 cattle. A convenient way of expressing this is to invent a new operation called **multiplication,** which is symbolized in one of the following ways: $3 \cdot 5$, $3(5)$, 3×5, or $(3)(5)$. Any one of these means

$$5 + 5 + 5$$

Similarly, $7 \cdot 4$, 7×4, $7(4)$, or $(7)(4)$ means

$$4 + 4 + 4 + 4 + 4 + 4 + 4.$$

In general, $a \cdot b$ [also written as ab, $a(b)$, $a \times b$, or $(a)(b)$] means

$$\underbrace{b + b + b + \cdots + b}_{a \text{ of them}}$$

Figure 4.1

Note that since we can only multiply two numbers at a time, **multiplication is also a binary operation.**

According to our definition of multiplication, $2 \cdot 3$ means 2 groups of 3 things each. This can be pictured as in Fig. 4.1. On the other hand, $3 \cdot 2$ means 3 groups of 2 things each. This we picture as in Fig. 4.2. In both cases we have 6 things altogether. Therefore

Figure 4.2

$$2 \cdot 3 = 3 \cdot 2$$

This is an example of the commutative law of multiplication.

Commutative Law of Multiplication   *If a and b are any natural numbers, then $a \cdot b = b \cdot a$.*

Multiplication of natural numbers is also associative. If you multiply $2 \times 4 \times 5$, this can be done in two ways (without changing the **order** of the numbers). One way is

$$2 \times (4 \times 5) \qquad 4 \text{ is grouped with } 5$$
$$2 \times 20$$
$$40$$

Another way is

$$(2 \times 4) \times 5 \qquad 4 \text{ is grouped with } 2$$
$$8 \times 5$$
$$40$$

In both cases our answer is, of course, the same. This shows that the grouping does not matter and that

$$2 \times (4 \times 5) = (2 \times 4) \times 5$$

In general terms, we have the following law.

Associative Law of Multiplication   *If a, b, and c are any natural numbers, then $a(bc) = (ab)c$.*

Comment   As with addition, since the parentheses do not matter, it is customary to leave them out and write *abc* instead of *a*(*bc*) or (*ab*)*c*.

If we multiply any two natural numbers, our result will also be a natural number, that is, the natural numbers are closed under multiplication. This we state as the following law.

Law of Closure for Multiplication   *If a and b are any natural numbers, then ab is also a natural number.*

How would you perform the following calculation?

$$(36 \times 57) + (36 \times 43)$$

Most likely, you would multiply 36×57, getting 2052, and then multiply 36×43, getting 1548. So you would have

$$(36 \times 57) + (36 \times 43)$$
$$= 2052 \quad + \quad 1548$$
$$= 3600$$

This calculation takes some time. It can be done more easily if you know the distributive law of multiplication over addition. Notice that 36 appears in both multiplications. We can "factor out" the 36 (put the 36 outside in front of the parentheses), getting $36(57 + 43)$. We can now add $57 + 43$, getting 100. Thus we have

$$(36 \times 57) + (36 \times 43)$$
$$= 36(57 + 43) \qquad \text{We factor the 36.}$$
$$= 36(100)$$
$$= 3600$$

Formally stated, we have the following law.

Distributive Law of Multiplication over Addition   *If a, b, and c are natural numbers, then a(b + c) = ab + ac. Also (b + c)a = ba + ca.*

Comment   This law says that whether we add $b + c$ first and then multiply by *a* or multiply by *a* first and then add, our result is the same.

EXAMPLE 4   Multiply $(63 \times 51) + (63 \times 49)$, using the ditributive law.

SOLUTION
$$(63 \times 51) + (63 \times 49)$$
$$= 63(51 + 49) \qquad \text{We factor the 63.}$$
$$= 63(100)$$
$$= 6300 \quad \square$$

EXAMPLE 5 Multiply 80×23, using the distributive law.

SOLUTION

$$80 \times 23$$
$$= 80(23)$$
$$= 80(20 + 3)$$
$$= (80 \times 20) + (80 \times 3) \qquad \text{This is where we use the distributive law.}$$
$$= 1600 + 240$$
$$= 1840 \quad \square$$

Subtraction Up to now we have mentioned subtraction without considering precisely what it means. If you spend \$3 in a store and give the salesclerk a \$10 bill, you will usually get your change as follows: The clerk says "three" and then, handing you the money, counts out the change, "four, five, six, seven, eight, nine, ten dollars." The clerk has to subtract 3 from 10 and does this by starting with 3. The clerk then figures how much change must be added to 3 to make 10. Thus the subtraction

$$10 - 3 = \text{change}$$

becomes

$$3 + \text{change} = 10$$

In other words, the change is the amount that must be added to 3 to make 10. This procedure shows us what **subtraction** really is.

> **Definition 4.3** *If a and b are natural numbers, then a − b (read as "a minus b") means the number that must be added to b to obtain a (if such a number exists). If we call this number x, we have a − b = x, if b + x = a.*

EXAMPLE 6 $4 - 1$ means some number x that must be added to 1 to make 4. Thus $1 + x = 4$. Therefore, $x = 3$. Then we have $4 - 1 = 3$. $\square$

EXAMPLE 7 $17 - 6$ means the number x that must be added to 6 to make 17. Thus $6 + x = 17$. And $x = 11$. So $17 - 6 = 11$. $\square$

Comment Subtraction is also a **binary operation,** since it involves *two* numbers, the number that we are subtracting and the number from which we are subtracting.

We have already pointed out that subtraction of natural numbers is *not* closed, commutative, or associative. The following examples should emphasize these facts.

EXAMPLE 8 $5 - 3 = 2$. However, $3 - 5$ is *not* equal to 2; $3 - 5$ equals a *negative number*, -2, which we will discuss further in Section 4.4. Thus

$$5 - 3 \quad \text{is } not \text{ equal to} \quad 3 - 5$$

which shows that the operation of subtraction is *not* commutative. □

EXAMPLE 9 In the above example we saw that $3 - 5 = -2$, which is *not* a natural number. Thus subtraction of natural numbers is *not closed*. (This means that when you subtract one natural number from another, the result is not always a natural number.) □

EXAMPLE 10 $\quad\quad 9 - (6 - 4) \quad$ equals

$\quad\quad\quad\quad 9 - 2 \quad\quad$ (remember, do what is in parentheses first)

which equals

$\quad\quad\quad\quad 7$

On the other hand,

$\quad\quad (9 - 6) - 4 \quad$ equals

$\quad\quad\quad\quad 3 - 4 \quad\quad$ which equals

$\quad\quad\quad\quad -1$

Therefore $9 - (6 - 4)$ is *not* equal to $(9 - 6) - 4$, so subtraction of natural numbers is *not* an associative operation. □

Division The final binary operation that we will consider is **division.** If you were asked to divide 6 by 2, you would immediately give the answer 3. How do you get this answer? Undoubtedly, you would ask yourself, "What number multiplied by 2 will give 6?" With this in mind, we define division in general as follows.

> **Definition 4.4**   *If a and b are natural numbers, then $a \div b$, also denoted as $\frac{a}{b}$ (read as "a divided by b"), means the natural number x, such that $b \cdot x = a$.*

EXAMPLE 11 $39 \div 13$ means a natural number x such that $13 \cdot x = 39$. Thus $x = 3$. We have, then, $39 \div 13 = 3$. □

EXAMPLE 12 $24 \div 6$ means a natural number x, such that $6 \cdot x = 24$. Thus $x = 4$. Therefore $24 \div 6 = 4$. □

EXAMPLE 13 $\frac{100}{20}$ means a natural number x, such that $20 \cdot x = 100$. Thus $x = 5$. Therefore, $\frac{100}{20} = 5$. □

It can easily be seen that division of natural numbers is not closed, commutative, or associative. This is illustrated by the following examples.

EXAMPLE 14 $3 \div 6$ equals $\frac{1}{2}$. This is not a natural number. Thus we see that when we divide one natural number by another, the result is *not* always a natural number. This shows that division of natural numbers is not closed.   ☐

EXAMPLE 15   We know that $6 \div 3 = 2$. However, $3 \div 6$ is not equal to 2. Thus $6 \div 3$ is *not* equal to $3 \div 6$, which shows that division is *not* commutative.   ☐

EXAMPLE 16

$$24 \div (6 \div 2) \quad \text{equals}$$
$$24 \div 3 \quad \text{which equals}$$
$$8.$$

However,

$$(24 \div 6) \div 2 \quad \text{equals}$$
$$4 \div 2 \quad \text{which equals}$$
$$2.$$

Thus we see that $24 \div (6 \div 2)$ is *not* equal to $(24 \div 6) \div 2$, which shows that division is *not* associative.   ☐

Comment   We have seen that subtraction and division are not associative operations. Therefore when we subtract or divide more than two numbers, we *must* insert parentheses to show which numbers are to be subtracted or divided first. For example, we must not write $24 \div 6 \div 2$, since (as we saw in Example 16) this can have two different answers, depending on where the parentheses are inserted.

EXERCISES

In Exercises 1–10, state the property for natural numbers that justifies the computation being performed.

1. $12 + 8 = 8 + 12$
2. $(6 \cdot 5)8 = 6(5 \cdot 8)$
3. $5(8 + 12) = 5 \cdot 8 + 5 \cdot 12$
4. $9 + 8$ is a natural number
5. $5 + (3 + 9) = (5 + 3) + 9$
6. $5 + (3 + 9) = 5 + (9 + 3)$

7. $6 \times 9 = 9 \times 6$
8. $5(9 \cdot 8) = (5 \cdot 9)8$
9. $(7 + 8)2 = 7 \cdot 2 + 8 \cdot 2$
10. $7 \cdot 6 = 42$

In Exercises 11 –14, perform the calculations by using the distributive law.

11. 17×102 12. $(16 \times 91) + (16 \times 9)$
13. 103×38 14. $(76 \times 21) + (24 \times 21)$

In Exercises 15–21, determine whether the given set of numbers is closed under the indicated operation.

Numbers	Operation
15. {2, 4, 6}	addition
16. {0}	multiplication
17. {even numbers}	multiplication
18. {odd numbers}	multiplication
19. {5, 10, 15, . . . }	addition
20. {natural numbers less than 17}	multiplication
21. {0, 1}	multiplication

Which of the activities given in Exercises 22–28 are commutative and which are not?

22. Taking a picture with a camera and putting film in the camera.

23. Applying for admission to college and taking the Scholastic Aptitude Test (SAT).

24. Getting on a plane and paying the fare.

25. Tuning up your car and rotating the tires.

26. Drinking alcohol and driving a car.

27. Learning computer programming and buying a computer.

28. Getting undressed and taking a shower.

29. Consider an operation that we shall call | given by the following table:

$\mid$	x	y	z
x	x	y	z
y	y	y	z
z	z	z	z

a) Is | a closed operation?

b) Is | a commutative operation?

c) Is | an associative operation?

30. Consider an operation that we shall call ↑ given by the following table:

↑	x	y	z
x	x	z	z
y	y	x	z
z	z	z	x

a) Is ↑ a closed operation?

b) Is ↑ a commutative operation?

c) Is ↑ an associative operation?

31. Explain how the distributive law can be applied in the following situation: Willy Martin is the manager of a Little League baseball team. Baseball gloves for a player on the team cost $19 apiece. On Wednesday, Willy orders 6 gloves, and on Thursday he orders 17 gloves. What is the total cost for all the gloves?

32. An operation called # is given by the rule

$$a \# b = a + b + 2.$$

In other words, $a \# b$ is obtained by first adding a and b and then adding 2 to the results. For example, if a and b are 5 and 7, respectively, then

$$a \# b = 5 + 7 + 2 = 14.$$

If a and b are 9 and 12 respectively, then

$$a \# b = 9 + 12 + 2 = 23.$$

a) Is the operation # commutative?

b) Is the operation # associative?

The distributive law can be generalized as follows:

$$a(b + c + d + \ldots) = ab + ac + ad + \ldots.$$

Use the generalized distributive law to evaluate each of the computations in Exercises 33–35.

33. $(19 \times 62) + (19 \times 12) + (19 \times 26)$

34. $(42 \times 93) + (27 \times 93) + (31 \times 93)$

35. 9×126

36. A **magic triangle** is an arrangement of natural numbers in the form of a triangle where the sum of the numbers on each of the sides (called the **magic sum**) is always the same. The amount of numbers on each side is called the **order**. For example,

represents a magic triangle whose order is 3 and whose magic sum is 11. Find a magic triangle of order 3 whose magic sum is 23.

37. The Pythagoreans investigated a special group of natural numbers known as **triangular numbers.** The numbers that have geometric forms that look like triangles are called triangular numbers. The first few triangular numbers are 1, 3, 6, 10, and 15, as can be seen from the diagram below:

Natural number   1   3   6   10   15

Determine the next three triangular numbers.

Complete each of the computations in Exercises 38–44. Can you generalize?

38. $1 \cdot 9 + 2 = ?$
      $12 \cdot 9 + 3 = ?$
      $123 \cdot 9 + 4 = ?$
      etc.

39. $9 \cdot 9 + 7 = ?$
      $98 \cdot 9 + 6 = ?$
      $987 \cdot 9 + 5 = ?$
      etc.

40. $1 \cdot 8 + 1 = ?$
      $12 \cdot 8 + 2 = ?$
      $123 \cdot 8 + 3 = ?$
      etc.

41. $3 \cdot 37 = 111$ and $1 + 1 + 1 = 3$
    $6 \cdot 37 = ?$ and   ?
    $9 \cdot 37 = ?$ and   ?
    etc.

42. $1 \cdot 1 \quad = ?$
      $11 \cdot 11 = ?$
      $111 \cdot 111 = ?$
      etc.

43. $7 \cdot 7 \quad = ?$
      $67 \cdot 67 \quad = ?$
      $667 \cdot 667 = ?$
      etc.

44. $7 \cdot 15873 = ?$
      $14 \cdot 15873 = ?$
      $21 \cdot 15873 = ?$
      etc.

45. Pick a 3-digit number, such as 273. Write the digits again in the same order to make the 6-digit number 273273. Divide 273273 by 13 and you will see that there is no remainder. Now pick another 3-digit number, rewrite the digits again to make a 6-digit number, and divide the result by 13. Again you will get no remainder.

   a) Generalize this result.

   b) Try to explain why this happens.

46. Follow the same procedure as in Exercise 45, except divide by 7 instead of 13. What happens? Can you generalize this result? Can you explain why this happens?

In Exercises 47–49, use the results of Example 2 in Section 4.1 to find the sum of the numbers.

47. $1 + 3 + 5 + \ldots + 29$

48. $1 + 3 + 5 + \ldots + 85$

49. $11 + 13 + 15 + \ldots + 81$

50. Let us calculate the following:

$$3 + 5 \cdot 6$$
$$3 + 30$$
$$33.$$

   Now compute

$$3(5 + 6)$$
$$3 \cdot 11$$
$$33.$$

   Thus $3 + 5 \cdot 6 = 3(5 + 6)$. From this we can conclude that if $a, b,$ and c are natural numbers, then

$$a + bc = a(b + c)?$$

   Explain.

51. Recall that when we discussed sets in Chapter 1, we had several operations involving two sets. Specifically, these were $\cup$, $\cap$, and $\times$. Although we did not state it at the time, these are, of course, binary operations.

   a) Which of these operations are commutative?

   b) Which of these operations are closed?

   c) Which of these operations are associative?

4.3
THE WHOLE NUMBERS

If ever a number was misunderstood, then zero is it. Ask any five people (nonmathematicians, of course) what zero means to them. When we did this, some of the responses we got were the following.

1. Zero is nothing.
2. Zero is not just plain nothing; it is a something nothing.
3. Zero is not a number at all.
4. Zero is the absence of anything.
5. Zero is the null set.
6. Zero is either the beginning or end of something.

Despite beliefs to the contrary, zero is a number just like any other number. However, it has certain properties that other numbers do not have. In this section and the following ones we will discuss these properties.

One source of confusion is the fact that zero plays different roles in different situations. In the number 405, zero acts as a "place holder." If you say sadly, "My bank balance is zero," then zero is used as a quantity indicating that there is nothing left.

Zero probably appeared first as a place holder. The Babylonians, who used a base 60 number system, at first had no symbol to represent empty places. Later in their history, they used the symbol ⟅ as a place holder. This is probably the first appearance of any symbol for zero. The Mayans also had invented zero symbols. Some of these were

Later, the Greeks invented their own symbol for zero and developed the *concept* of zero to represent nothingness.

Earlier civilizations do not appear to have used zero as a *number*. Later, the Hindus developed the concept of and a symbol for zero. Present evidence indicates that zero was used in India from the ninth century on and possibly earlier. The Hindus appear to have used zero not merely as a place holder or as a concept for nothing, but also as a *number*. India was invaded by the Arabs around 700 A.D. The Arabs later introduced the Hindu numerals to Europe. In the Western world, zero was not completely accepted and used as a number until much later, probably around the sixteenth century.

Let us investigate the role of zero as a number. We first have the following definition.

Definition 4.5 *The **whole numbers** are the natural numbers together with the number 0. Using set notation, this would be {0, 1, 2, 3, 4, . . .}.*

Comment The only difference between the set of natural numbers and the set of whole numbers is the number 0.

What can we say about 0? Perhaps the most important property that zero has is that when zero is added to any number, it does not do anything to that number. For example,

$$5 + 0 = 5,$$
$$0 + 4 = 4,$$
$$117 + 0 = 117, \quad \text{and}$$
$$0 + 0 = 0$$

In general, if *a* is any number, then $a + 0 = a$ and $0 + a = a$.

A "do nothing" number of this type is called an **identity.** Thus *zero is the identity for addition.* Would you say that zero is the "do-nothing" number for multiplication? Obviously not. Consider $5 \cdot 0$. This means

$$\underbrace{0 + 0 + 0 + 0 + 0}_{\text{5 of them}}$$

Thus we have $5 \cdot 0 = 0$. Generally, if *a* is *any* number, then $a \cdot 0 = 0$ and $0 \cdot a = 0$.

How about division? What can we say about division that involves zero? Very often, division by zero and division into zero are confused. To help us understand the difference, suppose your Uncle Sam dies and leaves a will dividing his money equally among 20 people. Upon investigation it is found that his entire fortune is in Confederate money and is therefore worthless. So there is really nothing to divide. How much money do these 20 people get? Obviously, nothing. Thus

$$\frac{0}{20} = 0$$

To look at it another way, let us say that 0 divided by 20 is "something." So

$$\frac{0}{20} = \text{something}$$

Now what could this something be? Remember what division means.

$$\frac{0}{20} = \text{something}$$

means

$$0 = 20 \cdot \text{something}$$

If the something is 1, then $0 = 20 \cdot 1$, which is obviously wrong. If the something is 2, the $0 = 20 \cdot 2$, which is again wrong. If the something is 3, then $0 = 20 \cdot 3$, which is clearly wrong. Obviously, if the something is anything other than 0, the statement

$$0 = 20 \cdot \text{something}$$

cannot be true. We conclude that the something must be zero.

In other words, we can say that if x is any nonzero number, then x divided into 0 is 0, that is,

$$\frac{0}{x} = 0$$

On the other hand, what would we mean by $\frac{20}{0}$? In a mathematics class recently, students were asked this question. Various answers were given. Some said that the answer was zero. Others believed that the answer was 20. Still others claimed that the answer was 1. Some even thought that the answer was "infinity."

To see who is right, let us check each of the suggested answers. If we claim that $\frac{6}{3} = 2$, then we can easily check this answer, since $6 = 3 \cdot 2$, so $\frac{6}{3} = 2$ is true.

For those who claimed that $\frac{20}{0} = 0$, is it true that

$$20 = 0 \cdot 0?$$

Obviously not, since we know that 0 times anything is 0 and not 20.

For those who claimed that $\frac{20}{0} = 20$, again this doesn't work, since it must follow that $20 = 0 \cdot 20$. But $0 \cdot 20$ is not 20.

What about those who thought that $\frac{20}{0} = 1$? Were they right? Again no, since 20 is not equal to $0 \cdot 1$.

For those who believed the answer to be infinity, we point out here only that the answer to a division problem (if an answer exists) must be a specific number. "Infinity" (usually denoted by ∞) is *not* a number.[2]

In fact, *it is not possible to divide by 0 at all*. Why? Suppose it were possible. There are two cases:

Case 1. A nonzero number divided by 0.

Case 2. Zero divided by 0.

For case 1, just to be specific, let the nonzero number be 20. Then as we have seen earlier, this is not possible. It would be the same had we used any other nonzero number besides 20. Try it to convince yourself.

---

2. For those who want to know more about ∞, we suggest that you consult any elementary calculus text or your teacher. The discussion of this symbol is beyond the scope of this book.

For case 2, suppose 0 divided by 0 equals something. Thus

$$\frac{0}{0} = \text{something}$$

This means $0 = 0 \cdot$ "something" (from our definition of division). What could this something be? Any number will do. For example, $0 = 0 \cdot 5$, or $0 = 0 \cdot 17$, or $0 = 0 \cdot 141$. Thus the something is not specific. The answer to a division problem *must* be a specific number.

Summarizing We can summarize division involving 0 as follows.

1. Division into 0 always gives 0 (assuming we are not dividing by 0).
2. Division by zero is not possible.

It is interesting to find that the question of dividing by zero was actually considered by Aristotle more than 2000 years ago. He concluded that division by zero was impossible. This is surprising because it is so much like the modern approach to the problem. Division by zero was also considered by the Indian mathematician Bhaskara. He believed that when you divided by zero the result was an "unchangeable infinity" that was almost religious in nature. In fact, he compared this infinity with the unchanging nature of God.

In the following section we will have more to say about zero and its properties.

EXERCISES

1. Is there an identity for multiplication? If yes, what is it? If no, why not?

2. Is zero the identity for subtraction?

3. Is one the identity for division? Explain your answer.

4. On p. 193 we showed that if a is any number, then $a \cdot 0 = 0$. We also stated that $0 \cdot a = 0$ without justifying it. Explain why $0 \cdot a = 0$.

5. Explain how the number 0 is used in our decimal system.

6. Suppose we define a new operation $\circ$ on the set of whole numbers as follows:

$a \circ b = 0$ if either a or b or both are odd
$a \circ b = 1$ if a and b are both even

where a and b are any whole numbers.

a) Is $\circ$ a commutative operation?

b) Is $\circ$ an associative operation?

Exercises 7 and 8 are suggested only for those who have had some algebra. If you have not had any algebra, then skip these problems and go on to the next section.

†7. What is wrong with the following?
Given: $a = b$.
Multiply both sides by b:

$$ab = b^2$$

Subtract a^2 from both sides:

$$ab - a^2 = b^2 - a^2$$

---

† The dagger throughout indicates those exercises involving algebra.

Factor $ab - a^2$ as $a(b - a)$.
Factor $b^2 - a^2$ as $(b + a)(b - a)$, so that

$$a(b - a) = (b + a)(b - a)$$

Divide both sides by $b - a$:

$$\frac{a(b - a)}{b - a} = \frac{(b + a)(b - a)}{b - a}$$

We then have $a = b + a$.
Since we know $a = b$, we then have $a = a + a$
or that $a = 2a$.
Divide both sides by a, and we have

$$\frac{a}{a} = \frac{2a}{a}$$

or that $1 = 2$.

†8. What is wrong with the following?
Let a be any number but 1.
Also let $a = b$.
Add 1 to both sides:

$$a + 1 = b + 1$$

Multiply both sides by $a - 1$ so that

$$(a + 1)(a - 1) = (b + 1)(a - 1).$$

Simplifying, this gives

$$a^2 - 1 = ab - b + a - 1$$

Adding $+1$ to both sides and simplifying, we get

$$a^2 = ab - b + a$$

This can be further simplified by subtracting ab from both sides. We get

$$a^2 - ab = a - b$$

Factoring, we have

$$a(a - b) = a - b$$

Divide both sides by $a - b$. We get $a = 1$.
This contradicts the assumption that a was any number but 1.

4.4

THE INTEGERS   The natural numbers that we introduced in Section 4.2 are also known as the **positive integers.** They are sometimes written as

$$+1, +2, +3, +4, +5, \ldots$$

instead of

$$1, 2, 3, 4, 5, \ldots.$$

Are the positive integers and 0 sufficient for all our everyday needs? Definitely not! To see why, recall that in Section 4.2 we showed that subtraction of natural numbers is not a closed operation. This means that when we subtract one natural number from another natural number, our answer is not always a natural number. For example, $5 - 2$ is the natural number 3. On the other hand, $2- 5$ is *not* a natural number. Thus if we had only the natural numbers to work with, we would not always be able to subtract one number from another.

This situation is very inconvenient, and we can remedy it by introducing new numbers, which are called **negative integers.** This can be done as follows: Consider a building that has many floors both above and below street level. (The ones below street level are the basement and subbasements.)

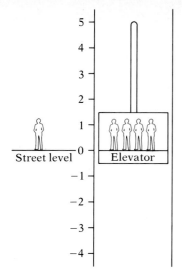

Figure 4.3

The street level floor is labeled 0 (see Fig. 4.3). Floors above street level will be labeled +1, +2, +3, etc. Floors below street level will be labeled −1, −2, −3, and so on. If you get in on the street level floor and push the + 5 button, you will go 5 floors up; if you push the −5 floor button, you will go 5 floors in the *opposite* direction, down.

Definition 4.6   *The **negative integers** are the numbers*

$$-1, -2, -3, -4, -5, \ldots.$$

This leads us to the following definition of integers.

Definition 4.7   *The set of **integers** consists of the whole numbers and the negative integers, that is, the set of integers is*

$$\{\ldots, -5, -4, -3, -2, -1, 0, +1, +2, +3, \ldots\}$$

Addition   We already know how to add positive integers (since these are just the natural numbers). How do we add negative numbers?

Let us go back to our elevator (see Fig. 4.4). Renée gets in on the tenth floor. She takes the elevator down two floors, picks up her paycheck there, and then goes down another eight floors to the street floor.

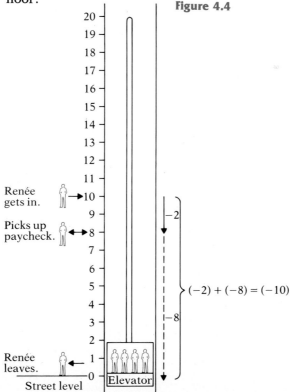

Figure 4.4

We can represent Renée's trip using negative numbers; (-2) stands for going down two floors. Similarly, (-8) represents going down eight floors. The total descent was (-10). Thus we see that $(-2) + (-8) = (-10)$.

Notice that $2 + 8$ is 10, so that

$$-(2 + 8) \text{ is } (-10)$$

Therefore

$$(-2) + (-8) = -(2 + 8)$$

To further illustrate the idea, consider Frances, who works on the twentieth floor. She first takes the elevator down to the seventh floor where she meets her friend Leo. Together they go down to the cafeteria on the fifth floor. Her descent was (-13) and then (-2), or a total of (-15). Another way of saying this is

$$(-13) + (-2) = (-15)$$

Notice again that $13 + 2$ equals 15, so that $-(13 + 2)$ equals (-15). This gives

$$(-13) + (-2) = -(13 + 2)$$

This shows us how to add two negative integers.

RULE 4.1   *If $(-a)$ and $(-b)$ represent two negative integers, then*

$$(-a) + (-b) = -(a + b).$$

How do we add a positive and a negative integer? To answer this, again consider Frances (who works on the twentieth floor). She goes down to visit her friend Leo on the seventh floor. They have an argument, and Frances goes up to the tenth floor to consult her friend Renée. Her trip from the twentieth floor to the seventh floor can be represented by (-13). Her trip from the seventh floor to the tenth floor can be represented by $(+3)$.

Now Frances starts on the twentieth floor and ends up on the tenth floor. The end result of her trip is (-10). Thus we see that

$$(-13) + (+3) = (-10).$$

While Frances is consulting with Renée, Leo goes looking for her. Since he goes from the seventh floor to the twentieth floor, this trip is $(+13)$. Upon hearing that Frances is with Renée, he goes down to the tenth floor. This trip is (-10). Since he starts from the seventh floor and ends up on the tenth floor, the end result of his trip is $(+3)$. Thus we see that

$$(+13) + (-10) = (+3).$$

We now give a general rule for adding a positive and a negative integer.

RULE 4.2 *Consider the numbers without their signs and*

i): *Select the larger.*

ii): *Subtract the smaller from the larger.*

iii): *Put the sign of the larger of step (i) in front of the answer of step (ii).*

EXAMPLE 1 Add $(+2)$ and (-5).

SOLUTION The numbers without their signs are 2 and 5.
Step (i): The larger is 5.
Step (ii): $5 - 2 = 3$.
Step (iii): Since the sign of the larger is $-$, our answer is (-3). Thus

$$(+2) + (-5) = (-3). \quad \square$$

EXAMPLE 2 Add $(-23) + (+50)$.

SOLUTION The numbers without their signs are 23 and 50.
Step (i): The larger is 50.
Step (ii): $50 - 23 = 27$.
Step (iii): Since the sign of the larger is $+$, our answer is $(+27)$. Thus

$$(-23) + (+50) = (+27). \quad \square$$

How much is $(+15)$ and (-15)? If we try to apply the rule, we notice that when we consider the numbers without their signs, neither is larger. What do we do now? To answer this question, we go back to the elevator. Dave gets in, goes up fifteen floors, and then goes down fifteen floors. This is represented by $(+15) + (-15)$. Since he returns to his starting point, the result of his trip is 0. Thus we have

$$(+15) + (-15) = 0.$$

In general, if b is any number, then $(+b) + (-b) = 0$.
We call (-15) the additive inverse of $(+15)$.

> **Definition 4.8**   *The **additive inverse** of a number b is the number which when added to b gives 0. This is $(-b)$.*

EXAMPLE 3   The additive inverse of $(+7)$ is (-7), since $(+7) + (-7) = 0$. ☐

EXAMPLE 4   The additive inverse of (-10) is $(+10)$, since $(-10) + (+10) = 0$. ☐

EXAMPLE 5   We mentioned that (-15) is the additive inverse of $(+15)$. Is $(+15)$ the additive inverse of (-15)? ☐

Subtraction   What do we mean by $2 - 5$? We can answer this by going back to our elevator. In this situation, one would normally interpret this to mean "go up 2 floors and then go down 5 floors." You would end up 3 floors below where you started, or (-3). Thus $2 - 5 = (-3)$.

If you look back at Example 1, you will notice that $(+2) + (-5)$ is also (-3). In other words, $2 - 5 = (+2) + (-5)$.

This suggests the following definition.

> **Definition 4.9**   *If a and b are any integers, then*
>
> $$a - b \qquad means \qquad a + (-b)$$
>
> *In other words, $a - b$ means "a plus the additive inverse of b."*

EXAMPLE 6 $4 - 7$ means $4 + (-7)$. Using the rule for addition, we get

$$4 + (-7) = (-3)$$

Thus

$$4 - 7 = (-3) ☐$$

EXAMPLE 7 $5 - (-3)$ means $5 + (+3)$. (Remember, $+3$ is the additive inverse of -3). Our result is $+8$. Therefore

$$5 - (-3) = (+8) ☐$$

Comment   In the line above, we have $5 - (-3)$. Although there are two minus signs, they have different meanings. The first minus represents subtraction. The second minus represents a negative integer.

Multiplication   Since positive integers are really natural numbers, multiplication of positive integers is exactly the same as multiplication of natural numbers. For example, $(+3) \cdot (+5)$ is the same as $3 \cdot 5$. Thus we really have $(+5) + (+5) + (+5)$. This sum is obviously $(+15)$.

Similarly, $(+7) \cdot (+4)$ means

$$(+4) + (+4) + (+4) + (+4) + (+4) + (+4) + (+4)$$

which equals $(+28)$.

Now what could we mean by $(+3) \cdot (-5)$? It seems reasonable to interpret this in a similar manner. Thus we say that $(+3) \cdot (-5)$ means

$$\underbrace{(-5) + (-5) + (-5)}$$
$$3 \text{ groups of } (-5)$$

which equals (-15).

Notice that $(+3) \cdot (-5) = -(3 \cdot 5)$.

Similarly, $(+4) \cdot (-6)$ means

$$\underbrace{(-6) + (-6) + (-6) + (-6)}$$
$$4 \text{ groups of } (-6)$$

which equals (-24).

Again notice that $(+4) \cdot (-6) = -(4 \cdot 6)$. This we can generalize as:

If $(+a)$ is a positive integer and $(-b)$ is a negative integer, then the product is

$$(+a) \cdot (-b) = -(a \cdot b)$$

What about $(-5) \cdot (+3)$? We would certainly like the commutative law to hold for multiplication. Thus, we want $(-5) \cdot (+3)$ to equal $(+3) \cdot (-5)$. Since $(+3) \cdot (-5)$ equals $-(3 \cdot 5)$, we have $(-5) \cdot (+3)$ equals $-(3 \cdot 5)$ or $-(5 \cdot 3)$. In both cases our answer is (-15).

In another example, $(-6) \cdot (+4)$ should equal $(+4) \cdot (-6)$. We therefore have

$$(-6) \cdot (+4)$$
$$= -(6 \cdot 4)$$
$$= (-24)$$

In general,

If $(-a)$ is a negative integer, and $(+b)$ is a positive integer, then the product is

$$(-a) \cdot (+b) = -(a \cdot b)$$

We can summarize our discussion as the following rule.

RULE 4.3 *The **product** of a negative integer and a positive integer is found by multiplying the numbers without their signs and putting a minus sign in front of the answer.*

A formal justification of this rule will be given in the exercises.

Finally, we must consider how to multiply two negative numbers. For example, what would $(-10) \cdot (-6)$ be? The answer is $(+60)$. To see why, consider Margaret, who has withdrawn from her bank account $10 a week for the past six weeks. Obviously, six weeks ago she was $60 richer.

Let $+1$ represent a deposit of $1.

Let -1 represent a withdrawal of $1.

Let $+1$ represent a week from now.

Let -1 represent a week ago.

Then (-10) represents Margaret's weekly withdrawal, and (-6) represents six weeks ago. Since six weeks ago she had $60 more, this situation can be written as $(-10) \cdot (-6) = (+60)$.

Now let us look at Rosalie, who has been on a diet for several weeks and has been steadily losing two pounds a week. If we let -4 represent four weeks ago and -2 represent a weight loss of two pounds, then four weeks ago she weighed $(-2) \cdot (-4)$ or $+8$ pounds more than she does now.

This leads us to the following rule.

> **RULE 4.4** *If $(-a)$ and $(-b)$ are negative integers, then the **product** is $(-a) \cdot (-b) = +(a \cdot b)$.*

EXAMPLE 8   The product of $(-5) \cdot (-4)$ is $+(5 \cdot 4)$, which equals $(+20)$.   ☐

EXAMPLE 9   The product of $(-20) \cdot (-3)$ is $+(20 \cdot 3)$. This equals $(+60)$.   ☐

EXERCISES

In Exercises 1–15, calculate the answer to the indicated problem.

1. $(+15) + (+9)$
2. $(-4) + (-7)$
3. $(+8) + (-3)$
4. $(-6) + (+7)$
5. $(-9) + (+4)$
6. $(-3) - (-2)$
7. $(-4) - (+9)$
8. $0 - (-8)$
9. $(-9) - (-8)$
10. $(-8)(+3)$
11. $(-8)(-3)$
12. $(+8)(-3)$
13. $(-7) - 0$
14. $(-7) \cdot 0$
15. $(+7) - (-7)$

16. What is the identity for addition of integers?

17. Are the integers closed under division? Explain your answer.

18. The commutative, associative, and closure laws hold for addition and multiplication of integers. For example, the commutative law for addition states:

    If x and y are any integers, then $x + y = y + x$.

    State, in symbols, each of the other laws.

19. For every signed number a and every signed number b, is $a + b$ a unique signed number?

20. The **distributive law of multiplication over addition** states that if a, b, and c are any inte-

gers, then

$$a(b + c) = ab + ac.$$

For example,

$$5(4 + 3) = 5 \cdot 4 + 5 \cdot 3$$
$$5(7) \quad = \quad 20 + 15$$
$$35 \quad = \quad 35.$$

Is it also true that $(a + b)c = ac + bc$? Give examples to support your answer.

21. The rule for multiplying a positive integer and a negative integer can be justified with the following argument. We will show that $(+4) \cdot (-16)$ must equal (-64):

$$4[16 + (-16)]$$
$$= 4(16) + 4(-16) \qquad \text{(Distributive law)}$$
$$= 64 + 4(-16). \qquad\qquad (1)$$

On the other hand, we know that

$$4[16 + (-16)] = 4(0) \text{ because } 16 + (-16) = 0.$$

Therefore $4[16 + (-16)] = 0$. So $64 + 4(-16) = 0$ in view of (1). From this it follows that $4(-16)$ is the additive inverse of 64. We already know that the additive inverse of 64 of -64. Thus

$$4(-16) = (-64).$$

Construct a similar argument showing that $(-a)(-b) = + (ab)$.

Which law justifies each of the computations performed in Exercises 22–30?

22. $(+8) + (+6) = (+6) + (+8)$

23. $(-5)[(-8) + (-3)] = (-5)(-8) + (-5)(-3)$

24. $(+6) + [(-7) + (-9)] = [(+6) + (-7)] + (-9)$

25. $(-5)(-7)$ is an integer

26. $(-7)[(-8)(-9)] = [(-7)(-8)](-9)$

27. $(-6)[8 + 2] = (-6)(8) + (-6)(2)$

28. $(-11)(0) = 0$

29. $(-5) + (+5) = 0$

30. $(+8)(-3) = (-3)(+8)$

Use the distributive law to find the value of each of the computations in Exercises 31–33.

31. $(-16) \cdot (+94) + (-16) \cdot (+6)$

32. $(-9) \cdot (+72) + (-9)(+21) + (-9)(+7)$

33. $(-19)(-24) + (-19)(-76)$

34. Show that $(-1) \cdot x = -x$.

35. The symbol a^n stands for

$$\underbrace{a^n = a \cdot a \cdot a \cdot a}_{a \text{ multiplied by itself } n \text{ times}}$$

where n is a positive integer. If a is a negative integer, will a^n be positive or negative when:

a) n is even? Explain.

b) n is odd? Explain.

36. An archaeologist discovered an ancient relic that dates back to the year 426 B.C. How old was the relic in the year 1985?

37. Aristotle was born in 384 B.C. and Euclid was born in 365 B.C. Who was born first?

For Exercises 38 and 39, let I represent the set of integers, W the set of whole numbers, and N the set of natural numbers.

38. Write a set relationship (using subsets) connecting each of the following:

a) W and I **b)** W and N **c)** N and I

39. Find each of the following:

a) $N \cup W$ **b)** $N \cap W$ **c)** $I \cap N$

d) $W \cap I$ **e)** $W \cup I$ **f)** $N \cup I$

The **absolute value** of an integer is defined as the undirected distance of that integer from zero on a number line. For example, the absolute value of -9 is 9. The absolute value of $+9$ is 9. The symbol for absolute value is $|\ \ |$. Thus $|-9|$ is read as "the absolute value of -9." The undirected distance (direction does not count) between 0 and -9 on a number line is 9.

Evaluate each of the absolute values given in Exercises 40–48.

40. $|+11|$ **41.** $|-8|$

42. $|3 - 10|$ **43.** $|0|$

44. $|+9 - 9|$ **45.** $|9 - 5|$

46. $|(7 - 6) - 4|$ **47.** $|7 - (6 - 4)|$

48. $|12 - 12|$

**49. The rule for addition of a positive signed number and a negative signed number can be stated formally as follows: If a is a positive signed number and if b is a negative signed number, then

$$\text{if } |a| \geq |b| \text{ then } a + b = |a| - |b|$$

and

$$\text{if } |b| > |a| \text{ then } a + b = -(|b| - |a|)$$

Illustrate this rule by selecting different values for a and b and calculating $a + b$. Check your answer by using the rule given on page 199.

4.5

DIVISIBILITY

A basic idea of number theory is that of one number dividing another number evenly. For example, 4 divides 12 evenly, but 4 does *not* divide 11 evenly. We therefore begin with a discussion of what we mean by **divisibility.**

We say that 4 divides 12 evenly because 12 consists of exactly three 4's. To put it another way, $4 \cdot 3 = 12$.

Similarly, we know that 4 does not divide 11 because 11 does not consist exactly of a whole number of 4's. In other words, there is no whole number m such that $4 \cdot m = 11$. This leads us to the following definition.

> **Definition 4.10**   *We say that x **divides** y whenever there is a whole number m such that x · m = y. If no such number m exists, then we say that x does not divide y. If x divides y, then we call x a **divisor** of y.*

Comment   In this definition we *never* let $x = 0$. Remember, you cannot divide by 0. However, y *can* equal 0.

EXAMPLE 1   7 divides 21 because $7 \cdot 3 = 21$.

9 does *not* divide 21, because there is no whole number m such that $9 \cdot m = 21$.
8 divides 200, because $8 \cdot 25 = 200$.
8 does *not* divide 100, because there is no whole number m such that $8 \cdot m = 100$. ▭

EXAMPLE 2   The divisors of 18 are 1, 2, 3, 6, 9, and 18 itself, since each of these divides 18. ▭

EXAMPLE 3   5 divides 50 and 5 also divides 15. Let us now add 50 and 15. We get 65. How about 65? It is obvious that 5 divides 65 also, since $5 \cdot 13 = 65$. This leads us to the following useful statement about divisibility. ▭

> **Statement 1**   *Suppose x divides y and x also divides z. Then x divides y + z.*

This just says that if x divides each of two numbers, then it also divides their sum.

†Proof of statement 1 If x divides y, this means that there is a whole number m such that

$$y = x \cdot m \qquad\qquad (1)$$

Similarly, if x divides z, this means that there is a whole number n such that

$$z = x \cdot n \qquad\qquad (2)$$

If we now add (1) and (2) together, we see that

$$\begin{aligned} y + z &= x \cdot m + x \cdot n \\ &= x(m + n) \qquad \text{(by the distributive law)}. \end{aligned}$$

Now $(m + n)$ is the sum of two whole numbers and is therefore itself a whole number. Why? Let us call it M. Thus $M = m + n$. Then

$$y + z = x(m + n)$$

or

$$y + z = xM \qquad\qquad (3)$$

This last line (3) says that there is a number M such that $x \cdot M = y + z$. By Definition 4.10, we therefore have that x divides $y + z$.

EXAMPLE 4 4 divides 8, and 4 also divides 20. Statement 1 tells us that 4 divides $8 + 20$, or that 4 divides 28. ☐

EXAMPLE 5 6 divides 30 and 6 divides 12. Statement 1 tells us that 6 divides $30 + 12$, or that 6 divides 42. ☐

EXAMPLE 6 3 divides 12; 3 does not divide 5. Does 3 divide $12 + 5$, or 17? Obviously not! ☐

The last example leads us to another useful statement.

> **Statement 2** *If x divides y, and x does not divide z, then x does not divide $y + z$.*

This just says that if x divides one part of a sum and x does not divide the second part, then it cannot divide the entire sum.

†Proof of statement 2 This proof is a proof by contradiction, so you should review this technique before reading further.

There are two possibilities:

a) x does divide $y + z$, or

b) x does not divide $y + z$.

Either (a) or (b) *must* be true.

Let us try possibility (a) and suppose that it is true. Then x divides $y + z$. By Definition 4.10 this means that there is a number m such that $x \cdot m = y + z$. Since we are given that x divides y, then this means that there is a number n such that $x \cdot n = y$. Now

$$x \cdot m = y + z$$
$$x \cdot m = (x \cdot n) + z \qquad \text{(We just substitute for } y.)$$

Let us subtract $x \cdot n$ from both sides of this equation. We get

$$(x \cdot m) - (x \cdot n) = (x \cdot n) + z - (x \cdot n)$$
$$(x \cdot m) - (x \cdot n) = z \qquad \text{(The } x \cdot n\text{'s on the right side cancel out.)}$$
$$x(m - n) = z \qquad \text{(Here we used the distributive law on the left side of the equation.)}$$

Now $m - n$ is the difference of two whole numbers, and m is larger than n. (Why?) Thus $m - n$ is also a whole number. Let us call it M. We then have

$$x(m - n) = z$$
$$x \cdot M = z \qquad \text{(We substituted } M \text{ for } m - n.)$$

But this last line says that there is a number M that, when multiplied by x, will give z. By Definition 4.10 this means that x *divides* z.

However, since we were told that x *does not divide* z, we therefore have a *contradiction*. This means that possibility (a) is wrong. Since possibility (a) cannot be correct, then we conclude that possibility (b), which says that x does *not* divide $y + z$, is correct.

EXAMPLE 7   6 divides 12. However, 6 does not divide 13. What about the sum of 12 and 13? Clearly, 6 does not divide $12 + 13$, or 25.   □

Tests for Divisibility   There are many situations in which we want to know whether one number divides another evenly. Take the following example. In a certain town the sanitation department employs 291 people. The department works in crews of 3. Can these 291 people be evenly divided into crews of 3? You can answer this question by actually dividing 3 into 291. If you do, you find that 3 goes into 291 exactly 97 times. However, sometimes we do not want to spend the time required to do this. It turns out that we can often answer such question without actually doing the division, but by using the following **divisibility tests**.

Test for Divisibility by 2   *A number is divisible by 2 when the ones digit is 0, 2, 4, 6, or 8.*

EXAMPLE 8   The number 4308 is divisible by 2, since the ones digit is 8.

The number 23456 is divisible by 2, since the ones digit is 6.

The number 51694 is divisible by 2, since the ones digit is 4. ☐

Test for Divisibility by 3

A number is divisible by 3 when the sum of its digits is divisible by 3.

EXAMPLE 9

The number 52341 is divisible by 3, since the sum of its digits is 5 + 2 + 3 + 4 + 1, or 15, and 15 is divisible by 3.

The number 291 is divisible by 3, since the sum of its digit is 2 + 9 + 1, or 12, and 12 is divisible by 3. ☐

Test for Divisibility by 4

A number is divisible by 4 when the number formed by the last two digits is divisible by 4.

EXAMPLE 10

The number 5344 is divisible by 4, since the last two digits form the number 44, and 44 is divisible by 4.

The number 6213 is *not* divisible by 4, since the last two digits form the number 13, and 13 is not divisible by 4. ☐

Test for Divisibility by 5

A number is divisible by 5 when the ones digit is 0 or 5.

EXAMPLE 11

The number 42805 is divisible by 5, since the ones digit is 5.

The number 28130 is divisible by 5, since the ones digit is 0. ☐

Test for Divisibility by 9

A number is divisible by 9 if the sum of it digits is divisible by 9.

EXAMPLE 12

The number 5346 is divisible by 9, since the sum of its digits is 5 + 3 + 4 + 6, or 18, and 18 is divisible by 9.

The number 3289 is *not* divisible by 9, since the sum of its digits is 3 + 2 + 8 + 9, or 22, and 22 is not divisible by 9. ☐

A very useful application of these divisibiity tests is in reducing fractions to lowest terms. For example, the fraction $\frac{1470}{21657}$ can be reduced. The divisibility test for 3 shows us that both the numerator and denominator can be divided by 3. Can the fraction be reduced by any other number?

These tests for divisibility can be proved by using divisibility theorems. Consult the suggested readings for these proofs.

EXERCISES

Find all the divisors of each of the numbers given in Exercises 1–10.

1. 56   **2.** 86   **3.** 98   **4.** 62

5. 132   **6.** 160   **7.** 149   **8.** 482

9. 376   **10.** 558

11. Find all the divisors of 0.

Test each of the numbers given in Exercises 12–21 to see whether it is divisible by 2, 3, 4, 5, or 6.

12. 324   **13.** 436   **14.** 864   **15.** 685

16. 293   **17.** 456   **18.** 890   **19.** 3402

20. 5678   **21.** 1396

22. a) Make up a test to determine whether a number is divisible by 6.

    b) Use this test to determine which of the numbers given in Exercises 12–21 are divisible by 6.

23. a) Make up a test to determine whether a number is divisible by 8.

    b) Use this test to determine which of the numbers given in Exercises 12–21 are divisible by 8.

24. Can you find a five-digit number that is divisible by 4, 5, 8, and 9?

25. Can you find a six-digit number that is divisible by 5, 6, 7, and 9?

26. Can you find a six-digit number that is divisible by 5 and 7 and not by 4, 6, or 9?

27. Find a rule for determining when a number is divisible by 50.

28. Can you find a three-digit number that is divisible by the product of its digits?

A leap year is a year whose date is divisible by 4. However, century years (that is, years that end in two zeros) are leap years only when their dates are divisible by 400. Which of the years given in Exercises 29–36 are leap years?

29. 1984   **30.** 1623   **31.** 1492   **32.** 1900

33. 1400   **34.** 1424   **35.** 1776   **36.** 1943

37. If the sum of the divisors of a number, other than itself, is equal to another number, and *vice versa*, then we say that the numbers are **amicable** or **friendly**. Show that 220 and 284 are friendly numbers.

38. If 1184 is one of the two numbers of a pair of friendly numbers, find the other number.

39. a) Multiply the numbers 5, 6, and 7 together. Divide the results by 6.

    b) Multiply the numbers 13, 14, and 15 together. Divide the results by 6.

    c) Multiply the numbers 25, 26, and 27 together. Divide the results by 6.

    d) On the basis of the results obtained in parts (a), (b), and (c), can you generalize about the product of any three numbers?

    e) What statement can be made about the product of any four numbers?

40. Determine whether the following statement is true: A number is divisible by 8 if and only if the difference between the sum of the digits in the odd places and the sum of the digits in the even places is divisible by 11.

4.6

THE RATIONAL NUMBERS

Division is not always possible if one has only the integers. For example, try to divide 2 by 7. The answer is 2/7, which, of course, is not an integer. To overcome this difficulty, the rational numbers were invented. We define these as follows.

> **Definition 4.11**   *A **rational number** is any number that can be written as $\frac{a}{b}$, where a and b are integers and b is not zero. Here a (the number on top) is called the numerator, and b (the number on the bottom) is called the denominator.*

Notice that in this definition we stated that the denominator cannot be zero. Why?

EXAMPLE 1 The following are examples of rational numbers:

$$\frac{2}{3}, \quad \frac{+5}{-3}, \quad \frac{-5}{+3}, \quad \frac{8}{2}, \quad \frac{-13}{1}$$

The integer 4 is a rational number since it can be written as $\frac{4}{1}$. Similarly, -17 is a rational number. It can be written as $\frac{-17}{1}$. Also, 0 is a rational number. Why?

As a matter of fact, all the integers (and therefore all the natural numbers also) are rational numbers, since we can write them as $\frac{\text{integer}}{1}$. ☐

Comment The rational numbers are so named not because they are in better mental health than other numbers, but because they can be written as the **ratio** of two integers.

Equal Rational Numbers

If you have two quarters and your friend has a half-dollar, then clearly you both have the same amount of money. This means that $\frac{2}{4}$ represents the same quantity as $\frac{1}{2}$ (the half-dollar). Consequently,

$$\frac{2}{4} = \frac{1}{2}$$

The rational numbers $\frac{3}{9}$ and $\frac{1}{3}$ are also equal. To see this, consider a pie divided into nine equal parts (Fig. 4.5). Clearly $\frac{3}{9}$ of the pie is a third of the entire pie.

Consider the statement $\frac{2}{4} = \frac{1}{2}$. Notice that if we "cross multiply"

$$\frac{2}{4} \diagup\!\!\!\!\diagdown \frac{1}{2}$$

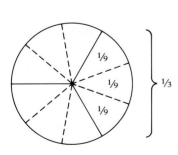

Figure 4.5

we have $2 \cdot 2 = 4 \cdot 1$. Similarly, in the statement $\frac{3}{9} = \frac{1}{3}$, if we cross multiply

$$\frac{3}{9} \diagup\!\!\!\!\diagdown \frac{1}{3}$$

we get $3 \cdot 3 = 9 \cdot 1$.

In general, we have the following definition.

Definition 4.12   *The rational numbers $\frac{a}{b}$ and $\frac{c}{d}$ are equal if, when we cross multiply*

$$\frac{a}{b} \diagdown\!\!\!\!\diagup \frac{c}{d}$$

we get $ad = bc$.

EXAMPLE 2   The rational numbers $\frac{3}{8}$ and $\frac{12}{32}$ are equal, since when we cross multiply

$$\frac{12}{32} \diagdown\!\!\!\!\diagup \frac{3}{8}$$

we get

$$12 \cdot 8 = 32 \cdot 3$$
$$96 = 96 \quad \square$$

EXAMPLE 3 $\frac{5}{11}$ is *not* equal to $\frac{4}{9}$, since if we cross multiply

$$\frac{5}{11} \diagdown\!\!\!\!\diagup \frac{4}{9}$$

we get $5 \cdot 9$, which is 45. This is not equal to $11 \cdot 4$, which is 44.
$\square$

It is easy to see that $\frac{2}{3} = \frac{8}{12}$, since $2 \cdot 12 = 3 \cdot 8$. Let us look more closely at the rational number $\frac{8}{12}$. It can be written as $\frac{2 \cdot 4}{3 \cdot 4}$. Since 4 appears in the numerator and in the denominator, we can **"cancel" out** (divide the numerator and denominator by) the 4. As a result we have

$$\frac{2 \cdot \cancel{4}}{3 \cdot \cancel{4}} = \frac{2}{3}$$

In general, if a, b, and c are any integers (with b and c not zero), then

$$\frac{ac}{bc} = \frac{a}{b} \quad \text{and} \quad \frac{ca}{cb} = \frac{a}{b}$$

To justify this, apply the definition of equal rational numbers. We cross multiply $\frac{ac}{bc} = \frac{a}{b}$, getting $(ac)b = (bc)a$. By using the commutative, associative, and closure laws for multiplication of integers, it can be shown that $(ac)b$ does equal $(bc)a$, so the rational numbers $\frac{ac}{bc}$ and $\frac{a}{b}$ are equal. Similarly, we can show that $\frac{ca}{cb}$ and $\frac{a}{b}$ are equal. Another justification of this cancellation principle will be given after we discuss multiplication of rational numbers.

The rational number $\frac{8}{12}$ can be written as $\frac{4 \cdot 2}{6 \cdot 2}$. If we cancel the 2, we are left with $\frac{4}{6}$. So we have another rational number $\frac{4}{6}$, which is equal to $\frac{8}{12}$. Also, $\frac{8}{12}$ can be written as $\frac{2}{3}$, as we saw above. The rational numbers $\frac{8}{12}, \frac{4}{6}$, and $\frac{2}{3}$ are all equal. However, $\frac{2}{3}$ is different from the others because it is not possible to cancel any more numbers from its numerator and denominator. When $\frac{8}{12}$ is written as $\frac{2}{3}$, we say that it has been **reduced to lowest terms.**

EXAMPLE 4   Reduce $\frac{15}{20}$ to lowest terms.

SOLUTION $\dfrac{15}{20} = \dfrac{3 \cdot \cancel{5}}{4 \cdot \cancel{5}} = \dfrac{3}{4}$ □

EXAMPLE 5   Reduce $\dfrac{-18}{21}$ to lowest terms.

SOLUTION $\dfrac{-18}{21} = \dfrac{(-6)\cancel{(+3)}}{(+7)\cancel{(+3)}} = \dfrac{-6}{7}$ □

EXAMPLE 6   A student was asked to reduce $\dfrac{360}{240}$ to lowest terms. He wrote

$$\frac{360}{240} = \frac{36 \cdot \cancel{10}}{24 \cdot \cancel{10}} = \frac{36}{24}$$

What is wrong?

SOLUTION   We see that this has not been reduced to lowest terms, since we can further reduce $\dfrac{36}{24}$ as

$$\frac{36}{24} = \frac{3 \cdot \cancel{12}}{2 \cdot \cancel{12}} = \frac{3}{2}$$

Therefore $\dfrac{360}{240}$ reduced to lowest terms is $\dfrac{3}{2}$.   □

EXAMPLE 7 $\dfrac{7}{7} = \dfrac{1 \cdot \cancel{7}}{1 \cdot \cancel{7}} = \dfrac{1}{1}$. This is written as 1.   □

Addition and Subtraction

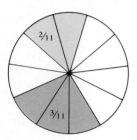

Figure 4.6

How do we add $\frac{3}{11}$ and $\frac{2}{11}$? Look at the pie in Fig. 4.6, which has been divided into eleven equal parts. Clearly $\frac{2}{11} + \frac{3}{11}$ equals $\frac{5}{11}$. Both denominators are the same. Since $2 + 3 = 5$, then we have

$$\frac{2}{11} + \frac{3}{11} = \frac{2 + 3}{11} = \frac{5}{11}$$

In general terms, we have the following rule.

RULE 4.5   *If $\frac{a}{b}$ and $\frac{c}{b}$ are rational numbers, then $\frac{a}{b} + \frac{c}{b} = \frac{a + c}{b}$.*

Now let us subtract $\frac{2}{11}$ from $\frac{3}{11}$, that is, $\frac{3}{11} - \frac{2}{11}$. Again referring back to the pie, we find that $\frac{3}{11} - \frac{2}{11} = \frac{1}{11}$. Since the denominators are the same and $3 - 2 = 1$, we have

$$\frac{3}{11} - \frac{2}{11} = \frac{3 - 2}{11} = \frac{1}{11}$$

which suggests the following rule.

RULE 4.6   *If $\frac{a}{b}$ and $\frac{c}{b}$ are rational numbers, then $\frac{a}{b} - \frac{c}{b} = \frac{a - c}{b}$.*

EXAMPLE 8 $\dfrac{3}{10} + \dfrac{4}{10} = \dfrac{3 + 4}{10} = \dfrac{7}{10}$ □

EXAMPLE 9 $\dfrac{3}{14} + \dfrac{5}{14} = \dfrac{3 + 5}{14} = \dfrac{8}{14}$. Since $\dfrac{8}{14}$ can be written as $\dfrac{4 \cdot 2}{7 \cdot 2}$, we can

cancel the 2's, getting $\dfrac{4}{7}$. Thus $\dfrac{3}{14} + \dfrac{5}{14} = \dfrac{4}{7}$.   □

EXAMPLE 10 $\dfrac{2}{5} - \dfrac{3}{5} = \dfrac{2 - 3}{5} = \dfrac{-1}{5}$ □

Now let us add $\frac{2}{3}$ and $\frac{4}{5}$. Since they do not have the same denominators, the previous rule does not appear to apply. You may think (or wish) that these numbers cannot be added. However, we have a way out of this difficulty.

$\dfrac{2}{3}$ is the same as $\dfrac{2 \cdot 5}{3 \cdot 5} = \dfrac{10}{15}$ (cancellation principle)

$\dfrac{4}{5}$ is the same as $\dfrac{4 \cdot 3}{5 \cdot 3} = \dfrac{12}{15}$ (cancellation principle)

Therefore

$$\frac{2}{3} + \frac{4}{5} \text{ is the same as } \frac{10}{15} + \frac{12}{15}$$

Now we can use our rule:

$$\frac{10}{15} + \frac{12}{15} = \frac{10 + 12}{15} = \frac{22}{15}$$

We conclude that $\frac{2}{3} + \frac{4}{5} = \frac{22}{15}$. □

Our procedure is generalized in the following rule.

RULE 4.7 *To add two rational numbers $\frac{a}{b}$ and $\frac{c}{d}$ that have different denominators,*

1. *rewrite each number so that they have the same denominator, and*
2. *add the resulting rational numbers by the rule on p. 212.*

Subtraction is done by a similar procedure.

EXAMPLE 11   Add $\frac{5}{6} + \frac{3}{4}$.

SOLUTION $\frac{5}{6} = \frac{5 \cdot 4}{6 \cdot 4} = \frac{20}{24}$ and $\frac{3}{4} = \frac{3 \cdot 6}{4 \cdot 6} = \frac{18}{24}$.

Therefore

$$\frac{5}{6} + \frac{3}{4} = \frac{20}{24} + \frac{18}{24} = \frac{20 + 18}{24}$$

$$= \frac{38}{24}$$

which can be reduced as

$$\frac{38}{24} = \frac{19 \cdot \cancel{2}}{12 \cdot \cancel{2}} = \frac{19}{12}$$

We have $\frac{5}{6} + \frac{3}{4} = \frac{19}{12}$. □

EXAMPLE 12   Add $\frac{2}{3} + \frac{1}{9}$.

SOLUTION $\frac{2}{3} = \frac{2 \cdot 3}{3 \cdot 3} = \frac{6}{9}$

We do not have to do anything to $\frac{1}{9}$, since the denominator is already

9. Thus we have the following:

$$\frac{2}{3} + \frac{1}{9} = \frac{6}{9} + \frac{1}{9}$$

$$= \frac{6+1}{9} = \frac{7}{9}$$

Our answer is $\frac{2}{3} + \frac{1}{9} = \frac{7}{9}$. $\square$

EXAMPLE 13   Subtract $\frac{5}{7} - \frac{1}{2}$.

SOLUTION $\frac{5}{7} = \frac{5 \cdot 2}{7 \cdot 2} = \frac{10}{14}$ and $\frac{1}{2} = \frac{1 \cdot 7}{2 \cdot 7} = \frac{7}{14}$.

Therefore

$$\frac{5}{7} - \frac{1}{2} = \frac{10}{14} - \frac{7}{14}$$

$$= \frac{10-7}{14} = \frac{3}{14}$$

Thus

$$\frac{5}{7} - \frac{1}{2} = \frac{3}{14} \quad \square$$

Multiplication   The area of a rectangle may be calculated by multiplying the length times the width. The following two diagrams illustrate this.

3 | Area is $3 \cdot 5 = 15$
   5

4 | Area is $4 \cdot 6 = 24$
   6

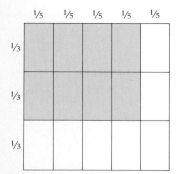

Figure 4.7

Now consider a rectangle measuring 1 by 1 that has been broken up into fifteen smaller rectangles of the same size, as shown in Fig. 4.7.

Each little rectangle measures $\frac{1}{3}$ by $\frac{1}{5}$. Suppose we wanted to find the area represented by the shaded portion of the diagram. From the area formula this would be $\frac{2}{3} \cdot \frac{4}{5}$. On the other hand, the shaded portion contains 8 little rectangles. Since there are 15 altogether, the area is $\frac{8}{15}$ of the total area. Thus we see that

$$\frac{2}{3} \cdot \frac{4}{5} = \frac{8}{15}$$

Notice that $2 \cdot 4 = 8$, and $3 \cdot 5 = 15$. So we have

$$\frac{2}{3} \cdot \frac{4}{5} = \frac{2 \cdot 4}{3 \cdot 5}$$

$$= \frac{8}{15}$$

What we have done is to multiply the numerators and also the denominators to get the product. This gives us the following rule.

RULE 4.8 *If $\frac{a}{b}$ and $\frac{c}{d}$ are rational numbers, then the product is*

$$\frac{a}{b} \cdot \frac{c}{d} = \frac{ac}{bd}.$$

EXAMPLE 14   Multiply $\frac{3}{7} \cdot \frac{5}{8}$.

SOLUTION $\dfrac{3}{7} \cdot \dfrac{5}{8} = \dfrac{3 \cdot 5}{7 \cdot 8} = \dfrac{15}{56}.$ $\square$

EXAMPLE 15   Multiply $\dfrac{-10}{3} \cdot \dfrac{5}{9}$.

SOLUTION $\dfrac{-10}{3} \cdot \dfrac{5}{9} = \dfrac{(-10) \cdot 5}{3 \cdot 9} = \dfrac{-50}{27}.$ $\square$

EXAMPLE 16   Multiply $4 \cdot \dfrac{6}{7}$.

SOLUTION   Since 4 is the same as $\dfrac{4}{1}$, we have

$$\frac{4}{1} \cdot \frac{6}{7} = \frac{4 \cdot 6}{1 \cdot 7} = \frac{24}{7} \quad \square$$

Comment   We now have another way of justifying the cancellation principle introduced earlier.

$$\frac{ac}{bc} = \frac{a \cdot c}{b \cdot c} \qquad (b \text{ and } c \text{ are not zero})$$

$$= \frac{a}{b} \cdot \frac{c}{c}$$

$$= \frac{a}{b} \cdot 1 = \frac{a}{b}$$

Therefore

$$\frac{ac}{bc} = \frac{a}{b}$$

which means that we can cancel the c's.

Division   In your previous mathematical studies you learned a rather peculiar procedure for dividing one rational number by another. It can be stated as follows.

RULE 4.9   *If $\frac{a}{b}$ and $\frac{c}{d}$ are rational numbers, then $\frac{a}{b} \div \frac{c}{d} = \frac{a}{b} \cdot \frac{d}{c}$.*

In words, this says that you flip over $\frac{c}{d}$ to make $\frac{d}{c}$ and multiply $\frac{d}{c}$ by $\frac{a}{b}$.

EXAMPLE 17   Divide $\frac{4}{7}$ by $\frac{3}{5}$.

SOLUTION $\frac{4}{7} \div \frac{3}{5} = \frac{4}{7} \cdot \frac{5}{3} = \frac{4 \cdot 5}{7 \cdot 3} = \frac{20}{21}.$ □

EXAMPLE 18   Divide $\frac{-2}{3}$ by $\frac{8}{9}$.

SOLUTION
$$\frac{-2}{3} \div \frac{8}{9} = \frac{-2}{3} \cdot \frac{9}{8}$$
$$= \frac{(-2) \cdot 9}{3 \cdot 8}$$
$$= \frac{-18}{24} \quad \text{which can be reduced as}$$
$$= \frac{-3 \cdot \cancel{6}}{4 \cdot \cancel{6}} = \frac{-3}{4} \quad □$$

EXAMPLE 19   Divide $\frac{2}{5}$ by 3.

SOLUTION $\frac{2}{5} \div 3 = \frac{2}{5} \div \frac{3}{1}$
$$= \frac{2}{5} \cdot \frac{1}{3} = \frac{2 \cdot 1}{5 \cdot 3} = \frac{2}{15}. \quad □$$

If you are like most students (including the authors), this rule probably seemed very mysterious when you first saw it. It was something you learned to humor the teacher, without really understanding why it works. The mystery will now be unraveled.

Remember that $6 \div 3 = 2$ means $6 = 3 \cdot 2$. This says that $6 \div 3$ is a number (namely 2) that when multiplied by 3 results in 6.

Similarly, $\frac{4}{7} \div \frac{3}{5}$ means a rational number that when multiplied by $\frac{3}{5}$ will give $\frac{4}{7}$. We claim that this number is $\frac{4}{7} \cdot \frac{5}{3}$, or $\frac{20}{21}$. So we want to multiply $\frac{3}{5}$ by some number and come up with an answer of

$\frac{4}{7}$. This can be accomplished by first multiplying $\frac{3}{5}$ by $\frac{5}{3}$. This yields

$$\frac{3}{5} \cdot \frac{5}{3} = \frac{3 \cdot 5}{5 \cdot 3} = \frac{15}{15} = 1$$

Now if we multiply the 1 by $\frac{4}{7}$, we get

$$1 \cdot \frac{4}{7} = \frac{1}{1} \cdot \frac{4}{7} = \frac{1 \cdot 4}{1 \cdot 7} = \frac{4}{7}$$

Therefore to get from $\frac{3}{5}$ to $\frac{4}{7}$, we multiply by $\frac{5}{3}$ and then by $\frac{4}{7}$ or by $\left(\frac{5}{3} \cdot \frac{4}{7}\right)$. To check that we are right, we have

$$\frac{4}{7} = \frac{3}{5} \cdot \left(\frac{4}{7} \cdot \frac{5}{3}\right) = \frac{3}{5} \cdot \left(\frac{4 \cdot 5}{7 \cdot 3}\right)$$
$$= \frac{3}{5} \cdot \frac{20}{21} = \frac{3 \cdot 20}{5 \cdot 21} = \frac{60}{105}$$
$$= \frac{4 \cdot 15}{7 \cdot 15} = \frac{4}{7}$$

Comment   The commutative, associative, and distributive laws hold for addition and multiplication of rational numbers.

The number 1 is the identity for multiplication.

The number 0 is the identity for addition.

Decimals   We know that the rational number $\frac{1}{2}$ can be written as 0.5 in **decimal** form. Similarly, 0.3333... is the decimal representation of the rational number $\frac{1}{3}$. The three dots indicate that there are infinitely many 3's following. Here the 3 forms a pattern that repeats forever. More generally, a decimal is a number written with a decimal point, such as 67.32, 8.1245, 0.5, or 0.33333.... Each digit to the right or left of the decimal point has a place value as shown below:

thousands	hundreds	tens	ones	·	tenths	hundredths	thousandths

Thus the number 563.482 is read as five hundred sixty-three and four hundred eighty-two thousandths, and the number 3.21 is read as three and twenty-one hundredths.

The decimal 0.5 is called a **terminating** decimal, since it ends after a specific number of places.

The decimal 0.3333... is called a **repeating** decimal. It does not end after a specific number of places. The same number repeats itself endlessly.

The decimal 0.434343. . . is also a repeating decimal. In this case the repeating pattern consists of the two numbers 43, which repeat themselves endlessly.

> **Definition 4.13**   *A **terminating decimal** is one that ends after a specific number of places. A **repeating decimal** is one in which (after a certain point) the same pattern of numbers repeats itself endlessly. A **nonterminating and nonrepeating decimal** is one that goes on forever without the same group of numbers repeating over and over in the same pattern.*

EXAMPLE 20

0.25 is a terminating decimal.

0.76914 is a terminating decimal.

0.484848. . . is a repeating decimal.

0.592592. . . is a repeating decimal.

0.1434343. . . is a repeating decimal.

0.12345678910111213. . . is a nonterminating and nonrepeating decimal (What would the next few places be?).

0.43794162. . . (where the numbers are chosen at random) is a nonterminating and nonrepeating decimal.   □

Often when dealing with numbers, we find it necessary to round. For example, if you live on the East Coast and your sister lives on the West Coast, 2003 miles away, you would probably say that she lives 2000 miles away. In this case you rounded 2003 to 2000. The same is true for decimals. We often round to the nearest tenth, hundredth, thousandth, etc., as specified. To round decimals, we use the following rule.

> **RULE 4.10**   **Rounding Decimals**
>
> **1.** *Underline the digit that appears in the position to which the number is to be rounded.*
>
> **2.** *Examine the first digit to the right of the underlined position.*
>
>   **a)** *If the digit is 0, 1, 2, 3, or 4, replace all digits to the right of the underlined position by zeros.*
>
>   **b)** *If the digit is 5, 6, 7, 8, or 9, add 1 to the digit in the underlined position and replace all digits to the right of the underlined position by zeros.*
>
> **3.** *If any of these zeros (from step 2) are to the right of the decimal point, omit them.*

Let us see how this rule is used.

EXAMPLE 21   Round 61.379 to the nearest hundredth.

SOLUTION   We underline the digit that appears in the position to which the number is to be rounded.

$$61.3\underline{7}$$

Since the digit to the right of the underlined position is 5 or more, we add 1 to the digit in the underlined position. Thus 61.379 rounded to the nearest hundredth is 61.38. □

EXAMPLE 22   0.0792 rounded to the nearest thousandth is 0.079.

36.746 rounded to the nearest whole number is 37. □

It can be shown that every rational number can be written as either a terminating decimal or a repeating decimal. It is also true that repeating decimals and terminating decimals represent rational numbers. To change a rational number to a decimal, simply divide the denominator of the fraction into the numerator.

EXAMPLE 23   The fraction $\frac{3}{5}$ can be converted to decimal form by the following procedure. We divide 5 into 3, getting

$$
\begin{array}{r}
0.6 \\
5\overline{)3.0} \\
\underline{3.0}
\end{array}
$$

Therefore $\frac{3}{5} = 0.6$. □

EXAMPLE 24   The fraction $\frac{1}{7}$ can be converted to decimal form by dividing 7 into 1:

$$
\begin{array}{r}
0.1428571\ldots \\
7\overline{)1.0000000} \\
\underline{7} \\
30 \\
\underline{28} \\
20 \\
\underline{14} \\
60 \\
\underline{56} \\
40 \\
\underline{35} \\
50 \\
\underline{49} \\
10 \\
\underline{7} \\
3
\end{array}
$$

Notice that as soon as we got a remainder of ① the pattern repeats.

Thus, $\frac{1}{7} = 0.1428571\ldots$ is a repeating decimal. □

Similarly, we can convert terminating decimals to rational numbers.

EXAMPLE 25   Convert 0.25 to a rational number.

SOLUTION   The decimal 0.25 stands for $\frac{25}{100}$ or $\frac{1}{4}$.   ☐

EXAMPLE 26   Convert 3.45 to a rational number.

SOLUTION   3.45 means $3\frac{45}{100}$ or

$$3 + \frac{45}{100} = \frac{300}{100} + \frac{45}{100} = \frac{345}{100}.$$

Of course, this reduces to $\frac{69}{20}$.   ☐

EXAMPLE 27   Convert 0.0059 to a rational number

SOLUTION   0.0059 means $\frac{59}{10,000}$.   ☐

Decimals can be added and subtracted in much the same way that we add and subtract whole numbers. The only exception is that we require that the decimal points be aligned. For example, to add 16.38 with 7.581 and 9.1, we first align the decimal points as shown:

$$\begin{array}{r} 16.38 \\ 7.581 \\ \underline{9.1} \\ 33.061 \end{array}$$

Most people prefer to add additional zeros to the right so that all the numbers contain the same number of decimal places. Thus the above problem could be rewritten as

$$\begin{array}{r} 16.380 \\ 7.581 \\ \underline{9.100} \\ 33.061 \end{array}$$

16.380 ⟵ 1 zero added

9.100 ⟵ 2 zeros added

EXAMPLE 28   Subtract 3.32 from 8.761.

SOLUTION   We first rewrite the problem with the decimals lined up and zeros added. We get

$$\begin{array}{r} 8.761 \\ \underline{-3.320} \\ 5.441 \end{array}$$

−3.320 ⟵ 1 zero added   5.441 ☐

EXAMPLE 29   Subtract 3.946 from 5.

SOLUTION   We have

$$5.000 \longleftarrow 3 \text{ zeros added}$$
$$\underline{-3.946}$$
$$1.054 \quad \square$$

Multiplying decimals is very much like multiplying whole numbers. *We simply multiply the numbers as if they were whole numbers and then locate the decimal point in the answer by* **adding** *the number of decimal points in each number being multiplied.*

EXAMPLE 30   Multiply 1.073 by 68.21.

SOLUTION

$$1.073 \longleftarrow 3 \text{ decimal places}$$
$$\underline{\times 68.21} \longleftarrow 2 \text{ decimal places}$$
$$1073$$
$$2146$$
$$8584$$
$$\underline{6438}$$
$$73.18933 \longleftarrow 3 + 2 = 5 \text{ decimal places} \quad \square$$

EXAMPLE 31   Multiply 4.1 by 0.0029.

SOLUTION

$$4.1 \qquad \longleftarrow 1 \text{ decimal place}$$
$$\underline{\times 0.0029} \longleftarrow 4 \text{ decimal places}$$
$$369$$
$$\underline{82}$$
$$0.01189 \longleftarrow 1 + 4 = 5 \text{ decimal places} \quad \square$$

Comment   In the preceding example it was necessary to place a zero to the left of one so that the decimal point could be placed in the correct position.

Division involving decimals is similar to division involving whole numbers. We just have to be sure that we are dividing by a whole number.

EXAMPLE 32   Divide 8.05 by 2.3.

SOLUTION   If we rewrite, we have $\dfrac{8.05}{2.3}$. We first move the decimal one place to the right in the numerator and the denominator as shown:

$$\frac{8.05}{2.3} = \frac{80.5}{23}$$

(This is actually accomplished by multiplying both the numerator and denominator by 10.) Now that the decimal point in the denominator has been repositioned, we can divide as usual. We have

$$
\begin{array}{r}
3.5 \\
23\overline{)80.5} \\
69 \\
\hline
115 \\
115 \\
\hline
00
\end{array}
$$

The decimal point in the answer is placed straight above where it was in the dividend. Thus 8.05 divided by 2.3 is 3.5. □

EXERCISES

Reduce each of the fractions given in Exercises 1–8 to lowest terms.

1. $\dfrac{70}{72}$ **2.** $\dfrac{-130}{180}$ **3.** $\dfrac{24}{144}$ **4.** $\dfrac{17}{153}$

5. $\dfrac{76}{38}$ **6.** $\dfrac{-12}{12}$ **7.** $\dfrac{15}{3}$ **8.** $\dfrac{30}{-25}$

Which of the pairs of rational numbers given in Exercises 9–16 are equal?

9. $\dfrac{5}{7}, \dfrac{15}{21}$ **10.** $\dfrac{12}{14}, \dfrac{5}{7}$ **11.** $\dfrac{13}{17}, \dfrac{91}{119}$

12. $\dfrac{8}{20}, \dfrac{56}{140}$ **13.** $\dfrac{9}{13}, \dfrac{8}{117}$ **14.** $\dfrac{9}{7}, \dfrac{4}{3}$

15. $\dfrac{6}{7}, \dfrac{7}{8}$ **16.** $\dfrac{8}{12}, \dfrac{12}{18}$

Write each of the rational numbers given in Exercises 17–20 in two different ways.

17. $\dfrac{8}{9}$ **18.** $\dfrac{4}{13}$ **19.** $\dfrac{5}{7}$ **20.** $\dfrac{4}{11}$

In Exercises 21–34, perform the indicated operations and simplify the results.

21. $\dfrac{3}{4} + \dfrac{9}{4}$ **22.** $\dfrac{8}{9} - \dfrac{3}{9}$ **23.** $\dfrac{2}{9} + \dfrac{4}{3}$

24. $\dfrac{7}{12} + \dfrac{8}{16}$ **25.** $\dfrac{8}{9} - \dfrac{5}{7}$ **26.** $\dfrac{5}{11} \cdot \dfrac{55}{12}$

27. $\dfrac{4}{9} \cdot \dfrac{27}{17}$ **28.** $\dfrac{-8}{11} \cdot \dfrac{13}{17}$ **29.** $\left(\dfrac{-8}{11}\right)\left(\dfrac{-11}{8}\right)$

30. $\dfrac{4}{3} \div \dfrac{5}{7}$ **31.** $\left(\dfrac{-8}{11}\right) \div \left(\dfrac{22}{16}\right)$

32. $16 \div \dfrac{1}{8}$ **33.** $\dfrac{7}{9}\left(\dfrac{1}{2} - \dfrac{1}{3}\right)$ **34.** $\dfrac{5}{8}\left(\dfrac{2}{3} - 4\right)$

†**35. If a, b, and c are integers, prove that $(ac)b = (bc)a$. (*Hint:* Use the closure, commutative, and associative laws.)

36. What is wrong with the following?

$$\frac{5}{8} = \frac{5 \cdot 0}{8 \cdot 0} = \frac{0}{0} \quad \text{and} \quad \frac{9}{10} = \frac{9 \cdot 0}{10 \cdot 0} = \frac{0}{0}.$$

Since $\dfrac{5}{8}$ and $\dfrac{9}{10}$ are both equal to $\dfrac{0}{0}$, they are equal to each other. That is, $\dfrac{5}{8} = \dfrac{9}{10}$.

†**37. Using the definition of multiplication of rational numbers, prove that $0 \cdot \dfrac{a}{b} = 0$ where $\dfrac{a}{b}$ is any rational number.

†**38. Prove each of the following:

a) $\dfrac{-a}{b} = \dfrac{a}{-b}$ **b)** $\dfrac{-a}{b} = -\dfrac{a}{b}$

This shows that $-\dfrac{a}{b} = \dfrac{a}{-b} = \dfrac{-a}{b}$. (*Hint:* Use the definition of equal rational numbers.)

†**39. Prove that $\dfrac{-a}{-b} = \dfrac{a}{b}$. (*Hint:* Use the definition of equal rational numbers.)

Identify each of the decimals given in Exercises 40–48 as either terminating, nonterminating but repeating, or nonterminating and nonrepeating.

40. 0.123   **41.** 0.878787...

42. 0.19191919

43. 0.893893

44. 0.76121212...

45. 0.86429571

46. 0.81828384...

47. 0.040040004...

48. 0.132134136138

49. When you add two repeating decimals, will your answer be a repeating decimal? Explain.

50. When you add two nonrepeating and nonterminating decimals, will your answer be a nonrepeating and nonterminating decimal? Explain.

Convert each of the rational numbers given in Exercises 51–58 to decimals.

51. $\frac{5}{8}$ **52.** $\frac{2}{9}$ **53.** $\frac{9}{14}$ **54.** $\frac{15}{12}$

55. $\frac{11}{13}$ **56.** $\frac{12}{7}$ **57.** $\frac{9}{11}$ **58.** $\frac{8}{24}$

Convert each of the decimals given in Exercises 59–62 to rational numbers.

59. 0.83

60. 0.2

61. 58.321

62. 0.00000029

In Exercise 63–68, perform the indicated operations and round off all answers to the nearest hundredth.

63. Subtract 6.69 from 9.723.

64. Subtract 5.432 from 8.

65. Multiply 2.096 by 79.37.

66. Multiply 8.532 by 0.0039.

67. Divide 7.32 by 4.5.

68. Divide 6.81 by 0.23.

The repeating decimal 0.838383... can be converted to a rational number as follows: Let N stand for the number:

$$\text{Thus } N = 0.838383\ldots$$

Multiply by 100. We get

$$100N = 100(0.838383\ldots) \quad \text{or}$$

$$100N = 83.8383\ldots$$

$$\text{Subtract } N: \quad \underline{N = 0.8383\ldots}$$

$$99N = 83$$

Dividing both sides by 99, we obtain

$$N = \frac{83}{99} \quad \text{or that} \quad \frac{83}{99} = 0.838383\ldots$$

By a similar procedure, convert each of the repeating decimals in Exercises 69–76 to rational numbers.

69. 0.282828...

70. 0.898989...

71. 0.191919...

72. 0.356356...

73. 0.832832...

74. 7.292929...

75. 6.878787...

76. 5.83121212...

**77. Can you find a fraction such that if you double $\frac{1}{8}$ of it and multiply the result by the original fraction, the answer will be $\frac{1}{9}$?

78. When Madeline started on her car trip, the odometer of the car read 41632.8. At the end of the trip it read 42591.2. For the entire trip she used 46.8 gallons of gas. How many miles per gallon (to the nearest hundredth) did the car average? (*Hint:* Divide the total number of miles traveled by the number of gallons of gas used.)

79. The electric meter on Don's house read 1453.6 kilowatt hours at the beginning of the month. At the end of the month it read 1710.4 kilowatt hours. What is Don's electric bill for that month if electricity in his area costs 12.38 cents per kilowatt hour?

80. Find two rational numbers whose product is a whole number.

81. Find two rational numbers whose sum is a whole number.

82. Find two rational numbers whose difference is a whole number.

83. Find two rational numbers whose quotient is a whole number.

84. The following is a series of steps that can be used to show that the commutative law of addition, $\frac{a}{b} + \frac{c}{d} = \frac{c}{d} + \frac{a}{b}$, is valid for rational numbers. State the reason for each of these steps.

a) $\frac{a}{b} + \frac{c}{d} = \frac{ad + bc}{bd}$

b) $\dfrac{a}{b} + \dfrac{c}{d} = \dfrac{da + cb}{db}$

c) $\dfrac{a}{b} + \dfrac{c}{d} = \dfrac{cb + da}{db}$

d) $\dfrac{a}{b} + \dfrac{c}{d} = \dfrac{c}{d} + \dfrac{a}{b}$

85. Consider the two rational numbers $\dfrac{a}{b}$ and $\dfrac{c}{d}$.

Find values of a, b, c, and d for which:

a) $\dfrac{a}{b} + \dfrac{c}{d} = \dfrac{a + c}{b + d}$

b) $\dfrac{a}{b} + \dfrac{c}{d} \neq \dfrac{a + c}{b + b}$

86. Is it ever true that $\dfrac{1}{a} + \dfrac{1}{b} = \dfrac{1}{a + b}$? Explain your answer.

4.7

IRRATIONAL NUMBERS

In the last section we pointed out that there are three different kinds of decimals:

1. terminating decimals,
2. repeating decimals, and
3. nonrepeating and nonterminating decimals.

The first two kinds represent rational numbers. On the other hand, *numbers that can be written as nonrepeating and nonterminating decimals are called* **irrational numbers.** The fact that they are called irrational does not mean that they do not make sense. They are as meaningful as any other numbers. We do not introduce them merely to make our discussion of decimals complete. They are important and interesting in themselves, and mathematicians could not work without them.

Some examples of irrational numbers are π (which you will remember if you have studied geometry), $\sqrt{2}$, $\sqrt{3}$, etc.[3]

The story of the discovery of irrational numbers is one of the more interesting chapters in the history of mathematics. Irrational numbers were first discovered around the sixth century B.C. by the Pythagoreans, a school of Greek mathematicians. This school was named after its founder, Pythagoras, a philosopher, mathematician, and mystic. We know nothing certain about him except that he was born in Samos in Greece and is believed to have traveled widely, as far as Egypt, Babylon, and even India. When Pythagoras returned from his travels, he settled at Croton in southern Italy. There he founded a school for the study of religion, philosophy, mathematics, and science. An interesting story is told about Pythagoras's attempt to get students

---

3. Read $\sqrt{2}$ as "the square root of 2." It means a number that when multiplied by itself gives 2. For example, $\sqrt{4} = 2$, since $2 \cdot 2 = 4$; and $\sqrt{9} = 3$, since $3 \cdot 3 = 9$.

for his school. He found a poor workman and offered to pay him to learn geometry. Pythagoras promised to give him a coin for each theorem that he learned. The workman happily accepted the challenge and earned many coins. Gradually, the workman became so interested in geometry that he wanted Pythagoras to teach him more and more. To persuade him, the workman now offered Pythagoras a coin for each theorem that he taught him. In the end, Pythagoras got back all his money.

The Pythagorean school was also a secret society. This society was in some ways like a modern commune. Men and women were equal, all property was common, and activities were communal. Even mathematical and scientific achievements were considered work of the entire community.

Many of the Pythagoreans' religious beliefs seem somewhat strange to us. They believed in the transmigration of souls and would not eat meat, being afraid that they might be dining off some departed friend. Among the things they considered sinful were

eating beans,

picking up anything that had fallen (a belief shared by many small children today),

eating from a whole loaf,

walking on highways (apparently Croton traffic was an earlier version of the Los Angeles Freeway),

letting swallows sit on one's roof.

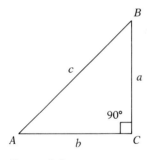

Figure 4.8

When they were not worrying about these matters, the Pythagoreans passed the time studying philosophy and mathematics. Pythagoras said, "All is number," which meant that the form of all things in the world can be explained in terms of numbers. The mathematical basis of music was one of his important discoveries.

Probably the greatest accomplishment of his school was its investigation of what is today known as the Pythagorean theorem. This theorem says: In a right triangle (labeled as shown in Fig. 4.8),

$$a^2 + b^2 = c^2.$$

It is said that Pythagoras was so pleased by this theorem that he sacrificed an ox to celebrate its discovery. Actually, various special cases of the theorem had been known for centuries before the Pythagoreans.

Unfortunately, this achievement led to the downfall of the society and its philosophy. Remember that they believed that the universe could be explained entirely in terms of *numbers*, which for them meant rational numbers. Now consider the right triangle shown in Fig. 4.9.

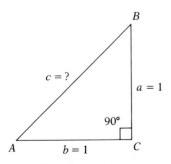

Figure 4.9

The Pythagorean theorem says

$$a^2 + b^2 = c^2$$
$$1^2 + 1^2 = c^2$$
$$1 + 1 = c^2$$
$$2 = c^2$$
$$2 = c \cdot c$$

Thus c is a number that when multiplied by itself gives 2. Symbolically (see the footnote on p. 224), $c = \sqrt{2}$. It was discovered by a Pythagorean named Hippasus (at least so the story goes) that $\sqrt{2}$ *is not a rational number.* Thus we have a physical thing, that is, the longest side of our triangle (the hypotenuse), that is *not* a rational number. This, of course, upset the whole Pythagorean philosophy.

The discovery that $\sqrt{2}$ is not a rational number was a terrible shock to the Pythagoreans. According to one report, they drowned Hippasus to prevent him from spreading the "bad" news. (This is one way to solve a mathematical dispute.) Another version states that Hippasus was shipwrecked by the gods for his wickedness. However, nothing the Pythagoreans could do was able to change mathematical fact: $\sqrt{2}$ was not, and is not, rational. It is irrational.

It turns out that $\sqrt{2}$ is not the only irrational number. There are infinitely many of them. Another important irrational number that you might have come across in other mathematics course is π (pronounced "pie"). In the Bible (I Kings 7:23 and II Chronicles 4:22), calculations indicate that 3 was used as the value of π. In recent years, π has been computed accurately to 500,000 places. The value of π correct to 14 places is 3.14159265358979. In the nineteenth century the legislature of Indiana attempted to pass a law establishing a fixed decimal value for π. They gave up when the idea was ridiculed in the press.

Fact: $\sqrt{2}$ was, and is, not rational. It is irrational.

Let us see why. In order to prove this, we need a few simple definitions and facts.

> **Definition 4.14**   *An integer is **even** if it can be divided exactly (no remainder) by 2. Thus every even number can be written as*
>
> $$2 \cdot (integer)$$

If you try to divide any number by 2, then either there is no remainder or the remainder is one. If there is no remainder, then the number is even. If the remainder is 1, then we have the following definition.

> **Definition 4.15**   *An integer is said to be **odd** if it can be written as*
>
> $$2 \cdot (integer) + 1$$

EXAMPLE 1   The integer 16 is even, since we can write it as

$$16 = 2 \cdot 8 \quad \square$$

EXAMPLE 2   The integer 17 is odd, since we can write it as

$$17 = 2 \cdot 8 + 1 \quad \square$$

Now consider the following:

The number 9 is an odd number. If we multiply it by itself, we get 9×9, or 9^2, which is 81. Notice that 81 is also an odd number.

Similarly, 7 is an odd number. And 7 multiplied by itself, or 7^2, gives 49, which is also an odd number.

In general, we have this statement.

Definition 1   *If a number, n, is odd, then n multiplied by itself is odd, or* **if n is odd, then n^2 is odd.**

Now look at the number 4, which is even: 4 is 2^2, and 2 is also even.

Similarly, 100 is also an even number; 100 is 10^2, and 10 is also even. This leads to the next statement.

Definition 2   *If the square of a number is even, then the number itself is even, or* **if n^2 is even, then n is even.**

We will now prove these two statements. The proof of Statement 1 requires a little algebra, so if you have had *no* algebra, go on to Example 3 below.

†Proof of statement 1   We are given an odd number, n.

We want to show that n^2 is also odd.

Since n is odd, we can write n as

$$2 \cdot (\text{integer}) + 1$$

Let us call the integer k. Then

$$n = 2k + 1$$

Therefore

$n^2 = n \cdot n$

$\quad = (2k + 1) \cdot (2k + 1)$

$\quad = 4k^2 + 4k + 1$ (since the product $(2k + 1)(2k + 1)$ equals $4k^2 + 4k + 1$)

$\quad = (4k^2 + 4k) + 1$ (Here we use the associative law to group $4k^2 + 4k$ together.)

$\quad = 2(2k^2 + 2k) + 1$ (The distributive law is used here to factor out the 2.)

Now by the closure laws for integers, $(2k^2 + 2k)$ is an integer also. Let us call it c. Then we have

$$n^2 = 2c + 1$$

That is,

$$n^2 = 2 \cdot (\text{integer}) + 1$$

This means that n^2 is an odd number, and our statement is proved.
Let us illustrate Statement 1 with some examples.

EXAMPLE 3   3 is odd. Therefore 3^2, which is 9, is also odd.   ☐

EXAMPLE 4   5 is odd. Thus 5^2, which equals 25, is also odd.   ☐

Before we prove Statement 2, we first need to review the idea of a **proof by contradiction.** Suppose you were invited to a party at a certain address. When you arrive there, you discover that the house has two apartments, A and B, and you do not know which of them is the right one. You would pick one apartment, say A, and ring the bell. If the occupants of apartment A said that the party was not there, you would then know that it was in apartment B.

The reasoning process you would use in the above situation is the same as the reasoning in a proof by contradiction. We can summarize it as follows:

1. You know that either possibility A or B must be true.

2. You try possibility A and find that it is wrong.

3. You conclude that possibility B is correct.

This method will be used in our proof of Statement 2.

†Proof of statement 2   We are given an even number n^2 and asked to prove that n is also even. We know that *either n is odd (possibility A) or n is even (possibility B).*

Let us consider whether possibility A can be right. Possibility A says that n is odd.

If n is odd, then Statement 1 tells us that n^2 is also odd.

But we are given that n^2 is even.

Thus n *cannot be odd.*

We conclude that possibility A is wrong. It then follows that possibility B is right. This means that n must be even. Thus Statement 2 is proved.

EXAMPLE 5   64, which is 8^2, is even. Therefore 8 is also even.   ☐

EXAMPLE 6 36, which is the same as 6^2, is even. Thus 6 is also even. □

Now we are ready to prove that $\sqrt{2}$ is not rational. The proof is again a proof by contradiction and uses a little algebra. If you have *never* studied algebra, omit it.

†Proof that $\sqrt{2}$ is not rational Either $\sqrt{2}$ is rational (possibility *A*) or $\sqrt{2}$ is not rational (possibility *B*.)

Suppose *A* is correct, so that $\sqrt{2}$ is rational. Then, by our definition of rational number, $\sqrt{2}$ can be written as $\frac{a}{b}$, where *a* and *b* are integers and *b* is not 0.

Now we know that every rational number can be reduced to lowest terms. So we can assume that

$$\sqrt{2} = \frac{a}{b},$$ where *a* and *b* have no common divisors (that is, $\frac{a}{b}$ is reduced to lowest terms.)

Square both sides of this equation. We get

$$(\sqrt{2})^2 = \left(\frac{a}{b}\right)^2$$

$$2 = \frac{a^2}{b^2}$$

Multiply both sides by b^2. We get

$$2b^2 = \frac{a^2}{b^2} \cdot b^2$$

or upon simplifying,

$$2b^2 = a^2$$

We see that

$$a^2 = 2 \cdot b^2 = 2 \cdot (\text{integer}) \tag{1}$$

which means that a^2 is even.

Since a^2 is even, Statement 2 tells us that *a* must be even.

Now that we know *a* is even, we can write it as

$$a = 2 \cdot (\text{integer})$$

If we call this integer *p*, we have

$$a = 2p$$

Squaring both sides, we have

$$a^2 = (2p)^2 = 4p^2$$

If we substitute $4p^2$ for the a^2 in equation (1) above, we will get $4p^2 = 2b^2$. Dividing both sides by 2, we get

$$2p^2 = b^2$$

or

$$2 \cdot (\text{integer}) = b^2$$

This means that b^2 is even. Since b^2 is even, Statement 2 tells us that b is even. Thus we can write b as

$$b = 2 \cdot (\text{integer})$$

Now we have

$$a = 2 \cdot (\text{integer})$$
$$b = 2 \cdot (\text{integer})$$

So *a and b have a common divisor of 2.* But when we started out, *a and b had no common divisors.* This is obviously a **contradiction.** We must conclude that possibility A ($\sqrt{2}$ is rational) is wrong. Since A is wrong, B (which says that $\sqrt{2}$ is not rational) must be correct. Finally, we conclude that $\sqrt{2}$ *is not rational.*

EXERCISES

†*1. Two examples of irrational numbers are $5\sqrt{3}$ and $4\sqrt{3}$.

     a) Is their product irrational? Explain.

     b) Is their quotient irrational? Explain.

     c) Is their sum irrational? Explain.

†*2. Show, by examples, that the product of two irrational numbers may be rational or may be irrational.

†*3. Construct a line segment which measures $\sqrt{5}$ inches long. (*Hint:* Use the Pythagorean theorem for triangles.)

†*4. Prove: If n is not divisible by 3, then n^2 is not divisible by 3.

†*5. Use Exercise 4 to prove: If n^2 is divisible by 3, then n is also divisible by 3.

†*6. Use Exercise 5 to prove that $\sqrt{3}$ is not rational.

†*7. Write a statement similar to that of Exercise 4 using 5 instead of 3.

†*8. Write a statement similar to that of Exercise 5 using 5 instead of 3.

†*9. Use Exercise 8 to prove that $\sqrt{5}$ is not rational.

†*10. Where, in our proof of the fact that $\sqrt{2}$ is not rational, did we use Statement 1?

†*11. We know that 1 is rational and $\sqrt{2}$ is not rational. Prove that the sum $1 + \sqrt{2}$ is not rational. (*Hint:* Use proof by contradiction.)

Solve Exercises 12 and 13 by using the Pythagorean theorem.

12. A 17-foot ladder leans against a building so that the base of the ladder is 15 feet away from the building as shown. How high up the building does the ladder reach?

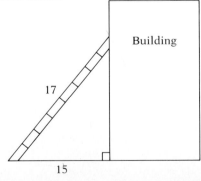

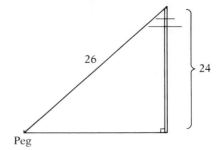

13. A 24-foot telephone pole is to be anchored by a 26-foot-long guy wire attached to a peg in the ground as shown. How far away from the base of the pole is the peg?

4.8

THE REAL NUMBERS—ORDER OF OPERATIONS

We started this chapter with the natural numbers. However, subtraction was not always possible using only the natural numbers, so we needed the integers. Although this made subtraction a closed operation, we still found that division was not always possible. To remedy this situation, we introduced the rational numbers. Then, we discovered that there exist certain numbers such as $\sqrt{2}$, $\sqrt{3}$, etc., that are not rational numbers but do occur frequently in mathematics. These we called the irrational numbers. All of these numbers together make up what we call the **real numbers.** These are the numbers used in most of elementary mathematics and in everyday situations. They are defined as follows.

> **Definition 4.16** *A real number is any number that is either a rational number or an irrational number. In the notation of sets we have*
>
> *{real numbers} = {rational numbers} ∪ {irrational numbers}.*

We can draw a diagram illustrating the relationship among the different kinds of numbers:

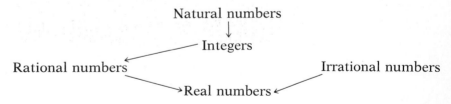

Another way of picturing the relationships between numbers is by means of a Venn diagram, as shown in Fig. 4.10.

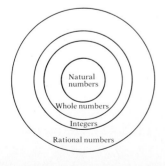

Real numbers

The shaded portion represents the irrational numbers

The real numbers include everything in all the circles.

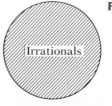

Figure 4.10

EXAMPLE 1   Let R = {real numbers}, I = {integers}, Q = {rational numbers}, and W = {irrationals}.

   a) Find $R \cap Q$.   **b)** Find $I \cap W$.   **c)** Is $W \subset I$ true?

SOLUTION   **a)** $R \cap Q = Q$, since the diagram shows that the only elements that are in both sets are the rational numbers.

   b) $I \cap W = \varnothing$, since the diagram shows that there is no number that is both an integer and an irrational number.

   c) If $W \subset I$ were true, then the circle for the irrationals would be inside the circle for the integers. Since it is not true, then the statement $W \subset I$ cannot be true.   □

The commutative, asssociative, distributive, and closure laws hold for addition and multiplication. Subtraction and division are closed for real numbers (with the exception of division by 0 which, of course, is not possible).

The identity for addition is 0.

The identity for multiplication is 1.

Thus we see that the real-number system is a very complete system.

A very important property of the real numbers is that every real number is either positive, negative, or zero. A convenient way of picturing the real numbers is by means of a *real-number line*. Such a line is shown in Fig. 4.11.

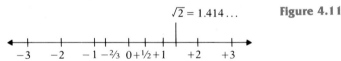

$\sqrt{2} = 1.414\ldots$ **Figure 4.11**

We pick a point on the line and label it 0. Then we pick another point to the right of 0 and label it +1. Next, we take a point to the right of +1 that is the same distance from +1 as +1 is from 0. We label this point +2. Similarly, we mark off +3, +4, +5, etc., ... Points to the left of 0 are labeled −1, −2, −3, etc.

Fractions are points between these numbers. (For example, see $+\frac{1}{2}$ and $-\frac{2}{3}$ in Fig. 4.11.)

Can we find a number like $\sqrt{2}$ on the number line? Look at the right triangle shown in Fig. 4.12. On p. 225 we showed that side c has length equal to $\sqrt{2}$. Now suppose we take this triangle and place point M at 0 on the number line, and place side c along the number line as shown in Fig. 4.13. Then N will be exactly $\sqrt{2}$ units to the right of 0. Thus the point on the number line that N touches is $\sqrt{2}$. It can be shown that $\sqrt{2}$ is approximately equal to 1.414. The number

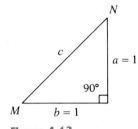

Figure 4.12

Figure 4.13

$\sqrt{2}$ cannot be represented exactly by a repeating or terminating decimal.

Every point on the line represents a real number. Similarly, every real number can be represented by a point on the line.

Pick any two numbers on the line. To be specific, let us take $+3$ and -2. The number $+3$ is to the right of -2. We say that $+3$ is larger than or greater than -2. This is symbolized by writing $(+3) > (-2)$. The symbol ">" stands for "is greater than."

We also see that -2 is to the left of $+3$. In this case we say that -2 is less than or smaller than $+3$. This is symbolized as $(-2) < (+3)$. The symbol "<" stands for "is less than."

Sometimes we know that one number is either less than another number or equal to it. But we do not know which one is the case. In this situation we use the symbol "$\leq$," which is read as "equal to or less than."

Similarly, the symbol "$\geq$" means "equal to or greater than."

We can summarize our discussion in the following definition.

Definition 4.17 *If a and b are any real numbers on a number line, then:*

$a < b$ means a is to the left of b,

$a > b$ means a is to the right of b,

$a \leq b$ means a is to the left of b or a is the same as b, and

$a \geq b$ means a is to the right of b or a is the same as b.

Comment Given any two numbers a and b, then *one and only one* of the following must be true:

i) $a = b$,

ii) $a > b$, or

iii) $a < b$.

This is sometimes called the **law of trichotomy.**

EXAMPLE 2 $2 < 3$ means 2 is to the left of 3, as shown below. ☐

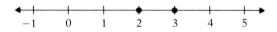

EXAMPLE 3 $5 > -3$ means 5 is to the right of -3, as shown below. ☐

EXAMPLE 4 $10 \leq b$ means 10 is to the left of point b or the same as point b. ☐

Order of Operations   Suppose we are asked to evaluate $2 + 3 \times 4$. There are two ways to approach this problem as follows:

One person might multiply first and then add, getting	Another person might add first and then multiply, getting
$2 + 3 \times 4 = 2 + 12$	$2 + 3 \times 4 = 5 \times 4$
$= 14$	$= 20$

Obviously, the answers are not the same. Which is right?

Often when working with expressions involving two or more operations, we must decide which operation is to be performed first. Mathematicians have agreed upon procedures that specify the exact order to follow. We have

RULE 4.11   Order of Operations   *When simplifying expressions involving several operations,*

1. *do all the multiplications and divisions first, performing them in order from left to right,*

2. *then do all the additions and subtractions, performing them in order from left to right.*

Applying Rule 4.11, we see that the value of $2 + 3 \times 4$ is 14, since we first multiply 3×4, getting 12, and then add $2 + 12$, getting 14.

EXAMPLE 5   Evaluate $8 \times 4 + 3 \times 5$.

SOLUTION   We do all the multiplications first and then add. Thus

$$8 \times 4 + 3 \times 5 = 32 + 15$$
$$= 47 \quad \square$$

EXAMPLE 6   Evaluate $18 + 0 \div 9 - 5 \times 0.2$.

SOLUTION   We first do all the multiplications and divisions from left to right and then add. Thus

$$18 + 0 \div 9 - 5 \times 0.2 = 18 + 0 - 1$$
$$= 17 \quad \square$$

Let us refer back to the example $2 + 3 \times 4$ given earlier. Rule 4.11 tells us that the answer is 14. Suppose we wanted to add the 2 and 3 together before multiplying by 4. In this case we would enclose the sum $2 + 3$ in parentheses and write this example as $(2 + 3) \times 4$. When written in this manner, we find the sum before doing the multiplication. Thus

$$(2 + 3) \times 4 = 5 \times 4$$
$$= 20$$

When $(2 + 3) \times 4$ is written with parentheses, then we do not follow Rule 4.11 as $(2 + 3) \times 4$ is *not* the same as $2 + 3 \times 4$. Thus when simplifying any expression involving parentheses (which act as a grouping symbol), we always perform the operations indicated on the numbers within the parentheses first.

Comment   Parentheses are not the only symbols used to indicate grouping. Brackets [ ] and braces { } are also used and have the same meaning as parentheses.

Comment   If an expression contains two or more grouping symbols, then we perform the operations on the numbers within the innermost symbols first.

We can now expand our rule for order of operations. We have

RULE 4.12   Order of Operations When Working with Parentheses
To simplify an expression involving parentheses (or any other grouping symbol) and several operations,

1. *first perform the operations within the parentheses starting with the innermost group,*
2. *then do all the multiplications and divisions, performing them in order from left to right, and*
3. *finally do all the additions and subtractions, performing them in order from left to right.*

EXAMPLE 7   Simplify the expression $4 + 5[8 + (4 - 2) \times 3)]$.

SOLUTION   We start with the innermost parentheses first:

$$4 + 5[8 + (4 - 2) \times 3] = 4 + 5[8 + 2 \times 3]$$

$$= 4 + 5[8 + 6] \qquad \text{(Inside the brackets we do multiplication first.)}$$

$$= 4 + 5(14) \qquad \text{(We evaluate the numbers inside the parentheses.)}$$

$$= 4 + 70 \qquad \text{(We multiply first.)}$$

$$= 74 \qquad \qquad \square$$

When simplifying expressions containing powers (exponents), we first evaluate the exponent and then follow the usual order for the other operations. Thus

$$2 \times 5^2 = 2 \times 25 \qquad \text{(since } 5^2 = 5 \times 5 = 25)$$
$$= 50$$

Our answer is 50 and *not* 100, which is obtained by first multiplying 2×5 getting 10 and then squaring. Exponents are evaluated first.

We summarize our discussions with the following general rule.

RULE 4.13   Order of Operations (General Case)   *To simplify an expression involving exponents, parentheses, and several operations,*

1. *simplify any expressions that are within parentheses (or any other grouping symbol), starting with the innermost grouping symbol,*
2. *evaluate any powers or roots,*
3. *do all multiplications and divisions, performing them in order from left to right,*
4. *do all additions and subtrations, performing them in order from left to right.*

EXAMPLE 8   Evaluate $5(7 - 3)^2 - 2$.

SOLUTION $5(7 - 3)^2 - 2 = 5(4)^2 - 2$ (We first simplify the expression within parentheses.)

$$= 5(16) - 2$$ (We evaluate the exponent.)

$$= 80 - 2$$ (We do the multiplication.)

$$= 78$$ (We do the subtraction.)   □

Comment   In the next chapter, when we discuss hand-held calculators, we will indicate how scientific calculators perform calculations according to Rule 4.13 as opposed to nonscientific calculators, which perform the various operations in the order in which they are written.

You may have the impression that there can be no numbers other than real numbers. This is not true. To see this, consider the innocent-looking $\sqrt{-1}$ (the square root of -1). What does this symbol really mean? The symbol $\sqrt{-1}$ means some number which when multiplied by itself gives -1. Let us call this number i. Then $i^2 = -1$. Now what kind of number is i? Is it negative, positive, or zero?

If i is negative, then i^2 would have to be positive (since a negative number times a negative number is a positive number). But i^2 is -1, which is negative. Therefore i cannot be negative.

If i is positive, then i^2 would have to be positive (since a positive number times a positive number is a positive number). Since i^2 is -1, then i cannot be positive.

If i is zero, then i^2 is 0 (since 0 times 0 is 0). This definitely is not -1.

So here we have a number, i (that is, $\sqrt{-1}$), that is neither negative, positive, nor zero. It follows that i cannot be a real number. (Remember that every real number is either negative, positive, or zero.)

Numbers like i are called **imaginary** or **complex numbers.** They play an important role in mathematics, physics, and technology. In particular, they are used in many branches of engineering, such as electrical engineering, heat conduction, elasticity, and aeronautical engineering. The references for this chapter in the Suggested Further Reading List contain a further discussion of imaginary numbers.

EXERCISES

1. Let R = {real numbers}, I = {integers}, Q = {rational numbers}, and W = {irrational numbers}. Find each of the following:

   a) $R \cap I$ b) $I \cap Q$ c) $W \cap I$
   d) $R \cup Q$ e) $Q \cup W$ f) $W \cap R$
   g) $W \cup R$ h) $I \cup Q$ i) $I \cap (Q \cup W)$

Using the notation of Exercise 1, which of the statements in Exercises 2–7 are true?

2. $Q \subset R$ 3. $Q \subset I$ 4. $I \subset Q$

5. $R \subset W$ 6. $W \subset R$ 7. $I \subset (Q \cup W)$

Represent each of the numbers in Exercises 8–13 on a number line.

8. 8   9. −9   10. $+\dfrac{1}{4}$

11. −0.7

12. 8.6

13. $\sqrt{7}$

Using a number line, determine which is the correct symbol (>, =, <) for each of the pairs of numbers in Exercises 14–21.

14. 4.5, 4.6   15. +5, −6

16. −7, −8   17. $\dfrac{1}{4}, \dfrac{1}{5}$

18. 4, 4   19. $\dfrac{1}{8}$, 0.125

20. $\sqrt{3}$, 1.7   21. $\dfrac{1}{6}$, 0.166 . . .

Complete the following chart.

Number	Natural number	Whole number	Integer	Rational number	Irrational number	Real number
−5	no	no	yes	yes	no	yes
$\sqrt{13}$	no	no	no	no	yes	yes
22. 5						
23. $-\dfrac{1}{8}$						
24. 3.021021 . . .						
25. 7.1						
26. 0						
27. $6\sqrt{3}$						
28. 2.13587 . . .						
29. $\dfrac{2}{\sqrt{3}}$						

30. If x is any real number, then $x > 3$ means x is to the right of 3 on the number line. We can picture this as shown at right:

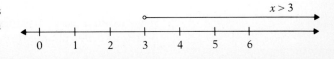

Similarly, $x \leq -1$ can be pictured as

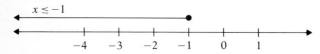

Picture each of the following, using a number line.

a) $x \leq 4$ **b)** $x > -6$ **c)** $x < \dfrac{3}{4}$

d) $x \geq -8$ **e)** $x < -3$ **f)** $x \geq 5$

g) $x > -\dfrac{3}{5}$ **h)** $x \leq 0.8$

†31. Let a, b, and x be any real numbers. Using the properties of real numbers, prove that if

$$a + x = b + x,$$

then

$$a = b.$$

This is called **cancellation law for addition.**

†32. Let a, b, and x be any real numbers, with x not equal to zero. The **cancellation law of multiplication** says that if

$$ax = bx,$$

then

$$a = b.$$

a) Prove this using the properties of real numbers.

b) Why do we require that x not be equal to zero?

†33. **a)** Find a rational number between $\dfrac{5}{8}$ and $\dfrac{7}{9}$.

b) Show that there is always a rational number between the rational numbers $\dfrac{a}{b}$ and $\dfrac{c}{d}$.

†34. **a)** Find a rational number between $\sqrt{3}$ and $\sqrt{5}$.

b) Find an irrational number between $\sqrt{3}$ and $\sqrt{5}$.

†35. **a)** Add the numbers $2.131313\ldots$ and 0.76767676.

b) Multiply the numbers $2.131313\ldots$ and 0.76767676.

Evaluate each of the expressions in Exercises 36–50.

36. $40 + \dfrac{1}{2} \times 20$ **37.** $16 - 4 \div \dfrac{1}{2}$

38. $5 \times 9 - 3 \times 8$ **39.** $40 + 40 \div 5 + 3$

40. $4\,(5 - 9) - 4$ **41.** $6 + 7\,(9 - 5)$

42. $8 + 3\,[7 + (6 - 2) \times 5]$

43. $5 + 9\,[8 + (6 - 3) \times 2]$

44. $(2^4)\left(\dfrac{1}{2}\right)^5$ **45.** $6^2 + 7^2$

46. $5\,(6^2) - 36$ **47.** $(12 + 6)^2$

48. $14 - 3\,(4 - 2)^3$ **49.** $9\,(4^2 - 3^2)$

50. $2 + (10^2 - 9^2)\,(8^2 + 7^2) - 7$

4.9

SUMMARY   In this chapter we discussed the many different types of numbers and their properties. We started with the natural numbers and worked our way through to the real number system. In the process, not only did we indicate the various properties that each of these systems satisfies, but we also reviewed how to convert rational numbers to decimals and how to analyze whether a decimal represents a rational or an irrational number. Rules for rounding decimals were also given.

Finally, when we discussed the real number system, we pointed out how we can use a number line to represent inequalities. We also specified a particular order of operations when performing calculations.

STUDY GUIDE You should now be able to demonstrate your knowledge of the following ideas by giving definitions, descriptions, or specific examples. Page references are given in parentheses.

Different Types of Numbers

Natural or counting numbers (p. 180)
Zero (p. 192)
Whole numbers (p. 192)
Positive integers (p. 196)
Negative integers (p. 197)
Integers (p. 197)
Additive inverse (p. 200)
Rational numbers (p. 208)
Repeating decimals (p. 218)

Terminating decimals (p. 218)
Nonterminating, nonrepeating decimals (p. 218)
Irrational numbers (p. 224)
Even numbers (p. 226)
Odd numbers (p. 227)
Real numbers (p. 231)
The number i (p. 236)
Complex numbers (p. 236)

Operations

Addition of natural numbers (p. 180)
Binary operation (p. 181)
Multiplication of natural numbers (p. 184)
Subtraction of natural numbers (p. 187)
Division of natural numbers (p. 188)

Rules

Commutative law of addition (p. 181)
Associative law of addition (p. 183)
Law of closure for addition (p. 184)
Commutative law of multiplication (p. 185)
Associative law of multiplication (p. 185)
Law of closure for multiplication (p. 186)
Distributive law of multiplication over addition (p. 186)
Identity for addition (p. 193)
Identity for multiplication (p. 195)

Other Important Ideas

Theory of numbers (p. 177)
Number patterns (p. 180)
Sum (p. 180)
Division into 0 is 0 (p. 194)
Division by 0 is impossible (p. 194)
Arithmetic (p. 199)
Absolute value (p. 203)
Divisibility (p. 204)
Divisor (p. 204)
Divisibility tests (p. 206)
Amicable or friendly numbers (p. 208)
Numerator (p. 208)
Denominator (p. 208)
Ratio (p. 209)
Equal rational numbers (p. 210)
Cancellation principle (p. 210)

Reducing fractions to lowest terms (p. 211)
Addition and subtraction of rational numbers (p. 212)
Multiplication and division of rational numbers (p. 214, 216)
Decimals (p. 217)
Rounding decimals (p. 218)
Performing operations with decimals (p. 220)
Number line (p. 232)
Greater than or equal to (p. 233)
Law of trichotomy (p. 233)
Order of operations (p. 234)
Cancellation law for addition (p. 238)
Cancellation law of multiplication (p. 238)

| FORMULAS TO REMEMBER | The following list summarizes all the formulas discussed in this chapter. |

Laws   If a and b are real numbers, then:

$a + b$ is a real number	Law of closure for addition
$a + b = b + a$	Commutative law of addition
$a + (b + c) = (a + b) + c$	Associative law of addition
$a \cdot b$ is a real number	Law of closure for multiplication
$a \cdot b = b \cdot a$	Commutative law of multiplication
$(a \cdot b)c = a(b \cdot c)$	Associative law of multiplication
$a(b + c) = ab + ac$	Distributive law of multiplication
$a + 0 = a$	Additive identity
$a + (-a) = 0$	Additive inverse

If $\dfrac{a}{b}$ and $\dfrac{c}{d}$ are rational numbers that are equal, then $ad = bc$.

If $\dfrac{a}{b}$ and $\dfrac{c}{b}$ are rational numbers, then $\dfrac{a}{b} \pm \dfrac{c}{b} = \dfrac{a \pm c}{b}$.

If $\dfrac{a}{b}$ and $\dfrac{c}{d}$ are rational numbers, then $\dfrac{a}{b} + \dfrac{c}{d} = \dfrac{ad + bc}{bd}$.

If $\dfrac{a}{b}$ and $\dfrac{c}{d}$ are rational numbers, then $\left(\dfrac{a}{b}\right) \cdot \left(\dfrac{c}{d}\right) = \dfrac{ac}{bd}$.

If $\dfrac{a}{b}$ and $\dfrac{c}{d}$ are rational numbers, then $\left(\dfrac{a}{b}\right) \div \left(\dfrac{c}{d}\right) = \dfrac{ad}{bc}$.

Rules for adding and multiplying signed numbers as given in Section 4.4.

Divisibility tests as given in Section 4.5.

Rules for rounding decimals as given in Section 4.6.

Rules for the order of operations as given in Section 4.8.

The Pythagorean theorem, which states that $a^2 + b^2 = c^2$ for the diagram at left.

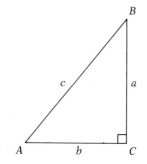

MASTERY TESTS

Form A

1. Find $\left(\dfrac{4}{9}\right) \cdot \left(\dfrac{-3}{11}\right)$.

2. Simplify: $\dfrac{7\,(-2) - 4\,(-8)}{2\,(-1) - (-6 - 3)}$.

3. What is $-7 - (-2)$?

4. Which computation illustrates the associative law of addition?

    a) $4 + (3 + 2) = 4 + (2 + 3)$ **b)** $5\,(4 + 3) = 5 \cdot 4 + 5 \cdot 3$

    c) $(4 + 3) + 2 = 4 + (3 + 2)$ **d)** $4\,(3 \cdot 2) = (4 \cdot 3)2$

    e) none of these

5. Is it true or false that $(-7) < (-8)$?

6. Are the rational numbers $\frac{5}{4}$ and $\frac{40}{32}$ equal?

7. Find the value of x such that the rational numbers $\frac{2}{8}$ and $\frac{x}{12}$ are equal.

8. Express $0.181818\ldots$ as a rational number.

9. Is it true or false that every rational number is an integer?

10. Simplify: $\frac{5}{7} - \frac{3}{8}$.

11. Simplify: $(-7) \div \frac{3}{5}$.

12. Simplify: $\frac{4}{3}\left(\frac{3}{8} + 1\right)$.

13. Convert $\frac{5}{12}$ to decimals.

14. Simplify: $8 \times 7 - 2^3$.

15. Evaluate: $[(-15) \div (5)] \div (-1)$.

Form B

1. Evaluate $10 \times 8 \div 2 \div 4^2$.

2. Refer back to the magazine clipping at the beginning of this chapter. Multiply 142857 by 7. What happens?

3. Find all the divisors of 1472.

4. On Monday a certain stock rose $\frac{1}{8}$ of a dollar per share. On Tuesday the stock dropped $\frac{3}{32}$ per share. On Wednesday the stock rose $\frac{7}{64}$ per share. On Thursday the stock dropped $\frac{3}{16}$ per share, and on Friday it rose $\frac{1}{4}$ per share. Find the net change per share of this stock over the five-day period.

5. A 13-foot ladder leans against a building so that the base of the ladder is 5 feet away from the building. How high up the building does the ladder reach?

6. Perform the following calculation by using the distributive law:

$$68 \times 102.$$

7. Is 1992 a leap year?

8. Combine: $7 + \frac{3}{5} - \frac{2}{3}$.

9. When simplified, $(-4)[(-7) - (-8)]$ becomes

    a) $+4$ **b)** -4 **c)** $+60$ **d)** -60 **e)** none of these

10. When 8 is multiplied by $\frac{8}{7}$, the product is

    a) 7 **b)** $\frac{1}{7}$ **c)** $\frac{64}{7}$ **d)** $\frac{64}{56}$ **e)** none of these

11. Nonterminating but repeating decimals are

    a) rational numbers   **b)** integers   **c)** irrational numbers

    d) whole numbers   **e)** none of these

12. Find a rational number between $\frac{3}{7}$ and $\frac{4}{7}$.

13. An archaeologist discovered an ancient relic that dates back to the year 728 B.C. How old was the relic in the year 1985?

14. The electric meter on Jan's house read 4376.2 kilowatt hours at the beginning of the month. At the end of the month it read 4624.9 kilowatt hours. What is Jan's electric bill for this month if electricity in her area cost 13.76 cents per kilowatt hour?

15. Find a rational number such that if you quadruple $\frac{1}{8}$ of it and multiply the result by the original number, the answer will be $\frac{32}{25}$.

5

CONSUMER MATHEMATICS

CHAPTER OBJECTIVES

To describe the different types of pocket calculators. (*Section 5.2*)

To study the different logical systems used by the various calculators. (*Section 5.2*)

To learn about ratios and proportions and how they can be used. (*Section 5.3*)

To understand percents and how they are used. (*Section 5.4*)

To convert percents to decimals and decimals to percents. (*Section 5.4*)

To distinguish between simple interest and compound interest, which is used by most banks. (*Section 5.5*)

To discuss installment buying and mortgages and how interest charges are calculated for each. (*Section 5.6*)

To point out how we use mortality tables in determining the premium for life insurance. (*Section 5.7*)

243

Are You Still Stuck
With A
14% Mortgage?
Cut Down to Today's Low Fixed Rates!
NOW
11½%
11.81% A.P.R.
NO INCOME VERIFICATION AVAILABLE
Yes, you can get rid of those higher payments by
REFINANCING YOUR HOME.

August 21, 1985

If you look closely at this newspaper advertisement, you will notice the 11.81% A.P.R. in small print. What does A.P.R. stand for? Shouldn't an $11\frac{1}{2}$% mortgage be an $11\frac{1}{2}$% mortgage?

HISTORICAL NOTE

Interest on money borrowed has had an unusual and often tainted history. Over the years, scoundrels have charged enormous interest rates for the use of their money. There also have been "legitimate" businesses that purposely disguised the true amount or percentage of interest charged. An interest charge of 6% may have been an actual rate of 18%.

In 1969 the U.S. Congress passed the Truth in Lending Law for the benefit of consumers who use credit. Under the provisions of the law, full dis-

closure must be made of any interest charged (finance charges) and of the method used to calculate the interest. Interest must be expressed as an annual percentage rate on the monthly statement received. Although these actual charges vary from state to state, great progress has been made in making the ideas of interest understandable by the everyday person. It should be noted, however, that the Truth in Lending Law is a consumer protection act and does not apply to business loans.

5.1

INTRODUCTION

Let us look in on Evelyn, who is analyzing the bills she just received in the mail. Among these are

her life insurance premium bill,

her mortgage payment bill,

her credit card bill, on which a new revolving interest charge appears, and

her real estate tax bill announcing that the assessed value of her house has been increased by 8%.

On the way to the bank to withdraw the interest from her savings account, Evelyn stops off at a local supermarket to buy food. With her calculator in hand, she is able to determine the unit costs of various items so that she gets the most for her money.

Mathematics is quickly becoming an important tool for the consumer. Whether you are interested in obtaining a loan for your college tuition or need to make a down payment on your car, a knowledge of percents and interest calculations is vital.

In this chapter we will discuss some of the tools that can be used to make you an intelligent and financially confident consumer.

5.2

USING A CALCULATOR

In the past few years, sales of hand-held calculators have been phenomenal. When these calculators first came on the market in the early 1950s, they were extremely expensive. Nowadays, as a result of improved technology, you can buy a calculator for less than $5, and it seems that nearly everyone has.

The major advantage in using these calculators is the speed and accuracy of their calculations. Since so many different models are currently on the market, we will merely discuss some of the features common to all of them. Generally speaking, the price you pay for a calculator will reflect its capabilities.

The general-purpose calculators, which are the cheapest ones, are capable of performing only the four basic arithmetic operations of addition $\boxed{+}$, subtraction $\boxed{-}$, multiplication $\boxed{\times}$, and division $\boxed{\div}$. Some general-purpose calculators have a memory register in which partial calculations can be stored for later use. These memory registers are usually indicated by $\boxed{M+}$ or $\boxed{STO}$ or $\boxed{M}$ buttons.

The scientific calculators have features that make them especially useful to engineers and mathematicians. In addition to the functions already mentioned, scientific calculators have buttons that enable them to perform special calculations such as square root $\boxed{\sqrt{}}$, trigonometric functions $\boxed{\text{SIN}}$, $\boxed{\text{COS}}$, and $\boxed{\text{TAN}}$ as well as logarithmic functions $\boxed{\text{LOG}}$. Most also have square $\boxed{x^2}$ and exponent $\boxed{y^x}$ buttons.

The programmable calculators allow the user to insert cards that have special programs on them enabling the calculators to perform complicated calculations for specific situations. There are also some special-purpose calculators currently on the market that are used primarily in business, medicine, and statistics.

All the calculators consist of two parts, the keyboard and the display panel. The keyboard has separate buttons for each of the numbers from 0 to 9 and buttons for the four basic operations $\boxed{+}$, $\boxed{-}$, $\boxed{\times}$, and $\boxed{\div}$. In addition there are $\boxed{\text{CE}}$ and $\boxed{\text{C}}$ buttons, the purpose of which we will explain shortly. A typical calculator is shown in Fig. 5.1. The more expensive calculators have additional buttons that are designed to perform special calculations.

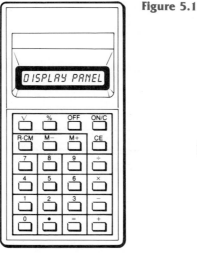

Figure 5.1

Your calculator may use either **arithmetic logic, algebraic logic,** or **Reverse Polish Notation** (RPN) logic. This will determine how you enter numbers.

Suppose we wish to evaluate $5 + 6 \times 7$ with a hand-held calculator. We notice that we have to perform the operations of addition and multiplication. In the last chapter (Section 4.8), we stated a rule for the proper order of performing multiplication, division, addition, and subtraction of real numbers. We also gave rules for dealing with exponents. Most scientific calculators are programmed so that they will automatically perform the operations in the order stated in our rule. (We suggest that you review the three rules given in Section 4.8 before reading further.)

If you have a calculator that uses arithmetic logic, then you will need to be careful about the order of operations. A calculator using the RPN logic requires you to push the ENTER and SAVE buttons. Such calculators do not have an equal = button. Instead the operation symbols are entered after the numbers have been entered. Since most calculators use algebraic logic, that is the system that we will use in the rest of this book.

The following examples show many calculations (using the algebraic logic system). We suggest that you turn on your calculator and push the appropriate buttons as you read on. We will assume that you are familiar with the keyboard of the calculator.

EXAMPLE 1   Using a calculator, add 23 and 62.

SOLUTION

What you do	*What appears on display panel*
1. Turn on/off button to on.	0.
2. Push 2 button.	2.
3. Push 3 button.	23.
4. Push + button.	23.
5. Push 6 button.	6.
6. Push 2 button.	62.
7. Push = button.	85. ▭

Comment   If you make a mistake while entering a number, push the CE (clear error) button. This will remove *only* the last entries and leave everything else. To clear the calculator completely of *everything* in it, push the C (clear) button.

EXAMPLE 2   Using a calculator, calculate 20,412 ÷ 28.

SOLUTION

What you do	*What appears on display panel*
1. Turn on/off button to on.	0.
2. In order, push the 2, 0, 4, 1, and 2 buttons.	20412.
3. Push ÷ button.	20412.
4. Push 2 and 8 buttons in order.	28.
5. Push = button.	729. ▭

EXAMPLE 3   Using a calculator, calculate 20,413 ÷ 28.

SOLUTION

What you do	*What appears on display panel*
1. Turn on/off button to on.	0.
2. In order, push the 2, 0, 4, 1, and 3 buttons.	20413.
3. Push ÷ button.	20413.
4. Push, in order, the 2 and 8 buttons.	28.
5. Push = button.	☐

Notice that in the last step of Example 3 we have not indicated what will appear on the display panel. That is because the display will depend on the particular calculator being used. Some calculators are programmed with a **fixed decimal point.** For example, a machine that has a fixed decimal point with two decimal places will give the answer to Example 3 as 729.04. The decimal point has been programmed to be in this "fixed" position and cannot be moved. All answers are rounded off to two decimal places. On the other hand, if the calculator has a **floating decimal point,** then the decimal point will "float" and appear where it belongs when calculations are performed. If Example 3 were done with a floating point calculator, the display panel would show 729.0357143 at the last step. (We are assuming here that the display panel has room for 10 digits.)

EXAMPLE 4   Using a calculator, find 3 × 4 + 5 × 6.

SOLUTION   In mathematics it is agreed that multiplications and divisions are always done before additions and subtractions. Thus this problem means: Multiply 3 × 4, getting 12. Then multiply 5 × 6, getting 30. Finally, add the results 12 and 30 to get 42. The scientific calculators are programmed to always perform multiplication and division before addition and subtraction. Therefore such calculators would treat this problem as if it were written as (3 × 4) + (5 × 6).

Calculators that are not programmed this way (usually the general-purpose models) will perform the calculations exactly as they are entered. This is shown below.

	What appears on display panel when using	
What you do	*scientific calculator programmed to do multiplication and division first*	*general-purpose calculator programmed to perform calculations as they are entered*
Turn on/off button to on.	0	0.
Push 3 button.	3	3.
Push × button.	3.	3.
Push 4 button.	4	4.
Push + button.	12.	12.
Push 5 button.	5	5.
Push × button.	5.	17.
Push 6 button.	6	6.
Push = button.	42.	102. ☐

Comment   To do the calculation above as $(3 \times 4) + (5 \times 6)$ on a general-purpose calculator, you would have to first compute 3×4 and record the result, which is 12. Then you would compute 5×6, getting 30. Finally, you would add 12 to the 30 to get 42.

Depending upon the logic used by your calculator, the problem discussed at the beginning of this section, $5 + 6 \times 7$, would be done as follows:

For calculators using

Arithmetic Logic		*Algebraic Logic*		*RPN Logic*	
Turn on/off button to on	on/off	Turn on/off button to on	on/off	Turn on/off button to on	on/off
Push the 6 button	6	Push the 5 button	5	Push the 5 button	5
Push the × button	×	Push the + button	+	Push the enter button	ENTER
Push the 7 button	7	Push the 6 button	6	Push the 6 button	6
Push the = button	=	Push the × button	×	Push the enter button	ENTER
Push the + button	+	Push the 7 button	7	Push the 7 button	7
Push the 5 button	5	Push the = button	=	Push the × button	×
Push the = button	=			Push the + button	+
Result 47		Result 47		Result 47	

EXAMPLE 5 Using a scientific calculator, evaluate 7×9^2.

SOLUTION

What you do	*What appears on display panel*
Turn on/off button to on	0.
Push the 7 button	7.
Push the × button	7.
Push the 9 button	9.
Push the x² button	81.
Push the = button	567. ▢

EXAMPLE 6 Using a scientific calculator, evaluate 8^3.

What you do	*What appears on display panel*
Turn on/off button to on	0.
Push the 8 button	8.
Push the yˣ button	8.
Push the 3 button	3.
Push the = button	512. ▢

Before deciding which particular calculator to buy, you should consider exactly what you will be using it for. In many cases the general-purpose model may be quite adequate. Buy a scientific model only if your work requires it.

EXERCISES

Explain how you can use a general-purpose hand-held calculator to obtain the correct answer to each of the calculations given in Exercises 1–8. Actually perform the calculations using any kind of calculator.

1. $(5 + 7)^3 - 3 \times 4^2$

2. $\dfrac{700}{400 + 30 + 70}$

3. $7^3 + 8^3$

4. $\dfrac{7^2}{5^2 + 6^2}$

5. $9^3 + 10^2$

6. $\dfrac{7 \times 9^2}{8^2 - 4}$

7. $\dfrac{56 \times 81}{73 \times 69 \times 45}$

8. $(7 + 8 \times 3)^5$

9. Try to perform the calculation $\frac{3}{0}$ on your calculator. What happens?

10. Try to perform the calculation $\sqrt{-4}$ on your calculator. What happens?

In Exercises 11–20, perform each of the calculations on a calculator and then round your answer to the nearest hundredth.

11. $78 - 61.63$

12. 567.123×723.791

13. $0.00723 \div 53.72$

14. 0.5791×0.6423

15. $5693.721 + 71.6 + 39.54763$

16. $7239.23 \div 41.61795$

17. $723.39 \times 71.38 \div 73.002$

18. $642.73 \times 8.002 \div 61.732$

19. $\dfrac{68.93 \times 53.987 \times 0.793}{18.3 \times 0.0079 \times 6.93}$

20. $\dfrac{19.83 \times 71.82 \times 0.0073}{51.76 \times 0.5863 \times 0.0078}$

21. Using a calculator, evaluate 7×9^3. Use the $\boxed{y^x}$ button. (Be careful!)

22. Using a calculator, evaluate $(-3)^4$. Can you use the $\boxed{y^x}$ button? (Be careful!)

23. Using a calculator, evaluate 52^0.

24. Enter the number 7734 into your calculator. Now turn the machine upside down. What do you get?

25. Enter the number 57738 into your calculator. Now turn the machine upside down. What do you get?

26. Convert the fraction $732\frac{17}{23}$ to decimals.

27. Pick any three-digit number. Make a six-digit number by repeating the digits again. Divide the six-digit number by 7. Divide the result by 11. Divide the new result by 13. What is your answer? Repeat this process for three different numbers.

28. Using a calculator, evaluate $2 \div 3$. (Some calculators will give the answer as 0.666666667. Other calculators will not round the answer but will give 0.66666666 as the answer.)

5.3

RATIO AND PROPORTION

Often we find ourselves comparing different quantities. Thus a business major may find that 30 credits out of the 128 credits required for a bachelor's degree must be in business courses. The student may be interested in determining what part of the degree consists of business courses. Ratios are used to make such comparisons.

> **Definition 5.1**   A **ratio** of one number to another (nonzero) number is the quotient of the first number divided by the second number. The ratio of a to b can be expressed as $\frac{a}{b}$ or $a \div b$ or $a:b$. The numbers a and b are called the **terms** of the ratio.

For example, the ratio of 12 to 6 is $12 \div 6$ or $\frac{12}{6}$. The quotient $\frac{12}{6}$ is equivalent to $\frac{2}{1}$ or 2. We can thus say that the first number is twice the second number.

A ratio is expressed in **simplest form** when both terms of the ratio are whole numbers and when there is no number (other than 1) that divides exactly into these terms.

In the example discussed earlier the ratio of the number of business courses to the total number of credits required for a degree is

$$\frac{30}{128} \quad \text{or} \quad \frac{15}{64}.$$

EXAMPLE 1   A certain cookie recipe calls for $1\frac{1}{3}$ cups of water to $1\frac{5}{6}$ cups of flour. What is the ratio of the number of cups of water to the number of cups of flour?

SOLUTION   The ratio of cups of water to the number of cups of flour is

$$\frac{1\frac{1}{3}}{1\frac{5}{6}} = \frac{\frac{4}{3}}{\frac{11}{6}}$$

$$= \frac{4}{3} \div \frac{11}{6}$$

$$= \frac{4}{3} \cdot \frac{6}{11}$$

$$= \frac{8}{11}. \quad \square$$

As shoppers in supermarkets, we often find similar items with different prices. Since the similar items may be packaged differently, **unit pricing** allows us to compare the costs and get the best buy. The following example shows us how this can be done.

EXAMPLE 2   Kim is in a supermarket comparing the prices of several brands of rice krispies. The national brand contains 567 grams and sells for $2.09. The store brand contains 539 grams and sells for $1.69. Which brand has the lower cost per gram?

SOLUTION   To determine which brand has the lower cost per gram, we set up a ratio between the cost of the item and the number of units in the package. This will give us the **unit cost** for each brand. In our case we have

	National brand		Store brand	
$\dfrac{\text{cost}}{\text{number of units}} =$	$\dfrac{\$2.09}{567 \text{ grams}} = \dfrac{209 \text{ cents}}{567 \text{ grams}}$		$\dfrac{\$1.69}{539 \text{ grams}} = \dfrac{169 \text{ cents}}{539 \text{ grams}}$	
	$= 0.37 \text{ cents/gram}$		$= 0.31 \text{ cents/gram}$	

Thus the store brand is $0.37 - 0.31$ or 0.06 cents per gram cheaper.
$\square$

EXAMPLE 3   Will Rogers runs 320 meters in 35 seconds. His brother Gill runs 250 meters in 26 seconds. Which runner is faster?

SOLUTION   To determine which runner is faster, we set up a ratio between the distance and the time. We get

	Will		Gill
	$\dfrac{320 \text{ meters}}{35 \text{ seconds}} = 9.14 \text{ meters/second}$		$\dfrac{250 \text{ meters}}{26 \text{ seconds}} = 9.62 \text{ meters/second}$

Thus Gill is the faster runner. $\square$

EXAMPLE 4   During a baseball season the ratio of the number of times that Daryl got a hit to the official number of times he was at bat is $4:11$. If Daryl had 440 official times at bat that season, how many hits did Daryl get?

SOLUTION   The ratio $4:11$ tells us that $\frac{4}{11}$ of Daryl's official times at bat resulted in a hit. Since Daryl was at bat 440 times, we know that $\frac{4}{11}$ of these times he got a hit. Thus Daryl got $\frac{4}{11} \times 440$ or 160 hits that season.

□

Consider the ratio $\frac{3}{12}$ and the ratio $\frac{1}{4}$. We notice that $\frac{3}{12} = \frac{1}{4}$. The equation $\frac{3}{12} = \frac{1}{4}$ is called a proportion.

> **Definition 5.2**   *A **proportion** is an equation stating that two ratios are equal.*

In the previous example the proportion $\frac{3}{12} = \frac{1}{4}$ can also be written as $3:12 = 1:4$. In both cases we read this as "3 is to 12 as 1 is to 4."

More generally, if we have the proportion $\frac{a}{b} = \frac{c}{d}$, then the first and fourth terms are called the **extremes,** and the second and third terms are called the **means** of the proportion.

$$
\overbrace{a:b \;=\; c:d}^{\text{Extremes}} \qquad \text{Extremes}
$$

Extremes

$$a:b = c:d$$

Means

Extremes

$$\frac{a}{b} = \frac{c}{d}$$

Means

Notice that in the proportion $3:12 = 1:4$, the product of the means 12×1 is equal to the product of the extreme 3×4. In both cases the answer is 12.

In Chapter 4 (Definition 4.12), we indicated that two rational numbers $\frac{a}{b}$ and $\frac{c}{d}$ are equal if when we cross-multiply

$$\frac{a}{b} = \frac{c}{d}$$

we get $ad = bc$.

We can now state this in words as follows.

Formula 1   *In the proportion $\frac{a}{b} = \frac{c}{d}$, the product of the means is equal to the product of the extremes: that is, ad = bc.*

Let us apply Formula 1.

EXAMPLE 5   Tell whether $\frac{2}{3} = \frac{10}{15}$ is a valid proportion.

SOLUTION   In the equation $\frac{2}{3} = \frac{10}{15}$ the product of the means (second and third terms) is 3×10 or 30. The product of the extremes (first and fourth terms) is 2×15 or 30. Since the answer is the same for both, we conclude that $\frac{2}{3} = \frac{10}{15}$ is a valid proportion.   □

EXAMPLE 6   Solve for x in the proportion $\frac{3}{5} = \frac{24}{x}$.

SOLUTION   We apply Formula 1:

$$\frac{3}{5} = \frac{24}{x}$$
$$3x = 24 \times 5$$
$$3x = 120$$
$$x = \frac{120}{3} \qquad \text{(We divide both sides of the equation by 3.)}$$
$$x = 40 \quad □$$

EXAMPLE 7   The real estate tax in Culver City in 1985 was $8.60 for every $100 of assessed value. If a house is assessed at $9000, what is the real estate tax?

SOLUTION   We can solve this problem by setting up a proportion as follows:

$$\frac{\text{Real estate tax}}{\text{Assessed value}} = \frac{\$8.60}{\$100} = \frac{\text{Real estate tax}}{\$9000}.$$

Since we wish to find the real estate tax for a house assessed at $9000, let us call this quantity x. The proportion then becomes

$$\frac{8.60}{100} = \frac{x}{9000}.$$

We must now solve for x. Applying Formula 1 gives

$$100x = 8.60 \times 9000$$
$$100x = 77400$$
$$x = \frac{77400}{100} \qquad \text{(We divide both sides of the equation by 100.)}$$
$$x = 774$$

The real estate tax on a house assessed at $9000 is $774. ☐

EXAMPLE 8   The United States is gradually converting from the English system to the metric system of measurement. In most states today the posted speed limit is 55 miles per hour. What will the posted speed limit be in kilometers per hour when the conversion is completed? (Assume that 1 kilometer is approximately 0.62 miles.)

SOLUTION   We can write a proportion to compare kilometers to miles. Let $x =$ the number of kilometers in 55 miles. We then have

$$\frac{1 \text{ km}}{0.62 \text{ miles}} = \frac{x \text{ km}}{55 \text{ miles}}.$$

We now apply Formula 1:

$$0.62x = 1 \times 55$$
$$0.62x = 55$$
$$x = \frac{55}{0.62} \quad \text{(We divide both sides of the equation by 0.62)}$$
$$x = 88.71$$

Thus the posted speed limit will be 89 kilometers per hour. ☐

EXERCISES

Express each of the ratios given in Exercises 1–6 in simplest form.

1. 80 meters to 45 meters

2. 16 cm to 64 cm

3. $60 to $80

4. $6 to 50 cents

5. 14 ounces to 2 pounds

6. 4 hours to 12 minutes

7. If 14 cars can be serviced by 7 mechanics in one day, how many cars can be serviced by 8 mechanics in one day?

8. Mark typed 1400 words in 35 minutes. Arlene typed 1200 words in 30 minutes. Who is the faster typist?

9. Howie replaced 3 of the 8 spark plugs in his car. What is the ratio of the number of spark plugs replaced to the number of spark plugs not replaced?

10. An obstetrician delivered 30 boys and 20 girls during the month of July. For every 3 boys delivered, how many girls were delivered?

11. In a supermarket a package containing 6 bars of soap sells for $1.39. At the same time, a package containing 10 bars of the same soap sells for $2.25. Which is the better buy?

12. Doug's monthly income is $1000. Doug spends this money on rent, food, education, entertainment and miscellaneous items in the ratio of 6:5:3:2:4. How much does Doug spend on each item?

13. A painter needs 25 gallons of a paint mixture. This mixture is obtained by mixing white paint and blue paint in the ratio of 3:2. How many gallons of each kind must the painter use?

14. In a magazine containing 192 pages the ratio of the number of pages devoted to advertisements to the number of pages devoted to arti-

cles is 13 to 3. How many pages of each are contained in the magazine?

15. In a certain bag of mixed fruit the ratio of the number of dried apricots to the number of dried apples is 8 : 7. If the bag contains 300 pieces of fruit, how many of them are apricots?

In Exercises 16–21, tell whether the given ratios form a valid proportion.

16. $\dfrac{5}{4} = \dfrac{15}{12}$ 　　　　**17.** $\dfrac{8}{7} = \dfrac{15}{14}$

18. $\dfrac{7}{8} = \dfrac{8}{7}$ 　　　　**19.** $\dfrac{7}{1} = \dfrac{14}{2}$

20. $\dfrac{120}{15} = \dfrac{16}{2}$ 　　　**21.** $\dfrac{18}{4} = \dfrac{27}{6}$

Solve for x in each of the proportions given in Exercises 22–27.

22. $\dfrac{x}{15} = \dfrac{3}{5}$ 　　**23.** $\dfrac{10}{3} = \dfrac{x}{12}$ 　　**24.** $\dfrac{20}{3x} = \dfrac{30}{4.5}$

25. $\dfrac{18}{x} = \dfrac{12}{9}$ 　　**26.** $\dfrac{2}{16} = \dfrac{2x}{40}$ 　　**27.** $\dfrac{7}{x} = \dfrac{28}{16}$

28. Refer back to Example 7. What is the real estate tax on a house that is assessed for $12,500?

29. If the posted speed limit on a highway is 40 miles per hour, what is the speed limit in kilometers per hour?

30. In a sporting store, 7 bats cost $56. At the same rate, what is the cost of 12 bats?

31. Leo drove 300 miles while using 14.5 gallons of gas. Assuming the same driving conditions, how many gallons of gas are needed to drive 425 miles?

32. Three ounces of a certain food contain 95 calories. How many calories are contained in seven ounces of the food?

33. Spencer owns 15 shares of stock of the Back Corp. He receives $4.95 in dividends. How much would he receive in dividends if he owned 25 shares of the stock?

34. Harley Smith made 234 photocopies of a poster and paid $14.04. At the same rate, what would 185 photocopies of the poster cost?

35. Gary stayed in the hospital for 8 days. The bill came to $1968. Assuming the same daily charges, what would be the cost of a 10-day stay at the hospital?

5.4
PERCENTS

We often read newspaper articles such as the one shown below or are involved in discussions in which the word "percent" is used. Actually, the word is derived from the Latin word *per centum* meaning "per hundred." A **percent** can be considered as the ratio of a number to 100. For example, 9%, which means $\dfrac{9}{100}$, is the ratio of 9 to 100. Similarly, 5.5% means $\dfrac{5.5}{100}$, which is the ratio of 5.5 to 100. Also, x% means $\dfrac{x}{100}$.

Higher Phone Bills

DOVER: The Public Service Commission yesterday authorized the telephone company to raise its rates by 3%. This represents the third increase in less than 5 years. Last year the telephone company raised its rates by 5.5%.

Business News, August 5, 1986

Percents are usually written with the percent symbol, as in the above examples. However, since they are often written in fraction form (where the denominator is 100) or in decimal form, let us pause for a moment to indicate how we change fractions to percents and vice versa.

EXAMPLE 1 Change $\frac{3}{5}$ to a percent.

SOLUTION Since percent means the ratio of a number to 100, we can change $\frac{3}{5}$ to a percent by setting up the following proportion:

$$\frac{3}{5} = \frac{x}{100}$$ (This is a proportion because it consists of two equal ratios.)

$$5x = 300$$ (In a proportion the product of the means equals the product of the extremes.)

$$x = \frac{300}{5}$$ (We divide both sides of the equation by 5.)

$$x = 60$$

Thus $\frac{3}{5} = \frac{60}{100}$ or 60%.

We can also change $\frac{3}{5}$ to a percent by dividing 5 (the denominator) into 3 (the numerator), then multiplying the quotient by 100 and adding a percent symbol. Thus in our case we get

$$\frac{3}{5} = 0.6$$ (First we divide 5 into 3.)

$$0.6 \times 100 = 60$$ (We multiply the quotient by 100.)

Therefore $\frac{3}{5} = 60\%$. We add the percent symbol. ☐

Comment If you are using a calculator, the second method of changing a fraction to a percent is much easier.

EXAMPLE 2 Convert $\frac{1}{3}$ to a percent.

SOLUTION We first change $\frac{1}{3}$ from a fraction into a decimal. What we get is $\frac{1}{3} = 0.3333\ldots$, a nonterminating but repeating decimal. Now we multiply by 100 and add the percent symbol. We get $\frac{1}{3} = 33.33\%$.

You will notice that we have rounded our answer to 2 decimal places. In this section we will *round* all our answers to 2 decimal places. ☐

If a number is already written in decimal form, then we change it to a percent by multiplying by 100 and then adding the percent symbol. Since multiplication by 100 is equivalent to moving the decimal point two places to the right, we can change a number from decimal form to a percent form very quickly by moving the decimal point two places to the right.

EXAMPLE 3   Change 0.792 to a percent.

SOLUTION   We can change 0.792 to a percent by first multiplying it by 100. Then we add the percent symbol. Thus $0.792 \times 100 = 79.2\%$.

Alternatively, we move the decimal point in 0.792 two places to the right and add the percent symbol as shown:

$$0.792 = 79.2\%. \quad \square$$

Finally, to convert a number written with a percent symbol to a decimal equivalent, we drop the percent symbol and divide the number by 100. This is really an application of the definition of percent, which means the ratio of a number to 100.

EXAMPLE 4   Convert 94% to a decimal.

SOLUTION   To convert 94% to a decimal, we drop the percent symbol and then divide the number by 100. We get

$$94\% = \frac{94}{100} = 0.94.$$

Thus $94\% = 0.94$. $\square$

Comment   Since division by 100 is equivalent to moving the decimal point two places to the left, we can change a number from percent form to decimal form very quickly by moving the decimal point two places to the left.

EXAMPLE 5   Many banks pay $5\frac{1}{4}\%$ interest on savings accounts. Express this percent as a decimal.

SOLUTION $5\frac{1}{4}\%$ is the same as 5.25% since $\frac{1}{4} = 0.25$. Then we move the decimal point two places to the left and drop the percent symbol. Thus 5.25% = 0.0525.

Notice that we have to add an extra zero. Why? $\square$

EXAMPLE 6   Of the 780 graduates of Stoneyville College receiving their bachelor's degree this year, 663 said that they would continue their education and go on to the master's degree. What percent of the graduates will continue their education?

SOLUTION   The percent of the graduates who will continue their education is the ratio of the number who will continue to the total number of graduates. Thus we divide the number who will continue by the total number of graduates and multiply the result by 100. At the end we add a percent symbol. In our case we get

$$\frac{663}{780} \times 100 = 0.85 \times 100 = 85\%.$$

Therefore 85% of the graduates will continue their education.   □

EXAMPLE 7   In the previous example, 40% of the graduates have a grade point average (GPA) above 3.3. How many students have a GPA above 3.3?

SOLUTION   We first write 40% in decimal form as

$$40\% = \frac{40}{100} = 0.40.$$

Since there are 780 graduates, 40% of whom had a GPA above 3.3, this means that there are 780×0.40 or 312 students with a GPA above 3.3.   □

EXAMPLE 8   Vicky purchased a new car for $7850. She made a down payment of 14% of the price of the car when she bought it. How much was her down payment?

SOLUTION   We first write 14% in decimal form as

$$14\% = \frac{14}{100} = 0.14.$$

Since the down payment was 14%, this means that it was

$$0.14 \times 7850 \text{ or } \$1099.   □$$

EXAMPLE 9   Mr. Jaskel, who earns $300 per week in wages, has just been informed by his boss that his salary will be increased to $375. Find the percent of increase in his salary.

SOLUTION   To determine the **percent increase** in his salary, we first find the actual increase. We then divide this by the original salary. Finally, we multiply our answer by 100 and add the percent symbol. Since Mr. Jaskel's salary went from $300 to $375, his actual increase was $75.

His original salary was $300. Thus the percent increase is

$$\frac{375 - 300}{300} \times 100 = \frac{75}{300} \times 100$$
$$= 0.25 \times 100$$
$$= 25\%. \quad \square$$

EXERCISES

Convert each of the fractions or numbers given in Exercises 1–10 into percents.

1. $\frac{4}{5}$ **2.** $\frac{1}{4}$ **3.** $\frac{2}{3}$ **4.** $\frac{5}{7}$

5. 2   **6.** 3   **7.** $1\frac{2}{5}$ **8.** $2\frac{1}{3}$

9. $1\frac{2}{3}$ **10.** $\frac{53}{200}$

Convert each of the decimals given in Exercises 11–18 into percents.

11. 0.12   **12.** 0.567   **13.** 0.01
14. 0.002   **15.** 5.1   **16.** 6.23
17. 1.01   **18.** 1.00

Convert each of the percents given in Exercises 19–26 into decimals.

19. $5\frac{1}{2}\%$ **20.** $2\frac{1}{4}\%$ **21.** $6\frac{2}{3}\%$ **22.** $12\frac{1}{2}\%$

23. 18%   **24.** $1\frac{1}{2}\%$ **25.** 10.1%   **26.** 16.23%

27. What is 7% of 95?

28. What is 12% of 426?

29. What is 8.2% of 578?

30. What is $9\frac{1}{4}\%$ of 360?

31. One hundred is 125% of what number?

32. Eleven is $5\frac{1}{2}\%$ of what number?

33. Gwendolyn just purchased a new dress for $49.95. The sales tax is $8\frac{1}{4}\%$. How much tax does Gwendolyn have to pay?

34. Bob just bought a new set of radial tires for $198. Sales tax is $7\frac{1}{2}\%$. What is the total cost of the tires (including tax)?

35. The general sales tax in a certain state is 7% of gross sales. One week a department store collected $36,820 in sales tax. What were the gross sales for the week?

36. Catherine rented a car for a week at a cost of $79 (car only). The sales tax was $4.59 extra. What is the percent of the sales tax?

37. *Cash or Charge.* Many gas stations give customers a discount on cash purchases for gasoline. At one gas station, credit card customers pay $1.119 per gallon of gas, whereas cash customers receive a discount of 6 cents per gallon of gas. What is the percent of discount for cash-paying customers?

38. At a recent sale a 35-mm camera, which costs the dealer $89.98 and usually lists for $124.49, was selling for 104.98.

    a) Find the regular percent markup.

    b) Find the percent decrease of the sale price.

39. After taking a 3% discount for early payment, Marty paid his credit card company bill with a check in the amount of $163.93. What was the original amount of the bill?

40. Alexis purchased a new car for $7800. At the end of the year, the value of the car had depreciated by $1560. By what percent did the value of the car decrease?

41. As a result of a new union contract, Joanne's annual salary will be increased from $42,000 to $45,500. Find the percent increase in her salary.

42. The police force of Ardsley consists of 78 police officers, 27 of whom are from minority groups. What percent of the police force is from minority groups?

43. A doctor tested 476 samples of blood and found that 6 of these samples had a certain virus in them. What percent of the blood samples had the virus in them?

44. Ninety-eight percent of all people with a certain form of skin cancer (when detected early) will be cured after appropriate medical treatment. If 412 new cases were reported in Shawnee last year, how many of these patients will be cured after appropriate medical treatment?

45. To generate sales, a dealer reduced the price of a camera by $67.25 for a week. This represented

a discount of 25% of the original price. On the last day of the sale the camera was further discounted to sell for 60% of its original price. What was the final selling price of the camera?

46. Pedro arranged for a $50,000 mortgage from a bank. The annual rate of interest is 0.14 on the *entire* amount, and the interest is payable monthly. What is the monthly interest charge?

5.5

SIMPLE AND COMPOUND INTEREST

At one time or another in our lives we must either borrow money or lend money. We may borrow money to help pay for college tuition, to help finance a car, or to buy a house that sells for $225,000 when we have only $30,000 for the down payment. In these situations we need money and are willing to pay **interest** to someone or some business for the use of the money.

On the other hand, we may lend money to someone or to some business or governmental agency. In this case the business or governmental agency may sell bonds, which in essence promise to repay the loan at a specified date. In the meantime, interest will be paid, usually at six-month intervals, for the use of the money.

Many of us have department store credit cards or general bank credit cards such as Master Card, Visa, and American Express. When we do not repay the monthly charges by the date specified, we are charged interest. This is usually labeled **finance charges** on the monthly statement that we receive.

As a matter of fact, a large part of the U.S. economy is based on borrowing and interest. It is for this reason that a knowledge of interest, whether on money borrowed or on money saved, is a valuable tool.[1]

Interest is the money paid for the use of money. The amount of money borrowed when an individual applies for a loan from a bank or when an individual lends money to a bank in the form of a savings account or to a governmental agency in the form of a bond is called **principal.**

When money is borrowed, the borrower agrees to pay the lender interest at a specified rate over a period of time—for example, 6% over 3 years or 10% over 4 years. In each case the rate of interest is

---

1. It is strongly suggested that the reader use a calculator to work through many of the examples and exercises in this section.

specified as a percentage, either by using the % symbol or by writing it in decimal form. Thus we can write an interest rate of 18 percent as 18% or as 0.18 and an interest rate of five percent as 5% or as 0.05.

Comment   Many savings banks pay $5\frac{1}{4}\%$ interest on savings accounts. This is written in decimal form as 0.0525 and *not* as 0.525. The latter represents 52.5%, a rather high interest rate.

The interest charged on a loan may be simple or compound. *Simple interest* is generally used for short-term loans of a year or less and is computed on the whole amount of money borrowed for the entire period of the loan. On the other hand, when a loan is for a longer period of time or for interest paid by a bank, the interest is added to the original loan at specified times (quarterly, semiannually, monthly, etc.), thereby increasing the amount of interest paid. This is known as **compound interest** and will be discussed later in this section.

Simple Interest   Suppose $6000 is borrowed for 1 year. If the rate of interest is 8% per year, then the interest is 8% of $6000 or

$$0.08 \times \$6000 = \$480.$$

If the money is borrowed for 2 years, then the total interest paid is

$$0.08 \times \$6000 \times 2 = \$960,$$

where 0.08 is the rate of interest, $6000 is the amount of money borrowed (the principal), and 2 is the time period.

More generally, we have the following simple-interest formula.

> **Formula 2   *Simple-Interest Formula***   *The total simple interest, I, in dollars on a loan of P dollars for n years, where the annual interest rate is r, is given by*
>
> $$I = P \cdot r \cdot n.$$

EXAMPLE 1   Find the simple interest on a loan of $900 for 4 years, given that the rate of interest is 6% per year.

SOLUTION   We use the simple-interest formula. Here $P = 900$, $r = 0.06$, and $n = 4$. Then

$$\begin{aligned} I &= P \cdot r \cdot n \\ &= 900(0.06)(4) \\ &= \$216. \end{aligned}$$

Thus the total simple interest is $216.   □

When working with the simple-interest formula, you must express both r (rate) and n (time) in the same unit of time—for example, days, months, years. If this is not the case, then one of them must be converted so that they are both in the same unit of time, as illustrated in the following examples.

EXAMPLE 2 *Friendly Loans.* Bill arranges to borrow $800 from his friend for 9 months. The simple interest rate is 30% per year. How much money does Bill have to repay to his friend?

SOLUTION We use the simple-interest formula (Formula 2). Here $P = 800$ and $r = 0.30$. Since the loan is for only 9 months, we must express this as a fraction of the year. Thus $n = 9$ months $= \frac{9}{12}$ year (n is usually expressed in years). Then

$$I = P \cdot r \cdot n$$
$$= 800\left(0.30\right)\left(\frac{9}{12}\right)$$
$$= \$180.$$

Thus the interest is $180. Since the loan was for $800, Bill must repay $980 to his friend. ☐

EXAMPLE 3 *Off-Track Betting (OTB).* Harry is in an OTB office in New York when he receives a tip on a particular horse. He borrows $940 from a friend and agrees to repay $1000 to his friend in two months. Find the annual rate of interest.

SOLUTION Again we will use the simple-interest formula. Here $P = 940$ and $n = 2$ months $= \frac{2}{12}$ year. Since Harry borrowed $940 and will repay $1000, the interest is $1000 - \$940 = \$60$. Then

$$I = P \cdot r \cdot n$$
$$60 = 940 \cdot r \cdot \left(\frac{2}{12}\right)$$

Solving this equation for r gives us

$$r = \frac{18}{47} \quad \text{or} \quad 0.3830$$

Thus the annual rate of interest is 38.30%. ☐

EXAMPLE 4 *Short-Term Loan.* Marie Cartright arranges for a short-term loan at an annual interest rate of 15%. The total interest charged is $300. If Marie repays the loan in 3 months, then how much money did she borrow?

SOLUTION We are interested in finding the amount of money borrowed, so we must find the value of P. We use Formula 2 with $r = 0.15$, $n = 3$ months $= \frac{3}{12}$ year, and $I = 300$. We have

$$I = P \cdot r \cdot n$$

$$300 = P(0.15)\left(\frac{3}{12}\right)$$

$$300 = 0.0375P.$$

Now we divide both sides of this equation by 0.0375. We get

$$\frac{300}{0.0375} = \frac{0.0375P}{0.0375}$$

$$8000 = P.$$

Thus Marie borrowed $8000. ☐

Compound Interest Suppose a person has some money to deposit in a savings bank that pays compound interest. Since some banks in the United States now pay interest *compounded continuously*, let us analyze what is meant by **compound interest.**

Suppose we deposit $1000 (called the principal) in a bank that pays 6% interest per year compounded annually. Then at the end of one year we would have $1060. This amount represents the $1000 principal plus the 1000(0.06), or $60, interest earned on the money.

During the second year the bank will pay 6% interest on $1060, so we will earn $1060(0.06), or an additional $63.60, in interest. Thus at the end of the second year we would have a total accumulation of $1123.60. The same thing will happen in succeeding years. Table 5.1 indicates the amount of money accumulated after several years, assuming the bank pays 6% interest compounded annually.

On the other hand, if the bank pays 6% interest compounded semiannually, then it is really paying only 3% interest for each six-month period. Similarly, when the bank pays 6% interest compounded quarterly, it is really paying only 1.5% interest for every three-month period. What happens when the bank pays 6% interest compounded continuously? Table 5.2 gives the amount of money that can be accumulated when different compounding periods are used. The values given in this table were obtained by using the **compound-interest formula,** to be discussed shortly.

Tables 5.1 and 5.2 suggest that as the number of compounding periods increases, the amount of money that we accumulate also increases. Yet the amount of money that we accumulate does not increase without bounds (unfortunately!). There is a limit. To find this limit, we can use the compound-interest formula, which follows directly from the fourth column of Table 5.1. We have the following.

Formula 3 **Compound-Interest Formula** *The compound amount A (principal + interest) that results when P dollars (the principal) is invested at a rate of r per period for n periods is given by the formula*

$$A = P(1 + r)^n.$$

When the money is compounded continuously, we have the following.

Formula 4 *If P dollars is invested in a bank that pays interest at the rate of r per year* **compounded continuously,** *the amount of money accumulated, A, after n years is*

$$A = Pe^{rn},$$

where the values of e to the appropriate power can be determined from Table 5.3.

TABLE 5.1

Amount of money accumulated after six years, assuming a 6% interest rate compounded annually

Year	Amount of money on deposit at beginning of year	Interest earned during year	Amount of money on deposit at end of year	Simplified form
1	$1000	1000(0.06)	1000 + 1000(0.06) = 1000(1 + 0.06)	$1060
2	$1060	1060(0.06)	1060 + 1060(0.06) = 1000(1 + 0.06)^2	$1123.60
3	$1123.60	1123.60(0.06)	1123.60 + 1123.60(0.06) = 1000(1 + 0.06)^3	$1191.02
4	$1191.02	1191.02(0.06)	1191.02 + 1191.02(0.06) = 1000(1 + 0.06)^4	$1262.48
5	$1262.48	1262.48(0.06)	1262.48 + 1262.48(0.06) = 1000(1 + 0.06)^5	$1338.23
6	$1338.23	1338.23(0.06)	1338.23 + 1338.23(0.06) = 1000(1 + 0.06)^6	$1418.52

TABLE 5.2

Amount of money accumulated after one year at 6% interest using different compounding periods

Principal	Compounded annually	Compounded semiannually	Compounded quarterly	Compounded monthly	Compounded continuously
$1000.	$1060.	$1060.90	$1061.36	$1061.68	$1061.84
$5000.	$5300.	$5304.50	$5306.82	$5308.39	$5309.18

TABLE 5.3
The exponential function

x	e^x	x	e^x	x	e^x	x	e^x	x	e^x
0.00	1.00000	0.50	1.64872	1.00	2.71828	1.50	4.48169	2.00	7.38906
0.01	1.01005	0.51	1.66529	1.01	2.74560	1.51	4.52673	2.01	7.46332
0.02	1.02020	0.52	1.68203	1.02	2.77319	1.52	4.57223	2.02	7.53832
0.03	1.03045	0.53	1.69893	1.03	2.80107	1.53	4.61818	2.03	7.61409
0.04	1.04081	0.54	1.71601	1.04	2.82922	1.54	4.66459	2.04	7.69061
0.05	1.05127	0.55	1.73325	1.05	2.85765	1.55	4.71147	2.05	7.76790
0.06	1.06184	0.56	1.75067	1.06	2.88637	1.56	4.75882	2.06	7.84597
0.07	1.07251	0.57	1.76827	1.07	2.91538	1.57	4.80665	2.07	7.92482
0.08	1.08329	0.58	1.78604	1.08	2.94468	1.58	4.85496	2.08	8.00447
0.09	1.09417	0.59	1.80399	1.09	2.97427	1.59	4.90375	2.09	8.08491
0.10	1.10517	0.60	1.82212	1.10	3.00417	1.60	4.95303	2.10	8.16617
0.11	1.11628	0.61	1.84043	1.11	3.03436	1.61	5.00281	2.11	8.24824
0.12	1.12750	0.62	1.85893	1.12	3.06485	1.62	5.05309	2.12	8.33114
0.13	1.13883	0.63	1.87761	1.13	3.09566	1.63	5.10387	2.13	8.41487
0.14	1.15027	0.64	1.89648	1.14	3.12677	1.64	5.15517	2.14	8.49944
0.15	1.16183	0.65	1.91554	1.15	3.15819	1.65	5.20698	2.15	8.58486
0.16	1.17351	0.66	1.93479	1.16	3.18993	1.66	5.25931	2.16	8.67114
0.17	1.18530	0.67	1.95424	1.17	3.22199	1.67	5.31217	2.17	8.75828
0.18	1.19722	0.68	1.97388	1.18	3.25437	1.68	5.36556	2.18	8.84631
0.19	1.20925	0.69	1.99372	1.19	3.28708	1.69	5.41948	2.19	8.93521
0.20	1.22140	0.70	2.01375	1.20	3.32012	1.70	5.47395	2.20	9.02501
0.21	1.23368	0.71	2.03399	1.21	3.35348	1.71	5.52896	2.21	9.11572
0.22	1.24608	0.72	2.05443	1.22	3.38719	1.72	5.58453	2.22	9.20733
0.23	1.25860	0.73	2.07508	1.23	3.42123	1.73	5.64065	2.23	9.29987
0.24	1.27125	0.74	2.09594	1.24	3.45561	1.74	5.69734	2.24	9.39333
0.25	1.28403	0.75	2.11700	1.25	3.49034	1.75	5.75460	2.25	9.48774
0.26	1.29693	0.76	2.13828	1.26	3.52542	1.76	5.81244	2.26	9.58309
0.27	1.30996	0.77	2.15977	1.27	3.56085	1.77	5.87085	2.27	9.67940
0.28	1.32313	0.78	2.18147	1.28	3.59664	1.78	5.92986	2.28	9.77668
0.29	1.33643	0.79	2.20340	1.29	3.63279	1.79	5.98945	2.29	9.87494
0.30	1.34986	0.80	2.22554	1.30	3.66930	1.80	6.04965	2.30	9.97418
0.31	1.36343	0.81	2.24791	1.31	3.70617	1.81	6.11045	2.31	10.07442
0.32	1.37713	0.82	2.27050	1.32	3.74342	1.82	6.17186	2.32	10.17567
0.33	1.39097	0.83	2.29332	1.33	3.78104	1.83	6.23389	2.33	10.27794
0.34	1.40495	0.84	2.31637	1.34	3.81904	1.84	6.29654	2.34	10.38124
0.35	1.41907	0.85	2.33965	1.35	3.85743	1.85	6.35982	2.35	10.48557
0.36	1.43333	0.86	2.36316	1.36	3.89619	1.86	6.42374	2.36	10.59095
0.37	1.44773	0.87	2.38691	1.37	3.93535	1.87	6.48830	2.37	10.69739
0.38	1.46228	0.88	2.41090	1.38	3.97490	1.88	6.55350	2.38	10.80490
0.39	1.47698	0.89	2.43513	1.39	4.01485	1.89	6.61937	2.39	10.91349
0.40	1.49182	0.90	2.45960	1.40	4.05520	1.90	6.68589	2.40	11.02318
0.41	1.50682	0.91	2.48432	1.41	4.09596	1.91	6.75309	2.41	11.13396
0.42	1.52196	0.92	2.50929	1.42	4.13712	1.92	6.82096	2.42	11.24586
0.43	1.53726	0.93	2.53451	1.43	4.17870	1.93	6.88951	2.43	11.35888
0.44	1.55271	0.94	2.55998	1.44	4.22070	1.94	6.95875	2.44	11.47304
0.45	1.56831	0.95	2.58571	1.45	4.26311	1.95	7.02869	2.45	11.58835
0.46	1.58407	0.96	2.61170	1.46	4.30596	1.96	7.09933	2.46	11.70481
0.47	1.59999	0.97	2.63794	1.47	4.34924	1.97	7.17068	2.47	11.82245
0.48	1.61607	0.98	2.66446	1.48	4.39295	1.98	7.24274	2.48	11.94126
0.49	1.63232	0.99	2.69123	1.49	4.43710	1.99	7.31553	2.49	12.06128

(*continued on page 268*)

x	e^x	x	e^x	x	e^x	x	e^x	x	e^x
2.50	12.18249	3.00	20.08554	3.50	33.11545	4.00	54.59815	4.50	90.01713
2.51	12.30493	3.01	20.28740	3.51	33.44827	4.01	55.14687	4.51	90.92182
2.52	12.42860	3.02	20.49129	3.52	33.78443	4.02	55.70110	4.52	91.83560
2.53	12.55351	3.03	20.69723	3.53	34.12397	4.03	56.26091	4.53	92.75856
2.54	12.67967	3.04	20.90524	3.54	34.46692	4.04	56.82634	4.54	93.69080
2.55	12.80710	3.05	21.11534	3.55	34.81332	4.05	57.39745	4.55	94.63240
2.56	12.93582	3.06	21.32756	3.56	35.16320	4.06	57.97431	4.56	95.58347
2.57	13.06582	3.07	21.54190	3.57	35.51659	4.07	58.55696	4.57	96.54411
2.58	13.19714	3.08	21.75840	3.58	35.87354	4.08	59.14547	4.58	97.51439
2.59	13.32977	3.09	21.97708	3.59	36.23408	4.09	59.73989	4.59	98.49443
2.60	13.46374	3.10	22.19795	3.60	36.59823	4.10	60.34029	4.60	99.48431
2.61	13.59905	3.11	22.42104	3.61	36.96605	4.11	60.94671	4.61	100.48415
2.62	13.73572	3.12	22.64638	3.62	37.33757	4.12	61.55924	4.62	101.49403
2.63	13.87377	3.13	22.87398	3.63	37.71282	4.13	62.17792	4.63	102.51406
2.64	14.01320	3.14	23.10387	3.64	38.09184	4.14	62.80282	4.64	103.54435
2.65	14.15404	3.15	23.33606	3.65	38.47467	4.15	63.43400	4.65	104.58498
2.66	14.29629	3.16	23.57060	3.66	38.86134	4.16	64.07152	4.66	105.63608
2.67	14.43997	3.17	23.80748	3.67	39.25191	4.17	64.71545	4.67	106.69774
2.68	14.58509	3.18	24.04675	3.68	39.64639	4.18	65.36585	4.68	107.77007
2.69	14.73168	3.19	24.28843	3.69	40.04485	4.19	66.02279	4.69	108.85318
2.70	14.87973	3.20	24.53253	3.70	40.44730	4.20	66.68633	4.70	109.94717
2.71	15.02928	3.21	24.77909	3.71	40.85381	4.21	67.35654	4.71	111.05216
2.72	15.18032	3.22	25.02812	3.72	41.26439	4.22	68.03348	4.72	112.16825
2.73	15.33289	3.23	25.27966	3.73	41.67911	4.23	68.71723	4.73	113.29556
2.74	15.48698	3.24	25.53372	3.74	42.09799	4.24	69.40785	4.74	114.43420
2.75	15.64263	3.25	25.79034	3.75	42.52108	4.25	70.10541	4.75	115.58428
2.76	15.79984	3.26	26.04954	3.76	42.94843	4.26	70.80998	4.76	116.74592
2.77	15.95863	3.27	26.31134	3.77	43.38006	4.27	71.52163	4.77	117.91924
2.78	16.11902	3.28	26.57577	3.78	43.81604	4.28	72.24044	4.78	119.10435
2.79	16.28102	3.29	26.84286	3.79	44.25640	4.29	72.96647	4.79	120.30136
2.80	16.44465	3.30	27.11264	3.80	44.70118	4.30	73.69979	4.80	121.51041
2.81	16.60992	3.31	27.38512	3.81	45.15044	4.31	74.44049	4.81	122.73161
2.82	16.77685	3.32	27.66035	3.82	45.60421	4.32	75.18863	4.82	123.96509
2.83	16.94546	3.33	27.93834	3.83	46.06254	4.33	75.94429	4.83	125.21096
2.84	17.11577	3.34	28.21913	3.84	46.52547	4.34	76.70754	4.84	126.46935
2.85	17.28778	3.35	28.50273	3.85	46.99306	4.35	77.47846	4.85	127.74039
2.86	17.46153	3.36	28.78919	3.86	47.46535	4.36	78.25713	4.86	129.02420
2.87	17.63702	3.37	29.07853	3.87	47.94238	4.37	79.04363	4.87	130.32091
2.88	17.81427	3.38	29.37077	3.88	48.42421	4.38	79.83803	4.88	131.63066
2.89	17.99331	3.39	29.66595	3.89	48.91089	4.39	80.64042	4.89	132.95357
2.90	18.17414	3.40	29.96410	3.90	49.40245	4.40	81.45087	4.90	134.28978
2.91	18.35680	3.41	30.26524	3.91	49.89895	4.41	82.26946	4.91	135.63941
2.92	18.54129	3.42	30.56941	3.92	50.40044	4.42	83.09628	4.92	137.00261
2.93	18.72763	3.43	30.87664	3.93	50.90698	4.43	83.93141	4.93	138.37951
2.94	18.91585	3.44	31.18696	3.94	51.41860	4.44	84.77494	4.94	139.77024
2.95	19.10595	3.45	31.50039	3.95	51.93537	4.45	85.62694	4.95	141.17496
2.96	19.29797	3.46	31.81698	3.96	52.45732	4.46	86.48751	4.96	142.59379
2.97	19.49192	3.47	32.13674	3.97	52.98453	4.47	87.35672	4.97	144.02688
2.98	19.68782	3.48	32.45972	3.98	53.51703	4.48	88.23467	4.98	145.47438
2.99	19.88568	3.49	32.78595	3.99	54.05489	4.49	89.12144	4.99	146.93642

TABLE 5.3
The exponential function (*continued*)

x	e^x	x	e^x	x	e^x	x	e^x
5.00	148.41316	7.00	1096.63309	9.00	8103.08295	11.00	59874.13477
5.10	164.02190	7.10	1211.96703	9.10	8955.29187	11.10	66171.15430
5.20	181.27224	7.20	1339.43076	9.20	9897.12830	11.20	73130.43652
5.30	200.33680	7.30	1480.29985	9.30	10938.01868	11.30	80821.63379
5.40	221.40641	7.40	1635.98439	9.40	12088.38409	11.40	89321.72168
5.50	244.69192	7.50	1808.04231	9.50	13359.72522	11.50	98715.75879
5.60	270.42640	7.60	1998.19582	9.60	14764.78015	11.60	109097.78906
5.70	298.86740	7.70	2208.34796	9.70	16317.60608	11.70	120571.70605
5.80	330.29955	7.80	2440.60187	9.80	18033.74414	11.80	133252.34570
5.90	365.03746	7.90	2697.28226	9.90	19930.36987	11.90	147266.62109
6.00	403.42877	8.00	2980.95779	10.00	22026.46313		
6.10	445.85775	8.10	3294.46777	10.10	24343.00708		
6.20	492.74903	8.20	3640.95004	10.20	26903.18408		
6.30	544.57188	8.30	4023.87219	10.30	29732.61743		
6.40	601.84502	8.40	4447.06665	10.40	32859.62500		
6.50	665.14159	8.50	4914.76886	10.50	36315.49854		
6.60	735.09516	8.60	5431.65906	10.60	40134.83350		
6.70	812.40582	8.70	6002.91180	10.70	44355.85205		
6.80	897.84725	8.80	6634.24371	10.80	49020.79883		
6.90	992.27469	8.90	7331.97339	10.90	54176.36230		

Comment The letter e given in Formula 4 is used often in mathematics to represent a number whose value is approximately 2.71828.... The values of e to the appropriate power can be determined from Table 5.3.

EXAMPLE 5 John has $4000 that he can deposit in one of three banks, all of which pay 9% yearly interest. However, one bank compounds the interest annually, one compounds it quarterly, and one compounds it continuously. Compute the different amounts of interest that can be earned in each of these banks in 3 years.

SOLUTION We will use the compound-interest formula, $A = P(1 + r)^n$. For the bank that compounds its interest annually, there will be 3 paying periods (one for each year), so that $n = 3$. Then

$$A = \$4000(1 + 0.09)^3$$
$$= 4000(1.09)^3$$
$$= \$5180.12$$

For the bank that compounds its interest quarterly there will be 12 paying periods (4 for each year), so that $n = 12$. Also the rate of interest is 9% per year or $\dfrac{9\%}{4} = 2.25\%$ per paying period. Then

$$A = \$4000(1 + 0.0225)^{12}$$
$$= 4000(1.0225)^{12}$$
$$= \$5224.20$$

For the bank that compounds its interest continuously, we use Formula 4, which is $A = Pe^{rn}$. The money will remain in the bank for 3
years, so that $n = 3$. (When using Formula 3, we express n in years.
The same is true for the interest rate, r.) Then

$$A = 4000e^{0.27}$$
$$= 4000(1.30996)$$
$$= \$5239.84$$

In this problem the difference between compounding continuously
and compounding annually amounts to $5239.84 − $5180.12, or
$59.72. ▢

Many banks now quote interest rates with the expression "effective interest rate." Using the above notation, we have the following.

Formula 5   *The **effective annual interest rate** corresponding to an interest rate of r compounded continuously is given by $e^r − 1$.*

Thus if a bank pays 6% interest compounded continuously, the effective annual interest rate is

$$e^{0.06} − 1 = 1.0618 − 1$$
$$= 0.0618, \text{ or } 6.18\%$$

This means that a 6% interest rate compounded continuously will
yield the same amount of money to the depositor as a 6.18% interest
rate compounded annually.

EXAMPLE 6   What is the effective annual yield of an 8% interest rate compounded
continuously?

SOLUTION   We will use Formula 5. We have $r = 0.08$, so

$$e^r − 1 = e^{0.08} − 1$$
$$= 1.08329 − 1$$
$$= 0.08329$$

Thus the effective annual interest is 0.08329, or 8.329%. ▢

EXERCISES

1. Find the simple interest on a loan of $8000 that
is borrowed for 3 years at an 18% annual interest rate.

2. Find the simple interest on a loan of $8800 that
is borrowed for 20 months at an 18% annual
interest rate.

3. Leslie borrowed $4800 from his friend and

repaid $5578 to his friend in six months. What
was the (simple) annual rate of interest?

4. Jennifer borrowed $4000 from a finance com-
pany. She repaid the loan in 5 months and was
charged $450 as interest. What was the annual
rate of interest?

5. Heather borrowed $6000 from a cousin and

repaid her $6560. The interest was 24% per year. How long did it take Heather to repay the loan?

6. If $3000 is deposited in a bank that pays 8% annual interest compounded quarterly, how much money will accumulate in 6 years?

7. What is the effective annual interest rate of a 7% interest rate compounded continuously?

8. What is the effective annual interest rate of a $6\frac{1}{4}$% interest rate compounded continuously?

9. Marlene deposits $5000 in a time deposit account of a bank that pays 10% interest compounded annually. If Marlene plans to leave the money in the bank for 11 years, how much money will she have in the account?

10. *Financing a College Education.* Charlotte deposits $3000 in a bank that pays 8% annual interest compounded quarterly. She plans to leave the money in the bank for 15 years, at which time she will use it to finance her son's college education. How much money will Charlotte have then?

11. *Individual Retirement Accounts (IRA).* Judy opens an Individual Retirement Account by making an initial deposit of $2000. Judy is $39\frac{1}{2}$ years old. According to government regulations, Judy may not withdraw any of the money until she reaches $59\frac{1}{2}$ years. If Judy makes no other deposits and the bank pays 9% interest compounded quarterly, how much money will she have when she reaches $59\frac{1}{2}$ years of age?

12. Loretta is about to deposit $8000 in a bank that pays 7% annual interest. If the money will be kept in the bank for 9 years, then find the total amount that will accumulate if the bank compounds its interest (a) annually; (b) quarterly; (c) daily; (d) continuously.

13. Scott borrowed some money from a finance company that charged him $480 interest. The loan was for 8 months. If the finance company charges simple interest at the rate of 22% per year, then how much money did Scott borrow from the company?

14. Richard bought a used car from a dealer who charges interest at the rate of 24% compounded per year. Richard repaid the loan in 8 months and was charged $320 interest. How much money did Richard borrow from the dealer?

15. Miguel invests $2000 in a bank that pays $8\frac{3}{4}$% annual interest compounded quarterly. Julia invests $2000 in a bank that pays 8.85% annual interest compounded semiannually. After one year who will have more money?

16. Is it wiser to invest money in a bank that pays $6\frac{3}{4}$% annual interest compounded quarterly or in a bank that pays 7% annual interest compounded semiannually?

5.6

INSTALLMENT BUYING AND MORTGAGES

Often people buy things on an installment plan. For example, suppose Matthew wishes to buy a television set that costs $500. Furthermore, suppose he has only $100 available. The store may agree to sell Matthew the television set on the **installment plan.** Under this plan, Matthew will pay $100 **down** (as a **down payment**) plus an additional charge in a series of regular payments (usually monthly). If the monthly charge is $36 and these payments are spread over a 20-month period, then the total cost of the television set can be found by multiplying the monthly charge by the number of payments and then adding the down payment. In our case we have

$$\$36 \times 20 = \$720 \qquad \text{Total amount of monthly payments}$$
$$+ \ \underline{\$100} \qquad \text{Down payment}$$
$$\$820 \qquad \text{Total cost of television set.}$$

Thus Matthew is paying \$500 for the television and \$820–\$500, or \$320, as interest.

EXAMPLE 1 Martha has just purchased a \$1500 piano. She has agreed to make a down payment of \$200 and to pay the balance on the installment plan by making 15 monthly payments of \$110. What is the total cost of the piano (including interest)?

SOLUTION Martha will make 15 monthly payments of \$110 each. Thus her monthly payments will amount to \$1650. To this we must add the initial down payment of \$200. Thus the total cost of the piano is \$1650 + 200 = \$1850. Of this amount, \$1500 is for the piano and \$350 is for interest. ☐

EXAMPLE 2 Bill has just purchased a used car for \$1500. He pays \$500 as a down payment and agrees to pay the \$1000 balance in 5 monthly installments. Interest is $1\frac{1}{2}\%$ per month on any unpaid balance. What is the total cost of the car including interest?

SOLUTION The first monthly payment is for the \$1000 balance. Since the interest is $1\frac{1}{2}\%$ per month on any unpaid balance, the interest is

$$1000 \times 0.015 = \$15$$

so Bill owes \$1015. If Bill makes a payment of \$215, the next bill will be for \$1015 − \$215, or \$800. The interest charge for the second month is $1\frac{1}{2}\%$ per month on the unpaid balance of \$800, or

$$800 \times 0.015 = \$12$$

so Bill owes \$812. If Bill makes a payment of \$215, the next bill will be for \$812 − \$215, or \$597. The interest charge for the third month is $1\frac{1}{2}\%$ per month on the unpaid balance of \$597, or

$$597 \times 0.015 = \$8.96$$

so Bill owes \$605.96. If Bill again makes a payment of \$215, the next bill will be for \$605.96 − \$215 = \$390.96. The interest charge for the fourth month is $1\frac{1}{2}\%$ per month on the unpaid balance of $390.96, or

$$390.96 \times 0.015 = \$5.86$$

so Bill owes $396.82. If Bill makes a payment of $215, the next bill will be for $396.82 - $215 = $181.82. The interest charge for the fifth month is $1\frac{1}{2}\%$ per month on the unpaid balance of $181.82, or

$$181.82 \times 0.015 = \$2.73$$

so Bill owes $184.55. Bill pays this completely. Bill has paid a total of $500 + $215 + $215 + $215 + $215 + $184.55, or $1544.55. Of this amount, $1500 is for the car and $44.55 is for interest.   □

Comment   Most banks and finance companies use a scheme similar to the one outlined in the preceding examples to compute the interest charges.

Often an individual who wishes to buy a home does not have enough money to pay for the house entirely. He or she then arranges for a long-term **mortgage** from a bank and agrees to repay the mortgage by making equal periodic payments to the bank. In this case, each periodic payment includes partial repayment on the principal plus interest payments on the declining balance of the principal. The process of making payments under these conditions is called **amortization.**

It should be noted that although each periodic payment is the same, the percentage of the periodic payments that is used to pay for the interest charge and the percentage of the amount that is used to repay the principal will change. Thus in the early years of a mortgage, most of the periodic payments are used to pay off the interest for the loan over the entire period, and relatively little is used to repay the principal. However, in the later years, almost all of each periodic payment is used to repay the principal. This fact is clearly illustrated in Table 5.4, in which we have given the amortization schedule for a mortgage loan of $130,000 at a 9.5% rate of interest to be paid over a 15-year period.

How do we determine the periodic payments necessary to amortize (pay off under the conditions described) a loan? Fortunately, mathematicians have compiled many charts that simplify the computations considerably. One such chart is given in Table 5.5. We illustrate the use of this chart with several examples.

EXAMPLE 3   *Paying a Mortgage.* Martha Galzen arranged for a 15-year mortgage for $34,000 with the Second National City Bank. She decided to amortize the loan by making equal payments to the bank every 3 months.

Interest is 8% a year compounded quarterly. How much money will Martha have to pay the bank every 3 months?

SOLUTION   Since the mortgage is to run for 15 years and there are to be 4 payments per year, there will be a total of 15 × 4, or 60, payment periods. Interest is 8% a year compounded quarterly, or $\frac{8\%}{4}$ = 2% per payment period. Now we use Table 5.5, with 60 payment periods and an interest rate of 2%. The chart value is 0.028768. We multiply the chart value by the amount of the mortgage, getting

$$34000(0.028768) = 978.11$$

Thus Martha will have to pay the bank $978.11 every 3 months.

EXAMPLE 4   *Buying a Television on Installment.*   Mack Jones bought a color TV for $489. He made a down payment of $89 and agreed to pay the $400 balance in equal monthly payments over a 3-year period. Interest is 18% a year compounded monthly. How much money must Mack pay monthly in order to amortize the loan over the 3-year period?

SOLUTION   Since Mack will repay the loan over a 3-year period, there will be 3 × 12, or 36, payment periods. The interest rate is 18% a year, or $\frac{18\%}{12}$ = 1.5% a month. Now we use Table 5.5. The chart value for 36 payment periods and a $1\frac{1}{2}\%$ rate of interest is 0.036152. We multiply the chart value by the amount of the balance getting

$$400(0.036152) = 14.46 \text{ (rounded off).}$$

Thus his monthly payment is $14.46. Over the 3 years, Mack will pay 36 × $14.46, or $520.56. The original loan was for $400. Therefore he will pay $120.56 in interest.

EXAMPLE 5   To combat the rise in shoplifting, the Rochelle Department Store purchased $400,000 worth of closed-circuit television monitors for its several stores. The department store chain agreed to pay for the equipment by making equal monthly payments to the manufacturer over a 5-year period. Interest is 9% a year, compounded monthly. What are the monthly payments?

SOLUTION   Since the loan will be repaid over a 5-year period, there will be a total of 12 × 5, or 60, payment periods. The interest rate is $\frac{9\%}{12}$, or $\frac{3}{4}\%$ per month. Now we use Table 5.5. The chart value is 0.020758. Since the loan was for $400,000, the monthly payments will be

$$400,000(0.020758) = \$8303.20$$

TABLE 5.4

Mortgage amortization program

Mortgage Amount	$130,000
Fixed Interest Rate	9.5%
Number of Years	15
Monthly Payments	$1357.50

Payment Number	Date	Principal	Interest	Balance	Payment Number	Date	Principal	Interest	Balance
1	7/1/86	$328.32	$1,029.18	$129,671.68	31	1/1/89	415.95	941.55	118,515.52
2	8/1/86	330.92	1,026.58	129,340.76	32	2/1/89	419.24	938.26	118,096.28
3	9/1/86	333.54	1,023.96	129,007.22	33	3/1/89	422.56	934.94	117,673.72
4	10/1/86	336.18	1,021.32	128,671.04	34	4/1/89	425.91	931.59	117,247.81
5	11/1/86	338.84	1,018.66	128,332.20	35	5/1/89	429.28	928.22	116,818.53
6	12/1/86	341.53	1,015.97	127,990.67	36	6/1/89	432.68	924.82	116,385.85
7	1/1/87	344.23	1,013.27	127,646.44	37	7/1/89	436.10	921.40	115,949.75
8	2/1/87	346.96	1,010.54	127,299.48	38	8/1/89	439.55	917.95	115,510.20
9	3/1/87	349.70	1,007.80	126,949.78	39	9/1/89	443.03	914.47	115,067.17
10	4/1/87	352.47	1,005.03	126,597.31	40	10/1/89	446.54	910.96	114,620.63
11	5/1/87	355.26	1,002.24	126,242.05	41	11/1/89	450.08	907.42	114,170.55
12	6/1/87	358.07	999.43	125,883.98	42	12/1/89	453.64	903.86	113,716.91
13	7/1/87	360.91	996.59	125,523.07	43	1/1/90	457.23	900.27	113,259.68
14	8/1/87	363.77	993.73	125,159.30	44	2/1/90	460.85	896.65	112,798.83
15	9/1/87	366.65	990.85	124,792.65	45	3/1/90	464.50	893.00	112,334.33
16	10/1/87	369.55	987.95	124,423.10	46	4/1/90	468.18	889.32	111,866.15
17	11/1/87	372.47	985.03	124,050.63	47	5/1/90	471.88	885.62	111,394.27
18	12/1/87	375.42	982.08	123,675.21	48	6/1/90	475.62	881.88	110,918.65
19	1/1/88	378.39	979.11	123,296.82	49	7/1/90	479.38	878.12	110,439.27
20	2/1/88	381.39	976.11	122,915.43	50	8/1/90	483.18	874.32	109,956.09
21	3/1/88	384.41	973.09	122,531.02	51	9/1/90	487.00	870.50	109,469.09
22	4/1/88	387.45	970.05	122,143.57	52	10/1/90	490.86	866.64	108,978.33
23	5/1/88	390.52	966.98	121,753.05	53	11/1/90	494.75	862.75	108,483.48
24	6/1/88	393.61	963.89	121,359.32	54	12/1/90	498.66	858.84	107,984.82
25	7/1/88	396.73	960.77	120,962.59	55	1/1/91	502.61	854.89	107,482.21
26	8/1/88	399.87	957.63	120,562.72	56	2/1/91	506.59	850.91	106,975.62
27	9/1/88	403.03	954.47	120,159.69	57	3/1/91	510.60	846.90	106,465.02
28	10/1/88	406.22	951.28	119,753.47	58	4/1/91	514.64	842.86	105,950.30
29	11/1/88	409.44	948.06	119,344.03	59	5/1/91	518.72	838.78	105,431.66
30	12/1/88	412.68	944.82	118,931.35	60	6/1/91	522.82	834.68	104,908.84

Payment Number	Date	Principal	Interest	Balance
61	7/1/91	526.96	830.54	104,381.88
62	8/1/91	531.13	826.37	103,850.75
63	9/1/91	535.34	822.16	103,315.41
64	10/1/91	539.58	817.92	102,775.83
65	11/1/91	543.85	813.65	102,231.98
66	12/1/91	548.15	809.35	101,683.83
67	1/1/92	552.49	805.01	101,131.34
68	2/1/92	556.87	800.63	100,574.47
69	3/1/92	561.28	796.22	100,013.19
70	4/1/92	565.72	791.78	99,447.47
71	5/1/92	570.20	787.30	98,877.27
72	6/1/92	574.71	782.79	98,302.56
73	7/1/92	579.26	778.24	97,723.30
74	8/1/92	583.85	773.65	97,139.45
75	9/1/92	588.47	769.03	96,550.98
76	10/1/92	593.13	764.37	95,957.85
77	11/1/92	597.82	759.68	95,360.03
78	12/1/92	602.56	754.94	94,757.47
79	1/1/93	607.33	750.17	94,150.14
80	2/1/93	612.13	745.37	93,538.01
81	3/1/93	616.98	740.52	92,921.03
82	4/1/93	621.87	735.63	92,299.16
83	5/1/93	626.79	730.71	91,672.37
84	6/1/93	631.75	725.75	91,040.62
85	7/1/93	636.75	720.75	90,403.87
86	8/1/93	641.79	715.71	89,762.08
87	9/1/93	646.87	710.63	89,115.21
88	10/1/93	651.99	705.51	88,463.22
89	11/1/93	657.16	700.34	87,806.06
90	12/1/93	662.36	695.14	87,143.70
91	1/1/94	667.60	689.90	86,475.96
92	2/1/94	672.89	684.61	85,803.21
93	3/1/94	678.21	679.29	85,125.00
94	4/1/94	683.58	673.92	84,441.42
95	5/1/94	689.00	668.50	83,752.42
96	6/1/94	694.45	663.05	83,057.97
97	7/1/94	699.95	657.55	82,358.02
98	8/1/94	705.49	652.01	81,652.53
99	9/1/94	711.07	646.43	80,941.46

Payment Number	Date	Principal	Interest	Balance
100	10/1/94	716.70	640.80	80,224.76
101	11/1/94	722.38	635.12	79,502.38
102	12/1/94	728.10	629.40	78,774.28
103	1/1/95	733.86	623.64	78,040.42
104	2/1/95	739.67	617.83	77,300.75
105	3/1/95	745.53	611.97	76,555.22
106	4/1/95	751.43	606.07	75,803.79
107	5/1/95	757.38	600.12	75,046.41
108	6/1/95	763.37	594.13	74,283.04
109	7/1/95	769.42	588.08	73,513.62
110	8/1/95	775.51	581.99	72,738.11
111	9/1/95	781.65	575.85	71,956.46
112	10/1/95	787.83	569.67	71,168.63
113	11/1/95	794.07	563.43	70,374.56
114	12/1/95	800.36	557.14	69,574.20
115	1/1/96	806.69	550.81	68,767.51
116	2/1/96	813.08	544.42	67,954.43
117	3/1/96	819.52	537.98	67,134.91
118	4/1/96	826.01	531.49	66,308.90
119	5/1/96	832.54	524.96	65,476.36
120	6/1/96	839.14	518.36	64,637.22
121	7/1/96	845.78	511.72	63,791.44
122	8/1/96	852.47	505.03	62,938.97
123	9/1/96	859.22	498.28	62,079.75
124	10/1/96	866.03	491.47	61,213.72
125	11/1/96	872.88	484.62	60,340.84
126	12/1/96	879.79	477.71	59,461.05
127	1/1/97	886.76	470.74	58,574.29
128	2/1/97	893.78	463.72	57,680.51
129	3/1/97	900.85	456.65	56,779.66
130	4/1/97	907.98	449.52	55,871.68
131	5/1/97	915.17	442.33	54,956.51
132	6/1/97	922.42	435.08	54,034.09
133	7/1/97	929.72	427.78	53,104.37
134	8/1/97	937.08	420.42	52,167.29
135	9/1/97	944.50	413.00	51,222.79
136	10/1/97	951.98	405.52	50,270.81
137	11/1/97	959.51	397.99	49,311.30
138	12/1/97	967.11	390.39	48,344.19

TABLE 5.4
Mortgage amortization program (continued)

Payment Number	Date	Principal	Interest	Balance	Payment Number	Date	Principal	Interest	Balance
139	1/1/98	974.77	382.73	47,369.42	160	10/1/99	1,150.32	207.18	25,019.03
140	2/1/98	982.48	375.02	46,386.94	161	11/1/99	1,159.42	198.08	23,859.61
141	3/1/98	990.26	367.24	45,396.68	162	12/1/99	1,168.60	188.90	22,691.01
142	4/1/98	998.10	359.40	44,398.58	163	1/1/00	1,177.85	179.65	21,513.16
143	5/1/98	1,006.00	351.50	43,392.58	164	2/1/00	1,187.18	170.32	20,325.98
144	6/1/98	1,013.97	343.53	42,378.61	165	3/1/00	1,196.58	160.92	19,129.40
145	7/1/98	1,021.99	335.51	41,356.62	166	4/1/00	1,206.05	151.45	17,923.35
146	8/1/98	1,030.08	327.42	40,326.54	167	5/1/00	1,215.60	141.90	16,707.75
147	9/1/98	1,038.24	319.26	39,288.30	168	6/1/00	1,225.22	132.28	15,482.53
148	10/1/98	1,046.46	311.04	38,241.84	169	7/1/00	1,234.92	122.58	14,247.61
149	11/1/98	1,054.74	302.76	37,187.10	170	8/1/00	1,244.70	112.80	13,002.91
150	12/1/98	1,063.09	294.41	36,124.01	171	9/1/00	1,254.55	102.95	11,748.36
151	1/1/99	1,071.51	285.99	35,052.50	172	10/1/00	1,264.48	93.02	10,483.88
152	2/1/99	1,079.99	277.51	33,972.51	173	11/1/00	1,274.49	83.01	9,209.39
153	3/1/99	1,088.54	268.96	32,883.97	174	12/1/00	1,284.58	72.92	7,924.81
154	4/1/99	1,097.16	260.34	31,786.81	175	1/1/01	1,294.75	62.75	6,630.06
155	5/1/99	1,105.84	251.66	30,680.97	176	2/1/01	1,305.00	52.50	5,325.06
156	6/1/99	1,114.60	242.90	29,566.37	177	3/1/01	1,315.33	42.17	4,009.73
157	7/1/99	1,123.42	234.08	28,442.95	178	4/1/01	1,325.75	31.75	2,683.98
158	8/1/99	1,132.32	225.18	27,310.63	179	5/1/01	1,336.24	21.26	1,347.74
159	9/1/99	1,141.28	216.22	26,169.35	180	6/1/01	1,346.82	10.68	0.92

TABLE 5.5
Amortization payments table

n	$\frac{1}{4}\%$	$\frac{1}{2}\%$	$\frac{3}{4}\%$	1%	$1\frac{1}{4}\%$	$1\frac{1}{2}\%$	$1\frac{3}{4}\%$	2%	$2\frac{1}{2}\%$	3%
1	1.002500	1.005000	1.007500	1.010000	1.012500	1.015000	1.017500	1.020000	1.025000	1.030000
2	0.501876	0.503753	0.505632	0.507512	0.509394	0.511278	0.513163	0.515050	0.518827	0.522611
3	0.335002	0.336672	0.338346	0.340022	0.341701	0.343383	0.345067	0.346755	0.350137	0.353530
4	0.251565	0.253133	0.254705	0.256281	0.257861	0.259445	0.261032	0.262624	0.265818	0.269027
5	0.201503	0.203010	0.204522	0.206040	0.207562	0.209089	0.210621	0.212158	0.215247	0.218355
6	0.168128	0.169595	0.171069	0.172548	0.174034	0.175525	0.177023	0.178526	0.181550	0.184598
7	0.144289	0.145729	0.147175	0.148628	0.150089	0.151556	0.153031	0.154512	0.157495	0.160506
8	0.126410	0.127829	0.129256	0.130690	0.132133	0.133584	0.135043	0.136510	0.139467	0.142456
9	0.112505	0.113907	0.115319	0.116740	0.118171	0.119610	0.121058	0.122515	0.125457	0.128434
10	0.101380	0.102771	0.104171	0.105582	0.107003	0.108434	0.109875	0.111327	0.114259	0.117231
11	0.092278	0.093659	0.095051	0.096454	0.097868	0.099294	0.100730	0.102178	0.105106	0.108077
12	0.084694	0.086066	0.087451	0.088849	0.090258	0.091680	0.093114	0.094560	0.097487	0.100462
13	0.078276	0.079642	0.081022	0.082415	0.083821	0.085240	0.086673	0.088118	0.091048	0.094030
14	0.072775	0.074136	0.075511	0.076901	0.078305	0.079723	0.081156	0.082602	0.085537	0.088526
15	0.068008	0.069364	0.070736	0.072124	0.073526	0.074944	0.076377	0.077825	0.080766	0.083767
16	0.063836	0.065189	0.066559	0.067945	0.069347	0.070765	0.072200	0.073650	0.076599	0.079611
17	0.060156	0.061506	0.062873	0.064258	0.065660	0.067080	0.068516	0.069970	0.072928	0.075953
18	0.056884	0.058232	0.059598	0.060982	0.062385	0.063806	0.065245	0.066702	0.069670	0.072709
19	0.053957	0.055303	0.056667	0.058052	0.059455	0.060878	0.062321	0.063782	0.066761	0.069814
20	0.051323	0.052666	0.054031	0.055415	0.056820	0.058246	0.059691	0.061157	0.064147	0.067216
21	0.048939	0.050282	0.051645	0.053031	0.054437	0.055865	0.057315	0.058785	0.061787	0.064872
22	0.046773	0.048114	0.049477	0.050864	0.052272	0.053703	0.055156	0.056631	0.059647	0.062747
23	0.044795	0.046135	0.047498	0.048886	0.050297	0.051731	0.053188	0.054668	0.057696	0.060814
24	0.042981	0.044321	0.045685	0.047073	0.048487	0.049924	0.051386	0.052871	0.055913	0.059047
25	0.041313	0.042652	0.044016	0.045407	0.046822	0.048263	0.049730	0.051220	0.054276	0.057428
26	0.039773	0.041112	0.042477	0.043869	0.045287	0.046732	0.048203	0.049699	0.052769	0.055938
27	0.038347	0.039686	0.041052	0.042446	0.043867	0.045315	0.046791	0.048293	0.051377	0.054564
28	0.037023	0.038362	0.039729	0.041124	0.042549	0.044001	0.045482	0.046990	0.050088	0.053293
29	0.035791	0.037129	0.038497	0.039895	0.041322	0.042779	0.044264	0.045778	0.048891	0.052115
30	0.034641	0.035979	0.037348	0.038748	0.041079	0.041639	0.043130	0.044650	0.047778	0.051019
31	0.033565	0.034903	0.036274	0.037676	0.039109	0.040574	0.042070	0.043596	0.046739	0.049999
32	0.032556	0.033895	0.035266	0.036671	0.038108	0.039577	0.041078	0.042611	0.045768	0.049047
33	0.031608	0.032947	0.034320	0.035727	0.037168	0.038641	0.040148	0.041687	0.044859	0.048156
34	0.030716	0.032056	0.033431	0.034840	0.036284	0.037762	0.039274	0.040819	0.044007	0.047322
35	0.029875	0.031215	0.032592	0.034004	0.035451	0.036934	0.038451	0.040002	0.043206	0.046539

(continued)

TABLE 5.5
Amortization payments table

n	$\frac{1}{4}\%$	$\frac{1}{2}\%$	$\frac{3}{4}\%$	1%	$1\frac{1}{4}\%$	$1\frac{1}{2}\%$	$1\frac{3}{4}\%$	2%	$2\frac{1}{2}\%$	3%
36	0.029081	0.030422	0.031800	0.033214	0.034665	0.036152	0.037675	0.039233	0.042452	0.045804
37	0.028330	0.029671	0.031051	0.032468	0.033923	0.035414	0.036943	0.038507	0.041741	0.045112
38	0.027618	0.028960	0.030342	0.031761	0.033220	0.034716	0.036250	0.037821	0.041070	0.044459
39	0.026943	0.028286	0.029669	0.031092	0.032554	0.034055	0.035594	0.037171	0.040436	0.043844
40	0.026302	0.027646	0.029030	0.030456	0.031921	0.033427	0.034972	0.036556	0.039836	0.043262
41	0.025692	0.027036	0.028423	0.029851	0.031321	0.032831	0.034382	0.035972	0.039268	0.042712
42	0.025111	0.026456	0.027845	0.029276	0.030749	0.032264	0.033821	0.035417	0.038729	0.042192
43	0.024557	0.025903	0.027293	0.028727	0.030205	0.031725	0.033287	0.034890	0.038217	0.041698
44	0.024029	0.025375	0.026768	0.028204	0.029686	0.031210	0.032778	0.034388	0.037730	0.041230
45	0.023523	0.024871	0.026265	0.027705	0.029190	0.030720	0.032293	0.033910	0.037268	0.040785
46	0.023040	0.024389	0.025785	0.027228	0.028717	0.030251	0.031830	0.033453	0.036827	0.040363
47	0.022578	0.023927	0.025325	0.026771	0.028264	0.029803	0.031388	0.033018	0.036407	0.039961
48	0.022134	0.023485	0.024885	0.026334	0.027831	0.029375	0.030966	0.032602	0.036006	0.039578
49	0.021709	0.023061	0.024463	0.025915	0.027416	0.028965	0.030561	0.032204	0.035623	0.039213
50	0.021301	0.022654	0.024058	0.025513	0.027018	0.028572	0.030174	0.031823	0.035258	0.038865
51	0.020909	0.022263	0.023669	0.025127	0.026636	0.028195	0.029803	0.031459	0.034909	0.038534
52	0.020532	0.021887	0.023295	0.024756	0.026269	0.027833	0.029447	0.031109	0.034574	0.038217
53	0.020169	0.021525	0.022935	0.024400	0.025917	0.027485	0.029105	0.030774	0.034254	0.037915
54	0.019820	0.021177	0.022589	0.024057	0.025578	0.027151	0.028777	0.030452	0.033948	0.037626
55	0.019483	0.020841	0.022256	0.023726	0.025251	0.026830	0.028461	0.030143	0.033654	0.037349
56	0.019159	0.020518	0.021935	0.023408	0.024937	0.026521	0.028158	0.029847	0.033372	0.037084
57	0.018845	0.020206	0.021625	0.023102	0.024635	0.026223	0.027866	0.029561	0.033102	0.036831
58	0.018543	0.019905	0.021326	0.022806	0.024343	0.025937	0.027585	0.029287	0.032842	0.036588
59	0.018251	0.019614	0.021037	0.022520	0.024062	0.025660	0.027314	0.029022	0.032593	0.036356
60	0.017969	0.019333	0.020758	0.022244	0.023790	0.025393	0.027053	0.028768	0.032353	0.036133
61	0.017696	0.019061	0.020489	0.021978	0.023528	0.025136	0.026802	0.028523	0.032123	0.035919
62	0.017431	0.018798	0.020228	0.021720	0.023274	0.024888	0.026559	0.028286	0.031901	0.035714
63	0.017176	0.018543	0.019976	0.021471	0.023029	0.024647	0.026325	0.028058	0.031688	0.035517
64	0.016928	0.018297	0.019731	0.021230	0.022792	0.024415	0.026098	0.027839	0.031482	0.035328
65	0.016688	0.018058	0.019495	0.020997	0.022563	0.024191	0.025880	0.027626	0.031285	0.035146

66	0.034971	0.031094	0.027421	0.025668	0.023974	0.022341	0.020771	0.019265	0.017826	0.016455
67	0.034803	0.030910	0.027223	0.025464	0.023764	0.022126	0.020551	0.019043	0.017602	0.016229
68	0.034642	0.030733	0.027032	0.025266	0.023560	0.021917	0.020339	0.018827	0.017384	0.016010
69	0.034486	0.030562	0.026847	0.025075	0.023363	0.021715	0.020133	0.018618	0.017172	0.015797
70	0.034337	0.030397	0.026668	0.024889	0.023172	0.021519	0.019933	0.018415	0.016967	0.015590
71	0.034193	0.030238	0.026494	0.024710	0.022987	0.021329	0.019739	0.018217	0.016767	0.015389
72	0.034054	0.030084	0.026327	0.024536	0.022808	0.021145	0.019550	0.018026	0.016573	0.015194
73	0.033921	0.029936	0.026165	0.024367	0.022634	0.020966	0.019367	0.017839	0.016384	0.015004
74	0.033792	0.029792	0.026007	0.024204	0.022465	0.020792	0.019189	0.017658	0.016201	0.014819
75	0.033668	0.029654	0.025855	0.024046	0.022301	0.020623	0.019016	0.017482	0.016022	0.014639
76	0.033548	0.029520	0.025708	0.023892	0.022141	0.020459	0.018848	0.017310	0.015848	0.014464
77	0.033433	0.029390	0.025564	0.023743	0.021987	0.020300	0.018684	0.017143	0.015679	0.014293
78	0.033322	0.029265	0.025426	0.023598	0.021836	0.020144	0.018525	0.016981	0.015514	0.014127
79	0.033215	0.029143	0.025291	0.023457	0.021690	0.019993	0.018370	0.016822	0.015354	0.013965
80	0.033112	0.029026	0.025161	0.023321	0.021548	0.019847	0.018219	0.016668	0.015197	0.013807
81	0.033012	0.028912	0.025034	0.023188	0.021410	0.019704	0.018072	0.016518	0.015044	0.013653
82	0.032916	0.028803	0.024911	0.023059	0.021276	0.019564	0.017929	0.016371	0.014896	0.013503
83	0.032823	0.028696	0.024792	0.022934	0.021145	0.019429	0.017789	0.016228	0.014750	0.013356
84	0.032733	0.028593	0.024676	0.022812	0.021018	0.019297	0.017653	0.016089	0.014609	0.013213
85	0.032647	0.028493	0.024563	0.022694	0.020894	0.019168	0.017520	0.015953	0.014470	0.013074
86	0.032563	0.028396	0.024454	0.022578	0.020773	0.019043	0.017391	0.015820	0.014335	0.012937
87	0.032482	0.028303	0.024347	0.022466	0.020656	0.018920	0.017264	0.015691	0.014203	0.012804
88	0.032404	0.028212	0.024244	0.022357	0.020541	0.018801	0.017141	0.015564	0.014074	0.012674
89	0.032328	0.028124	0.024144	0.022251	0.020430	0.018685	0.017021	0.015441	0.013948	0.012546
90	0.032256	0.028038	0.024046	0.022148	0.020321	0.018571	0.016903	0.015320	0.013825	0.012422
91	0.032185	0.027955	0.023951	0.022047	0.020215	0.018461	0.016788	0.015202	0.013705	0.012300
92	0.032117	0.027875	0.023859	0.021949	0.020112	0.018353	0.016676	0.015087	0.013587	0.012181
93	0.032051	0.027797	0.023769	0.021853	0.020011	0.018247	0.016567	0.014974	0.013472	0.012064
94	0.031987	0.027721	0.023681	0.021760	0.019913	0.018144	0.016460	0.014864	0.013360	0.011950
95	0.031926	0.027648	0.023596	0.021669	0.019817	0.018044	0.016355	0.014756	0.013249	0.011839
96	0.031866	0.027577	0.023513	0.021581	0.019723	0.017945	0.016253	0.014650	0.013141	0.011730
97	0.031809	0.027507	0.023432	0.021495	0.019632	0.017849	0.016153	0.014547	0.013036	0.011623
98	0.031753	0.027440	0.023354	0.021411	0.019543	0.017756	0.016055	0.014446	0.012932	0.011518
99	0.031699	0.027375	0.023277	0.021329	0.019456	0.017664	0.015959	0.014347	0.012831	0.011415
100	0.031647	0.027312	0.023203	0.021249	0.019371	0.017574	0.015866	0.014250	0.012732	0.011314

(continued)

TABLE 5.5
Amortization payments table *(continued)*

n	$3\frac{1}{2}\%$	4%	$4\frac{1}{2}\%$	5%	$5\frac{1}{2}\%$	6%	$6\frac{1}{2}\%$	7%	$7\frac{1}{2}\%$	8%
1	1.035000	1.040000	1.045000	1.050000	1.055000	1.060000	1.065000	1.070000	1.075000	1.080000
2	0.526401	0.530196	0.533998	0.537805	0.541618	0.545437	0.549262	0.553092	0.556928	0.560769
3	0.356934	0.360349	0.363773	0.367209	0.370654	0.374110	0.377576	0.381052	0.384538	0.388034
4	0.272251	0.275490	0.278744	0.282012	0.285294	0.288591	0.291903	0.295228	0.298568	0.301921
5	0.221481	0.224627	0.227792	0.230975	0.234176	0.237396	0.240635	0.243891	0.247165	0.250456
6	0.187668	0.190762	0.193878	0.197017	0.200179	0.203363	0.206568	0.209796	0.213045	0.216315
7	0.163544	0.166610	0.169701	0.172820	0.175964	0.179135	0.182331	0.185553	0.188800	0.192072
8	0.145477	0.148528	0.151610	0.154722	0.157864	0.161036	0.164237	0.167468	0.170727	0.174015
9	0.131446	0.134493	0.137574	0.140690	0.143839	0.147022	0.150238	0.153486	0.156767	0.160080
10	0.120241	0.123291	0.126379	0.129505	0.132668	0.135868	0.139105	0.142378	0.145686	0.149029
11	0.111092	0.114149	0.117248	0.120389	0.123571	0.126793	0.130055	0.133357	0.136697	0.140076
12	0.103484	0.106552	0.109666	0.112825	0.116029	0.119277	0.122568	0.125902	0.129278	0.132695
13	0.097062	0.100144	0.103275	0.106456	0.109684	0.112960	0.116283	0.119651	0.123064	0.126522
14	0.091571	0.094669	0.097820	0.101024	0.104279	0.107585	0.110940	0.114345	0.117797	0.121297
15	0.086825	0.089941	0.093114	0.096342	0.099626	0.102963	0.106353	0.109795	0.113287	0.116830
16	0.082685	0.085820	0.089015	0.092270	0.095583	0.098952	0.102378	0.105858	0.109391	0.112977
17	0.079043	0.082199	0.085418	0.088699	0.092042	0.095445	0.098906	0.102425	0.106000	0.109629
18	0.075817	0.078993	0.082237	0.085546	0.088920	0.092357	0.095855	0.099413	0.103029	0.106702
19	0.072940	0.076139	0.079407	0.082745	0.086150	0.089621	0.093156	0.096753	0.100411	0.104128
20	0.070361	0.073582	0.076876	0.080243	0.083679	0.087185	0.090756	0.094393	0.098092	0.101852
21	0.068037	0.071280	0.074601	0.077996	0.081465	0.085005	0.088613	0.092289	0.096029	0.099832
22	0.065932	0.069199	0.072546	0.075971	0.079471	0.083046	0.086691	0.090406	0.094187	0.098032
23	0.064019	0.067309	0.070682	0.074137	0.077670	0.081278	0.084961	0.088714	0.092535	0.096422
24	0.062273	0.065587	0.068987	0.072471	0.076036	0.079679	0.083398	0.087189	0.091050	0.094978
25	0.060674	0.064012	0.067439	0.070952	0.074549	0.078227	0.081981	0.085811	0.089711	0.093679
26	0.059205	0.062567	0.066021	0.069564	0.073193	0.076904	0.080695	0.084561	0.088500	0.092507
27	0.057852	0.061239	0.064719	0.068292	0.071952	0.075697	0.079523	0.083426	0.087402	0.091448
28	0.056603	0.060013	0.063521	0.067123	0.070814	0.074593	0.078453	0.082392	0.086405	0.090489
29	0.055445	0.058880	0.062415	0.066046	0.069769	0.073580	0.077474	0.081449	0.085498	0.089619
30	0.054371	0.057830	0.061392	0.065051	0.068805	0.072649	0.076577	0.080586	0.084671	0.088827
31	0.053372	0.056855	0.060443	0.064132	0.067917	0.071792	0.075754	0.079797	0.083916	0.088107
32	0.052442	0.055949	0.059563	0.063280	0.067095	0.071002	0.074997	0.079073	0.083226	0.087451
33	0.051572	0.055104	0.058745	0.062490	0.066335	0.070273	0.074299	0.078408	0.082594	0.086852
34	0.050760	0.054315	0.057982	0.061755	0.065630	0.069598	0.073656	0.077797	0.082015	0.086304
35	0.049998	0.053577	0.057270	0.061072	0.064975	0.068974	0.073062	0.077234	0.081483	0.085803

36	0.049284	0.052887	0.056606	0.060434	0.064366	0.068395	0.072513	0.076715	0.080994	0.085345
37	0.048613	0.052240	0.055984	0.059840	0.063800	0.067857	0.072005	0.076237	0.080545	0.084924
38	0.047982	0.051632	0.055402	0.059284	0.063272	0.067358	0.071535	0.075795	0.080132	0.084539
39	0.047388	0.051061	0.054856	0.058765	0.062780	0.066894	0.071099	0.075387	0.079751	0.084185
40	0.046827	0.050523	0.054343	0.058278	0.062320	0.066462	0.070694	0.075009	0.079400	0.083860
41	0.046298	0.050017	0.053862	0.057822	0.061891	0.066059	0.070318	0.074660	0.079077	0.083561
42	0.045798	0.049540	0.053409	0.057395	0.061489	0.065683	0.069968	0.074336	0.078778	0.083287
43	0.045325	0.049090	0.052982	0.056993	0.061113	0.065333	0.069644	0.074036	0.078502	0.083034
44	0.044878	0.048665	0.052581	0.056616	0.060761	0.065006	0.069341	0.073758	0.078247	0.082802
45	0.044453	0.048262	0.052202	0.056262	0.060431	0.064700	0.069060	0.073500	0.078011	0.082587
46	0.044051	0.047882	0.051845	0.055928	0.060122	0.064415	0.068797	0.073260	0.077794	0.082390
47	0.043669	0.047522	0.051507	0.055614	0.059831	0.064148	0.068553	0.073037	0.077592	0.082208
48	0.043306	0.047181	0.051189	0.055318	0.059559	0.063898	0.068325	0.072831	0.077405	0.082040
49	0.042962	0.046857	0.050887	0.055040	0.059302	0.063664	0.068112	0.072639	0.077232	0.081886
50	0.042634	0.046550	0.050602	0.054777	0.059061	0.063444	0.067914	0.072460	0.077072	0.081743
51	0.042322	0.046259	0.050332	0.054529	0.058835	0.063239	0.067729	0.072294	0.076924	0.081611
52	0.042024	0.045982	0.050077	0.054294	0.058622	0.063046	0.067556	0.072139	0.076787	0.081490
53	0.041741	0.045719	0.049835	0.054073	0.058421	0.062866	0.067394	0.071995	0.076659	0.081377
54	0.041471	0.045469	0.049605	0.053864	0.058232	0.062696	0.067243	0.071861	0.076541	0.081274
55	0.041213	0.045231	0.049388	0.053667	0.058055	0.062537	0.067101	0.071736	0.076432	0.081178
56	0.040967	0.045005	0.049181	0.053480	0.057887	0.062388	0.066969	0.071620	0.076330	0.081090
57	0.040732	0.044789	0.048985	0.053303	0.057729	0.062247	0.066846	0.071512	0.076236	0.081008
58	0.040508	0.044584	0.048799	0.053136	0.057580	0.062116	0.066730	0.071411	0.076148	0.080932
59	0.040294	0.044388	0.048622	0.052978	0.057440	0.061992	0.066622	0.071317	0.076067	0.080862
60	0.040089	0.044202	0.048454	0.052828	0.057307	0.061876	0.066520	0.071229	0.075991	0.080798
61	0.039892	0.044024	0.048295	0.052686	0.057182	0.061766	0.066426	0.071147	0.075921	0.080738
62	0.039705	0.043854	0.048143	0.052552	0.057064	0.061664	0.066337	0.071071	0.075856	0.080683
63	0.039525	0.043692	0.047998	0.052424	0.056953	0.061567	0.066254	0.071000	0.075796	0.080632
64	0.039353	0.043538	0.047861	0.052304	0.056847	0.061476	0.066176	0.070934	0.075740	0.080585
65	0.039188	0.043390	0.047730	0.052189	0.056748	0.061391	0.066103	0.070872	0.075688	0.080541
66	0.039030	0.043249	0.047606	0.052081	0.056654	0.061310	0.066034	0.070814	0.075639	0.080501
67	0.038879	0.043115	0.047488	0.051978	0.056565	0.061235	0.065970	0.070760	0.075594	0.080464
68	0.038734	0.042986	0.047375	0.051880	0.056482	0.061163	0.065910	0.070710	0.075553	0.080429
69	0.038595	0.042863	0.047267	0.051787	0.056402	0.061096	0.065854	0.070663	0.075514	0.080397
70	0.038461	0.042745	0.047165	0.051699	0.056328	0.061033	0.065801	0.070620	0.075478	0.080368
71	0.038333	0.042633	0.047068	0.051616	0.056257	0.060974	0.065752	0.070579	0.075444	0.080340
72	0.038210	0.042525	0.046975	0.051536	0.056190	0.060918	0.065705	0.070541	0.075413	0.080315
73	0.038092	0.042422	0.046886	0.051461	0.056127	0.060865	0.065662	0.070505	0.075384	0.080292
74	0.037978	0.042323	0.046802	0.051390	0.056067	0.060815	0.065621	0.070472	0.075357	0.080270

(continued)

TABLE 5.5
Amortization payments table (*continued*)

n	$3\frac{1}{2}\%$	4%	$4\frac{1}{2}\%$	5%	$5\frac{1}{2}\%$	6%	$6\frac{1}{2}\%$	7%	$7\frac{1}{2}\%$	8%
75	0.037869	0.042229	0.046721	0.051322	0.056010	0.060769	0.065583	0.070441	0.075332	0.080250
76	0.037765	0.042139	0.046644	0.051257	0.055956	0.060725	0.065547	0.070412	0.075304	0.080231
77	0.037664	0.042052	0.046571	0.051196	0.055906	0.060683	0.065513	0.070385	0.075287	0.082214
78	0.037567	0.041969	0.046501	0.051138	0.055858	0.060644	0.065482	0.070359	0.075267	0.080198
79	0.037474	0.041890	0.046434	0.051082	0.055812	0.060607	0.065452	0.070336	0.075248	0.080183
80	0.037385	0.041814	0.046371	0.051030	0.055769	0.060573	0.065424	0.070314	0.075231	0.080170
81	0.037299	0.041741	0.046310	0.050980	0.055729	0.060540	0.065398	0.070293	0.075215	0.080157
82	0.037216	0.041671	0.046252	0.050932	0.055690	0.060509	0.065374	0.070274	0.075200	0.080146
83	0.037137	0.041605	0.046197	0.050887	0.055654	0.060480	0.065351	0.070256	0.075186	0.080135
84	0.037060	0.041541	0.046144	0.050844	0.055619	0.060453	0.065329	0.070239	0.075173	0.080125
85	0.036987	0.041479	0.046093	0.050803	0.055587	0.060427	0.065309	0.070223	0.075161	0.080116
86	0.036916	0.041420	0.046045	0.050764	0.055556	0.060402	0.065290	0.070209	0.075150	0.080107
87	0.036848	0.041364	0.045999	0.050727	0.055527	0.060380	0.065272	0.070195	0.075139	0.080099
88	0.036782	0.041310	0.045955	0.050692	0.055499	0.060358	0.065256	0.070182	0.075129	0.090092
89	0.036719	0.041258	0.045913	0.050659	0.055473	0.060338	0.065240	0.070170	0.075120	0.080085
90	0.036658	0.041208	0.045873	0.050627	0.055448	0.060318	0.065225	0.070159	0.075112	0.080079
91	0.036599	0.041160	0.045835	0.050597	0.055424	0.060300	0.065212	0.070149	0.075104	0.080073
92	0.036543	0.041114	0.045798	0.050568	0.055402	0.060283	0.065199	0.070139	0.075097	0.080067
93	0.036488	0.041070	0.045763	0.050541	0.055381	0.060267	0.065186	0.070130	0.075090	0.080062
94	0.036436	0.041028	0.045730	0.050515	0.055361	0.060252	0.065175	0.070121	0.075084	0.080058
95	0.036385	0.040987	0.045698	0.050490	0.055342	0.060238	0.065164	0.070113	0.075078	0.080053
96	0.036337	0.040949	0.045667	0.050466	0.055324	0.060224	0.065154	0.070106	0.075072	0.080050
97	0.036290	0.040911	0.045638	0.050444	0.055307	0.060211	0.065145	0.070097	0.075067	0.080046
98	0.036245	0.040875	0.045610	0.050423	0.055291	0.060199	0.065136	0.070092	0.075063	0.080042
99	0.036201	0.040841	0.045584	0.050402	0.055276	0.060188	0.065128	0.070086	0.075058	0.080039
100	0.036159	0.040808	0.045558	0.050383	0.055261	0.060177	0.065120	0.070081	0.075054	0.080036

All of the loans and mortgages discussed up to this point are referred to as **conventional loans** or **conventional mortgages.** The money is borrowed from a commercial bank or a savings and loan association.

Over the last few years, as money has become very "tight," interest rates for home mortgage loans have risen dramatically. Many prospective home buyers, particularly younger people, have become discouraged from buying homes because of the high interest rates. To promote more home mortgage loans, lending institutions have developed several alternative forms of mortgages to suit the home buyer's pocketbook. Among some of these alternatives are the following.

Adjustable Rate Mortgages (ARM)

In this type of mortgage loan, the rate of interest is adjusted up or down at a specific interval (usually every six months). The adjusted rate depends upon the yield of one-year U.S. treasury bills. The initial interest rate is much less than that of regular (conventional) mortgages. However, the rates can be raised at every adjustment period. Some adjustable rate mortgages have special clauses that place a cap on the maximum rate of interest over the life of the mortgage.

Variable Rate Mortgages

In this type of mortgage the rate of interest is adjustable up or down and is based upon an index that reflects the cost of funds to the bank.

Points

Many lending institutions now require the prospective home buyer to pay one or more points when the house is bought, that is, at closing time. One point equals 1% of the amount of the mortgage. Thus the monthly mortgage payments will be lower, since the rate of interest on the mortgage will be lower.

Since there are so many different types of mortgages available, the prospective home buyer should shop around carefully before signing on the dotted line.

On the basis of what we have seen until now, we notice that when the interest charges are added to any item purchased on the installment plan, the true interest rate may be considerably higher than what it is initially specified by the seller. To make it easier for the consumer to determine the true interest rate, Congress passed the Truth in Lending Act in 1969. This law requires *all* lenders to accurately state the **annual percentage rate** (APR). Each lender must specify the APR, whether the lender is a local merchant, a bank, or a credit card issuer.

Lengthy and extensive tables have been computed by the Federal Reserve System for determining the APR. In Table 5.6 we present part of the APR table. We use the numbers in this table as follows.

TABLE 5.6 Annual percentage rates

Annual percentage rate / Number of Payments	6	12	18	24	30	36	42	48	60
10.00%	2.94	5.50	8.10	10.75	13.43	16.16	18.93	21.74	27.48
10.25%	3.01	5.64	8.31	11.02	13.78	16.58	19.43	22.32	28.22
10.50%	3.08	5.78	8.52	11.30	14.13	17.01	19.93	22.90	28.96
10.75%	3.16	5.92	8.73	11.58	14.48	17.43	20.43	23.48	29.71
11.00%	3.23	6.06	8.93	11.86	14.83	17.86	20.93	24.06	30.45
11.25%	3.31	6.20	9.14	12.14	15.19	18.29	21.44	24.64	31.20
11.50%	3.38	6.34	9.35	12.42	15.54	18.71	21.94	25.23	31.96
11.75%	3.45	6.48	9.56	12.70	15.89	19.14	22.45	25.81	32.71
12.00%	3.53	6.62	9.77	12.98	16.24	19.57	22.96	26.40	33.47
12.25%	3.60	6.76	9.98	13.26	16.60	20.00	23.47	26.99	34.23
12.50%	3.68	6.90	10.19	13.54	16.95	20.43	23.98	27.58	34.99
12.75%	3.75	7.04	10.40	13.82	17.31	20.87	24.49	28.18	35.77
13.00%	3.83	7.18	10.61	14.10	17.66	21.30	25.00	28.77	36.52
13.25%	3.90	7.32	10.82	14.38	18.02	21.73	25.51	29.37	37.29
13.50%	3.97	7.46	11.03	14.66	18.38	22.17	26.03	29.97	38.06
13.75%	4.05	7.60	11.24	14.95	18.74	22.60	26.55	30.57	38.83
14.00%	4.12	7.74	11.45	15.23	19.10	23.04	27.06	31.17	39.61
14.25%	4.20	7.89	11.66	15.51	19.45	23.48	27.58	31.77	40.39
14.50%	4.27	8.03	11.87	15.80	19.81	23.92	28.10	32.37	41.17
14.75%	4.35	8.17	12.08	16.08	20.17	24.35	28.62	32.98	41.95
15.00%	4.42	8.31	12.29	16.37	20.54	24.80	29.15	33.59	42.74
15.25%	4.49	8.45	12.50	16.65	20.90	25.24	29.67	34.20	43.53
15.50%	4.57	8.59	12.72	16.94	21.26	25.68	30.19	34.81	44.32
15.75%	4.64	8.74	12.93	17.22	21.62	26.12	30.72	35.42	45.11
16.00%	4.72	8.88	13.14	17.51	21.99	26.57	31.25	36.03	45.91
16.25%	4.79	9.02	13.35	17.80	22.35	27.01	31.78	36.65	46.71
16.50%	4.87	9.16	13.57	18.09	22.72	27.46	32.31	37.27	47.51
16.75%	4.94	9.30	13.78	18.37	23.08	27.90	32.84	37.88	48.31
17.00%	5.02	9.45	13.99	18.66	23.45	28.35	33.37	38.50	49.12
17.25%	5.09	9.59	14.21	18.95	23.81	28.80	33.90	39.13	49.92
17.50%	5.17	9.73	14.42	19.24	24.18	29.25	34.44	39.75	50.73
17.75%	5.24	9.87	14.64	19.53	24.55	29.70	34.97	40.37	51.55
18.00%	5.32	10.02	14.85	19.82	24.92	30.15	35.51	41.00	52.36
20.00%	5.91	11.16	16.52	22.15	27.89	33.79	39.85	46.07	58.96
25.00%	7.42	14.05	20.95	28.09	35.49	43.14	51.03	59.15	76.11
30.00%	8.93	16.98	25.41	34.19	43.33	52.83	62.66	72.83	94.12
35.00%	10.45	19.96	29.96	40.44	51.41	62.85	74.74	87.06	112.94
40.00%	11.99	22.97	34.59	46.85	59.73	73.20	87.24	101.82	132.51

RULE 5.1 **Rule for Finding the Annual Percentage Rate** *To find the actual APR, we first divide the actual interest charge (or finance charge) by the amount borrowed. We then multiply the result by 100. This gives us the interest (finance) charge per $100 of the amount being financed. Using this number, look in Table 5.6 to find the actual APR for the appropriate number of payments.*

EXAMPLE 6 Hugh Carson purchased a $7000 car with a down payment of $1200. The balance is being financed by the dealer. Hugh has agreed to make 48 monthly payments of $168.86 each. What is the APR?

SOLUTION Since Hugh made a down payment of $1200, the amount borrowed is $7000 − $1200 or $5800. He has agreed to make 48 monthly payments of $168.86 each. This amounts to 48 × $168.86 or $8105.28. Since only $5800 was borrowed, the total finance charge is

$$\$8105.28 - \$5800 = \$2305.28.$$

The actual interest (finance charge) is $2305.28, and the amount borrowed is $5800, so

$$\frac{\text{Actual interest}}{\text{Amount borrowed}} \times 100 = \frac{2305.28}{5800} \times 100 = \$39.75.$$

Thus Hugh is paying $39.75 for every $100 borrowed. We now look in Table 5.6. We look for 48 in the number of payments row. Then we move down this column until we find the number closest to 39.75. In this case, we find that 39.75 is listed exactly in this table. We look in the left column and find that the APR is 17.50%. ☐

EXAMPLE 7 Yolanda purchased a stereophonic cable-ready video cassette recorder for $945. She made a down payment of $150 and paid the balance in 18 equal installments of $49.70 each. What was the APR?

SOLUTION The actual amount borrowed is $945 − $150 or $795. Yolanda made 18 installment payments of $49.70 each. This amounts to 18 × 49.70 or $894.60. Since only $795 was borrowed, the actual interest charge is $99.60. Thus

$$\frac{\text{Actual interest}}{\text{Amount borrowed}} \times 100 = \frac{99.60}{795} \times 100 = \$12.53.$$

Thus Yolanda is paying $12.53 for every $100 borrowed. Now we look in Table 5.6. We look for 18 in the number of payments row. Then we move down this column until we find the number closest to 12.53. This number is between 12.50 and 12.72 but closer to 12.50. Thus the APR was approximately 15.25%. ☐

EXERCISES

1. Kathy wishes to purchase some stereo equipment that costs $1800. She agrees to pay $800 down and to pay the $1000 balance over a 15-month period by making monthly payments of $90. How much is Kathy really paying for the stereo equipment?

2. Steve works in a jewelry store. He wishes to buy a $3000 engagement ring for his fiancée. His boss agrees to sell him the ring by deducting $35 from Steve's weekly paycheck over a two-year period. No down payment is required. How much does the ring really cost Steve?

3. Jack has an outstanding balance of $385 with a major credit card company that charges $1\frac{1}{2}\%$ interest per month on any unpaid balance. Jack will repay the credit card company by making monthly payments of $45 until the balance is completely paid. How many months does it take Jack to repay the loan?

4. In Exercise 3, how much money does Jack end up paying the company?

5. Bruce can repay a $600 loan by making monthly payments of $60 each, where the interest is 3% per month on any unpaid balance, or by making monthly payments of $80 each, where interest is 2% per month on any unpaid balance. Which of the two methods results in a lower total interest charge?

6. Trudy joined a health club that charges $385 annual membership dues. Trudy decides to pay the $385 on the installment plan by making monthly payments of $50. Interest is $3\frac{1}{2}\%$ on any unpaid balance. How much does Trudy end up paying for the membership dues?

7. Al Jordan arranged for a 8-year conventional mortgage for $35,000 with a bank that charges 12% annual interest compounded monthly. Al agrees to amortize the loan by making equal monthly payments to the bank. What are Al's monthly payments?

8. What are the quarterly payments for a $25,000 conventional mortgage from a bank that charges 10% annual interest compounded quarterly, given that the loan is for 6 years?

9. Michael has just renovated his house. To accomplish this, he borrowed $5000 from a bank that charges interest at an annual rate of 14% compounded quarterly. Michael wants to repay the loan by making equal quarterly payments over a 5-year period. What are his quarterly payments?

10. What are the monthly payments for a $45,000 conventional mortgage from a bank that charges 12% annual interest compounded monthly given that the loan is to run for 6 years?

11. In Exercise 10, what are the periodic payments if they are to be made every month and the loan is to run for 8 years?

12. David borrowed $3000 from a finance company that charges 18% annual interest compounded monthly. David agrees to repay the loan over a 30-month period. What are his monthly payments?

13. Juan Rodrigues arranges with his college to pay for his $1400 tuition bill by making monthly payments over a 4-month period. The college will not allow Juan to register next semester until this semester's bills are paid completely. Furthermore, the college charges 9% annual interest compounded monthly. What are Juan's monthly payments?

14. David bought Christmas presents for his family amounting to $427.88. He paid the bill with his credit card. The credit card company's revolving plan requires that he pay a minimum monthly installment of $50. Interest is 18% a year ($1\frac{1}{2}\%$ a month) on any unpaid balance. David intends to pay the minimum amount until the total bill is paid. What is the *total* finance charge?

15. Linda borrowed $15,000 from her company pension plan for 3 years. As a benefit to employees, the plan charges interest at the rate of 8% a year compounded quarterly. Linda plans to repay the loan and accumulated interest charges by making quarterly payments. In the

meantime, Linda takes the $15,000 and lends it to her cousin, Jeff, as a conventional mortgage for 3 years. Jeff agrees to pay interest at the rate of 12% a year compounded quarterly and to amortize the loan by making quarterly payments to Linda. After Linda repays her loan plus interest charges, what is her profit?

16. Bridget Connors purchased a complete home computer system for $1600. She made a down payment of $400 and paid the balance over a 12-month period in 12 equal installments of $107.32 each. What was the APR?

17. *Paying Your Tuition Can Be Costly.* Jerry Bates paid $\frac{1}{8}$ of his $2400 college tuition bill at registration and agreed to pay the balance in 6 equal installments of $364.95 each.
    a) What was the finance charge?
    b) What was the APR?

18. *Bad Credit Rating.* Connie Johnson had a very bad credit rating and had to borrow $2000 from a local finance company. She repaid the loan in 12 equal installments of $200 each. What was the APR?

5.7

LIFE INSURANCE   Almost every consumer buys some form of insurance in his or her lifetime. Today, there are companies providing health insurance, liability insurance, cancer (only) insurance, fire insurance, and life insurance, to name a few. Since the basic concept of insurance is the sharing of risks or losses and insurance companies are not in business to lose money, it is important to see how companies make decisions regarding the premium to charge.

Let us see how this is done by analyzing one particular type of insurance available to the consumer: life insurance. **Mortality tables,** one of which is shown in Table 5.7, list the number of individuals living at various ages, the (average) number of deaths per 1000 at these ages, and the average number of years people of a given age can expect to live (**life expectancy**). These tables are periodically reviewed and revised by each of the insurance companies using them in order to reflect changing (favorable or unfavorable) company experiences, medical and technological advances, etc.

When an individual applies for life insurance, the company must determine the probability that the individual will die (or live) while the insurance policy is in effect. This can be easily calculated by using the data of Table 5.7.

We will illustrate the technique with several examples.

EXAMPLE 1   Find the probability that a 50-year-old individual will *not* live to be 51 years old.

SOLUTION   From Table 5.7 we find that there were 8,762,306 people alive at age 50 and 8,689,404 alive at age 51. Thus 8,762,306 − 8,689,404, or 72,902

TABLE 5.7

Commissioner's standard ordinary mortality table

Age	Number living	Deaths per 1000	Life expectancy (Number of years people of a given age can expect to live)
0	10,000,000	1.08	68.30
1	9,929,200	1.76	67.78
2	9,911,725	1.52	66.90
3	9,896,659	1.46	66.00
4	9,882,210	1.40	65.10
5	9,868,375	1.35	64.19
6	9,855,053	1.30	63.27
7	9,842,241	1.26	62.35
8	9,829,840	1.23	61.43
9	9,817,749	1.21	60.51
10	9,805,870	1.21	59.58
11	9,794,005	1.23	58.65
12	9,781,958	1.26	57.72
13	9,769,633	1.32	56.80
14	9,756,737	1.39	55.87
15	9,743,175	1.46	54.95
16	9,728,950	1.54	54.03
17	9,713,967	1.62	53.11
18	9,698,230	1.69	52.19
19	9,681,840	1.74	51.28
20	9,664,994	1.79	50.37
21	9,647,694	1.83	49.46
22	9,630,039	1.86	48.55
23	9,612,127	1.89	47.64
24	9,593,960	1.91	46.73
25	9,575,636	1.93	45.82
26	9,557,155	1.96	44.90
27	9,538,423	1.99	43.99
28	9,519,442	2.03	43.08
29	9,500,118	2.08	42.16
30	9,480,358	2.13	41.25
31	9,460,165	2.19	40.34
32	9,439,447	2.25	39.43
33	9,418,208	2.32	38.51
34	9,396,358	2.40	37.60

Age	Number living	Deaths per 1000	Life expectancy (Number of years people of a given age can expect to live)
35	9,373,807	2.51	36.69
36	9,350,279	2.64	35.78
37	9,325,594	2.80	34.88
38	9,299,482	3.01	33.97
39	9,271,491	3.25	33.07
40	9,241,359	3.53	32.18
41	9,208,737	3.84	31.29
42	9,173,375	4.17	30.41
43	9,135,122	4.53	29.54
44	9,093,740	4.92	28.67
45	9,048,999	5.35	27.81
46	9,000,587	5.83	26.95
47	8,948,114	6.36	26.11
48	8,891,204	6.95	25.27
49	8,829,410	7.60	24.45
50	8,762,306	8.32	23.63
51	8,689,404	9.11	22.82
52	8,610,244	9.96	22.03
53	8,524,486	10.89	21.25
54	8,431,654	11.90	20.47
55	8,331,317	13.00	19.71
56	8,223,010	14.21	18.97
57	8,106,161	15.54	18.23
58	7,980,191	17.00	17.51
59	7,844,528	18.59	16.81
60	7,698,698	20.34	16.12
61	7,542,106	22.24	15.44
62	7,374,370	24.31	14.78
63	7,195,099	26.57	14.14
64	7,003,925	29.04	13.51
65	6,800,531	31.75	12.90
66	6,584,614	34.74	12.31
67	6,355,865	38.04	11.73
68	6,114,088	41.68	11.17
69	5,859,253	45.61	10.64

(continued)

TABLE 5.7
Commissioner's standard ordinary mortality table (*continued*)

Age	Number living	Deaths per 1000	Life expectancy (Number of years people of a given age can expect to live)
70	5,592,012	49.79	10.12
71	5,313,586	54.15	9.63
72	5,025,855	58.65	9.15
73	4,731,089	63.26	8.69
74	4,431,800	68.12	8.24
75	4,129,906	73.37	7.81
76	3,826,895	79.18	7.39
77	3,523,881	85.70	6.98
78	3,221,884	93.06	6.59
79	2,922,055	101.19	6.21
80	2,626,372	109.98	5.85
81	2,337,524	119.35	5.51
82	2,058,541	129.17	5.19
83	1,792,639	139.38	4.89
84	1,542,781	150.01	4.60
85	1,311,348	161.14	4.32
86	1,100,037	172.82	4.06
87	909,929	185.13	3.80
88	741,474	198.25	3.55
89	594,477	212.46	3.31
90	468,174	228.14	3.06
91	361,365	245.77	2.82
92	272,552	265.93	2.58
93	200,072	289.30	2.38
94	142,191	316.66	2.07
95	97,165	351.24	1.80
96	63,037	400.56	1.51
97	37,787	488.42	1.18
98	19,331	668.15	0.83
99	6,415	1000.00	0.50

people died between their 50th and 51st birthdays. Thus the probability that a 50-year-old individual will not live to be 51 years old is

$$\frac{72,902}{8,762,306} = 0.00832$$

The same probability can be obtained by referring to the "Deaths per 1000" column of Table 5.7. According to the table, there were 8.32 deaths per 1000 for an individual aged 50 years old. □

EXAMPLE 2   Find the probability that a 45-year-old person will live to be 60 years old.

SOLUTION   From Table 5.7 we find that there were 9,048,999 people alive at age 45 and 7,698,698 alive at age 60. Thus 7,698,698 of the original 9,048,999 remained alive, so that the probability that a 45-year-old individual will live to be 60 years old is

$$\frac{7,698,698}{9,048,999} = 0.8508$$ □

EXAMPLE 3   George Smith is 85 years old. Find (a) George's life expectancy; (b) the probability that George will live to be 90 years old.

SOLUTION   **a)** By referring to the fourth column of Table 5.7 we find that the life expectancy of an 85-year-old individual is 4.32 years. Thus a person of George's age can expect to live, on the average,

$$85 + 4.32 \quad \text{or} \quad 89.32 \text{ years}.$$

b) From Table 5.7 we find that there were 1,311,348 people living at age 85 and 468,174 of them at age 90. Thus the probability that 85-year-old George will live to be 90 years old is

$$\frac{468,174}{1,311,348} = 0.3570$$ □

Comment   It is important to properly interpret the data given in Table 5.7. For example, according to Table 5.7, the life expectancy of a 30-year-old individual is 41.25 years. This *does not guarantee* that a 30-year-old individual will definitely live an additional 41.25 years. On the contrary, the individual may die within a year. However, in the long run, when many individuals are considered, then a 30-year-old person will live, on the average, an additional 41.25 years.

We can summarize the results of the previous examples in the following formula.

Formula 6   *Let N_a be the number of individuals alive at age a years old and let N_b be the number of individuals alive at age b years old. Then the probability that an individual who is a years old will*

i) *live to be b years old is*

$$\frac{N_b}{N_a}$$

ii) *not live to be b years old is*

$$\frac{N_a - N_b}{N_a}$$

where the values of N_a and N_b can be obtained by referring to Table 5.7.

Once an insurance company knows the probability that an individual who is *a* years old will live to be *b* years old, the next thing to determine is the premium to charge for life insurance.

In our discussion we will neglect any overhead costs, costs associated with building up a reserve, and any other incidental costs. In reality, the premium actually charged to an individual must cover both mortality costs and all other costs.

To calculate the premium charges, let us assume that Hiram Kirkland who is 47 years old applies to a life insurance company for a one-year $10,000 life insurance policy. Thus if he dies within the year, the company will pay his beneficiary $10,000. How much should the company charge for such a policy? It should be obvious that if Hiram *does not die* within the year, then the premium paid will be the profit to the company. On the other hand, if Hiram does die within the year, then the company will have to pay $10,000 to his beneficiary. Obviously, the company will then lose money since the premium will be considerably less than the amount the company must pay out.

Since Hiram will live or die within the year, we can calculate the company's expected profit as follows: Let *p* be the premium charged. Then we have the following.

If Hiram lives, then the profit to the company is	If Hiram dies within a year, then the profit to the company is
p	$p - 10,000$

Comment   If Hiram dies within a year, then $p - 10,000$ will be a negative number. This represents the company's loss.

Let us now calculate the probability that a 47-year-old person will live to be 48 years old. Using Formula 6 and Table 5.7, we have

probability that a 47-year-old person will live to be 48 years old $= \dfrac{N_{48}}{N_{47}} = \dfrac{8,891,204}{8,948,114} = 0.99364$

and

$$\begin{pmatrix}\text{probability that a 47-year-}\\ \text{old person will } not \text{ live to be}\\ \text{48 years old}\end{pmatrix} = \frac{N_{47} - N_{48}}{N_{47}}$$

$$= \frac{8,948,114 - 8,891,204}{8,948,114}$$

$$= 0.00636$$

Then the expected profit for the insurance company is

$$\begin{pmatrix}\text{Profit if}\\ \text{Hiram lives}\end{pmatrix} \cdot \begin{pmatrix}\text{Probability that}\\ \text{Hiram will live}\end{pmatrix} + \begin{pmatrix}\text{Profit if}\\ \text{Hiram dies}\end{pmatrix} \cdot \begin{pmatrix}\text{Probability that}\\ \text{Hiram will not live}\end{pmatrix}$$

$$= p(0.99364) + (p - 10,000) \cdot (0.00636)$$

$$= 0.99364p + 0.00636p - (10,000) \cdot (0.00636) \quad \text{(by the distributive law)}$$

$$= p - 63.60. \quad \text{(combining terms)}$$

If the company wants to break even, then its expected profit is 0. Thus we solve the equation

$$0 = p - 63.60$$

for p. We get

$$p = \$63.60$$

This means that if the company charges $63.60 as its premium for every 47-year-old person who applies for a one-year $10,000 life insurance policy, then in *the long run* the company will break even. In order to make a profit, the premium charged by the company must be more than $63.60

Comment   As we indicated earlier, our calculations do not take into consideration any overhead costs, costs associated with building up a reserve fund, and any other incidental costs. These costs are added to the premium that an individual must pay.

EXAMPLE 4   *Living to a Ripe Old Age.* Marilyn Black is 88 years old and has applied to an insurance company for a $5000 one-year life insurance policy. On the basis of the mortality tables only, what premium should the company charge Marilyn?

SOLUTION   We let p be the premium that the company should charge Marilyn. Then we have the following.

If Marilyn lives, then the profit to the company is	If Marilyn dies within a year, then the profit to the company is
p	$p - 5000$

Using Formula 6 and Table 5.7, we calculate the probability that Marilyn will live or die within the year. Thus

$$\text{probability that an 88-year-old person will live to be 89 years old} = \frac{N_{89}}{N_{88}} = \frac{594{,}477}{741{,}474} = 0.80175$$

and

$$\text{probability that an 88-year-old person will } not \text{ live to be 89 years old} = \frac{N_{88} - N_{89}}{N_{88}}$$

$$= \frac{741{,}474 - 594{,}477}{741{,}474}$$

$$= 0.19825$$

The expected profit for the insurance company is

$$\begin{pmatrix} \text{Profit if} \\ \text{Marilyn lives} \end{pmatrix} \cdot \begin{pmatrix} \text{Probability that} \\ \text{Marilyn will live} \end{pmatrix} + \begin{pmatrix} \text{Profit if} \\ \text{Marilyn dies} \end{pmatrix} \cdot \begin{pmatrix} \text{Probability that} \\ \text{Marilyn will die} \end{pmatrix}$$

$$= p(0.80175) + (p - 5000) \cdot (0.19825)$$

$$= 0.80175p + 0.19825p - (5000) \cdot (0.19825)$$

$$= p - 991.25$$

Then we solve the equation

$$0 = p - 991.25$$

for p and get $p = \$991.25$. Thus the company should charge Marilyn at least $991.25 for the one-year $5000 life insurance policy. □

EXERCISES

1. Find the probability that a 63-year-old person will live to be 65 years old.

2. Felicia is 49 years old and plans to retire when she reaches 65 years of age so that she can begin collecting Social Security. Find the probability that she *will not* live to be 65 years old.

3. Ernesto is 49 years old and has invested a lot of money in a business enterprise that promises to make him a millionaire at the end of six years. Find the probability that Ernesto will be alive then.

4. Corey Johnson is 42 years old and has just purchased a house with a 25-year mortgage from the bank. Find the probability that Corey *will* not be alive when the mortgage is completely paid off, that is, after 25 years.

5. Sharon Gallo has just opened a new dentist's office. Sharon is 38 years old. Find the probability that Sharon will live to be 62 years old.

6. Gary Thompson is 32 years old. Each day he jogs 6 miles. Find the probability that Gary *will not* live to be 52 years old.

7. Marilyn is 93 years old. Find (a) her life expectancy; (b) the probability that she will live to be 95 years old.

8. Sonya Washington is 66 years old. Find (a) her life expectancy; (b) the probability that she *will not* live to be 68 years old.

9. Howard Zehna is 59 years old and has applied to an insurance company for a one-year $100,000 life insurance policy. What premium should the company charge Howard?

10. Edith Bressler is 61 years old and wishes to obtain a $75,000 one-year life insurance policy. What premium should an insurance company charge her?

11. Douglas Lake is 63 years old. He wishes to obtain a one-year $200,000 life insurance policy to provide protection for his pregnant wife should he die within the year. What premium should an insurance company charge him?

** 12. Marjorie Suthers is 47 years old. She wishes to obtain a five-year $100,000 life insurance pol-

icy. What premium should the insurance company charge her each year that the policy is in effect? (*Hint:* The premium will change each year as Marjorie gets older.)

** 13. Alexis Taylor is 38 years old. She wishes to obtain a six-year $85,000 life insurance policy. What premium should the insurance company charge her each year that the policy is in effect?

** 14. Maurice Bauman is 62 years old. He wishes to obtain a thirteen-year $125,000 life insurance policy. However, he wishes to pay the same premium each year. What premium should the company charge him? (*Hint:* First find the premium for each year and then find the average for the 13 years.)

5.8

SUMMARY   In this chapter we presented several topics to provide the information you need to become a more intelligent shopper and a wiser consumer. Our initial discussion dealt with hand-held pocket calculators, which are fast becoming a rather common item in most households. Depending upon your particular needs, you can purchase a calculator with any one of three different types of logic. Each of these logics requires that the data be entered in a different format.

We also discussed percents, ratio, and proportion as well as applications of these ideas to everyday life situations. It is suggested that you use a calculator for the exercises in this chapter.

We then analyzed simple and compound interest as well as installment buying and mortgages, both of which use interest calculations. Particular care must be exercised when using Table 5.5 (the amortization payments table). You should now be able to understand how the finance charges are determined when you receive your monthly bill from the credit card company or how monthly mortgage payments are computed.

Finally, we studied life insurance and mortality tables. With a knowledge of the meaning of such tables we can better understand how life insurance premiums are calculated.

STUDY GUIDE   You should now be able to demonstrate your knowledge of the following ideas by giving definitions, descriptions, or specific examples. Page references are given in parentheses.

Arithmetic logic (p. 246)   Reverse Polish Notation (RPN) logic (p. 246)
Algebraic logic (p. 246)   Fixed decimal point (p. 248)

Floating decimal point (p. 248)
Ratio (p. 251)
Terms of a ratio (p. 251)
Simplest form of a ratio (p. 251)
Unit pricing (p. 252)
Unit cost (p. 252)
Proportion (p. 253)
Extremes of a proportion (p. 253)
Means of a proportion (p. 253)
Percent (per centum) (p. 256)
Interest (p. 261)
Finance charges (p. 261)
Principal (p. 261)
Simple interest (p. 262)
Compound interest (p. 264)
Compound interest formula
   (p. 265)

Compounded continuously
   (p. 265)
Effective interest rate (p. 269)
Installment plan (p. 270)
Mortgage (p. 272)
Amortization (p. 272)
Conventional loan or mortgage (p.283)
Adjustable rate mortgage (p. 283)
Variable rate mortgage (p. 283)
Points (p. 283)
Annual percentage rate (APR)
   (p. 283)
Mortality tables (p. 287)
Life expectancy (p. 287)

FORMULAS TO REMEMBER

The following list summarizes all the formulas discussed in this chapter.

1. In the proportion $\frac{a}{b} = \frac{c}{d}$, the product of the means is equal to the product of the extremes, that is, $ad = bc$.

2. Changing fractions to percents: To change a fraction to a percent, divide the denominator into the numerator. Then multiply the quotient obtained by 100 and add a percent symbol. If a number is already written in decimal form, we change it to a percent by multiplying by 100. Then we add the percent symbol.

3. Changing percents to fractions: To change a number written with a percent symbol to a decimal equivalent, drop the percent symbol and divide the number by 100.

4. Simple interest formula: $I = P \cdot r \cdot n$ where I = total simple interest, P = principal, r = annual rate of interest, and n = number of years.

5. Compound interest formula: $A = P(1 + r)^n$, where A = compound amount (principal + interest).

6. For continuously compounded interest: $A = Pe^{rn}$, where the value of e to the appropriate power can be obtained from Table 5.3.

7. The effective interest rate: $e^r - 1$.

8. Probability that an individual who is a years old will

  a) live to be b years old is

$$\frac{N_b}{N_a}$$

  b) not live to be b years old is

$$\frac{N_a - N_b}{N_a}$$

where the values of N_a and N_b are obtained from Table 5.7.

MASTERY TESTS

Form A

1. Do the ratios $\frac{5}{7}$ and $\frac{30}{42}$ form a valid proportion?

2. Solve for x in the proportion $\frac{5}{6} = \frac{x}{24}$.

3. Convert $\frac{8}{9}$ to a percent.

4. Convert 0.532 into a percent.

5. Change $2\frac{3}{4}\%$ into decimals.

6. What is $4\frac{1}{2}\%$ of 360?

7. Nine is $7\frac{1}{2}\%$ of what number?

8. Maurice picked $\frac{2}{3}$ of a quart of strawberries and found that $\frac{1}{5}$ of those picked were not edible. What is the ratio of the number of those picked that were not edible to the total number of strawberries picked?

9. Augustine and Lena decided to divide their $100,000 winning lottery ticket in the ratio of $13:7$. How much did each person get?

10. The Board of Commissioners of the Tareltown school district has just approved the expenditure of funds for computer hardware and computer software in the ratio of $11:3$. If they plan to spend $15,000 for computer software, how much will be spent on computer hardware?

11. Find the simple interest on a loan of $7500 for $2\frac{1}{2}$ years given that the rate of interest is 9% per year.

12. Alan borrows $1500 from his friend for 8 months. The simple interest rate is 25% per year. How much money does Alan have to repay to his friend?

13. Zelda borrows $525 from her boss and agrees to repay $600 to her boss in three months. What is the annual simple rate of interest?

14. Gazelle arranges for a short term loan with the mechanic to repair her car. The annual rate of interest is 15%. If the total interest charged is $25 and Gazelle repays the loan in 2 months, how much money did she borrow?

15. What is the effective annual interest yield of a 9% annual interest rate compounded continuously?

Form B

1. Gregory invests his son's gifts from his first birthday, which amount to $743.00, in a certificate of deposit that pays interest at an annual rate of 12% compounded quarterly. How much will this money be worth when Gregory's son starts college at age 18?

2. Helen, who is $45\frac{1}{2}$ years old, opens an Individual Retirement Account (IRA) by depositing $2000. If Helen makes no other deposits and the bank pays 10% interest a year compounded quarterly, how much money will

Helen have when she reaches $59\frac{1}{2}$ years of age and is eligible to withdraw the money?

3. Karen has just purchased a used car for $3000. She makes a down payment of 20% and agrees to pay the balance on the installment plan by making 40 weekly payments of $80. What is the total cost of the car (including interest)?

4. In a certain state, the personal income tax rate is $825 plus $12\frac{1}{4}$% for any earnings over $38,000. If Jane Rodgers earns $43,712 a year, what is her personal tax liability in her state?

5. Reggie has just renovated his house at a cost of $10,000. He paid the contractor $4000 as a down payment and agreed to pay the $6000 balance in 6 monthly installments with a minimum payment of $1000 per month. Interest is $1\frac{1}{2}$% per month on any unpaid balance. Reggie wishes to pay the minimum amount. What is the total cost of the renovations including interest?

6. Rich Jones bought a complete home entertainment center for $1289. He made a down payment of $200 and agreed to pay the $1089 balance in equal monthly payments over a 2-year period. Interest is 18% a year compounded monthly. How much money must Rich pay monthly in order to amortize the loan over the 2-year period?

7. Terry Raskin arranged for a $65,000 conventional mortgage from a bank that charges 12% annual interest compounded quarterly. The loan is to run for 11 years. What are Terry's quarterly payments?

8. What is the likelihood (probability) that a 56-year-old person will live to be 69 years old?

9. Wesley Eagleton is 51 years old and wishes to obtain a $88,000 one-year life insurance policy. What premium should the company charge him?

10. Thomas Blair is 63 years old. Find the probability that he will *not* live to be 88 years old.

**11. Lance Soka is 60 years old. He wishes to obtain a 10-year $150,000 life insurance policy. However, he wishes to pay the same premium each year. What premium should the company charge him?

12. The average number of scheduled daily flights at an airport increased from 153 in August 1984 to 188 in August 1985. What is the percent increase in the average number of scheduled flights?

13. The number of reported cases of measles in Bugleville decreased from 196 in 1984 to 147 in 1985. What was the percent decrease in the number of reported cases of measles?

14. A 3.3-ounce bottle of perfume costs $8.49. A 4.4-ounce bottle of the same perfume costs $10.23. Which is a better buy?

15. A 1.4-ounce serving of a certain brand of bran flakes contains 130 calories. How many calories are contained in a 1.9-ounce serving of this brand of bran flakes?

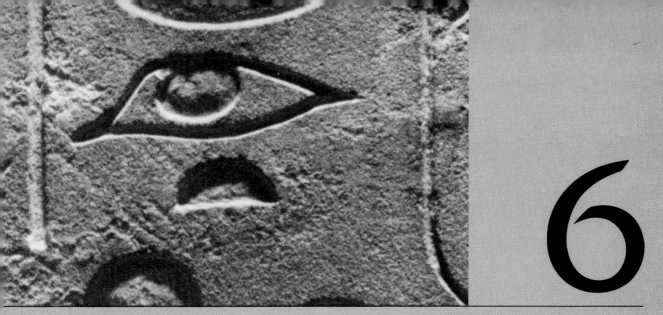

6

MATHEMATICAL SYSTEMS

CHAPTER OBJECTIVES

To describe what is meant by clock arithmetic and how to add using different clocks. (*Section 6.2*)

To discuss modular arithmetics, which deal with the remainders that are obtained when we divide. (*Section 6.3*)

To study how we can check the answers for addition, multiplication, and subtraction problems by casting out 9's and by casting out 11's. (*Sections 6.4 and 6.5*)

To determine what a group is and which properties must be satisfied by a system in order to be classified as a group. (*Section 6.6*)

To apply the concept of groups to various symmetries. (*Section 6.7*)

To discuss games that involve two players and in which there is no kitty—that is, what one player wins, the other loses. (*Section 6.8*)

To analyze games that are strictly determined and to indicate how to find optimal strategies or best moves for each player in such games. (*Sections 6.8 and 6.9*)

To study games that are not strictly determined and to develop a formula for finding optimal strategies for such games. We can also find the value of a game which will alert us to its fairness. (*Sections 6.8 and 6.9*)

To use the ideas of game theory in social science situations. (*Sections 6.8 and 6.9*)

Top Fruit to Install New Computer Scales

SPRINGFIELD: In an effort to make sure that the consumer is not being cheated, the management of the Top Fruit Chain Stores has announced that it would begin replacing the scales in all of its 100 stores by new computer scales. The new scales print out on a display panel the weight of an object correct to the nearest hundredth. This should eliminate the numerous complaints received by the Consumer Fraud Department that the clerks at several of the stores were cheating customers by reading the scales incorrectly.

The total cost to the company was estimated to be $100,000.

THE CHRONICLE, April 17, 1984

Electronic Games Win Again

by Gwen Jackson

HAMILTON: So you think that you are smart? Have you seen the latest electronic games? On my recent visit to Macy's in New York City, I discovered well over 150 electronic games ranging from electronic backgammon and checkers to poker, baseball, hockey, and football simulation.

The number of computer games on the market is also growing constantly. In each case the game or gadget has been programmed to respond with a best move to any move made by a player. It is very difficult, or almost impossible, to outwit these "little brains."

The games that I surveyed ranged in price from a low of $22 to a high of over $200.

THE TIMES, Saturday, December 13, 1984

All of us have seen an old-fashioned scale such as the one discussed in the first newspaper article. When using such a scale, we must determine two things: where the pointer stops and how many times the pointer passes the 0. Although the Top Fruit Store Chain mentioned in the article is replacing such scales with the more modern computer scales, these older scales do illustrate the ideas of clock and modular arithmetics to be discussed in this chapter.

Now consider the second newspaper article. As part of our natural way of living we often play different types of games. When playing a game with an opponent, we try to make a move that will outwit the opponent. As the newspaper article indicates, there are many computer games on the market today. Playing against them is not an easy task, since in each case the gadget has been programmed to analyze all the possible moves and to respond with the optimal strategies (best moves).

How do we determine optimal strategies?

6.1
INTRODUCTION

Throughout our lives we play various games. Some of these games are played for fun, or to pass time, or for money. Card games, chess, dominoes, checkers, and tic-tac-toe are of this type. There are other situations (such as the stock market, politics, war, bargaining between a union and an employer, and business activities) that so closely resemble games that we frequently use the language of games in discussing them. We say, for example, "I made a good deal," "That was a good move," "She plays the market," "He's bluffing." Because these activities are similar to games in many ways, we can use the mathematical theory of games in analyzing them. For this reason, even nonmathematicians such as economists and sociologists find this branch of mathematics of great interest.

For example, when unions and employers sit down to negotiate a new contract, each side pushes for certain items that will result in benefits to itself. In trying to obtain the best terms, each side will use various strategies. By a combination of bluffing, arm-twisting, and "friendly persuasion," each will try to get the other side to agree to its terms. In doing so, not only must each side plan its own moves, but it must also anticipate the other's moves and act accordingly.

This is the same type of situation that occurs when two or more people are playing a game such as chess, Monopoly, or football. Each player tries to figure out his or her opponents' moves and act accordingly.

Similar situations arise in many areas of business. For example, suppose a large hamburger chain is considering opening a new restaurant in one of two towns. In deciding which town to open in, it must consider the reactions of its competitors (if any).

There are certain games that have "best" moves for a player, no matter what his or her opponent does. Then again, there are other games that have no single best possible strategy. Most military situations are of the latter type. In this case a player must first decide on one or more strategies. There is no "perfect" or "best" strategy for all situations. Obviously, a player must hide his or her own strategy so that the opponent does not figure it out. The first person to develop a method for finding best moves under these circumstances was the Hungarian mathematician John Von Neumann (1903–1957).

In mathematical game theory we deal only with **games of strategy.** These are games in which two or more players are competing against each other. Each player has several alternative moves available. He or she must decide which is the best move to make. All of the games mentioned above are games of strategy. On the other hand, such games as Russian roulette, dice, pitching pennies, and lotteries are not games of strategy. In them, no plan is involved. Some of them involve physical skill. Others are just betting against the odds.

In this chapter we deal only with two-person, zero-sum games. A **two-person game** is any game that involves only two players. A game is called a **zero-sum game** if on each move, one player wins what the other player loses. A game in which there is a bank or kitty is not a zero-sum game and will not be discussed in this book.

In many different areas of mathematics the same patterns occur over and over again. It is this property that helps make mathematics interesting and valuable. For example, we mentioned earlier that the relationship between sets were very much like the relationships between statements that occur in logic (see p. 106).

HISTORICAL NOTE

The term "group" was first used by the brilliant young French mathematician Évariste Galois in 1830. Although he was a mathematical genius, his abilities were not recognized by his teachers. In fact, he was twice denied admission to the famous École Polytechnique, *the* school for mathematicians and scientists in Paris. Twice he submitted papers containing his important discoveries to the French Academy, and each time they were lost.

At the age of 20 he became involved in a quarrel over a woman and was challenged to a duel. He spent the entire night before the duel writing down his ideas. The next morning he was killed. It is interesting to think about what he might have contributed to mathematics had he lived a normal life span.

Galois studied groups in order to solve certain problems in algebra. His discoveries greatly expanded the field of algebra. Furthermore, his ideas have also been applied to physics. They have been particularly valuable in the study of quantum mechanics.

Évariste Galois (1811–1832). Above is a page from Galois' papers.

In this chapter we will examine in detail one very important system that occurs in practically all branches of mathematics. This system is called a **group.**

6.2
CLOCK ARITHMETIC

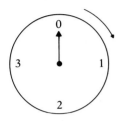

Figure 6.1

To introduce the idea of a group, let us consider a clock that has only four numbers on it, as shown in Fig. 6.1.

Starting at 0, the clock hand will point to 1, 2, and 3 in order, and then return to 0. This cycle repeats itself over and over. Let $\oplus$ represent the turning of the hand of the clock in the direction of the arrow. Then $2 \oplus 3$ means that the clock is at the 2 position and then moves through 3 positions. It stops at the 1 position, so we say that

$$2 \oplus 3 = 1.$$

Similarly, $1 \oplus 2$ would mean that the clock is first in the 1 position and then moves 2 more places, ending up in the 3 position. Thus we have

$$1 \oplus 2 = 3.$$

We can make up a table showing all possible starting positions and all possible ending positions (Table 6.1).

TABLE 6.1

	$\oplus$	0	1	$\downarrow$ 2	3
	0	0	1	2	3
	1	1	2	3	0
	2	2	3	0	1
$\rightarrow$	3	3	0	(1)	2

This chart is read as follows: To find $3 \oplus 2$, for example, go to row 3 then over to column 2. We have indicated this by means of arrows in Table 6.1. We find the answer to be 1. Thus

$$3 \oplus 2 = 1.$$

In a similar manner we read from the table that

$$3 \oplus 3 = 2.$$

We can think of the different positions of this clock as a set G, whose members are 0, 1, 2, and 3. So,

$$G = \{0, 1, 2, 3\}.$$

The operation of turning the hand, which we denoted by $\oplus$, can be considered a **binary operation.** (Recall that a binary operation involves combining any two elements of a set to get a third element.)

We can observe a number of interesting things about this system. First of all, Table 6.1 indicates that no matter where we start, we will always end up at one of four positions 0, 1, 2, or 3. This means that the operation $\oplus$ is **closed.** (Remember, an operation is closed if, when we perform the operation, the result is always within the set we started with.)

Now notice the following:

$$0 \oplus 0 = 0, \qquad 0 \oplus 0 = 0,$$
$$0 \oplus 1 = 1, \qquad 1 \oplus 0 = 1,$$
$$0 \oplus 2 = 2, \qquad 2 \oplus 0 = 2,$$
$$0 \oplus 3 = 3, \qquad 3 \oplus 0 = 3.$$

Thus when we perform the operation $\oplus$ with 0 on any element, nothing happens. Here 0 is the **identity element** (see p. 193) for the operation $\oplus$.

Now suppose we are at any position on the clock. Can we always get back to 0 by the operation $\oplus$? The answer is clearly yes, as the following results indicate:

$$0 \oplus 0 = 0,$$
$$1 \oplus 3 = 0,$$
$$2 \oplus 2 = 0,$$
$$3 \oplus 1 = 0.$$

We call 3 the **inverse** of 1 for the operation $\oplus$ because when we perform $3 \oplus 1$, we get 0. Similarly, 2 is the inverse of 2 for the operation $\oplus$, 1 is the inverse of 3, and 0 is its own inverse.

Every element of this system has an inverse. In other words we can always get back to 0.

Consider the expression $2 \oplus 3 \oplus 1$. This can be interpreted in two different ways. One way is to first do $2 \oplus 3$, getting 1. Then do $1 \oplus 1$, getting 2. In other words,

$$\begin{aligned}
&(2 \oplus 3) \oplus 1 \\
= \quad &1 \oplus 1 \\
= \quad &2.
\end{aligned}$$

Another way is to first do $3 \oplus 1$, getting 0. Then do $2 \oplus 0$, which gives 2. That is,

$$\begin{aligned}
&2 \oplus (3 \oplus 1) \\
= \quad &2 \oplus 0 \\
= \quad &2.
\end{aligned}$$

In both cases our answer is 2. We conclude that $(2 \oplus 3) \oplus 1 = 2 \oplus (3 \oplus 1)$. This shows that the associative law holds for these three numbers. In a similar manner we can show that *the associative law holds for any three numbers in this system.*

This clock is called a **mod 4 clock** because of its four positions.

Next, let us look at a familiar object, a three-way switch on a table lamp. Such a switch can be pictured as shown in Fig 6.2. We turn the switch in a clockwise direction as indicated by the arrow. If the switch is in the "off" position, turning it once puts it in "low." Another turn moves it to "medium." Still another turn moves it to "high." The next turn moves it to the off position again. If the switch is in the low position, turning it twice moves it to the high position. Turning it three times puts it in the off position. This cycle can be repeated as often as we like.

If we start at the off position, then we need 0 turns to get to the off position, 1 turn to get to low, 2 turns to medium, and 3 turns to high. Thus we let

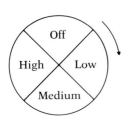

Figure 6.2

0 stand for the off position,

1 stand for the low position,

2 stand for the medium position,

3 stand for the high position, and

$\oplus$ stand for turning the switch clockwise.

Table 6.2 shows the final position of the switch, depending on where you start and the number of turns you make.

TABLE 6.2

$\oplus$	0	1	2	3
0	0	1	2	3
1	1	2	3	0
2	2	3	0	1
3	3	0	1	2

Compare this table with the table for the mod 4 clock (Table 6.1). It is exactly the same. What this means is that both situations have the same mathematical structure, although they appear to be completely different.

EXAMPLE 1 Let us construct a table for a mod 3 clock (that is, a clock with 3 positions). The clock is shown in Fig. 6.3. As before, let $\oplus$ represent turning the clock. We then have the results shown in Table 6.3.

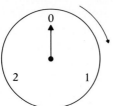

Figure 6.3

TABLE 6.3

$\oplus$	0	1	2
0	0	1	2
1	1	2	0
2	2	0	1

EXAMPLE 2   What would $6 \oplus 5$ be in a mod 10 clock? in a mod 8 clock?

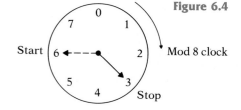

Figure 6.4

Mod 10 clock

Mod 8 clock

SOLUTION   We draw both clocks as shown in Fig. 6.4. We start at 6 and move through 5 additional positions. In the mod 10 clock we stop at 1 as indicated in the diagram. Thus

$$6 \oplus 5 = 1 \text{ in mod 10.}$$

In the mod 8 clock we stop at 3, again as indicated in Fig. 6.4. Thus

$$6 \oplus 5 = 3 \text{ in mod 8.}$$

EXAMPLE 3   In what clock is $4 \oplus 3 = 1$?

SOLUTION   The number 4 shows us that the clock must be *at least* a mod 5 clock. (Why?) Let us try mod 5. A quick check (which the reader should verify by actually drawing such a clock) will show that

$$4 \oplus 3 = 2 \text{ in mod 5.}$$

So mod 5 is wrong. Next we try mod 6. This works. (Try it, to convince yourself.)

$$4 \oplus 3 = 1 \text{ in mod 6.}$$

EXAMPLE 4   Let us define "subtraction" on a clock as "turning the clock backwards." We will denote this as $\ominus$. Let us find $3 \ominus 5$ in a mod 8 clock. (See the mod 8 clock in Fig. 6.4.) We start at 3 and move back 5 positions, stopping at 6. Thus

$$3 \ominus 5 = 6 \text{ in mod 8.}$$

Comment Some readers may say that $3 \ominus 5 = -2$. However, there is no -2 on the clock. Nevertheless, if we interpret -2 to mean "start at 0 and go back 2," then, of course, we stop at 6. We will always give our answer as 6 rather than as -2.

EXAMPLE 5 In a mod 7 clock, $4 \ominus x = 6$. Find x.

SOLUTION We start at 4 and move back x places until we get to 6. From the clock in Fig. 6.5 we find that

$$4 \ominus 5 = 6.$$

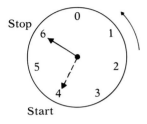

Figure 6.5

Thus $x = 5$. $\square$

Comment We define "multiplication" on a clock as moving the hands of the clock through an appropriate number of positions. We will denote this as $\otimes$. Thus $4 \otimes 3$ on a mod 5 clock means that we move through 12 positions. Of course, we stop at the 2 position. Thus,

$$4 \otimes 3 = 2 \text{ in mod 5}.$$

This idea will be further illustrated in the exercises for this section.

EXERCISES

1. Make up a table for the operation $\oplus$ on the following clocks:

  a) mod 7   **b)** mod 8

  c) mod 2   **d)** mod 5

2. Make up a table for the operation $\ominus$ on the following clocks:

  a) mod 5   **b)** mod 9

Determine the answer to each of the calculations in Exercises 3–14.

3. $4 \oplus 6$ in mod 9   **4.** $2 \oplus 3$ in mod 5

5. $5 \ominus 9$ in mod 10   **6.** $4 \ominus 7$ in mod 11

7. $4 \oplus 4$ in mod 6   **8.** $4 \oplus 4$ in mod 7

9. $8 \ominus 8$ in mod 15   **10.** $7 \ominus 11$ in mod 13

11. $5 \otimes 4$ in mod 6   **12.** $6 \otimes 7$ in mod 8

13. $9 \otimes 0$ in mod 10   **14.** $3 \ominus 14$ in mod 9

In Exercises 15–20, find the value of x.

15. $2 \oplus x = 7$ in mod 8   **16.** $2 \ominus x = 4$ in mod 5

17. $3 \oplus x = 1$ in mod 4   **18.** $x \ominus 5 = 4$ in mod 9

19. $4 \oplus x = 2$ in mod 7   **20.** $1 \ominus x = 3$ in mod 12

In what clock are the calculations performed in Exercises 21–26 true?

21. $4 \oplus 5 = 3$ **22.** $2 \ominus 5 = 4$

23. $3 \oplus 3 = 1$ **24.** $4 \ominus 5 = 6$

25. $7 \oplus 8 = 3$ **26.** $3 \ominus 4 = 7$

27. In a mod 6 clock, name the inverse of each element.

28. In a mod 9 clock, name the inverse of each element.

29. Make up a table for the operation $\oplus$ on a mod 5 clock, and determine whether the associative law

$$a \oplus (b \oplus c) = (a \oplus b) \oplus c$$

is valid.

Each of the equations given in Exercises 30–37 is valid for a mod 11 clock. Solve for x.

30. $4 \otimes 5 = x$ **31.** $5 \ominus x = 9$

32. $4 \otimes (5 \oplus x) = 3$ **33.** $\dfrac{x}{3} = 5$

34. $x \otimes (2 \oplus 5) = 3$ **35.** $5 \otimes 5 = x$
36. $9 \oplus x = 3$ **37.** $3 \otimes x = 4$

**38. *Tiles.* In designing tiles we can often use the addition or multiplication tables of the various clock arithmetics to obtain different patterns. Thus, in mod 3 we have the following addition and multiplication tables:

$\oplus$	0	1	2		$\otimes$	0	1	2
0	0	1	2		0	0	0	0
1	1	2	0		1	0	1	2
2	2	0	1		2	0	2	1

If we represent the number 0 as ▱ , the number 1 as a box with three vertical lines in it ⊞⊞ , and the number 2 as a box with two semicircles in it ⋈ , then we get the following tile design based on the addition table.

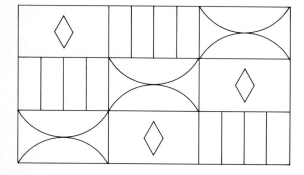

A design based on the multiplication table might be pictured as follows (using the same representations).

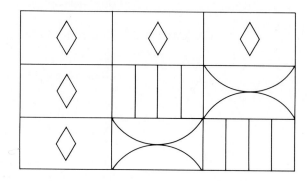

Using the addition and multiplication tables for the mod 5 and 6 clocks, construct tiles with appropriate designs based on these tables.

39. Read the article "Using Mathematical Structures to Generate Artistic Designs" by Sonia Forseth and Andria Price Troutman in *Mathematics Teacher* (May 1984, pp. 393–398) for an interesting discussion on how modular arithmetic can be used to create interesting patterns.

6.3

MODULAR ARITHMETIC

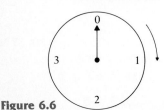

Figure 6.6

Let us look again at the mod 4 clock of the last section, which is shown in Fig. 6.6.

Suppose we were to start at 0 and move through 11 places. We would make 2 complete turns and then stop at 3. Notice that if you divide 11 by 4, you get a remainder of 3. This is no coincidence. Every time we make a complete turn on the clock, we go through 4 positions. Thus the number of complete turns is the number of times that 4 goes into 11 (that is, 2). We then have to go through the remaining 3 positions. This is the remainder when 11 is divided by 4.

Similarly, if we start at 0 and move through 18 positions, we would stop at 2. If we divide 18 by 4, the remainder would also be 2.

In other words, each position on the clock represents the remainder that we get when we divide a number by 4. This gives us a new way of looking at this clock. We can think of it as just the remainders we get when we divide by 4. From this point of view this system is called a **modulo 4 arithmetic.** The number 4 is called the **modulus.**

EXAMPLE 1 In a modulo 5 arithmetic the numbers would be the remainders we get when we divide by 5. These are 0, 1, 2, 3, 4. In this system the modulus is 5. In a mod 5 clock, moving through 17 positions, starting at 0, leaves us at 2. We get the same result if we divide 17 by 5.

In a modulo 4 arithmetic, suppose we were to add $3 + 2$. We know that $3 + 2 = 5$. However, we are only interested in the remainder when we divide by 4, so our answer would be 1. We write this answer as $3 + 2 \equiv 1$ in modulo 4 arithmetic. The three-lined equals sign indicates that we are working only with remainders. This corresponds to $3 \oplus 2 = 1$ of the table for the mod 4 clock (see Table 6.2). □

Notation   Instead of writing out the words "in modulo arithmetic" for the previous example, we abbreviate this as "(mod 4)." We write the answer as $3 + 2 \equiv 1$ (mod 4). This is read as

3 plus 2 is **congruent** to 1 modulo 4.

EXAMPLE 2   Add $4 + 5$ (mod 6).

SOLUTION $4 + 5$ is 9. If we divide 9 by 6, our remainder will be 3. Therefore
$$4 + 5 \equiv 3 \text{ (mod 6)}. \quad \square$$

EXAMPLE 3   Add $7 + 9$ (mod 10).

SOLUTION $7 + 9 = 16$. Dividing 16 by 10, we get a remainder of 6. Therefore
$$7 + 9 \equiv 6 \text{ (mod 10)}. \quad \square$$

EXAMPLE 4   Add $5 + 5$ (mod 7).

SOLUTION $5 + 5 = 10$. Dividing 10 by 7, the remainder is 3. Our answer is
$$5 + 5 \equiv 3 \text{ (mod 7)}. \quad \square$$

EXAMPLE 5   Multiply $7 \cdot 3$ (mod 5).

SOLUTION $7 \cdot 3 = 21$. If we divide 21 by 5 we get a remainder of 1. Thus
$$7 \cdot 3 \equiv 1 \text{ (mod 5)}.$$

On a mod 5 clock this means that we move through 21 positions, stopping at the 1 position. ☐

EXAMPLE 6   Multiply $5 \cdot 9$ (mod 15).

SOLUTION $5 \cdot 9 = 45$. Dividing 45 by 15, we get a remainder of 0. Thus

$$5 \cdot 9 \equiv 0 \text{ (mod 15)}. \ \square$$

EXAMPLE 7   Subtract $5 - 2$ (mod 6).

SOLUTION $5 - 2 = 3$. Dividing 3 by 6, we get a remainder of 3. Therefore

$$5 - 2 \equiv 3 \text{ (mod 6)}. \ \square$$

EXAMPLE 8   Subtract $3 - 6$ (mod 9).

SOLUTION   To do this problem, we look at a mod 9 clock (see Fig. 6.7). We start at 3 and go back 6, stopping at 6. Thus

$$3 - 6 \equiv 6 \text{ (mod 9)}. \ \square$$

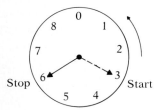

Figure 6.7

In a modular arithmetic a number corresponds to its remainder when divided by the modulus. Thus in a mod 5 system the number 38 corresponds to 3, since 38 divided by 5 gives a remainder of 3. We write this as $38 \equiv 3$ (mod 5).

Notice that $38 - 3$ is exactly divisible by 5. This leads us to a useful way of rephrasing the idea of modular arithmetics.

> **Definition 6.1**   *We say that two integers a and b* **are congruent modulo** *m, if a − b can be divided exactly by m (m is a natural number). We write this as*
>
> $$a \equiv b \text{ (mod } m).$$

EXAMPLE 9 $7 \equiv 3$ (mod 2) means 7 is congruent to 3 modulo 2. This is true because $7 - 3$ can be divided exactly by 2. ☐

Notice that if we divide 7 by 2, the remainder is 1, and if we divide 3 by 2, the remainder is also 1. *If any two numbers are congruent mod m, then their remainders when divided by m are the same.*

EXAMPLE 10 $10 \equiv 15$ (mod 5) because $10 - 15$, which is −5, is exactly divisible by 5. ☐

EXAMPLE 11 $12 \equiv 7$ (mod 3) is *not* true because $12 - 7$, which is 5, cannot be exactly divided by 3. ☐

EXAMPLE 12 $25 \equiv 7$ (mod 9) is true because $25 - 7$, which equals 18, is divisible by 9. ☐

EXAMPLE 13 $38 \equiv 5 \pmod{11}$ is true because $38 - 5 = 33$, and 33 is exactly divisible by 11. ☐

EXAMPLE 14 $-5 \equiv 6 \pmod{11}$ is true because $-5 - 6$, which is -11, can be divided exactly by 11. ☐

EXAMPLE 15 $5 \equiv -13 \pmod 9$ because $5 - (-13)$, which equals $5 + 13$ or 18, is exactly divisible by 9. ☐

We wish to emphasize a comment made in Example 9. We pointed out that if $a \equiv b \pmod m$, then when we divide a by m or b by m, the remainder is the same in both cases. Thus whether two numbers are congruent modulo m depends on whether they have the same remainders when we divide them by m. To put it another way, if $a \equiv b \pmod m$, the positions of a and b on the mod m clock are the same.

EXAMPLE 16   If New Year's Day is Thursday, on what day of the week will February 1 fall?

SOLUTION   January has 31 days. Since there are 7 days in a week, we can consider this as a modulo 7 system, where Sunday is 0, Monday is 1, and so forth. If we now divide 31 by 7, our remainder is 3. Thus from January 1 until February 1 there are exactly 4 weeks and 3 days. Therefore, February 1 will occur three days after a Thursday. It will occur on a Sunday. ☐

EXERCISES

In Exercises 1–10, perform the indicated operations in the specified modular arithmetics.

1. $4 + 8 \pmod 3$　　**2.** $6 + 9 \pmod 9$

3. $5 \cdot 6 \pmod{11}$　　**4.** $8 \cdot 3 \pmod 7$

5. $5 - 8 \pmod 3$　　**6.** $1 - 6 \pmod 8$

7. $0 - 5 \pmod 5$　　**8.** $-4 \cdot 3 \pmod 7$

9. $-8 \cdot 5 \pmod 7$　　**10.** $-4 \cdot 8 \pmod 9$

Which of the statements given in Exercises 11–20 are true?

11. $18 \equiv 3 \pmod 5$　　**12.** $16 \equiv 2 \pmod 7$

13. $14 \equiv -1 \pmod 5$　　**14.** $-6 \equiv 14 \pmod{10}$

15. $8 \cdot 7 \equiv 6 \pmod 9$　　**16.** $15 \equiv 11 \pmod 8$

17. $10 \equiv 10 \pmod 3$　　**18.** $-8 \equiv -5 \pmod 3$

19. $18 \equiv 20 \pmod 2$　　**20.** $6 \cdot 8 \equiv -2 \pmod 5$

In Exercises 21–30, find one possible replacement for x, so that the resulting statement will be true. (More than one answer may be possible.)

21. $x \equiv 3 \pmod 5$　　**22.** $2x \equiv 1 \pmod 7$

23. $x \equiv 3 \pmod{14}$　　**24.** $x + 5 \equiv 2 \pmod 6$

25. $4 - x \equiv 7 \pmod{11}$　　**26.** $x - 4 \equiv 7 \pmod 8$

27. $x + 6 \equiv 3 \pmod{12}$　　**28.** $7x + 2 \equiv 1 \pmod{15}$

29. $3x \equiv 6 \pmod 5$　　**30.** $4x \equiv 2 \pmod{10}$

In Exercises 31–40, find one possible replacement for m, so that the resulting statement will be true. (More than one answer may be possible.)

31. $8 \equiv 3 \pmod m$　　**32.** $2 \equiv -9 \pmod m$

33. $13 \equiv 3 \pmod m$　　**34.** $-5 \equiv -5 \pmod m$

35. $-6 \equiv -8 \pmod m$　　**36.** $4 \equiv -6 \pmod m$

37. $8 \equiv -6 \pmod m$ **38.** $-3 \equiv -7 \pmod m$

39. $12 \equiv -8 \pmod m$ **40.** $-5 \equiv 20 \pmod m$

41. In a certain year, April 15 falls on a Thursday. In that same year, on what day of the week will July 4 fall?

42. August 15, 1985, fell on a Thursday. On what day of the week will August 15, 1991, fall?

43. If George Washington's birthday (February 22) falls on a Tuesday this year, on what day of the week will Independence Day (July 4) fall?

**44. July 4, 1985, occurred on a Thursday. On what day of the week was the Declaration of Independence (July 4, 1776) signed?

**45. *Friday, the Thirteenth.* Thomas Jefferson, the third president of the United States, is the only president who was born on the 13th of the month. His birthday was April 13, 1743. Was he born on Friday the 13th?

46. A computer programmer is rearranging his computer discs. He knows that he has fewer than 100 discs in his collection. When he arranges the discs in piles of 5 each, he has 4 left over. When he arranges the discs in piles of 7 each, he has 5 left over. When he arranges the discs in piles of 4 each, he has 1 left over. How many discs does the programmer have?

47. A caterer has to launder identical soiled tablecloths. She knows that there are fewer than 100. However, she does not know exactly how many tablecloths there are. In an effort to determine how many there are, she piles the tablecloths into stacks of 10 each. When she does this, she has 8 left over. When she arranges the tablecloths in piles of 7 each, she has 1 left over. When she arranges the tablecloths in piles of 4 each, she has 2 tablecloths left over. How many soiled tablecloths does the caterer have?

6.4

CASTING OUT 9'S Whenever we perform a computation, we like to check our answer. One way of doing this is by **casting out 9's.** It works this way.

$$\text{Add:} \quad \begin{array}{r} 476 \\ +237 \\ \hline 713 \end{array}$$

We check this by adding the digits in each number as shown. Thus we have

$$
\left.
\begin{array}{l}
476 \rightarrow 4 + 7 + 6 = 17 \rightarrow 1 + 7 = 8 \\
+237 \rightarrow 2 + 3 + 7 = 12 \rightarrow 1 + 2 = 3
\end{array}
\right\} \text{We add these.}
$$

$$11 \rightarrow 1 + 1 = ②$$

$$713 \rightarrow 7 + 1 + 3 = 11 \rightarrow 1 + 1 = ②$$

Let us examine this procedure. We first add the digits in 476. This gives 17. Now we add the digits in 17. This gives 8.

We repeat the process on 237. First we get 12. Then adding the digits of 12, we get 3.

Next we *add* the results 8 and 3, getting 11. Finally, adding the digits of 11, we get 2.

Now we do the same thing to the total 713. Again we end up with 2.

Since in both cases the result is 2, our answer is *probably* correct. There is a *slight* possibility that the answer is wrong.

If the results are not the same, then we know definitely that we have made a computational error. To illustrate this, suppose we add 2437 and 5617 and obtain the incorrect answer 8044. Let us check.

$$2437 \rightarrow 2 + 4 + 3 + 7 = 16 \rightarrow 1 + 6 = \qquad\qquad 7 \Big\} \text{ We}$$
$$+5617 \rightarrow 5 + 6 + 1 + 7 = 19 \rightarrow 1 + 9 = 10 \rightarrow 1 + 0 = 1 \Big\} \begin{array}{l}\text{add}\\\text{these.}\end{array}$$
$$\qquad\qquad\qquad\qquad\qquad\qquad\qquad\qquad\qquad\qquad ⑧$$
$$8044 \rightarrow 8 + 0 + 4 + 4 = 16 \rightarrow 1 + 6 = ⑦$$

The fact that one result is 8 and the other is 7 tells us that we made a mistake. Find the error.

EXAMPLE 1 Add 368, 47, and 5928, and check by casting out 9's.

SOLUTION

$$368 \rightarrow 3 + 6 + 8 \qquad = 17 \rightarrow 1 + 7 = 8 \Big\}$$
$$47 \rightarrow 4 + 7 \qquad\quad = 11 \rightarrow 1 + 1 = 2 \Big\} \text{ We add these.}$$
$$+5928 \rightarrow 5 + 9 + 2 + 8 = 24 \rightarrow 2 + 4 = \underline{6} \Big\}$$
$$\qquad\qquad\qquad\qquad\qquad\qquad 16 \rightarrow 1 + 6 = ⑦$$
$$6343 \rightarrow 6 + 3 + 4 + 3 = 16 \rightarrow 1 + 6 = ⑦$$

Since we get 7 in both cases, the addition is *probably* correct. □

EXAMPLE 2 Multiply 731 by 26 and check by casting out 9's.

SOLUTION

$$731 \rightarrow 7 + 3 + 1 = 11 \rightarrow 1 + 1 = 2 \Big\}$$
$$\times\ 26 \rightarrow 2 + 6 = \qquad\qquad = \underline{8} \Big\} \text{ We multiply these.}$$
$$\overline{4386} \qquad\qquad\qquad\qquad 16 \rightarrow 1 + 6 = ⑦$$
$$\underline{1462}$$
$$\overline{19006} \rightarrow 1 + 9 + 0 + 0 + 6 = 16 \rightarrow 1 + 6 = ⑦$$

This time, because it is a multiplication problem, we multiply the 2 and the 8, as we indicated. Since the result in both cases is 7, our answer is *probably* correct. □

EXAMPLE 3 Multiply 68 by 281 and check by casting out 9's.

SOLUTION

$$68 \rightarrow 6 + 8 \qquad = 14 \rightarrow 1 + 4 = 5 \Big\}$$
$$\times 281 \rightarrow 2 + 8 + 1 = 11 \rightarrow 1 + 1 = \underline{2} \Big\} \text{ We multiply these.}$$
$$\overline{68} \qquad\qquad\qquad\qquad\qquad 10 \rightarrow 1 + 0 = ①$$
$$564$$
$$\underline{136}$$
$$\overline{19308} \rightarrow 1 + 9 + 3 + 0 + 8 = 21 \rightarrow 2 + 1 = ③$$

The difference in the results tells us that we have made a mistake. Can you find it? ☐

EXAMPLE 4   Subtract 256 from 468 and check by casting out 9's.

SOLUTION

$$468 \rightarrow 4 + 6 + 8 = 18 \rightarrow 1 + 8 = 9$$
$$-256 \rightarrow 2 + 5 + 6 = 13 \rightarrow 1 + 3 = \underline{4}$$
We subtract.

→ ⑤

$$212 \rightarrow 2 + 1 + 2 = ⑤ \leftarrow$$ ☐

EXAMPLE 5   Subtract 273 from 465 and check by casting out 9's.

SOLUTION

$$465 \rightarrow 4 + 6 + 5 = 15 \rightarrow 1 + 5 = 6$$
$$-273 \rightarrow 2 + 7 + 3 = 12 \rightarrow 1 + 2 = \underline{3}$$
We subtract.

③

$$192 \rightarrow 1 + 9 + 2 = 12 \rightarrow 1 + 2 = ③$$ ☐

EXAMPLE 6   Subtract 213 from 778 and check by casting out 9's.

SOLUTION

$$778 \rightarrow 7 + 7 + 8 = 22 \rightarrow 2 + 2 = 4$$
$$-213 \rightarrow 2 + 1 + 3 \quad\quad = \underline{6}$$
We subtract.

(-2)

$$565 \rightarrow 5 + 6 + 5 = 16 \rightarrow 1 + 6 = ⑦$$

On first thought it would appear that our answer is incorrect. However, note that 7 is congruent to -2 (mod 9) because $7 - (-2) = 7 + 2 = 9$ is exactly divisible by 9. Thus in mod 9 arithmetic, -2 and 7 are really the same, so that our answer is *probably* right. ☐

You are probably wondering why this mysterious procedure works. The secret will now be revealed.

Consider the number 3221, which can be written as

$$3(\text{thousands}) + 2(\text{hundreds}) + 2(\text{tens}) + 1$$
$$3(1000) \quad + 2(100) \quad + 2(10) \quad + 1.$$

If you divide 1000 by 9, the remainder is 1. Thus if you divide 3(1000) by 9, the remainder is 3(1), or 3.

Similarly, if we divide 100 by 9, the remainder is 1. Therefore if we divide 2(100) by 9, the remainder is 2(1) = 2.

Also if we divide 10 by 9, the remainder is 1. So if you divide 2(10) by 9, the remainder is 2(1), which is 2.

But 1 divided by 9 doesn't go. We just have a remainder of 1.

Putting this all together, if we divide 3221 by 9, we will get a total remainder of $3 + 2 + 2 + 1$, or 8. (You should verify this by actually dividing 3221 by 9.) Notice that the remainder is exactly the sum of the digits.

In general, if we divide any number by 9, the remainder will be the sum of the digits. (If the sum of the digits is greater than 9, repeat the procedure.)

Now let us look at an example of casting out 9's.

$$436 \rightarrow 4 + 3 + 6 = 13 \rightarrow 1 + 3 = 4$$
$$+237 \rightarrow 2 + 3 + 7 = 12 \rightarrow 1 + 2 = 3 \Big\} \text{ We add.}$$

$$\textcircled{7}$$

$$673 \rightarrow 6 + 7 + 3 = 16 \rightarrow 1 + 6 = \textcircled{7}$$

The numbers 4 and 3 are the remainders we get when we divide 436 and 237, respectively, by 9. If the addition is correct, then the sum of these remainders should equal the remainder we get when we divide 673 by 9. It is the same in this case, so we are *probably* correct.

Comment We stated that if the remainders check, then our answer is *probably* correct. You cannot be 100% sure that your answer is correct because if you make a mistake that is a multiple of 9, then the remainders will not be affected by the error. The remainders will still check.

EXERCISES

In Exercises 1–12, perform the indicated operations and check your result by casting out 9's.

1. 768
   +423

2. 7893
   5329
   +6458

3. 582
   −293

4. 56834
   −37945

5. 4567
   −3672

6. 327
   ×645

7. 231
   × 68

8. 6038
   × 246

9. 5138
   6257
   +1894

10. 5462
    −2837

11. 2234
    ×5683

12. 1234
    ×1234

13. A teacher making up an arithmetic exam accidentally wrote the problem

$$\begin{array}{c} 2357 \\ +1245 \end{array} \quad \text{as} \quad \begin{array}{c} 2537 \\ +1245 \end{array}$$

She has prepared an answer sheet in advance and decides to check by casting out 9's. She does not find her error. Why?

14. What does casting out 9's have to do with congruence and modular arithmetic?

15. Is there any way to tell whether a number is divisible by 9 (without actually dividing by 9)?

16. By looking at the last digit only, can we tell whether a number is divisible by 9?

17. In base 3 the method of checking by casting out 9's will *not* work, since there are no 9's. Can you think of a method similar to that of casting out 9's that would work for base 3?

6.5
CASTING OUT 11'S

Another method commonly used to check computations is **casting out 11's.** In this method, for each number, *we start at the right* and move to the left. We put a + in front of the first digit, then a − in front of the second digit, and continue alternating the signs. We add the results. This is illustrated below.

$$2741 \rightarrow + 1 - 4 + 7 - 2 = 2$$
$$5673 \rightarrow + 3 - 7 + 6 - 5 = -3$$
$$781 \rightarrow + 1 - 8 + 7 \qquad = 0$$

From here on, the procedure is the same as that for casting out 9's.

EXAMPLE 1   Add 2741 and 781 and check the result by casting out 11's.

SOLUTION

$$2741 \rightarrow + 1 - 4 + 7 - 2 = 2$$
$$+\ 781 \rightarrow + 1 - 8 + 7 \qquad = 0$$
$$\left.\right\} \text{We add.}$$

$$3522 \rightarrow + 2 - 2 + 5 - 3 = ②$$

Since in both cases we get 2, our answer is probably correct.   ☐

EXAMPLE 2   Add 142936 and 782225 and check the result by casting out 11's.

SOLUTION

$$142936 \rightarrow + 6 - 3 + 9 - 2 + 4 - 1 = 13 \rightarrow + 3 - 1 = 2$$
$$+782225 \rightarrow + 5 - 2 + 2 - 2 + 8 - 7 = \qquad\qquad\qquad 4$$
$$\left.\right\} \begin{array}{l}\text{We} \\ \text{add.}\end{array}$$

$$925161 \rightarrow + 1 - 6 + 1 - 5 + 2 - 9 = \boxed{-16} \longleftarrow ⑥$$

−16 is congruent to 6 (mod 11) because −16 − 6, which is −22, is exactly divisible by 11. Hence our answer is probably correct.   ☐

EXAMPLE 3   Multiply 321 and 68 and check by casting out 11's.

SOLUTION

$$321 \rightarrow + 1 - 2 + 3 \qquad\qquad = 2$$
$$\times\ 68 \rightarrow + 8 - 6 \qquad\qquad\qquad = 2$$
$$\left.\right\} \text{We multiply.}$$

$$\begin{array}{r} 2568 \\ \hline 1926 \\ \hline \end{array}$$

$$④$$

$$21828 \rightarrow + 8 - 2 + 8 - 1 + 2 = 15 \rightarrow + 5 - 1 = ④$$

Hence our answer is probably correct.   ☐

EXERCISES

In Exercises 1–12, perform the indicated operations and check your result by casting out 11's.

1.   278
        +587

2.   394
        +685

3.   4642
        +8975

4.   6832
        7614
        +8329

5.   321
        −263

6.   7123
        −3567

7.   8234
        −1762

8.   23
        ×69

9.   523
        × 65

10.   823
        ×746

11.   6942
        × 837

12.   7432
        ×6578

13. Can you explain why the method of casting out 11's works?

14. Several students were asked to multiply 69 and 73. One answer given was 5048. Check by casting out 9's and by casting out 11's. What happens? Can you explain why this happens?

15. Perform each of the following calculations in the indicated base.

a) $212_{(8)}$
       $+627_{(8)}$

b) $27_{(8)}$
       $×45_{(8)}$

c) $376_{(8)}$
       $×127_{(8)}$

d) Check each of the answers obtained in parts (a), (b), and (c) by casting out by a number one less than the base. (This method is similar to the method of casting out 9's.)

6.6

GROUPS

In Section 6.2 we pointed out that the mod 4 clock (or modulo 4 arithmetic, which is the same thing) has the following important properties.

1. The binary operation $\oplus$ is closed.

2. $\oplus$ is associative.

3. 0 is an identity for $\oplus$.

4. Every number on the clock has an inverse for $\oplus$.

Clearly, any other clock besides the mod 4 clock has these four properties also. Many other systems found in widely varying branches of mathematics also have these properties. Such systems are called **groups.** Because they occur so often, mathematicians devote a good deal of attention to them. We will first give a formal definition of a group and then illustrate it with several examples.

Definition 6.2   *A **group** is a set of elements G, together with a binary operation "∘", with the following properties.*

Property 1.   ∘ is closed.

Property 2.   ∘ is associative.

Property 3.   There is an identity element in G for "∘." If we call this identity element i, then this means that for any element a in G we have a ∘ i = a and i ∘ a = a. In other words, i is the "do-nothing" element for "∘." In the mod 4 clock, 0 was the identity for the operation ⊕.

Property 4.   Every element in G has an inverse for "∘." This means that for any element a, we can find another element which operating on a with "∘" gives i. For the mod 4 clock we actually listed the inverse for each element on p. 304.

Comment   The operation ∘ may vary from one situation to another, as we shall see. It may sometimes be ordinary addition or multiplication. Sometimes it will be the turning of a clock hand as in clock arithmetic. In other situations it will be a totally different operation.

The best way to understand groups is to look at some examples.

EXAMPLE 1   Let G = {integers} and let ∘ stand for ordinary addition. This system is a group. Let us see why.

In Chapter 4 we saw that addition of integers is closed and associative. We also saw that 0 is the identity element. So the first three properties are satisfied.

Let us verify Property 4. Consider the integer 3. Its inverse is -3, since $3 + (-3) = 0$, which is the identity. Similarly the inverse of -11 is $+11$, since $(-11) + (+11) = 0$. It is clear that *every* element has an inverse. Thus Property 4 is satisfied.

Since all four properties are satisfied, it follows that this system is a group. ▢

EXAMPLE 2   Let G = {integers} and let ∘ be ordinary multiplication. Let us see if this system is a group.

Again we know from our work in Chapter 4 that multiplication of integers is closed and associative. We also know that 1 is the identity for multiplication. So Properties 1, 2, and 3 are satisfied. What about Property 4? Consider the integer 3. Does it have an inverse for multiplication? We are asking if there is an *integer* such that

$$3 \cdot (\text{integer}) = 1.$$

The only number which when multiplied by 3 gives 1 is $\frac{1}{3}$. However, $\frac{1}{3}$ is not an integer. Thus *there is no element in G that is an inverse for 3*. Property 4, which says that *every* element must have an inverse, is *not* true in this case. So this system is *not* a group. ▢

EXAMPLE 3   In Example 2 we saw that the set of integers with the operation of multiplication did not form a group. The problem was that not every element had an inverse. The only numbers that could be inverses were rational numbers like $\frac{1}{3}$.

If we change G from the integers to the rational numbers, then we *do* have numbers like $\frac{1}{3}$. Thus if we let G = {rational numbers} and let ∘ be ordinary multiplication, it looks as if we have a group. Unfortunately, this is not the case; 0 has no inverse. If you doubt this, try to find a number which when multiplied by 0 will give you 1. It can't be done! However, if we let G = {rational numbers, without 0} and let ∘ be multiplication, then we *do* have a group. ☐

EXAMPLE 4   Let G = {−1, 1} and let ∘ be ordinary multiplication. We will verify that this is a group.

Property 1.

$$1 \cdot 1 = 1$$
$$1 \cdot (-1) = -1$$
$$(-1) \cdot 1 \doteq -1$$
$$(-1) \cdot (-1) = +1$$

These results are summarized in the following table:

∘	−1	1
−1	1	−1
1	−1	1

All possible ways of multiplying elements of G result in some element of G. Thus ∘ is a closed operation.

Property 2.   Since ∘ is just ordinary multiplication, we already know that the associative law holds for ∘.

Property 3.   1 is the identity.

Property 4.   The inverse of 1 is 1, since $1 \cdot 1 = 1$. The inverse of −1 is −1, since $(-1) \cdot (-1) = 1$.

Therefore this system is a group. ☐

EXAMPLE 5   Let G = {1, −1} and let ∘ be ordinary addition. Then this system is *not* a group. For one thing, ∘ is not a closed operation, since $1 + 1$ is 2, and 2 is not an element of G. Moreover, there is no identity. Why? ☐

EXAMPLE 6   Consider a set G whose elements are a, b, and c. Let the operation ∘ be given by Table 6.4. Does this system form a group?

TABLE 6.4

∘	a	b	c
a	a	b	c
b	b	c	a
c	c	a	b

SOLUTION   We must verify that the four properties of a group hold.

Property 1.   Since all the elements in the table are elements of G, ∘ is a closed operation.

Property 2.   The associative law holds. We will verify it for one case. Readers should verify it for a few others to convince themselves.

$$
\begin{array}{rcl|rcl}
\multicolumn{3}{c|}{a \circ (b \circ c)} & \multicolumn{3}{c}{(a \circ b) \circ c} \\
 & = & a \circ a & & = & b \circ c \\
 & = & a & & = & a
\end{array}
$$

Property 3.   We see from the table that a is the identity, since a applied to any element results in the same element.

Property 4.   Every element has an inverse.
The inverse of a is a because $a \circ a =$ the identity which is a.
The inverse of b is c because $b \circ c =$ the identity which is a.
The inverse of c is b because $c \circ b =$ the identity which is a.

Since all four properties are satisfied, this system forms a group.   ▭

Some groups have an additional property called the *commutative* property. This is defined as follows.

Definition 6.3   *A group is called an **Abelian group** if the commutative property holds. This means that if a and b are any elements of G, then $a \circ b = b \circ a$.*

All the groups we have discussed so far are commutative.

EXERCISES

Determine whether each of the sets described in Exercises 1–11 forms a group under the specified operation.

1. The set of odd integers with the operation of addition.

2. The set of odd integers with the operation of multiplication.

3. The set of even integers with the operation of multiplication.

4. The set of negative integers with the operation of addition.

5. The set of negative integers with the operation of multiplication.

6. Mod 4 arithmetic, that is, 0, 1, 2, 3 (mod 4), with the operation of multiplication.

7. Mod 4 arithmetic, that is, 1, 2, 3 (mod 4), without zero with the operation of multiplication.

8. The set of multiples of 5 with the operation of addition.

9. The set of multiples of 5 with the operation of multiplication.

10. The set of real numbers with the operation of addition.

11. The set of real numbers with the operation of multiplication.

In each of Exercises 12–15 an operation $\circ$ is given by the table shown. Determine whether each is a group.

12.

$\circ$	a	b
a	a	b
b	b	a

13.

$\circ$	c	d
c	c	d
d	e	c

14.

$\circ$	d	e	f
d	d	e	g
e	e	d	f
f	f	e	d

15.

$\circ$	d	e	f
d	d	d	d
e	d	e	f
f	d	e	f

16. Which of the groups given in Exercises 1–15 are Abelian (or commutative) groups?

17. Let $G = \{a, b, c\}$ and let $\circ$ be the operation given by the following table:

$\circ$	a	b	c
a	a	b	c
b	b	a	c
c	c	b	a

a) Find the identity element.

b) Find the inverse of each element.

c) Is G an Abelian group?

18. Let $G = \{0, 5, 8, 9\}$ and let $\circ$ be defined by the following table:

$\circ$	0	5	8	9
0	0	0	0	0
5	0	5	0	8
8	0	8	5	9
9	0	8	9	8

Does this system form a group? Explain your answer.

19. Let $G = \{\alpha, \beta, \gamma\}$ and let $\circ$ be defined by the following table:

$\circ$	α	β	γ
α	α	β	γ
β	β	γ	α
γ	γ	α	β

Does this system form a group?

20. Let $G = \{1, 2, 3, 4, 5, 6\}$ and let $\circ$ be defined by

the following table:

∘	1	2	3	4	5	6
1	1	1	1	1	1	1
2	1	2	3	4	5	6
3	1	3	5	1	3	5
4	1	4	1	4	1	4
5	1	5	3	1	5	3
6	1	6	5	4	3	2

a) Is ∘ a closed operation?

b) Is there an identity? If yes, find it.

c) Is there an inverse for 3? If yes, find it.

d) Is there an inverse for 6? If yes, find it.

e) Is ∘ commutative?

f) Does this system form a group?

21. Let $G = \{\alpha, \beta, \gamma, \Delta\}$ and let ∘ be given the following table:

∘	α	β	γ	Δ
α	α	β	γ	Δ
β	β	α	Δ	γ
γ	γ	Δ	α	β
Δ	Δ	γ	β	α

a) Is ∘ a closed operation?

b) Is ∘ a commutative operation?

c) Is there an identity element? If yes, find it.

d) Find the inverse of each element.

e) Does this system form a group?

22. Let $G = \{x, y, z\}$ and let ∘ be given the following table:

∘	x	y	z
x	x	y	z
y	x	z	y
z	x	y	z

a) Find $(x \circ y) \circ (z \circ y)$.

b) Find $(x \circ y) \circ z$ and $x \circ (y \circ z)$.

c) Is the set closed under the operation of ∘? Explain.

d) Does this set form a group?

**23. Let $G = \{$all integers$\}$ and let ∘ be defined as follows: If a and b are in G, then $a \circ b = (a \cdot b) - (a + b)$. For example, if a and b are 6 and 7, then $a \circ b = (6 \cdot 7) - (6 + 7) = 42 - 13 = 29$. Does this system form a group?

**24. Let $G = \{$all rational numbers$\}$ and let ∘ be defined as follows: If a and b are in G, then $a \circ b = (a \cdot b) - \frac{a}{b}$. For example, if a and b are 3 and 2, then $a \circ b = (3 \cdot 2) - \frac{3}{2}$ or $4\frac{1}{2}$. Does this system form a group? Explain.

**25. Consider the following operations on $G = \{$integers$\}$.

i) □ is defined as $a \,\square\, b = a + 2b$.

ii) ⊙ is defined as $a \odot b = a$.

iii) △ is defined as $a \,\triangle\, b = a(a + b)$.

iv) ∗ is defined as $a * b = a + b - ab$.

For each of the above, answer the following questions.

a) Is the operation associative?

b) Is the operation commutative?

c) Is there an identity? If so, what is it?

d) Does the system form a group?

**26. Let $G = \{$rational numbers between 0 and 1 inclusive$\}$ and let ∘ be defined as $x \circ y = x + y - xy$, where x and y are any elements in G.

a) Is this system a group?

b) Is this system an Abelian group?

**27. Let $G = \{$integers$\}$ and let ∘ be multiplication modulo m, where m is a composite number. Does this system form a group?

28. Let $G = \{1, 2, 3, 4, 5\}$ and let ∘ be the operation of addition modulo 6. Does this system form a group? Explain.

29. Let $G = \{1, 2, 3, 4, 5, 6\}$ and let ∘ be the operation of multiplication modulo 8. Does this system form a group? Explain.

6.7

* **MORE EXAMPLES OF GROUPS**

In this section we will examine some interesting groups that arise when we move around geometric shapes such as triangles, squares, etc. (A background in geometry is not needed. You only have to know what triangles, squares, and rectangles are.)

EXAMPLE 1

B

A   *C*   **Figure 6.8**

Let us look at an **equilateral triangle.** This is just a triangle with three equal sides and three equal angles as shown (Fig. 6.8).

Suppose we cut such a triangle out of a wooden block. We ask the following question: How many ways can we put the triangle back into the wooden block without turning the triangle over? (We suggest that the reader cut such a triangle out of a piece of paper or cardboard and refer to it as he or she reads along.) The triangle and block are shown in Fig. 6.9.

Original position   **Figure 6.9**

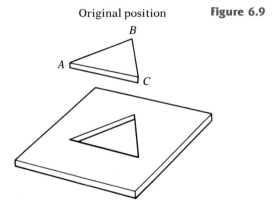

One way of putting the triangle back into the block is to put it back exactly as it was taken out. This we will call rotation *X*. (See Fig. 6.10.)

Another way of putting the triangle back into the block is to rotate it one-third of the way around (120°). This we call rotation *Y*.

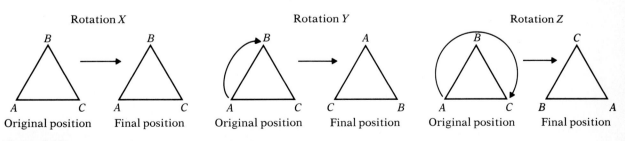

Rotation *X*   Rotation *Y*   Rotation *Z*

Figure 6.10

The only other way of putting the triangle back into the block is to rotate it two-thirds of the way around (240°). This we will call rotation Z.

These illustrations show the rotation performed on the triangles starting in the original positions only. However, we may perform the rotation from any starting position that fits into the block. One such possibility is shown in the rotation Y of Fig. 6.11.

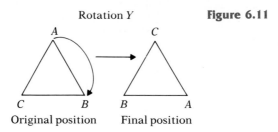

Figure 6.11

Rotation Y

Original position   Final position

Let G be the set containing these three rotations, that is, $G = \{X, Y, Z\}$. Now we define an operation $\circ$ on G in the following way: $Y \circ Z$ means "first apply rotation Y and then apply rotation Z to the result." If we start with the triangle in the original position, then $Y \circ Z$ is as illustrated in Fig. 6.12. $Y \circ Z$ leaves the triangle in the original position. This is the same as rotation X. Thus $Y \circ Z = X$.

Similarly, we define $X \circ Z$ as first applying X and then Z. If we start in the original position, then the result is as shown in Fig. 6.13. Thus $X \circ Z = Z$.

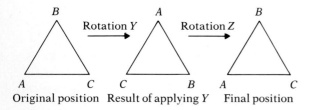

Rotation Y Rotation Z

Original position   Result of applying Y Final position

Figure 6.12

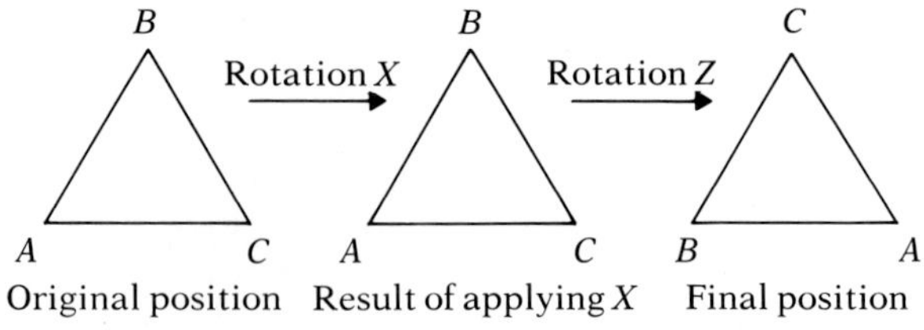

Rotation X Rotation Z

Original position   Result of applying X Final position

Figure 6.13

In the same way we define $\circ$ for other elements of G. We can summarize all possibilities by the following table:

$\circ$	X	Y	Z
X	X	Y	Z
Y	Y	Z	X
Z	Z	X	Y

Now we ask: Is this system a group? We must verify the four group properties.

Property 1.   From the table we see that ∘ is a closed operation.

Property 2.   It is easy to verify that the associative law holds. Try it.

Property 3.   The identity is X.

Property 4.   The inverse of X is X.
The inverse of Y is Z.
The inverse of Z is Y.   ☐

Hence this system is a group. As a matter of fact, it is even a commutative or Abelian group.

EXAMPLE 2   We consider the same situation as in the last example. However, we now allow the triangle to be turned over as well. We still have the three rotations as in the preceding example, but we rename them *symmetries X, Y,* and *Z.* In addition, we have three new ways of putting the triangle back into the block. These are shown in Fig. 6.14 and are called symmetries *P, Q,* and *R.* In each case we flip the triangle over around the dotted line.

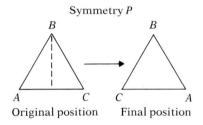
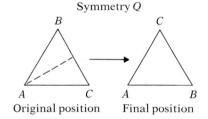
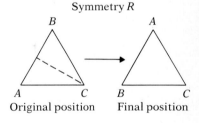

Figure 6.14

The operation ∘ is defined as in Example 1. Thus $X \circ P$ means "first apply X and then apply P." See Fig. 6.15.

Similarly, $Q \circ Y$ means "first apply Q and then apply Y." See Fig. 6.16.

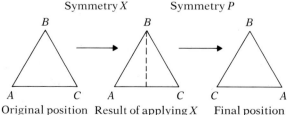

Figure 6.15

Figure 6.16

All possible results are summarized in the following table:

∘	X	Y	Z	P	Q	R
X	X	Y	Z	P	Q	R
Y	Y	Z	X	R	P	Q
Z	Z	X	Y	Q	R	P
P	P	Q	R	X	Y	Z
Q	Q	R	P	Z	X	Y
R	R	P	Q	Y	Z	X □

This system also forms a group called the **symmetries of a triangle.**
It is *not* a commutative group. (See p. 320.)

EXAMPLE 3   Suppose we now have a square cut out of a block of wood, as in the previous two examples. Let us find all possible ways of replacing the square in the block of wood. Flipping over is allowed. There are eight possible ways of doing this, as shown in Fig. 6.17. These are called **symmetries.**

    We define ∘ as we did for the triangle. For example, $r \circ v$ means "first apply r and then apply v to the result." The table for ∘ is shown below. This system forms a group called the **symmetries of the square.** It is clear from the table that ∘ is a closed operation.

∘	p	q	r	s	t	u	v	w
p	p	q	r	s	t	u	v	w
q	q	r	s	p	w	v	t	u
r	r	s	p	q	u	t	w	v
s	s	p	q	r	v	w	u	t
t	t	v	u	w	p	r	q	s
u	u	w	t	v	r	p	s	q
v	v	u	w	t	s	q	p	r
w	w	t	v	u	q	s	r	p

The associative law holds. Since there are 512 possible combinations, we will check just one of them:

$$
\begin{array}{c|c}
t \circ (q \circ r) & (t \circ q) \circ r \\
= \quad t \circ s & = \quad v \circ r \\
= \quad w & = \quad w
\end{array}
$$

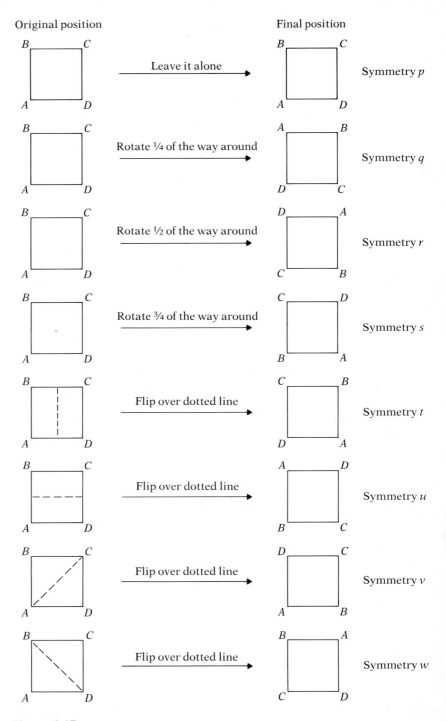

Figure 6.17

The identity element is obviously p. Every element has an inverse. For example, the inverse of s is q. □

It can be shown that the identity for a group is unique. This means that there cannot be more than one identity. As an illustration, if we look at the group of symmetries of the square (Fig. 6.17) we notice that the identity is p. There is no other identity. Now if we were to study *any* other group, we would immediately know that there is only one identity. We would not have to prove it.

This process is typical of mathematical activity. Mathematicians try to discover general properties and then apply them to specific situations as they arise. Of course, in establishing general concepts they are usually motivated by specific examples. (Refer back to the discussion of inductive and deductive reasoning in Chapter 2.)

EXERCISES

*1. Verify all cases of the associative law for Example 1.

*2. In Example 3, verify the associative law for three different cases.

*3. In Example 2, find the inverse of each element.

*4. a) Find all the symmetries (possible ways of putting it back into the wood block) for any rectangle that is not a square. See Fig. 6.18. (*Hint*: There are four of them.)

b) Show that this system is a group. It is called the **Klein 4 group.**

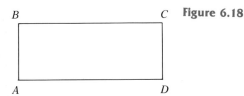

B   *C*   **Figure 6.18**

A   *D*

*5. An isosceles triangle is a triangle with two equal sides, as in Fig. 6.19 where side a = side

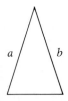

Figure 6.19

a b

b. (The third side does not necessarily have to equal the other two sides.)

a) Find all the symmetries of this isosceles triangle.

b) Do these symmetries form a group?

*6. A regular pentagon has five equal sides, as shown in Fig. 6.20.

a) Find all the symmetries of a regular pentagon.

b) Do these symmetries form a group?

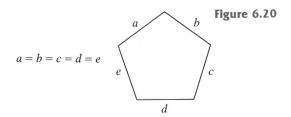

Figure 6.20

$a = b = c = d = e$

*7. Private Bond is standing at attention. Sergeant Stone comes along and gives him orders such as "right turn," "left turn," "about face," or "as you were." Let

> r stand for right turn,
> l stand for left turn,
> a stand for about face, and
> s stand for as you were.

Let ∘ represent any combination of these orders. For example, $l \circ r$ would mean "first make a left turn and then a right turn." Your result is, of course, the starting position. So $l \circ r = s$.

a) Make up a table for ∘.

b) Is ∘ closed?

c) Is ∘ commutative?

d) Is there an identity element? If so, find it.

e) Find the inverse of each element.

f) Find $l \circ (s \circ a)$ and $(l \circ s) \circ a$.

g) Does this system form a group?

*8. Show by example that the group of Example 2 is not commutative.

6.8

**STRICTLY
DETERMINED GAMES**

Let us look in on Jack Row and Jill Column, who are playing a game. Each player has two cards. Row has a black 7 and a red 7. Column has a black 7 and a red 4. The rules of the game are as follows.

Each player puts out one card at the same time. If the colors match, then Row is the winner. If the colors do not match, then Column is the winner. In each case the amount of money won is the difference of the numbers (the larger minus the smaller) appearing on the cards. All possible plays and the resulting payoffs are given in the following table, which is called the **payoff matrix.** Each entry represents Row's winnings or losses.

		Player Column	
		Bk 7	Rd 4
Player Row	Bk 7	0	−3
	Rd 7	0	3

If both players play black 7, then the colors match. Row wins, and the amount of money won is the difference of the numbers appearing on the cards. So Row wins $7 - 7$, or 0. We put 0 in the first row and first column.

If Row plays a black 7 and Column plays a red 4, then the colors do not match. Therefore Column wins. The amount of money won is the difference between the numbers, or $7 - 4$, which is 3. Since Column wins 3, Row loses 3. This we indicate by −3 in the chart.

If Row plays red 7 and Column plays black 7, then Column wins $7 - 7$, or 0.

If Row plays red 7 and Column plays red 4, then Row wins $7 - 4$, or 3.

What is the best move for each player? Row reasons as follows: If he plays black 7, then he can either break even or lose 3, depending on Column's play. He can never win. On the other hand, if he plays red 7, then he can either break even or win 3. (Remember that the entries in the chart always represent Row's winnings, which are the same as Column's losses.) Clearly, the best move for Row is a red 7. He can never lose any money with this play, no matter what Column does, and he could win 3.

What should Column do? She knows that Row will always play red 7. If Column plays black 7, then she will break even. On the other hand, if she plays red 4, then she will lose 3. Her best move would then be to play black 7. We circle each player's best move as shown:

Player Column

	Bk 7	Rd 4
Bk 7	0	−3
Rd 7	0	3

Player Row

Notice that one number has been circled twice. It is 0. This number is called the **value** of the game. If the value of a game is zero, then we say that it is a **fair** game. This game is obviously a fair game. What this means is that in the long run, the amount of money won by either Row or Column is 0 (assuming that both players do not make any foolish moves).

Perhaps the thought has crossed your mind that there is no point in playing this game, since no money can be won. In a sense you are right. However, you do not know that this is the case until you have analyzed it mathematically. In more complicated games it is usually impossible to tell without mathematical analysis whether a game is worth playing at all. Some games are too complicated to be analyzed, even mathematically.

HISTORICAL NOTE

John Von Neumann was born in Budapest, Hungary, on December 28, 1903. He was a brilliant mathematician and received the Ph.D. degree in mathematics from the University of Budapest in 1923 when he was only twenty years old.

He left Europe just before World War II to come to the United States. For many years he worked at the Institute for Advanced Study at Princeton University, as well as for the United States government. He was one of the scientists who developed the atomic bomb.

Together with Oskar Morgenstern in 1944, Von Neumann wrote a book called *Theory of Games and Economic Behaviour*, in which he developed the theory that enables the economist to study economic behavior by using methods of game theory. He proved that there is a "rational course of action" or best strategy in all games, even those that are not strictly determined. This important result allows the theory to be applied to many different subjects.

The game of *Go-Bang* in Japan is an ancient war game. A heavy block of wood is used as the board and the surface is divided into squares by cross lines. Smooth elliptically shaped stones, called "Ishi" in Japanese, are used for "men."

Peabody Museum of Salem, photo by Mark Sexton

Let us refer back to the game on page 329. The payoff matrix again is

	Bk 7	Rd 4
Bk 7	0	−3
Rd 7	0	3

What is the smallest number in the first row? It is −3. (Remember, −3 is smaller than 0.) Is −3 the largest number of the column that it is in? Clearly not, since 3 is larger.

In the second row the smallest number is 0. Is 0 the maximum (largest number) in that column? Since there is no number that is larger than 0, then 0 is the maximum. In other words, *this number is at the same time the smallest in its row and the largest in its column.* Whenever a number like this exists in a game, then we say the game is **strictly determined.** We restate as definitions the important ideas given above.

Definition 6.4   *The **payoff matrix** of a game is a table of numbers representing the amount of money won or lost by each player as a result of any move. Each entry represents player Row's winnings or losses.*

Definition 6.5   *A game is said to be **strictly determined** if there is a number in the payoff matrix that is **at the same time** the smallest number of the row it's in and the largest number of its column.*

Definition 6.6   *If we can find a number satisfying the conditions described by Definition 6.5, then this number is called the **value** of the game. We will refer to the value of a game as v.*

Definition 6.7   *A game is called a **fair game** if its value is zero.*

Definition 6.8   *By an **optimal strategy** for a player we mean a best move. This will mean either winning the most amount of money or losing the least amount.*

In any game a player is interested in finding his or her optimal strategy (best move). If we know that a game is strictly determined, then it is very easy to find the optimal strategy for each player. This is done as follows:

Player Row: Play the row that contains the value v. (If there is more than one such row, play either.)

Player Column: Play the column that contains the value v. (If there is more than one such column, play either.)

The following examples illustrate all of these ideas.

EXAMPLE 1   Row and Column are playing a card game. Each player has only three cards, which are a 2, a 4, and a 5. The rules are as follows. At a given signal, each player puts out a card. If the numbers match, then nobody wins. If the numbers don't match, the person with the higher number wins. The amount won is the sum of the numbers shown. The payoff matrix is the following.

		Player Column		
		2	4	5
	2	0	−6	−7
Player Row	4	6	0	−9
	5	7	9	0

The smallest number in the first row is −7. This is *not* the largest in its column; 0 is larger.

The smallest number in the second row is −9. This is *not* the largest in its column.

The smallest number in the third row is 0. This *is* the largest in its column.

This means that the game is strictly determined and its value is 0. Since the value, v, is 0, the game is *fair*. The value v is in the third row and third column. The optimal strategies are:

Player Row: Play row 3, or the card with 5 on it.

Player Column: Play column 3, or the card with 5 on it.   ☐

EXAMPLE 2   In a card game, each player has three cards, a 2, a 4, and a 5. Each player puts out one card at the same time. If the numbers match, then player Row wins the sum of the numbers. If the numbers don't match, then player Column wins the sum of the numbers. The payoff matrix is the following.

Player Column

		2	4	5
	2	4	−6	−7
Player Row	4	−6	8	−9
	5	−7	−9	10

Is this game strictly determined? To answer this, we notice that the smallest number in row 1 is −7. This is not the largest number in the third column.

The smallest number in row 2 is −9. This again is not the largest in the third column.

In row 3 the smallest number, −9, is not the largest in the second column. Since there is no number that is the smallest of the row and the largest of its column, the game is *not* strictly determined. ☐

EXAMPLE 3

Choosing. Two players, Sue and Pete, are "choosing." Each player shows 1 or 2 fingers at the same time. If they both show the same number of fingers, then Sue wins the sum of the fingers shown. If they both do not show the same number of fingers, then Pete wins the sum of the fingers shown. The payoff matrix is shown at the left.

Pete

		1	2
	1	2	−3
Sue	2	−3	4

During and after the period of the Crusades the cultural exchange between Christians and Moslems included, along with such things as Arabic numerals and the works of Aristotle, the game of chess. From a thirteenth century manuscript of rules comes this miniature in which a Christian and a Moslem play at chess. In 1985 and again in 1986, Anatoly Karpov and Gary Kasparov competed for the world chess championship.

Time Magazine

Is this game strictly determined? The smallest number in row 1 is −3. This is not the largest in the second column. Similarly, −3, which is the smallest of the second row, is not the largest of the first column, so the game is not strictly determined. ☐

EXAMPLE 4   Suppose we have a game whose payoff matrix is given by

		Player C			
		1	2	3	4
	1	1	4	5	9
Player R	2	0	−2	−7	−8
	3	1	3	2	8

The smallest number of the first row is 1. This is also the largest number in the column. Similarly, the minimum of the third row is 1. This is again the largest number in the column.

The game is strictly determined. The value of the game is 1. This means that the game is not fair. It is in favor of Row. The optimal strategies are as follows.

Player R: Play *either* row 1 or row 3.

Player C: Play column 1. ☐

EXAMPLE 5   *Stone, Paper, and Scissors.* Mrs. Jones has told her two children, Pat and Ned, to play the game of stone, paper, and scissors. The game works as follows: Each player says the word "stone" or "scissors" or "paper." Scissors cuts paper, paper wraps stone, and stone breaks scissors. Therefore if one says "stone" and the other says "scissors," then "stone" wins a lollipop. If one says "scissors" and the other says "paper," then "scissors" wins a lollipop. Similarly, if one says "stone" and the other says "paper," then "paper" wins a lollipop. If they both say the same thing, then the game is a tie. The payoff matrix is shown below.

		Ned		
		Scissors	Stone	Paper
	Scissors	0	−1	+1
Pat	Stone	+1	0	−1
	Paper	−1	+1	0

Is this game strictly determined? If so, find the optimal strategies for each player. ☐

EXAMPLE 6 Two car rental agencies offer weekly discounts to customers in certain weeks. If they both offer the discount in the same week, nobody gets any points. If one company offers a discount and the other company offers this discount a week later, credit the first company with 5 points. If one company offers a discount and the other company offers it two weeks later, credit the first company with 3 points. The payoff matrix is as shown.

		Company 2 offers discount		
		Week 1	Week 2	Week 3
	Week 1	0	+5	+3
Company 1 offers discount	Week 2	−5	0	+5
	Week 3	−3	−5	0

This chart is read as follows: The entry in the second row and third column indicates that company 1 offers the discount in week 2 and company 2 offers the discount in week 3. Thus we credit company 1 with 5 points. The +5 entry represents this.

The business contest between the two companies is a strictly determined game. The value is 0. The optimal strategy for each company is to offer the discount in week 1. ☐

EXERCISES

1. What do you think are the optimal strategies for each player in Example 2?

2. **a)** If you were Sue, how would you play the game in Example 3?

   b) How would you play if you were Pete?

3. Why is the game of Example 4 in favor of Player R?

4. **a)** Is the game of Example 5 strictly determined?

   b) Are there optimal strategies? Explain.

5. In Example 6, why is the contest a strictly determined game?

Each of the payoff matrices given in Exercises 6–13 represents some game. Determine whether each game is strictly determined or not. If the game is strictly determined, find the value of the game and the optimal strategies for each player. Are any of the games fair?

6.
C

R	4	10
	3	−2

7.
C

R	8	0
	0	−4

8.
C

R	4	−4
	−4	4

9.
C

R	0	0
	0	8

10.

	C	
R	0	0
	0	-3

11.

	C		
	0	-2	8
R	14	-16	18
	8	-12	6

12.

	C			
	0	2	-2	4
	-6	2	8	-4
R	2	0	-12	0
	-16	6	0	8

13.

	C			
	4	10	2	8
	6	2	-10	0
R	8	0	-6	2
	0	4	-8	0

14. Lisa and Felix have each been given three cards as follows: Lisa has a black 4, a red 8, and a red 9. Felix has a black 3, a red 7, and a black 10. They each put out a card at the same time. If the colors match, then Lisa wins the sum of the numbers appearing on the cards. If they don't match, then Felix wins the sum of the numbers.

    a) Set up the payoff matrix for this game.

    b) Is the game strictly determined?

15. Karen and Frank have each hidden one or two single dollar bills under a hat. At a given signal, both hats are raised and the number of dollar bills hidden under each is revealed. If they both have hidden the same number of dollar bills, then Karen wins an amount of money that represents the sum. Otherwise, Frank wins an amount that represents the product of the number of dollars shown.

    a) Set up the payoff matrix for this game.

    b) Is the game strictly determined?

16. Two competing taxi services, Yellow Cab Co. and Expresso Cab Co., are considering expanding their operations into two nearby cities, Blakeville and Bushtown. Credit the Yellow Cab Co. with 10 points if it decides on Blakeville and the Expresso Cab Co. decides on Bushtown. Credit the Expresso Cab Co. with 8 points if it decides on Blakeville and the Yellow Cab Co. decides on Bushtown. Credit both companies with 0 if they both decide on the same city.

    a) Set up the payoff matrix.

    b) Is the game strictly determined?

    c) Find the value of the game.

    d) Is it fair?

    e) Find the optimal strategies for each company.

17. *Promotions.* Bill and Gail are salespeople for a large publishing company. The company is considering promoting either or both of them to district managers. Bill has promised to pay Gail $100 if they both are promoted. Otherwise, Gail will pay him $40.

    a) Set up the payoff matrix.

    b) Find the value of the game (if it exists).

6.9

NON-STRICTLY DETERMINED GAMES   In Section 6.8 we discussed how to find the optimal strategies for strictly determined games. We have come across many games that are not strictly determined (that is, there exists no number that is the smallest of its row and also the largest of its column). In this section we will consider how to find the optimal strategies for such games. For simplicity we will consider only 2 × 2 games. These are games in which each player has only two possible moves.

EXAMPLE 1   Consider a game whose payoff matrix is as shown below.

		C Bk 2	Rd 4
R	Bk 1	0	6
	Rd 3	6	0

This is a non-strictly determined game. However, it is not obvious what R should do. If R plays black 1, then he either breaks even or wins 6. If R plays red 3, then he again will win 6 or break even. It would seem that R should play either black 1 or red 3, since in both cases, the amount of money won is the same. If R always plays black 1, then C will undoubtedly notice this. C will then play black 2 to break even. Similarly, if R always plays red 3, then C will again notice this and play red 4 to break even. R may decide to alternate his play between black 1 and red 3. This is also dangerous, since C is sure to notice this and play accordingly. The best strategy for R is to play black 1 sometimes and red 3 at other times, but not according to any pattern that C will recognize. One way of doing this is to flip a coin (without C seeing what is being done) and play black 1 if heads comes up and play red 3 if tails comes up. In the long run, heads and tails will each come up approximately half the time. Thus R will end up playing black 1 half the time and red 3 half the time.

C should follow a strategy similar to that of R. He should toss a coin and play black 2 or red 4 depending on whether heads or tails comes up. Why is this the best strategy for C?   □

EXAMPLE 2   Mabel and Bob are playing the following game. Mabel hides either 1 penny or 2 pennies in her hand. Bob has to guess 1 or 2 and wins the number of pennies if he guesses correctly. Otherwise he wins nothing. The payoff matrix of this game is as follows:

		Bob guesses 1	2
Mabel hides	1 penny	−1	0
	2 pennies	0	−2

This game is not strictly determined. If Mabel always hides 1 penny and Bob observes this pattern, then Bob will always say 1.

Mabel will always lose 1 penny. Similarly, if Mabel always hides 2 pennies and Bob figures this out, then Bob will always say 2. Mabel will always lose 2 pennies.

It is obvious that Mabel should hide 1 penny sometimes and 2 pennies other times, but according to no particular pattern. She should *mix* her strategies. How is this to be done?

You may think that Mabel could do what R did in the last example. In that situation, no matter what R did, his payoff was either 0 or 6. This meant that any move he made would give him the same amount of money. In this example, Mabel loses different amounts of money, depending on her move. Thus just flipping a coin will not work. For this example and others like it, we need the following results. □

It can be shown[2] that a non-strictly determined game whose payoff matrix is

$$C$$

	Column 1	Column 2
Row 1	a	b
Row 2	c	d

R

has the following optimal strategies. (The formulas may look frightening, but don't let them scare you off; they are really quite simple to use.)

RULE 6.1 *Let*

$$R_1 = \frac{d - c}{a + d - b - c} \qquad R_2 = \frac{a - b}{a + d - b - c}$$

$$C_1 = \frac{d - b}{a + d - b - c} \qquad C_2 = \frac{a - c}{a + d - b - c}$$

Player R: *Depending on the total number of games played, play row 1, R_1 of the time and row 2, R_2 of the time.*

Player C: *Depending on the total number of games played, play column 1, C_1 of the time and column 2, C_2 of the time.*

The value of the game is

$$v = \frac{ad - bc}{a + d - b - c}.$$

---

2. See, for example, J. Kemeny, J. Snell, and G. Thompson, *Introduction To Finite Mathematics*, pp. 343–350. Englewood Cliffs, N.J.: Prentice-Hall, 1966.

Comment The value for a non-strictly determined game is not defined as it was for strictly determined games. In this case the value is given by a formula.

Let us now return to Example 2. The payoff matrix was as follows.

		Bob guesses	
		1	2
Mabel hides	1 penny	−1	0
	2 pennies	0	−2

Here $a = -1$, $b = 0$, $c = 0$, and $d = -2$. Therefore

$$R_1 = \frac{d - c}{a + d - b - c} = \frac{(-2) - 0}{(-1) + (-2) - 0 - 0} = \frac{-2}{-3} = \frac{2}{3}$$

$$R_2 = \frac{a - b}{a + d - b - c} = \frac{(-1) - 0}{(-1) + (-2) - 0 - 0} = \frac{-1}{-3} = \frac{1}{3}$$

$$C_1 = \frac{d - b}{a + d - b - c} = \frac{(-2) - 0}{(-1) + (-2) - 0 - 0} = \frac{-2}{-3} = \frac{2}{3}$$

$$C_2 = \frac{a - c}{a + d - b - c} = \frac{(-1) - 0}{(-1) + (-2) - 0 - 0} = \frac{-1}{-3} = \frac{1}{3}$$

$$v = \frac{ad - bc}{a + d - b - c} = \frac{(-1)(-2) - 0 \cdot 0}{(-1) + (-2) - 0 - 0} = \frac{+2}{-3} = \frac{-2}{3}$$

Thus the optimal strategies are for R (or Mabel) to play row 1 (1 penny) $\frac{2}{3}$ of the time and to play row 2 (2 pennies) $\frac{1}{3}$ of the time.

Similarly, the optimal strategies for C (or Bob) are to play column 1 (guess 1) $\frac{2}{3}$ of the time and to play column 2 (guess 2) $\frac{1}{3}$ of the time.

The value of the game is $-\frac{2}{3}$. Since the value is negative, this means that the game is in favor of C (Bob). We could have guessed this, since the payoff matrix indicates that R can never win.

R now knows that she should play row 1, $\frac{2}{3}$ of the time, and row 2, $\frac{1}{3}$ of the time. Again she does not want to play row 1 twice and then row 2 once, constantly repeating this pattern, since C is bound to catch on and will play accordingly. She is therefore looking for some way of playing row 1, $\frac{2}{3}$ of the time, and row 2, $\frac{1}{3}$ of the time, but without C being able to figure out what R is going to do next. One way of doing this is to use a spinner such as the one shown in Fig. 6.21.

Figure 6.21

It is obvious that the pointer will land in the shaded portion $\frac{2}{3}$ of the time and in the unshaded portion $\frac{1}{3}$ of the time. R should spin the pointer. If it lands in the shaded area, she should play row 1 (hide 1 penny). If the pointer stops in the unshaded area, then she should play row 2 (hide 2 pennies). By using this scheme, R will end up playing row 1, $\frac{2}{3}$ of the time, and row 2, $\frac{1}{3}$ of the time.

C can use a similar spinner to determine his moves.

EXAMPLE 3 Row and Column each have two cards. Row has a black 4 and a red 4. Column has a black 2 and a red 3. Each player puts out a card at the same time. If the colors match, Row wins the difference. Otherwise Column wins the difference. The payoff matrix is shown at the left.

		Column	
		Bk 2	Rd 3
	Bk 4	2	−1
Row	Rd 4	−2	1

This game is not strictly determined. If Row plays black 4, then he can win 2 but lose only 1. If he plays red 4, then he can win 1 but lose 2. His best move would seem to be to play black 4. However, if Row always plays black 4, then Column will undoubtedly notice this. She will then play red 3 all the time, and Row will wind up losing 1 each time.

The formula tells us how often Row should play black 4 and red 4. It also will tell us Column's best strategy. We have the following:

$$a = 2, \quad b = -1, \quad c = -2, \quad d = 1;$$

$$R_1 = \frac{d - c}{a + d - b - c} = \frac{1 - (-2)}{2 + 1 - (-1) - (-2)}$$

$$= \frac{1 + 2}{2 + 1 + 1 + 2} = \frac{3}{6} = \frac{1}{2};$$

$$R_2 = \frac{a - b}{a + d - b - c} = \frac{2 - (-1)}{2 + 1 - (-1) - (-2)}$$

$$= \frac{2 + 1}{2 + 1 + 1 + 2} = \frac{3}{6} = \frac{1}{2};$$

$$C_1 = \frac{d - b}{a + d - b - c} = \frac{1 - (-1)}{2 + 1 - (-1) - (-2)}$$

$$= \frac{1 + 1}{2 + 1 + 1 + 2} = \frac{2}{6} = \frac{1}{3};$$

$$C_2 = \frac{a - c}{a + d - b - c} = \frac{2 - (-2)}{2 + 1 - (-1) - (-2)}$$

$$= \frac{2 + 2}{2 + 1 + 1 + 2} = \frac{4}{6} = \frac{2}{3};$$

$$v = \frac{ad - bc}{a + d - b - c} = \frac{2(1) - (-1)(-2)}{2 + 1 - (-1) - (-2)}$$

$$= \frac{2 - 2}{2 + 1 + 1 + 2} = \frac{0}{6} = 0.$$

The optimal strategies are

$\quad$ *Player Row:* Play row 1 (black 4) $\frac{1}{2}$ of the time.
$\qquad\qquad\quad$ Play row 2 (red 4) $\frac{1}{2}$ of the time.
$\quad$ *Player Column:* Play column 1 (black 2) $\frac{1}{3}$ of the time.
$\qquad\qquad\qquad$ Play column 2 (red 3) $\frac{2}{3}$ of the time.

The value of the game is zero, so that the game is fair.

$\quad$ Row can flip a coin each time and play black 4 if heads comes up and red 4 if tails comes up. In the long run (assuming the coin is fair), approximately $\frac{1}{2}$ of the tosses will show heads, and $\frac{1}{2}$ will show tails. Thus Row will end up playing black 4, $\frac{1}{2}$ of the time and red 4, $\frac{1}{2}$ of the time.

$\quad$ In order for Column to play black 2, $\frac{1}{3}$ of the time and red 3, $\frac{2}{3}$ of the time, she should use a spinner such as the one shown in Fig. 6.21. $\qquad\square$

Comment After you have computed the values of $R_1, R_2, C_1,$ and $C_2,$ you can check your answers by using the fact that $R_1 + R_2 = 1$ and $C_1 + C_2 = 1.$

EXERCISES

Each of the payoff matrices in Exercises 1–8 is for a non-strictly determined game. In each case, find the values of $R_1, R_2, C_1, C_2,$ and $v.$

1.

$$C$$
$$R \begin{array}{|c|c|} \hline 2 & 8 \\ \hline 6 & 4 \\ \hline \end{array}$$

2.

$$C$$
$$R \begin{array}{|c|c|} \hline 4 & -4 \\ \hline 0 & 2 \\ \hline \end{array}$$

3.

$$C$$
$$R \begin{array}{|c|c|} \hline -12 & 10 \\ \hline 14 & 8 \\ \hline \end{array}$$

4.

$$C$$
$$R \begin{array}{|c|c|} \hline 16 & 20 \\ \hline 18 & -10 \\ \hline \end{array}$$

5.

$$C$$
$$R \begin{array}{|c|c|} \hline -16 & 6 \\ \hline 12 & -8 \\ \hline \end{array}$$

6.

$$C$$
$$R \begin{array}{|c|c|} \hline -10 & 0 \\ \hline 0 & -16 \\ \hline \end{array}$$

7.

$$C$$
$$R \begin{array}{|c|c|} \hline -4 & 3 \\ \hline 1 & -1 \\ \hline \end{array}$$

8.

$$C$$
$$R \begin{array}{|c|c|} \hline 5 & -3 \\ \hline -2 & 8 \\ \hline \end{array}$$

9. Diana and Jane are playing the following game. Diana hides either a $5 bill or a $10 bill under a hat. Jane must guess which bill has been hidden. If she guesses correctly, then she gets the amount guessed. Otherwise, she must give Diana the amount guessed.

$\quad$ **a)** Set up the payoff matrix.

$\quad$ **b)** Verify that the game is not strictly determined.

$\quad$ **c)** Find the value of the game.

$\quad$ **d)** Is the game fair?

$\quad$ **e)** Find the optimal strategies for each player.

10. Sam and Jim are choosing by each showing one or two fingers at the same time. If they both show the same number of fingers, then Sam wins an amount equal to the *product* of the number of fingers shown. Otherwise Jim wins an amount equal to the difference of the number of fingers shown (largest minus smallest).

$\quad$ **a)** Set up the payoff matrix.

b) Find the value of game.

c) Is the game fair? If not, in whose favor is the game?

d) Find the optimal strategies for each player.

11. Sam and Jim again are choosing by each showing one, two, or three fingers. If the sum of the fingers shown is even, Sam gets the sum; if odd, Jim gets the sum.

a) Set up the payoff matrix.

b) Can you find the value of the game?

c) Using the techniques discussed in this section, can you find the optimal strategies for each player? Explain.

12. *War Games.* The following is a war game. A navy ship is to shell a coastal town. If the ship comes in close to the shore, the shelling will of course be more accurate. It is also more dangerous for the ship, since the enemy has coastal gunners whose accuracy is greater at short range. The coastal guns may be adjusted in one of two positions: short range and long range. Credit the ship with 20 points if it avoids enemy gunfire and −8 points if it is within range of the guns. Also credit the ship with 9 extra points for accurate shelling if it comes in close.

a) Set up the payoff matrix.

b) Find the optimal strategies for each side.

c) What device should each side use to carry out this strategy?

13. Two competing businessmen, John Grimes and Peter Sharp, have decided to open a computer supplies store in either of two sections of a city. The following payoff matrix indicates the differences in projected weekly sales (in thousands of dollars) depending upon which section of the city each one chooses.

		Peter chooses	
		North side	South side
John chooses	North side	16	15
	South side	14	18

a) Find the optimal strategies for each businessman.

b) What device should each businessman use to carry out this strategy?

14. Why, in the above games, is it necessary for a player to use a spinner or a coin to decide on a move? Why can't the player decide what move to make the appropriate number of times without such a device?

A professional war game specialist in London, England, prepares his moves, like a general, on his operations table.

Bandphoto

15. *Sherlock Holmes Escaping Death.* In Arthur Conan Doyle's famous story "The Final Problem," Sherlock Holmes was being pursued by his archenemy Professor Moriarty. In the story, Holmes boards a train traveling from London to Dover in an attempt to make good his escape to the Continent. However, as the train speeds out of London Station, Holmes and Professor Moriarty spot each other. Professor Moriarty decides to charter a special train and pursue Holmes. There is only one stop between London and Dover. This is at Canterbury. What should each person do? Should Sherlock or Professor Moriarty get off at Canterbury or go all the way to Dover?

If both get off at Dover or both get off at Canterbury, then Moriarty is sure to kill Holmes. If Professor Moriarty gets off at Canterbury and Holmes goes on to Dover then he will escape the professor (at least temporarily). If Holmes gets off at Canterbury and Professor Moriarty goes on to Dover, the chase is not over. It can be considered a temporary draw. Assign a value of +300 if Professor Moriarty catches

Holmes and kills him. Assign a value of $+150$ to Holmes if he escapes to the Continent and a value of $+40$ if he gets off at Canterbury whereas Moriarty gets off at Dover.

a) Set up the payoff matrix.

b) Find the value of the game. Is the game fair?

c) What should Sherlock Holmes do?

16. *Prisoner's Dilemma.* Police are interrogating two suspects, Al and Nina, in an attempt to find out more information on the latest terrorist bombings that have plagued the city. The suspects are being questioned separately. The prosecuting attorney knows that he does not have enough evidence to convict either of them but nevertheless approaches each prisoner separately with the following deal: "If you confess, then you will be set free and your partner will be given a 30-year jail sentence. If both you and your partner do not confess, then both of you will be given a 26-year jail sentence. If both of you do confess, then each of you will get a light jail sentence of 2 years.

a) Set up the payoff matrix.

b) Find the optimal strategy for each prisoner.

17. *Escaping Bank Robber.* A bank robber is being chased by a police car when he comes to the end of the road. He can turn either right or left. If he turns right and the pursuing police car turns right also, then there is an 80% chance that he will be captured. If he turns right and the police car turns left, then there is a 20% chance that he will be captured by other police cars. If he turns left, then there is a 50% chance that he will evade being captured, since the area is covered by heavy brush.

a) Set up the payoff matrix.

b) Find the optimal strategy for the escaping bank robber and for the pursuing police officer.

18. Read the article "Military Decision and Game Theory" by O. G. Haywood, which appeared in the *Journal of the Operations Research Society of America* (Volume 2, pp. 365–385, 1954) to see how game theory can be used in actual war games.

19. Read the article "Aircraft Landing Fees: A Game-Theory Approach" by S. C. Littlechild and G. F. Thompson, which appeared in the *Bell Journal of Economics* (Volume 8, no. 1, pp. 186–204, 1977) to see how the officials of Birmingham Airport in England used game theory techniques to determine airport landing fees as well as the optimal runway size.

6.10

CONCLUDING REMARKS

In this chapter we have barely scratched the surface of game theory. We have discussed very simple games only. However, far more complicated games can also be analyzed mathematically. This, of course, does not mean that game theory will provide optimal strategies for every game.

A simplified form of poker has been studied by mathematicians, and techniques have been devised for finding optimal strategies. However, even this version of the game (which is played with a deck of only three cards) requires a payoff matrix that has 1728 entries. To calculate the optimal strategies for each player with an accuracy of ten percent, we would have to perform about two billion multiplications and additions. (Of course, part of the strategy is that while

Modern computers that have been programmed with the optimal strategies can never lose when playing the game tic-tac-toe.

you are doing these computations, your opponent will get tired of waiting and give up.)

Very complicated games, such as regular poker, chess, or bridge, involve so many computations that no formula exists that will give us the optimal strategies for each move in a reasonable amount of time. It has been estimated that even in a simple game where one player has 100 possible moves and his or her opponent has only 200 possible moves, it would take a modern high-speed computer a rather long time (many months) to compute the optimal strategies. For a human being to attempt this would therefore be ridiculous.

HISTORICAL NOTE

Computers have been programmed to play chess. One of the first people to do this was the mathematician A. M. Turing, who designed a chess-playing machine called *MADAM*. His machine was a very poor chess player and made foolish moves. After several moves the machine would be forced to give up. Many advances have been made since Turing's machine. Today it is possible to play a fairly advanced game of chess with a computer.

However, no machine has been designed that analyzes every possible strategy corresponding to any move. It would take 10^{108} years to play all the possible games if we had a machine that could play a million games a second.

Computers have been programmed to play such a game as tic-tac-toe. The computer can never lose because it has been programmed with the optimal strategies.

A. I. Laboratory, Massachusetts Institute of Technology

Computers can not only write music and compose poetry; they can also play the intellectual game of chess.

6.11

SUMMARY

In this chapter we discussed clock arithmetic and how it is used. Our method of telling time on our familiar clock is a modular system. We applied the ideas of a modular system to checking calculations by either casting out 9's or casting out 11's.

Finally, we studied groups that occur in many areas of mathematics. By studying groups in general we can discover the properties that *all* groups have. Then when we work with a specific group, we know that *this* group will have all those properties. We do not have to reestablish these properties for this specific group.

We also applied the ideas of groups to the various symmetries.

Additionally, we studied two-person, zero-sum games. It turns out that when we set up the payoff matrix for any game, the game may be strictly determined or non-strictly determined.

To find the optimal strategies for strictly determined games, we find a number that is both the minimum of the row that it is in and also the maximum of its column. We then play the row or column that contains that number.

If the game is not strictly determined, we apply the rule given on p. 338 to find the optimal strategies for each player.

The above rules also enable us to determine the value of a game, which, in turn, will tell us if a game is fair.

STUDY GUIDE

You should now be able to demonstrate your knowledge of the following ideas by giving definitions, descriptions, or specific examples. Page references are given in parentheses.

Systems	Clock arithmetic (p. 303)	Symmetries of a square (p. 326)
	Modular arithmetic (p. 308)	Klein 4 group (p. 328)
	Group (p. 317)	Strictly determined games (p. 329)
	Abelian group (p. 320)	Non-strictly determined games
	Symmetries of a triangle (p. 326)	(p. 336)

Properties of Groups	Closure property (p. 317)	Inverses (p. 317)
	Identities (p. 317)	Commutative law (p. 320)
	Associative law (p. 317)	

Applications of Modular Arithmetic Were Given For	Casting out 9's (p. 312)	Casting out 11's (p. 316)

Other Ideas Discussed	Games of strategy (p. 302)	Congruence modulo a natural
	Two-person games (p. 302)	number (p. 310)
	Zero-sum games (p. 302)	Strictly determined games (p. 331)
	Closed operation (p. 304)	Payoff matrix (p. 331)
	Identity element (p. 304)	Value of a game (p. 331)
	Inverse element (p. 304)	Fair game (p. 331)
	Binary operation (p. 304)	Optimal strategy (p. 332)
	Modulus (p. 309)	Non-strictly determined games
		(p. 336)

FORMULAS TO REMEMBER

The following list summarizes all the formulas discussed in this chapter.

1. $a \equiv b \pmod{m}$ means that $a - b$ is exactly divisible by m.

2. A group is a set of elements G, together with a binary operation $\circ$, with the following properties:

a) $\circ$ is closed.

b) $\circ$ is associative.

c) There is an identity element in G for $\circ$.

d) Every element in G has an inverse under $\circ$.

For strictly determined games whose value is v, the optimal strategies are as follows.

Player R: Play the row that contains v.

Player C: Play the column that contains v.

For non-strictly determined games whose payoff matrix is

$$
\begin{array}{c}
\quad C \\
R\ \begin{array}{|c|c|}
\hline
a & b \\
\hline
c & d \\
\hline
\end{array}
\end{array}
$$

the optimal strategies are as follows:

Player R: Play row 1, R_1 of the time and row 2, R_2 of the time;

Player C: Play column 1, C_1 of the time and column 2, C_2 of the time;

where

$$
R_1 = \frac{d-c}{a+d-b-c}, \qquad R_2 = \frac{a-b}{a+d-b-c},
$$

$$
C_1 = \frac{d-b}{a+d-b-c}, \qquad C_2 = \frac{a-c}{a+d-b-c}.
$$

Also, for non-strictly determined games the value v is given by the formula

$$
v = \frac{ad-bc}{a+d-b-c}.
$$

MASTERY TESTS

Form A   1. If $x \equiv 4 \pmod 6$, then one replacement for x is

   a) 0   **b)** 4   **c)** 2   **d)** 7   **e)** none of these

2. $2 \oplus 3 \pmod 5 = ?$

   a) 0   **b)** 1   **c)** 2   **d)** 3   **e)** none of these

3. $4 \otimes 3 \pmod 6 = ?$

   a) 0   **b)** 12   **c)** 3   **d)** 5   **e)** none of these

4. Let $G = \{0, 1, 2, 3, 4, 5\}$ under addition modulo 6. If $4 \oplus x = 3$, then $x = ?$

   a) 5   **b)** 1   **c)** 2   **d)** 4   **e)** none of these

5. If $9 \equiv 6 \pmod m$ is true, then one possible value of m is

   a) 0   **b)** 2   **c)** 6   **d)** 3   **e)** none of these

6. If Washington's Birthday (February 22) is on a Friday, on what day of the week will July 4th occur? (Assume no leap year.)

   a) Sunday   **b)** Monday   **c)** Wednesday   **d)** Tuesday
   e) none of these

7. If $x \oplus 4 \equiv 0 \pmod 9$, then $x = ?$

   a) 3   **b)** 4   **c)** 5   **d)** 7   **e)** none of these

For questions 8–10, refer to the following table, which represents some operation $\circ$.

$\circ$	x	y	z
x	x	y	z
y	y	z	x
z	z	x	y

8. What is the identity element?

   a) x **b)** y **c)** z **d)** none of these

9. What is the inverse of x?

   a) x **b)** y **c)** z **d)** none of these

10. What is the inverse of z?

   a) x **b)** y **c)** z **d)** none of these

11. If $-5 \equiv -8 \pmod x$, then $x = ?$

   a) -3 **b)** 3   **c)** 8   **d)** 5   **e)** none of these

12. If $5 \ominus x \equiv 2 \pmod 8$, find x.

   a) 2   **b)** 3   **c)** 4   **d)** 5   **e)** none of these

For questions 13–14, refer to the following chart:

$\circ$	a	b	c	d	e	f
a	a	a	c	d	f	b
b	a	b	c	d	e	f
c	b	c	a	e	f	d
d	a	d	c	b	e	f
e	d	e	b	a	c	f
f	e	f	a	d	b	c

13. Find $a \circ (d \circ e)$.

   a) a **b)** b **c)** c **d)** d **e)** none of these

14. Find $(c \circ b) \circ (a \circ f)$.

   a) a **b)** b **c)** c **d)** d **e)** none of these

15. A math teacher is rearranging the test papers from one of her lecture classes to make it easier for her to grade them. She knows that there are fewer than 100 students in the class. When she arranges the papers in piles of 5 each, she has 1 left over. When she arranges the papers in piles of 7 each, she has 6 left over, and when she arranges the papers in piles of 4 each, she has 0 left over. How many test papers does the teacher have to grade?

Form B

1. Which of the following payoff matrices are strictly determined?

a)
2	4
3	1

b)
8	4
3	2

c)
-3	-2
-4	0

d)
3	5
4	9

e) none of these

2. For the non-strictly determined game whose payoff matrix is

-5	-3
-4	-6

, what is the value of R_2?

3. Refer back to Exercise 2. What is the value of v?

4. Perform each of the indicated operations and check your result by casting out 9's and 11's.

a)
$$\begin{array}{r} 368 \\ +527 \\ \hline \end{array}$$

b)
$$\begin{array}{r} 872 \\ -583 \\ \hline \end{array}$$

c)
$$\begin{array}{r} 732 \\ \times 345 \\ \hline \end{array}$$

For questions 5–8, refer to the following payoff matrix for a certain game.

$$C$$

R
16	3
8	10

5. What is the value of the game?

6. What is the value of R_1?

7. What is the value of C_2?

8. What is the value of R_2?

9. A game whose value is $-\frac{2}{3}$ favors

a) player R

b) player C

c) neither player R nor player C

d) none of these

10. In what clock is $4 \ominus 6 \equiv 8$ true?

a) mod 7 **b)** mod 8 **c)** mod 9 **d)** mod 10 **e)** none of these

11. August 20, 1985, occurred on a Tuesday. In that same year, on what day of the week did November 25 occur?

12. A coin box contains only nickels. Leslie knows that there are fewer than 100 nickels in the box. However, she does not know exactly how many there are. When she arranges the coins in piles of 8, she has 5 left. When she arranges them in piles of 9 each, she also has 5 left. When she arranges them in piles of 12 each, she again has 5 left. How many coins does Leslie have?

13. What are the optimal strategies for a game whose payoff matrix is as follows?

$$C$$

$$R \quad \begin{array}{|c|c|} \hline 3 & 8 \\ \hline 1 & -7 \\ \hline \end{array}$$

14. *True or False?* If a game is not strictly determined, then its value can never be 0.

15. If a group is Abelian, this means that

    a) every element has an inverse   **b)** there is an identity

    c) the operation is associative   **d)** the operation is commutative

    e) none of these.

7

THE NATURE OF ALGEBRA

CHAPTER OBJECTIVES

To summarize rules for working with exponents. (*Section 7.2*)

To understand what a polynomial is. (*Section 7.3*)

To demonstrate how to perform the four basic operations involving polynomials. (*Section 7.3*)

To study functions and the notation used for them. (*Section 7.4*)

To review the techniques that are used to solve linear equations. (*Section 7.5*)

To solve a system of simultaneous linear equations algebraically. (*Section 7.6*)

To show how we can use linear inequalities to solve verbal problems. (*Section 7.7*)

To discuss three different methods of factoring polynomials. (*Section 7.8*)

To learn how to solve quadratic equations by factoring or by using the quadratic formula. (*Section 7.9*)

To apply the different algebraic techniques to a variety of verbal problems. (*Sections 7.4, 7.5, 7.6, 7.7, and 7.9*)

351

Massive Cleanup Campaign to Begin

DOVER: State officials announced this morning that they would begin a massive cleanup on the site of the former ABC Chemical Corp. The rectangular plot of land, located off State Highway 28, was occupied by the chemical company during the late 1960s before the company declared bankruptcy. Tons of dangerous chemicals, which were allowed to seep into the ground, now present an environ- mental hazard to the residents of the neighboring communities.

The officials estimate that the cleanup campaign will take several months and will cost millions of dollars. It is anticipated that the federal government will contribute some funds from its Superfund to the cleanup campaign.

The Evening Ledger, April 17, 1985

Can we use mathematical techniques to analyze the above article and determine the largest area that can be cleaned under certain conditions? Fortunately, the answer is yes.

7.1

INTRODUCTION

Often when applying algebra to business decision models, we encounter both linear and quadratic profit equations. For example, suppose that the S & W Knitting Corp. finds that its profit per week y, in hundreds of dollars, is given by $y = 10x - 0.43x^2$, where x is the number of racks of knit dresses sold per week. This is a quadratic expression.

In your later work you will often need to factor as well as solve both linear and quadratic equations and inequalities. In this chapter we will review the ideas that you are most likely to need in your later analysis of such expressions. Our discussion is merely intended as a review of some of the basic ideas. A more detailed analysis can be found in any algebra book.

7.2

EXPONENTS

Throughout the remainder of this book we will be studying the real number system. Let us then pause for a moment to review some of its properties that we will use in analyzing the nature of algebra.

When working with real numbers, we must often deal with exponents, which are defined as follows:

Definition 7.1   *If x is a real nonzero number and n is any integer, then*

$$x^n = \underbrace{x \cdot x \cdot x \cdot \ldots \cdot x}_{n \text{ of them}}$$

*x is called the **base,** and n is called the **exponent.***

HISTORICAL NOTE

The ancient Babylonians were the first to find formulas for what we would today describe as the solutions of quadratic equations. Since the Babylonians had no conception of what a negative square root is, their approach was completely different from ours. Certain tablets that have been found seem to point in the direction of logarithms.[1]

The Greeks, on the other hand, who did not have the advantages and notations of present-day algebra, still could solve many quadratic equations by geometric methods. For example, to solve the quadratic equation $x^2 + 8x = 9$, the Greeks would use a geometric dissection method as follows: First draw rectangular regions whose area is $x^2 + 8x$ as shown in Figure 7.1.

Now draw a square whose area is 9. This is shown in Figure 7.2.

Since $x^2 + 8x = 9$, we have the diagram shown in Figure 7.3, so that

$$x^2 + 8x = 9.$$

Add a square whose area is 16 to each side of Figure 7.3. We get Figure 7.4. so that

$$(x + 4)^2 = 25,$$
$$x + 4 = 5,$$
$$\text{and} \quad x = 1.$$

Although the Greeks had much more sophisticated methods for solving quadratic equations, the above illustration does show us what their methods were like.

1. S. Gandz, "Origin and development of the quadratic equation," *Osiris*, Volume III (1937), pp. 405–557.

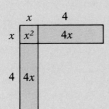

Figure 7.1
The shaded portion represents
$x^2 + 8x$

Figure 7.2
The shaded portion represents 9

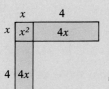

Figure 7.3

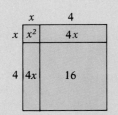

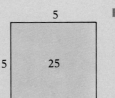

Figure 7.4

For example, x^3 represents $x \cdot x \cdot x$, and 5^4 represents $5 \cdot 5 \cdot 5 \cdot 5$. If $n = 0$, then the value of x^n is defined to be 1. Thus $5^0 = 1$. Also, $x^{-n} = \frac{1}{x^n}$. For example,

$$x^{-7} = \frac{1}{x^7} \quad \text{and} \quad 7^{-3} = \frac{1}{7^3}.$$

Similarly, $(100,000)^0 = 1$.

If your calculator has a $\boxed{y^x}$ button, then you can evaluate any expression involving exponents quite simply. For example, to evaluate 7^3 using a calculator, we proceed as follows:

What you do	*What appears on display panel*
Turn on machine	0.
Push 7 button	7.
Push y^x button	7.
Push 3 button	3.
Push = button	343.

Comment   On most calculators the base must be a positive number when using the $\boxed{y^x}$ button. Otherwise you will get an error message.

Comment   Great care must be exercised when using the $\boxed{y^x}$ button on the calculator. Try to evaluate 2×5^3 using the $\boxed{y^x}$ button. What happens?

When dealing with expressions involving exponents, we use certain properties. Since our discussion is intended only as a review, we will merely present the rules. A complete discussion and justification can be found in any algebra book.

Multiplication Rule $x^a \cdot x^b = x^{a+b}$ (*We simply add the exponents.*)

Power Rule $(x^a)^b = x^{ab}$ (*We simply multiply the exponents.*)

Quotient Rule $\frac{x^a}{x^b} = x^{a-b} \, (x \neq 0)$ (*We simply subtract the exponents.*)

EXAMPLE 1   **a)** $x^7 \cdot x^5 = x^{7+5} = x^{12}$

b) $3^5 \cdot 3^9 = 3^{5+9} = 3^{14}$ (*We do* not *multiply the 3's. We only add the exponents.*)

c) $(5^7)^4 = 5^{7 \cdot 4} = 5^{28}$

d) $\frac{x^{15}}{x^5} = x^{15-5} = x^{10}$ (*We do* not *divide the exponents.*)

e) $\frac{10^5}{10^2} = 10^{5-2} = 10^3$ (*We do* not *divide the 10's. We only subtract the exponents.*) ☐

EXAMPLE 2 $\dfrac{16^2}{4^3}$ cannot be evaluated in its present form by using the quotient rule. The reason is that the bases are not the same. Of course, we can evaluate 16^2, getting 256, and we can evaluate 4^3, getting 64. We then divide 256 by 64. Our answer is 4. $\square$

EXAMPLE 3 Refer back to Example 2. We notice that $16 = 4^2$. Thus

$$\frac{16^2}{4^3} = \frac{(16)^2}{4^3} = \frac{(4^2)^2}{4^3}$$

$$= \frac{4^{2 \cdot 2}}{4^3} \qquad \text{(by the power rule)}$$

$$= \frac{4^4}{4^3}$$

$$= 4^1. \qquad \text{(by the quotient rule)} \quad \square$$

Comment If the exponent of a number is 1, it is customary to omit it. Thus 4^1 will be written as 4.

Until now we have assumed that the exponent was an integer. Of course, exponents need not necessarily be integers. A number can have a rational number as an exponent also. Thus if x is a positive number, $x^{\frac{1}{2}}$ is defined as $\sqrt{x}$, and $x^{\frac{1}{3}}$ is defined as $\sqrt[3]{x}$. Similarly, $x^{\frac{1}{4}}$ is defined as $\sqrt[4]{x}$. Thus $25^{\frac{1}{2}} = \sqrt{25}$ or the square root of 25. This means we are looking for a number that when multiplied by itself gives 25. One answer is 5, so that $25^{\frac{1}{2}} = 5$. Also, $8^{\frac{1}{3}} = \sqrt[3]{8}$ or the cube root of 8. This means that we are looking for a number that when multiplied by itself 3 times gives 8. One answer is 2, so that $8^{\frac{1}{3}} = 2$.

More generally, $x^{\frac{1}{n}} = \sqrt[n]{x}$, where $n \neq 0$. (If x is a negative number, then n may be an odd integer only.) This means that we are looking for a number that when multiplied by itself n times gives x. Applying this definition gives $\sqrt[n]{x^m} = (x^m)^{\frac{1}{n}} = x^{\frac{m}{n}}$. Also, $(\sqrt[n]{x})^m = (x^{\frac{1}{n}})^m = x^{\frac{m}{n}}$. Therefore

RULE

$$x^{\frac{m}{n}} = \sqrt[n]{x^m} \quad \text{or} \quad x^{\frac{m}{n}} = (\sqrt[n]{x})^m \text{ where } n \neq 0 \text{ and } x > 0.$$

The above rule is easy to use, as the following examples will illustrate.

EXAMPLE 4
a) $27^{\frac{2}{3}}$ means $(\sqrt[3]{27})^2$. Since $\sqrt[3]{27}$ represents the cube root of 27, or 3, we must evaluate 3^2. This, of course, equals 9. Thus $27^{\frac{2}{3}} = 9$.

b) $32^{\frac{3}{5}}$ means $(\sqrt[5]{32})^3$. Since $\sqrt[5]{32}$ means that we are looking for a number that when multiplied by itself 5 times gives 32, its value must be 2. Thus we must evaluate 2^3. This, of course, equals 8. Therefore $32^{\frac{3}{5}} = 8$. $\square$

EXAMPLE 5   Evaluate $8^{-\frac{2}{3}}$.

SOLUTION   We first eliminate the negative exponent by rewriting

$$8^{-\frac{2}{3}} \quad \text{as} \quad \frac{1}{8^{\frac{2}{3}}}.$$

Then we apply the above rule. We get

$$8^{-\frac{2}{3}} = \frac{1}{8^{\frac{2}{3}}} \qquad \text{(definition of negative exponents)}$$

$$= \frac{1}{(\sqrt[3]{8})^2} \qquad \text{(rule for fractional exponents)}$$

$$= \frac{1}{2^2}$$

$$= \frac{1}{4}. \quad \square$$

We can summarize our discussion of exponents in the following table:

Notation	Name
$b^x = \underbrace{b \cdot b \cdot b \dots \cdot b}_{x \text{ of them}}$ where x is an integer	Definition of exponents
$b^x \cdot b^y = b^{x+y}$	Multiplication rule
$(b^x)^y = b^{xy}$	Power rule
$b^{-x} = \dfrac{1}{b^x} \qquad (b \neq 0)$	Negative exponents
$b^0 = 1$	Zero exponents
$\dfrac{b^x}{b^y} = b^{x-y} \qquad (b \neq 0)$	Quotient rule
$b^{\frac{x}{y}} = (\sqrt[y]{b})^x = \sqrt[y]{b^x} \qquad (b > 0)$	Fractional exponents

EXERCISES

Simplify each of the expressions given in Exercises 1–13, which involve exponents.

1. $16^{\frac{3}{2}}$

2. $64^{-\frac{3}{2}}$

3. $-8^{\frac{5}{3}}$

4. $4^{-\frac{1}{2}}$

5. $\left(\dfrac{1}{2}\right)^{-3}$

6. $(5^2)^3$

7. 10^0

8. $4^{\frac{3}{2}} \cdot 4^{\frac{7}{2}}$

9. $\dfrac{16^4}{16^3}$

10. $(2^3)(2^2)(2^7)$

11. $49^{-\frac{3}{2}}$

12. $2 \cdot 3^2$

13. $(2^3 \cdot 3^3)^3$

14. Evaluate $25^{\frac{1}{2}}$ using a calculator.

15. Evaluate $5 \cdot 4^3$ using a calculator.

16. Evaluate $16^{-\frac{3}{4}}$ using a calculator.

7.3

BRIEF ALGEBRA REVIEW

In this section we review some ideas from algebra. The material is intended only as a *review* of the ideas needed to solve some verbal problems. It is by no means a complete discussion of all the ideas of algebra. A complete discussion can be found in the references given for this chapter at the end of this book.

Variables and Constants

In algebra, two kinds of symbols are used to represent numbers: **variables** and **constants.**

A **constant** has a fixed value, such as 2, −4, $\frac{1}{2}$, π, etc. Sometimes we know that a number is fixed, but we do not know what its value is. Then we denote it by a letter at the beginning of the alphabet, such as *a*, *b*, *c*, etc.

A **variable** is a symbol that may be replaced by more than one number. We denote variables by letters at the end of the alphabet, such as *x*, *y*, *z*, etc. For example, in the expression "*x* is less than 4," *x* is not a fixed number, but can represent *any* number less than 4.

In algebra, multiplication is denoted in various ways. For example, "2 multiplied by *x*" can be denoted as $2 \cdot x$, $2(x)$, $(2)(x)$, or just $2x$. "2 multiplied by 4" can be denoted as $2 \cdot 4$, $2(4)$, or $(2)(4)$. It cannot be written as 24 because it would then be confused with the number twenty-four. Whenever such confusion is possible, the multiplication dot or parentheses *must* be used. They may be omitted, as in $2x$, if no other meaning is possible.

EXAMPLE 1   In the expression $2x + 3y$, 2 and 3 are constants, and *x* and *y* are variables. ☐

EXAMPLE 2   In the equation $3s - 2.9t = 6$, the constants are 3, 2.9, and 6. The variables are *s* and *t*. ☐

EXAMPLE 3   In the equation $x - 2y = 4$, the constants are 2 and 4. (There is a constant 1 in front of the *x*, which is understood.) The variables are *x* and *y*, and they can be replaced by many different values. For example, if *x* is 8, then *y* is 2; if *x* is 4, then *y* is 0; and if *x* is 10, then *y* is 3. There are actually infinitely many *pairs* of values that can replace *x* and *y* in this example. (Note that once we give a value to *x*, the value of *y* is specifically determined.) ☐

EXAMPLE 4   In the expression $1/t$, 1 is a constant, and *t* is a variable that can represent any value except 0. (See p. 194.) ☐

EXAMPLE 5   Let us evaluate the expression $2x - y + 3z$ for several different values of *x*, *y*, and *z*. If *x* is 1, *y* is 0, and *z* is 3, we have $2x - y + 3z$

$= 2(1) - 0 + 3(3)$, which equals 11. If x is -2, y is 4, and z is $\frac{1}{3}$, we have $2x - y + 3z = 2(-2) - 4 + 3(\frac{1}{3})$, which equals -7. $\square$

Polynomials   Let us review some of the basic ideas of polynomials. A **polynomial** is any expression that is formed by adding, subtracting, and multiplying constants and variables. Only natural numbers may appear as exponents of variables. Also, no variable may appear in a denominator. The parts of the expression that are added (or subtracted) are called **terms.** The expression $7x$, which has only one term, is called a **monomial.** The polynomial $8x + 3$, which has two terms, is called a **binomial.** The polynomial $8x^2 + 9x - 7$, which has three terms, is called a **trinomial.**

EXAMPLE 6   The following algebraic expressions are polynomials:

a) $7x$

b) $\dfrac{8x^2}{3} - \dfrac{7y^3}{10}$

c) $12x^2 + 8x^2 - 3x + 4$

d) 0

e) $\dfrac{5}{9}x - \dfrac{2}{3} + \pi$

f) $7x^2y^3z^4 + 83$ $\square$

EXAMPLE 7   The following algebraic expressions are *not* polynomials:

a) $\dfrac{10}{y}$ is *not* a polynomial, since the variable y appears in the denominator.

b) $x^{\frac{3}{2}} + 7$ is *not* a polynomial, since the exponent of x is not a natural number. $\square$

The **degree of a polynomial** in one variable is the same as the greatest exponent that appears in it. For example, $10x^3 - 8x^2 + 7x + 3$ is a polynomial whose degree is 3. Also, $10x^2 - 12x + 1$ is a polynomial whose degree is 2, and $12x + 1$ is a polynomial whose degree is 1. A polynomial may involve more than one variable. For example, $7x + 3y$ and $x^2 + 5xy + 7y^3$ are polynomials that involve two variables. The polynomial $7x^2y^3z$ has degree 6, since the exponents of the variables are 2, 3, and 1.

When dealing with polynomials it is often advisable to arrange the terms so that the term with the greatest exponent of the variable is written first and the exponents in the other terms appear in **descending order.** Thus the polynomials $8x^3 - 7x^2 + 3x - 9$, $18x^2 + 7x - 3$, and $2x + 7$ are all written in **standard form.** These polynomials are said to be arranged in **descending powers** of x. (Of course, the variable in a polynomial need not be x; it may be any other letter.)

To simplify a polynomial that has several terms, we simply combine like terms by using the commutative, associative, and distributive properties as needed. The terms $7xy$ and $3xy$ are called **like terms,** since they both have the same variables as factors and the variables have the same exponent. Also, $5x^2$ and $10x^2$ are like terms, and $12(x + y)^3$ and $4(x + y)^3$ are pairs of like terms. On the other hand, $3x$ and $4y$ are **unlike terms,** since the variables are different. Also $7xy$ and $5ab$ are unlike terms. Although the terms $8x^2$ and $12x^3$ have the same variables as factors, they are still unlike terms, since they do not have the same exponents.

To add polynomials, we simply combine like terms by adding the numerical coefficients. It is suggested that the polynomials be arranged in descending or ascending powers of a particular variable so that like terms are arranged in vertical columns. We can then add or subtract each column separately.

EXAMPLE 8   Add the polynomials $8x + 7y - 3z$, $2x - y + 5z$, and $-7x + 2y + 2z$.

SOLUTION   We arrange the polynomials so that like terms are in vertical columns as shown. We then add each column separately.

$$
\begin{array}{r}
8x + 7y - 3z \\
2x - y + 5z \\
-7x + 2y + 2z \\
\hline
3x + 8y + 4z
\end{array}
$$

Thus our answer is $3x + 8y + 4z$. $\square$

EXAMPLE 9   Add $12x^2 - 9xy + 2y^2$, $4xy - 3x^2$, $-8y^2 + 17xy$.

SOLUTION   We arrange the polynomials as shown:

$$
\begin{array}{r}
12x^2 - 9xy + 2y^2 \\
- 3x^2 + 4xy \\
+ 17xy - 8y^2 \\
\hline
9x^2 + 12xy - 6y^2
\end{array}
$$

Our answer is $9x^2 + 12xy - 6y^2$. $\square$

EXAMPLE 10   Simplify $3x + [8x + (9 - 3x)]$.

SOLUTION   We first remove the innermost grouping symbol. We have

$$3x + [8x + (9 - 3x)] = 3x + [8x + 9 - 3x]$$
$$= 3x + [5x + 9]$$
$$= 3x + 5x + 9$$
$$= 8x + 9.$$

Our answer is $8x + 9$.   ☐

   To subtract one polynomial from another, we first arrange the polynomials so that like terms are in vertical columns. Then we subtract the like terms in each column separately by adding the opposite of the subtrahend to the minuend. This procedure is illustrated in the following examples.

EXAMPLE 11   Subtract $8x^2 - 3x + 2$ from $12x^2 + 9x - 1$.

SOLUTION   We arrange the polynomials so that like terms are in vertical columns as shown. We then *subtract* each column separately.

$$12x^2 + 9x - 1 \leftarrow \text{minuend}$$
$$\underline{8x^2 - 3x + 2} \leftarrow \text{subtrahend}$$
$$4x^2 + 12x - 3$$

Our answer is $4x^2 + 12x - 3$.   ☐

EXAMPLE 12   Simplify $4x - [3 - (-8 + 9x)]$

SOLUTION

$$4x - [3 - (-8 + 9x)] = 4x - [3 + 8 - 9x]$$
$$= 4x - [11 - 9x]$$
$$= 4x - 11 + 9x$$
$$= 13x - 11$$

Our answer is $13x - 11$.   ☐

   What about multiplication? To multiply a polynomial by a monomial, we simply use the distributive property and multiply each term of the polynomial by the monomial. We then combine the resulting products.

EXAMPLE 13   Simplify:

a) $-7x(3x^2 + 7x - 3)$
b) $-5x^2y^3(12xy^2 - 3y^5)$

SOLUTION We apply the distributive property. This gives

a) $-7x(3x^2 + 7x - 3) = -21x^3 - 49x^2 + 21x$

b) $-5x^2y^3(12xy^2 - 3y^5) = -60x^3y^5 + 15x^2y^8$ $\square$

To multiply one polynomial by another polynomial, we apply the distributive property. If we treat one of the polynomials as a single expression, then the distributive law is easy to apply.

EXAMPLE 14 Multiply: $(7x + 3)(2x - 3)$.

SOLUTION $(7x + 3)(2x - 3)$

$\quad = 7x(2x - 3) + 3(2x - 3) \quad$ (by the distributive property)

$\quad = 14x^2 - 21x + 6x - 9 \quad$ (by the distributive property)

$\quad = 14x^2 - 15x - 9 \quad$ (combining like terms) $\square$

Actually, to multiply two polynomials, we multiply each term in one polynomial by each term in the other. This is best accomplished by arranging the multiplier and the multiplicand according to descending or ascending order and then applying the distributive law so that each term of the multiplicand is multiplied by each term of the multiplier. We then combine like terms.

EXAMPLE 15 Multiply: $(2x - 7y)(3x - 5y)$.

SOLUTION

$$
\begin{array}{r}
2x - 7y \leftarrow \text{multiplier} \\
3x - 5y \leftarrow \text{multiplicand} \\
\hline
-\,10xy + 35y^2 \\
6x^2 - 21xy \\
\hline
6x^2 - 31xy + 35y^2
\end{array}
$$

Our answer is $6x^2 - 31xy + 35y^2$. $\square$

EXAMPLE 16 Multiply: $(2x - 1)(8x^3 - 9x^2 + 7x - 3)$.

SOLUTION

$$
\begin{array}{r}
8x^3 - 9x^2 + 7x - 3 \\
2x - 1 \\
\hline
-\,8x^3 + 9x^2 - 7x + 3 \\
16x^4 - 18x^3 + 14x^2 - 6x \\
\hline
16x^4 - 26x^3 + 23x^2 - 13x + 3
\end{array}
$$

Our answer is $16x^4 - 26x^3 + 23x^2 - 13x + 3$. $\square$

To divide one polynomial by a monomial, we simply divide each term of the polynomial by the monomial. This is illustrated in the following examples.

EXAMPLE 17   Divide $18x^6 + 12x^4$ by $-2x^2$.

SOLUTION $\dfrac{18x^6 + 12x^4}{-2x^2} = \left(\dfrac{18x^6}{-2x^2}\right) + \left(\dfrac{12x^4}{-2x^2}\right)$

$$= -9x^4 - 6x^2$$

Our answer is $-9x^4 - 6x^2$.   ☐

EXAMPLE 18   Divide $20x^4y^4 + 15x^3y^3 - 5xy$ by $-5xy$.

SOLUTION $\dfrac{20x^4y^4 + 15x^3y^3 - 5xy}{-5xy} = \left(\dfrac{20x^4y^4}{-5xy}\right) + \left(\dfrac{15x^3y^3}{-5xy}\right) + \left(\dfrac{-5xy}{-5xy}\right)$

$$= -4x^3y^3 - 3x^2y^2 + 1$$

Our answer is $-4x^3y^3 - 3x^2y^2 + 1$.   ☐

To divide one polynomial by another polynomial, we follow the same pattern as long division in arithmetic. We proceed as follows:

a) Arrange the polynomials in standard form so that the terms of the divisor and the terms of the dividend are in descending powers of one variable.

b) Divide the first term of the dividend by the first term of the divisor to get the first term of the quotient.

c) Multiply the first term of the quotient by the whole divisor.

d) Subtract this product from the dividend to obtain the new dividend.

e) Repeat steps b–d until the remainder is 0 or until the degree of the remainder is less than the degree of the divisor.

f) Check the answer by using the relationship

quotient × divisor + remainder = dividend.

EXAMPLE 19   Divide $6y^3 - 28y + 3y^2 + 15$ by $2y - 3$.

SOLUTION   In this case the dividend is $6y^3 - 28y + 3y^2 + 15$, and the divisor is $2y - 3$. We arrange the terms in descending order. We get

$$6y^3 + 3y^2 - 28y + 15.$$

Now we divide by following the same pattern that is used in arithmetic. We have

$$
\begin{array}{r}
3y^2 + 6y - 5 \\
2y - 3 \overline{)6y^3 + 3y^2 - 28y + 15} \\
\underline{6y^3 - 9y^2} \\
12y^2 - 28y \\
\underline{12y^2 - 18y} \\
-10y + 15 \\
\underline{-10y + 15}
\end{array}
$$

Our answer is $3y^2 + 6y - 5$. ☐

EXAMPLE 20   Divide $6x^3 + 11x^2 - 1$ by $3x + 1$.

SOLUTION   We first arrange both polynomials in descending order. We add zeros as the coefficients of any missing terms so that like terms will be properly aligned. We have

$$
\begin{array}{r}
2x^2 + 3x - 1 \\
3x + 1 \overline{)6x^3 + 11x^2 + 0x - 1} \\
\underline{6x^3 + 2x^2} \\
9x^2 + 0x \\
\underline{9x^2 + 3x} \\
-3x - 1 \\
\underline{-3x - 1}
\end{array}
$$

Our answer is $2x^2 + 3x - 1$. ☐

EXAMPLE 21   Divide $6x^3 + 5 + 11x^2$ by $3x + 1$.

SOLUTION   After arranging the terms in descending order and adding zeros as coefficients of any missing terms, we get

$$
\begin{array}{r}
2x^2 + 3x - 1 \\
3x + 1 \overline{)6x^3 + 11x^2 + 0x + 5} \\
\underline{6x^3 + 2x^2} \\
9x^2 + 0x \\
\underline{9x^2 + 3x} \\
-3x + 5 \\
\underline{-3x - 1} \\
6
\end{array}
$$

Thus the quotient is $2x^2 + 3x - 1$ with a remainder of 6. ☐

EXERCISES

In Exercises 1–42, perform the indicated operation and simplify your answer as much as possible.

1. Add: $x^3 - 7x^2 + 3x, 4x^3 - 5 + 2x^2, 9x^3 + 2x^2 + 7x - 3, 2x^3 - 3x + 7$

2. Add: $-7x, 3x + 2y, -3x^2 + 8y + 9x, -5x^2 + 3y - 2x, -9y$

3. Add: $2x^3 - 3x^2y + 8xy^2 - 2y^3, -5xy^2 - 3x^3 + 2x^2y, -8y^3 + 2xy^2 - 5x^2y$

4. Simplify: $-3x + (10 - 5x) + 8x$

5. Simplify: $(2x^2 + 8x - 10) + (-3x^2 - 4x - 5)$

6. Simplify: $(2x^3 + 8x - 1) + (-3x^2 - 2x - 5)$

7. Simplify: $10 - [-3 + (9x - 2)]$

8. Simplify: $(5x^2 + 8x - 3) - (2x^2 - 2x - 9)$

9. Simplify: $5x^2 - [9x - (3x - x^2) + 8]$

10. Simplify: $5x^2 - 9x + 5x^2 - (8x + 3) + 8$

11. Simplify: $2x - 8x^2 - (4 + 5x - 8x^2)$

12. From the sum of $x^2 + 5x - 3$ and $2x^2 - 3x + 9$, subtract $12x^2 - 8x + 17$.

13. Multiply: $-8x(5x^2 - 3x + 7)$

14. Multiply: $-5x^2y^2z^3(7x^2y^3 - 3xy + 8)$

15. Multiply: $-\dfrac{5}{3}(15x^2 - 21xy^3 + 36y^3)$

16. Multiply: $10x^4y(-5x^5y^8 - 3xy + 17)$

17. Multiply: $-2x^2y^5(1 - 3xy + y^5)$

18. Multiply: $(5x + 8)(3x - 2)$

19. Multiply: $(x - 5y)(x + 3y)$

20. Multiply: $(2x + 3y)(5x - 7y)$

21. Multiply: $(8x^2 - 5xy + 3y^2)(3x - 5)$

22. Multiply: $(x^2 - 3x + 17)(2x + 3)$

23. Multiply: $(5xy + x^2 - 3y^2)(2x + y)$

24. Multiply: $(5x^2 - 3xy + 7y^2)(2x - 5y)$

25. Simplify: $5x(2x + 3) - (x - 3)(2x + 1)$

26. Simplify: $(3x + 7)(2x - 3) - (x + 4)(6x - 5)$

27. Simplify: $(2x - 3)^2 - (5x + 1)^2$

28. Divide: $\dfrac{6a^2b - 18a^3b^2}{-3ab}$

29. Divide: $\dfrac{-15x^3y^5 + 27x^3y}{-3xy}$

30. Divide: $\dfrac{48x^6y^4 - 24x^2y^2}{-24x^2y^2}$

31. Divide: $\dfrac{x^4y^4 - x^2y^2 - xy}{-xy}$

32. Divide: $\dfrac{-3.2x^2y^5 - 2.4x^3y^2 + 1.6x^2y^3}{-0.4x^2y^2}$

33. Divide $x^2 - 8x + 7$ by $x - 1$.

34. Divide $2x^2 - 5x - 3$ by $x - 3$.

35. Divide $x^3 - 8x^2 + 17x - 10$ by $x - 5$.

36. Divide $x^3 + 6x + 4$ by $x + 2$.

37. Divide $8x^3 - 24x^2 + 16x + 7$ by $x - 5$.

38. Divide $x^2 - 28y^2 + 3xy$ by $x - 6y$.

39. Divide $x^3 + 64$ by $x + 4$.

40. Divide: $\dfrac{2x^5 + 4x^4 - x^3 - 5x + 10}{x^2 + 2x + 1}$

41. Divide: $\dfrac{15x^4 - 1}{x - 1}$

42. Divide: $\dfrac{2x^5 - x^4 - x^2 - 8x + 6}{x^2 - 2x + 1}$

43. If José buys $5x + 1$ suits and pays $200x + 3$ dollars for each suit, find the total cost for all the suits.

44. The length of a parking lot is $15x + 7$ meters, and its width is $21x + 3$ meters. Find its area.

45. The Acme Trucking Company finds that its daily operating expenses for its x trucks is given by $300x^2 + 2yx + 9$ dollars. Represent, in simplest form, the daily operating expenses for 1 truck.

7.4

FUNCTIONS AND FUNCTION NOTATION

Let us consider the charts of numbers given in the following examples.

EXAMPLE 1   A neighborhood movie theater owner finds that the number of people coming to the movies per evening is dependent on the charge for admission, as shown in this table.

Admission charge	$0.50	$0.75	$1.00	$2.00	$3.00	$4.00	$5.00
Number of people attending per evening	3000	2700	2500	2000	1700	1200	600

EXAMPLE 2 A baseball team owner finds that the attendance at baseball games at home depends on who the opposing team is, as shown in this table.

Opposing team	Team A	Team B	Team C	Team D	Team E	Team F
Attendance at game	10,000	25,000	30,000	37,000	46,000	55,000

EXAMPLE 3 A homeowner finds that the number of gallons of oil used in heating the house for the entire winter season depends on the setting of the thermostat, as shown in this table.

Thermostat (°Fahrenheit)	74°	72°	70°	68°	66°	64°
Number of gallons of oil needed to heat house	2600	2300	2000	1700	1400	1100

In each of the above examples we notice that there is a correspondence between the entries in the top row and the entries in the bottom row. Thus in Example 1 we see that when the admission charge is 50¢, then 3000 people will attend the movie, and when the admission charge is $5.00, only 600 people will attend the movie. We notice that to each number in the top row we associate one and only one number. Mathematicians say that the number of people attending the movie is a **function of** or **depends on** the admission charge.

> **Definition 7.2** A *function* is a rule that assigns to each number in a set of numbers a unique second number. If the two members of the set are x and y, then we can say that a function is a relationship between the two variables x and y (any letters may be used) such that for each value substituted for x, there is obtained a unique value of y.

Since we will be studying functions a great deal, we introduce some special notation.

Notation   We abbreviate the words "y is a function of x" as $y = f(x)$. This is read "y equals f of x."

EXAMPLE 4   The equation $y = x^2$ represents a function. If we replace x by 1, we get $y = 1$. If we replace x by -2, we get $y = 4$, etc. For each value of x we get a unique value of y. In this case, y is a function of x, and we write $y = f(x) = x^2$.   ☐

EXAMPLE 5   The equation $y = 3x + 2$ represents a function, since if we replace x by any number, we get a unique value of y. We write $y = f(x) = 3x + 2$.   ☐

EXAMPLE 6   The equation $y = \dfrac{1}{x}$ represents a function, and we write $y = f(x) = \dfrac{1}{x}$, provided that $x \neq 0$.   ☐

EXAMPLE 7   The equation $y = \pm\sqrt{x}$ (which means y equals either the positive or negative square root of x) cannot represent a function. The reason for this is that if we replace x by any number, we do *not* get a *unique* value for y. Thus if x is 4, we get $y = 2$ and $y = -2$.   ☐

EXAMPLE 8   The equation $y = -\sqrt{x}$ represents a function, since now there *is* a unique value of y for each value of x. Thus if x is 4, y can have only the value -2.   ☐

   A function is like a coin-operated machine: Each time you put in a coin, you get an item out. We put x-values into the "function machine." For each x-value we put in, we get a unique y-value out of the machine.
   Like any coin-operated machine, the function will "accept" some values of x, but may reject others. For example, in the function $y = 1/x$, we cannot put in the value $x = 0$. (Why not?) This leads us to the following definition.

> **Definition 7.3**   *When using the function notation, the set of all values of x that can be substituted in the function is called the **domain** of the function. The set of all values of y that these substitutions create is called the **range** of the function.*

EXAMPLE 9   In the function $y = 1/x$ we can put in any value for x except 0. So the domain of this function is all real numbers except 0.   ☐

EXAMPLE 10   In the function $y = -\sqrt{x}$ we can put in any positive number or 0 for x. We cannot put in negative values for x because these would give us complex numbers. Thus the domain of this function is $x \geq 0$.   ☐

EXAMPLE 11   In the function $y = 2x - 3$ we can put in *any* value for x. So the domain is all real numbers.   ☐

EXAMPLE 12   In the function $y = -\sqrt{x}$, only negative values of y or 0 come out, so the range is $y \le 0$. ☐

EXAMPLE 13   In the function $y = x^2$, only positive values of y or 0 come out, so the range is $y \ge 0$. ☐

EXAMPLE 14   In the function $y = |x|$ (y is the absolute value of x), only positive values of y or 0 come out, so the range is $y \ge 0$. ☐

Notation   If we have a function $y = f(x)$, then the symbol $f(a)$ means the value of the function when $x = a$.

EXAMPLE 15   For the function $y = 2x - 3$, $f(1) = 2(1) - 3$ or -1.
For the function $y = 2x - 3$, $f(0) = 2(0) - 3$ or -3.
For the function $y = 2x - 3$, $f(-1) = 2(-1) - 3$ or -5.
For the function $y = 2x - 3$, $f(a) = 2a - 3$. ☐

EXAMPLE 16   For the function $y = x^2$, $f(3) = 3^2$ or 9.
For the function $y = x^2$, $f(1) = 1^2$ or 1.
For the function $y = x^2$, $f(t) = t^2$. ☐

EXAMPLE 17   For the function $\dfrac{3x - 4}{5}$, $f(2) = \dfrac{3(2) - 4}{5}$ or $\dfrac{2}{5}$.

For the function $\dfrac{3x - 4}{5}$, $f(-6) = \dfrac{3(-6) - 4}{5}$ or $\dfrac{-22}{5}$.

For the function $\dfrac{3x - 4}{5}$, $f(0) = \dfrac{3(0) - 4}{5}$ or $\dfrac{-4}{5}$. ☐

EXERCISES

For each of the situations described in Exercises 1–2, determine whether y is a function of x. If not, explain why.

1.

Weight of an envelope, x	Postage required (in 1985), y
Up to 1 ounce	22¢
Between 1 and 2 ounces	39¢
Between 2 and 3 ounces	56¢
Between 3 and 4 ounces	73¢
Between 4 and 5 ounces	90¢

2.

Number of college credits being taken this semester, x	Student fee payable, y
Between 0 and 3 credits	$12
Between 3 and 6 credits	$17
Between 6 and 9 credits	$25
Between 9 and 12 credits	$32
More than 12 credits	$44

For each of the functions given in Exercises 3–11, find $f(1)$, $f(0)$, $f(-3)$, $f\left(\dfrac{1}{2}\right)$, $f(a)$, and $f(x + h)$.

3. $f(x) = 4x - 3$ **4.** $f(x) = 3x + 2$

5. $f(x) = x^2 + 3$ **6.** $f(x) = \dfrac{3x + 1}{5x + 2}$

7. $f(x) = 3x^2$ **8.** $f(x) = (2 - x)(7 + x)$

9. $f(x) = x^2 + 7x + 6$

10. $f(x) = 2x^3 + 3x^2 + 7x + 5$

11. $f(x) = \dfrac{5x^2 + 7x - 3}{2x - 1}$

12. The Brookhaven Camp provides door-to-door luggage pickup service for its campers. The charges are as follows: $15 for a camp trunk and one duffel bag. There is a $6 charge for each additional duffel bag. Thus if a camper has x duffel bags, the cost is y dollars.

   a) Is y a function of x?

   b) If y is a function of x, can you find an equation expressing this relationship?

   c) If Linda has 4 duffel bags altogether (in addition to the trunk), what is the cost for the luggage pickup service?

13. A lottery dealer makes a profit of 6¢ on each state lottery ticket that she sells. Furthermore, if she sells more than 25,000 tickets in a month, the state gives her a rebate, which results in an extra 2¢ on each ticket over 25,000 sold. If she sells x tickets, her profit is y dollars.

   a) Is y a function of x?

   b) If y is a function of x, can you find an equation expressing this relationship?

   c) For each of the months shown in the table, find the dealer's profit.

Month	Number of lottery tickets sold
January	24,228
February	17,368
March	26,014
April	28,132
May	29,485
June	38,692

14. Melissa's monthly rent is $428. She has signed a long-term lease with her landlord whereby her rent will be raised 1% every six months. Thus x months from now, her monthly rent will be y dollars.

   a) Is y a function of x?

   **b) If y is a function of x, can you find an equation expressing this relationship?

   c) Find her monthly rent five years from now.

Find the domain and range of each of the functions given in Exercises 15–18.

15. $f(x) = 2x^2 + 5$

16. $f(x) = \sqrt{7x + 2}$

17. $f(x) = \dfrac{4x - 3}{2x + 1}$

18. $f(x) = |2x + 2|$

7.5

LINEAR EQUATIONS IN ONE VARIABLE AND THEIR APPLICATION

Equations and Their Solution

Much of the beauty of mathematics lies in the fact that many real-world situations can be translated into mathematical equations. Such mathematical descriptions of real-world situations are called **mathematical models.** These mathematical models are often given in the form of **equations,** which may be simple or complicated.

Basically, an equation is a mathematical statement that consists of two expressions joined together by an equal sign. It specifies that the expression on the left side of the equality sign is equal to the expression on the right side. Any unknown in an equation is represented by a letter and is known as a **variable.** A **solution** to an equation is a number which when substituted for the variable results in a true statement.

EXAMPLE 1 The following are examples of equations:

    a) $5x = 20$ is an equation.

    b) $9x + 3 = 6x + 8$ is an equation.

    c) $8x^2 + 3x + 9 = 0$ is an equation.

    d) $8x + 27 - 3 = 0$ is an equation.

    e) $4x + 1$ is *not* an equation. Can you see why? ☐

    Any equation containing x (the unknown) raised only to the exponent 1 and to no higher or lower exponent is called a **linear equation in x.**

EXAMPLE 2 **a)** $7x - 3 = 18$ is a linear equation in x.

    b) $5x^2 - 3x + 7 = 0$ is *not* a linear equation in x because it contains an x^2 term.

    c) $7x - 3y + 8z^2 = 10$ is a linear equation in x. It is also a linear equation in y. It is *not* a linear equation in z. ☐

    To **solve** an equation in x means to find a value of x that makes the equation true.

EXAMPLE 3 **a)** The equation $5x = 45$ has as its solution $x = 9$, since $5(9) = 45$ is true.

    b) $x^2 - 16 = 0$ has as its solutions $x = 4$ and $x = -4$, since $4^2 - 16 = 0$ is true and $(-4)^2 - 16 = 0$ is true.

    c) $x + 8 = x$ has no solution, since there is no value of x that will make the equation true. ☐

    To solve an equation for x, we must get the equation into the form "$x =$ some expression." To do this, we use the following rules.

RULES FOR SOLVING LINEAR EQUATIONS

    1. Any number may be added to or subtracted from both sides of an equation.

    2. Both sides of an equation may be multiplied or divided by the same number (except that we cannot divide by 0).

The following examples illustrate the use of these rules.

EXAMPLE 4 Solve the equation $8x = 12x - 7$ for x.

SOLUTION Subtract $12x$ from both sides of the equation. This gives

$$
\begin{array}{rcr}
8x & = & 12x - 7 \\
-12x & = & -12x \\
\hline
-4x & = & -7
\end{array}
$$

We get

Now we divide both sides of the equation by -4.

$$\frac{-4x}{-4} = \frac{-7}{-4}$$

So the solution is $x = \frac{7}{4}$. $\square$

EXAMPLE 5   Solve the equation $8x - 5 = 3x + 10$ for x.

SOLUTION   Add 5 to both sides of the equation.

$$\begin{array}{rcl}
8x - 5 &=& 3x + 10 \\
+ 5 && + 5 \\
\hline
8x &=& 3x + 15
\end{array}$$

Subtract $3x$ from both sides of the equation.

$$\begin{array}{rcl}
8x &=& 3x + 15 \\
-3x && -3x \\
\hline
5x &=& 15
\end{array}$$

Divide both sides of the equation by 5.

$$\frac{\cancel{5}x}{\cancel{5}} = \frac{15}{5}$$

So the solution is $x = 3$. $\square$

EXAMPLE 6   Solve the equation $2ax - 5 + 3z = 7z$ for x.

SOLUTION   Add $5 - 3z$ to both sides of the equation.

$$\begin{array}{rcl}
2ax - 5 + 3z &=& 7z \\
+ 5 - 3z && 5 - 3z \\
\hline
2ax &=& 5 + 4z
\end{array}$$

We get

Divide both sides of the equation by $2a$.

$$\frac{\cancel{2}a x}{\cancel{2}a} = \frac{5 + 4z}{2a}$$

Therefore

$$x = \frac{5 + 4z}{2a}. \quad \square$$

EXAMPLE 7   Solve the equation $2ax - 5 + 3z = 7z$ for z.

SOLUTION   Subtract $2ax$ from both sides of the equation.

$$\begin{array}{rcl}
2ax - 5 + 3z &=& 7z \\
-2ax && - 2ax \\
\hline
- 5 + 3z &=& 7z - 2ax
\end{array}$$

Add 5 to both sides of the equation.

$$-5 + 3z = 7z - 2ax$$
$$+5 \qquad\qquad\qquad + 5$$
$$\overline{\qquad 3z = 7z - 2ax + 5}$$

Subtract $7z$ from both sides of the equation.

$$3z = \quad 7z - 2ax + 5$$
$$-7z \quad -7z$$
$$\overline{-4z = -2ax + 5}$$

Divide both sides of the equation by -4.

$$\frac{\cancel{-4}z}{\cancel{-4}} = \frac{-2ax + 5}{-4}$$

Therefore

$$z = \frac{-2ax + 5}{-4}. \quad \square$$

Comment   We could have also obtained the same answer for the previous example if we had first subtracted $3z$ from both sides of the equation, getting $2ax - 5 = 4z$, and then divided both sides of the equation by 4. Our answer would then be $z = \dfrac{2ax - 5}{4}$. Is this the same answer that we obtained previously?

Forming Algebraic Equations   Since we must often form an equation that is an adequate model of some real-world situation, it is important to practice the formation of such equations. This is always the first step in building such mathematical models. The procedure is illustrated in the following examples. The reader should analyze each example carefully.

EXAMPLE 8   The cost of a certain computer and its software programs is $36,000. The computer costs 8 times as much as the software programs. Find the cost of each.

SOLUTION   Let x = the cost of the software programs; then $8x$ = the cost of the computer. We then have the following equation based upon the given information.

$$8x + x = 36{,}000$$

Now we solve for x:

$$8x + x = 36{,}000$$

We combine terms:

$$9x = 36{,}000 \qquad \text{(Remember that } x = 1x.\text{)}$$

We divide both sides of the equation by 9, getting

$$\frac{\cancel{9}x}{\cancel{9}} = \frac{36,000}{9}$$

so that

$$x = 4000$$

Thus the software programs cost $4000, and the computer costs $32,000. ▢

EXAMPLE 9   Evelyn and Margaret are stockbrokers. One week, they opened 89 new customer accounts. Evelyn opened 1 more than 3 times the number of new accounts opened by Margaret. Find the number of new customer accounts opened by each.

SOLUTION   Let x = the number of new customer accounts opened by Margaret. Then $3x + 1$ = the number of new customer accounts opened by Evelyn. Since Evelyn and Margaret opened 89 new accounts, we have

$$x + 3x + 1 = 89$$

so that

$$4x + 1 = 89$$

We subtract 1 from both sides of the equation, getting

$$
\begin{array}{rcr}
4x + 1 & = & 89 \\
-1 & & -1 \\
\hline
4x & = & 88
\end{array}
$$

Now we divide both sides of the equation by 4. This gives

$$\frac{\cancel{4}x}{\cancel{4}} = \frac{88}{4}$$

so that

$$x = 22.$$

Thus, Margaret opened 22 new customer accounts, and Evelyn opened 67 new customer accounts. ▢

EXAMPLE 10   A supermarket has just received a shipment of 65 cases of corn flakes. The supermarket wishes to divide the 65 cases so that they can be stored in 2 different warehouses, with the larger warehouse getting more than the smaller warehouse. If 3 times the number of cases that the larger will receive is 6 more than 6 times the number of cases that the smaller warehouse will receive, how many cases will each warehouse receive?

SOLUTION   Let x = the number of cases assigned to the larger warehouse. Then $65 - x$ = the number of cases to be stored in the smaller warehouse. On the basis of the given information, we have

$$3x = 6(65 - x) + 6$$

or

$$3x = 390 - 6x + 6$$

so that

$$3x = 396 - 6x$$

Add $6x$ to both sides of the equation. We get

$$\begin{array}{rcl} 3x & = & 396 - 6x \\ +6x & & + 6x \\ \hline 9x & = & 396 \end{array}$$

Now we divide both sides of the equation by 9. This gives

$$\frac{\cancel{9}x}{\cancel{9}} = \frac{396}{9}$$

or

$$x = 44.$$

Thus the larger warehouse should receive 44 cases, and the smaller warehouse should receive $65 - 44$ or 21 cases.   □

EXERCISES

Solve each of the equations given in Exercises 1–26 for x.

1. $8x = 24$

2. $-6x = 24$

3. $10x + 4x = 28$

4. $9x + 3 = 21$

5. $4x + 3x - 8 = 62$

6. $7x - 14 - 3x = -38$

7. $9x + 7 = 5x + 47$

8. $9x - 2 = 3x - 14$

9. $\frac{x}{3} = 7$

10. $\frac{4}{7}x = \frac{16}{49}$

11. $8x - 4 + 7 = 6x + x + 9$

12. $6x - 12 - x = 9x + 53$

13. $5(3x - 2) = 5$

14. $15x - 3(x + 6) = 6$

15. $2 - 7(x - 1) = 3(x - 2) - 5(x + 3)$

16. $2(x - 3) - 17 = 13 - 3(x + 2)$

17. $bx - 5 = c$

18. $m^2x - 3m^2 = 12m^2$

19. $a = bx + 6$

20. $8bx - 6b^2 = 4bx + 12b^2$

21. $2ax = 10a^2 - 3ax$

22. $22x - 3(5x + 4) = 16$

23. $3x + (2x - 5) = 13 - 2(x + 2)$

24. $15x - 8(x + 6) = 14x + 3$

25. $\dfrac{7x + 5}{8} - \dfrac{3x + 15}{10} = 2$

26. $\dfrac{3}{2}(2x + 1) - \dfrac{1}{3}(4x - 1) = -3\dfrac{1}{6}$

27. Let n represent a number. Then match each of the sentences given in column A with an equation that represents it from column B.

Column A	Column B
a) Four times a number is 64.	**i.** $7 - 5n = 58$
b) A number increased by 7 is 28.	**ii.** $8n = 72$
c) One number is $\frac{1}{5}$ of another, and the difference is 58.	**iii.** $\frac{1}{8}n - 7 = 58$
d) One-eighth of a number decreased by 7 is 58.	**iv.** $5n - 7 = 58$
e) One number is $\frac{1}{5}$ of another, and their sum is 103.	**v.** $n + \frac{1}{5}n = 103$
f) The product of 8 and a number is 72.	**vi.** $n - \frac{1}{5}n = 58$
g) When 7 is subtracted from 5 times a number, the result is 58.	**vii.** $\frac{1}{5}n - n = 58$
h) When 5 times a number is subtracted from 7, the result is 58.	**viii.** $4n = 64$
	ix. $n + 7 = 28$

Use an algebraic equation to solve each of Exercises 28–39.

28. A merchant sold a home video system for $450, which was 25% above its cost to him. What was the cost of the home video system to the merchant?

29. During a sale, the cost of a television set was decreased from $99 to $66. What was the percent decrease in price?

30. One evening, a movie theater sold 450 tickets and received $1650 in proceeds for the performance. Children's tickets were $3.00, and adult tickets were $4.00. How many children's and how many adult tickets were sold?

31. Marilyn deposited $160 in her bank. The number of $5 bills she deposited was 3 more than the number of $10 bills she deposited, and the number of $1 bills she deposited was 30 more than the number of $5 bills she deposited. How many bills of each type did Marilyn deposit?

32. Phyllis and Madeline are account executives for a large Wall Street brokerage company. During the first two business days of January, they completed 162 transactions. If the number of transactions completed by Phyllis is 2 more than 3 times the number completed by Madeline, find the number completed by each.

33. Gail is analyzing the number of credit card sales completed. On Tuesday the number of credit card sales completed was 2 more than Monday's number of credit card sales completed. Twice as many credit card sales were completed on Wednesday as on Monday. The total number of credit card sales completed on Monday and Wednesday exceeds the number of credit card sales completed on Tuesday by 2. How many credit card sales were completed on each day?

34. George purchased a new car for $8784 and paid $724.68 as sales tax. What is the percent of the sales tax?

35. Mark receives a shipment of transistor parts. The bill is $47,392.68. The shipper allows a discount of $3\frac{1}{2}$% if the total bill is paid C.O.D. If Mark decides to pay C.O.D., how much does he have to pay?

36. Kay Parsons invests a sum of money in municipal bonds paying 6% annual interest. She then decides to invest $500 more than the sum originally invested in a real estate venture paying 10% annual interest. Her total annual income from both investments is $210. How much did she invest in each venture?

37. The personnel director of the Calyn Company is analyzing the age of the president of the company and the age of her secretary. This is being studied so that a new pension plan can be started. The director finds that the president is 3 times as old as her secretary. Eight years from now, the president's age will exceed twice her secretary's age at that time by 14 years. How old is each one now?

** **38.** The photocopying machine on the eighth floor of a large office building is situated on a platform whose length is 8 feet more than its width. If the length of the platform is increased by 4 feet and the width is decreased by 1 foot, the area will remain unchanged. Find the dimensions of the platform.

39. Bill, Mag, and Chris have each sent out résumés to numerous companies seeking employment. The number of résumés sent out by Mag is 1 less than the number sent out by Bill. The number of résumés sent out by Chris is 5 less than twice the number sent out by Mag. If the number of résumés sent out by Chris exceeds the number sent out by Bill by 12, find the number of résumés sent out by each.

7.6

ALGEBRAIC SOLUTION OF A SYSTEM OF SIMULTANEOUS LINEAR EQUATIONS

Let us consider Michael, who knows that he has 12 coins in his piggy bank, which amount to 80¢. He also knows that there are only nickels and dimes in the piggy bank. How many coins of each kind does Michael have? Do we have enough information to solve the problem?

Let n = the number of nickels in the piggy bank and let d = the number of dimes in the piggy bank. Since there are 12 coins in the bank altogether, we must have $n + d = 12$. Also, each nickel is worth 5 cents, so n nickels are worth $5n$ cents. Similarly, each dime is worth 10 cents, so d dimes are worth $10d$ cents. The value of all the coins together is 80 cents. Thus $5n + 10d = 80$. Here the two equations $n + d = 12$ and $5n + 10d = 80$ impose two conditions on the variables at the same time. We call the two equations a **system of simultaneous linear equations.**

To **solve** such a system, that is, to find a **solution** of a system of simultaneous linear equations in two variables, means that we are looking for a pair of numbers that satisfies both equations. While there are numerous techniques for solving such a system, at this point we will merely discuss an algebraic solution. A system of linear equations may have exactly one solution, no solution, or infinitely many solutions.

EXAMPLE 1 Solve the following system of equations:

$$n + d = 12 \qquad [A]$$
$$5n + 10d = 80 \qquad [B]$$

SOLUTION Let us multiply both sides of equation [A] by -5. This yields an equivalent equation [C], in which the number in front of n is exactly the same as the number in front of n in equation [B] but is of the opposite sign. We get

	Original equation	*New equation*	
[A]	$n + d = 12$	$-5n - 5d = -60$	[C]
[B]	$5n + 10d = 80$	$\underline{5n + 10d = 80}$	[B]

Now we add the corresponding members of equations [B] and [C] to eliminate the variable n. We get

$$5d = 20.$$

We now solve for the value of d by dividing both sides of the resulting equation by 5, getting

$$d = 4.$$

We then replace d by its value in any equation involving both variables and solve for n. We get

$$n + d = 12,$$
$$n + 4 = 12,$$
$$n = 8.$$

Thus we have $n = 8$ and $d = 4$. Referring back to the problem given at the beginning of this section, we conclude that Michael has 8 nickels and 4 dimes in his piggy bank.   □

Comment   We can easily check that 8 nickels and 4 dimes is indeed the solution, since 8 nickels are worth 40¢ and 4 dimes are also worth 40¢. The total value of the 12 coins is 80¢.

EXAMPLE 2   Solve the following system of equations:

$$3x - 7 = 7y \qquad \text{[A]}$$
$$4x = 3y + 22 \qquad \text{[B]}$$

SOLUTION   We will first transform each of the given equations [A] and [B] into two equivalent equations [C] and [D], in which the terms containing the variables appear on one side and the constant appears on the other side of the equation. We get

$$3x - 7y = 7 \qquad \text{[C]}$$
$$4x - 3y = 22 \qquad \text{[D]}$$

Now we eliminate one of the variables, say y, from both equations. To accomplish this, let us multiply both sides of equation [C] by 3 and both sides of equation [D] by -7. This gives

$$9x - 21y = 21 \qquad \text{[E]}$$
$$-28x + 21y = -154 \qquad \text{[F]}$$

We notice that in both of the equations [E] and [F] the numbers in front of y have the same numerical value but are of opposite sign. Adding these two equations eliminates the variable y. We get

$$-19x = -133.$$

Dividing both sides of this last equation by -19 gives

$$x = 7.$$

Replacing x by its value in any equation containing both variables gives

$$3x - 7y = 7$$
$$3(7) - 7y = 7$$
$$21 - 7y = 7$$

$$-7y = -14 \qquad \text{(We subtract 21 from both sides of the equation.)}$$

$$y = 2 \qquad \text{(We divide both sides of the equation by } -7.)$$

Our answer is $x = 7$ and $y = 2$. The reader should check that these values satisfy *both* equations. ☐

EXAMPLE 3 Yogi is the manager of a Little League baseball team. Yesterday he purchased 8 bats and 4 gloves for the team. The total cost was \$156. Today he purchased at the same prices an additional 3 bats and 7 gloves. The total cost was \$108. Find the cost of a bat and the cost of a glove.

SOLUTION Let $b = $ the cost of a bat, and let $g = $ the cost of a glove. On the basis of the information given in the problem, 8 bats and 4 gloves cost \$156. This means $8b + 4g = 156$. Also, 3 bats and 7 gloves cost \$108, or $3b + 7g = 108$. Thus we must solve the following set of equations:

$$8b + 4g = 156$$
$$3b + 7g = 108$$

Multiply both sides of the first equation by 7 and both sides of the second equation by -4. We get

$$56b + 28g = 1092$$
$$\underline{-12b - 28g = -432}$$

Adding these two equations will eliminate the variable g. Our result is

$$44b = 660.$$

Dividing both sides of this equation by 44 gives $b = 15$. Replacing b

by its value in any equation containing both variables gives

$$8b + 4g = 156$$
$$8(15) + 4g = 156$$
$$120 + 4g = 156$$
$$4g = 36$$
$$g = 9$$

Thus $b = 15$ and $g = 9$. This means that a bat costs \$15 and a glove costs \$9. □

EXERCISES

Use the techniques of this section to solve each of the sets of equations given in Exercises 1–13.

1. $4x + 3y = 29$
$2x - 3y = 1$

2. $4x + 5y = -7$
$2x - 3y = 13$

3. $3x + 4y = -5$
$4x + 5y = -7$

4. $y = x + 12$
$10y + 5x = 360$

5. $3x + 7y = 33$
$2x + 5y = 23$

6. $3x - y = 3$
$x + 3y = 11$

7. $4x + 5y = 22$
$-4x + 3y = -6$

8. $3x + 2y = 41$
$4x - y = 40$

9. $4y - 6x = -26$
$6y - 4x = -24$

10. $3x + 2y = 6$
$x + y = 1$

11. $6x + 10y = 7$
$15x - 4y = 3$

12. $x - y = 11$
$3x + 2y = 3$

13. $3x + 6y = 15$
$-x - 2y = 12$

14. *Admission Charges.* A family of 2 adults and 5 children paid \$34 for admission to the Animal Town Amusement Park. Another family of 3 adults and 2 children paid \$29 for admission. (The admission price includes all rides.) Find the admission price for an adult and the price for a child.

15. *Check Cashing.* Jessica is a bank teller. Yesterday she cashed a \$200 check and gave the customer 25 bills in \$10 and \$5 bills. How many \$5 bills and how many \$10 bills did she give the customer?

16. *Stamp Collecting.* Ken is a stamp collector. In one group of 75 stamps he has only 20¢ and 22¢ stamps with a total face value of \$16.00. How many stamps of each kind does he have?

17. *Little League Baseball.* The coach of a Little League baseball team purchased 6 balls and 4 bats for the team at a cost of \$72. The following day the coach purchased at the same prices 8 additional balls and 6 additional bats at a cost of \$104. Find the cost of a bat and the cost of a ball.

18. *Wage Discrimination.* A farmer in the West has been accused of paying female sharecroppers lower wages than male sharecroppers for identical jobs. Records show that during one week the farmer paid out \$550 in wages for a work force consisting of four men and five women. In another week the farmer paid out \$1150 for a work force consisting of ten men and eight women. Find the weekly wages paid by the farmer for a male sharecropper and for a female sharecropper.

19. *College Admissions.* A large southern university received 6000 applications for admission to the premedical or the prelaw program. The number of applications for the premedical program was three times the number of applications for the prelaw program. How many students applied to the premedical program and how many applied to the prelaw program?

20. The Outboard Amusement Company rents out motorboats and rowboats for the afternoon. The rental fee received for 7 rowboats and 6 motorboats is \$63. The rental fee for 4 rowboats and 3 motorboats is \$33. Find the rental fee for a motorboat and for a rowboat.

21. *Anthropology.* In 1953, anthropologists discovered a group of 2200 people living on a small island in the Pacific. Of these 2200 people, 400 were considered functionally illiterate. This represented 20% of the male population and 16% of the female population. How many males and how many females were in the group?

22. The owners of the White Lake Recreation Area wish to spray the shores of the lake with a chemical to kill the mosquitoes, which bother the sunbathers. Local environmental regulations require that the chemical spray contain only 20% Dursban. The owners have two chemical solutions, one that contains 4% Dursban and one that contains 40% Dursban. How many gallons of each should the owners use to produce 36 gallons of a solution that contains 20% Dursban?

23. Alexis is on a special diet. Her doctor has told her to limit her daily intake of calcium and vitamin C to 800.08 and 65.395 milligrams, respectively. (These are the minimum daily requirements.) Medical research indicates that each cup of skim milk contains 284 milligrams of calcium and 1.25 milligrams of vitamin C. A cup of orange juice contains 22 milligrams of calcium and 129 milligrams of vitamin C. How much of each of these foods should Alexis consume to maintain her special diet?

7.7

SOLVING VERBAL PROBLEMS BY USING LINEAR INEQUALITIES

Until now we have been dealing with equations. Of course, on many occasions we must deal with mathematical expressions in which one expression is not equal to another expression. Such statements are known as **inequalities.** When dealing with inequalities the following symbols of inequality are used.

Symbol	Meaning
$x < y$	x is less than y
$x \leq y$	x is less than or equal to y
$x > y$	x is greater than y
$x \geq y$	x is greater than or equal to y

As with equations, we can have linear inequalities in one variable or in many variables.

EXAMPLE 1 The following are examples of linear inequalities.

a) $3x > 12$ is a linear inequality in x.

b) $9x + 6 \leq 4x + 3$ is a linear inequality in x.

c) $4x + 5y > 3$ is a linear inequality in x and y.

In Section 4.8 of Chapter 4 we indicated how we picture all real numbers that are less than or equal to some number on a number line. We suggest that you review the discussion before reading further.

Let us solve the inequality $2x > 10$ algebraically, where x is a real number. We can find the solution to an inequality in much the same way that we solve equations. However, there are some exceptions.

We may add or subtract the same number to both sides of an inequality, a procedure that is similar to what we did with equalities. When we multiply or divide both sides of an inequality by the same number, we must be careful. When both sides of an inequality are multiplied or divided by a positive number, then this is allowed, as we do when we work with equalities. On the other hand, when both sides of an inequality are multiplied or divided by a negative number, then the order of the inequality is reversed. Thus if $2x > 10$ and we divide both sides of this inequality by 2 (a positive number), we get $x > 5$. On the other hand, if $-2x > +8$ and we divide both sides of this inequality by -2 (a negative number), we get $x < -4$. The order of the inequality sign is reversed.

EXAMPLE 2   Find and graph the solution set of the inequality

$$x - 8 > 1.$$

SOLUTION   We add 8 to both sides of the inequality. This will leave us with x alone on one side of the inequality. We get

$$
\begin{array}{rl}
x - 8 & > 1 \\
+\ 8 & +\ 8 \\
\hline
x & > 9
\end{array}
$$

The graph of the solution set is shown in Figure 7.5. Note that 9 is not included in the graph.

Figure 7.5

EXAMPLE 3   Find and graph the solution set of the inequality

$$3x + 2 \le 12 - 2x.$$

SOLUTION   We first add $+2x$ to both sides of the inequality. We get

$$
\begin{array}{rl}
3x + 2 & \le 12 - 2x \\
+\ 2x & \quad\quad +\ 2x \\
\hline
5x + 2 & \le 12
\end{array}
$$

Now we subtract 2 from both sides of the inequality. This gives

$$
\begin{array}{rl}
5x + 2 & \le 12 \\
-\ 2 & -\ 2 \\
\hline
5x & \le 10
\end{array}
$$

Finally, we divide both sides of this inequality by 5 (a positive number). Our result is

$$\frac{\cancel{5}x}{\cancel{5}} \le \frac{10}{5} \quad \text{or} \quad x \le 2.$$

Figure 7.6

The graph of the solution set is shown in Figure 7.6. Note that 2 is included in the graph. ☐

EXAMPLE 4   Find and graph the solution set of the inequality

$$4(2x - 3) - 12x \le 0.$$

SOLUTION   Using the distributive law, we remove parentheses.

$$
\begin{array}{rll}
4(2x - 3) - 12x & \le 0 & \\
8x - 12 - 12x & \le 0 & \\
-4x - 12 & \le 0 & \\
\underline{+ 12 \qquad\quad + 12} & & \text{(We add $+12$ to both sides.)} \\
\dfrac{-4x}{-4} & \le \dfrac{12}{-4} & \text{(We divide both sides by -4,} \\
& & \text{a negative number. The sense} \\
x & \ge -3 & \text{of the inequality is reversed.)}
\end{array}
$$

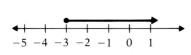

Figure 7.7

The graph of the solution set is shown in Figure 7.7. Note that -3 is included in the graph. ☐

EXAMPLE 5   Christopher needs $80 to fix his car. His boss has agreed to pay him $6 an hour for shoveling snow in addition to his regular part-time salary of $50. What is the minimum number of hours that Christopher must work to earn the $80 to fix his car?

SOLUTION   Let $x =$ the number of hours that Christopher must work shoveling snow. Since he earns $6 an hour, his total extra money earned from shoveling the snow for x hours is $6x$. His weekly part-time salary is $50. Since he needs at least $80, we have

$$6x + 50 \ge 80.$$

Now we solve this inequality for x:

$$
\begin{array}{rll}
6x + 50 & \ge \quad 80 & \\
\underline{-50 \quad -50} & & \text{(We subtract 50 from both sides.)} \\
\dfrac{6x}{6} & \ge \dfrac{30}{6} & \text{(We divide both sides by 6,} \\
& & \text{a positive number.)} \\
x & \ge \quad 5 &
\end{array}
$$

Thus Christopher must work at least 5 hours shoveling snow. ☐

EXAMPLE 6   Allison received grades of 98, 83, 84, and 86 on her first four math exams. What grade must she receive on the next exam so that her average will be at least 90% for all five exams?

SOLUTION   Let x = the fifth exam grade. An average is found by adding the grades together and dividing the sum by the number of exams. Thus her average is

$$\frac{98 + 83 + 84 + 86 + x}{5}$$

Since this must be at least 90%, we must have

$$\frac{98 + 83 + 84 + 86 + x}{5} \geq 90$$

or

$$\frac{351 + x}{5} \geq 90$$

$$5\left(\frac{351 + x}{5}\right) \geq 90(5) \quad \text{(We multiply both sides by 5, a positive number.)}$$

$$351 + x \geq 450$$

$$
\begin{array}{r}
351 + x \geq 450 \\
-351 \qquad -351 \\
\hline
x \geq 99
\end{array}
\quad
\begin{array}{l}
\text{(We subtract 351 from} \\
\text{both sides.)}
\end{array}
$$

Thus Allison must receive at least a 99 on her next exam for her to have a 90% average on all five exams.   □

EXERCISES

Find and graph the solution set for each of the inequalities given in Exercises 1–20.

1. $x - 5 > 7$

2. $x + 2 > 10$

3. $x - 2.5 \leq 3.5$

4. $x + 4 < 6$

5. $12 < x + 3$

6. $16 \geq 4y$

7. $-6x \leq 24$

8. $-10x \geq -30$

9. $\frac{x}{2} \leq 3$

10. $\frac{x}{3} \geq -4$

11. $\frac{2}{3}x \leq 30$

12. $\frac{-4}{9}x \geq -36$

13. $3x + 1 \geq -31$

14. $6x + 1 \geq -17$

15. $14x - 8 < 12 + 4x$

16. $12x + 4 - 16x < 28$

17. $24x - 6(2x + 3) \geq 0$

18. $6x - 12 \leq 6(7 + 2x)$

19. $10x \leq 20 + 4(3x - 4)$

20. $-3(2x + 4) - 2 \geq -5(x - 3)$

21. Steve saves at least \$18 a week from his salary. He already has accumulated \$324. In how many weeks will Steve accumulate at least \$468?

22. Mary has at least 6 times as many cassette tapes as does her sister Stephanie. If together, Mary and Stephanie have at least 98 tapes, find the least number of cassette tapes that each girl has.

23. Suzanne has joined a weight watchers' club and has agreed to lose at least 54 pounds in equal amounts per week over a 9-week period. What is the minimum amount that she must lose each week to accomplish this goal?

24. Hector has been told by his supplier that his franchise will be canceled unless he sells at least 90,000 gallons of gas over a 4-month period. During the first month, Hector sold 21,000 gallons of gas. At least how many gallons of gas must Hector sell in each of the remaining 3 months (assuming equal monthly sales) so that his franchise will not be canceled?

25. Gina purchased 4 computer discs and handed the clerk a $20 bill. She received less than $8 change. What is the minimum cost of a disc?

26. Lester purchased a set of tires for his car. The tires are guaranteed to provide at least 30,000 miles of driving (under normal conditions). Lester plans to drive from the East Coast to the West Coast and back, a distance of at most 6000 miles. Assuming that he makes the trip, what is the minimum number of miles remaining for the tires to be guaranteed?

27. Cecile Robinson is a guest speaker. For each lecture delivered she receives at least $175 in compensation. Last year she earned a minimum of $2000 from her speaking engagements. What is the least number of guest lectures that she delivered?

7.8

FACTORING

Factoring a polynomial means to express it as a product of other expressions. Although there are several methods of factoring that we can use, we suggest that you first try to find any **greatest common factor.** This involves using the distributive law in reverse. Thus to factor a polynomial whose terms have a common monomial factor, we first find any greatest monomial that is a factor of each term of the polynomial. Then we divide the polynomial by the monomial factor. The answer that we get is the other factor. Finally, we express the polynomial as a product of the two factors thus obtained.

EXAMPLE 1 Factor $10x - 10y$.

SOLUTION We first notice that 10 is the greatest common factor of $10x$ and $10y$. Thus it will be one of the factors. To find the other factor, we divide $10x - 10y$ by 10, getting $x - y$. This represents the other factor. Therefore

$$10x - 10y = 10(x - y). \quad \square$$

EXAMPLE 2 Factor $3xy^2 + 6x^2y$.

SOLUTION We notice that $3xy$ is the greatest common factor of $3xy^2$ and $6x^2y$. Thus it will be one of the factors. To find the other factor, we divide $3xy^2 + 6x^2y$ by $3xy$, getting $y + 2x$. This represents the other factor. Therefore

$$3xy^2 + 6x^2y = 3xy(y + 2x). \quad \square$$

EXAMPLE 3   Factor $6x^4 + 24x^3 - 15x^2$.

SOLUTION   We notice that $3x^2$ is the greatest common factor of $6x^4$, $24x^3$, and $-15x^2$. Thus it will be one of the factors. To find the other factor, we divide $6x^4 + 24x^3 - 15x^2$ by $3x^2$, getting $2x^2 + 8x - 5$. This represents the other factor. Therefore

$$6x^4 + 24x^3 - 15x^2 = 3x^2(2x^2 + 8x - 5). \quad \square$$

     A second method of factoring involves expressions of the form $x^2 - y^2$. Since $x + y$ multiplied by $x - y$ gives $x^2 - y^2$, the factors of $x^2 - y^2$ are $x + y$ and $x - y$. Any expression of the form $x^2 - y^2$ is called the **difference of two squares.** To factor any polynomial that is the difference of two squares, we simply express each of the terms as the square of a monomial and then use the fact that $x^2 - y^2 = (x + y)(x - y)$.

EXAMPLE 4   Factor **a)** $x^2 - 16$ **b)** $9x^2 - 25y^2$ **c)** $\dfrac{16}{49}x^2y^2 - 0.09$

SOLUTION   **a)** $x^2 - 16 = (x)^2 - (4)^2 = (x + 4)(x - 4)$

             b) $9x^2 - 25y^2 = (3x)^2 - (5y)^2 = (3x + 5y)(3x - 5y)$

             c) $\dfrac{16}{49}x^2y^2 - 0.09 = \left(\dfrac{4}{7}xy\right)^2 - (0.3)^2 = \left(\dfrac{4}{7}xy + 0.3\right)\left(\dfrac{4}{7}xy - 0.3\right)$ $\square$

     A third method of factoring involves factoring trinomials of the form $ax^2 + bx + c$. Recall that in Section 7.3 we discussed how to multiply polynomials. Let us analyze the procedure that we used so that we can find the product of two binomials mentally. Suppose we want to multiply $5x - 2$ by $2x - 3$. We can use both a horizontal and a vertical format as shown below:

Horizontal format: $(5x - 2)(2x - 3) = 5x(2x - 3) - 2(2x - 3)$

$$= 10x^2 - 15x - 4x + 6$$

$$= 10x^2 - 19x + 6$$

Vertical format:

$$
\begin{array}{r}
5x - 2 \\
2x - 3 \\
\hline
-15x + 6 \\
10x^2 - 4x \\
\hline
10x^2 - 19x + 6
\end{array}
$$

We notice the following: The first term of our answer, $10x^2$, is equal to the product of $5x$ and $2x$, the first term of each of the binomials. Also, $+6$, the last term, is equal to the product of -2 and -3, which are the last terms of each of the binomials. Finally, the middle term of $-19x$ is obtained by multiplying the first term of each binomial by the second term of the other and adding these products. Thus $(-15x) + (-4x) = -19x$. We can indicate this by using the following schematic diagram:

$$-4x$$
$$(5x - 2) \qquad (2x - 3)$$
$$-15x$$

We can summarize the procedure for multiplying (mentally) two binomials of the form $ax + b$ and $cx + d$: We first multiply the first terms of the binomials. Then we multiply the first term of each binomial by the last term of the other binomial and add these products. Finally, we multiply the last terms of the binomials and add all the terms together.

EXAMPLE 5 Multiply $(5x - y)(2x - y)$.

SOLUTION $(5x - y)(2x - y)$ ($10x^2$ is the product of the first two terms.)

$$-5xy$$
$(5x - y)(2x - y)$ ($-7xy$ is the sum of the cross products of
$$-2xy$$ $-2xy$ and $-5xy$.)

$(5x - y)(2x - y)$ ($+y^2$ is the product of the last terms.)

Thus

$$(5x - y)(2x - y) = 10x^2 - 7xy + y^2. \quad \square$$

EXAMPLE 6 Multiply $(-5x + 3)(-8x + 7)$.

SOLUTION $(-5x + 3)(-8x + 7)$ ($40x^2$ is the product of the first two terms.)

$$-35x$$
$(-5x + 3)(-8x + 7)$ ($-59x$ is the sum of the cross products of
$$-24x$$ $-35x$ and $-24x$.)

$(-5x + 3)(-8x + 7)$ ($+21$ is the product of the last terms.)

Thus

$$(-5x + 3)(-8x + 7) = 40x^2 - 59x + 21. \quad \square$$

To factor any trinomial of the form $ax^2 + bx + c$ means to find two binomials of the form $(dx + e)$ and $(fx + g)$ such that their product is the trinomial that we started with. The procedure involves a trial-and-error process, since we must try all possible pairs of factors until we find the one that works. Thus we must find two binomials such that (1) the product of the first terms of both binomials is equal to the first term of the trinomial (ax^2); (2) the product of the last terms of both binomials is equal to the last term of the trinomial (c); and finally, (3) when the first term of each binomial is multiplied by the second term of the other and the sum of these products is determined, the result obtained must equal the middle term of the trinomial (bx).

EXAMPLE 7   Factor $x^2 + 5x + 6$.

SOLUTION   The product of the first terms of the binomials must be x^2. Thus each first term of the binomial factors must be x. Therefore the factors are of the form

$$x^2 + 5x + 6 = (x \quad)(x \quad).$$

Since the product of the last terms of both binomial factors must be $+6$, both of these terms must be either positive or negative. As the middle term of the trinomial to be factored is $+5x$, the last terms of both binomials must be positive. Several pairs of possible integers whose product is $(+6)$ are $(+6)$ and $(+1)$ and $(+3)$ and $(+2)$. We then have the following possibilities: $(x + 6)(x + 1)$ and $(x + 3)(x + 2)$. We now try each of these possibilities until we find the one that works.

$$\overset{\displaystyle +1x}{\overbrace{(x + 6)(x + 1)}}$$
$$+6x$$

is *not* correct because the middle term is $(+1x) + (+6x) = +7x$ and not $+5x$.

$$\overset{\displaystyle +2x}{\overbrace{(x + 3)(x + 2)}}$$
$$+3x$$

is correct because the middle term is $(+3x) + (+2x) = +5x$. Therefore

$$x^2 + 5x + 6 = (x + 3)(x + 2). \quad \square$$

EXAMPLE 8   Factor $x^2 - 9x + 18$.

SOLUTION   The product of the first terms of the binomials must be x^2. Thus each first term of the binomial factors must be x. Therefore the factors are of the form

$$x^2 - 9x + 18 = (x \quad)(x \quad).$$

Since the product of the last terms of both binomial factors must be +18, both of these terms must be either positive or negative. As the middle term of the trinomial to be factored is $-9x$, the last term of both binomials must be negative. Several pairs of possible integers whose product is +18 are (-6) and (-3), (-9) and (-2), and (-18) and (-1). We then have the following possibilities: $(x - 6)(x - 3)$, $(x - 9)(x - 2)$, and $(x - 18)(x - 1)$. We now try each of these possibilities until we find the one that works.

$$\overset{-2x}{(x - 9)(x - 2)}$$
$$-9x$$

is *not* correct because the middle term is $(-9x) + (-2x) = -11x$ and not $-9x$.

$$\overset{-1x}{(x - 18)(x - 1)}$$
$$-18x$$

is *not* correct because the middle term is $(-18x) + (-1x) = -19x$ and not $-9x$.

$$\overset{-3x}{(x - 6)(x - 3)}$$
$$-6x$$

is correct because $(-3x) + (-6x) = -9x$. Therefore

$$(x^2 - 9x + 18) = (x - 6)(x - 3). \quad \square$$

EXAMPLE 9 Factor $3x^2 - 5x - 2$.

SOLUTION Since the product of the first terms of the binomials must be $3x^2$, one of the terms must be $3x$, and the other must be x. Thus the factors are of the form

$$3x^2 - 5x - 2 = (3x \quad)(x \quad).$$

Since the product of the last terms of both binomial factors must be -2, one of the terms must be positive, and the other must be negative. The pairs of possible integers whose product is -2 are (-2) and $(+1)$ and $(+2)$ and (-1). We then have the following possibilities: $(3x - 2)(x + 1)$, $(3x - 1)(x + 2)$, $(3x + 2)(x - 1)$, and $(3x + 1)(x - 2)$. We now try each of these possibilities until we find the one that works.

$$\overset{+3x}{\overbrace{(3x - 2)(x + 1)}}$$
$$\underset{-2x}{\underbrace{}}$$

is *not* correct because the middle term is $(+3x) + (-2x) = +1x$ and not $-5x$.

$$\overset{+6x}{\overbrace{(3x - 1)(x + 2)}}$$
$$\underset{-1x}{\underbrace{}}$$

is *not* correct because the middle term is $(+6x) + (-1x) = +5x$ and not $-5x$.

$$\overset{-3x}{\overbrace{(3x + 2)(x - 1)}}$$
$$\underset{+2x}{\underbrace{}}$$

is *not* correct because the middle term is $(-3x) + (+2x) = -1x$ and not $-5x$.

$$\overset{-6x}{\overbrace{(3x + 1)(x - 2)}}$$
$$\underset{+1x}{\underbrace{}}$$

is correct because $(-6x) + (+1x) = -5x$. Therefore
$$3x^2 - 5x - 2 = (3x + 1)(x - 2). \quad \square$$

EXERCISES

Factor each of the following, using any of the techniques or combination of the techniques discussed in this section.

1. $10x + 20$

2. $3x - 6y$

3. $12 - 18y$

4. $6x^2 - 12x$

5. $33x^3y^2 - 44x^2y$

6. $80x^4y^5z^3 + 64x^2y^3z$

7. $24xy^3 - 18x^2y^2 + 9xy$

8. $48x^3y^2 - 60x^2y^5 + 54xy$

9. $144 - x^2$

10. $16 - x^4$

11. $x^4 - 64$

12. $144x^2 - 169y^2$

13. $x^2y^2 - 49$

14. $\frac{1}{64} - x^2$

15. $\frac{16}{25}x^2y^8 - \frac{4}{9}$

16. $16x^2 - 16$

17. $3x^2 - 27y^2$

18. $x^3 - x$

19. $x^4 - 81$

20. $x^6 - x^2$

21. $4x^2 - 36$

22. $49x^2 - 64y^2$

23. $x^2 + 7x + 6$

24. $x^2 + 2x - 24$

25. $x^2 - 10x + 21$

26. $x^2 - x - 6$

27. $x^2 + x - 6$

28. $x^2 + 9x - 36$

29. $2x^2 + 7x + 6$ **30.** $3x^2 + 10x + 8$ **31.** $6x^2 + 5x - 4$ **32.** $4x^2 - 12x + 5$

33. $10x^2 - 9x + 2$ **34.** $4x^2 - 5xy - 6y^2$ **35.** $4x^2 - 8x - 32$ **36.** $x^4 - 20x^2 + 64$

37. $3x^2 + 15x + 18$ **38.** $4x^2 - 32x + 48$ **39.** $x^3 + 10x^2 + 24x$ **40.** $-3x^2 + 21x + 54$

7.9

QUADRATIC EQUATIONS AND THEIR USE IN SOLVING VERBAL PROBLEMS

Any equation of the second degree in one variable such as $x^2 - 5x + 6 = 0$ is called a **quadratic equation**. A quadratic equation in one variable is said to be arranged in **standard form** when it is written in the form $ax^2 + bx + c = 0$, where a, b, and c are real numbers and $a \neq 0$.

Often an equation may appear to be a quadratic equation. However, when we rearrange and combine like terms on one side and 0 appears on the other side of the equation, we no longer have a quadratic equation. For example, the equation $5x^2 + 8x = 5x(x + 7) - 8$ appears to be a quadratic equation. When we remove parentheses and collect all terms to one side of the equation, we get

$$5x^2 + 8x = 5x^2 + 35x - 8 \qquad \text{(by the distributive law)}$$

or

$$-27x + 8 = 0.$$

This is a linear equation.

While there are numerous methods for solving quadratic equations, we will discuss only two methods: factoring and the quadratic formula. We will assume that all of our quadratic equations are arranged in standard form. If this is not the case, we must transform the equation into an equivalent quadratic equation that is in the standard form $ax^2 + bx + c = 0$. Every quadratic equation has two roots (solutions).

RULE   *To solve a quadratic equation by factoring, we:*

1. *First rearrange the equation into standard form by removing parentheses, eliminating fractions, combining terms as needed, and then placing all terms on the left side of the equation and leaving 0 on the right side.*

2. *Factor the left side.*

3. *Set each of the factors obtained in Step 2 equal to 0 and solve the resulting equations for the unknown. Here we use the fact that if the product of two real numbers is 0, then at least one of the numbers is 0.*

4. *Check each answer obtained in Step 3 by substituting each value into the original equation.*

We illustrate the procedure with several examples.

EXAMPLE 1   Solve $x^2 + 9x + 18 = 0$ for x.

SOLUTION   The equation is already in standard form, so we factor. We have

$$x^2 + 9x + 18 = 0$$
$$(x + 6)(x + 3) = 0$$

Setting each of these factors equal to 0 and solving for x gives

$$x + 6 = 0 \quad \text{or} \quad x = -6$$

and

$$x + 3 = 0 \quad \text{or} \quad x = -3.$$

Thus our solution is $x = -6$ and $x = -3$. The reader should check that these are indeed solutions. ☐

EXAMPLE 2   Solve the equation $2x(x - 6) - 20 = -17(x + 4) + 60$ for x.

SOLUTION   We must first rearrange the equation so that it is in standard form. We have

$$2x(x - 6) - 20 = -17(x + 4) + 60$$
$$2x^2 - 12x - 20 = -17x - 68 + 60$$
$$2x^2 - 12x - 20 = -17x - 8$$
$$2x^2 + 5x - 12 = 0$$

Now we factor $2x^2 + 5x - 12$ and set each of the factors equal to 0. We get

$$2x^2 + 5x - 12 = (2x - 3)(x + 4).$$

Then

$$2x - 3 = 0 \qquad \qquad x + 4 = 0$$
$$2x = 3 \qquad \qquad \quad x = -4$$
$$x = \frac{3}{2}$$

Therefore our solution is $x = \frac{3}{2}$ and $x = -4$. ☐

Quadratic Formula   *When a quadratic equation is written in standard form, its roots can be obtained by using the following formula, which is known as the **quadratic formula.***

$$Root\ 1 = \frac{-b + \sqrt{b^2 - 4ac}}{2a}$$

and

$$Root\ 2 = \frac{-b - \sqrt{b^2 - 4ac}}{2a}$$

Comment When we use the quadratic formula, the value of a is the coefficient of x^2, the value of b is the coefficient of the x term, and the value of c is the remaining number (assuming that the equation is arranged in standard form).

EXAMPLE 3 Solve the equation $x^2 + 3x - 7 = 0$.

SOLUTION In this case the equation is in standard form. Now we determine the values of a, b, and c. We have $a = 1$, $b = 3$, and $c = -7$. Substituting these values into the quadratic formula gives

$$\text{Root 1} = \frac{-3 + \sqrt{3^2 - 4(1)(-7)}}{2(1)}$$

$$= \frac{-3 + \sqrt{9 + 28}}{2}$$

$$= \frac{-3 + \sqrt{37}}{2} \qquad \text{(Note that } \sqrt{37} \approx 6.0828.\text{)}$$

$$= \frac{-3 + 6.0828}{2} = 1.5414 \quad \text{or} \quad 1.5 \text{ to the nearest tenth}$$

Also

$$\text{Root 2} = \frac{-3 - \sqrt{3^2 - 4(1)(-7)}}{2(1)}$$

$$= \frac{-3 - \sqrt{9 + 28}}{2}$$

$$= \frac{-3 - \sqrt{37}}{2}$$

$$= \frac{-3 - 6.0828}{2} = -4.5414 \quad \text{or} \quad -4.5 \text{ to the nearest tenth}$$

Thus the two roots are 1.5 and -4.5. ☐

Comment The quadratic formula is often written in combined form as $\dfrac{-b \pm \sqrt{b^2 - 4ac}}{2a}$.

EXAMPLE 4 Solve the equation $3x^2 = 5x + 7$ for x.

SOLUTION We first rearrange the equation so that it is in standard form. We get

$$3x^2 = 5x + 7$$

or

$$3x^2 - 5x - 7 = 0.$$

Now we determine the values of a, b, and c. We have $a = 3$, $b = -5$, and $c = -7$. Substituting these values into the quadratic formula gives

$$\frac{-b \pm \sqrt{b^2 - 4ac}}{2a} = \frac{-(-5) \pm \sqrt{(-5)^2 - 4(3)(-7)}}{2(3)}$$

$$= \frac{5 \pm \sqrt{25 + 84}}{6}$$

$$= \frac{5 \pm \sqrt{109}}{6} \qquad \text{(Note that } \sqrt{109} \approx 10.4403.\text{)}$$

$$= \frac{5 \pm 10.4403}{6}$$

$$\text{Root 1} = \frac{5 + 10.4403}{6} \qquad \text{Root 2} = \frac{5 - 10.4403}{6}$$

$$= \frac{15.4403}{6} \qquad\qquad = \frac{-5.4403}{6}$$

$$= 2.5734 \qquad\qquad\quad = -0.9067$$

$$= 2.6 \text{ to the} \qquad\quad = -0.9 \text{ to the}$$
$$\text{nearest tenth} \qquad\qquad \text{nearest tenth}$$

Thus the two roots are 2.6 and −0.9. ▢

Often the solutions of some verbal problems involve solving quadratic equations. We use the most convenient method for solving such equations.

EXAMPLE 5   In the game of baseball, the baseball diamond is shaped in the form of a square 90 feet on a side as shown in Figure 7.8. Baseball regulations require that the pitcher's mound be 60.5 feet from home plate along a straight line joining home plate to second base (as indicated in Figure 7.8). What is the distance between the pitcher's mound and second base to the nearest tenth of a foot?

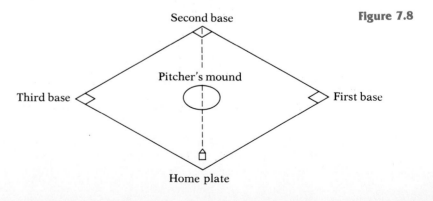

Figure 7.8

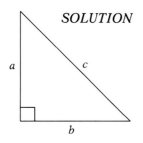

Figure 7.9

SOLUTION We can solve this problem by applying the Pythagorean theorem. Recall that in a right triangle labeled as in Figure 7.9 the Pythagorean theorem states that $a^2 + b^2 = c^2$.

Since the baseball diamond is shaped in the form of a square, we have right angles at all the bases. Thus

$$\left(\begin{array}{c}\text{distance from} \\ \text{second base to} \\ \text{third base}\end{array}\right)^2 + \left(\begin{array}{c}\text{distance from} \\ \text{third base to} \\ \text{home plate}\end{array}\right)^2 = \left(\begin{array}{c}\text{distance from} \\ \text{second base to} \\ \text{home plate}\end{array}\right)^2$$

Let $x =$ the entire distance from second base to home plate. We then have

$$90^2 + 90^2 = x^2$$
$$8100 + 8100 = x^2$$
$$16{,}200 = x^2$$

or

$$x^2 - 16{,}200 = 0.$$

We now have a quadratic equation in standard form where $a = 1$, $b = 0$, and $c = -16{,}200$. Substituting these values into the quadratic formula gives

$$\frac{-0 \pm \sqrt{0^2 - 4(1)(-16200)}}{2(1)}$$
$$= \frac{\pm \sqrt{64800}}{2}$$
$$= \frac{\pm 254.5584}{2} \qquad (\text{From a calculator, } \sqrt{64800} \approx 254.5584.)$$

$$\text{Root 1} = \frac{+254.5584}{2} \qquad\qquad \text{Root 2} = \frac{-254.5584}{2}$$
$$= 127.2792 \qquad\qquad\qquad = -127.2792$$
$$= 127.3 \quad \text{to the nearest} \qquad \text{We do not consider this}$$
$$\text{tenth} \qquad\qquad\qquad \text{negative value, since}$$

We do not consider this negative value, since distance cannot be a negative number.

Thus the entire distance between home plate and second base is 127.3 feet. However, we already know that the distance between home plate and the pitcher's mound is 60.5 feet, so the distance between the pitcher's mound and second base is

$$127.3 - 60.5 \quad \text{or} \quad 66.8 \text{ feet.}$$

EXERCISES

Solve each of the quadratic equations given in Exercises 1–30 by factoring or by using the quadratic formula. Use the square root button $\boxed{\sqrt{x}}$ on your calculator to express all irrational roots to the nearest tenth.

1. $x^2 + x - 72 = 0$
2. $2x^2 + x - 10 = 0$
3. $2x^2 + 7x = -6$
4. $3x^2 - 10x + 3 = 0$
5. $3x^2 - 8x + 4 = 0$
6. $2x^2 + x - 10 = 0$
7. $\dfrac{x}{3} + \dfrac{9}{x} = 4$
8. $x(x - 3) = 4$
9. $x(x - 2) = 25$
10. $x(x + 3) = 17$
11. $x^2 + 7x = 0$
12. $x^2 + 12x = 0$
13. $x^2 = 9$
14. $x^2 - 25 = 0$
15. $x^2 - 16x + 64 = 0$
16. $x^2 + 5x - 7 = 0$
17. $x^2 - 12x + 2 = 0$
18. $x^2 - 8x + 12 = 0$
19. $2x^2 - 11x = 4$
20. $x = 4 + \dfrac{3}{x}$
21. $2x^2 + x = 5$
22. $2x^2 = 8x - 3$
23. $2x^2 - 11x + 1 = 0$
24. $3x^2 - 5x + 2 = 0$
25. $2x^2 - 3x - 1 = 0$
26. $5x^2 - x - 2 = 0$
**27. $2x^2 - 5x + a = 0$
**28. $5x^2 + x - c = 0$
29. $2x = 4 + \dfrac{3}{x}$
30. $9x^2 + 4x - 3 = 0$

31. Heather and Jason start from the same point and travel along straight roads that are at right angles to each other. Heather and Jason are traveling at rates of 30 and 40 miles per hour, respectively. In how many hours will they be 200 miles apart?

**32. Dawn Prescott purchased shares of stock in the Mellow Corporation, for which she paid $1350. Had the price per share been $5 less, Dawn would have been able to purchase 3 extra shares of stock for the same total expenditure of $1350. How many shares of stock did Dawn purchase, and what was the price per share paid?

33. The product of two consecutive odd integers is 195. Find the integers.

34. Yvonne Brezinsky has a 30-foot ladder. She wants to lean the ladder against her house so that she can reach the ledge of a window that is 16 feet above the ground. How close to the house should the foot of the ladder be placed so that she will be able to reach the ledge of the window?

35. The ratio of the length and width of a rectangular conference room in an office building is $3:4$. The area of the room is 1200 square meters. Find the length and width of the conference room.

36. Edwin wishes to purchase carpeting for his living room. He knows that the length of the room is 4 more than twice the width of the room. Furthermore, the carpet installer tells him that he needs 198 square feet of carpet. How long and how wide is Edwin's living room?

37. To make it easier for disabled students who are confined to wheelchairs to enter the main administration building, the carpenter at River College constructed a ramp as shown in Figure 7.10. The length of the ramp is 3 more than twice the height of the ramp, and the distance from the bottom of the ramp to the base of the building is 2 more than twice the height of the ramp, as shown below. How long is the ramp?

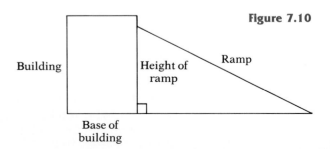

Figure 7.10

**38. Several students in a science laboratory have constructed a miniature rocket ship. When the rocket ship is launched into the air, its height h at any time t is given by the formula $h = -16t^2 + 80t$. At what time will the rocket ship reach a height of 40 feet?

7.10

SUMMARY

In this chapter we reviewed some important ideas from algebra. We first reviewed the definition of, and rules for working with, exponents. We also reviewed how to solve equations and inequalities for the unknown variables. After defining what a polynomial is, we analyzed the four basic operations that can be performed with polynomials.

We studied one technique for solving a system of simultaneous linear equations. Other techniques will be studied in later chapters. Finally, we analyzed the quadratic equation. We pointed out (after analyzing the different factoring techniques) how to solve such equations either by factoring them or by using the quadratic formula. The reader should be aware that quadratic equations can be solved by other techniques such as completing the square and graphically.

Again, we remind the reader that our discussion of the various topics from algebra is intended to serve only as a review of these ideas. A complete and thorough analysis can be found in any algebra book.

STUDY GUIDE

You should now be able to demonstrate your knowledge of the following ideas presented in this chapter by giving definitions, descriptions, or specific examples. Page references (in parentheses) are given for each term so that you can check your answer.

Exponent (p. 352)
Base (p. 352)
Variables (p. 357)
Constants (p. 357)
Polynomials (p. 358)
Terms (p. 358)
Monomial (p. 358)
Binomial (p. 358)
Trinomial (p. 358)
Degree of a polynomial (p. 358)
Descending order (p. 359)
Standard form of a polynomial (p. 359)
Like terms (p. 359)
Unlike terms (p. 359)
Function (p. 365)
Domain (p. 366)
Range (p. 366)

Function notation (p. 367)
Mathematical models (p. 368)
Equations (p. 368)
Variables (p. 368)
Solution (p. 368)
Solving linear equations (p. 368)
Systems of simultaneous linear equations (p. 375)
Inequalities in one variable (p. 379)
Factoring polynomials (p. 383)
Greatest common factor (p. 383)
Difference of two squares (p. 384)
Quadratic equation (p. 389)
Standard form of a quadratic equation (p. 389)
Solving quadratic equations by factoring (p. 389)
Quadratic formula (p. 390)

FORMULAS   The following list summarizes all of the formulas given in this chapter.
TO REMEMBER
1. Rules for exponents

$$b^x = \underbrace{b \cdot b \cdot b \cdot b \ldots \cdot b}_{x \text{ of them}}$$ Definition of exponents

$$b^x \cdot b^y = b^{x+y}$$ Multiplication rule

$$(b^x)^y = b^{xy}$$ Power rule

$$b^{-x} = \frac{1}{b^x} \quad b \neq 0$$ Negative exponent

$$b^0 = 1$$ Zero exponents

$$\frac{b^x}{b^y} = b^{x-y} \quad b \neq 0$$ Quotient rule

$$b^{\frac{x}{y}} = (\sqrt[y]{b})^x = \sqrt[y]{b^x} \quad b > 0$$ Fractional exponents

2. Rules for solving linear equations

a. Any number may be added to or subtracted from both sides of an equation.

b. *Both* sides of an equation may be multiplied by or divided by the same number (except that we cannot divide by 0).

3. Rules for solving linear inequalities

a. Any number may be added to or subtracted from both sides of an inequality.

b. *Both* sides of an inequality may be multiplied or divided by the same positive number. The sense of the inequality is not changed.

c. *Both* sides of an inequality may be multiplied or divided by the same negative number. The sense of the inequality is reversed.

4. Quadratic equation

When a quadratic equation is arranged in the standard form $ax^2 + bx + c = 0$, where $a \neq 0$, the roots are

$$\text{Root } 1 = \frac{-b + \sqrt{b^2 - 4ac}}{2a} \qquad \text{Root } 2 = \frac{-b - \sqrt{b^2 - 4ac}}{2a}$$

MASTERY TESTS

Form A   **1.** Factor: $x^2 - 0.64y^2$.

2. Solve for y: $\dfrac{2y + 1}{5} - \dfrac{3y - 7}{2} = 7$.

3. If $f(x) = 2x^{\frac{3}{2}} - 4x^0$, find $f(4)$.

4. In the temperature formula $F = \dfrac{9}{5}C + 32$, at what temperature is $F = C$?

  a) $+ 8°$ **b)** $-8°$ **c)** $+40°$ **d)** $-40°$ **e)** none of these

5. The quotient of $(2x^3 - 3x^2) \div x^2$ is

    a) $2x^3 - 1$ **b)** $2x - 3$ **c)** $2x - 3x^2$ **d)** $2x^5 - 3x^4$

    e) none of these

6. If $x + 2$ is a factor of $x^2 + Kx - 8$, find the numerical value of K.

7. Solve the following inequality for x: $-2x + 7(3x - 1) > -4(9x + 3)$.

8. If $x \neq 0$, then $x^{-\frac{1}{2}}$ is equal to

    a) $\sqrt{x}$ **b)** 2 **c)** $\dfrac{1}{x^2}$ **d)** $\dfrac{1}{\sqrt{x}}$ **e)** none of these

9. Solve for x: $x^2 - 5x - 24 = 0$.

10. Factor: $2x^2 - 13x + 18$.

11. If $2x^2 + 7x - 15$ is divided by $2x - 3$, the remainder is

    a) -30 **b)** 0 **c)** -9 **d)** -21 **e)** none of these

12. Solve for x: $\dfrac{3}{x} + \dfrac{5}{2x} = 1$.

13. Solve for d: $t = a + (n - 1)d$.

14. If a and b are real numbers and $ab > 0$, which is *never* true?

    a) $a > b$ **b)** $a > 0$ and $b > 0$ **c)** $a > 0$ and $b < 0$

    d) $a < 0$ and $b < 0$ **e)** none of these

15. Solve the following inequality for x: $7x - 3 \geq 10x + 21$.

Form B

1. A concrete mixture requires 2 parts of gravel, 3 parts of cement, and 4 parts of sand by weight. How many pounds of cement are required for $4\frac{1}{2}$ tons of this mixture? (1 ton = 2000 pounds)

    a) 2000   **b)** 1000   **c)** 3000   **d)** 4000   **e)** none of these

2. Find, to the nearest tenth, the roots of the equation $2x + \dfrac{4}{x} = 7$.

3. The Blais Paper Company has determined that x truckloads of bleached paper costs $7x^3 + 2x^2 + 5x + 9$ hundreds of dollars. Find the cost of one truckload of bleached paper.

4. Find, to the nearest tenth, the values of x that are solutions to the equation $x^2 + 2x - 1 = 0$.

5. Find, to the nearest tenth, the roots of the equation $\dfrac{2x}{x - 1} = \dfrac{3}{x + 9}$.

6. A mechanic's hourly wage is 4 times her helper's. They were paid a total of $78 for a job on which the mechanic worked 5 hours and the helper worked 6 hours. Find the hourly wage of the helper.

7. A soda-vending machine in the student cafeteria contains 20 coins. Some of the coins are nickels, and the rest are quarters. If the value of the coins is $4.40, find the number of coins of each kind. (Solve by algebraic techniques only.)

8. Working alone, Maria can draw the architectural plans for a building in 10 hours, and Adolf can draw the plans in 12 hours. How long will it take them to do the job if they work together?

9. Express as a single fraction in lowest terms $\dfrac{2x + 1}{8} - \dfrac{x + 2}{6}$.

10. Jennifer invested $5000, part at 6% and the remainder at 8%. The annual income from the 8% investment is $260 greater than the annual income from the 6% investment. Find the amount of money invested at each rate.

11. One bricklayer takes twice as long as a second bricklayer to build a certain wall. Together they can build the wall in 6 hours. How long would it take each bricklayer to build the wall alone?

** 12. Refer back to the newspaper article given at the beginning of this chapter. If the state officials have 10,000 yards of special material available for the job, what is the largest area that can be cleansed from the chemical pollutants?

13. The supply S of a certain computer component and the demand D for the computer component are given by $S = 50{,}000p - 70{,}000$ and $D = \dfrac{50{,}000}{p}$, where p is the price. Find the price p at which the supply will be equal to the demand. (This price is known as the **equilibrium price**.)

14. A scientist is experimenting with certain mosquitoes whose population P at any given time t is given by $P = -15t^2 + 40t + 7000$. Assuming that the time is measured in hours after the experiment begins, find the time at which the scientist will have a mosquito population of 5000.

15. The laws of supply and demand tell us that the price of a particular video cassette recorder (VCR) depends upon the quantity demanded. If x video cassette recorders are demanded, then the price of a VCR is given by the equation $P = 50{,}000 - 300x - x^2$. If VCRs are currently selling for $400 apiece, find the demand, x.

8

GRAPHING FUNCTIONS

CHAPTER OBJECTIVES

To review the rectangular coordinate system and how we plot points in this system. (*Section 8.2*)

To learn how we draw the graph of a linear equation. (*Section 8.3*)

To study the different forms of an equation of a straight line. (*Section 8.3*)

To indicate how we find the x-intercept and y-intercept of a line. (*Section 8.4*)

To understand what is meant by the slope of a line. (*Section 8.4*)

To demonstrate how we draw the graph of linear inequalities. (*Sections 8.5 and 8.7*)

To point out how we can solve a system of linear equations in two variables graphically. (*Section 8.6*)

To introduce you to linear programming. (*Section 8.8*)

To apply graphical techniques to solve linear programming problems. (*Section 8.9*)

Teachers and Board of Education at an Impasse

NEW YORK: Despite round-the-clock negotiations, Board of Education officials announced yesterday that they were unable to reach any agreement with the teacher's union. The main stumbling block is not salary raises but working conditions. The Board is attempting to maximize the educational opportunities available to students under such constraints as curtailed teaching personnel, a severely reduced budget, and diminished classroom space and supplies.

    The teachers vowed to go out on strike unless their demands for more favorable working conditions and smaller class sizes are met.

DAILY PRESS September 1, 1985

In recent years, linear programming has been applied to analysis in many different areas of study, including education and the energy shortage. Is it possible to determine how to maximize educational opportunity when we have certain restrictions?

In this chapter we indicate how we can apply graphical techniques to solve some of the linear programming problems associated with the ideas discussed in the newspaper article.

8.1

INTRODUCTION

During World War II the United States found itself facing urgent military needs but had limited resources with which to meet them. For example, the Navy needed submarines, aircraft carriers, battleships, cruisers, destroyers, and PT boats. To produce any one of these required a certain amount of labor, money, and raw materials (metal, rubber, etc.). Each of these ships had various advantages and limitations. Of course, the cost varied greatly. Manufacturing a PT boat is much cheaper than producing a battleship, but a battleship has greater destructive power than a PT boat.

Similarly, an aircraft carrier is very expensive to build. However, with its planes it can inflict enormous damage on the enemy. On the other hand, the loss of an aircraft carrier may mean losing some planes also. The cost would then be very great. The government had to decide how to use its limited money and materials so as to maximize the destructive power and minimize losses.

The theory of linear programming was developed to answer this and similar questions. It turns out that in addition to military problems, linear programming can be applied to a wide variety of business and economic situations as well.

British and American transport planes carried supplies to West Berlin during the Berlin airlift in 1948. Planes landed at frequent and regular intervals, scheduled by linear programming.

Wide World Photos, Inc.

Linear programming problems can be solved either by graphical techniques or by algebraic techniques. The algebraic technique is known as the *simplex method*. In this chapter we indicate how we can apply graphical techniques to solve linear programming problems. However, to solve such problems, we must first study the ideas of functions and graphs.

8.2
THE RECTANGULAR COORDINATE SYSTEM

Most of us are familiar with the concept of a graph, although we might not use this name specifically. For example, consider the map of part of San Francisco shown in Figure 8.1. Chinatown has a map

---

HISTORICAL NOTE

The idea of graphs can be traced back to the ancient Egyptians and Greeks. Many Greek mathematicians, including Euclid and Apollonius, studied graphs. *The Elements*, written by Euclid, contains a detailed treatment of figures formed by straight lines and circles. *Conic Sections*, written by Apollonius, presents a thorough discussion of the graphs of conic sections, which he named ellipses, parabolas, and hyperbolas. In the seventeenth century the French mathematician René Descartes (1596–1650) developed an algebraic method for analyzing curves. This is the basis of the coordinate geometry and graphs that we study in this chapter.

So successful was Descartes's work that he soon was invited to the courts of many kings and queens. He routinely declined most of these invitations. In 1649, however, he was invited to the court of Queen Christina of Sweden to give the queen instruction in philosophy. He accepted this invitation. Queen Christina preferred to work in the mornings in the unheated castle, whereas Descartes was accustomed to staying in bed until noon, when it warmed up a bit. Several months after arriving in Stockholm, Descartes contracted pneumonia. In 1650, Descartes died at the age of 54, apparently as a result of his inability to adapt to the rigors of the Swedish winter.

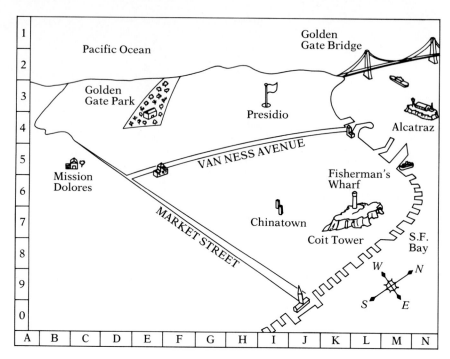

Figure 8.1

location of approximately I7. To find it, we move across the bottom scale until we arrive at the I position. We draw a vertical line through this location. Next, we find 7 on the side of the map and draw a horizontal line through this point. The place where the horizontal and vertical lines meet (intersect) is the approximate location of Chinatown. What is the map location of Alcatraz? Alcatraz has a map location of N3. What is the map location of Fisherman's Wharf?

Since most of the functions that we discuss in this book are given by means of an equation, we often find it convenient to draw a "picture," called the **graph** of the equation. It is for this reason that we study the rectangular coordinate system.

To get started, let us first discuss a convenient way of representing numbers, that is, by means of a **number line.** A horizontal line is drawn, and any point on it is selected as the starting point. This point is labeled 0 and is called the **origin** (see Figure 8.2). Since a line can be extended indefinitely in either direction, we indicate this by putting arrows at the left and at the right.

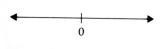

Figure 8.2

Next we select any convenient unit of length and mark off points in succession both to the right and left of zero. When we move in a direction that is to the right of 0, we say that we are moving in the **positive direction.** When we move to the left, we say that we are moving in the **negative direction.** We label the points as shown in

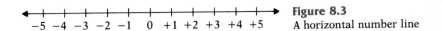

Figure 8.3
A horizontal number line

Figure 8.3. The number that names a point is called the **coordinate** of the point.

Number lines do not necessarily have to be drawn horizontally. They may also be drawn vertically, as shown in Figure 8.4.

In the rectangular coordinate system we draw two lines (one vertical and one horizontal) that are perpendicular to each other. The horizontal line is called the **x-axis,** and the vertical line is called the **y-axis.** The point where the x-axis and the y-axis meet is called the **origin.** Both of the axes are labeled with a **number scale,** as shown in Figure 8.5. Starting at the origin, if we move to the right, we are going in the positive direction on the x-axis as opposed to moving to the left, which is the negative direction. Similarly, if we start at the origin and move up, we are going in the positive direction on the y-axis as opposed to moving down, which is the negative direction.

Look at Figure 8.5. To get to point P, we start at the origin and move 4 units to the right and then, from that point, move up 3 units. We call 4 the **x-coordinate** or **abscissa** of the point and 3 the **y-coordinate** or **ordinate** of the point. We label the point as (4, 3), always being careful to write the x-coordinate first and then the y-coordinate. The numbers (4, 3) are called the **coordinates** of point P. We always enclose the x- and y-coordinates within parentheses and separate them by a comma.

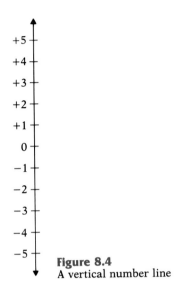

Figure 8.4
A vertical number line

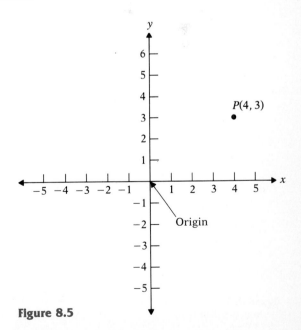

Figure 8.5

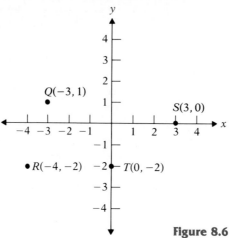

Figure 8.6

In Figure 8.6 the coordinates of point Q are $(-3, 1)$. This means that we start at the origin, move 3 units to the left, and then, from that point, move up 1 unit.

The coordinates of point R are $(-4, -2)$. We start at the origin, move 4 units to the left, and then, from that point, move down 2 units.

The coordinates of point S are $(3, 0)$. We start at the origin, move 3 units to the right, and then move 0 units up or down.

The coordinates of point T are $(0, -2)$. We start at the origin and move 0 units to the right or left. Then, from that point, we move down 2 units.

Notice that the x-axis and y-axis divide the plane (paper) into four regions. Each region is called a **quadrant,** and these are labeled counterclockwise as shown in Figure 8.7. The values of x and y for any point may be positive or negative depending upon the quadrant in which they are located. The different possibilities are summarized below:

y

Quadrant II	Quadrant I
Quadrant III	Quadrant IV

x

Figure 8.7
The four quadrants

If point is in quadrant	Then x is	Then y is
I	positive	positive
II	negative	positive
III	negative	negative
IV	positive	negative

What are the values of x and y when the point is on the line between the quadrants?

The system that we have just described is called a ***rectangular coordinate system.*** It is the one that is most commonly used in mathematics. It should be noted that other systems are possible. In this text we use only a rectangular coordinate system.

EXERCISES

For Exercises 1–8, refer to the graph shown in Figure 8.8. Find the coordinates of the points indicated.

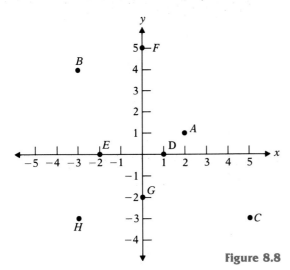

Figure 8.8

1. Point A 2. Point B 3. Point C
4. Point D 5. Point E 6. Point F
7. Point G 8. Point H

For Exercises 9–20, draw a pair of coordinate axes and graph the specified points.

9. $(7, 3)$ 10. $(4, -3)$ 11. $(-2, -7)$
12. $(3, 0)$ 13. $(-4, 0)$ 14. $(0, 3)$
15. $(0, -8)$ 16. $(0, 0)$ 17. $(|-3|, 7)$
18. $(|-4|, |-3|)$ 19. $(-4, |-3|)$ 20. $(|5|, |-3|)$

For Exercises 21–24, determine the quadrant in which the point lies.

21. $(3, -7)$ 22. $(-2, -4)$
23. $(-7, 5)$ 24. $(|-2|, |-1|)$

25. Graph the points $(1, -4)$ and $(4, 5)$. Join them together with a straight line. Where does this line cut the y-axis? In other words, find the y-value at this point.

26. Graph the points $(10, -3)$ and $(-5, 6)$. Join them together with a straight line. Where does this line cut the x-axis? In other words, find the x-value at this point.

27. Locate several points on the x-axis. What is the y-value for every point on the x-axis?

28. Locate several points on the y-axis. What is the x-value for every point on the y-axis?

29. What are the coordinates of the origin?

30. Graph the points $(3, 5)$ and $(7, 9)$. Join them together with a straight line. What is the y-value of the point on this line when $x = 6$?

8.3

THE GRAPH AND EQUATION OF A STRAIGHT LINE

Consider the equation $y = 2x + 1$. One solution of this equation is $x = 2$ and $y = 5$. We can write this solution in abbreviated form as $(2, 5)$. Another solution is $x = 1$ and $y = 3$, which we write as $(1, 3)$. Other solutions are $(3, 7)$, $(-1, -1)$, $(0, 1)$, and so on. Actually, there are infinitely many such pairs of values that are solutions for this equation. To find other solutions, we let x be any convenient number and solve for y as shown in the following table.

x	$y = 2x + 1$	y	*Point*
4	$2(4) + 1$	9	$(4, 9)$
3	$2(3) + 1$	7	$(3, 7)$
2	$2(2) + 1$	5	$(2, 5)$
1	$2(1) + 1$	3	$(1, 3)$
0	$2(0) + 1$	1	$(0, 1)$
-1	$2(-1) + 1$	-1	$(-1, -1)$
-2	$2(-2) + 1$	-3	$(-2, -3)$
-3	$2(-3) + 1$	-5	$(-3, -5)$
-4	$2(-4) + 1$	-7	$(-4, -7)$

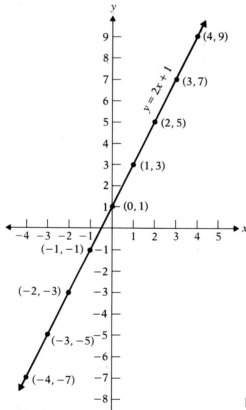

Figure 8.9
Graph of $y = 2x + 1$

Let us plot these points on a rectangular coordinate system, as shown in Figure 8.9. If we join these points, we get a straight line. This line is called the *graph* of the equation. Actually, the graph of the equation consists of *all* of the points (x, y) that are solutions of the equation.

EXAMPLE 1 Draw the graph of $y = 5x + 3$.

SOLUTION We set up a table of values by finding several pairs of numbers that are solutions to the equation, as shown below. Then we plot the points and join them. The resulting straight line is the graph of $y = 5x + 3$. See Figure 8.10.

x	$y = 5x + 3$	y
2	$5(2) + 3$	13
1	$5(1) + 3$	8
0	$5(0) + 3$	3
-1	$5(-1) + 3$	-2
-2	$5(-2) + 3$	-7

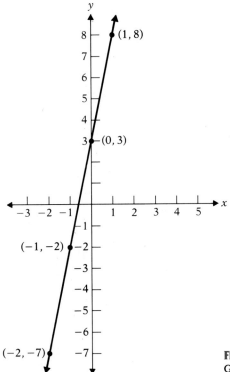

Figure 8.10
Graph of $y = 5x + 3$.

EXAMPLE 2 Draw the graph of $y = -3x + 2$.

SOLUTION We set up a table of values by finding several pairs of numbers that are solutions to the equation. The graph is given in Figure 8.11.

x	$y = -3x + 2$	y
2	$-3(2) + 2$	-4
1	$-3(1) + 2$	-1
0	$-3(0) + 2$	2
-1	$-3(-1) + 2$	5
-2	$-3(-2) + 2$	8

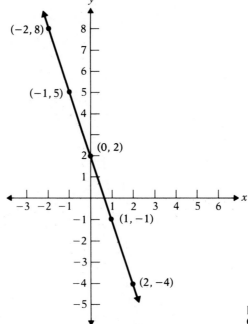

Figure 8.11
Graph of $y = -3x + 2$ ☐

EXAMPLE 3   Draw the graph of $y = 3$.

SOLUTION   We set up a table of values as we did in the previous examples. In this case, we notice they y must always be 3, no matter what value x has. The graph of $y = 3$ is given in Figure 8.12. It is a line parallel to the x-axis.

x	y
2	3
1	3
0	3
-1	3
-2	3

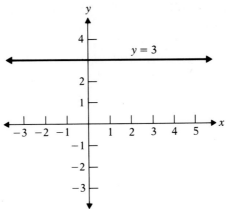

Figure 8.12 ☐

EXAMPLE 4   Draw the graph of $x = 4$.

SOLUTION   We set up a table of values. In this case we notice that x must always be 4, no matter what value y has. The graph of $x = 4$ is given in Figure 8.13. It is a line parallel to the y-axis.

x	y
4	-2
4	-1
4	0
4	1
4	2

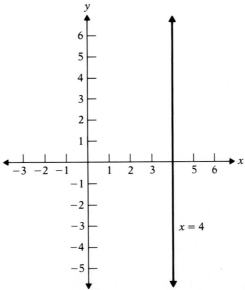

Figure 8.13 ☐

All of the graphs that we have drawn so far are straight lines. Actually, the graph of any equation of the form $Ax + By + C = 0$ (where A, B, and C are constants and A and B are not both 0) will be a straight line. An equation whose graph is a straight line is called a *linear equation.*

The results of the previous examples can be summarized as follows:

Summary

1. The graph of $Ax + By + C = 0$ (where A, B, and C are constants and A and B are not both 0) is called a straight line.
2. The graph of $x = C$ is a straight line parallel to the y-axis.
3. The graph of $y = B$ is a straight line parallel to the x-axis.

EXAMPLE 5   Is the point (2, 3) on the graph of $3x + 5y = 21$?

SOLUTION   The point (2, 3) is on the graph of $3x + 5y = 21$ if and only if $x = 2$, $y = 3$ is a solution of the equation. Substituting, we find that $3(2) + 5(3) = 21$. Thus the point (2, 3) is on the graph of $3x + 5y = 21$. □

Graphing equations involving absolute values is also done by using a table of values. Thus to draw the graph of $y = |x|$, we first find several solutions of $y = |x|$ by replacing x with several signed numbers and finding the corresponding y-values. Of course, y must always be a nonnegative number. We have the following table of values and its graph in Figure 8.14.

x	y
−3	3
−2	2
−1	1
0	0
1	1
2	2
3	3

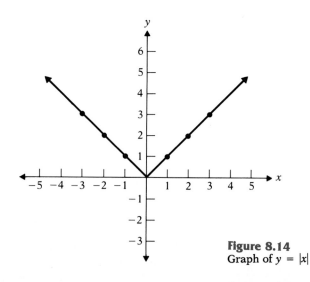

Figure 8.14
Graph of $y = |x|$

EXERCISES

Draw the graph of each of the equations given in Exercises 1–11.

1. $y = 3x + 7$

2. $y = x + 4$

3. $y = -3x + 7$

4. $y = -5x + 3$

5. $y = -3x$

6. $4x = 8$

7. $5y = -15$

8. $4x + 3y = 12$

9. $5x + 3y = 15$

10. $y = \dfrac{x}{4} + 3$

11. $y = -\dfrac{x}{3} - 4$

**12. Draw the graph of $y = |2x - 3|$.

**13. Draw the graph of $|x - 2| = 9$.

**14. Draw the graph of $|y - 3| = 6$.

**15. Draw the graph of $|x - y| = 3$.

**16. Draw the graph of $|x| - |y| = 3$.

**17. Draw the graph of $|y| - |x| = 3$.

8.4

INTERCEPTS AND SLOPE OF A STRAIGHT LINE

Consider the graph of $2x + 3y = 6$, which is given formally below in Figure 8.15. Notice that this graph crosses the x-axis at the point $(3, 0)$, and then crosses the y-axis at the point $(0, 2)$. Mathematicians refer to these crossing points as the x-intercept and the y-intercept, respectively.

> **Definition 8.1**   *The point at which a graph crosses the x-axis is called the **x-intercept**. To find the x-value of this point, set y = 0, and then solve for x.*

> **Definition 8.2**   *The point at which a graph crosses the y-axis is called the **y-intercept**. To find the y-value of this point, set x = 0, and then solve for y.*

EXAMPLE 1   Find the x-intercept and the y-intercept of the line $4x + 3y = 12$.

SOLUTION   We apply Definitions 8.1 and 8.2. To find the x-intercept, we let $y = 0$ and solve for x, getting

$$4x + 3(0) = 12$$
$$4x = 12$$
$$x = 3$$

Thus the x-intercept is at $(3, 0)$. To find the y-intercept, we let $x = 0$ and solve for y, getting

$$4(0) + 3y = 12$$
$$3y = 12$$
$$y = 4$$

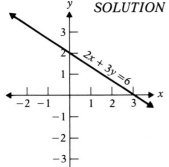

Figure 8.15
Graph of $2x + 3y = 6$

Thus the *y*-intercept is at (0, 4). The graph of this equation is given in Figure 8.16.

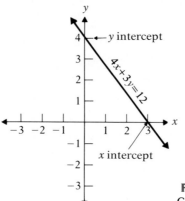

Figure 8.16
Graph of $4x + 3y = 12$ ☐

EXAMPLE 2   Draw the graph of a straight line whose *x*-intercept is at $(-3, 0)$ and whose *y*-intercept is at $(0, -2)$.

SOLUTION   Since we are told that the *x*-intercept is at $(-3, 0)$, we know that the graph crosses the *x*-axis at the point $(-3, 0)$. Also, since the *y*-intercept is at $(0, -2)$, the graph crosses the *y*-axis at the point $(0, -2)$. Thus we plot these two points on a coordinate axis and join them to get the straight line shown in Figure 8.17.   ☐

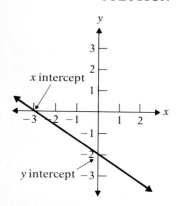

Figure 8.17
Graph of a line whose *x* intercept is at $(-3, 0)$ and whose *y* intercept is at $(0, -2)$

Another important number associated with the graph of a straight line is its *slope*. We define the slope of a line as the change in the vertical distance divided by the change in the horizontal distance. To understand this concept, consider the roof shown in Figure 8.18. To climb to the top, a person would cover a horizontal distance of 365 centimeters and a vertical distance of 730 centimeters. Therefore in this case the slope of the roof is $\frac{730}{365}$ or 2. The slope of the roof shown in Figure 8.19 is $\frac{365}{1095}$ or $\frac{1}{3}$. The first roof has a greater slope and hence is steeper.

Figure 8.18

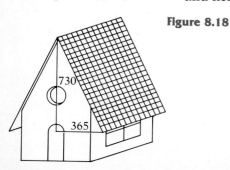

Figure 8.19

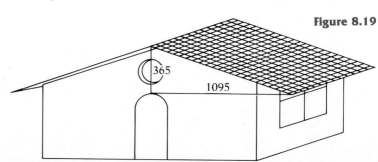

(Source: United Press International)

Figure 8.20

Now consider the photograph shown in Figure 8.20. On August 7, 1974, Philippe Petit walked across a guy wire connecting the twin towers of the World Trade Center in New York City. In accomplishing this feat the only change that occurred was the horizontal distance covered. There was no vertical distance covered. Such a line has zero slope.

On May 26, 1977, George Willig scaled the side of the World Trade Center in New York City. In his climb up the 110 floors the only change that occurred was the vertical distance. There was no horizontal distance covered. Such a line has no slope, since we would have to divide by zero. (See Figure 8.21.)

(Source: United Press International)

Figure 8.21

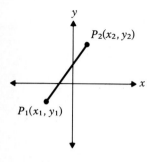

Figure 8.22

More formally, suppose that we are given two distinct points, P_1 and P_2, in a rectangular coordinate plane as shown in Figure 8.22. The subscripts are used merely to distinguish the points. Let the coordinates of these points be $P_1(x_1, y_1)$ and $P_2(x_2, y_2)$. These two points determine a unique number called the slope of the line.

Formula 8.1   *The **slope** of the line passing through the points (x_1, y_1) and (x_2, y_2) is given by*

$$\frac{y_2 - y_1}{x_2 - x_1},$$

provided that $x_2 \neq x_1$. If $x_2 = x_1$, then $x_2 - x_1 = 0$, and

$$\frac{y_2 - y_1}{x_2 - x_1}$$

does not exist.[1] We usually use the letter m to represent the slope.

EXAMPLE 3   Find the slope of the line passing through the points $(2, 5)$ and $(7, 8)$.

SOLUTION   We apply Formula 8.1. Let $x_1 = 2$, $y_1 = 5$, $x_2 = 7$, and $y_2 = 8$. Then

$$x_2 - x_1 = 7 - 2 = 5$$
$$y_2 - y_1 = 8 - 5 = 3$$

so that

$$\frac{y_2 - y_1}{x_2 - x_1} = \frac{3}{5}$$

Thus the slope of the line passing through the points $(2, 5)$ and $(7, 8)$ is $\frac{3}{5}$. □

Comment   In applying Formula 8.1 it does not make a difference which point is called $P_1(x_1, y_1)$ and which point is called $P_2(x_2, y_2)$. In both cases we get the same answer. To illustrate, suppose that in Example 3 we let $x_1 = 7$, $y_1 = 8$, $x_2 = 2$, and $y_2 = 5$. Then the slope of the line is

$$\frac{y_2 - y_1}{x_2 - x_1} = \frac{5 - 8}{2 - 7} = \frac{-3}{-5} = \frac{3}{5}$$

This is exactly the same answer we obtained in Example 3.

---

1. In mathematics, division by zero is not defined. For a detailed discussion of division involving zero, see pages 194–195.

EXAMPLE 4   Find the slope of the line passing through the points.

   a) (3, 4) and (5, 1)

   b) (7, 6) and (9, 6)

   c) (4, 5) and (4, −2)

SOLUTION   We apply Formula 8.1.

a) $\dfrac{y_2 - y_1}{x_2 - x_1} = \dfrac{1 - 4}{5 - 3} = \dfrac{-3}{2} = -\dfrac{3}{2}$

b) $\dfrac{y_2 - y_1}{x_2 - x_1} = \dfrac{6 - 6}{9 - 7} = \dfrac{0}{2} = 0$

c) $\dfrac{y_2 - y_1}{x_2 - x_1} = \dfrac{-2 - 5}{4 - 4} = \dfrac{-7}{0}$ (This cannot be done. Therefore there is no slope; see footnote 1.)

It can be shown that if the slope of a line is positive, then the line rises as we go from left to right. If the slope of a line is negative, then the line falls as we go from left to right. If the slope of a line is zero, then the line is parallel to the x-axis. If a line has no slope, then the line is parallel to the y-axis. Thus the slope of a line tells us the "direction" of the line. These possibilities are pictured in Figures 8.23–8.26.

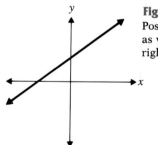

Figure 8.23
Positive slope (line rises as we move from left to right)

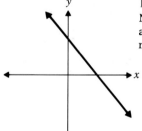

Figure 8.24
Negative slope (line falls as we move from left to right)

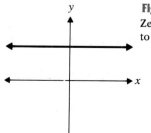

Figure 8.25
Zero slope (line is parallel to x-axis)

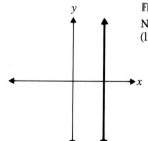

Figure 8.26
No slope
(line is parallel to y-axis)

Lines that are parallel have the same slope. Can you see why? (See Figure 8.27.)

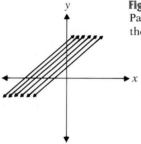

Figure 8.27
Parallel lines (all lines have the same slope)

Let us again analyze the graph of $y = 3x + 2$. Several pairs of values that satisfy this equation are shown in the accompanying table.

x	y
2	8
1	5
0	2
−1	−1
−2	−4

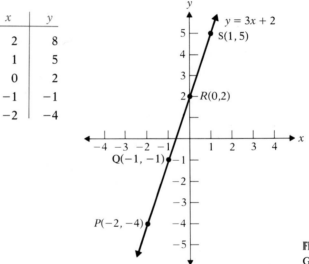

Figure 8.28
Graph of $y = 3x + 2$

The graph is given in Figure 8.28. To calculate the slope of this line we pick any two points on the line, say P and Q, and find the slope of the line passing through these points. We have

$$\text{Slope} = m = \frac{-4 - (-1)}{-2 - (-1)} = \frac{-4 + 1}{-2 + 1} = \frac{-3}{-1} = +3.$$

If we had selected points R and S, the slope of the line would still be 3. In both cases we get the same value for the slope; it is 3. More generally, if we select any two points on this line, the slope will always be 3. We say that *the slope of a line is constant.* Notice that 3 is the coefficient of x in the equation $y = 3x + 2$. Next we observe that the line crosses the y-axis at $(0, 2)$. In the equation $y = 3x + 2$, the 2 is the number that stands alone. We generalize our results in the following formula.

Formula 8.2 *The equation $y = mx + b$ represents a straight line whose slope is m and whose y-intercept is at (0, b).*

The accompanying table gives several examples of slope and y-intercept.

Equation	Slope	y-intercept at
$y = 2x + 5$	2	(0, 5)
$y = -3x + 4$	-3	(0, 4)
$y = 5x$	5	(0, 0)
$y = -x$	-1	(0, 0)
$y = 4$	0	(0, 4)
$y = x + 1$	1	(0, 1)
$x = 2$	Undefined	There is none

EXAMPLE 5   Find the equation of a line with a slope of 3 and whose y intercept is at (0, 14).

SOLUTION   Using Formula 8.2, we find that the equation is $y = 3x + 14$. ☐

EXAMPLE 6   Find the equation of a line that has a slope of -2 and passes through the origin.

SOLUTION   Since the line passes through the origin, its y-intercept is at (0, 0). Thus by using Formula 8.2 the equation is $y = -2x + 0$, or $y = -2x$. ☐

Since we will be interested in finding the equation of a line under many different conditions, we list several different forms of the equation, depending on the information given.

Name	Given information	Equation
slope-intercept form	slope $= m$, y-intercept $= (0,b)$	$y = mx + b$
point-slope form	slope $= m$, point $= (x_1, y_1)$	$y - y_1 = m(x - x_1)$
two-point form	point $= (x_1, y_1)$, point $= (x_2, y_2)$	$y - y_1 = \dfrac{y_2 - y_1}{x_2 - x_1}(x - x_1)$ $(x_1 \neq x_2)$
two-intercept form	x-intercept $= (a, 0)$, y-intercept $= (0, b)$	$\dfrac{x}{a} + \dfrac{y}{b} = 1$

EXAMPLE 7   Find the equation of a line whose slope is 3 and that passes through the point $(-1, -2)$.

SOLUTION   Since we are given the slope of the line and a point on the line, we use the point-slope form. We have $m = 3$ and $x_1 = -1$, $y_1 = -2$, so when we substitute into $y - y_1 = m(x - x_1)$, we get

$$y - (-2) = 3[x - (-1)]$$
$$y + 2 = 3(x + 1)$$
$$y = 3x + 1$$

The equation of the line then is $y = 3x + 1$.   □

EXAMPLE 8   Find the equation of the line that passes through the points $(3, 4)$ and $(6, 10)$.

SOLUTION   Since we are given two points, we use the two-point form. We let $x_1 = 3$, $y_1 = 4$, $x_2 = 6$, and $y_2 = 10$, so when we substitute these values into

$$y - y_1 = \frac{y_2 - y_1}{x_2 - x_1}(x - x_1),$$

we get

$$y - 4 = \frac{10 - 4}{6 - 3}(x - 3)$$

Simplifying, we get

$$y - 4 = 2(x - 3)$$
$$y = 2x - 2$$

Thus the equation of the line is $y = 2x - 2$.   □

EXAMPLE 9   Find the equation of the line whose x-intercept is 4 and whose y-intercept is 12.

SOLUTION   Since we are given both the x-intercept and the y-intercept, we use the two-intercept form. We let $a = 4$ and $b = 12$, so on substitution we get

$$\frac{x}{a} + \frac{y}{b} = 1$$

$$\frac{x}{4} + \frac{y}{12} = 1$$

Simplifying this equation gives $3x + y = 12$. Therefore the equation of the line is $3x + y = 12$.   □

EXAMPLE 10   Find the equation of the line that passes through the point $(1, 9)$ and that is parallel to the line whose equation is $y = 4x + 7$.

SOLUTION The line that we are interested in is parallel to the line whose equation is $y = 4x + 7$. Therefore the slope of both lines is 4. A point on the line is $(1, 9)$. Now that we know the slope of the line and a point on the line, we use the point-slope form with $m = 4$ and $x_1 = 1$, $y_1 = 9$. We have

$$y - y_1 = m(x - x_1)$$
$$y - 9 = 4(x - 1)$$
$$y - 9 = 4x - 4$$
$$y = 4x + 5$$

Therefore the equation of the line is $y = 4x + 5$. ☐

EXERCISES

1. Find the slope of each of the following situations.

a)

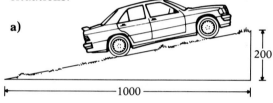

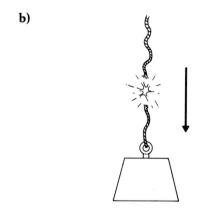

b)

c)

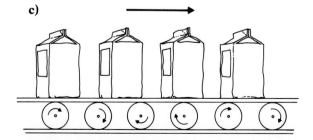

2. Find the slope of the line passing through the following points.

   a) $(3, 5)$ and $(5, 9)$
   b) $(6, 1)$ and $(3, 0)$
   c) $(3, -2)$ and $(-6, 5)$
   d) $(5, 3)$ and $(0, -4)$
   e) $(3, 9)$ and $(-3, -5)$
   f) $(-4, -5)$ and $(-3, -2)$
   g) $(-4, 0)$ and $(0, -5)$
   h) $(4, 3)$ and $(3, 4)$
   i) $(4, 4)$ and $(8, 9)$
   j) $(5, 6)$ and $(2, 6)$

3. Find the slope and the y-intercept of the line whose equation is as follows.

   a) $y = 3x - 1$ **b)** $y = \frac{x}{3} + 4$
   c) $5y - 3x = 15$ **d)** $8x - 5y = 40$
   e) $y = x$ **f)** $y = 7$
   g) $x = 1$

4. Is the point $(-3, -2)$ on the graph of $5x - 3y = -21$?

5. Find the x-intercept and the y-intercept of the lines whose equations are as follows.

   a) $4x - 3y = 12$ **b)** $-7x + 14y = 21$
   c) $y = 3x$ **d)** $6x - 3y - 15 = 0$
   e) $8x = 24$ **f)** $-3y = 14$

6. Find the equation of the lines satisfying the indicated conditions.

    a) has slope -3 and y-intercept is at $(0, -8)$.

    b) passes through the points $(4, 5)$ and $(8, 10)$.

    c) has slope $-\dfrac{2}{3}$ and passes through the point $(5, 8)$.

    d) passes through points $(-3, 7)$ and $(-2, -1)$.

    e) has slope $-\dfrac{5}{7}$ and passes through the origin.

7. Find the equation of the line that passes through the point $(4, 2)$ and is parallel to the line whose equation is $y = 4x + 5$.

8. Write an equation of the line that is parallel to $y = 2x + 3$ and has the same y-intercept as $y + 5 = 4x$.

** 9. Write an equation of the line that is perpendicular to the line $x + 3y = 9$ and that passes through the origin.

10. Are the following sets of lines parallel?

    a) $y = 3x + 1$ **b)** $y = -5x$

        $y = 3x - 8$ $3y + 15x = 11$

    c) $4x = 8y - 3$

        $24y = 12x + 17$

11. The vertices of parallelogram $ABCD$ are at $A(-2, 4)$, $B(2, 6)$, $C(7, 2)$, and $D(x, 0)$.

    a) Find the slope of line $\overline{AB}$.

    b) Express the slope of line $\overline{CD}$ in terms of x.

    c) Using the results in answer to parts (a) and (b), find the value of x.

    d) Write an equation of line $\overline{BD}$.

8.5

LINEAR INEQUALITIES AND THEIR GRAPHS

In linear programming we are usually concerned with **inequalities.** These are expressions such as

$$2x - y < 4, \qquad x + y > 5, \qquad 4x \geq 2, \qquad y + 3 \leq 4.$$

Inequalities contain "less than" ($<$) or "greater than" ($>$) signs instead of only the equals sign. In this chapter we will deal only with inequalities in which the sign in front of the y-term is positive. If this is not the case, we apply the rules for inequalities to change the sign in front of the y-term to a positive sign.

The graph of the type of inequality to be discussed in this chapter can easily be drawn by the following procedure:

a) Change the inequality sign to an equals sign.

b) Graph the line given by step (a).

c) If the inequality is "less than," shade in the area below the line. If the inequality is "greater than," shade in the area above the line.

d) If the inequality is "$\leq$" or "$\geq$," include the line. Otherwise do not include the line in the shaded area. This situation is usually indicated by a dashed line.

e) If the line is vertical and the inequality is "greater than," shade the area to the right of the line.

f) If the line is vertical and the inequality is "less than," shade the area to the left of the line.

EXAMPLE 1 Graph the inequality $x + y > 5$.

SOLUTION We first change the ">" sign to an "$=$" sign, getting $x + y = 5$. We graph this line, using the table of values below.

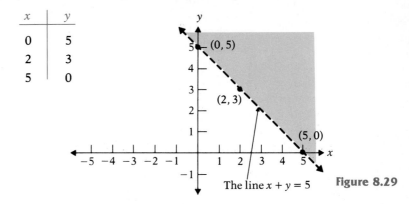

x	y
0	5
2	3
5	0

The line $x + y = 5$

Figure 8.29

Since the inequality is ">," we shade the area above the line. The line itself is not included. We indicate this by drawing it as a dashed line. (See Fig. 8.29.) The line is called a **plane divider.** The plane divider always divides the plane into two **half-planes,** one of which is shaded and the other which is not. ☐

EXAMPLE 2 Graph the inequality $2x + 3y \leq 6$.

SOLUTION We first change the "$\leq$" sign to an "$=$" sign, getting $2x + 3y = 6$. We graph this line as shown in Fig. 8.30. Since the inequality is "$\leq$," we shade in the area below the line and include the line as well. The half-plane below the line as well as the line itself represents the graph of $2x + 3y \leq 6$.

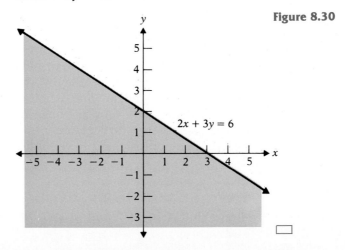

Figure 8.30

$2x + 3y = 6$

EXAMPLE 3   Draw the graphs of the following inequalities:

a) $x \geq 1$ **b)** $x < 1$ **c)** $y \geq 2$ **d)** $y < 2$

SOLUTION   The graphs of these inequalities are given in the Figures 8.31–8.34. Notice that since the inequality is $\geq$ in parts (a) and (c), we shade in the area that is to the right of or above the line and include the line as well. On the other hand, in parts (b) and (d) the inequality is "$<$." We shade in only the region that is to the left of or below the line. The line itself is not included. We indicate this by drawing it as a dashed line.

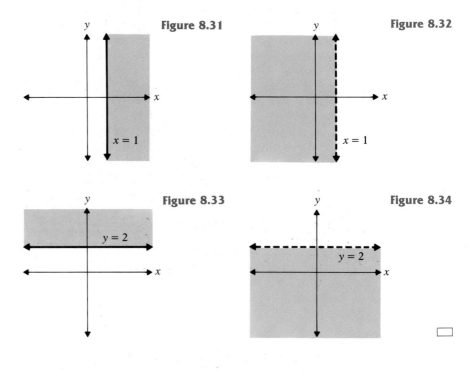

Figure 8.31   Figure 8.32

Figure 8.33   Figure 8.34

There is an alternative, and often more convenient, way of determining whether the region above or below a line or the region to the right of or left of a line should be shaded. We simply select any convenient point, and determine whether the coordinates of the point satisfy the inequality. If they do, then the region containing that point is shaded. Otherwise, the region containing the point is not shaded. This technique is illustrated in the following example.

EXAMPLE 4   Graph the inequality $3x + 4y \leq 12$.

SOLUTION   We first change the "$\leq$" sign to an "$=$" sign, getting $3x + 4y = 12$.

We graph this line, using the table of values below.

x	y
4	0
0	3
8	−3

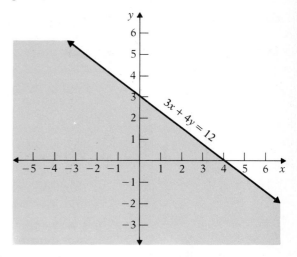

Figure 8.35

Now we must determine whether we shade the region above the line or below the line. Let us select (0, 0) as a convenient point. Substituting $x = 0$ and $y = 0$ into the inequality $3x + 4y \leq 12$ gives

$$3(0) + 4(0) \leq 12$$

or $0 \leq 12$, which is a true statement. Thus (0, 0) is a point in the solution set. We thus shade the region (half-plane) below the line. The line itself is also included as shown in Figure 8.35. ☐

Comment　When using this alternative technique for determining which region to shade, it is not necessary to make sure that the coefficient of y is positive or not.

EXERCISES

Draw the graph of each of the inequalities given in Exercises 1–20.

1. $x > 3$

2. $x \leq -5$

3. $y \geq 5$

4. $y < -2$

5. $x \geq 0$

6. $y \leq -1$

7. $y > 3x$

8. $y > x + 1$

9. $2y - 3x > 12$

10. $2x + 5y \leq 10$

11. $2x + 3y > 6$

12. $2x - 3y \leq 6$

13. $y - x \leq -4$

14. $x - y \geq 2$

15. $x \leq -2y$

16. $3y - 6x \geq 9$

17. $2x + y - 3 \geq 0$

18. $7x - 3y \leq -21$

19. $5x - 4y \leq 20$

20. $15 \leq 3x - 5y$

Rearrange each of the inequalities given in Exercises 21–26 so that the number in front of the y-term is positive.

21. $5x - 6y \leq 12$

22. $7x - 3y \leq 8$

23. $x - y \leq 0$

24. $2x - y \geq 0$

25. $3x - 2y \geq -4$

26. $15x - 18y \leq -10$

8.6

GRAPHICAL SOLUTION OF A SYSTEM OF LINEAR EQUATIONS IN TWO VARIABLES

In Section 7.6 we studied an algebraic technique for solving a system of two equations in two unknowns. Now that we have learned how to draw graphs, we can solve such a system graphically. To find the solution of a system of linear equations in two unknowns graphically, we simply draw the graphs of both equations on the same set of axes. The coordinates of the point of intersection represent the one common solution that satisfies both equations. The following situations may arise:

a) If a system of linear equations has *one* common solution, it is called a **system of consistent independent equations.** Graphically, this will occur when the two straight lines have unequal slopes and intersect in one point.

b) If a system of linear equations has *no* common solution, it is called a **system of inconsistent equations.** Graphically, this will occur when two straight lines have equal slopes but different y-intercepts and are thus parallel.

c) If a system of linear equations is of such a nature that every solution of either one of the equations is also a solution of the other, it is called a **system of consistent dependent equations.** Graphically, this will occur when two linear equations are graphed in a coordinate plane and turn out to be the same line.

We illustrate the above ideas with several examples.

EXAMPLE 1   Find the solution to the following system of linear equations graphically:

$$y + x = 6$$
$$2y - 3x = 2$$

SOLUTION   We first draw the graph of $y + x = 6$. Several pairs of values that satisfy this equation are shown below. The graph is given in Figure 8.36.

Table of values for
$y + x = 6$

x	y
3	3
5	1
6	0

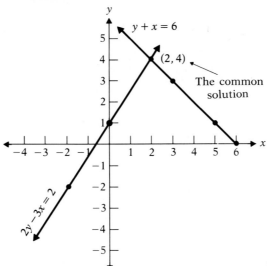

Figure 8.36

On the same set of axes we draw the graph of $2y - 3x = 2$. Several pairs of values that satisfy this equation are shown below. The graph is also given in Figure 8.36.

Table of values for 2y − 3x = 2

x	y
4	7
0	1
−2	−2

The coordinates of the point of intersection of the two lines is $(2, 4)$. This represents the common solution. It is the *only* point that satisfies both equations. Thus our answer is $x = 2$ and $y = 4$. The system of equations is consistent. ☐

EXAMPLE 2 Solve the following system of linear equations graphically:

$$y - x = 2$$
$$2y = 2x + 10$$

SOLUTION We first draw the graph of $y - x = 2$ by finding several pairs of values that satisfy this equation. The graph is given in Figure 8.37.

Table of values for $y - x = 2$

x	y
3	5
0	2
−1	1

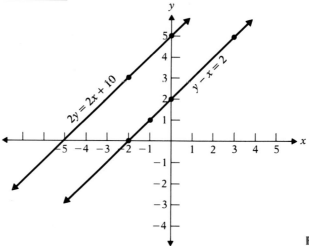

Figure 8.37

On the same set of axes we draw the graph of $2y = 2x + 10$ by finding several pairs of values that satisfy this equation.

Table of values for $2y = 2x + 10$

x	y
3	8
0	5
−2	3

We notice that the two lines do not intersect; they are parallel. Both lines have the same slope of 1. (Verify this.) There is no common solution, and the system is said to be inconsistent. ☐

EXAMPLE 3   Solve the following system of linear equations graphically:

$$2x - \frac{1}{3}y = 4$$

$$y = 6x - 12$$

8.6 Graphical Solution of a System of Linear Equations in Two Variables

SOLUTION   We first draw the graph of $2x - \frac{1}{3}y = 4$ by finding several pairs of values that satisfy this equation. The graph is given in Figure 8.38.

Figure 8.38

Table of values for

$$2x - \frac{1}{3}y = 4$$

x	y
2	0
3	6
−1	−18

$y = 6x - 12$

$2x - \frac{1}{3}y = 4$

$(-1, 18)$

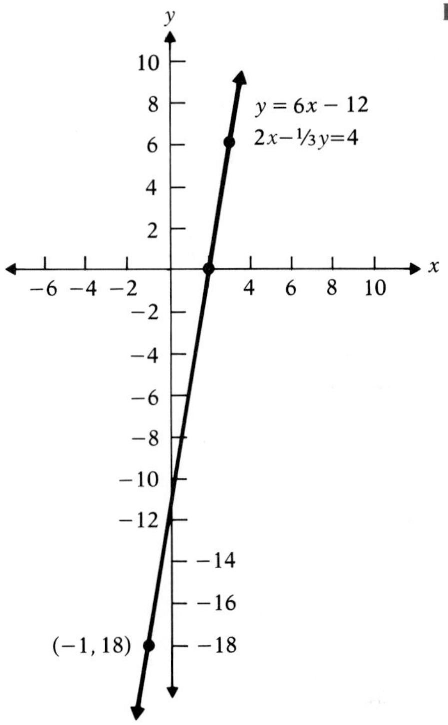

On the same set of axes we draw the graph of $y = 6x - 12$ by finding several pairs of values that satisfy this equation.

Table of values for

$$y = 6x - 12$$

x	y
2	0
3	6
−1	−18

We notice that all the points lie on the same line. The two equations represent the same line, so any solution for one of the equations will also be the solution for the second equation. The system of equations is **dependent.** It has an infinite number of solutions.

EXERCISES

Solve the system of linear equations given in Exercises 1–14 graphically.

1. $\begin{cases} y = 4x \\ y = 2x + 6 \end{cases}$ 2. $\begin{cases} y = x + 3 \\ y = 2x + 5 \end{cases}$

3. $\begin{cases} y = -2x + 1 \\ y = \frac{1}{2}x + 1 \end{cases}$ 4. $\begin{cases} y = 2x \\ 3x - y = 3 \end{cases}$

5. $\begin{cases} 3x + y = 5 \\ 4x - y = 9 \end{cases}$ 6. $\begin{cases} 5x - 3y = 11 \\ 2x + 3y = 17 \end{cases}$

7. $\begin{cases} x = 1 \\ y = 2 \end{cases}$ 8. $\begin{cases} x = -3 \\ y = -1 \end{cases}$

9. $\begin{cases} 2x = y + 1 \\ 3x = 5y - 10 \end{cases}$ 10. $\begin{cases} 2x + 3y = 11 \\ 3x - 2y = 10 \end{cases}$

11. $\begin{cases} x + y + 3 = 0 \\ 3x = -7y + 3 \end{cases}$ 12. $\begin{cases} 5x - 4y = 5 \\ 2x - 3y = -5 \end{cases}$

13. $\begin{cases} 4x - 3y = -6 \\ 3x - 4y = -1 \end{cases}$ 14. $\begin{cases} x = y + 2 \\ 3y = 5x - 8 \end{cases}$

In Exercises 15–20, determine whether the system is consistent, inconsistent, or dependent.

15. $7x - 2y = 20$
    $4y = 14x + 6$

16. $3x - 2y = 12$
    $2x - 3y = -6$

17. $2x - 5y = 10$
    $y = \frac{2}{5}x - 2$

18. $x + 6y - 3 = 0$
    $x = 3$

19. $\quad y = 2x + 1$
    $-6x = 3 - 3y$

20. $2x + 3y = 2$
    $5x + 3y = 14$

8.7

GRAPHING SYSTEMS OF LINEAR INEQUALITIES

In Section 8.5 we indicated how to draw the graph of an inequality. In the last section we showed how we can solve a system of equations graphically. Now we will combine these ideas so that we will be able to solve a system of linear inequalities graphically.

To solve a system of linear inequalities graphically, we simply draw the graphs of both inequalities on the same set of axes. We find the ordered pairs that satisfy all the inequalities. These are the solutions to the system. There may be an infinite number of such ordered pairs.

EXAMPLE 1   Graph the solution set for the following system of inequalities:

$$x < 3$$
$$y > -1$$

SOLUTION   We first draw the graph of $x < 3$ by drawing the plane divider $x = 3$ as indicated in Figure 8.39. The line itself is drawn as a broken line. The half-plane to the left of this line represents the graph of the solution set of $x < 3$. This has vertical shading.

On the same set of axes we draw the graph of $y > -1$ by drawing the plane divider $y = -1$ as indicated in Figure 8.39. Again, the line itself is drawn as a broken line. The half-plane above this line, which has been horizontally shaded, represents the graph of the solution set of $y > -1$. Notice that in one region the shading overlaps. This shaded region, which is the intersection of both of the graphs $y > -1$

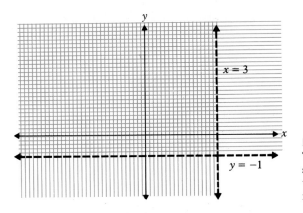

Figure 8.39
The region where the shading overlaps represents the solution of both inequalities.

and $x < 3$, is the graph of the solution set of these inequalities. The coordinates of any point in this region satisfy both inequalities. The coordinates of no other point can satisfy both inequalities. □

EXAMPLE 2   Graph the solution set for the following system of inequalities:

$$x + y \geq 1$$
$$y \leq 3x - 1$$

SOLUTION   We draw the graph of $x + y \geq 1$ by first graphing the plane divider $x + y = 1$. The half-plane *above* the line as well as the line itself form the graph of the solution set of $x + y \geq 1$. (See Figure 8.40.) Using the same set of axes, we draw the graph of $y \leq 3x - 1$ by first graphing the plane divider $y = 3x - 1$. The half-plane *below* the line as well as the line itself form the graph of the solution set of $y \leq 3x - 1$. In the diagram, one region has been shaded twice. This region, which

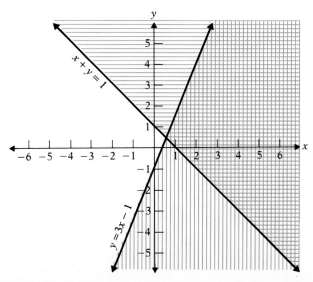

Figure 8.40
The region where the shading overlaps represents the solution set of both inequalities.

is the intersection of both graphs $x + y \geq 1$ and $y \leq 3x - 1$, is the graph of the solution set of these inequalities. Again, the coordinates of any point in this region satisfy both inequalities. ☐

EXAMPLE 3   Graph the solution set for the following system of inequalities:

$$3x + 2y \geq 12$$
$$2x + 3y \leq 15$$
$$x \geq 0$$
$$y \geq 0$$

SOLUTION   We first draw the graph of $3x + 2y \geq 12$ by graphing the plane divider $3x + 2y = 12$. The half-plane above the line as well as the line itself form the graph of the solution set of $3x + 2y \geq 12$. On the same set of axes we draw the graph of $2x + 3y \leq 15$ by graphing the plane divider $2x + 3y = 15$. The half-plane below the line as well as the line itself form the graph of the solution set of $2x + 3y \leq 15$. (See Figure 8.41.) On the same set of axes we draw the graphs of $x \geq 0$

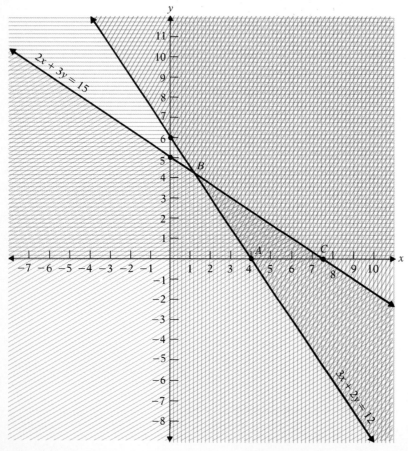

Figure 8.41
Triangle *ABC*, the region that has all four types of shading, represents the solution set of all four inequalities.

and $y \geq 0$ by first drawing the plane dividers $x = 0$ and $y = 0$. The half-plane to the right of the line together with the line $x = 0$ forms the graph of the solution of $x \geq 0$. Similarly, the half-plane above the line together with the line $y = 0$ forms the graph of the solution of $y \geq 0$.

The region with all four types of shading is the graph of the solution set of the inequalities $3x + 2y \geq 12$, $2x + 3y \leq 15$, $x \geq 0$, and $y \geq 0$. This region includes all the points that are in the interior of triangle ABC as well as those points that are on lines AB, BC, and AC, which are the sides of this triangle. Any point in this region satisfies all four inequalities. ☐

EXERCISES

In Exercises 1–16, graph the solution set of each system of inequalities.

1. $\begin{cases} x \geq 5 \\ y < 7 \end{cases}$

2. $\begin{cases} x \leq 3 \\ y \geq -1 \end{cases}$

3. $\begin{cases} x > 1 \\ y < 2 \end{cases}$

4. $\begin{cases} y \leq x \\ x < 4 \end{cases}$

5. $\begin{cases} x \geq 1 \\ y \leq 3x \end{cases}$

6. $\begin{cases} x + y > 5 \\ x - y < 8 \end{cases}$

7. $\begin{cases} 2x - y \leq -6 \\ x + y > 3 \end{cases}$

8. $\begin{cases} 3x + y \geq -6 \\ 2x - 3y < 6 \end{cases}$

9. $\begin{cases} 3x + 2y \leq 6 \\ x - y > 5 \end{cases}$

10. $\begin{cases} 4x - 7y \leq 28 \\ 2x - y > 1 \end{cases}$

11. $\begin{cases} x \geq 2 \\ y \leq 3 \\ x + y \geq 1 \end{cases}$

12. $\begin{cases} x + 3y \leq 6 \\ 2x > 5 \\ 3y < 8 \end{cases}$

13. $\begin{cases} 5x - 2y < 10 \\ 2x + 3y \geq -1 \\ x > 0 \end{cases}$

14. $\begin{cases} 4x - 3y \leq 12 \\ 2x + y \leq 7 \\ y < 10 \\ x > 3 \end{cases}$

15. $\begin{cases} 2x + y \leq 8 \\ 3x - 4y \leq 9 \\ x \geq 0 \\ y \geq 0 \end{cases}$

16. $\begin{cases} 3x + 4y \leq 12 \\ 2x + 3y \leq 6 \\ x + y < 1 \\ x \geq 0 \\ y \geq 0 \end{cases}$

8.8

LINEAR PROGRAMMING

In this section we will discuss some different types of problems that can be solved by linear programming methods. We will solve some of them. Others have solutions that are beyond the scope of this text. We include them merely to indicate the range of applicability.

EXAMPLE 1 The Mark Toy Company has two best-selling toys, the *Clue* and the *Ringer*. To produce one case of *Clues* requires 3 hr on machine I and 2 hr on machine II and yields a profit of $14 per case. On the other hand, to produce one case of *Ringers* requires 2 hr on machine I and 4 hr on machine II. *Ringers* yield a profit of $12 a case. How many of each should the company produce in order to maximize its profit if no machine can be used more than 8 hr per day?

SOLUTION   Here we will merely translate the problem from words into mathematical expressions. The solution will be completed in Section 8.9. The important expressions are marked with • and are summarized at the end.

Let x = the number of cases of *Clues* to be produced.

Let y = the number of cases of *Ringers* to be produced.

We list all the information given, in the form of a box.

		Hours needed		Profit from each product (in dollars)
		Machine I	Machine II	
Products	Clue	3	2	14
	Ringer	2	4	12
	Limit on machines	8	8	

Let us look at machine I. It gives 3 hr for each case of *Clues* and 2 hr for each case of *Ringers*. It cannot give more than 8 hr in one day. It could, of course, give less, since the machine does not have to operate all day long. The total time spent on *Clues* by machine I is

$$\text{(time per case)} \quad \cdot \quad \text{(number of cases)}$$
$$= \quad 3 \qquad\qquad \text{times} \qquad x$$
$$= \quad 3x.$$

Similarly, the total time spent on *Ringers* by machine I is

$$\text{(time per case)} \quad \cdot \quad \text{(number of cases)}$$
$$= \quad 2 \qquad\qquad \text{times} \qquad y$$
$$= \quad 2y.$$

In producing both toys, machine I has given (in one day) $3x + 2y$ hr. We know that machine I works at most 8 hr, and possibly less. Thus the number of hours that machine I works is equal to or less than 8. We use the symbol $\leq$ to indicate "less than or equal to." Therefore for machine I we have

• $3x + 2y \leq 8.$ (Machine I)

Similarly, the total time spent on *Clues* by machine II is

$$\text{(time per case)} \quad \cdot \quad \text{(number of cases)}$$
$$= \quad 2 \qquad\qquad \text{times} \qquad x$$
$$= \quad 2x.$$

Also, the total time spent by machine II on *Ringers* is

$$\text{(time per case)} \quad \cdot \quad \text{(number of cases)}$$
$$= \quad 4 \qquad \text{times} \qquad y$$
$$= \quad 4y.$$

In producing both toys, machine II works $2x + 4y$ hr. Since machine II also works 8 hr or less, this gives

- $2x + 4y \leq 8.$ (Machine II)

Now let's consider the profit. Mark Toy Company makes $14 on each case of *Clues* and $12 on each case of *Ringers*. Its profit on *Clues* is

$$\text{(profit per case)} \quad \cdot \quad \text{(number of cases)}$$
$$= \quad 14 \qquad \text{times} \qquad x$$
$$= \quad 14x.$$

Similarly, its profit on *Ringers* is

$$\text{(profit per case)} \quad \cdot \quad \text{(number of cases)}$$
$$= \quad 12 \qquad \text{times} \qquad y$$
$$= \quad 12y.$$

The total profit from both products, then, is $14x + 12y$. We call this P. Obviously, the manufacturer wishes to maximize the profit. So we want to maximize $P = 14x + 12y$. We will abbreviate this as

- $\max P = 14x + 12y.$ (Profit)

It should be obvious that x and y cannot be negative. They must be 0 or larger. Why?

We state this mathematically as

- $x \geq 0$ and $y \geq 0.$

The mathematical formulation of this problem can be summarized as:

$$\max P = 14x + 12y$$
$$\text{subject to} \begin{cases} 3x + 2y \leq 8, \\ 2x + 4y \leq 8, \\ x \geq 0 \text{ and } y \geq 0 \end{cases} \quad \square$$

EXAMPLE 2 John has asked his instructor, Professor Hutton, if he can submit problems for extra credit. Professor Hutton says that he can submit up to 100 problems of three types, A, B, and C. Type A problems are worth 5 points each, type B problems are worth 4 points each, and type C problems are worth 6 points each. John finds that type A problems require 3 minutes; type B, 2 minutes; and type C, 4 minutes. John can spend at most $3\frac{1}{2}$ hr (210 minutes) on the assignment. Furthermore, problems of types A and B involve computations that give him a headache. So he cannot bear to spend more than $2\frac{1}{2}$ hr (150 minutes) on these. How many problems of each type should he do to maximize his credit?

MATHEMATICAL
FORMULATION

Let x = the number of type A problems to be done.

Let y = the number of type B problems to be done.

Let z = the number of type C problems to be done.

Since John was asked to submit no more than 100 problems altogether, we have

- $x + y + z \leq 100$.

The time spent on type A problems is $3x$. The time spent on type B problems is $2y$. The time spent on type C problems is $4z$. The total time spent on all problems is $3x + 2y + 4z$.

Since John can devote at most 210 minutes to his mathematics assignment, we have

- $3x + 2y + 4z \leq 210$.

In a multilayered network of air routes, planes fly in layers 1000 feet apart and are protected on all sides by five miles of air space.

SEQ.	LOCAL TIME	GMT	DAYS	ID	EQUIP.	DEP. AP.	ARR. AP.	FIX	BEACON CODE	EFF. DATE	DISC. DATE
0216	22-00	3-00	1111 1	LH421	707	POS	FRA	7ZI	1104		3 31
0217	22-05	3-05	1 1 1	PA160	707	JFK	BOS	HTM	2614		
0220	22-05	3-05	1	PA162	737	JFK	BOS	HTM	2615		
0221	22-06	3-06	1 1111	AA848	B3F	JFK	BOS	HTM	2616		
0222	22-10	1-10		AL923	09S	BOS	PHL	7DJ	1105		
0223	22-10	3-10	1	AL947	09S	POS	PVD	HTM	1106		
	22-14	3-14	1111111	TW76	B7F	SFO	BOS	BED		3 15	
0224	22-15	3-15	111111	AL 805	09S	POS	PIT	9WE	1107		
0225	22-15	3-15	111111	EA917	72S	BOS	EWR	9WE	1110		
	22-18	3-18	111111	AA288	727	OCA	BOS	HTM		3 2	
0226	22-18	3-18	111111	UA492	08F	ORD	BOS	RED	2617	3 2	3 1
	22-20	3-20	111111	UA146	D10	ORD	BOS	RED			
0227	22-24	3-24	1111111	EA215	09S	LGA	POS	7IL	2620		
0230	22-25	3-25	1111111	PL912	B80	JFK	BOS	HTM	2621		
0231	22-29	3-29	1111111	TW76	B7F	SFO	BOS	BED	2622		3 14
0232	22-30	3-30	11	EA535	09S	BOS	EWR	9WE	1111		
0233	22-31	3-31	1111111	AL996	09S	PHL	BOS	HTM	2623		
0234	22-32	3-32	1111111	AL436	BAC	BGM	BOS	RED	2624		
0235	22-44	3-44	1111111	AA578	727	ALB	BOS	RED	2625		
0236	22-47	3-47	1111111	EA184	727	PVD	BOS	HTM	2626		
0237	22-48	3-48	1111111	AA288	727	OCA	BOS	HTM	2627		3 1
0240	22-54	3-54	1111111	EA990	727	JFK	BOS	HTM	2630		
0241	22-55	3-55	111	HA68	707	DTW	BOS	RED	2631		
0242	22-55	3-55	1	1N114	707	BOS	SNN	7ZI	1112		3 17
	22-55	3-55	1	1N114	707	BOS	SNN	7ZI		3 18	3 31
	22-55	3-55		1N116	707	BOS	SNN	7ZI		3 18	3 17
0243	22-55	3-55	1	1N116	707	BOS	SNN	7ZI	1113		
0244	22-58	3-58	1111111	AL454	BAC	BOC	BOS	RFD	2632		
0245	22-58	3-58	1111111	AL528	BAC	SYR	BOS	RED	2633		
0246	22-59	3-59	1111111	AL414	BAC	ALP	BOS	9ED	2634		
0247	23-00	4-00	1	1B308	08F	BOS	FRA	7ZI	1114		
0250	23-05	4-05	1	PA163	707	LIS	BOS	7ZI	2635		
0251	23-07	4-07	111111	DL226	72S	PAL	BOS	HTM	2636		
0252	23-10	4-10	111111	FT243	78S	SYR	BOS	BFD	2637		
0253	23-14	4-14	111111	EA166	09S	PVD	BOS	HTM	2640		
0254	23-15	4-15	1 1 1	PA163	707	ORY	BOS	7ZI	2641		
0255	23-23	4-23	1111111	AA52	B7F	JRO	BOS	BED	2642		3 1

A computer printout schedules arriving and departing flights. Linear programming is used to efficiently assign airlines to scheduled routes.

Furthermore, since he can spend at most 150 minutes on type A and B problems, we have

- $3x + 2y \leq 150$.

The credit John can receive for all the problems is

$$P = 5x + 4y + 6z.$$

Of course, John wants to maximize P.

Again notice that x, y, and z cannot be negative. Why? We denote this as

- $x \geq 0$, $y \geq 0$, and $z \geq 0$.

Finally, we summarize the formulation of the problem as:

$$\max P = 5x + 4y + 6z$$

$$\text{subject to} \begin{cases} x + y + z \leq 100, \\ 3x + 2y + 4z \leq 210, \\ 3x + 2y \leq 150, \\ x \geq 0, \ y \geq 0, \ \text{and} \ z \geq 0 \end{cases}$$ □

EXAMPLE 3 Mr. Burns, the cook at Camp McGregor, has decided to serve foods A and B for dinner. He knows that each unit of food A contains 20 grams of protein and 25 grams of carbohydrates and costs 20¢. Each unit of food B contains 10 grams of protein and 20 grams of carbohydrates and costs 15¢. He wants the meal to contain at least 500 grams of protein and 800 grams of carbohydrates per person. The camp director has ordered him to save as much money as possible. What is the cheapest meal he can serve using these two foods only?

MATHEMATICAL
FORMULATION

Let x = the number of units of food A to be prepared.

Let y = the number of units of food B to be prepared.

The following box summarizes the information given in the problem.

| | Nutritional content | | Cost |
	Protein	Carbohydrates	(*in cents*)
Food A	20	25	20
Food B	10	20	15
Minimum needed	500	800	

A camper can get 20 grams of protein from each unit of food A and 10 grams of protein from each unit of food B. The total protein she gets from these foods is then $20x + 10y$. Since she must have *at least* 500 grams of protein, we then have

- $20x + 10y \geq 500$.

Similarly, she gets 25 grams of carbohydrates from each unit of food A and 20 grams of carbohydrates from each unit of food B. Thus her total carbohydrate intake is

$25x + 20y$.

Again, she must have *at least* 800 grams of carbohydrates. Therefore

- $25x + 20y \geq 800$.

The cost of food A is

(cost for one unit) · (number of units)

=   20   times x

= $20x$.

The cost of food B is

(cost for one unit) · (number of units)

=   15   times y

= $15y$.

The total cost, then, is $20x + 15y$. Of course, we would like to minimize this cost. We indicate this by

- min $P = 20x + 15y$.

Again, x and y cannot be negative. Thus

- $x \geq 0$ and $y \geq 0$.

The problem can now be expressed as

$$\min P = 20x + 15y$$

$$\text{subject to} \begin{cases} 20x + 10y \geq 500, \\ 25x + 20y \geq 800, \\ x \geq 0 \quad \text{and} \quad y \geq 0 \end{cases} \quad \square$$

EXAMPLE 4 The Helena Perfume Company markets two types of perfumes, *Michelle* and *Sweet Scent*. To produce 1 gal of *Michelle* requires 5 units of ingredient A and 3 units of ingredient B. It yields a profit of $300. In producing 1 gal of *Sweet Scent*, 2 units of ingredient A and 3 of ingredient B are used. The profit is $200. The Helena Perfume Company has available 180 units of ingredient A and 135 units of ingredient B. How many gallons of each perfume should they make in order to maximize profit?

MATHEMATICAL
FORMULATION

Let $x =$ the number of gallons of *Michelle* to be made.

Let $y =$ the number of gallons of *Sweet Scent* to be made.

We summarize the information given as follows:

	Amount used of		Profit
	Ingredient A	Ingredient B	(in dollars)
Michelle	5	3	300
Sweet Scent	2	3	200
Amount available	180	135	

In producing *Michelle* and *Sweet Scent* we use $5x + 2y$ units of ingredient A. Since 180 units is the most that can be used, we have

- $5x + 2y \leq 180$.

Similarly, in producing the two perfumes we use $3x + 3y$ units of ingredient B. Since, in this case, we have at most 135 units available, we get

- $3x + 3y \leq 135$.

The profit is

- $300x + 200y$.

Again, we have the obviously necessary restriction that x and y cannot be negative. Thus

- $x \geq 0$ and $y \geq 0$.

The problem can now be written as

$$\max P = 300x + 200y$$

$$\text{subject to} \quad \begin{cases} 5x + 2y \le 180, \\ 3x + 3y \le 135, \\ x \ge 0 \quad \text{and} \quad y \ge 0 \end{cases} \square$$

In the following section we will complete the solution of Example 1 by using graphical techniques.

EXERCISES

Formulate each of the following problems mathematically. Do not try to actually solve them.

1. *Television Sets.* A manufacturer produces 12-inch and 19-inch television sets that require assembly, testing and packaging. Each 12-inch set produced requires 1 hour for assembly and 2 hours for testing and packaging. It yields a profit of $38 per set. Each 19-inch set produced requires 3 hours for assembly and only 1 hour for testing and packaging. It yields a profit of $48 per set. If the assembly line operates for at most 12 hours per day and the testing and packaging facility operates for at most 18 hours per day, how many of each kind of set should be produced so as to maximize profit?

2. *Armaments.* The ABC Munitions Corporation makes jeeps and tanks in its two factories. Factory I performs the basic assembly operations and operates for at most 120 hours per week. Factory II performs the finishing operations and operates for at most 90 hours per week. Each jeep manufactured requires 3 hours in Factory I and 4 hours in Factory II and yields a profit of $360. Each tank manufactured requires 4 hours in Factory I and 3 hours in Factory II and yields a profit of $480. How many of each should the company manufacture so as to maximize profit?

3. The Back Corporation manufactures suits and sport jackets. Manufacturing a suit requires 4 hours of work by operator A and 2 hours of work by operator B and yields a profit of $40 per suit manufactured. On the other hand, making a sport jacket requires 2 hours by operator A and 4 hours by operator B and yields a profit of $24 per jacket manufactured. No operator can work more than 12 hours per day. How many of each should be manufactured so as to maximize profit?

4. *Budget Cuts.* George Manville is a mechanic for the All-Boro Transit Corp. Because of budget cuts, George is required to service buses and vans at both of the company's garages. Since the equipment is different at each location, it takes George 3 hours to service a van and 4 hours to service a bus at the Bushwick location and 1 hour to service a van and 2 hours to service a bus at the Wyatt location. The company requires George to spend at most 24 hours per week at the Bushwick location and at most 16 hours per week at the Wyatt location. If George earns $40 for each van serviced and $60 for each bus serviced, how many of each should he service so as to maximize his earnings?

5. The Academic Sportswear Company manufactures numerous items for tennis players. It has three workers: Bill, Mary, and Lucille. Bill works at most 30 hours per week; Mary and Lucille each work at most 48 hours per week. Each worker's weekly production for several items is as follows.

Item	Bill	Mary	Lucille
Shoes	2	5	3
Racquets	3	4	1
Shorts	6	6	2

If the profits are $4 for a pair of shoes, $3 for a racquet, and $2 for a pair of shorts, how many of each should be produced so as to maximize profit?

6. Howie, Willie, and Vinnie do home and small business renovations. Howie is an electrician, Willie is a carpenter, and Vinnie is a plumber. They always work as a team on any renovation project. The number of hours needed by each worker to complete his task depends on the project involved and is given in the following chart:

		Project		
		1-family house	2-family house	small business
	Howie	12	10	28
Worker	Willie	10	14	26
	Vinnie	8	12	20

The profits from each type of job are approximately $200, $300, and $500 respectively. If no worker works more than 40 hours per week, how many projects of each type must be completed so as to maximize profit?

7. *Mixed Nuts.* A wholesale grocer has just received a shipment of 4000 ounces of Brazil nuts, 5000 ounces of cashew nuts, and 6000 ounces of pecan nuts. The grocer plans to package the nuts in one of three possible combinations of mixed nuts as shown in the accompanying table.

Product	Ounces of Brazil nuts	Ounces of cashew nuts	Ounces of pecan nuts	Profit
Package A	4	5	4	24¢
Package B	5	3	3	30¢
Package C	3	4	5	36¢

How many of each should be produced so as to maximize profit?

8. *Paper Production.* The Marro Lumber Company operates two paper mills, each of which produces three different qualities of print paper. The quantity produced weekly by each mill is shown in the accompanying chart:

	Low-grade paper	Medium-grade paper	High-grade paper	Profit
Mill A	20 tons	10 tons	14 tons	$4000
Mill B	14 tons	12 tons	18 tons	$6000

However, owing to a strike at one of its chemical suppliers, the company finds that it can produce at most 300 tons of low-grade paper, 400 tons of medium-grade paper, and 360 tons of high-grade paper per week. How many days should each mill operate so as to maximize profit?

9. Consider the newspaper article below. One of the protesting farmers owns a 300-acre farm on

Farmers Protest Rising Costs and Low Farm Food Prices

WASHINGTON: Many of the nation's farmers staged a demonstration for the second day in a row to protest rising costs. For two hours, traffic on heavily traveled Constitution Avenue was at a standstill as farmers parked their tractors on the roads and stopped the engines.

The farmers are protesting the rising prices that they are being forced to pay for their farm supplies and equipment while their profit margin decreases. Said one protesting farmer from Wisconsin, "The rising costs of labor and materials together with a decrease in the amount received for these farm foods are driving me out of business. Although I am attempting to maximize the yields from my farm, I still cannot make ends meet."

The farmers vowed not to leave Washington until they meet with President Reagan to voice their plight.

DAILY PRESS January 19, 1982

which he can plant wheat or corn. The farmer knows that on his field, wheat requires three man-days of labor and $60 of capital for each acre planted and corn requires 4 man-days of labor and $80 of capital for each acre planted. The farmer finds that he has only $4000 of capital and 200 man-days of labor available for the job. Because of decreasing prices, wheat produces $400 profit for each acre planted, and corn produces $300 profit for each acre planted. How many acres of each crop should the farmer plant so as to maximize profit?

10. *Flu Vaccine.* To combat the outbreak of a flu epidemic, health officials have decided to vaccinate as many senior citizens as possible. They have enlisted the services of 50 doctors and 120 nurses who can be used as teams of 1 doctor and 3 nurses (usually called a full team) or 1 doctor and 2 nurses (usually called a half team). On the basis of past experience, health officials estimate that a full team can vaccinate 300 senior citizens per day, whereas a half team can only vaccinate 200 senior citizens per day. How many teams of each type should be used so as to maximize the number of people receiving their flu vaccination?

8.9

SOLVING LINEAR PROGRAMMING PROBLEMS GRAPHICALLY

In this section we discuss one technique for finding maximum or minimum values of linear functions subject to certain conditions called **constraints.** This geometric approach to solving linear programming problems can be used primarily when there are two unknowns (x and y). However, if there are three or more unknowns (x, y, and z) or if there are many inequalities, this method becomes either very complicated or totally impossible. A more general approach for solving linear programming problems that can be used in every case is known as the **simplex method.**

Comment The term "linear programming" is derived from the fact that the problems to which we apply this theory can be translated mathematically (as was done in Section 8.8) into the form of linear equations and inequalities. The graphs of these equations are straight lines. Of course, when the problems translated mathematically are not linear, then we are dealing with nonlinear programming. We will not deal with such problems in this text.

Solution of Linear Programming Problems by the Graphical Method

Now we will see how to solve some linear programming problems by using the graphical method. Let us look again at Example 1 of Section 8.8. The mathematical formulation was

$$3x + 2y \le 8,$$
$$2x + 4y \le 8,$$
$$x \ge 0 \quad \text{and} \quad y \ge 0,$$
$$\max P = 14x + 12y.$$

We first graph $3x + 2y \le 8$ and shade its solution vertically. (See Figure 8.42.) The third line of the mathematical formulation tells us

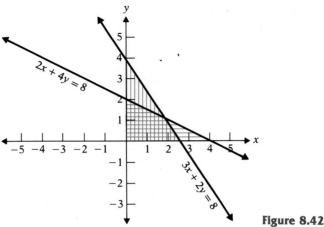

Figure 8.42

that both x and y must be positive. Therefore we shade nothing below the x-axis (since y cannot be negative). Similarly, we shade nothing to the left of the y-axis. On the same set of axes we graph $2x + 4y \leq 8$ and shade its solution horizontally. This is also shown in Figure 8.42.

Notice that in one part of the diagram the shading overlaps. This region is the part that satisfies *both* inequalities. Since we are interested in this region only, we copy over just that part in Figure 8.43. The region has four corners, which we have labeled A, B, C, and D.

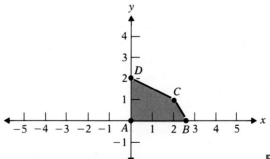

Figure 8.43

It can be proved in more advanced mathematics courses that a profit expression such as the P in our example will always have a maximum or minimum value at a *corner* of the region.

In our example the corners are at $(0, 0)$, $(2\frac{2}{3}, 0)$, $(2, 1)$, and $(0, 2)$. These are the coordinates of the corners. They can be read from the graph or by using algebra. Let us now substitute each of these values into the profit expression, $P = 14x + 12y$.

At $x = 0$ and $y = 0$ the profit is

$$P = 14(0) + 12(0)$$
$$P = 0 + 0 = 0.$$

This profit is clearly a minimum.

At $x = 2\frac{2}{3}$ and $y = 0$ the profit is

$$P = 14(2\frac{2}{3}) + 12 \cdot 0$$
$$P = 37\frac{1}{3} + 0 = 37\frac{1}{3}.$$

At $x = 2$ and $y = 1$ the profit is

$$P = 14(2) + 12(1)$$
$$P = 28 + 12 = 40.$$

At $x = 0$ and $y = 2$ the profit is

$$P = 14(0) + 12(2)$$
$$P = 0 + 24 = 24.$$

Thus the maximum profit occurs when $x = 2$ and $y = 1$, and it is 40. Referring back to Example 1 of Section 8.8, this means that the company should produce 2 cases of *Clues* and 1 case of *Ringers*. This results in a maximum profit of $40.

We will do one more example to illustrate the technique.

EXAMPLE 1   A manufacturer makes AM and FM radios. To produce an AM radio requires 2 hr in plant A and 3 hr in plant B. To produce an FM radio requires 3 hr in plant A and 1 hr in plant B. Plant A can operate for at most 15 hr a day, and plant B can operate for at most 12 hr a day. If the manufacturer makes a profit of $4 on an AM radio and $12 on an FM radio, how many of each should be produced in order to maximize the profit?

SOLUTION   Let $x =$ the number of AM radios to be produced.

Let $y =$ the number of FM radios to be produced.

As in the previous section, we make a box that shows all the information given in the problem.

	Plant A	Plant B	Profit (in dollars)
AM radio	2	3	4
FM radio	3	1	12
Limitations	15	12	

Crewmen in the combat information center aboard an assault ship keep track of the location of other ships cruising with them during a naval operation. The theory of linear programming is a valuable tool in this process.

Courtesy of U.S. Navy

The mathematical formulation, then, is

$$2x + 3y \leq 15,$$
$$3x + y \leq 12,$$
$$x \geq 0 \quad \text{and} \quad y \geq 0,$$
$$\max P = 4x + 12y.$$

Figure 8.44

We first graph $2x + 3y \leq 15$ and shade its solution horizontally. The third line of the mathematical formulation tells us that both x and y must be positive. Therefore we shade nothing below the x-axis and nothing to the left of the y-axis. We then draw, on the same set of axes, the graph of $3x + y \leq 12$ and shade its solution vertically. This is shown in Figure 8.44.

Again we redraw the doubly shaded region as shown in Figure 8.45. There are four corners, which we have labeled A, B, C, and D. These points are at $(0, 0)$, $(0, 5)$, $(3, 3)$, and $(4, 0)$, respectively.

Figure 8.45

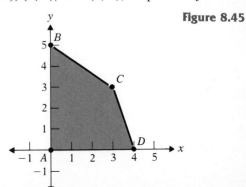

We now substitute each of these values into the profit expression, $P = 4x + 12y$.

At $x = 0$ and $y = 0$ the profit is

$$P = 4(0) + 12(0)$$
$$P = 0 + 0$$
$$P = 0.$$

At $x = 0$ and $y = 5$ the profit is

$$P = 4(0) + 12(5)$$
$$P = 0 + 60 = 60.$$

At $x = 3$ and $y = 3$ the profit is

$$P = 4(3) + 12(3)$$
$$P = 12 + 36 = 48.$$

At $x = 4$ and $y = 0$ the profit is

$$P = 4(4) + 12(0)$$
$$P = 16 + 0 = 16.$$

Thus the maximum profit occurs when $x = 0$ and $y = 5$, and it is $60. The manufacturer should produce no AM radios and 5 FM radios. This will result in a maximum profit of $60. □

This technique for solving linear programming problems is very simple and convenient to use. However, it does not work in all cases. In particular, if there are more than three unknowns (letters x, y, and z) or if there are many inequalities, then this method becomes either very complicated or totally impossible.

EXERCISES

In Exercises 1–12, use the graphical method discussed in this section to maximize or minimize each function as indicated.

1. Maximize $z = 5x + 3y$

subject to $\begin{cases} 2x + 3y \le 12 \\ 2x + y \ge 2 \\ x \ge 0, \quad y \ge 0 \end{cases}$

2. Maximize $z = 4x + 3y$

subject to $\begin{cases} -3x + 2y \le 6 \\ 4x + 3y \le 12 \\ x \ge 0, \quad y \ge 0 \end{cases}$

3. Maximize $z = 3x + 7y$

subject to $\begin{cases} 4x + 5y \ge 20 \\ 5x + 7y \le 35 \\ x \ge 0, \quad y \ge 0 \end{cases}$

4. Maximize $z = 8x + 2y$

subject to $\begin{cases} 2x + y \le 6 \\ x + y \le 5 \\ x \ge 0, \quad y \ge 0 \end{cases}$

5. Minimize $z = 3x + 6y$

subject to $\begin{cases} x + y \le 4 \\ y \le 2x - 7 \\ x \ge 0, \quad y \ge 0 \end{cases}$

6. Maximize $z = 8x + 3y$

subject to $\begin{cases} x + y \le 4 \\ x - y \ge 1 \\ x \ge 0, \quad y \ge 0 \end{cases}$

7. Minimize $z = 7x + 8y$

subject to $\begin{cases} 3x + 4y \ge 12 \\ 5x + 3y \le 15 \\ x \ge 0, \quad y \ge 0 \end{cases}$

8. Minimize $z = 7x + 9y$

subject to $\begin{cases} 8x + 3y \le 24 \\ 5x + 4y \ge 20 \\ x \ge 0, \quad y \ge 0 \end{cases}$

9. Minimize $z = 8x + 13y$

subject to $\begin{cases} 2x + 3y \ge 18 \\ x + 3y \ge 12 \\ x \ge 0, \quad y \ge 0 \end{cases}$

10. Minimize $z = 2x + 9y$

subject to $\begin{cases} 2x + 3y \ge 18 \\ x + 3y \ge 12 \\ 4x + 3y \ge 24 \\ x \ge 0, \quad y \ge 0 \end{cases}$

11. Maximize $z = 3x - 5y$

subject to $\begin{cases} x + y \ge 4 \\ 2x + 3y \le 12 \\ 4x + 5y \le 20 \\ x \ge 0, \quad y \ge 0 \end{cases}$

** 12. Minimize $z = 12x + 14y$

subject to $\begin{cases} x + y \le 5 \\ 3x + 4y \ge 12 \\ 2 \le x \le 5 \\ x \ge 0, \quad y \ge 0 \end{cases}$

13. A nutritionist at a hospital is considering one of two possible meal patterns for several patients in the intensive care unit of the hospital. The nutritional content of the two alternative meals is as follows:

Meal pattern	Protein content per unit	Carbohydrate content per unit
A	2	2
B	3	2

The doctors insist that the meals must provide at least 2 units of protein and at least 2 units of carbohydrates. The costs for preparing each of the meals A and B are 73¢ and 85¢, respectively. Assuming that each patient can be served either of these meals, how many of each should be prepared so as to minimize cost?

14. Dr. Achari, a cardiologist, has told Sylvia, one of his patients, to restrict her caloric intake from bread and cheese to at most 1200 calories per day. It is known that each slice of bread costs 12¢ and contains 110 calories, whereas each ounce of cheese costs 20¢ and contains 100 calories. How much of each can Sylvia eat so as to maximize her food consumption and yet satisfy the above requirements assuming that she wishes to spend at most $2.40 on these items?

15. *Mopeds.* The George Company manufactures two types of mopeds in its two factories. Factory A performs the basic assembly operations and operates for at most 144 hours per month. Factory B performs the finishing operations and operates for at most 180 hours per month. Each cheaper moped manufactured requires 3 hours in factory A and 4 hours in factory B and produces a profit of $81. Each expensive moped manufactured requires 6 hours in factory A and 5 hours in factory B and produces a profit of $104. How many of each should be manufactured so as to maximize profit?

16. *Ice Cream or Yogurt?* A dairy company has 1000 gallons of milk and 1300 gallons of cream from which it wishes to make yogurt and ice cream. Each gallon of ice cream manufactured requires 0.35 gallon of milk and 0.45 gallon of cream and produces a profit of 79¢. Each gallon of yogurt produced requires 0.30 gallon of milk and 0.42 gallon of cream and produces a profit of 90¢. How many of each should be manufactured so as to maximize profit?

17. *Part-Time Help.* A large farming cooperative in the West employs both part-time and full-time farm workers to harvest the crops. On any one day there is enough money to pay for at most 100 part-time and 150 full-time farm workers to harvest the crops. Each part-time farm worker is paid $4 per hour, and each full-time worker is paid $6 per hour. If officials of the cooperative know that at least 60 workers must be hired daily to prevent the crops from rotting, how many full-time and how many part-time farm workers must be hired so as to minimize costs?

18. The Ace Maintenance Corporation operates two laundromats. It is planning to install new coin-operated dry-cleaning machines and new coin-operated dryers at either or at both locations. The company has at most 24 square yards of usable space at location A and at most 36 square yards of usable space at location B for the proposed new equipment. However, fire department regulations require that each dry-cleaning machine have 2 square yards of space at location A and 3 square yards of space at location B. On the other hand, for the dryers, fire department regulations require 4 square yards of space at location A and 5 square yards of space at location B. If the company expects a profit of 75¢ from each use of a dry-cleaning machine and a profit of 50¢ from each use of a dryer, how many machines of each type should be installed?

**19. Spoduk College has just established a $100,000 trust fund to be used for yearly college scholarships. The money was obtained from a rich philanthropist who specified that college officials invest it in stocks and bonds as follows:

a) No more than half of the money may be invested in stocks.

b) At most $70,000 may be invested in bonds

c) The amount of money invested in bonds may be at most twice the amount of money invested in stocks.

If college officials estimate that they can receive 15% a year from money invested in stocks and 10% from money invested in bonds, how much money should be invested in each so as to maximize the return?

8.10

SUMMARY   In this chapter we introduced the rectangular coordinate system and the procedure used for graphing functions.

When we draw the graph of a linear function, the intercepts and slope of the line are quite helpful. The intercepts and slope can be determined from the equation of the line. Different versions of the equation of a line were discussed, and we indicated when each one is used.

We also indicated how to find the common solution for two equations graphically (if there is one). We simply draw the graphs on the same set of axes and read off the coordinates of the point of intersection.

Then we introduced what is meant by linear programming and learned how to set up certain problems as linear programming problems. Since linear inequalities are needed when working with linear programming, we reviewed them and explained how we draw their graphs. Finally, we studied the graphical procedure for solving linear programming problems.

STUDY GUIDE You should now be able to demonstrate your knowledge of the following ideas presented in this chapter by giving definitions, descriptions, or specific examples. Page references (in parentheses) are given for each term so that you can check your answer.

Simplex method (p. 401)
Graph (p. 402)
Number line (p. 402)
Origin (p. 402)
Positive direction (p. 402)
Negative direction (p. 402)
Coordinate (p. 403)
x-axis (p. 403)
y-axis (p. 403)
x-coordinate or abscissa (p. 403)
y-coordinate or ordinate (p. 403)
Quadrant (p. 404)
Rectangular coordinate system (p. 405)
Linear equation (p. 410)
x-intercept (p. 411)
y-intercept (p. 411)

Slope (p. 414)
Equation of a line (p. 417)
Plane divider (p. 421)
Half-plane (p. 421)
System of consistent equations (p. 424)
System of inconsistent equations (p. 424)
System of dependent equations (p. 424)
Graphing systems of linear inequalities (p. 428)
Linear programming (p. 431)
Solving linear programming problems graphically (p. 440)
Constraints (p. 440)

FORMULAS TO REMEMBER The following list summarizes all of the formulas given in this chapter.

1. The graph of $Ax + By + C = 0$ where A, B, and C are constants with A and B both not 0 is a straight line.

2. The graph of $x = C$ is a straight line parallel to y-axis.

3. The graph of $y = B$ is a straight line parallel to x-axis.

4. To find x-intercept of a line, set $y = 0$ and solve for x.

5. To find y-intercept of a line, set $x = 0$ and solve for y.

6. The slope of the line passing through the points (x_1, y_1) and (x_2, y_2) is
$$m = \frac{y_2 - y_1}{x_2 - x_1} \text{ provided that } x_1 \neq x_2.$$

7. The equation of the line whose slope is m and whose y-intercept is at $(0, b)$ is $y = mx + b$.

8. Other forms of the equation of a line:

$$y - y_1 = m(x - x_1) \qquad \text{Slope} = m \qquad \text{Point} = (x_1, y_1)$$

$$y - y_1 = \frac{y_2 - y_1}{x_2 - x_1}(x - x_1) \qquad \text{Point} = (x_1, y_1) \qquad \text{Point} = (x_2, y_2)$$

$$\frac{x}{a} + \frac{y}{b} = 1 \qquad x\text{-intercept} = (a, 0) \qquad y\text{-intercept} = (0, b)$$

9. The procedure for graphing linear inequalities as given on p. 420.

MASTERY TESTS

Form A

1. What is the slope of the line whose equation is $y - 3x + 7 = 0$?

 a) -7 **b)** -3 **c)** 3   **d)** 7   **e)** none of these

2. On the same set of coordinate axes, graph the following system of inequalities:

$$y < x - 2$$
$$y \geq -2x + 4$$

3. Write the coordinates of a point that is *not* in the solution set of the system of inequalities graphed in question 2.

4. Which graph represents the inequality $y < 4$?

a)

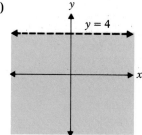

b)

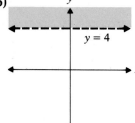

c)

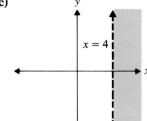

d)

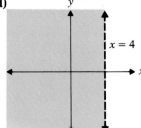

e) none of these

5. The graph of the equation $3x + 2y = 12$ intersects the y-axis at the point whose coordinates are

 a) $(0, 6)$ **b)** $(6, 0)$ **c)** $(0, 4)$ **d)** $(4, 0)$ **e)** none of these

6. Solve graphically and check:

$$y - 2x = 5$$
$$x + 2y = 0$$

7. What are the coordinates of a point that lies on both axes of a rectangular coordinate system?

 a) $(0, 0)$ **b)** $(0, 1)$ **c)** $(1, 0)$ **d)** $(1, 1)$ **e)** none of these

8. On the same set of coordinate axes, graph the lines whose equations are:

$$\begin{cases} y = 2x + 1 \\ y = 1 \\ x = 2 \end{cases}$$

9. Write the coordinates of the three vertices of the triangle formed by the lines graphed in question 8.

10. The point whose coordinates are $(3, K)$ lies on the graph of $2x + y = 5$. What is the value of K?

11. Find the slope of the line passing through the points $(-2, 3)$ and $(5, 9)$.

12. Find the equation of the line passing through the points $(4, 5)$ and $(8, 11)$

13. Find the equation of the line that passes through the point $(1, 2)$ and that is parallel to the line whose equation is $y - 8x = 5$.

14. Find the y-intercept of the line whose equation is $3x + 4y = 9$.

15. Write an equation of the line that is parallel to the line $2y - 6x = 9$ and that passes through the point $(-2, 1)$.

Form B

1. Which coordinates represent a point on the graph of $y = 4x - 14$?
   a) $(-2, 6)$ b) $(-6, 2)$ c) $(2, -6)$ d) $(6, -2)$
   e) none of these

2. On the same set of coordinate axes, graph the following system of inequalities and label the solution set S.

$$2x + y < 6$$
$$y \geq 2$$

3. What is the slope of the line whose equation is $y = 3$?

4. What is the slope of the line whose equation is $x = 1$?

5. The graph of which equation contains the point $(3, 1)$?
   a) $y = x$ b) $y = 3$ c) $y - 3x = 0$ d) $3y = x$
   e) none of these

6. Maximize $5x + 3y$

subject to $\begin{cases} x - 3y \leq 9 \\ 2x + y \leq 4 \\ x \geq 0 \\ y \geq 0 \end{cases}$

7. Minimize $8x + 2y$

subject to $\begin{cases} 2y \geq x - 6 \\ x + y \geq 2 \\ x \geq 0 \\ y \geq 0 \end{cases}$

8. The Acme Component Corp. manufactures ski poles and skateboards. Manufacturing a ski pole requires 2 hours of work by operator A and 4 hours of work by operator B and yields a profit of $3 per pole. On the other hand, making a skateboard requires 4 hours by operator A and 2 hours by operator B and yields a profit of $8 per skateboard. No operator can work more than 12 hours per day. How many of each product should be produced so as to maximize profit?

9. The Eastwood Munitions Company assembles jeeps and tanks in its two factories. Factory 1 performs the basic assembly operation. Factory 2 performs the finishing operations. For financial reasons, factory 1 has 185 man-days available per week. Factory 2 has 135 man-days available. Factory 1 needs 2 man-days on each jeep and 5 man-days on each tank. Factory 2 needs 3 man-days for each jeep or tank. How many of each should the company produce if its profit is $600 on a jeep and $900 on a tank?

10. The American Exterminating Company produces two kinds of mouse-traps, the *Normal* trap and the *Big* trap. For these traps to be produced, each must be processed by two factories, Y and Z. Factory Y assembles the traps and remains open at most 9 hours a day. Factory Z, which paints and packages the traps, can stay open at most 8 hours a day. Each *Normal* trap requires 1 hour in factory Y and 2 hours in factory Z. It yields a profit of $1. Each *Big* trap requires 3 hours in factory Y and only 1 hour in factory Z. It yields a profit of $4. How many of each kind of trap should the company manufacture so as to maximize profit?

11. A company produces two types of computer disc drives. Each drive of the first type requires twice as much labor time as does each drive of the second type. If only disc drives of the second type are produced, then the company can produce a total of 500 drives a day. Fluctuating market conditions limit daily sales of the first and second types to 150 and 200 drives, respectively. If the profit is $8 from each type I disc drive and $5 from each type II drive produced, how many of each type should be produced so as to maximize profit?

12. A chemical company plans to introduce a new mouthwash. It will advertise the product on both local radio and television stations. It plans to limit its advertisement expenses to $1000 a month. Each minute of radio advertisement costs $5 and each minute of television advertisement costs $100. The company intends to use radio at least twice as much as television. Past experience indicates that each minute of television advertisement usually generates 25 times as many sales as each minute of radio advertisement. How should the company allocate the monthly budget for radio and television advertisements so as to maximize sales?

13. Maximize $P = 3x + 2y$

$$\text{subject to} \quad \begin{cases} 2x + y \le 6 \\ 3x + 4y \le 12 \\ x \ge 0 \\ y \ge 0 \end{cases}$$

**14. Maximize $P = 5x + 8y$

$$\text{subject to} \begin{cases} x + y \leq 4 \\ 4x + 3y \leq 12 \\ -x + y \geq 1 \\ x + y \leq 6 \\ x \geq 0 \\ y \geq 0 \end{cases}$$

15. Name an ordered pair that is in the solution set of $y > 2x$ but not in the solution set of $y - 4 \leq 0$. (Use graphical techniques.)

9

GEOMETRY AND THE METRIC SYSTEM

CHAPTER OBJECTIVES

To review the basic geometrical ideas of points, lines, and planes. (*Section 9.2*)

To define what an angle is and to discuss two different systems of measuring angles: degree measure and radian measure and to learn how to convert angles measured in one system to measurements in the other system. (*Section 9.3*)

To study curves, the differences between similar and congruent triangles, and polygons as well as their properties. (*Sections 9.4 and 9.5*)

To introduce you to the metric system of measurement and learn how to convert from the British system to the metric system. (*Section 9.6*)

To compute the area and perimeter of various geometric figures. (*Section 9.7*)

To analyze the difference between Euclidean and non-Euclidean geometries. (*Sections 9.8 and 9.9*)

To apply the ideas discussed in this chapter to real-life situations. (*Throughout chapter*)

To discuss some of the basic ideas of topology. (*Sections 9.10, 9.11, and 9.12*)

To examine the Königsberg Bridge problem, the conveyor belt problem, the Möbius Strip, etc. (*Sections 9.10, 9.11 and 9.12*)

We're Going Metric

ALBANY: The New York State Thruway Authority announced yesterday that it would begin changing all of the destination marker signs and speed limit signs on the length of the New York Thruway to metric units in the spring. Not only will speed limit signs such as the one shown be changed, but signs indicating the distances between cities and towns will eventually be changed also.

The cost of the changeover is expected to run into millions of dollars and will be accomplished gradually over a period of several years.

THE NEWS, February 20, 1985

The above article indicates that America is gradually changing over to the metric system. Industries will have to replace machinery and tools as well as maintain dual inventories during this transition period. Furthermore, the public will have to be reeducated in how to use this new system and how to convert measurements from our present system to the metric system. In this chapter we will explain how this is done.

9.1
INTRODUCTION

The importance of geometry can be seen in the wide-ranging and ever-growing application of its concepts. For example, when the U.S. Mint produces dimes, it tries to make them all the same size and shape. (This idea is known as **congruence,** and we will have more to say about it later.) If you decide to carpet your home, you need some basic geometry in computing the number of square feet of carpet to buy. When railroad tracks are laid, we see geometry at work in the plan for them. It is necessary that they be the same distance apart everywhere. (When they are, they are then said to be **parallel.**)

Because geometry is so useful and necessary, it was one of the first branches of mathematics to be seriously studied. The science of geometry dates back at least to the ancient Babylonians and Egyptians, who needed it for the measurement of land for taxation purposes and for building. The precision with which the pyramids were built clearly indicates the Egyptians' skill with geometry. Geometry was also used in studying astronomy, which was, in turn, important in constructing calendars.

During the eighteenth century in the city of Königsberg (formerly in Prussia) there was an island surrounded by the river Pregel. This river was crossed by seven bridges as shown in Figure 9.1. On pleasant afternoons the citizens of Königsberg would amuse themselves with

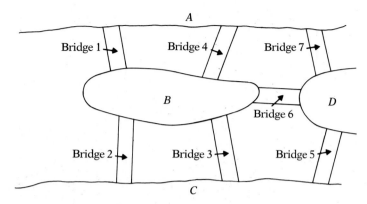

Figure 9.1

the following problem. Could anyone walk across *all* seven bridges without crossing any of the bridges twice? No one had ever been able to do it. Whenever anyone attempted it, either one bridge was not crossed at all or else one bridge had to be crossed more than once. Eventually, the citizens of Königsberg came to believe that it was impossible to do this, but they did not know why. The Swiss mathematician Leonhard Euler heard about the problem. Not only did he solve it, but in doing so he laid the basis for a whole new branch of mathematics called **topology.** Today topology is an important and growing field of mathematics with wide-ranging applications.

In this chapter we will discuss some of the basic concepts of geometry, the work of Euclid, and also two other interesting geometries that differ from Euclid's in some of the basic assumptions. Also, we will study the metric system of measurement. Finally, we will examine the basic ideas of topology and discuss several interesting applications of these ideas, such as the Königsberg Bridge problem, the conveyor belt problem, and the Möbius strip.

HISTORICAL NOTE

The Babylonians and Egyptians developed considerable skill in practical geometry, that is, in calculating areas, volumes, etc. They also knew such facts as certain special cases of the Pythagorean theorem. However, the geometric knowledge of the Babylonians and Egyptians was just a collection of facts accumulated through practical experience. The Greeks were the first people to undertake a formal study of geometry.

While there were many great Greek mathematicians, one of the best known was Euclid (about 300 B.C.). Although Euclid was a Greek, he was a mathematics professor at the University of Alexandria in Egypt. The story is told of a student who asked Euclid of what use it was to study geometry (a question quite familiar to all mathematics teachers). In reply, Euclid gave the student three pennies, since "he must make gain of what he learns."

Euclid's greatness lies not so much in his discovering new truths of geometry, but rather in his showing that all the known facts could be obtained from a few simple assumptions, using deductive logic. The modern high school geometry course is based on Euclid's ***Elements.***

9.2
POINTS, LINES, AND PLANES

The word **geometry** comes from the Greek *ge* meaning "earth" and *metria* meaning "measurement." Early humans observed that certain shapes, such as triangles, rectangles, circles, and surfaces, occurred frequently in nature. The study of geometry began with people's need to measure and understand the properties of these shapes.

Suppose you were asked, "What is a triangle?" You would probably say that a triangle is a figure bounded by three straight lines. This definition depends on your knowing what a line is. What exactly is a line? Everyone, of course, has an idea of what a line is. You could draw one if asked to do so, but could you describe a line in words?

Line

A line is often said to be a collection of points. However, understanding such a definition depends on your knowing what a point is. Moreover, the definition does not distinguish between lines and other geometric figures that can be thought of as collections of points. For example, a plane (flat surface) is also a collection of points.

Another definition of a line that is often given is that it is the shortest distance between two points. Again this depends on your knowing what a point is. It also depends on what is meant by "distance" and how it is measured. This definition is "circular."

No matter how we try to define a line, we encounter similar problems. Therefore although we all know what a line is, we do not attempt to define it formally. We accept it as an *undefined term.*

Point

Now let us consider the term **point.** What is a point? The following definitions have been suggested.

1. A point is a dot.

2. A point is a location in space.

3. A point is something that has no length, breadth, or thickness.

If we analyze these definitions, we see that none of them is really satisfactory. Dots have varying thicknesses and can be measured with a precise instrument (and a magnifying glass). Thus when speaking of a point, we would have to specify what size dot we mean. This would involve measurement and length. Obviously, this is not what we want.

We can make similar objections to the other definitions. Therefore we accept point as another *undefined term.*

In mathematics, as in language in general, when we try to define a term, we find that the definition depends on other terms, which in turn must also be defined. Ultimately, we see that we must start with some basic terms that we do not define. These are called **undefined**

A computer-drawn geometric figure.

terms. All other definitions are based on these undefined terms. In geometry, point and line are undefined terms, as we have stated.

Mathematicians did not always recognize the need for undefined terms. In fact, the most famous geometry book of the ancient world, Euclid's *Elements*, opens with twenty-three definitions. Many of these definitions involve other terms that have not been defined. As a result, they do not really define anything. For example, Euclid defines "point" as "that which has no part." However, no definition of what is meant by "part" is given. The modern mathematician recognizes the fact that we cannot define everything, and so we must start with some undefined terms, such as "point" or "line."

Because pictures are so useful in mathematics, a *point* is represented by a dot. The smaller we make the dot, the better it will represent the mathematical idea of a point.

Although the term *line* (by line, we will always mean a straight line) is undefined, it has certain important properties, which we now state.

1. *A line is a set of points in space.* The points of the line are said to lie on the line, and the line is said to contain them or pass through them. Points that lie on the same line are said to be **collinear.** In Figure 9.2 the line *l* is drawn with arrows pointing in both directions. This indicates that the line can be extended indefinitely in either direction.

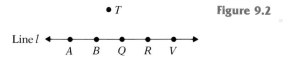

Figure 9.2

The points *A, B, Q, R,* and *V* all lie on the line *l*. They are collinear. Point *T* does not lie on the line *l*.

Notation   Lines will be denoted by lowercase letters such as *l, m, n,* . . . . Points are denoted by capital letters. A line containing two points, say *A* and *B*, can be denoted as $\overleftrightarrow{AB}$. The line of Figure 9.2 can be denoted as line *l*, $\overleftrightarrow{AB}$, $\overleftrightarrow{AQ}$, $\overleftrightarrow{QV}$, $\overleftrightarrow{AV}$, etc.

2. *Given any two different points in space, there is exactly one line containing these two points.*

3. *Any point on a line divides a line into two parts.* Each of these parts is called a **half-line.** The dividing point is known as the **end-point** of the half-line. The dividing point is not included in either of the half-lines. (See Figure 9.3.)

   Definition 9.1   *A half-line, together with its dividing point, is known as a **ray.***

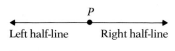

Figure 9.3

There will be many times when we will be interested in only certain finite parts of a line rather than the whole line or ray. For this reason we introduce the following definition.

Definition 9.2   *Any two points A and B on a line, together with all the points on the line that lie between them, are called a **line segment**. We denote it by $\overline{AB}$.*

EXAMPLE 1   The different terms and notations are illustrated below.

Description	Picture	Symbol
Line *AB*		$\overleftrightarrow{AB}$
Ray *AB*		$\overrightarrow{AB}$
Ray *BA*		$\overleftarrow{BA}$
Line segment *AB*		$\overline{AB}$

A solid dot, •, means that the point is to be included. An open dot, ◦, means that the point is not to be included.

Notice the difference between ray $\overrightarrow{AB}$ and ray $\overrightarrow{BA}$. Ray $\overrightarrow{AB}$ starts at point *A* and extends in the direction of point *B*. Ray $\overrightarrow{BA}$ starts at point *B* and extends in the direction of point *A*.   □

Two different lines that contain the same point are said to **intersect** at the point. Lines *l* and *m* of Figure 9.4 intersect at point *P*. Both lines contain this point.

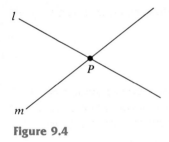

Figure 9.4

Plane   A **plane** can be thought of as a flat surface, such as the page of this book or the floor in your room. Mathematically, the word *plane* is an undefined term. Planes also have certain important properties.

1. *A plane is a set of points.* We say that the points are on the plane and that the plane contains them. Points that are on the same plane are said to be **coplanar.**

2. *Any three noncollinear points determine one and only one plane.* The plane of Figure 9.5 contains the three points *A*, *P*, and *Q*. This is the only plane that can contain these three points.

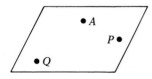

Figure 9.5

We have all seen four-legged tables that wobble because one of the legs is longer or shorter than the other three. This happens because the table can only be steady if all four feet are on the plane of the floor. When one leg is longer or shorter than the others, then the four feet do not lie in the same plane (the floor) and the table wobbles. A three-legged table is much more likely to be steady because it has only three feet and these three feet will always lie on some plane.

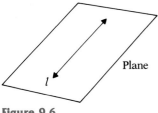

Figure 9.6

3. *If two points of a line are on a plane, then the whole line is on the plane.*

4. *A line on a plane divides the plane into two parts called* **half-planes.** The line does not belong to either of the half-planes. In Figure 9.6, line *l* divides the plane into two half-planes.

5. *If we are given two planes, then they either meet in a line or do not meet at all.* Planes that do not meet at all are called **parallel planes.**

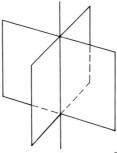

l   **Figure 9.7**

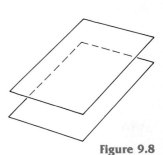

Figure 9.8

Figure 9.7 shows two planes intersecting in line *l*. Figure 9.8 shows two parallel (nonintersecting) planes. The ceiling and floor of your room are examples of parallel planes. The ceiling and walls of your room are examples of intersecting planes.

Parallel Lines   If we are given two lines in the same plane that *never* meet, then such lines are called **parallel lines.** Railroad tracks are an example of parallel lines. The edges of the pages of this book are another example of parallel lines. If line *l* and line *m* are parallel, we write this as *l* ∥ *m*; that is, the symbol "∥" stands for "is parallel to."

Parallel lines have played an important role in the development of geometry, as we shall see later.

EXERCISES

For Exercises 1–8, refer to the line shown in Figure 9.9 with the indicated points. Find each of the following.

For Exercises 9–18, refer to the line shown in Figure 9.10 with the indicated points. Find each of the following.

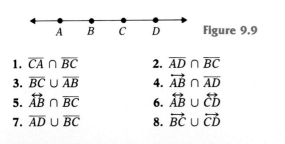

Figure 9.9

Figure 9.10

1. $\overline{CA} \cap \overline{BC}$ **2.** $\overline{AD} \cap \overline{BC}$

3. $\overline{BC} \cup \overline{AB}$ **4.** $\overrightarrow{AB} \cap \overline{AD}$

5. $\overleftrightarrow{AB} \cap \overline{BC}$ **6.** $\overleftrightarrow{AB} \cup \overleftrightarrow{CD}$

7. $\overline{AD} \cup \overline{BC}$ **8.** $\overrightarrow{BC} \cup \overrightarrow{CD}$

9. $\overline{AB} \cap \overline{BC}$ **10.** $\overline{AB} \cup \overline{BC}$

11. $\overrightarrow{AB} \cap \overline{CD}$ **12.** $\overrightarrow{DC} \cup \overrightarrow{DE}$

13. $\overleftrightarrow{DC} \cap \overrightarrow{DE}$ **14.** $\overleftrightarrow{AB} \cap \overline{CD}$

15. $\overrightarrow{FE} \cap \overrightarrow{EA}$ **16.** $\overline{EF} \cap \overline{AB}$

17. $\overline{BC} \cup \overline{DE}$ **18.** $\overrightarrow{BC} \cup \overline{EF}$

For Exercises 19–26, refer to Figure 9.11 with the indicated points. Find each of the following:

19. $\overline{BC} \cup \overline{CD}$ **20.** $\overline{BC} \cup (\overline{CD} \cup \overrightarrow{BD})$

21. $\overline{BC} \cap \overline{CD}$ **22.** $\overrightarrow{BC} \cap \overrightarrow{AD}$

23. $(\overline{AB} \cup \overline{BC}) \cup (\overline{CD} \cup \overline{DA})$

24. $\overrightarrow{BC} \cap \overrightarrow{AC}$

25. $(\overline{AB} \cup \overline{BC}) \cap \overline{BD}$ **26.** $\overline{AD} \cup \overline{DC}$

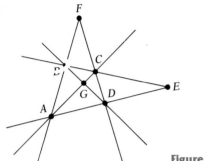

Figure 9.11

27. How many different lines can be drawn connecting three different noncollinear points?

28. How many different lines can be drawn connecting four different noncollinear points that lie on the same plane?

**29. How many different lines can be drawn connecting four different points that do not lie on the same plane if no three of the points are collinear?

**30. Is it possible to have two lines on the same plane that are not parallel and that do not intersect? Explain your answer. What if the lines are on different planes?

**31. If l is a line in plane P, m is a line in plane Q, and plane P is parallel to plane Q, then is line l parallel to line m?

**32. Suppose we are given four different points, A, B, C, and D, which do not all lie on the same plane. How many different planes can be drawn containing any three of the points if no three of the points are collinear?

9.3

ANGLES   What is an angle? We can all draw one as shown in Figure 9.12. We can define an angle formally as follows.

> **Definition 9.3**   An **angle** is the union of two rays that have a common endpoint. The rays are called the **sides** of the angle, and the endpoint is called the **vertex** of the angle.

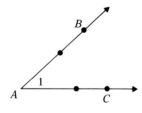

Figure 9.12

In Figure 9.12 the angle is formed by the rays $\overrightarrow{AB}$ and $\overrightarrow{AC}$ and the vertex is A. We identify this angle in one of the following ways:

1. By writing $\angle A$. In this notation, $\angle$ stands for angle, and A represents the vertex.

2. By writing $\angle BAC$ or $\angle CAB$. In this notation the vertex is in the middle, and the other two points are points on either side of the angle.

3. By inserting a "1," a "2," etc., and calling it $\angle 1$ or $\angle 2$, as we have done in Figure 9.12.

An angle divides the points of a plane that are not on the angle into two sets of points called **regions.** One region is called **the interior of the angle,** and the other is called the **exterior of the angle.** These are labeled in the diagram below. The region that contains all points *P* is called the interior of the angle. All other points on the plane (except those in the angle itself) are called the exterior of the angle.

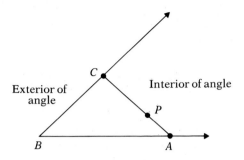

Angles are measured by an instrument called a **protractor,** such as the one shown in Figure 9.13. Angles are usually measured in units called **degrees,** and these are denoted by the symbol °. The scale on the protractor is marked from 0 to 180 degrees as shown. To measure an angle, the point marked O on the protractor is placed at the vertex, and the 0° line is placed along one of the rays. The size of the angle is determined by the position of the second ray on the protractor. Thus ∠AOB of Figure 9.14 measures 40°. We denote this as $m(\angle AOB) = 40°$ where *m* stands for "the measure of."

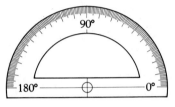

Figure 9.13

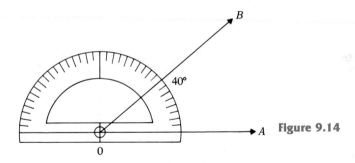

Figure 9.14

Each degree is divided into 60 smaller units known as **minutes.** The symbol for minute is ′. Thus 1° = 60′. Each minute is further divided in 60 smaller units known as **seconds,** which are denoted as ″. Therefore we have

$$1° = 60'$$

and

$$1' = 60''.$$

This system of angular measurement can be traced back to the Babylonians, who used a base-60 number system (see Chapter 3).

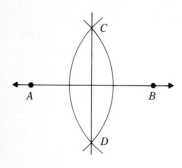

Figure 9.15

Consider the following construction. Draw a straight line $\overleftrightarrow{AB}$ as shown in Figure 9.15. Now take a compass, place the steel point on A, and draw an arc. Then put the steel point on B and, keeping the compass open the same amount, draw another arc that intersects the first arc in two places, C and D, as shown. (If they do not intersect, repeat the above procedure using larger compass settings.) Now join the two intersecting points.

Lines $\overleftrightarrow{CD}$ and $\overleftrightarrow{AB}$ form an angle. Measure it! It measures 90°. Such an angle is called a **right angle,** and the lines $\overleftrightarrow{AB}$ and $\overleftrightarrow{CD}$ that form the right angle are said to be **perpendicular.**

> **Definition 9.4**   *Two lines are said to be **perpendicular** if the angle at which they intersect is 90°, or a right angle.*

We denote this by using the symbol "⊥." Thus if line $\overleftrightarrow{AB}$ is perpendicular to line $\overleftrightarrow{CD}$, we write $\overleftrightarrow{AB} \perp \overleftrightarrow{CD}$.

It is convenient to distinguish among the different kinds of angles, depending on their measure. We do this in the following definition.

> **Definition 9.5**   *A **right angle** is an angle whose measure is 90°. A **straight angle** is an angle whose measure is 180°. An **acute angle** is an angle whose measure is between 0° and 90°. An **obtuse angle** is an angle whose measure is between 90° and 180°.*

EXAMPLE 1   An angle of 35° is an acute angle.

An angle of 138° 12′ 16″ is an obtuse angle.   ☐

Comment   According to our definition, an angle is merely a set of points. It consists of the points on the two rays (and the vertex, of course).

Comment   The measurement of an angle as described above requires that the measure of any angle be between 0° and 180°. Why?

It is not always convenient to restrict the measure of an angle to between 0° and 180°.

To extend angular measure beyond the 0°–180° restriction, we can think of an angle as being formed in the following way. Draw any ray and call this the **initial side.** Keeping its endpoint fixed, rotate this side a certain amount and stop. The place where we stop is another ray, which we call the **terminal side.** Some angles formed in this manner are shown in Figure 9.16. The arrow indicates the direction and amount of the rotation. In each case the vertex is at point A.

In Figure 9.16 (a) the rotation is counterclockwise, and its measure is considered to be positive.

In Figure 9.16 (b) the rotation is clockwise, and its measure is thought of as being negative.

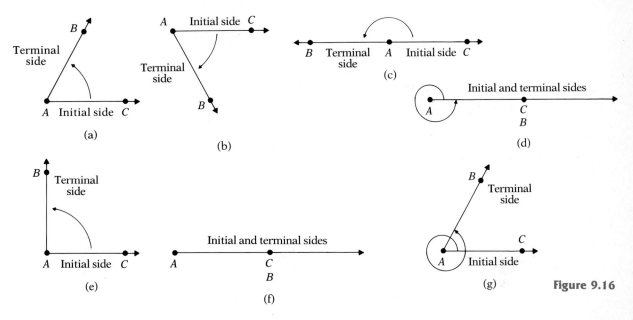

(a)   (b)   (c)   (d)

(e)   (f)   (g)   **Figure 9.16**

Notice that the angle of Figure 9.16 (d) represents one complete revolution, and therefore the terminal side is the same as the initial side. It is agreed that in one complete revolution there are 360 degrees or 360°. This standard of measurement also comes from the Babylonians and was related to their studies in astronomy.

Notice that the angle of Figure 9.16 (c) contains one-half of a complete revolution, or measures 180°. This conforms to the method first discussed for measuring angles using protractors.

In Figure 9.16 (g), we have an angle whose measure is larger than 360°. Using rotations, we can have angles of any size, both positive and negative, as shown in Figure 9.17.

Figure 9.17

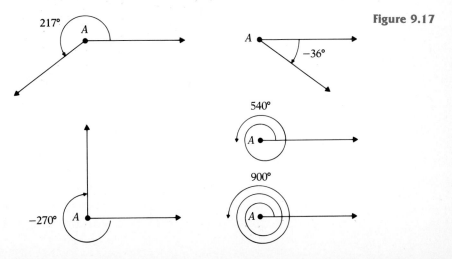

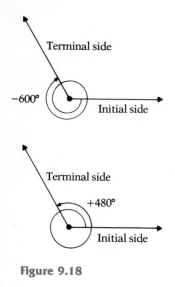

Terminal side

−600°

Initial side

Terminal side

+480°

Initial side

Figure 9.18

Such **rotational angles** frequently occur in everyday situations. For example, the rotational angles +480° and −600° have the same initial and terminal sides, as shown in Figure 9.18. If we consider these angles in terms of the initial and terminal sides, Definition 9.3 tells us that the two angles are the same, since they are determined by the same rays and the same vertex. However, from the rotational point of view they are obviously different. Suppose your car is stuck near the edge of a cliff and someone is giving you directions on how to proceed safely. It makes a big difference whether he tells you to rotate the wheels of the car + 480° or −600° (see Figure 9.19).

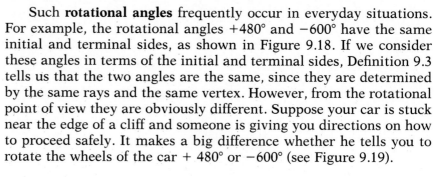

+480°

−600°

Figure 9.19

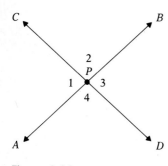

C   B

2
P
1   3
4

A   D

Figure 9.20

When any two lines intersect, four angles are formed, as shown in Figure 9.20. Angles 1 and 2 have a common ray, $\overrightarrow{PC}$, and we call them **adjacent angles.** Similarly, angles 2 and 3 have a common ray, $\overrightarrow{PB}$, and they are also adjacent angles. There are two other pairs of adjacent angles in this diagram. Name them.

In the same illustration, angles 2 and 4 are nonadjacent. We call them **vertical angles.** Similarly, angles 1 and 3 are called vertical angles. This leads us to the following definitions.

Definition 9.6   *Two angles are said to be* ***adjacent*** *if they have a common ray and a common vertex but do not have any interior points in common.*

Definition 9.7   *When two lines intersect, the nonadjacent angles formed are said to be* ***vertical*** *angles.*

Radian Measure   We know that we can measure something—for instance, the length of this piece of paper—with standard rulers of different units. The length can be given in either inches or centimeters. Similarly, weight can be measured in different units—in either pounds or grams. So far in this chapter, we have been using degrees as the unit of measurement for angles, but there is another unit that is often used in mathematics, science, and engineering. This is called the **radian.**

Consider a circle whose radius is 1 in. (a radius is any line segment drawn from the center of the circle to the circle). Draw a radius in this circle (see Figure 9.21). Now take a piece of string 1 in. long and place it *along the circle* with one end at the point where the radius

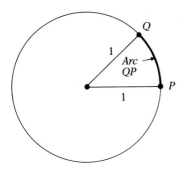

Figure 9.21

drawn meets the circle (point *P*). Put the other end of the string at point *Q*. Draw another radius from the center to point *Q*. The arc *QP* (portion of the circle between points *P* and *Q*) obviously has length equal to 1 in. The angle between the two radii (plural of radius) is assigned a measure of one **radian.**

We know that in one complete rotation there are 360°. How many radians are there in one complete rotation? Let us see. It can be shown that if a string were to be placed around the entire circle, it would measure 2π in. As you know, π is an irrational number that cannot be written exactly as a decimal. It is approximately equal to 3.1415. . . . An arc of 1 in. gives an angle of 1 radian. Therefore an arc of 2π in. (the whole circle) gives an angle of 2π radians.

Thus in one complete rotation there are 2π radians or 360°, depending on which unit you are using.

Formula 1 2π *radians* $= 360°$ (1)

Since 2π radians is 360°, we can find how many degrees there are in 1 radian by dividing by 2π. We get

$$\frac{2\pi}{2\pi} \text{ radians} = \frac{360°}{2\pi}. \qquad \left(\frac{2\pi}{2\pi} \text{ equals } 1.\right)$$

Formula 2   1 *radian* $= \dfrac{180°}{\pi}$ (2)

Similarly, if we divide (1) by 360, we get

$$\frac{2\pi}{360} \text{ radians} = \frac{360°}{360}.$$

Formula 3 $\dfrac{\pi}{180}$ *radians* $= 1$ *degree* (3)

If we use 3.14 as an approximation for π, we get

$$1 \text{ radian} = \frac{180°}{\pi}$$
$$= \frac{180°}{3.14},$$

which is approximately 57°. Also,

$$1 \text{ degree} = \frac{\pi}{180} \text{ radians}$$
$$= \frac{3.14}{180} \text{ radians},$$

which is approximately 0.0174 radian. *Thus 1 radian is approximately 57 degrees, and 1 degree is approximately 0.0174 radian.*

Formulas 2 and 3 enable us to convert from degrees to radians and from radians to degrees, as shown in the following examples.

EXAMPLE 2   Convert $\frac{\pi}{4}$ radians to degree measure.

SOLUTION   We use Formula 2, which says that

$$1 \text{ radian} = \frac{180}{\pi} \text{ degrees.}$$

We multiply both sides by $\frac{\pi}{4}$ and get

$$\frac{\pi}{4}(1 \text{ radian}) = \frac{\pi}{4}\left(\frac{180}{\pi} \text{ degrees}\right)$$

$$\frac{\pi}{4} \text{ radians} \quad = \frac{180}{4} \text{ degrees}$$

$$\frac{\pi}{4} \text{ radians} \quad = 45°. \quad \square$$

EXAMPLE 3   Convert 60° to radian measure.

SOLUTION   We use Formula 3, which says that

$$1 \text{ degree} = \frac{\pi}{180} \text{ radians.}$$

We multiply both sides by 60 and get

$$60(1 \text{ degree}) = 60\left(\frac{\pi}{180} \text{ radians}\right)$$

$$60 \text{ degrees} = \frac{60\pi}{180} \text{ radians}$$

$$60 \text{ degrees} = \frac{\pi}{3} \text{ radians.} \quad \square$$

Complementary and Supplementary Angles   Consider the following angle *ABC*. The measure of angle *a* is 35°, and the measure of angle *b* is 55°. The sum of the measures of these two angles is 90°. Angles *a* and *b* are said to be **complementary,** and each angle is the complement of the other. Formally, we have that *two angles are complementary angles if the sum of their measures is 90°.*

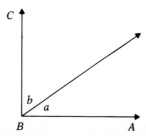

Now consider angle *IJK* shown below. The measure of angle c is 60°, and the measure of angle d is 120°. The sum of the measures of these two angles is 180°. Angles c and d are said to be **supplementary,** and each angle is the supplement of the other. Formally, we have that *two angles are supplementary angles if the sum of their measures is* 180°.

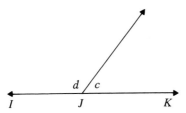

Comment   We can represent the measure of the complement of an angle whose measure is x degrees by $(90 - x)$, since $x + (90 - x) = 90$.

Comment   We can represent the measure of the supplement of an angle whose measure is x degrees by $(180 - x)$, since $x + (180 - x) = 180$.

EXAMPLE 4   The measure of the complement of an angle is eight times the measure of the angle. Find the measure of the angle.

SOLUTION   Let $x =$ the measure of the angle. Then $8x =$ the measure of the complement of the angle. Since the angles are complementary, the sum of their measures is 90°, so

$$x + 8x = 90$$
$$9x = 90 \quad \text{(We combine the } x \text{ and the } 8x.)$$
$$x = 10 \quad \text{(We divide both sides of the equation by 9.)}$$

Thus the measure of the angle is 10°.   □

EXAMPLE 5   Find the measure of an angle whose measure is 80° more than the measure of its supplement.

SOLUTION   Let $x =$ the measure of the supplement of the angle. Then $x + 80°$ represents the measure of the angle. Since the angles are supplementary, the sum of their measures is 180°, so

$$x + x + 80 = 180$$
$$2x + 80 = 180 \quad \text{(combining terms)}$$
$$2x = 100 \quad \text{(We subtract 80 from both sides of the equation.)}$$
$$x = 50$$

Thus the measure of the supplement is 50°, and the measure of the angle is 50° + 80°, or 130°. □

Parallel Lines Cut by a Transversal

Often we are given two parallel lines that are cut by a transversal. A **transversal** is a line that intersects two other lines in two different points as shown below, where line $\overleftrightarrow{EF}$ is a transversal that intersects the two parallel lines $\overleftrightarrow{AB}$ and $\overleftrightarrow{CD}$. Angles 3, 4, 5, and 6 are called **interior angles,** and angles 1, 2, 7, and 8 are called **exterior angles.** Furthermore, angles 4 and 5 as well as angles 3 and 6 are called **alternate interior angles.** They are interior angles on opposite sides of the transversal and do not have the same vertex.

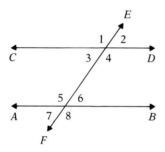

Angles 1 and 8 are called **alternate exterior angles.** They are exterior angles on opposite sides of the transversal and do not have the same vertex. Can you find another pair of alternate exterior angles?

Angles 4 and 6 are called **interior angles on the same side of the transversal.** Can you find another pair of interior angles on the same side of the transversal?

Angles 1 and 5 are called **corresponding angles.** One is an exterior angle and the other is an interior angle, both being on the same side of the transversal. There are three other pairs of corresponding angles. Can you find them? One such pair is angles 2 and 6.

In the above parallel lines it can be shown that the alternate interior angles 3 and 6 have the same measure. The same is true for the alternate interior angles 4 and 5. More generally, if two lines cut by a transversal are parallel, then their alternate interior angles have the same measure.

Also in the above parallel lines the interior angles on the same side of the transversal, angles 4 and 6 as well as angles 3 and 5, are supplementary.

Finally, the corresponding angles 1 and 5 have the same measure. The same is true for the corresponding angles 2 and 6, the corresponding angles 3 and 7, and the corresponding angles 4 and 8. Whenever two parallel lines are cut by a transversal, their corresponding angles always have the same measure.

EXERCISES

In Exercises 1–4, use a protractor to construct an angle having the indicated measures.

1. 50°   **2.** 65°   **3.** 130°   **4.** 95°

In Exercises 5–10, use a protractor to construct the following rotational angles.

5. 210°   **6.** 370°   **7.** 450°

8. −190°   **9.** −360°   **10.** −780°

In Exercises 11–15, find all the adjacent and all the vertical angles for the indicated diagrams.

11.
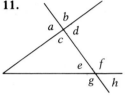

12.
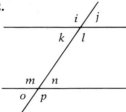

13.
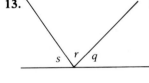

14.
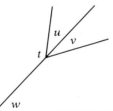

15.
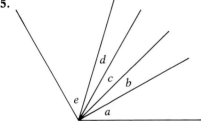

Each of the angles in Exercises 16–27 is given in degrees. Convert each to radians.

16. 10°   **17.** 210°   **18.** 315°   **19.** −45°

20. 190°   **21.** −720°   **22.** 1040°   **23.** −570°

24. 900°   **25.** −60°   **26.** −330°   **27.** 1600°

Each of the angles in Exercises 28–36 is given in radian measure. Convert each to degree measure.

28. $\dfrac{3\pi}{4}$ **29.** $\dfrac{\pi}{2}$ **30.** $\dfrac{7\pi}{6}$

31. $\dfrac{11\pi}{12}$ **32.** $-\dfrac{8\pi}{9}$ **33.** $-\dfrac{12\pi}{5}$

34. 10   **35.** 5π **36.** $-\dfrac{7\pi}{15}$

For Exercises 37–40, refer to Figure 9.22 and the indicated points. Find each of the following.

37. $\angle DAB \cap \angle BAC$

38. $\angle EAB \cap \angle DAE$

39. $\angle DAE \cap \angle BAC$

40. $\angle BAD \cap \angle CAE$

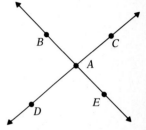

Figure 9.22

Classify each of the statements given in Exercises 41–43 as true or false.

41. In Figure 9.23, angles *ABC* and *ABD* are adjacent angles.

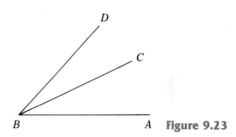

Figure 9.23

42. In Figure 9.24, angles *DBE* and *ABC* are vertical angles.

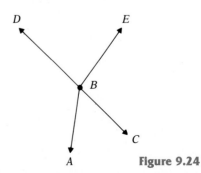

Figure 9.24

43. One degree measures approximately 0.0174 radians.

44. In the figure below, $\overleftrightarrow{AB} \parallel \overleftrightarrow{CD}$. Find the measure of

a) angle 5 when the measure of angle 3 is 150°.

b) angle 2 when the measure of angle 6 is 140°.

c) angle 4 when the measure of angle 5 is 60°.

d) angle 8 when the measure of angle 3 is 130°.

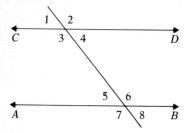

45. In the figure below, $\overleftrightarrow{AB} \parallel \overleftrightarrow{CD}$. If the measures

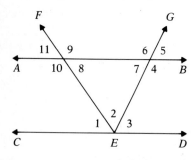

of angles 7 and 8 are 60° and 70°, respectively, find the measures of the remaining angles in the figure.

46. Find the measure of an angle that is 36° more than the measure of its complement.

47. Two angles are supplementary. The measure of the larger angle is twice the measure of the smaller angle. Find the measure of the smaller angle.

48. The measures of two complementary angles are in the ratio 7 : 2. Find the measure of each angle.

49. The measure of the supplement of an angle exceeds 4 times the measure of the angle by 30. Find the measure of the angle.

50. The measure of an angle is 40° less than the measure of its supplement. Find the measure of the supplement.

9.4

CURVES, TRIANGLES, AND POLYGONS   Using straight lines, we can construct figures in a plane that are known as **polygons.** Several examples of polygons are shown in Figure 9.25.

Figure 9.25

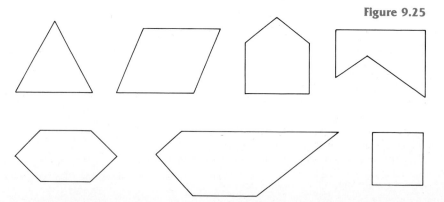

To define a polygon formally, we first introduce the idea of a simple closed curve.

Definition 9.8   A ***simple closed curve*** *is any curve that can be drawn without lifting the pencil (or other writing instrument) and that has the following properties:*

1. *The drawing starts and stops at the same point.*

2. *No point is touched twice (with the exception of the starting point).*

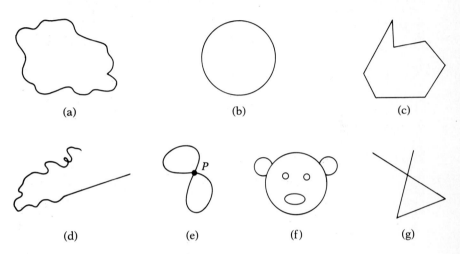

(a)   (b)   (c)

(d)   (e)   (f)   (g)

Figure 9.26

The examples of Figure 9.26 illustrate this definition. In Figure 9.26, (a), (b), and (c) are simple closed curves. Figure (d) is not a simple closed curve because it does not start and stop at the same point. (It is not closed.) Figure (e) is not a simple closed curve because point *P* is touched twice in drawing the curve. (To see this, try to draw it.) Figures (f) and (g) are not simple closed curves. Why not?

We can now define what we mean by a polygon.

Definition 9.9   *A **polygon** is a simple closed curve consisting of straight-line segments.*

The simplest kind of polygon is a **triangle,** which consists of *three* lines called **sides.** Triangles are very important in the study of geometry, since any polygon can be broken up into triangles, such as the figures shown in Figure 9.27. The Babylonians and Egyptians often used this idea in measuring land that was in the shape of a polygon. They divided the land into triangles and measured each part individually.

There are many different kinds of triangles. Three types that are important are equilateral, isosceles, and right triangles. These are defined as follows.

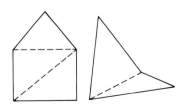

Figure 9.27

a)

b)

Alinari/Editorial Photocolor Archives

c)

Dennis Stock, Magnum Photos

d)

Dennis Stock, Magnum Photos

e)

Rene Burri, Magnum Photos

f)

Geometric shapes in architecture: (a) Transamerica Building (courtesy of Transamerica Corporation); (b) Giotto's Campanile, Florence; (c) Solomon Guggenheim Museum; (d) a commune building; (e) habitat; (f) the Pentagon (courtesy of the U.S. Navy).

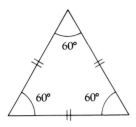

Figure 9.28

Definition 9.10 *An **equilateral triangle** is a triangle with three sides whose measures are equal. An **isosceles triangle** is a triangle with two sides whose measures are equal. A **right triangle** is a triangle that contains a 90° angle.*

It can be shown that if a triangle is equilateral, then all the angles are equal and each will measure exactly 60° (Figure 9.28).

It can also be shown that if a triangle is isosceles, then the measures of the two angles opposite the equal sides are also equal (Figure 9.29).

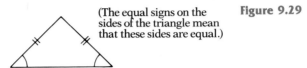

(The equal signs on the sides of the triangle mean that these sides are equal.)

Figure 9.29

The Pythagorean theorem states that in a right triangle, such as the one in Figure 9.30, $a^2 + b^2 = c^2$.

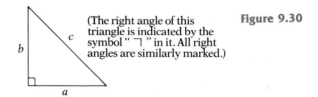

(The right angle of this triangle is indicated by the symbol " ⌐ " in it. All right angles are similarly marked.)

Figure 9.30

Consider the triangle shown in Figure 9.31. Trace each of the angles on separate pieces of paper and cut them out. Now place the cutout angles with the vertices (plural of vertex) together and the sides adjacent to each other, as shown in Figure 9.32.

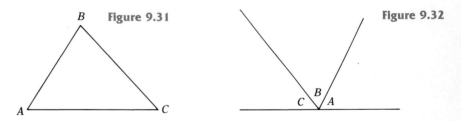

Figure 9.31

Figure 9.32

Notice that the three angles together add up to a straight angle (180°). This will be true for any triangle. What this means is that the sum of the measures of the angles of any triangle is 180°. This is a basic idea in the geometry of Euclid. We will see later that it is not true in non-Euclidean geometries.

Comment In discussing a right triangle the two sides that form the right angle are called the legs of the right triangle. The third side, which is opposite the right angle, is called the hypotenuse. These are shown below.

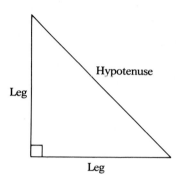

EXERCISES

Which of the figures given in Exercises 1–6 are simple closed curves?

1.

2.

3.

4.

5.

6.

Which of the figures given in Exercises 7–12 are polygons?

7.

8.

9.

10.

11.

STOP

12.

ONE WAY

13. Draw a triangle whose sides are 6 cm, 9 cm, and 12 cm. What happens?

14. Draw a triangle whose sides are 6 cm, 6 cm, and 12 cm. What happens?

15. Find the length of the hypotenuse of a right triangle *ABC* whose legs measure 16 and 30 cm as shown in Figure 9.33.

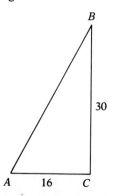

Figure 9.33

16. Find the length of the third side of a right triangle if the hypotenuse measures 34 cm and one of the legs measures 30 cm.

17. Draw a triangle whose angles are 50°, 80°, and 45°. What happens?

18. Draw two triangles that intersect at exactly one point.

19. Draw two triangles that intersect at exactly two points.

20. Draw two triangles that intersect at exactly three points.

21. Is an angle a simple closed curve? Explain.

22. Draw any four-sided polygon (known as a **quadrilateral**). Using tracing paper, copy each of the angles and cut them out. Now place the cutout angles with the vertices together and sides adjacent.

    a) What is the sum of these four angles?

    b) What can you say about the sum of the angles of any quadrilateral?

23. Draw a quadrilateral and a triangle such that the intersection is exactly

    a) one point

    b) two points

    c) three points

24. State the precise meaning of each of the following terms. (If you do not know, look them up in a dictionary or geometry book.)

    a) Parallelogram   **b)** Rhombus

    c) Rectangle   **d)** Square

25. Classify each of the following as true or false.

    a) Every parallelogram is a square.

    b) Every trapezoid is a parallelogram.

    c) The sum of the angles of any quadrilateral is 180°.

    d) A square is a rhombus.

    e) A triangle can be drawn when the sides measure 17 cm, 18 cm, and 35 cm.

    f) If a triangle is equilateral, then it is equiangular.

    g) An obtuse triangle may have two obtuse angles.

In Exercises 26–31, use the Pythagorean theorem to calculate the length of the missing side in each diagram.

26.

27.

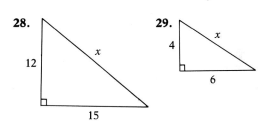

28.

29.

30.

31.

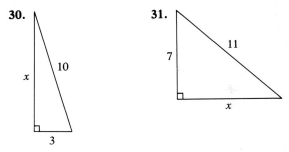

Use the Pythagorean theorem to solve Exercises 32 and 33.

32. As shown in Figure 9.34, a 25-foot ladder leans against the side of a house. The base of the ladder is 12 feet from the house on level ground.

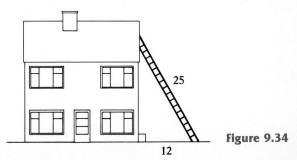

Figure 9.34

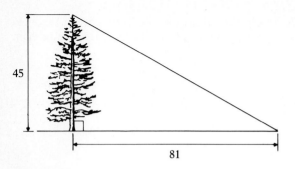

Figure 9.35

Find, to the *nearest foot*, the distance from the top of the ladder to the ground.

33. As shown in Figure 9.35, a tree 45 feet high on level ground casts a shadow 81 feet long. Find, to the nearest foot, the distance from the top of the tree to the end of the shadow on the ground.

34. Draw a right triangle that is isosceles.

35. Draw an obtuse triangle that is isosceles.

36. Can a triangle have two right angles? Explain.

37. Can a triangle have a right angle and an obtuse angle? Explain.

9.5

SIMILAR AND CONGRUENT TRIANGLES

Consider the three triangles shown in Figure 9.36. The triangles of parts (a) and (c) have exactly the same size and shape. If you cut out $\triangle LMN$[1] and place it on top of $\triangle FGH$ (with $\angle L$ falling on top of $\angle F$, and $\angle M$ falling on top of $\angle G$), then $\triangle LMN$ will fit *exactly* over $\triangle FGH$. Each can be considered a carbon copy of the other. We say that these two triangles are **congruent.** The symbol "≅" will stand for the words "is congruent to." Thus we write $\triangle LMN \cong \triangle FGH$, and we read this as "triangle LMN is congruent to triangle FGH." Although we have described what we mean by congruent triangles, in modern textbooks on geometry, "congruent" is considered an undefined term.

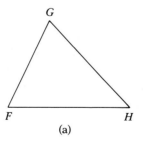 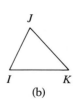 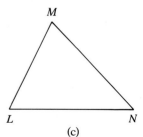

(a)   (b)   (c)

Figure 9.36

Now consider $\triangle FGH$ and $\triangle IJK$. Clearly, they are not congruent, since they are of different size. However, $\triangle IJK$ has exactly the same shape as $\triangle FGH$. Triangle FGH is an enlarged version of $\triangle IJK$.

---

1. The symbol "$\triangle LMN$" stands for "triangle LMN."

If all the angles of one triangle are congruent respectively to the angles of another triangle and the triangles have the same shape (but not necessarily the same size), they are called **similar triangles.**

Comment Congruent triangles are also similar triangles because they have the same shape. Similar triangles are not necessarily congruent because, although they have the same shape, their sizes may differ.

Congruent and similar triangles have many important properties, which we now state.

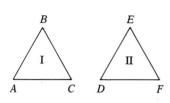

Figure 9.37

In Figure 9.37, triangles I and II are congruent. If we compare them, we find that side $\overline{AB}$ is equal in length to the **corresponding** side $\overline{DE}$. If $\triangle ABC$ is congruent to $\triangle DEF$, then side $\overline{AC}$ fits exactly on side $\overline{DF}$. Similarly, side $\overline{BC}$ fits exactly on side $\overline{EF}$, and side $\overline{AB}$ fits exactly on side $\overline{DE}$. Sides $\overline{AB}$ and $\overline{DE}$ are called **corresponding sides.** Sides $\overline{AC}$ and $\overline{DF}$ are corresponding sides, and sides $\overline{BC}$ and $\overline{EF}$ are also corresponding sides. The length of a side of a triangle is called its **measure.** Thus the length of side $\overline{AB}$ is called the measure of side $\overline{AB}$ and is denoted by the symbol $m(\overline{AB})$. Similarly, side $(\overline{BC})$ is equal in measure to corresponding side $\overline{EF}$. Also, side $\overline{AC}$ corresponds to side $\overline{DF}$, and they are equal in measure.

Property 1 *When two triangles are congruent, their corresponding sides are congruent (equal in measure).*

Now measure angle A and the corresponding angle D. They are equal. Similarly, $m(\angle C) = m(\angle F)$ and $m(\angle B) = m(\angle E)$.

Property 2 *When two triangles are congruent, their corresponding angles are congruent (equal in measure).*

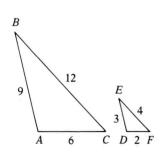

Figure 9.38

In Figure 9.38, side $\overline{DE}$ = 3 units, side $\overline{EF}$ = 4 units, and side $\overline{DF}$ = 2 units. Now measure sides $\overline{AB}$, $\overline{BC}$, and $\overline{AC}$. You should get 9, 12, and 6 units, respectively. Note that side $\overline{AB}$ is 3 times side $\overline{DE}$. Similarly, side $\overline{BC}$ is 3 times side $\overline{EF}$, and side $\overline{AC}$ is 3 times side $\overline{DF}$. Each side of $\triangle ABC$ is 3 times its corresponding side in $\triangle DEF$. We express this fact by saying that all three sides are **proportional.** In this example we have

$$\frac{m(\overline{AB})}{m(\overline{DE})} = \frac{m(\overline{BC})}{m(\overline{EF})} = \frac{m(\overline{AC})}{m(\overline{DF})}$$

$$\frac{9}{3} = \frac{12}{4} = \frac{6}{2} = 3.$$

This leads us to the following property of similar triangles.

Property 3 *If two triangles are similar, then the corresponding sides are proportional.*

Measure angles A and D with a protractor. You should find that $m(\angle A) = m(\angle D)$. Similarly, if you measure the other angles, you will find that $m(\angle B) = m(\angle E)$ and $m(\angle C) = m(\angle F)$. This leads us to the following property.

Property 4   *If two triangles are similar, then the corresponding angles are equal in measure. The converse of this statement is also true.*

EXAMPLE 1   In the similar triangles shown in Figure 9.39, find the measure of the lengths of the unmarked sides.

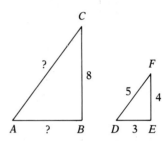

Figure 9.39

SOLUTION   Since the triangles are similar, we know by Property 3 that the sides are proportional. Thus

$$\frac{m(\overline{AB})}{m(\overline{DE})} = \frac{m(\overline{BC})}{m(\overline{EF})}$$

$$\frac{m(\overline{AB})}{3} = \frac{8}{4}.$$

Multiplying both sides by 3 gives

$$3\,\frac{m(\overline{AB})}{3} = 3 \cdot \frac{8}{4}$$

$$m(\overline{AB}) = 3 \cdot \frac{8}{4}$$

$$= 3 \cdot 2 = 6.$$

Similarly, we have

$$\frac{m(\overline{AC})}{m(\overline{DF})} = \frac{m(\overline{CB})}{m(\overline{FE})}$$

$$\frac{m(\overline{AC})}{5} = \frac{8}{4}.$$

Multiplying both sides by 5 gives

$$5\,\frac{m(\overline{AC})}{5} = 5 \cdot \frac{8}{4}$$

$$m(\overline{AC}) = 5 \cdot \frac{8}{4}$$

$$= 5 \cdot 2 = 10. \quad \square$$

EXAMPLE 2   In the similar triangles shown below, find the measure of the unmarked angles.

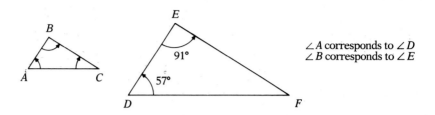

SOLUTION   Since the sum of the angles of a triangle is 180°, we know that

$$m(\angle D) + m(\angle E) + m(\angle F) = 180°$$
$$57° + 91° + m(\angle F) = 180°$$
$$148° + m(\angle F) = 180°.$$

Thus

$$m(\angle F) = 32°.$$

By property 4 the angles of $\triangle ABC$ must be equal in measure to those of $\triangle DEF$. Thus

$$m(\angle A) = m(\angle D) = 57°, \quad m(\angle B) = m(\angle E) = 91°, \quad \text{and}$$
$$m(\angle C) = m(\angle F) = 32°. \ \square$$

EXAMPLE 3   If a lamppost 30 ft high casts a shadow of 40 ft, how high is a pole next to it if the pole's shadow is 10 ft long?

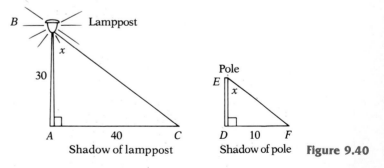

Figure 9.40

SOLUTION   First we draw a diagram (see Figure 9.40). We will assume that the lamppost and the pole stand vertically and that the sun strikes each at the same angle. These angles are marked with an "x." In the figure, each triangle has two equal angles, the right angle and the angle marked x. Thus the third angles are equal. (Why?) Since the triangles have their corresponding angles equal, the triangles are similar. It follows that their sides are proportional. Thus we have

$$\frac{m(\overline{DE})}{m(\overline{AB})} = \frac{m(\overline{DF})}{m(\overline{AC})}$$

$$\frac{m(\overline{DE})}{30} = \frac{10}{40}.$$

Multiplying both sides by 30, we get

$$m(\overline{DE}) = 30 \cdot \frac{10}{40} = 7\frac{1}{2} \text{ ft.} \quad \square$$

There are many other interesting and useful properties of congruent and similar triangles, some of which will be discussed in Section 9.8.

EXERCISES

In Exercises 1–6, the pairs of triangles or rectangles are similar. Find the sides or angles labeled with a question mark.

1.

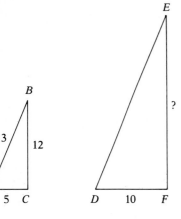

2.

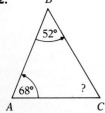
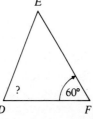

3.

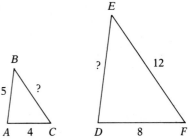

4.

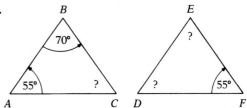

5.

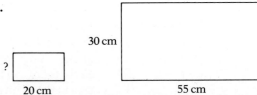

6.

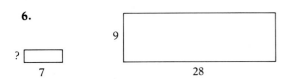

The idea of congruent triangles can be extended to any geometric figures. With this idea in mind, state whether each of the pairs of figures given in Exercises 7–15 appear to be congruent.

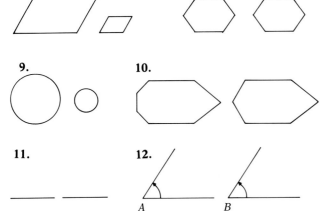

7.

8.

9.

10.

11.

12.

13.

14.

15.

16. Are all equilateral triangles congruent? Explain.

17. Are all equilateral triangles similar? Explain.

18. Are all right triangles similar? Explain.

19. Are all isosceles right triangles similar? Explain.

20. Are all isosceles right triangles congruent? Explain.

**21. Are all rhombuses similar? Explain.

22. Consider the figures shown below. Are the figures similar to each other?

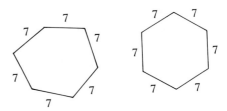

23. Can a quadrilateral be similar to a hexagon? Explain.

24. Consider the diagram below. Is parallelogram *ABCD* similar to rectangle *MNOP*? Explain.

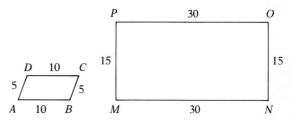

25. The following pairs of polygons are similar. Find the length of every side in each polygon.

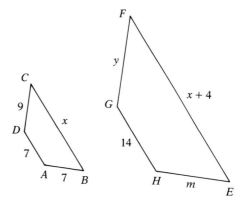

26. A picture 18 cm long and 12 cm wide is to be enlarged so that its width will be 15 cm. How long will the enlarged picture be?

27. Trudy has a wallet-size picture of her baby that is $2\frac{1}{2}$ cm wide by $4\frac{1}{4}$ cm long. She wishes to enlarge it for a wall poster that will be 42.5 cm long. How wide will the enlargement be?

28. In right triangles *ABC* and *DEF*, $m(\angle A) = m(\angle D)$ as shown.

   a) Is triangle *ABC* similar to triangle *DEF*?

   b) Find *x*.

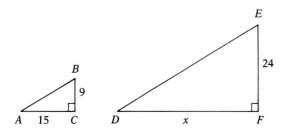

29. A certain flagpole casts a shadow 7 m long. At the same time a nearby boy who is $1\frac{1}{2}$ m tall casts a shadow 4 m long. Find the height of the flagpole.

30. A statue casts a shadow 20 ft long. At the same time an observer $4\frac{1}{2}$ ft tall casts a shadow 9 ft long. How tall is the statue?

31. In the accompanying diagram $\triangle ABC$ and $\triangle DEC$ are similar. Find $\overline{AB}$ and $\overline{DE}$.

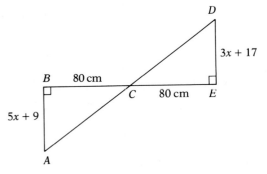

32. Consider $\triangle ABC$ and $\triangle ADE$. These triangles are similar. Can you explain why?

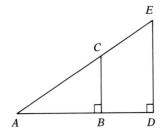

9.6

THE METRIC SYSTEM

Measure the line segment $\overline{AB}$ shown in Figure 9.41 with a ruler. You should find that it is three inches long. In measuring the line segment you compared it to the markings on a ruler. How did the company that made the ruler know how to mark it? Obviously, they had some standard measure, but where did that come from? Units of measurement, whether of weight, volume, or other measure, have been determined in the following way.

A   B **Figure 9.41**

    Scientists agree on a basic unit of measuring a particular quantity. For example, in the United States at present the standard unit of measurement of length is the **yard.** The standard yardstick is kept at the National Bureau of Standards in Washington, D.C. It is made of metal and is maintained at a constant temperature to prevent the metal from expanding. All other yardsticks are copies of this standard yardstick.

There are two standard systems of measurement in use today. One of these is known as the **metric system** and is used throughout most of the civilized world. The United States still uses the older **foot-pound,** or **British system** in which the standard unit of length is the foot. Many industries in the United States are opposed to converting to the metric system because of the cost of a changeover. After much debate, the U.S. Congress passed the Metric Conversion Act of 1975. In effect, this established a board to oversee the "voluntary" conversion. Thus in due time, many of the standard measures with which we are familiar will become metric units. Indeed, we are already beginning to "think metric," since many companies are now packaging their items in metric units.

One major advantage of using the metric system is that all measurements are expressed in powers of 10 (that is, 10, 100, 1000, etc.). The same is not true in the British system. For example, in the British system there are 3 feet in a yard, 12 inches in a foot, 5280 feet in a mile, etc.

Standard prefixes (which are powers of 10) are added to the various metric units to indicate multiples or submultiples of the basic units. These are shown in Table 9.1.

TABLE 9.1

Prefix	Symbol	Meaning
kilo-	k	one thousand times
hecto-	h	one hundred times
deka-	da	ten times
deci-	d	one tenth of
centi-	c	one hundredth of
milli-	m	one thousandth of

Metric Unit of Length

In the metric system of measurement *the meter is accepted as the unit of length.* Originally, the meter was defined as one ten-millionth of the distance measured along a meridian through Paris from the North Pole to the equator. A standard meter bar was constructed, but in fact it differed slightly (about 0.023%) from its intended length. Copies of the standard meter are kept in the major cities of the world. In 1961 the General Conference of Weights and Measures defined the meter in terms of the wavelength of a particular isotope of krypton. A meter measures approximately 39.37 inches. Similar standard measures have been agreed on for other quantities such as weight and liquid measure.

Some of the lengths used in the metric system, as well as their abbreviations, are shown in Table 9.2.

TABLE 9.2

Metric unit	Commonly used abbreviation	How many meters?
Kilometer	km	1000
Hectometer	hm	100
Dekameter	dam	10
Meter	m	1
Decimeter	dm	1/10 or 0.1
Centimeter	cm	1/100 or 0.01
Millimeter	mm	1/1000 or 0.001

The following examples show how we convert from one unit to another in the metric system.

EXAMPLE 1   How many centimeters are there in 325 meters?

SOLUTION   Since 1 m equals 100 cm, we have 1 m = 100 cm. Therefore

$$325 \times 1 \text{ m} = 325 \times 100 \text{ cm}$$
$$325 \text{ m} = 32{,}500 \text{ cm}.$$

Thus 325 m = 32,500 cm. ☐

EXAMPLE 2   How many kilometers are there in 7342 millimeters?

SOLUTION   Since 1 mm equals 0.001 m, we have 1 mm = 0.001 m. Therefore

$$7342 \times 1 \text{ mm} = 7342 \times 0.001 \text{ m}$$
$$7342 \text{ mm} = 7.342 \text{ m}.$$

Now

$$1 \text{ km} = 1000 \text{ m}.$$

This means that

$$1 \text{ m} = 1/1000 \text{ km or } 0.001 \text{ km}.$$

Thus

$$7.342 \times 1 \text{ m} = 7.342 \times 0.001 \text{ km} \quad \text{or} \quad 0.007342 \text{ km}.$$

Therefore

$$7342 \text{ mm} = 0.007342 \text{ km}. ☐$$

Since both the metric and British systems are still widely used, we should know how to convert from one system to the other. Table 9.3 shows the approximate relationships between the two systems.

TABLE 9.3

Converting from the metric system to the British system		Converting from the British system to the metric system	
1 millimeter	0.04 inch	1 inch	2.54 centimeters
1 centimeter	0.4 inch	1 foot	30.48 centimeters
1 meter	1.1 yards	1 yard	0.914 meter
1 kilometer	0.62 mile	1 mile	1.6 kilometers

Using the relationships given in Table 9.3, we can convert from the metric system to the British system and vice versa. This is illustrated in the following examples.

EXAMPLE 3 Convert 55 miles to kilometers.

SOLUTION From Table 9.3 we see that 1 mile is 1.6 km. Therefore 55 miles = 55 × 1.6 km or 88 km. ☐

EXAMPLE 4 Convert 75 centimeters to inches.

SOLUTION From Table 9.3 we see that 1 centimeter is 0.4 inch. Therefore 75 cm = 75 × 0.4 inch or 30 inches. ☐

EXAMPLE 5 The wheelbase of Bill's car measures 112 inches. How many centimeters does this measure?

SOLUTION From Table 9.3 we know that each inch is equivalent to 2.54 centimeters. Thus 112 inches is equivalent to 112 × 2.54 or 284.48 centimeters. ☐

EXAMPLE 6 Melissa's dining room is 12 feet by 9 feet. Express the measurement of her dining room in meters.

SOLUTION From Table 9.3 we know that 1 foot is 30.48 centimeters, so that 12 feet = 12 × 30.48 or 365.76 centimeters; also 9 feet = 9 × 30.48 or 274.32 centimeters. Since each centimeter is 1/100 of a meter, we have

$$365.76 \text{ cm} = \frac{365.76}{100}\text{m} \quad \text{or} \quad 3.6576 \text{ m}$$

and

$$274.32 \text{ cm} = \frac{274.32}{100}\text{m} \quad \text{or} \quad 2.7432 \text{ meters.}$$

Thus Melissa's room measures 3.6576 by 2.7432 meters. ☐

Metric Unit of Weight

In the metric system of measurement the standard unit of weight is the **gram,** as opposed to our pound. There are many abbreviations used when working with weights in the metric system. These are given in Table 9.4.

TABLE 9.4

Unit of weight	Commonly used abbreviation	How many grams?
Kilogram	kg	1000
Hectogram	hg	100
Dekagram	dag	10
Gram	gm	1
Decigram	dg	1/10 or 0.1
Centigram	cg	1/100 or 0.01
Milligram	mg	1/1000 or 0.001

To convert from the metric unit of weight to the British system or vice versa, we proceed as follows.

To convert from the metric unit of kilograms to the British unit of pounds, multiply by 2.2; that is, 1 kg ≈ 2.2 lb.

To convert from the British unit of pounds to the metric unit of kilograms, multiply by 0.45; that is, 1 lb ≈ 0.45 kg.

EXAMPLE 7   Convert 763 centigrams to kilograms.

SOLUTION   From Table 9.4 we find that 1 cg equals 0.01 gm. Therefore 1 cg = 0.01 gm, so that 763×1 cg = 763×0.01 gm, or 763 cg = 7.63 gm. Now we must convert 7.63 gm to kilograms. Since 1000 gm = 1 kg, we have 1 gm = $\frac{1}{1000}$ kg or 0.001 kg. Thus 7.63 gm = 7.63×0.001 kg or 0.00763 kg. ☐

EXAMPLE 8   Convert 2133 pounds to kilograms.

SOLUTION   Using the above conversion rule, we find that 1 lb = 0.45 kg. Therefore 2133 pounds = 2133×0.45 kg, or 959.85 kg. ☐

EXAMPLE 9   In a supermarket, Vera Nelson is examining a box of cookies that was manufactured in Europe. The label on the package reads "Net weight 355 grams." A similar box of cookies manufactured in the United States reads "Net weight 12 ounces." If both boxes of cookies sell for the same price, which box gives Vera the most for her money?

SOLUTION   In order to compare them we must express both weights in the same unit. We will convert grams to ounces. Using the conversion rule given above, the reader should verify that 1 gram equals approximately 0.035 ounce, so that 355 grams equals 355 × 0.035 or 12.425 ounces. Thus the box made in Europe gives Vera the most for her money.

Metric Unit of Volume

In the metric system the standard unit of liquid measure (volume) is the **liter,** as opposed to our quart. Some of the abbreviations commonly encountered when working with the metric system are shown in Table 9.5.

TABLE 9.5

Unit of volume	Commonly used abbreviation	How many liters?
Kiloliter	kL	1000
Hectoliter	hL	100
Dekaliter	daL	10
Liter	L	1
Deciliter	dL	1/10 or 0.1
Centiliter	cL	1/100 or 0.01
Milliliter	mL	1/1000 or 0.001

To convert from one system to the other, we can use the relationships given in Table 9.6.

TABLE 9.6

Converting from the metric system to the British system		Converting from the British system to the metric system	
1 liter	1.05 liquid quarts	1 liquid quart	0.95 liters
1 liter	0.91 dry quarts	1 dry quart	1.1 liters
1 kilogram	2.2 pounds	1 pound	0.45 kilograms

EXAMPLE 10   Convert 17.3 liters to liquid quarts.

SOLUTION   From Table 9.6 we find that 1 liter = 1.05 liquid quarts, so that 17.3 liters = 17.3 × 1.05 liquid quarts, or 17.3 liters = 18.165 liquid quarts.

For the benefit of the reader we present in Table 9.7 a chart that makes it easier to convert from one system to the other.

TABLE 9.7

	To convert from	to	multiply by
Length	inches	millimeters	25
	inches	centimeters	2.54
	feet	centimeters	30.48
	feet	meters	0.3
	yards	meters	0.914
	miles	kilometers	1.6
	millimeters	inches	0.04
	centimeters	inches	0.4
	meters	feet	3.28
	meters	yards	1.1
	kilometers	miles	0.62
Weight	ounces	grams	28.3
	pounds	kilograms	0.45
	grams	ounces	0.035
	kilograms	pounds	2.2
Liquid Measure	ounces	milliliters	29.76
	pints	liters	0.476
	quarts	liters (liquid)	0.95
	gallons	liters	3.81
	milliliters	ounces	0.034
	liters	pints	2.1
	liters	quarts (liquid)	1.05
	liters	gallons	0.26

EXAMPLE 11   Patrick Michaelson owns a car that has a 22-gallon gas tank. On a recent trip to Canada, Patrick stopped at a gas station to fill up. To his amazement, he was able to fill up his car with 47 liters. How many gallons of gas did Patrick actually purchase?

SOLUTION   We must convert 47 liters to gallons. From Table 9.7 we know that 1 liter equals 0.26 gallon, so that 47 liters equals 47 × 0.26 or 12.22 gallons. Thus Patrick purchased only 12.22 gallons of gas.   ☐

EXERCISES

In Exercises 1–26, convert each measurement to the units indicated.

1. 9 yards to meters

2. 553 pounds to grams

3. 7 liters to quarts

4. 2.3 kg to lb

5. 384 lb to cg

6. 3.7 L to cL

7. 12 ft to cm

8. 396 mm to inches

9. 1760 mL to gal

10. 77 km to cm

11. 275 kg to gm

12. 17.5 ft to m

13. 364 m to yd

14. 286 L to liquid quarts

15. 67 ft to mm

16. 83 mL to L

17. 54 kg to lb

18. 21 m to ft

19. 16 L to dry quarts

20. 62 dm to cm

21. 38 m, 4 dm to ft

22. 726 in. to mm

23. 354 dry quarts to L

24. 17 m, 5 cm to yd

25. 0.7 kg to cg

26. 138 m, 8 dm, 6 cm to miles

27. Consider the sign shown in Figure 9.42. How would this sign read in the metric system?

SPEED
LIMIT
35

Figure 9.42

28. Consider the sign shown in Figure 9.43. How would this sign read in the metric system?

BRIDGE
WEIGHT
LIMIT
7500 LB

Figure 9.43

29. A certain football player is 6′5″ tall, weighs 195 lbs, and has a waist of 38″. What are these measurements in the metric system?

30. A plane's normal cruising altitude is 10,000 meters. Express this altitude in feet.

31. Jason supplements his normal food intake with 1200 milligrams of vitamin C. Express this weight in ounces.

32. The oil tank in Heather's two-family house has a capacity of 275 gallons. Express the tank's capacity in liters.

33. On January 1, 1986, unleaded gasoline sold in New York City for $1.39 per gallon of premium gas. Find the cost of one liter.

In the metric system, temperature is measured in degrees **Celsius** (formerly centigrade). On the Celsius scale, water freezes at 0° and boils at 100°, whereas on the Fahrenheit scale it freezes at 32° and boils at 212°. A formula for converting degrees Fahrenheit to degrees Celsius is $C = \dfrac{F - 32}{1.8}$, where C and F are degrees Celsius and degrees Fahrenheit, respectively. Use this formula to answer Exercises 34–37.

34. To conserve energy, government officials recommend a thermostat setting of 68°F during the heating season. What is this in degrees Celsius?

35. At what temperature would you most likely go swimming?

   a) 40°C   **b)** 20°C   **c)** 5°C

36. The temperature on a cold winter day at the North Pole was −23°C. What was the temperature in degrees Fahrenheit?

37. The coldest temperature ever recorded was −126.9°F (recorded in the Antarctic). What is this temperature in degrees Celsius?

9.7

MEASURE AND AREA: USING THE METRIC SYSTEM

In dealing with figures in a plane we often want to measure quantities other than just length of line segments. For example, we may want to measure the area of a polygon. We all have some idea of what is meant by area. If someone were to ask you to find the area of this page, you could do so with little difficulty. What do we mean by area? How do we measure it?

To answer these questions, imagine that we have a room that measures 9 m by 6 m, as shown in Figure 9.44.

We wish to cover the floor with square tiles that measure 1 m by 1 m. It seems reasonable to say that each tile has unit area. If you are measuring length in meters, then the unit used for measuring area is square meters. Thus each of these tiles has area equal to 1 sq m.

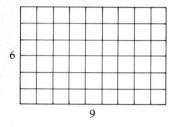

6

9

Figure 9.44

How many of the tiles will we need to cover the floor, assuming that no tiles will overlap? To answer this, we place 9 of these along the longer side of the floor and 6 along the shorter side of the floor, as shown. We continue to place tiles next to each other until the entire floor is covered. We see that it takes 54 tiles to completely cover the floor. Since one tile has an area of 1 sq m, it makes sense to say that the entire floor will have an area of 54 sq m. We note that $9 \times 6 = 54$, so we know that the area of the floor is its length times its width.

If we wanted to measure the area of any other rectangle (for example, a polygon with four sides and four right angles), then we could follow the same procedure. In each case it would lead us to the conclusion that the area is length times width. Thus for rectangles we agree that

$$\text{Area of rectangle} = \text{length} \times \text{width}.$$

In symbols we have

$$A = l \cdot w.$$

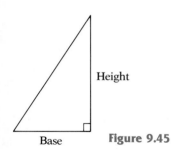

Height

Base

Figure 9.45

In our previous example we did not actually have to place tiles on the floor to determine its area. We could simply have multiplied the length by the width, getting 9×6, or 54, square meters.

How do we measure the area of a right triangle such as the one shown in Figure 9.45? In this case we cannot place square tiles over the triangle and expect to cover it exactly. The unit squares will cover either too much or too little. Instead, we can use the following procedure. Take tracing paper, copy the triangle, and then cut it out. Place the cutout triangle (the dotted one) alongside the original triangle as shown in Figure 9.46. We see that the two triangles together

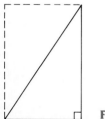

Figure 9.46

make a rectangle whose area we already know how to calculate by using the formula Area = length × width. Since our rectangle is two triangles, we see that the area of the original triangle is half the area of the rectangle. Thus

$$\text{Area of triangle} = \frac{1}{2} \text{ length} \times \text{width of rectangle.}$$

Note that the length of the rectangle equals the base of the triangle and that the width of the rectangle equals the height of the triangle. Therefore the area of the triangle can be written as

$$\text{Area of triangle} = \frac{1}{2} \text{ base} \times \text{height.}$$

In symbols,

$$A = \frac{1}{2}\, b \cdot h.$$

Now suppose that we are given any triangle (not necessarily a right triangle) and we wish to measure its area. From one of the vertices we can draw a perpendicular to the opposite side, as shown in Figure 9.47. The resulting figure will consist of two right triangles whose area we already know how to calculate. In this case the area of triangle I is $\frac{1}{2}b_1h$, and the area of triangle II is $\frac{1}{2}b_2h$. Thus the total area is

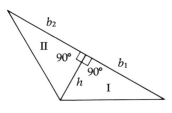

Figure 9.47

$$\frac{1}{2}b_1h + \frac{1}{2}b_2h$$

$$= \frac{1}{2}(b_1 + b_2)h \qquad \text{(by the distributive property)}$$

$$= \frac{1}{2}(\text{entire base})(\text{height})$$

$$= \frac{1}{2}\, b \cdot h.$$

Therefore the area of *any* triangle is given by $A = \frac{1}{2}bh$, where h is the perpendicular drawn from a vertex to the opposite side, b (see Figure 9.48).

Comment The area of any polygon can be calculated by dividing it into a collection of triangles, finding the area of each triangle individually, and then adding the areas together.

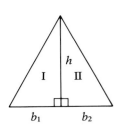

Figure 9.48

What about figures such as circles that cannot be broken up into triangles? It still seems reasonable to talk about the area of such figures, but what exactly do we mean by it? Let us consider a circle.

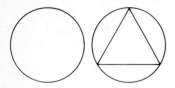

Figure 9.49

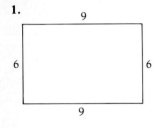

Figure 9.50

We can draw a triangle inside it as shown in Figure 9.49, which of course does not cover the circle completely. We can now replace the triangle with a square. This will cover still more of the circle. Now we replace the square with a pentagon (a five-sided polygon). This will cover still more of the circle (see Figure 9.50). If we continue in this way, drawing polygons with more and more sides in the circle, we find that each time, we cover more and more of the circle. (We can, of course, find the area of each polygon by using the techniques discussed earlier.)

It should be clear that as the number of sides of the polygon increases, the area of the polygon gets closer to the area of the circle. By using this procedure it can be shown that the area of the circle is equal to πr^2, where r is the radius of the circle.

Comment   Area is one measure of a closed plane figure. Another measure of such a figure is its **perimeter,** which for a polygon is just the sum of the length of its sides. The perimeter of a circle is called the **circumference,** and it can be shown that the measure of the circumference of a circle is equal to $2\pi r$.

EXERCISES

Find the perimeter of each of the closed plane figures given in Exercises 1–5.

1.

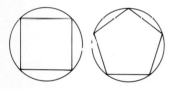

2.

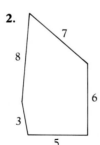

5.

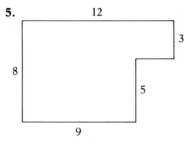

In Exercises 6–13, find the area of each of the closed plane figures.

6.

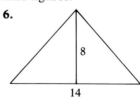

3.   4.

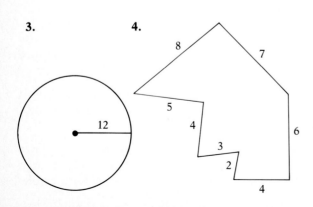

7.

8.

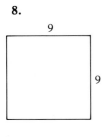

9

9

9.

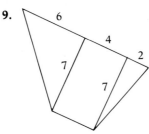

6

4

2

7

7

10.

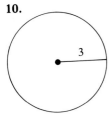

3

11.

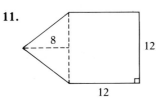

8

12

12

12.

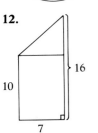

16

10

7

13.

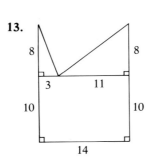

8

8

3

11

10

10

14

14. Two rooms have area equal to 600 square feet. One room measures 20 ft in width, and the other measures 40 ft in length. Which room is wider and why?

**15. By breaking up a parallelogram as shown in Figure 9.51, try to derive a formula for its area.

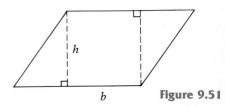

h

b

Figure 9.51

16. A **trapezoid is a four-sided figure, two of whose sides are parallel, as shown in Figure 9.52. By breaking up a trapezoid as indicated, try to derive a formula for its area.

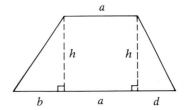

a

h h

b a d

Figure 9.52

9.8

EUCLIDEAN GEOMETRY Euclid's great work was his *Elements,* which was written around 300 B.C. It was a collection of all elementary mathematics, including geometry, and was actually an introductory textbook. When Euclid wrote the *Elements,* all the geometry in it was already known. His contribution consisted of organizing the available material into a mathematical system.

The *Elements* had an enormous influence on Western civilization. For 2000 years it was considered the best and most thorough example of mathematical reasoning. Early editions of the *Elements* were hand-copied in Greek, Latin, and Arabic, and the first printed version appeared in 1482. Since then, about 1000 different editions have been published. Most of our modern editions of Euclid are based on an edition of Euclid's *Elements* revised by Theon, a fourth century mathematician. Theon is also known as the father of the first important woman mathematician, Hypatia.

Euclid started (and so we have too) with certain basic terms such as *point* and *line*. However, unlike modern mathematicians, he did not see the need for undefined terms. He believed that it was possible to define all terms, and the *Elements* begins with twenty-three definitions. Unfortunately, these definitions are not adequate because they depend on other terms that have not been previously defined. For example, *point* is defined as "that which has no part." In order to understand this definition, you must first know what "part" is. Again, *line* is defined as "breadthless length." To understand this, you have to know what "breadth" and "length" mean. These were never defined by Euclid. His definitions are circular in the sense discussed in Section 9.2. Apparently, Euclid did not see this, and he considered his definitions satisfactory.

Although Euclid did not see that there were certain terms that had to be undefined, he did recognize that there were certain statements that had to be accepted without proof. Euclid intended to give deductive proofs of known mathematical facts such as the Pythagorean theorem. These proofs depend on other facts, which in turn depend on still others, and so on. This can go on forever unless you agree to stop somewhere. Therefore it is necessary to start with certain statements that must be accepted without proof. These are called **postulates** or **axioms,** and Euclid assumed ten. They are sometimes thought of as statements that are self-evident, statements that must be true because they are obvious. There is some indication that Euclid may have regarded at least some of his postulates in this way.

However, it often turns out that "obvious" truths are false. A good example of this is that it was once obvious to almost everyone that the sun revolved around the earth; one could watch it moving through the sky from east to west as the day passed. Today we all know that this obvious truth is false. Furthermore, in recent years, scientists and philosophers have shown that it is very difficult to be certain about anything at all. Thus present-day mathematicians do not claim that the postulates they use are true. They merely say that these postulates are what they are assuming. Postulates are just a starting point. If they turn out to be true statements about the physical world, so much the better. If not, it is still interesting to see what follows logically from them.

Euclid's ten postulates[2] or axioms were the following.

1. A straight line may be drawn connecting any two points.

2. A line segment can be extended indefinitely to form a line.

3. A circle may be drawn with any center and any radius. (The *radius* of the circle is a line segment drawn from the center to the circle.)

---

2. See Carl Boyer, *A History of Mathematics*, pp. 116–117. New York: John Wiley, 1968.

4. All right angles are equal.

5. Given a line and a point not on the line, then only one line can be drawn parallel to the first line passing through the given point. (Remember, parallel lines are lines that do not intersect.) This version of Euclid's fifth postulate, which was popularized by John Playfair, is the one that appears in many high school geometry texts. There are other versions of this postulate.

6. Things equal to the same things are equal to each other.

7. If equals are added to equals, then the sums are equal.

8. If equals are subtracted from equals, then the differences are equal.

9. Things that coincide with one another are equal to one another.

10. The whole is greater than any of its parts.

Starting from these postulates, Euclid was able to prove deductively many important and interesting statements called **theorems.** Some of the important theorems are the following.

1. Vertical angles are equal.

2. The base angles of an isosceles triangle are equal.

3. If the base angles of a triangle are equal, then the triangle is isosceles. (This is the converse of Theorem 2.)

4. The sum of the measures of the angles of any triangle equals 180°.

5. The Pythagorean theorem.

Although, as we pointed out earlier, Euclid's work was long considered a perfect example of deductive reasoning, within the past century it has been discovered that his reasoning is often incomplete. These logical gaps occur because he makes certain unstated assumptions based on diagrams. These assumptions cannot always be justified logically. To fill in the logical gaps, additional postulates must be introduced. This can and has been done, notably by the German mathematician David Hilbert (1862–1943).

One example of the kind of gap that occurs in Euclid's reasoning is the following "proof" that there exists a triangle with two right angles. (Of course this is ridiculous in Euclidean geometry, since the sum of the three angles of a triangle must be 180°. If two angles of a triangle are each 90°, then their sum alone is 180°, and when we add the third angle, the total is more than 180°.)

Proof that there exists a triangle with two right angles
Take two circles that meet in points *A* and *B* as in Figure 9.53. Let $\overline{AC}$ and $\overline{AD}$ be their diameters drawn from *A*. (A *diameter* is a line segment through the center of the circle that bisects the circle.) Draw

David Hilbert (1862–1943).

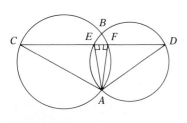

Figure 9.53

a)

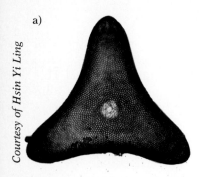

Courtesy of Hsin Yi Ling

b)

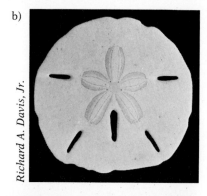

Richard A. Davis, Jr.

c)

William H. Amos

d)

William H. Amos

e)

Lee H. Somers

Geometric shapes in nature:
(a) Photomicrograph of a
diatom; (b) sand dollar;
(c) sea urchin; (d) jellyfish;
(e) starfish.

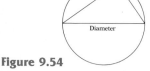

Figure 9.54

line segment $\overline{CD}$ meeting the circles at points E and F, as given. It can be shown in Euclidean geometry that $\angle AFC$ is a right angle. (Any angle inscribed in a semicircle is a right angle.)[3] Similarly, $\angle AED$ is a right angle (because it also is inscribed in a semicircle).

We now have triangle AEF with two right angles. What is wrong with this proof?

---

3. "An angle inscribed in a semicircle" is an angle such as $\angle A$ shown in Figure 9.54. Its vertex is on the circle, and its rays (sides) pass through the endpoints of a diameter.

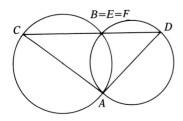

Figure 9.55

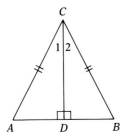

Figure 9.56

If you draw your own diagram very carefully, you will discover that $\overline{CD}$ passes through point B, as shown in Figure 9.55. Thus we see that points E and F are exactly the same as point B. So triangle AEF does not even exist. The carelessly drawn diagram of Figure 9.53 was misleading.

You may think that the above proof was rigged and that Euclid himself would never have made such an error. However, many of the proofs in the *Elements* and in high school geometry texts (which are based on Euclid) contain similar faults. The following proof is taken from a geometry text.

Theorem *The base angles of an isosceles triangle (Figure 9.56) are congruent.*

Given: $\triangle ABC$ with $\overline{AC} \cong \overline{BC}$.

To prove: $\angle A \cong \angle B$.

Proof

Statements	Reason
1. Draw the bisector of $\angle C$ (this is the line that divides $\angle C$ into two congruent angles).	1. Every angle has a bisector.
2. Extend the bisector of $\angle C$ to meet line segment $\overline{AB}$ at point D.	2. A line segment may be extended.
3. In $\triangle ACD$ and $\triangle BCD$, $AC \cong BC$.	3. Given.
4. $\angle 1 \cong \angle 2$.	4. An angle bisector divides the angle into two congruent angles.
5. $\overline{CD} \cong \overline{CD}$.	5. Anything is congruent to itself.
6. $\triangle ACD$ is congruent to $\triangle BCD$.	6. Two triangles are congruent when two sides and the included angle of one are equal, respectively, to two sides and the included angle of the other.
7. $\angle A \cong \angle B$.	7. If two triangles are congruent, then the corresponding parts are congruent.

This proof contains an error similar to the error in the previous proof. What is wrong? The problem is in the second step. Our reason for this step is that a line may be extended. This is Postulate 2 (p. 494). However, this postulate does not tell us that when this line is actually extended, it will meet line segment *AB*. The diagram certainly suggests that it does. But there is nothing that logically forces us to conclude that the bisector will meet $\overline{AB}$ at point *D* as shown. In Figure 9.57 the bisector does *not* meet line segment $\overline{AB}$.

You may say that straight lines simply do not behave like line $\overline{CD}$ of Figure 9.57. But what do you mean by a straight line? If you mean a straight pencil mark on paper or a straight chalk mark on the blackboard, then you are right. However, we are not discussing chalk or pencil marks. We are discussing lines, which have only the properties assumed in the postulates and no others. It does not follow *logically* from the postulates that a line cannot behave as $\overline{CD}$ does in Figure 9.57. Mathematicians are concerned with what follows *logically* from the postulates and not with what *appears* to be true in a picture.

Some of Euclid's other proofs contain similar faults. However, by adding suitable postulates to Euclid's original ten, modern mathematicians have been able to corrrect the faults. Let us not underestimate the work of Euclid because of these gaps. It took 2000 years for critics to discover and correct them.

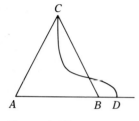

Figure 9.57

9.9

NON-EUCLIDEAN GEOMETRY

Euclid may have considered some of his postulates as obviously true. There is evidence, however, that Euclid was not entirely convinced of the truth of at least one postulate, namely, the fifth. This is called the parallel postulate[4] and it states:

> **Euclid's parallel postulate**   *Given a line and a point not on the line, then only one line can be drawn parallel to the first line passing through the given point.*

Euclid did not use the parallel postulate in proving theorems until he could not continue any further without it, and it is this apparent hesitation to use Postulate 5 that suggests that he was not completely satisified with it. Why not? Well, let us look at the postulate closely (see Figure 9.58).

The postulate states that it is possible to construct a line parallel to a given line. This means that the two lines will *never* meet, no

---

4. Euclid stated this postulate differently. The version given here is that of the English mathematician Playfair.

Point *P*

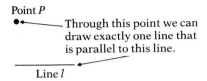

Through this point we can draw exactly one line that is parallel to this line.

Line *l*

Figure 9.58

matter how far they are extended. It is not humanly possible to go on extending lines forever. So how can we really know that these two lines will *never* meet at some point? Perhaps they would meet billions and billions of miles away from the starting point.

Other mathematicians were also doubtful about the truth of the parallel postulate; for 2000 years after the *Elements* first appeared, several attempts were made to *prove* this postulate rather than accept it as an unproved statement. However, these attempts all involved other unproved assumptions that were actually the same as the parallel postulate but were stated differently.

In the seventeenth century an Italian monk named Girolamo Saccheri (1667–1733) approached the problem in a new way. He used the method called *proof by contradiction*. Saccheri wanted to prove that Euclid's parallel postulate was true. Obviously, there are only two possibilities.

Possibility 1. Euclid's parallel postulate is false.

Possibility 2. Euclid's parallel postulate is true.

He assumed that Euclid's parallel postulate was false (Possibility 1) and tried to arrive at a contradiction. Believing that he had actually reached the contradiction, he concluded that Possibility 2 is correct, that is, that the parallel postulate is true. He published his results in *Euclides Vindicatus*, which means "Euclid vindicated," or "Euclid proved true." Apparently, Saccheri had great faith in Euclid and was very anxious to show that the parallel postulate was true. In fact, Saccheri made an error and did not really obtain a contradiction at all. Thus he did not prove the parallel postulate as he thought he had.

Saccheri's work is of great interest because he actually proved many theorems in the non-Euclidean geometries developed in the next century by Bolyai and Lobachevsky. So, although he failed to prove Euclid's geometry true, he paved the way for a new and important approach to geometry.

Saccheri's approach, as we have said, was to assume that the parallel postulate was false. If it is indeed false, then one of the following two situations is possible.

Possibility 1. Given a line *l* and a point *P* not on *l*, *at least two lines* can be drawn parallel to it passing through the given point (see Figure 9.59).

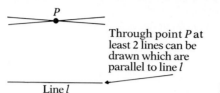

Through point *P* at
least 2 lines can be
drawn which are
parallel to line *l*

Line *l*

Figure 9.59

Possibility 2. Given a line *l* and a point *P* not on *l*, *no* lines can be drawn parallel to *l* that pass through the given point.

J. Bolyai (1802–1860), a Hungarian army officer, and N. I. Lobachevsky (1793–1856), a Russian mathematician, each independently developed a geometry based on Possibility 1. They used all the postulates of Euclid except the fifth, and in its place they substituted Possibility 1. This is known as the *Lobachevskian parallel postulate.* Then from this new set of postulates they proceeded to prove theorems. The geometry that they developed in this way is called *Lobachevskian geometry*, and it is different from Euclid's geometry in many startling ways. Some of the theorems of this geometry are as follows.

1. Given a line *l* and a point *P* not on *l*, then through point *P* many lines can be drawn parallel to it (compare this with the Euclidean parallel postulate).

2. The sum of the measures of the angles of any triangle is *less than* 180°.

3. Different triangles have different angle sums.

4. The sum of the measures of the angles of any quadrilateral is *less than* 360° (compare this with Exercise 22 in Section 9.4).

5. There are no rectangles. (This follows from the last theorem. Why?)

6. If two triangles are similar, then they must also be congruent. (This means that if two figures have the same shape, then they must also have the same size. Thus in a Lobachevskian world, miniature or enlarged copies of objects would be impossible to produce without distortion. To be accurate, all photographs would have to be life-size!)

7. Parallel lines are not spaced an equal distance apart.

It is interesting to note that the great German mathematician K. F. Gauss (1777–1855) also developed a geometry that is based on the Lobachevskian parallel postulate, but he did not want to publish it at the time because the ideas of Euclid were so widely accepted.

In 1854, B. Riemann (1826–1866), a German mathematician, introduced a different non-Euclidean geometry. He replaced Euclid's parallel postulate by Possibility 2, which is now called the *Riemann*

B. Riemann (1826–1866).

parallel postulate. When the parallel postulate is replaced by Possibility 2, it is also necessary to give up some of Euclid's other assumptions. There is actually a choice as to which assumptions can be abandoned. You can give up Postulate 1 (p. 494), or you can give up the principle, discussed in Section 9.2, that a line separates a plane into two half-planes.

The theorems of Riemann's geometry are also surprising and interesting. Some of them are as follows.

1. Parallel lines do not exist.

2. The sum of the measures of the angles of any triangle is greater than 180°.

3. There are no rectangles.

4. If two triangles are similar, then they must also be congruent. (This means that if the two figures have the same shape, then they also have the same size.)

5. A line is not separated by a point into two half-lines.

6. Two different lines intersect in *two* points. (This theorem is true only if you make the choice to abandon Postulate 1.)

Geometries make statements about physical objects—figures, shapes, areas, distances, etc. Thus they can be used to explain the physical world in which we live. The theorems of Riemannian and Lobachevskian geometry seem very strange to us. On first seeing these theorems it is natural to think that they cannot possibly be true in the real world. In fact, some of them actually seem to contradict our own experiences. (For example, haven't we all seen rectangles with our own eyes? Yet these don't exist in either Riemannian or Lobachevskian geometry.) It was partly this feeling that convinced Saccheri that he had proved Euclid's geometry to be true.

However, it has been shown that if Euclid's geometry is *logically* correct, then so are Riemann's and Lobachevsky's. Thus none of these three has any "logical superiority." Nevertheless, for 2000 years, Euclid's geometry had been accepted as an absolutely accurate description of the physical world. So it was difficult for mathematicians and scientists to accept the possibility that it might not be correct and that one of the non-Euclidean geometries might describe the world more accurately. Thus for a long time, most mathematicians believed that the non-Euclidean geometries were interesting logical works but could not have any application to the real world.

In this connection it is interesting to note that the great German mathematician Karl Friedrich Gauss (1777–1855) was actually the first to realize that Euclid's parallel postulate was not necessarily true. He created a non-Euclidean geometry. However, he did not publish his results, partly because he was afraid of being ridiculed.

Gauss tried to test the "truth" of his geometry in the following way. In Euclidean geometry the sum of the angles of a triangle is *exactly* 180°. In non-Euclidean geometry it is either less than or greater than 180°. (In Gauss's version it was less than 180°.) So he tried to measure the angle sums of triangles to see whether they would turn out to be exactly 180° or less than 180°. Now if you draw a triangle on a piece of paper, measure the angles, and add them up, you will find that the angle sum seems to be 180° (if you do it carefully). But measurements are only approximate, no matter how carefully they are made. Even worse, in Gauss's geometry the smaller the triangle is, the closer the angle sum is to 180°. For example, in Gauss's geometry a small triangle such as the one in Figure 9.60 might have angle sum equal to 179.99999999999999°. This is so close to 180° that it would be impossible to measure any difference between this triangle and one that was exactly 180°. Thus Gauss needed very large triangles. He got them by putting three people on three different mountains. Each one measured the angle between the lines of sight from himself to the other two observers, as shown in Figure 9.61.

Figure 9.60

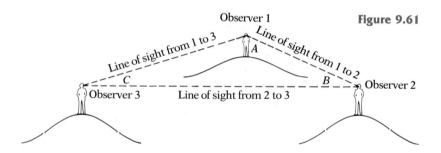

Figure 9.61

In this figure, observer 1 measured angle *A*, observer 2 measured angle *B*, and observer 3 measured angle *C*. The sum of these three angles turned out to be 179°59′ 58″. Did this mean that Gauss's geometry was correct because the result was less than 180°? No, the difference between his result and 180° was too small to be conclusive and might have been due to a measuring error. What was needed was an enormously large triangle, such as is found in astronomy.

In fact, the discoveries of twentieth century physics gave this kind of support to non-Euclidean geometry. In his work on relativity, Einstein used non-Euclidean geometry and obtained far better results than with Euclidean geometry. That is, predictions made on the basis of Einstein's theory agreed more closely with observed facts if non-Euclidean geometry was used. Then even the strongest supporters of Euclid had to acknowledge the applicability of non-Euclidean geometry to the real world.

Now the greater usefulness of non-Euclidean geometry in relativity theory is certainly a strong argument for its truth. But it does not *prove* that it *is* true.

Perhaps you are wondering why Euclidean geometry is still taught in all schools and used in engineering and practical applications if it is probably not true. The reason is simple—it works! Why does it work? As we have said with regard to Gauss's triangle experiment, in small areas there is no significant measurable difference between the results of Euclidean and non-Euclidean geometries. The earth is a very small region when compared to the whole universe. So on earth, Euclidean geometry is as "true" as any of the others for practical use. Since it is familiar and the easiest to use, we do so with perfectly good results.

The discovery and acceptance of non-Euclidean geometry has had an impact even outside mathematics. Most of science uses Euclidean geometry, and the doubt about its truth necessarily led to doubt about the truth of the scientific conclusions based on it. In fact, scientists began to doubt whether it was possible to ever find absolute scientific truths. The modern view is that such truths are not possible and that scientific "laws" are just approximate descriptions of the way we see the physical world. When the "absolutely true" geometry of Euclid turned out to be not as true as people had thought, scholars in other areas began to question their "truths." In fact, philosophers began to ask whether we can ever discover truths in general. This has led to a reexamination of what we can "know" in all areas of human knowledge—history, economics, law, ethics, etc. The debate about what we can know is still in progress. It will probably continue for many years to come.

9.10
THE KÖNIGSBERG BRIDGE PROBLEM

Our aim in this section is to "solve" the Königsberg Bridge problem mentioned at the beginning of this chapter. We will examine some of the basic facts needed for the solution (without proving them) and then show how these facts can be used to solve the problem.

The ideas needed have to do with **network theory.** A network begins with some points called **vertices.** Some of these may be connected with lines or curves that do not cross each other. (Arcs that meet at the same vertex are not considered to cross each other.) These connecting lines are called **simple arcs.** A **network** consists of these vertices and the connecting areas. Some examples of networks are shown in Figures 9.62–9.68.

In a network a vertex is **even** if it has an even number of arcs going to or from it. A vertex is **odd** if it has an odd number of arcs going to or from it. *For the purpose of determining whether a vertex is*

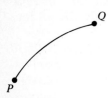

Figure 9.62
A network with two vertices, P
and Q, and one arc

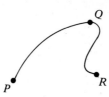

Figure 9.63
A network with three vertices,
P, Q, and R, and two arcs

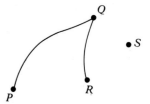

Figure 9.64
A network with four vertices,
P, Q, R and S, and two arcs

Figure 9.65
A network with two vertices, P
and Q, and two arcs

Figure 9.66
A network with one vertex, P.
There is one arc that connects
with point P twice. This arc is
called a **loop.**

Figure 9.67
This is *not* a network, since the
arc is not simple. It crosses
itself.

even or odd, a loop is counted twice. In Figures 9.62–9.68, some of the
vertices are even and some are odd, as shown in Table 9.8.

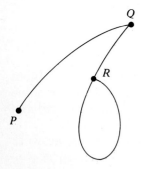

Figure 9.68
A network with three vertices,
P, Q and R, and three arcs.

TABLE 9.8

Figure number	Even vertices		Odd vertices
9.62	none		P, Q
9.63	Q		P, R
9.64	Q, S	(0 is an even number)	P, R
9.65	P, Q		none
9.66	P	(since the loop is counted twice)	none
9.68	Q		P, R

Now we are ready to solve the Königsberg Bridge problem. Look
at the Königsberg bridges, which we have redrawn in Figure 9.69.
This picture can be thought of as a network by taking the locations

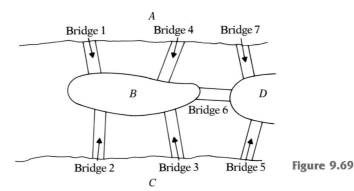

Figure 9.69

A, B, C, and D as the vertices and by taking the bridges as arcs. This is shown in Figure 9.70.

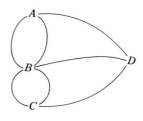

Figure 9.70

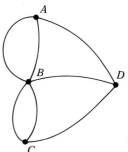

Figure 9.71

The picture can be further simplified as shown in Figure 9.71.

We can now restate the Königsberg Bridge problem as follows: *Can we trace the complete network exactly once without going over any one point twice and without lifting the pencil from the paper?* Before reading further, see if you can do this.

Euler proved that this could be done only if the number of odd vertices is exactly 0 or 2. Look at Figure 9.71 again. We see that A, B, C, and D are *all* odd vertices. Thus there is a total of 4 odd vertices in this network. Since the total number of odd vertices is *not* 0 or 2, the bridges can never be crossed as required. This does not mean that it is difficult to cross the bridges as specified, but that it is *impossible*.

EXERCISES

1. Draw a network that has

   a) 3 vertices and 1 arc.

   b) 1 vertex and 4 arcs.

   c) 5 vertices and 2 arcs.

   d) 4 vertices and 4 arcs.

   e) 3 vertices and 3 arcs.

   f) 3 vertices and 5 arcs.

2. For each of the networks drawn in Exercise 1, state whether the vertices are even or odd.

Which of the networks given in Exercises 3–12 can

be drawn without lifting the pencil from the paper and without drawing any line more than once? If possible, do it.

3.

4.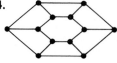

5.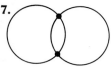

6.

7.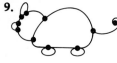

8.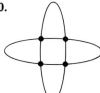

9.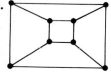

10.

11.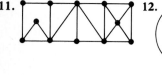

12.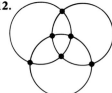

13. After Euler solved the Königsberg Bridge problem, an eighth bridge was built, as shown in Figure 9.72. With this extra bridge added, can a person cross all eight bridges without crossing any one bridge twice?

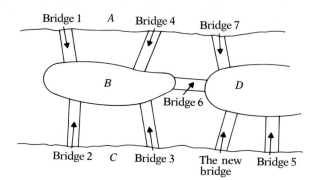

Figure 9.72

14. Figure 9.73 shows a map of part of the New York City area.

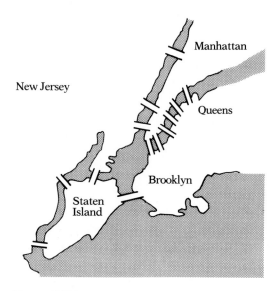

Figure 9.73

a) Redraw the map as a network.

b) Is it possible to cross every bridge and tunnel once without crossing any one more than once?

15. The following is a floor plan of one of the authors' houses.

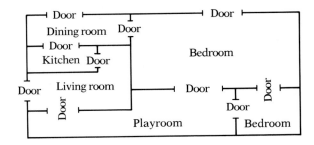

A thief has just broken into the house and wants to ransack every room as quickly as possible. Can the thief go from one room to another without ever going through any door twice?

16. Having found nothing in the first author's house, the thief has now broken into the second author's house. The floor plan for this house is shown below.

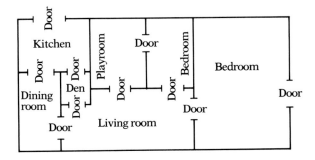

For this author's house, can the thief go from one room to another without going through any one door twice?

17. In the nineteenth century the mathematician William Rowan Hamilton invented the following puzzle. Consider the picture shown in Figure 9.74. Can we find a path along the edges of

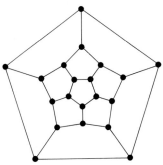

Figure 9.74
The Hamiltonian puzzle

the picture that passes through each vertex once and yet returns to its starting point? (*Hint*: Not every arc has to be crossed.)

18. A sanitation crew must collect garbage from every street in the territory shown in Figure 9.75.

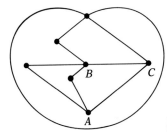

Figure 9.75

a) How can this be done so that no street is crossed twice?

b) A new road is being built between points A and B. When this road is built, will the sanitation crew still be able to collect the garbage from every street without covering any one street twice?

9.11

THE CONVEYOR BELT PROBLEM

Another interesting problem that can be solved by using network theory is the **conveyor belt problem.** Suppose that in a factory there are three different inspection stations and three different loading platforms as shown in Figure 9.76.

The manager wishes to connect each of the inspection stations to all three loading platforms by means of conveyor belts. For safety

| Inspection Station 1 | Inspection Station 2 | Inspection Station 3 |

| Loading Platform 1 | Loading Platform 2 | Loading Platform 3 |

Figure 9.76

reasons, none of the belts may cross each other. Can this be done? If so, how? We suggest that the reader try to draw a diagram connecting the stations to the platforms under the given conditions before reading further.

This problem can be solved by using the idea of a **connected network.** *A network is said to be connected if every two of its vertices are connected by one or more arcs in succession.* In Figure 9.77 the networks shown in parts (a), (b), (d), (f), (g), and (h) are connected. The network shown in part (c) is not connected because vertex *S* is not connected to any other arc. To put it another way, a network is connected if you can start at any vertex and get to any other vertex by moving a pencil along arcs without lifting the pencil. Similarly, the network shown in part (e) is not connected.

Figure 9.77

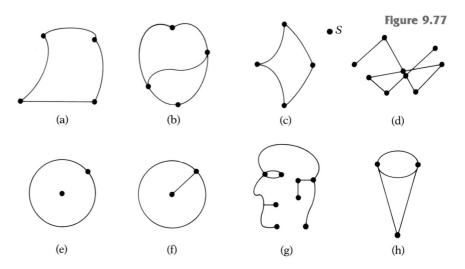

(a)   (b)   (c)   (d)

(e)   (f)   (g)   (h)

In discussing connecting networks the following symbols are used.

V = the number of vertices,

E = the number of arcs,   and

F = the number of regions into which the network divides the page.

Since this notation is important for the solution of the conveyor belt problem, we will illustrate it with some examples.

EXAMPLE 1   In Figure 9.78 there are three vertices, A, B, and C. So $V = 3$. There are three arcs, so $E = 3$. The network divides the page into two parts, the inside (labeled I) and all of the outside (labeled II), so $F = 2$.

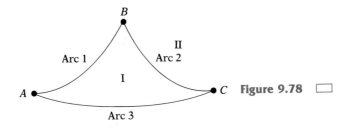

Figure 9.78 ▭

EXAMPLE 2   In Figure 9.79 there are three vertices, so $V = 3$. There are four arcs, so $E = 4$. And we see there are three regions labeled I, II, and III, so $F = 3$.

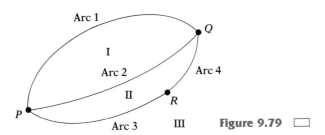

Figure 9.79 ▭

Euler proved that for any connected network the following must always be true.

Euler's Formula $V - E + F = 2$

Comment   In using Euler's formula a loop is considered to be *one* arc.

EXAMPLE 3   Verify Euler's formula for Example 1 above.

SOLUTION   In Example 1 we saw that $V = 3$, $E = 3$, and $F = 2$. Thus,

$$V - E + F = 3 - 3 + 2 = 2.$$

Therefore $V - E + F = 2$, which is exactly what Euler's formula states. ▭

EXAMPLE 4   Verify Euler's formula for Example 2 above.

SOLUTION   In Example 2 we saw that $V = 3$, $E = 4$, and $F = 3$. Thus

$$V - E + F = 3 - 4 + 3 = 2,$$

so that again we have $V - E + F = 2$, which is what Euler's formula states. □

EXAMPLE 5   Verify Euler's formula for the network shown in Figure 9.80.

Figure 9.80

SOLUTION   For this network, $V = 3$ and $E = 2$. How many regions are there for this network? From the diagram we see that the page is not divided into two or more parts by any of the arcs. Thus $F = 1$. Therefore

$$V - E + F = 3 - 2 + 1 = 2.$$

Again, Euler's formula is true. □

Let us now return to the conveyor belt problem described at the beginning of this section. We will solve it by using a proof by contradiction (see page 228). Suppose that the problem can be solved. That is, suppose that it *is* possible to connect the inspection stations with the loading platforms by belts that do not cross. In Figure 9.81 we have drawn the three stations and the three platforms connected by belts. Pretend that the belts do not cross, even though in the diagram they appear to do so.

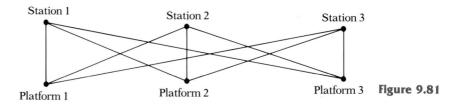

Figure 9.81

Let the inspection stations and loading platforms be vertices. There are six altogether. We can think of the belts as arcs. There are a total of nine such arcs. Since the problem specifies that the belts are not permitted to cross each other, these are actually simple arcs. This makes the diagram a network. Furthermore, it is a connected network. Therefore we can use Euler's formula. According to the formula, $V - E + F = 2$. Since $V = 6$ and $E = 9$, we have $6 - 9 + F = 2$. By trial and error, or by algebra, we find that $F = 5$. Remember that F represents the number of regions into which the network divides the page. Now look carefully at Figure 9.81. There are no regions enclosed by exactly three arcs, like the ones shown in Figures 9.82 and 9.83. Thus each separate region in the network must be enclosed by at least four arcs. Each individual arc borders on two regions, as shown in Figure 9.84.

Figure 9.82

Figure 9.83

Figure 9.84
Arc A borders on both region I and region II

Now F (which is the number of regions) is five, and each of these has at least four arcs on its border. This gives us 5×4 or 20 arcs. However, since each arc is shared by two regions, we can divide the 20 in half, getting at least 10 arcs.

Thus the network must have at least 10 arcs. However, we previously saw that E (the number of arcs) is exactly 9 for this network. This is a contradiction, since E cannot be exactly 9 and, at the same time, at least 10. Thus our assumption that it is possible to connect the stations to the platforms as specified leads to a contradiction. By the principle of the proof by contradiction our assumption is wrong. It follows that it is not possible to connect the stations with the platforms as specified.

EXERCISES

For each of the networks given in Exercises 1–9, determine whether the network is connected. If it is, find the values of V, E, and F, and verify Euler's formula.

3.

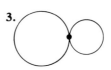

1.

2.

4.

5.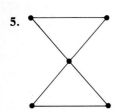

6.

9.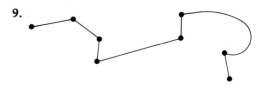

7.

8.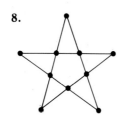

10. Suppose that one of the inspection stations in the conveyor belt problem has to be connected to only two of the loading platforms. All the other conditions remain the same. It is now possible to solve the problem. Draw a diagram showing how this can be done.

9.12

THE MÖBIUS STRIP

Figure 9.85

Figure 9.86

Figure 9.87
"Halving" a Möbius strip

Consider the page you are now reading. It has two sides, one called page 512 and the other called page 511. If you wish, you can paint one side blue and the other side red. The edge of the page separates the two sides.

Most surfaces have two sides. For example, take off one of your socks or stockings. There is an inside and an outside. If some paint fell on your sock while you were painting this page, you could turn it inside out, and the paint would not show because it would now be on the inside. Manufacturers of belts sometimes make use of this fact by making reversible belts. One side is one color, and the other side is another color.

The German mathematician Augustus Ferdinand Möbius (1790–1868) was able to create a surface that had only one side! Actually, it is not difficult to make such a surface. You can make one by taking off your belt, giving it a half-twist and then fastening it, as shown in Figure 9.85.

You now have a surface that is called a **Möbius strip.** If you do not have a belt, you can make a Möbius strip by taking a strip of paper, giving one end a half-twist, and then gluing the ends together, as shown in Figure 9.86.

Now take a Möbius strip made out of paper and start painting or coloring one side. What happens should convince you that this surface has only one side! Now take another Möbius strip and cut it along the middle of the strip as shown in Figure 9.87. What happens?

Next take another Möbius strip and cut it, this time one-third of the way in from an edge as shown in Figure 9.88. What happens this time?

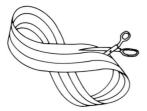

Figure 9.88
A Möbius strip cut in thirds

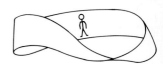

Figure 9.89

On a one-sided surface, such as a Möbius strip, many unusual things can happen. For example, imagine that we have a two-dimensional man who lives on this page, as shown above in Figure 9.89. This man lives entirely within this page and cannot come out. Suppose that a friend comes to visit him and walks up behind him. Our man wants to turn around to greet him. However, to do this, he must come out of the page, which he cannot do. If he tries to turn around to face his friend, he will end up standing on his head as shown in Figure 9.90.

Figure 9.90

However, if our man lived in a Möbius strip, while he would still be a two-dimensional man, he would be able to face his friend without standing on his head. All he has to do is simply walk around the strip once. To see how he does this, see Figure 9.91.

While Möbius strips are very entertaining, they are more than that. In recent years the Goodrich Rubber Company applied for a patent on a conveyer belt that was constructed as a Möbius strip. (See Figure 9.92.) What would be the advantage of using such a conveyor belt?

Figure 9.91

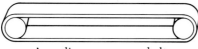

An ordinary conveyor belt   A conveyor belt using a Möbius strip

Figure 9.92

EXERCISES

1. Take a strip of paper and give it three half twists. Now glue the edges together. This is another type of Möbius strip.

   a) Cut it down the middle and see what happens.

   b) Cut in on a line one-third of the way from one edge. What happens?

2. Cut an ordinary Möbius strip one-fourth of the way from the edge. What happens?

3. Another interesting one-sided surface is the **Klein bottle**, which is shown below. The German mathematician Felix Klein (1849–1925) was the first to discover such a bottle. Consult the references in a library to find out all you can about Klein bottles.

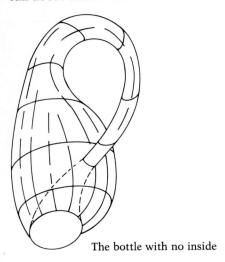

The bottle with no inside

*4. Refer back to the four-color map problem discussed in Chapter 2 (page 57). On a Möbius strip it is not true that every map can be colored with at most four colors. As many as six colors may be needed. Make a Möbius strip and

   a) draw a map that requires only 4 colors.

   b) draw a map that requires at least 5 colors.

   c) draw a map that requires at least 6 colors.

9.13

SUMMARY   In this chapter we studied some of the basic ideas of Euclidean geometry. Starting with the undefined terms *point*, *line*, and *plane*, we went on to study angles and their measurement. We discussed both degree and radian measure and indicated how to convert from one system of measurement to the other. We also analyzed curves and polygons and distinguished between similar and congruent polygons and the properties of each.

We then studied the metric system and the advantages of using it. We specified how to convert from our familiar way of measuring things to this new (to us, at least) system. We reviewed how to calculate the perimeter and the areas of various geometric figures. We also pointed out the difference between Euclidean and non-Euclidean geometries.

Finally, we discussed some ideas from network theory and how these ideas can be applied to solve the Königsberg Bridge problem and the conveyor belt problem as well as to analyze the Möbius strip.

STUDY GUIDE   You should now be able to demonstrate your knowledge of the following ideas presented in this chapter by giving definitions, descriptions, or specific examples. Page references (in parentheses) are given for each term so that you can check your answer.

Geometry (p. 456)	Undefined term (p. 456)
Line (p. 456)	Point (p. 456)

Collinear (p. 457)
Half-line (p. 457)
Endpoint (p. 457)
Ray (p. 457)
Line segment (p. 458)
Intersect (p. 458)
Plane (p. 458)
Coplanar (p. 458)
Half-plane (p. 459)
Parallel lines (p. 459)
Angle (p. 460)
Side of angle (p. 460)
Vertex (p. 460)
Region (p. 460)
Interior of an angle (p. 460)
Exterior of an angle (p. 460)
Protractor (p. 461)
Degrees (p. 461)
Minutes (p. 461)
Right angle (p. 462)
Perpendicular (p. 462)
Straight angle (p. 462)
Acute angle (p. 462)
Obtuse angle (p. 462)
Initial side (p. 462)
Terminal side (p. 462)
Rotational angles (p. 464)
Adjacent angles (p. 464)
Vertical angles (p. 464)
Radian measure (p. 464)
Complementary angles (p. 466)
Supplementary angles (p. 467)
Transversal (p. 468)
Interior angles (p. 468)
Exterior angles (p. 468)
Alternate interior angles (p. 468)
Interior angles on same side of
   transversal (p. 468)
Corresponding angles (p. 468)
Polygon (p. 471)
Simple close curve (p. 471)

Triangle (p. 471)
Equilateral triangle (p. 473)
Isosceles triangle (p. 473)
Right triangle (p. 473)
Quadrilateral (p. 475)
Congruent triangles (p. 476)
Similar triangles (p. 477)
Corresponding sides (p. 477)
Proportional sides (p. 477)
Metric system (p. 483)
Foot-pound or British system
   (p. 483)
Meter (p. 483)
Gram (p. 486)
Liter (p. 487)
Degrees Celsius (p. 489)
Perimeter (p. 492)
Circumference (p. 492)
Trapezoid (p. 493)
Euclid's *Elements* (p. 493)
Postulates or axioms (p. 494)
Theorems (p. 495)
Euclid's parallel postulate (p. 498)
Lobachevskian geometry (p. 500)
Riemannian geometry (p. 501)
Königsberg Bridge problem (p. 503)
Network theory (p. 503)
Vertices (p. 503)
Simple arcs (p. 503)
Even vertices (p. 503)
Odd vertices (p. 503)
Loop (p. 504)
Solution to Königsberg Bridge
   problem (p. 505)
Conveyor belt problem (p. 507)
Connected network (p. 508)
Euler's formula (p. 509)
Solution to conveyor belt problem
   (p. 510)
Möbius strip (p. 512)
Klein bottle (p. 513)

**FORMULAS
TO REMEMBER**

The following list summarizes the formulas given in this chapter.

1. To convert from an angle measured in degrees to radian measure, multiply by $\frac{\pi}{180}$.

2. To convert from an angle measured in radians to degree measure, multiply by $\frac{180}{\pi}$.

3. To convert from the British system to the metric system of measurement, use the charts shown on page 488.

4. The area of a triangle is $\frac{1}{2}bh$.

5. The area of a rectangle is bh.

6. The area of a circle is πr^2.

7. The circumference of a circle is $2\pi r$.

8. The sum of the measures of the angles of a triangle is $180°$.

9. The Pythagorean theorem: $a^2 + b^2 = c^2$.

10. Euler's formula for networks: $V - E + F = 2$.

MASTERY TESTS

Form A

1. If the length of each side of a square is represented by $2x$, which expression represents its area?

   a) $8x$ **b)** $4x$ **c)** $4x^2$ **d)** $2x^2$ **e)** none of these

2. If the length of the hypotenuse of a right triangle is 13 and the length of one leg is 12, find the length of the other leg.

3. The lengths of the sides of triangle *ABC* are 5, 6, and 7. Triangle *RST* is similar to △*ABC*. The longest side of △*RST* is 21. Find the length of the *shortest* side of △*RST*.

4. Two angles of a triangle are equal in measure, and the third angle is 150°. Find the number of degrees in one of the two equal angles.

5. If the perimeter of a polygon with five sides of equal measure is represented by $(15x - 20)$, express the length of one side of the polygon in terms of x.

6. The area of square *ABCD* is 36 square centimeters. What is the number of centimeters in the perimeter of the square?

7. If the circumference of a circle is doubled, the diameter of the circle

  a) increases by 2   **b)** is doubled   **c)** is multiplied by 4

  d) remains the same   **e)** none of these

8. In the accompanying figure, triangle *ABC* is similar to triangle *DEF*. If $m(\overline{AC}) = 6$, $m(\overline{AB}) = 7$, $m(\overline{DE}) = 28$, $\angle B \cong \angle E$, and $\angle C \cong \angle F$, find $m(\overline{DF})$.

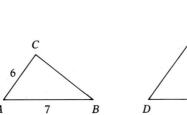 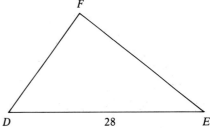

9. The length of a rectangle is represented by $(x + 3)$, and the width of the rectangle is represented by $(x - 8)$. Express the area of the rectangle as a trinomial in terms of x.

10. An angle of 75° measures how many radians?

11. Convert $\frac{7\pi}{12}$ radians to degree measure.

12. The circumference of a circle is 12π. What is the radius of the circle?

13. If p represents the perimeter of a square, represent the length of a side of the square in terms of p.

14. A tree 10 meters in height casts a shadow 25 meters long. At the same time a person casts a shadow 5 meters long. What is the number of meters in the height of the person?

15. The measures of the angles of a triangle are represented by x, $2x$, and $x + 20$. Find the number of degrees in the measure of the *smallest* angle of the triangle.

Form B **1.** The lengths of two legs of a right triangle are 3 and 5. Find, in radical form, the length of the hypotenuse.

For questions 2–5 refer to trapezoid $ABCD$, shown below, where $\overleftrightarrow{AB} \parallel \overleftrightarrow{DC}$, $\overleftrightarrow{AD} \perp \overleftrightarrow{AB}$ and $\overleftrightarrow{DB} \perp \overleftrightarrow{BC}$, $m(\overline{AB}) = 4$, $m(\overline{AD}) = 4$, and $m(\overline{DC}) = 8$.

2. Find $m(\overline{DB})$ in radical form.

3. Find the area of $\triangle ABD$.

4. Find the area of trapezoid $ABCD$.

5. Find the area of $\triangle DBC$.

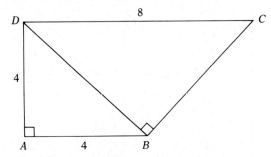

6. Consider a 2-liter bottle of soda. Does it contain more or less than a 64-ounce bottle of the same soda?

7. Draw a network that has five vertices and three arcs.

8. Determine whether the following network is connected.

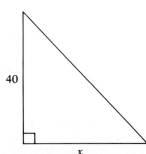

9. Convert 195 kg to lb.

10. The posted speed limit in a certain village is 30 mph. What is the speed limit in the metric system?

11. Convert 75°F to degrees Celsius.

12. A certain liquor sells for $7.95 per quart. Find the cost of one liter.

13. A chemical tank contains 7 kiloliters of a chemical. Express its contents in liters.

14. A rectangular room measures 13 meters by 84 centimeters. Find its area.

15. Consider the following triangles, which are similar. Find x.

10

PROBABILITY

CHAPTER OBJECTIVES

To discuss a counting principle and tree diagrams that can be used to find the number of possible outcomes of an experiment. (*Section 10.2*)

To distinguish between permutations and combinations which are arrangements of objects in which order does and does not count, respectively. (*Section 10.3*)

To study formulas for calculating the number of possible permutations, $_nP_r$, and the number of possible combinations, $_nC_r$, by using factorial notation (a shorthand notation for multiplication) or by using Pascal's triangle. (*Section 10.3*)

To define what is meant by probability (*Section 10.4*)

To indicate that mutually exclusive events are events that cannot occur at the same time. (*Section 10.5*)

To analyze independent events in which the occurrence of one event does not affect the occurrence of a second event. (*Section 10.6*)

To use probability to determine the odds in favor of or the odds against some event. (*Section 10.7*)

To apply probability and to determine the amount of money to be won or lost in the long run (mathematical expectation). (*Section 10.7*)

To look at a procedure for obtaining random numbers. (*Section 10.9*)

519

Bank Failure Sets Record

ROSWELL, N.M.: Moncor Bank of Roswell today became the 80th American bank to fail this year, eclipsing the record 79 bank failures in 1984, a Federal Deposit Insurance Corporation official said.

The bank, closed after being declared insolvent by the Deputy United States Comptroller of the Currency, Michael Patriarca, was the second Moncor bank in southeastern New Mexico to fail in the last two weeks. Moncor Bank of Hobbs was closed Aug. 30. The Roswell bank was the last of six banks held by Moncor Inc., which filed for reorganization under Chapter 11.

Mr. Patriarca said the F.D.I.C. was appointed receiver of the Roswell bank.

BUSINESS NEWS, September 12, 1985

Consider the lottery ticket. Does a person really have a "reasonable" chance (or probability) of winning when buying such a ticket? If a person buys many such tickets or plays the lottery on a regular basis, will this affect the person's expected amount of money to be won?

Now consider the clipping. Despite beliefs to the contrary, the probability of a bank failure in the United States is not zero. As the article indicates, the number of bank failures has been increasing over the past few years.

10.1
INTRODUCTION

Many people start the day by listening to the weather forecast. The announcer may say: "There is a 90% *probability* of rain today." What is meant by this statement? Either it will rain or it won't rain.

We also frequently hear expressions such as "I'll *probably* get an A in this course," or "I'll *probably* call her for a date," or "In all *probability*, you are right."

Historically, probability theory had its beginnings in the gambling halls and was used by gamblers to succeed in cheating. However, as the subject of statistics developed, it was discovered that a knowledge of probability is also essential to the statistician. Today, the use of probability in gambling is just one of its minor applications. The importance of probability lies in its wide range of application to such nonmathematical fields as medicine, psychology, economics, and business, to name a few.

In this chapter we will be discussing the meaning of probability and how it is used.

The mathematical study of probability can be traced back to the mathematician Jerome Cardan (1501–1576). The illegitimate child of a distinguished lawyer, Cardan became a famous doctor, who treated many prominent people throughout Europe. On various occasions he was also a professor of medicine at several Italian universities. While practicing as a doctor, he also studied, taught, and wrote mathematics.

Although he was extremely talented, Cardan's personality and personal life appear to have been less than perfect. He was very hot-tempered. In fact, he is said to have cut off one of his son's ears in a fit of rage. (His sons seem to have followed their father's example. One of them poisoned his own wife.)

Cardan was also an astrologer. There is a legend that claims that he predicted the date of his death astrologically and, to guarantee its accuracy, he drank poison on that day. That's one way of being right!

Cardan suffered from many illnesses that prevented him from enjoying life. To forget his troubles, he gambled daily for many years. His intense interest in gambling led him to write a book on the subject. This work, called *The Book on Games of Chance*, is really a textbook for gamblers, complete with tips on how to succeed in cheating. In this book we find the beginnings of the study of probability.

The development of mathematical probability was further helped along its way by the Chevalier de Méré. Like Cardan, he was a gambler. He

Jerome Cardan (1501–1576).

was also an amateur mathematician and was interested in the following problem: Suppose a gambling game must be interrupted before it is finished. How should the players divide up the money that is on the table? He sent the problem to his friend, the mathematical genius Blaise Pascal (1623–1662).

When Pascal received the Chevalier de Méré's gambling problem, he sent it to his friend, the great amateur mathematician Pierre Fermat (1602–1665). The two men wrote to each other on this subject. This correspondence was the starting point for the modern theory of probability. Many other gifted mathematicians were attracted by the work that Pascal and Fermat had begun.

10.2

COUNTING PROBLEMS

Suppose we toss two coins. What are the possible outcomes? There are four possibilities.

Coin 1	Coin 2
Head	Head
Head	Tail
Tail	Head
Tail	Tail

If we let H stand for head and T stand for tail, then the set of these outcomes can be written as {HH, HT, TH, and TT}.

If we were to flip a coin three times, then we would have eight possible outcomes. These form the set {HHH, HHT, HTH, HTT, THH, THT, TTH, and TTT}.

In these two examples, and in other similar problems, it is rather simple to list and count all the possible outcomes. In other situations there may be so many possible outcomes that it may be impractical or impossible to list all the possibilities. For problems like that, we will introduce an easy rule that can be used. The following examples will illustrate the above ideas.

EXAMPLE 1   A die (the plural is dice) is tossed once. The possible outcomes are 1, 2, 3, 4, 5, and 6.   ☐

EXAMPLE 2   If two dice are tossed, then there are 36 possible outcomes. These are

1, 1	1, 2	1, 3	1, 4	1, 5	1, 6
2, 1	2, 2	2, 3	2, 4	2, 5	2, 6
3, 1	3, 2	3, 3	3, 4	3, 5	3, 6
4, 1	4, 2	4, 3	4, 4	4, 5	4, 6
5, 1	5, 2	5, 3	5, 4	5, 5	5, 6
6, 1	6, 2	6, 3	6, 4	6, 5	6, 6 ☐

EXAMPLE 3   Four men and five women have signed up for mixed doubles at the Norfolk Tennis Club. (In mixed doubles two teams compete against each other, and each team consists of one man and one woman.) The men are Stu, Drew, Lou, and Hugh. The women are Nell, Adele, Anabel, Clarabel, and Maybelle. How many different teams can be arranged?

SOLUTION   There are 20 possible teams. These are listed below.

Men	Women	Men	Women
Stu	Nell	Lou	Nell
Stu	Adele	Lou	Adele
Stu	Anabel	Lou	Anabel
Stu	Clarabel	Lou	Clarabel
Stu	Maybelle	Lou	Maybelle
Drew	Nell	Hugh	Nell
Drew	Adele	Hugh	Adele
Drew	Anabel	Hugh	Anabel
Drew	Clarabel	Hugh	Clarabel
Drew	Maybelle	Hugh	Maybelle

There are four men and five women. Norfolk can select either Stu, Drew, Lou, or Hugh as the man for any team. If they select Stu, then they can select any one of the five women to be his partner. Thus there are five possible teams on which Stu can be the male partner. Similarly, there are five teams on which Drew can be the male partner, five teams for Lou, and five for Hugh. This makes a total of 4 × 5, or 20, teams. If there were 5 men and 6 women, then each of the 5 men would have 6 possible partners, and there would be 5 × 6, or 30, possible teams. ☐

This leads us to the following useful rule.

RULE   *If one thing can be done in m ways, and if, after this is done, something else can be done in n ways, then there are a total of m · n possible ways of doing both things (in the stated order).*

Comment   This rule is often known as **The Fundamental Principle of Counting.**

EXAMPLE 4   Sally is planning to go away for a week and is taking with her three blouses and four skirts. Sally will wear any of the blouses with any of the skirts. How many different outfits will Sally have if the colors of the clothes are as shown below?

Blouses	Skirts
Beige	Beige
White	Blue denim
Black	Gray
	Red

SOLUTION   For each outfit, Sally can select any one of three blouses and any one of four skirts. This gives her a total of 3 × 4, or 12, possible outfits. ☐

EXAMPLE 5   Assume that we have a deck of cards that consists of only four aces, four kings, four queens, and four jacks. We first select one card from this deck and then, without replacing it, select another. How many different outcomes are there?

SOLUTION   On the first draw, any one of 16 cards may be selected. There are now only 15 cards left for the second draw. This gives us a total of

$$16 \times 15 = 240 \text{ possible outcomes.} \quad \square$$

EXAMPLE 6   How many different three-digit numbers can be formed by using only the numbers 5, 7, 8, or 9, if

    a) repetitions are allowed?

    b) repetitions are not allowed?

    c) you can't start with 5, but repetitions are allowed?

SOLUTION   **a)** The first digit can be 5, 7, 8, or 9; that is, it can be chosen in four different ways. Similarly (since repetition is permitted), the second digit can be chosen in four different ways. The same is true for the third digit. This gives us four possibilities for the first digit, four possibilities for the second digit, and four possibilities for the third digit, so that we have

$$4 \times 4 \times 4 = 64$$

possible three-digit numbers. Notice that we are using the same rule as given above, but we have extended it to three possible things. The same rule can obviously be extended to any number of possible things.

    b) Again there are four different ways of selecting the first digit. Once we select a digit (whatever it is), it can no longer be used. Thus for the second digit we have only *three* choices. For the last digit there are only *two* choices. Why? Therefore we have a total of

$$4 \times 3 \times 2 = 24 \text{ possible three-digit numbers.}$$

    c) Since 5 cannot be used as the first digit, there are *three* choices for the first digit. However, *any* number, including 5, may be used for both the second and third digits. Thus for each of these there are four possible choices. This gives us a total of

$$3 \times 4 \times 4 = 48 \text{ possible three-digit numbers} \quad \square$$

EXAMPLE 7   In Example 4, Sally could choose any one of three blouses and any one of four skirts. The solution to this problem can be pictured in a diagram as shown in Figure 10.1.

    This diagram shows each blouse paired with all possible skirts. Such a diagram is called a **tree diagram.** We construct it as follows.

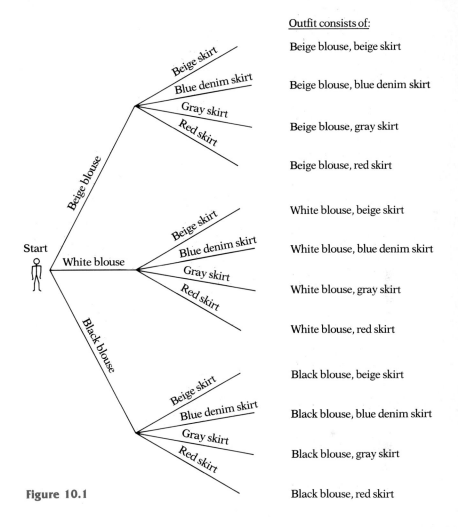

Outfit consists of:

Beige blouse, beige skirt

Beige blouse, blue denim skirt

Beige blouse, gray skirt

Beige blouse, red skirt

White blouse, beige skirt

White blouse, blue denim skirt

White blouse, gray skirt

White blouse, red skirt

Black blouse, beige skirt

Black blouse, blue denim skirt

Black blouse, gray skirt

Black blouse, red skirt

Figure 10.1

We draw a branch for each blouse. Each branch then breaks up into four smaller branches corresponding to the four skirts. The number of possible outcomes is obtained by counting the total number of smaller branches on the right. ☐

EXAMPLE 8 A coin is tossed four times. Using a tree diagram, find the total number of possible outcomes.

SOLUTION There are 16 little branches on the right of the diagram as shown in Figure 10.2. So there is a total of 16 possible outcomes. ☐

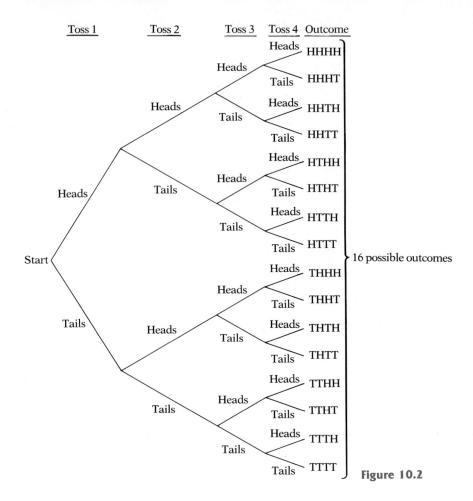

Figure 10.2

EXERCISES

1. A television network has the following list of possible shows: four quiz shows, three variety shows, three movies, and two newscasts. In how many different ways can it schedule these shows, assuming that any other factors can be neglected and assuming that order counts?

2. A restaurant offers the following menu:

Main Course	Dessert	Beverage
Shrimp	Tapioca	Soda
Lamb chop	Rice pudding	Tea
Ham	Ice cream	Coffee
Steak	Jelly tart	
Fish and chips		
Chicken		

In how many different ways can a meal be ordered?

3. In the first row of a jury box there are 6 vacant seats. In how many different ways can 6 jurors be seated in this row of 6 seats?

4. In a certain state all domestic animals (dogs and cats) must have a coded identification tag around their neck. These identification tags are of the following type: The first place of each identification code must be a 3, the second and third places must be letters, and the fourth and fifth places can be any numbers with repetition allowed. How many different identification tags can be made?

5. Seven people are waiting at a motor vehicle office to have their photos taken for new driver's licenses. In how many different ways can these people stand in line?

6. Consider the article below. Eight depositors are anxiously standing in line waiting to withdraw their money. In how many different ways can they stand in line?

Police Quell Depositor Unrest

NEW YORK: Local police had to be called in yesterday to restore order at the Golden Pacific Bank following the announcement by state banking officials that the bank was insolvent and that depositors would be allowed to withdraw only $100 per person pending further action by the Federal Deposit Insurance Corporation.

THE LOCAL TIMES, June 10, 1985

---

7. A jewelry salesman carries all of his samples in his attaché case. There is a combination lock on either side of the case. Each lock has three dials that have to be rotated independently so that any number from 0 to 9 inclusive shows on each dial. Both locks can be opened only if the correct number shows on each dial displayed. How many combinations does a thief have to try before the locks open and the thief can steal the sample jewels from the attaché case?

**8. In the baseball World Series the first team that wins four out of seven games is the winner. Construct a tree diagram showing the possible ways in which the World Series can end.

9. There are 9 approach roads leading to an airport. Because of heavy traffic, a taxi driver decides to go to the airport by one road and to leave by another road. In how many different ways can this be done?

10. How many different numbers greater than 2000 can be formed with the digits 1, 3, 4, and 7 if no repetitions are allowed?

11. The *New York Daily News* has as a daily feature a five-letter or a six-letter word that the reader has to unscramble to make a meaningful word. In how many different ways can the letters of the word "MODERN" be arranged? (*Note:* Each arrangement does not necessarily have to form a word.)

12. In the base 5 number system of counting, how many different three-digit numbers are there? (*Note:* A three-digit number cannot have zeros as the first digit or the first two digits. Thus 001 is not a three-digit number.)

13. In how many different ways can the letters of each of the following words be arranged if i) repetition is allowed and ii) repetition is not allowed?

   a) GOAL

   b) GREAT

   c) DESTROY

(*Note:* Each arrangement does not necessarily have to form a word.)

14. *Environmental Protection.* In an effort to protect the environment, the Nuclear Regulatory Commission of a certain state requires that all nuclear or chemical wastes be placed in special drums. These drums must then be stamped with numbers or letters as follows:

   a) The company number as assigned by the commission; there are 6 companies within the state.

   b) The plant location within the state; the state is divided into 8 geographical areas labeled A, B, C, D, E, F, G, or H.

   c) The date of sealing: month and last two digits of the year.

For example, a drum with the code 3A0784 stamped on it means that it was sealed by company 3 located in area A during July 1984. Using the above scheme, how many different codes are possible for drums sealed during the years 1980–1986?

15. Using tree diagrams, determine the total number of possible ways that a family can consist of four children.

10.3

PERMUTATIONS AND COMBINATIONS

In some of the exercises of the previous section, order was important, whereas in others it was not. For example, in Exercise 6, order *was* important. On the other hand, in Exercise 2, order was not important. This leads us to the useful idea of **permutations.** We state this as a definition.

> **Definition 10.1**   A **permutation** is any arrangement of objects **in a certain order.**

EXAMPLE 1   How many permutations are there of the letters in the word "cat"?

SOLUTION   There are 6. They are {cat, cta, act, atc, tca, tac}. ▢

EXAMPLE 2   How many different three-letter permutations can be formed by using the letters of the word "drug"?

SOLUTION   There are 24. They are

drg	dgu	grd	gud	rug	rdg	urg	udr
dgr	dru	gdr	gru	rgu	rdu	ugr	ugd
dug	dur	gdu	gur	rgd	rud	urd	udg ▢

In Example 2 we were interested in the number of possible permutations of 3 things that can be formed out of a possible 4 things. The symbol we use for this is $_4P_3$. We read this as "the number of permutations of 4 things taken 3 at a time."

If we were interested in the number of possible permutations of 2 things that can be formed out of a possible 4 things, then we would write this as $_4P_2$. More generally, we have the following.

Notation   The symbol $_nP_r$ means the number of permutations of n things taken r at a time. The symbol $_nP_n$ means the number of permutations of n things taken n at a time. This, of course, simply represents the number of different ways of arranging these n things.

There is a simple formula that allows us to calculate $_nP_r$ for any values of n and r. Before giving this formula we introduce the symbolism $n!$, read as "n factorial."

For example, $4!$, read as "4 factorial," means $4 \cdot 3 \cdot 2 \cdot 1$. Thus

$$4! = 4 \cdot 3 \cdot 2 \cdot 1 = 24.$$

Also,

$$5! = 5 \cdot 4 \cdot 3 \cdot 2 \cdot 1 = 120,$$

and

$$7! = 7 \cdot 6 \cdot 5 \cdot 4 \cdot 3 \cdot 2 \cdot 1 = 5040,$$
$$1! = 1.$$

The symbol 0! is taken to be equal to 1.

Now we are ready for the formula for the number of permutations of n things taken r at a time.

Formula 1 $\quad {}_nP_r = \dfrac{n!}{(n-r)!}$

EXAMPLE 3 Find ${}_5P_3$.

SOLUTION The symbol ${}_5P_3$ means the number of permutations of 5 things taken 3 at a time. Using the above formula, we have $n = 5$ and $r = 3$, so that we get

$$\begin{aligned}
{}_5P_3 &= \frac{5!}{(5-3)!} = \frac{5!}{2!} \\
&= \frac{5 \cdot 4 \cdot 3 \cdot 2 \cdot 1}{2 \cdot 1} \\
&= \frac{5 \cdot 4 \cdot 3 \cdot \cancel{2} \cdot \cancel{1}}{\cancel{2} \cdot \cancel{1}} \\
&= 5 \cdot 4 \cdot 3 = 60.
\end{aligned}$$

Thus ${}_5P_3 = 60$. ☐

EXAMPLE 4 Find ${}_6P_2$.

SOLUTION The symbol ${}_6P_2$ means the number of permutations of 6 things taken 2 at a time. Using Formula 1, we see that $n = 6$ and $r = 2$. Thus

$$\begin{aligned}
{}_6P_2 &= \frac{6!}{(6-2)!} \\
&= \frac{6!}{4!} \\
&= \frac{6 \cdot 5 \cdot 4 \cdot 3 \cdot 2 \cdot 1}{4 \cdot 3 \cdot 2 \cdot 1} \\
&= \frac{6 \cdot 5 \cdot \cancel{4} \cdot \cancel{3} \cdot \cancel{2} \cdot \cancel{1}}{\cancel{4} \cdot \cancel{3} \cdot \cancel{2} \cdot \cancel{1}} \\
&= 6 \cdot 5 = 30.
\end{aligned}$$

Therefore ${}_6P_2 = 30$. ☐

EXAMPLE 5 In Example 2 we found all the three-letter permutations of the four-letter word "drug." There were 24 of them. We could have used Formula 1 to obtain this answer. Since we have 4 letters to start with, $n = 4$. We are selecting three-letter permutations, so $r = 3$. Thus we want $_4P_3$, which equals

$$\frac{4!}{(4-3)!} = \frac{4!}{1!}$$
$$= \frac{4 \cdot 3 \cdot 2 \cdot \cancel{1}}{\cancel{1}}$$
$$= 24.$$

This confirms our previous answer, which we found by listing the permutations. ☐

EXAMPLE 6 Philip has just typed 5 letters and 5 envelopes. Before he can insert the letters into the envelopes, he drops them on the floor and they get all mixed up. When Philip picks them up, he inserts the letters into the envelopes without looking at them. In how many different ways can this be done?

SOLUTION Since the problem involves the ordering of envelopes, it is a permutation problem. When Philip picks up a letter, he has to select 1 out of 5 envelopes into which to put it. Thus $n = 5$ and $r = 5$. The total number of different ways that he can do this is

$$_5P_5 = \frac{5!}{0!}$$
$$= \frac{5 \cdot 4 \cdot 3 \cdot 2 \cdot 1}{1} \qquad \text{(Remember that } 0! = 1.\text{)}$$
$$= 120.$$

Philip can then insert the letters into the envelopes in 120 different ways. ☐

Next we consider a slightly different permutation problem. How many different four-letter words can be formed from the word "GURU"? Since there are two U's and we cannot tell them apart, Formula 1 has to be changed somewhat. Let us first list all the possible permutations. There are twelve of them, as shown.

GURU	UURG
GRUU	UUGR
GUUR	URGU
UGRU	RGUU
URUG	RUGU
UGUR	RUUG

Had we used Formula 1, we would have obtained

$$_4P_4 = \frac{4!}{(4-4)!} = \frac{4!}{0!}$$

$$= \frac{4 \cdot 3 \cdot 2 \cdot 1}{1} \qquad \text{(Remember that } 0! = 1.)$$

$$= 24.$$

Why did we get only 12 when Formula 1 gives 24? A little thought shows us that since we cannot tell the two U's apart, half of the 24 permutations of the formula will be repetitions. We therefore do not count them. So we end up with half of 24, or 12, different permutations. For example, if we label the two U's as U_1 and U_2, then two possible permutations given by Formula 1 are U_1U_2GR and U_2U_1GR. However, we cannot tell these apart (since when writing these, we do not really label the U's with 1 and 2). Thus we count these two possibilities as just one permutation.

This example leads us to the following formula for the number of permutations of n things, when some of them are alike.

Formula 2 *Suppose we have n things of which p are alike, q are alike, r are alike, etc. Then the number of different permutations is*

$$\frac{n!}{p!q!r! \ldots}.$$

(It is understood that $p + q + r + \cdots = n$.)

EXAMPLE 7 How many different permutations are there of the word (a) coffee? (b) Tennessee?

SOLUTION **a)** Since "coffee" has 6 letters, then $n = 6$. The "f" is repeated twice, and so is the "e." So $p = 2$ and $q = 2$. Formula 2 then tells us that the number of permutations is

$$\frac{6!}{2!2!1!1!} = \frac{6 \cdot 5 \cdot 4 \cdot 3 \cdot 2 \cdot 1}{2 \cdot 1 \cdot 2 \cdot 1 \cdot 1 \cdot 1}$$

$$= \frac{6 \cdot 5 \cdot \overset{2}{\cancel{4}} \cdot 3 \cdot \cancel{2} \cdot \cancel{1}}{\cancel{2} \cdot \cancel{1} \cdot \cancel{2} \cdot \cancel{1} \cdot \cancel{1} \cdot \cancel{1}} = 180.$$

There are 180 permutations.

b) "Tennessee" has nine letters, so n is 9. There are 4 e's, 2 n's, and 2 s's, so p is 4, q is 2, and r is 2. Formula 2 tells us that the number of permutations is

$$\frac{9!}{4!2!2!1!} = \frac{9 \cdot 8 \cdot 7 \cdot 6 \cdot 5 \cdot 4 \cdot 3 \cdot 2 \cdot 1}{4 \cdot 3 \cdot 2 \cdot 1 \cdot 2 \cdot 1 \cdot 2 \cdot 1 \cdot 1}$$

$$= \frac{9 \cdot \overset{2}{\cancel{8}} \cdot 7 \cdot 6 \cdot 5 \cdot \cancel{4} \cdot \cancel{3} \cdot \cancel{2} \cdot \cancel{1}}{\cancel{4} \cdot \cancel{3} \cdot 2 \cdot \cancel{1} \cdot \cancel{2} \cdot \cancel{1} \cdot \cancel{2} \cdot \cancel{1} \cdot \cancel{1}} = 3780.$$

So there are 3780 permutations. ☐

Combinations Suppose that Mike is in a record shop. He has enough money to buy only 3 records by the latest popular singing group, the Rockheads. The store has 5 different records by this group. In how many ways can Mike make his selection?

In this situation we are again interested in selecting 3 out of 5 things. However, this time we are *not* interested in the order in which the selection is made. We call a selection of this kind a **combination**.

> **Definition 10.2**   A *combination* is any selection of things where the order is not important.

Notation   The number of combinations of n things taken r at a time will be denoted as $_nC_r$.

Let us go back to Mike in the record shop. He must select 3 out of 5 records. This can be done in $_5C_3$ ways. We want to calculate $_5C_3$. If the records are labeled as A, B, C, D, and E and if order counts, then there are $_5P_3$ possible ways of selecting 3 records out of a total of 5. This gives

$$_5P_3 = \frac{5!}{(5-3)!}$$

$$= \frac{5!}{2!}$$

$$= 60.$$

This figure takes order into account, since it is the number of permutations. In our case we do not care about the order. Thus if he selects records A, B, and C, then all of the following permutations represent the same purchase: ABC, CAB, ACB, BAC, BCA, CBA. These six permutations are thus considered *one* combination. The same is true for any other combination of 3 records. Therefore to get the correct number of combinations, we divide the 60 by 6 and obtain 10. Notice that 6 is 3!. Thus $_5C_3$ is

$$\frac{_5P_3}{3!} = \frac{5!}{(5-3)!3!}$$

$$= 10.$$

Thus we have the following formula.

> **Formula 3** $_nC_r = \dfrac{n!}{(n-r)!r!}$

The following examples illustrate how the formula is used.

EXAMPLE 8   Eight workers at the Excelsior Music Corporation are unhappy about their working conditions. They wish to complain to the management. If management will listen to a committee of only 3 people, in how many ways can such a committee be formed?

SOLUTION   Since the order of selecting people for the committee is not important, the answer is the number combinations of 8 things taken 3 at a time. We thus want $_8C_3$, which is

$$_8C_3 = \frac{8!}{(8-3)!3!} = \frac{8!}{5!3!}$$
$$= \frac{8 \cdot 7 \cdot 6 \cdot 5 \cdot 4 \cdot 3 \cdot 2 \cdot 1}{5 \cdot 4 \cdot 3 \cdot 2 \cdot 1 \cdot 3 \cdot 2 \cdot 1}$$
$$= 56.$$

Thus 56 different committees can be formed. ☐

EXAMPLE 9   Tom is going to the supermarket to buy 4 pints of ice cream. The store sells 28 different flavors of ice cream, and the smallest amount they will sell of any one flavor is one pint. Tom wants to try as many different flavors as he can. In how many different ways can he buy 4 different flavors?

SOLUTION   Since order is not important, we want $_{28}C_4$. Formula 3 tells us that this is

$$\frac{28!}{(28-4)! \cdot 4!} = \frac{28!}{24! \cdot 4!} = 20{,}475$$

Thus Tom can select the 4 different flavors in 20,475 ways. ☐

EXAMPLE 10   In Playland Amusement Park there is a game that consists of throwing 3 balls into 6 baskets. You win if you get 1 ball each into 3 different baskets. In how many different ways can someone win?

SOLUTION   Since order is not important, we are interested in $_6C_3$. Formula 3 tells us that this is

$$\frac{6!}{(6-3)!3!} = \frac{6!}{3!3!}$$
$$= 20.$$

There are 20 winning combinations. ☐

EXAMPLE 11   In how many different ways can a committee of 3 men and 4 women be formed from a group of 8 men and 6 women?

SOLUTION   We must select any 3 men from a possible 8, and order does not matter. This can be done in $_8C_3$ ways.

```
                1
             1     1
          1     2     1
       1     3     3     1
    1     4     6     4     1
 1     5    10    10     5     1     Figure 10.3
```

Then we must select any 4 women from a possible 6, again where order does not count. This is $_6C_4$.

Since any group of men can be combined with any group of women to form the entire committee, then by the rule given in Section 10.2 (page 523) we have a total of

$$_8C_3 \cdot {_6C_4} = \frac{8!}{(8-3)!3!} \cdot \frac{6!}{(6-4)!4!}$$

$$= \frac{8!}{5!3!} \cdot \frac{6!}{2!4!}$$

$$= 56 \cdot 15 = 840.$$

Thus 840 committees can be formed. ☐

Pascal's Triangle Another useful technique for computing the number of possible combinations is by means of **Pascal's triangle.** The triangle is shown in Figure 10.3. It is not hard to see how this triangle is constructed. Each row has a 1 on either end. All the other entries are obtained by adding the numbers immediately above it directly to the right and left as shown by the arrows in Figure 10.4. Thus to get the entries for the sixth row, we add 1 and 5 to get 6. Then we add 5 and 10 to get 15. We next add 10 and 10 to get 20, and so on. To complete the row, we add 1's on each end. The numbers must be lined up exactly as shown in the diagram.

A triangle of numbers such as this is called Pascal's triangle. It can have as many rows as you want. This triangle was known to the Chinese for several centuries before Pascal's time. However, it is

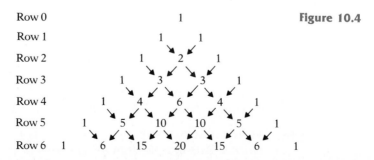

Figure 10.4

Row 0	1
Row 1	1 1
Row 2	1 2 1
Row 3	1 3 3 1
Row 4	1 4 6 4 1
Row 5	1 5 10 10 5 1
Row 6	1 6 15 20 15 6 1

HISTORICAL NOTE

Blaise Pascal had demonstrated his mathematical talent at an early age. He proved a very important theorem in geometry when he was only 16 years old. His interest in mathematics was not limited to geometry. When he was about 18 years old, he built the first successful computing machine. He also did valuable work in physics, notably confirming the fact that air has weight.

Throughout his life, Pascal suffered from severe illness, and hardly a day passed without pain. He was deeply religious, and when he was almost killed by a runaway horse in 1654, he regarded his narrow escape as a sign from God. As a result, he devoted himself even more than ever to religious meditation and writing. His great work is *Pensées*, which deals largely with philosophy and religion. Pascal died in 1662 at the age of only 39. He is famous both as a mathematician and philosopher.

named for Pascal because of the many interesting applications he found for it. (See Figure 10.5.)

Let us now see how Pascal's triangle can be used to solve problems in combinations.

EXAMPLE 12 In a recent mining accident a volunteer rescue squad consisting of 3 people was needed. Seven people volunteered. In how many different ways could a rescue squad be formed?

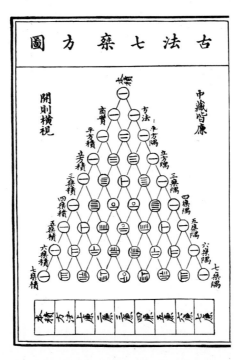

FIGURE 10.5 The Pascal triangle as depicted in 1303 at the front of Chu Shih-Chieh's *Ssu Yuan Yii Chien*. It is entitled "The Old Method Chart of the Seven Multiplying Squares" and tabulates the binomial coefficients up to the eighth power. *Reproduced with permission from Joseph Needham,* Science and Civilization in China, *III, 135 (New York: Cambridge University Press)*

SOLUTION Since order is not important, we want $_7C_3$. First let us evaluate $_7C_3$ by using Formula 3. We have

$$_7C_3 = \frac{7!}{(7-3)!3!}$$

$$= 35.$$

Thus there are 35 possible rescue squads that can be formed.

Now let us evaluate $_7C_3$ by using Pascal's triangle. We have 7 people to select from, so we write the first 7 rows of Pascal's triangle as shown in Figure 10.6. Look at Row 7. Since we must select 3 people, we go to the third entry (from the left) *after* the end 1. This entry is 35, and this is our answer. Thus again we see that $_7C_3$ is 35. We used the third entry (after the end 1) in row 7 because we wanted $_7C_3$. If we had wanted $_7C_5$, we would have used the fifth entry (after the end 1) in row 7. This entry is 21. Thus $_7C_5$ is 21. □

Row 0							1						**Figure 10.6**
Row 1						1		1					
Row 2					1		2		1				
Row 3				1		3		3		1			
Row 4			1		4		6		4		1		
Row 5		1		5		10		10		5		1	
Row 6	1		6		15		20		15		6		1
Row 7	1	7		21		35		35		21		7	1

In general, to find the value of $_nC_r$, we go to row n. Then we select the rth number (after the end 1) from the left. This entry is $_nC_r$. In this procedure *we always label the first row as row 0*.

EXAMPLE 13 Using Pascal's triangle, find (a) $_6C_4$; (b) $_6C_0$; and (c) $_5C_5$.

SOLUTION We will use the Pascal triangle shown in Figure 10.6.

a) To find $_6C_4$, go to row 6. Then go across to the fourth entry from the left (after the end 1). This entry is 15. Thus $_6C_4 = 15$.

b) To find $_6C_0$, go to row 6. Then go across to the "0'th entry" from the left (after the end 1). This means that we must remain at the 1. Thus $_6C_0 = 1$.

c) To find $_5C_5$, we go to row 5. Then we go across to the fifth entry from the left (after the end 1). This entry is 1. Thus $_5C_5 = 1$. □

Although we have used Pascal's triangle to evaluate $_nC_r$, there are many other interesting and important applications of this triangle. Consult the suggested further readings given for such applications.

EXERCISES

Evaluate each of the symbols given in Exercises 1–26.

1. $\dfrac{5!}{3!}$ **2.** $\dfrac{7!}{5!}$ **3.** $\dfrac{8!}{8!}$ **4.** $\dfrac{0!}{7}$

5. $_8P_6$ **6.** $_7P_7$ **7.** $_6P_4$ **8.** $_{11}P_8$

9. $_6P_6$ **10.** $_6P_0$ **11.** $_9P_4$ **12.** $_0P_0$

13. $_7C_3$ **14.** $_8C_5$ **15.** $_6C_5$ **16.** $_6P_5$

17. $_9C_9$ **18.** $_8C_7$ **19.** $_8C_1$ **20.** $_{10}C_9$

21. $_{10}C_1$ **22.** $_7C_0$ **23.** $_8C_8$ **24.** $_9C_4$

25. $_{10}C_{11}$ **26.** $_0C_0$

How many different permutations are there of the letters of the words given in Exercises 27–30?

27. DIFFERENCES   **28.** FLAMMABLE

29. SUCCESSION   **30.** MISSISSIPPI

31. A cosmetics company has 10 members on its board of directors. In how many different ways can it elect a president, vice-president, secretary, and treasurer?

32. In how many different ways can a police department arrange suspects in a police lineup if each lineup consists of 7 people?

33. A movie critic is asked to list, in order of preference, the ten best movies that she had seen in 1986. If she saw 30 movies during the year, in how many ways can she select the 10 best movies?

34. In how many different ways can a jury of 12 people be selected from a panel of 18 prospective jurors?

35. A drill sergeant enters the barracks, where there are 12 soldiers playing cards. He needs four "volunteers": one to mop the floor, one to peel potatoes, one to scrub the walls, and one to wash dishes. In how many different ways can he get his group of volunteers?

36. A stock clerk is arranging four distinguishable cases of corn flakes, three distinguishable cases of toilet tissue, and two distinguishable cases of paper towels on a loading platform. In how many different ways can these cases be arranged if

a) the cases can be arranged in any order?

b) the corn flakes are to be placed together, the toilet tissue together, and the paper towels together?

37. How many committees of seven people from a group of eight Blacks and nine Orientals can be formed to investigate discrimination charges if each committee must have

a) four Blacks?

b) at least four Blacks?

38. How many different seven-card rummy hands can be formed from a deck of 52 cards?

39. A television network president is arranging next month's schedule of shows. In how many different ways can eight shows be arranged for one evening's telecast if

a) the news special (which is one of the eight shows) must be the last show?

b) any show can be telecast at any time?

40. At a baseball game the commissioner and six other guests are to be seated in seven box seats along the third base line. In how many different ways can this be done if

a) anyone can sit anywhere?

b) the commissioner must sit in the middle?

c) the commissioner must sit in the middle and his press secretary must sit at the extreme left to answer reporters' questions?

41. A jewelry designer is making a necklace out of precious stones. She has 11 pearls, 8 rubies, and 5 emeralds. The necklace is to contain 4 pearls, 5 rubies, and 2 emeralds. In how many different ways can she select the jewels for the necklace? (Assume that the order in which the jewels appear on the necklace is not important.)

42. Dr. Bergen, a medical researcher, needs 5 human volunteers to test the effectiveness of a new pain-relieving arthritis drug. If 17 people have volunteered, in how many different ways can Dr. Bergen select the 5 volunteers to test the effectiveness of the new drug?

43. The Federal Savings Bank employs 16 full-time and 9 part-time bank tellers. As an economy move, the management decides to lay off 2 part-time and 3 full-time bank tellers. Neglecting seniority and any other considerations, in how many different ways can the workers to be laid off be selected?

44. Each day, a sample of 5 of the 12 microwave ovens produced on production line A are checked for radiation leakage before being shipped to the consumer.

a) In how many different ways can the 5 microwave ovens selected to be checked be chosen?

b) In how many different ways can the 7 microwave ovens selected *not* to be checked be chosen?

c) How do the answers in parts (a) and (b) compare? Explain your answer.

45. Look at the diagonals in the Pascal's triangle shown in Figure 10.7.

a) Find the sum of the number(s) in diagonal 1.

b) Find the sum of the numbers in diagonal 2.

c) Find the sum of the numbers in diagonal 3.

d) Find the sum of the numbers in diagonal 4, and so on throughout.

e) What do you notice about the results?

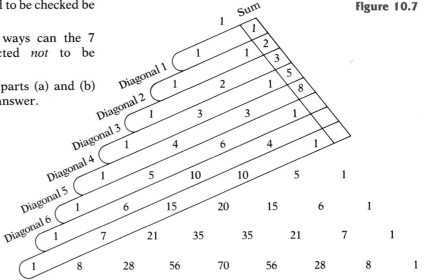

Figure 10.7

10.4

DEFINITION OF PROBABILITY

We are now ready to define what is meant by the concept of probability.

Suppose we toss an honest coin many times and observe the number of heads that appear. The results of several such experiments are summarized in the chart below.

Number of heads appearing	4	26	49	250	498	5,001
Number of tosses	10	50	100	500	1,000	10,000

We see that in each case the number of heads appearing is approximately $\frac{1}{2}$ the number of tosses. If we flip the coin one million times, we would expect to get approximately 500,000 heads. If we now toss a coin once, we would say that *the probability of getting a head is* $\frac{1}{2}$.

Note that when we flip the coin, there are two possible outcomes, heads and tails, both of which are equally likely. We are interested in only one of these outcomes—namely, heads. If heads occurs, we will call this a *favorable* outcome. The probability in this case is the number of favorable outcomes divided by the total number of possible outcomes, that is, it is 1 divided by 2, or $\frac{1}{2}$.

Suppose we were now to roll a die once. What is the probability of getting a 4? There are six possible outcomes, all of which are equally likely. These are 1, 2, 3, 4, 5, and 6. If the die is fair, then we would expect to get a 4 approximately $\frac{1}{6}$ of the time, since there are 6 possible outcomes and only 1 of these, namely the 4, is favorable. We would then say that *the probability of getting a 4 is* $\frac{1}{6}$.

We can now make this concept of probability more specific. Before doing so, however, we will give a definition that makes it easier to talk about probability.

Definition 10.3   *The set of all possible outcomes of an experiment is called the* **sample space.** *We will usually not be concerned with the entire sample space but rather with only some of these outcomes. Such a collection will be referred to as an* **event.** *(In other words, an event is a subset of the sample space.)*

EXAMPLE 1   If we toss a coin once, then the possible outcomes are H and T. Thus the sample space is {H, T}.

In this Japanese wine-drinking game, players use a marked die and boxes with matching marks. When the die is rolled, would you say the winner or the loser drinks the wine?

Peabody Museum of Salem, Photo by Mark Sexton

If we were to toss the coin twice, then the sample space would be {HT, TH, HH, TT}. The event "no heads" would be {TT}.   □

EXAMPLE 2   If a die is tossed once, then the sample space is {1, 2, 3, 4, 5, 6}.

The event "even number" is {2, 4, 6}.

The event "odd number" is {1, 3, 5}.

The event "number greater than 4" is {5, 6}.

The event "number divisible by 3" is {3, 6}.   □

We are now ready to define probability.

> **Definition 10.4**   *If an event can occur in any one of n equally likely ways and if f of these are considered as favorable outcomes, then the **probability** of getting a favorable outcome is*
>
> $$\frac{number\ of\ favorable\ outcomes}{total\ number\ of\ outcomes} = \frac{f}{n}.$$
>
> *Symbolically, we write*
>
> $$prob\ (favorable\ event) = \frac{f}{n}.$$

EXAMPLE 3   A card is drawn from a 52-card deck. What is the probability of getting

a) a heart?   **b)** a black card?   **c)** an ace?   **d)** the ace of spades?

SOLUTION   Since there are 52 cards in the deck, the total number of outcomes is 52.

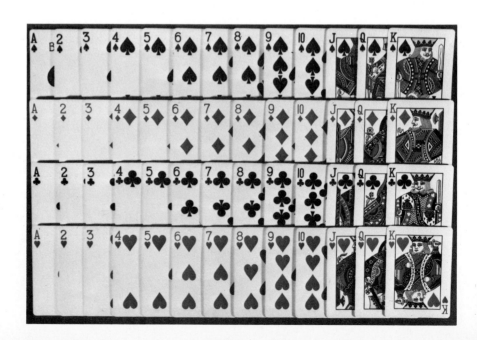

a) There are 13 hearts in a deck, so there are 13 favorable outcomes. Using Definition 10.4, we get

$$\text{prob (hearts)} = \frac{13}{52} = \frac{1}{4}.$$

b) Half of the deck consists of black cards, so there are 26 favorable outcomes. Hence

$$\text{prob (black card)} = \frac{26}{52} = \frac{1}{2}.$$

c) There are 4 aces in a deck, so there are 4 favorable outcomes. We then have

$$\text{prob (ace)} = \frac{4}{52} = \frac{1}{13}.$$

d) There is only one ace of spades in a deck. We therefore have only one favorable outcome. Thus

$$\text{prob (ace of spades)} = \frac{1}{52}. \quad \square$$

EXAMPLE 4   A fair die is tossed once. What is the probability of getting

a) an odd number larger than 1?
b) a number larger than 3?
c) a prime number? (Remember, a prime number is any number larger than 1 that is exactly divisible only by itself and 1.)
d) a number larger than 6?

SOLUTION   There are six possible outcomes. These are 1, 2, 3, 4, 5, and 6.

a) We know that 1, 3, and 5 are the possible odd numbers; 3 and 5 are larger than 1. So there are two possible favorable outcomes. Hence

$$\text{prob (odd number larger than 1)} = \frac{2}{6} = \frac{1}{3}.$$

b) There are three favorable outcomes 4, 5, and 6, so that

$$\text{prob (number larger than 3)} = \frac{3}{6} = \frac{1}{2}.$$

c) The prime numbers between 1 and 6 are 2, 3, and 5. There are three of them. Thus

$$\text{prob (prime number)} = \frac{3}{6} = \frac{1}{2}.$$

d) Since there are no numbers larger than 6 on a die, the number of favorable outcomes is 0. So

$$\text{prob (number larger than 6)} = \frac{0}{6} = 0. \quad \square$$

EXAMPLE 5   In a state lottery, each ticket has 5 numbers. If you get all 5 numbers right, you win $50,000. If you get the last 4 numbers in the correct order, you win $10,000. If you get the last 3 numbers in the correct order, you win $5000. If you get the last 2 numbers in the correct order, you win $1000. What is the probability of winning

a) $50,000?   **b)** $10,000?   **c)** $5000?

SOLUTION   Since there are 5 numbers on each ticket, ranging from 00001 to 99,999, there is a total of 99,999 possible outcomes.

a) Only 1 ticket has the winning number, so there is just 1 favorable outcome. Thus

$$\text{prob (winning \$50,000)} = \frac{1}{99,999}$$

b) To win $10,000, we must calculate how many tickets have the last 4 numbers in the correct order. We do not care about the first number. So the first number could be any one of 10 possible numbers. Hence there are 10 tickets that have the last 4 digits in the correct order. However, one of these (the one with the correct first number) is the big winner. So we don't count it. Thus there are only 9 favorable outcomes. Since there are 99,999 total possible outcomes we get

$$\text{prob (winning \$10,000)} = \frac{9}{99,999} = \frac{1}{11,111}$$

c) To win $5000, we must calculate how many tickets have the last 3 numbers in the correct order. This time we don't care about the first 2 numbers. The first place can be filled in 10 ways, and so can the second. So there is a total of 10 × 10, or 100, possible tickets with the last 3 digits correct. We must disregard 10 of these since 1 is the big winner and 9 are $10,000 winners. This then leaves us with 90 favorable outcomes out of a total of 99,999 possible outcomes. Therefore

$$\text{prob (winning \$5,000)} = \frac{90}{99,999} = \frac{10}{11,111} \quad \square$$

EXAMPLE 6   In Example 5 on p. 51 we stated that doctors in the United States have been treating patients who are suffering from mental depression with the drug lithium. They have found that approximately 80% of all such patients treated with lithium reported feeling better. What is the probability that a person who is suffering from mental depression and is treated with lithium will feel better?

SOLUTION   Since the doctors claim an 80% improvement rate when treated with lithium, this means that out of every 100 people treated, 80 will improve. Thus

$$\text{prob (improvement)} = \frac{80}{100} = \frac{4}{5} \quad \square$$

EXAMPLE 7   Greg, Rita, William, Frank, Yolando, and Dawn are six students who are enrolled in a math honors course at State University. The departmental policy is to award a $100 prize to each of the top two students. What is the probability that Dawn and William will receive the prize?

SOLUTION   We first find the total number of different ways in which the two winners can be selected. This is $_6C_2$ (the number of ways of selecting two out of six people where order does not count). Using Formula 3 given in Section 10.3, we get

$$_6C_2 = \frac{6!}{(6-2)!2!} = \frac{6!}{4!2!} = 15.$$

Of these 15 ways of selecting the two winners, only one consists of Dawn and William. Thus

$$\text{prob (Dawn and William win prize)} = \frac{1}{15} \quad \square$$

EXAMPLE 8   What is the probability that your math teacher will be fired on April 31?

SOLUTION   Since April has exactly 30 days, your math teacher cannot be fired on April 31. There are no favorable outcomes. Thus

$$\text{prob (your math teacher gets fired on April 31)} = 0. \quad \square$$

Something that can never happen is called the **null** event. Its probability is 0.

EXAMPLE 9   Mary has just been admitted to the maternity ward at a hospital to have a baby. What is the probability that the baby is a boy or a girl?

SOLUTION   There are only two possible outcomes (boy or girl), so that $n = 2$. Both of these are favorable, so $f = 2$. Thus

$$\text{prob (boy or girl)} = \frac{2}{2} = 1.$$

It is obvious that a favorable outcome *must* occur in this case. $\square$

*Something that is certain to occur is called the **definite** event. Its probability is 1.*

Comment   An event may never occur, in which case its probability is 0. An event may occur for certain, in which case its probability is 1. There are events that may or may not occur, and these will have probability between 0 and 1. Thus *the probability of any event is always somewhere between 0 and 1 and possibly including 0 or 1.*

Now that we have computed the probability of several different events, we can go back to the question we raised earlier: "What do we mean by probability?"

Let us analyze the weather forecaster's prediction that the probability of rain is 90%. We first point out that 90% can also be written as $\frac{90}{100}$. The weather forecaster means that, in the past, when the clouds and winds have been as they are today, then it has rained 90 times out of 100. In other words, on 100 days, when conditions have been as they are today, the event of rain has occurred on 90 of these days. Thus on the basis of past experience, he or she predicts rain for today with a probability of $\frac{90}{100}$, or 90%.

In a similar manner, when the doctor tells you that you have a 50–50 chance of surviving an operation, it means that on the basis of past experience, out of every 100 patients that the doctor has operated on, 50 pulled through and 50 didn't. Thus the probability of surviving is 50 out of 100, or $\frac{1}{2}$.

In general, if the probability of any event is $\frac{f}{n}$, this means that, in the long run, out of every n trials there will be f favorable outcomes. Thus probability represents the percentage of the time that the event will happen in the long run. This is sometimes called the **relative frequency** of the event.

This definition of probability is based on the number of favorable events occurring in many repeated trials. Obviously, this involves collecting statistical data about the number of events. For this reason this approach is often called **statistical probability.** Some books use another approach and define probability using axioms. This approach is often called **axiomatic probability.** The interested reader can consult any standard text on probability for a more detailed discussion of axiomatic probability.

EXERCISES

1. Two dice are rolled. What is the probability
   a) of getting a sum of 1?
   b) that the same number appears on both dice?

c) that the sum of the number of dots appearing on both dice is larger than 9?

d) that the sum of the number of dots appearing on both dice is an even number?

e) that two ones appear?

2. A card is drawn from an ordinary deck of 52 cards. What is the probability that it is

a) a club?

b) a two?

c) a picture card?

d) a card higher than 7? (Aces are considered to be 1's.)

3. A man and a woman who do not know each other board a plane in Paris that is bound for St. Louis, with stopoffs in London, New York, and Chicago. What is the probability that they both get off at the same airport?

4. A coin box has 9 nickels, 8 pennies, 6 dimes, and 4 quarters in it. A young child shakes the box, and a coin falls out. What is the probability that the coin that falls out is

a) a penny?   **b)** a nickel?

c) a dime?   **d)** a quarter?

5. The following information is available on the 418 members of the Beach Resort Health Club:

		Sex	
		Male	Female
Age	Under 30 years	126	109
	30 years and over	89	94

What is the probability that a club member selected at random

a) will be a female under 30 years of age?

b) will be a female?

c) will be under 30 years of age?

6. A computer programmer prepares 3 different computer discs (with different programs on each) and 3 different identifying labels. Before attaching each label to the appropriate disc, she drops everything on the floor. If she picks up all the discs and labels, and randomly attaches a label to a disc, what is the probability that each disc will be labeled correctly?

7. Which of the following numbers cannot be the probability of some event?

a) $-\dfrac{3}{4}$ **b)** 1.47   **c)** $\dfrac{11}{12}$

d) 0   **e)** 1   **f)** 0.99

8. The following is a breakdown of the different types of credit card sales of a large department store for a particular day.

Type of Credit Card Used

Amount of purchase	Master Card	VISA	American Express	Store Card
under $25	78	82	32	53
between $25 and $75	62	53	18	28
over $75	37	29	11	17

What is the probability that a credit card sale selected at random will be

a) a Master Card sale?

b) under $25?

c) a Master Card sale that is under $25?

9. *Mercy Killing.* A random survey[1] of 1000 adults in California was taken. Each person was asked to indicate his or her opinion on mercy killing of terminally ill patients. The following results were obtained:

	Were in favor of mercy killing	*Were opposed to mercy killing*	*Had no opinion*
Male	326	212	37
Female	207	199	19

What is the probability that an individual selected at random from this group

a) is a male who had no opinion?

b) is opposed to mercy killing?

c) is a female?

d) is a female who is in favor of mercy killing?

10. A local blood bank reports that 223 pints of blood were collected at a shopping center on

1. *Source:* Crescent and Bowes, Los Angeles, California.

Monday. After analysis it is determined that 7 of these pints are contaminated. The bottles are not labeled. A nurse selects a bottle at random. What is the probability that the bottle selected is *not* contaminated?

11. A computer company manufactured 1000 computer chip boards during one week, each numbered with a number from 1 to 1000. A few months later the company decides to recall those chip boards whose last digit is 9. Anton Jones owns one of the chip boards. What is the probability that he owns one of the chip boards that is being recalled?

12. A slot machine in a gambling casino has three wheels, and each wheel has a picture of a lemon, a cherry, and an apple on it. When the appropriate amount of money is inserted and the button is pushed, each wheel will rotate and then display a picture of one of the items mentioned. Each wheel operates independently. When all three wheels show the same item, then the player wins $5000.

    a) List all the possible outcomes for this machine.

    b) Find the probability of a player winning $5000 when playing this slot machine.

13. *Friday, the Thirteenth.* This is Exercise 13. Many people believe that 13 is an unlucky number. There are numerous hotels that do not have a thirteenth floor. Furthermore, some people believe that when the thirteenth day of a month falls on a Friday, this represents an unlucky event. Furthermore, they believe that this does not occur too often. To check the truth of this belief, let us consider the following facts. The calendar changes every year. By this we mean that if your birthday falls on a Monday this year, then next year it will fall on a Tuesday or Wednesday (depending on whether it is a leap year or not). However, our calendar repeats itself every 400 years. There are 4800 months during this period. The thirteenth day of the month in each of these 4800 months occurs on the different days of the week according to the following chart.

Day of week	Sun.	Mon.	Tues.	Wed.
How often the 13th day of month occurs on this day	687	685	685	687
	684	688	684	

a) Using the above chart, find the probability that the thirteenth day of the month will occur on a Friday.

b) Is this probability greater than, less than, or equal to the probability of its falling on any other day of the week?

10.5

RULES OF PROBABILITY

Consider the following problem. A card is selected from an ordinary deck of 52 cards. What is the probability that it is either a heart *or* a black card? Obviously, a card cannot be both a heart and a black card at the same time. We say that the events of drawing a heart and of drawing a black card are **mutually exclusive**. If A represents the event of drawing a heart and if B represents the event of drawing a black card, then $A \cap B = \varnothing$.

The probability of getting a heart is $\frac{13}{52}$, or $\frac{1}{4}$, since there are 13 hearts out of a possible 52 cards.

The probability of getting a black card is $\frac{26}{52}$, or $\frac{1}{2}$.

Patty Chock

It would require the use of a computer to determine the probability of a row on any one of these cards being completed ahead of the others, even though we are given the fact that some numbers have already been called. The fascination of Bingo lies in its complete unpredictability (to the average person).

Since there are 13 hearts and 26 black cards, this gives us a total of 39 favorable outcomes. As there are 52 cards in the deck, our answer is

$$\text{prob (heart or black card)} = \frac{39}{52} = \frac{3}{4}.$$

Note that if we add the probability of a heart and the probability of a black card, we get the following:

$$\text{prob (heart)} + \text{prob (black card)} = \frac{1}{4} + \frac{1}{2}$$
$$= \frac{3}{4}.$$

Thus we see that

   prob (heart or black card) = prob (heart) + prob (black card).

This leads us to the following.

Definition 10.5 *Two events, A and B, are said to be **mutually exclusive** if both A and B cannot occur at the same time. In terms of sets this means $A \cap B = \varnothing$.*

Formula 4 *If A and B are mutually exclusive events, then*

$$\text{prob (A or B)} = \text{prob (A)} + \text{prob (B)}.$$

EXAMPLE 1 A card is drawn from a deck of 52 cards. What is the probability of getting a 7 or a picture card?

SOLUTION   The events "getting a 7" and "getting a picture card" are mutually exclusive. We can therefore use Formula 4. We first calculate prob (getting a 7). Since there are four 7's out of a total of 52 cards,

$$\text{prob (7)} = \frac{4}{52} = \frac{1}{13}.$$

Also, we know that there are 12 picture cards in a 52-card deck. Thus

$$\text{prob (picture card)} = \frac{12}{52} = \frac{3}{13}.$$

By Formula 4 we get

$$\text{prob (7 or picture card)} = \text{prob (7)} + \text{prob (picture card)}$$
$$= \frac{1}{13} + \frac{3}{13}$$
$$= \frac{4}{13}. \quad \square$$

EXAMPLE 2   At the Fresh Air Fund Charity Bazaar, there is a table at which 30 unmarked surprise packages are being sold. Six of the packages contain transistor radios, 3 contain perfume, 10 contain wallets, 5 contain ashtrays, and 6 contain shavers. No package contains more than one item, and all the packages are wrapped identically. What is the probability that Ann, who buys the first package, gets either a radio or perfume?

SOLUTION   Since Ann is buying only one package, the events "getting a radio" and "getting perfume" are mutually exclusive. Thus Formula 4 can be applied. Since there are 30 packages altogether, 6 of which are radios and 3 of which are perfume, then

$$\text{prob (gets a radio)} = \frac{6}{30},$$

and

$$\text{prob (gets perfume)} = \frac{3}{30}.$$

Therefore

$$\text{prob (gets a radio or gets perfume)} = \text{prob (gets a radio)}$$
$$+ \text{prob (gets perfume)}$$
$$= \frac{6}{30} + \frac{3}{30} = \frac{9}{30} = \frac{3}{10}.$$

Therefore the probability that Ann gets a radio or perfume is $\frac{3}{10}$.

□

EXAMPLE 3   A mailman cannot read the address on a letter. He is not sure but thinks that the address is either 390 Main Street or 890 Main Street. The probability that he will deliver it to 390 Main Street is $\frac{1}{3}$, and the probability that he will deliver it to 890 Main Street is $\frac{2}{5}$. What is the probability that he will deliver the letter to 890 Main Street or 390 Main Street?

SOLUTION   Since the mailman cannot deliver the letter to both addresses (at the same time) we are dealing with mutually exclusive events. Formula 4 can be used. Therefore

$$\text{prob (390 or 890 Main St.)} = \text{prob (390)} + \text{prob (890)}$$
$$= \frac{1}{3} + \frac{2}{5} = \frac{5}{15} + \frac{6}{15} = \frac{11}{15}.$$

The probability that he delivers it to one of these addresses is $\frac{11}{15}$.

EXAMPLE 4   John drives his car over a nail. If the probability that he gets a flat tire is $\frac{4}{7}$, what is the probability that he does not get a flat tire?

SOLUTION   Since the events "getting a flat" and "not getting a flat" are mutually exclusive, we can use Formula 4. Clearly, one of these events *must* happen. Thus the event "flat or no flat" is the certain event and has probability of 1. By Formula 4 we then have

$$\text{prob (flat or no flat)} = \text{prob (flat)} + \text{prob (no flat)}$$
$$1 = \frac{4}{7} + \text{prob (no flat)}.$$

Subtracting $\frac{4}{7}$ from both sides, we get

$$1 - \frac{4}{7} = \text{prob (no flat)}$$
$$\frac{3}{7} = \text{prob (no flat)}.$$

Therefore John's chance of not getting a flat is $\frac{3}{7}$.

Events That Are Not Mutually Exclusive   Now let us consider the following problem. What is the probability of drawing from a deck of cards an ace *or* a spade? We first notice that the events "drawing an ace" and "drawing a spade" are *not* mutually exclusive, since the ace of spades is both an ace and a spade.

If we let A stand for the event of drawing an ace and let B stand for the event of drawing a spade, then $A \cap B \neq \varnothing$. Thus Formula 4 cannot be used. For situations of this type, we introduce the following.

> **Formula 5**   *If A and B are any events, then*
>
> $$prob\ (A\ or\ B) = prob\ (A) + prob\ (B) - prob\ (A\ and\ B).$$

Let us apply this formula to the above problem. We know that

$$prob\ (A) = \frac{4}{52} \quad \text{and} \quad prob\ (B) = \frac{13}{52}.$$

We now calculate prob $(A$ and $B)$. This event occurs only when the card drawn is an ace of spades. This had already been calculated on p. 541. Thus prob $(A$ and $B) = \frac{1}{52}$. Using Formula 5, we get

$$prob\ (A\ or\ B) = prob\ (A) + prob\ (B) - prob\ (A\ and\ B)$$
$$= \frac{4}{52} + \frac{13}{52} - \frac{1}{52} = \frac{16}{52} = \frac{4}{13}.$$

The probability of drawing an ace or a spade is $\frac{4}{13}$. $\square$

EXAMPLE 5   Sarah has just bought a new car. The probability that the body of the car will rust within six months is $\frac{1}{6}$. The probability that the oil will leak is $\frac{3}{8}$. The probability of both of these disasters happening is $\frac{1}{24}$. (We are assuming that these events are not mutually exclusive.) What is the probability that the body will rust *or* that the oil will leak?

prob (rust or oil leak)

$$= prob\ (rust) + prob\ (oil\ leak) - prob\ (rust\ and\ oil\ leak)$$
$$= \frac{1}{6} + \frac{3}{8} - \frac{1}{24} = \frac{4}{24} + \frac{9}{24} - \frac{1}{24} = \frac{12}{24} = \frac{1}{2}.$$

Thus the probability that Sarah's car will rust or have an oil leak is $\frac{1}{2}$. $\square$

EXAMPLE 6   Betty is anxious to know her grades in two courses, History 2.3 and Math 86. The probabilities of her getting an A in these courses are $\frac{1}{3}$ and $\frac{2}{9}$, respectively. Furthermore the probability of her getting an

Bruce Anderson

What would you think is the probability of two people in a crowd having the same birthday? Mathematically, the probability that *at least* two people in a crowd of 25 have a common birthday is greater than $\frac{1}{2}$. The probability increases to about 1 (almost a certainty) in a crowd of 60 people.

Peabody Museum of Salem, Photo by Mark Sexton

In the Japanese shell game, which originated around the twelfth century, the "right" halves of the shells are scattered on the floor, and the other halves are divided among the players. The players in turn present one shell at a time and compete to find the shell's counterpart among the halves on the floor. The order of matching halves does not matter; the player with the largest total count wins.

A in both courses is $\frac{1}{27}$. What is the probability that Betty will get an A in either course?

SOLUTION   Applying Formula 5 (since the events are not mutually exclusive), we get

prob (A in history or in math)

$$= \text{prob (A in history)} + \text{prob (A in math)} - \text{prob (A in both)}$$

$$= \frac{1}{3} + \frac{2}{9} - \frac{1}{27}$$

$$= \frac{9}{27} + \frac{6}{27} - \frac{1}{27}$$

$$= \frac{14}{27}$$

The chance of Betty getting an A in either course is $\frac{14}{27}$.   □

EXAMPLE 7   On a certain day the probability of rain is $\frac{4}{5}$, the probability of thunder is $\frac{2}{5}$, and the probability of both is $\frac{3}{5}$. What is the probability that it will rain *or* thunder?

SOLUTION   We use Formula 5.

prob (rain or thunder)

$$= \text{prob (rain)} + \text{prob (thunder)} - \text{prob (rain and thunder)}$$

$$= \frac{4}{5} + \frac{2}{5} - \frac{3}{5} = \frac{3}{5}$$

The chances of it raining or thundering are $\frac{3}{5}$.   □

Comment   Formula 4 is just a special case of Formula 5. Formula 5 applies to *any* events *A* and *B*. If these events happen to be mutually exclusive, then *A* and *B* cannot happen together. Thus prob (*A* and *B*) is 0. In this case, Formula 5 becomes

$$\text{prob } (A \text{ or } B) = \text{prob } (A) + \text{prob } (B) - \text{prob } (A \text{ and } B),$$
$$\text{prob } (A \text{ or } B) = \text{prob } (A) + \text{prob } (B) - 0,$$
$$\text{prob } (A \text{ or } B) = \text{prob } (A) + \text{prob } (B).$$

This is exactly the same as Formula 4.

EXERCISES

Determine which of the events given in Exercises 1–6 are mutually exclusive and which are not.

1. Catching a cold and getting a headache.

2. Getting a 3 and a 6 on one throw of a die.

3. Heating a home with gas heat and heating a home with oil heat.

4. Going on a skiing vacation in the Swiss Alps and going on a skiing vacation in Denver during the winter intersession period.

5. Smoking cigarettes and chewing tobacco.

6. Subscribing to the *Wall Street Journal* and subscribing to the *New York Times*.

7. A random survey of the members of the Lucky Health Club disclosed the following: the probability that a member uses the whirlpool is $\frac{3}{8}$, and the probability that a member uses the sauna is $\frac{1}{4}$. What is the probability that a member uses the whirlpool or the sauna if the probability that the member uses both is $\frac{1}{32}$?

8. According to local auto club officials, the probability that a gas station in the city has a functioning air pump is $\frac{2}{11}$, and the probability that the gas station attendant will check your oil is $\frac{5}{13}$. If the probability both that the gas station has a functioning air pump and that the attendant will check your oil is $\frac{2}{143}$, what is the probability of either event happening?

9. Insurance company statistics indicate that the probability that a married man in Rego Park has major medical insurance is 0.87 and the probability that he has disability insurance is 0.21. If the probability that he has both forms of insurance is 0.16, what is the probability that he has at least one of those forms of insurance?

10. How many credit cards are you carrying? A recent survey by a group of banks found that the probability that a person in Boulton has a Master Card issued in his or her name is 0.67 and the probability that the person has a VISA credit card is 0.71. Furthermore, the probability that the person has either of these credit cards is 0.92. What is the probability that a person has *both* a Master Card and a VISA credit card issued in his or her name in Boulton?

11. A computer club found that 59% of the people who own home computers use it for word processing, 31% of the owners use it for entertainment (games), and 27% of the owners use it for both word processing and entertainment. What is the probability that a home computer owner will use it either for word processing or for entertainment?

12. Willie walks into a bank. The probability that a teller is a part-time employee is $\frac{9}{17}$, and the probability that the teller is a male is $\frac{9}{34}$. If the probability of finding either a male teller or a part-time teller is $\frac{53}{68}$, what is the probability of finding a part-time male teller at the bank?

13. Motor vehicle records indicate that the probability that an applicant for a driver's license can pass the written exam on the first attempt is $\frac{2}{3}$. Furthermore, the probability is $\frac{2}{7}$ that the applicant can pass the road test on the first attempt, and the probability is $\frac{2}{21}$ that the applicant can pass both the written test and the road test on the first attempt. What is the probability that the applicant can pass either test on the first attempt?

14. Pete is in charge of the mail room of a stock brokerage office. The probability that he will send an important package by Federal Express is $\frac{5}{9}$, and the probability that he will send it by Purolator is $\frac{2}{9}$. What is the probability that he will *not* send an important package by Federal Express or Purolator?

15. Professor Gonzales teaches different types of computing courses. The probability that he will

teach Computing 106 (Introduction to BASIC) next semester is 0.59, and the probability that he will teach Computing 327 (Data Base Management) next semester is 0.62. If the probability that he will teach either of these courses is 0.86, what is the probability that he will teach both of these courses next semester?

16. Government records indicate that 74% of all mothers in San Piedro work (at least part-time) to supplement their family income, and 65% of all mothers in San Piedro send their young children to nursery school. If 44% of all working mothers send their children to nursery school, what is the probability that a mother will work or send her child to nursery school?

For any three events A, B, and C the probability of A or B or C is given by

$$\text{Prob } (A \text{ or } B \text{ or } C) = \text{prob } (A)$$
$$+ \text{ prob } (B) + \text{prob } (C) - \text{prob } (A \text{ and } B)$$
$$- \text{ prob } (A \text{ and } C) - \text{prob } (B \text{ and } C) +$$
$$\text{prob } (A \text{ and } B \text{ and } C).$$

Use this formula to solve the following problems:

17. Marilyn is the manager of a local fast-food restaurant. Over the past few years she has determined the following probabilities on the items that a customer will order:

Item(s)	Probability
Steak sandwich	$\frac{3}{7}$
French fries	$\frac{1}{2}$
Malted	$\frac{5}{11}$
Steak sandwich and French fries	$\frac{2}{9}$
French fries and malted	$\frac{1}{7}$
Steak sandwich and malted	$\frac{1}{4}$
Steak sandwich, French fries, and malted	$\frac{1}{11}$

What is the probability that a customer will order either a steak sandwich, French fries, or a malted?

18. Dave is putting on a going-away party for Sherry. He will definitely buy and serve scotch, bourbon, or rye whisky. The probability that he serves scotch is $\frac{2}{3}$, the probability that he serves bourbon is $\frac{1}{2}$, the probability that he serves rye is $\frac{25}{56}$, the probability that he serves scotch and rye is $\frac{71}{336}$, the probability that he serves rye and bourbon is $\frac{17}{84}$, and the probability that he serves scotch and bourbon is $\frac{23}{112}$. Find the probability that he serves all three whiskies.

19. An obstetrician is analyzing some facts about her patients and their recent deliveries. She has determined the following probabilities:

Fact	Probability
Mother over 35 years of age	0.27
First child for mother	0.32
Mother had a well-paying career job	0.42
First child for mother and over 35 years of age	0.17
First child for mother and mother had a well-paying career job	0.09
Mother over 35 years of age and had a well-paying career job	0.07
First child for mother and mother over 35 years of age who had a well-paying career job	0.02

What is the probability that the mother is either over 35 years of age or had a well-paying career job or that the child is a first child for the mother?

10.6

CONDITIONAL PROBABILITY

In the last section we considered problems in which more than one event occurred. The formulas given there do not apply to all situations, as the following examples illustrate.

EXAMPLE 1 Joan is at a stand in an amusement park where there are three identical boxes, two of which contain one red marble each. The third box has a white marble in it. To win, Joan must guess the color of the marble in each box. She has guessed that the first box has a red marble, and she is right. Joan claims that the marble in the second box is red. What is the probability that she is correct?

SOLUTION From the information given, we know that there is a red marble in box 1. This means that only one red marble and one white marble remain. Since either of them can be in box 2, the probability that she is correct if she guesses a red marble is $\frac{1}{2}$. □

EXAMPLE 2 Bob has drawn a card from a 52-card deck. What is the probability that it is a jack if we know that it is a picture card?

SOLUTION There are 12 picture cards: 4 jacks, 4 queens, and 4 kings and thus 12 outcomes, of which 4 are favorable. Thus the probability that the card is a jack, given that it is a picture card, is $\frac{4}{12}$, or $\frac{1}{3}$. □

EXAMPLE 3 Bill is getting dressed. He reaches into a drawer where he has 3 black and 2 gray socks (these are not pairs, but individual socks). He selects one sock and then, without replacing it, selects another. What is the probability that both are black?

SOLUTION An easy way to solve this problem is to list all the possible outcomes and then to count all the favorable ones. To do this, we label the black socks as b_1, b_2, b_3 and the gray socks as g_1, g_2. The possible outcomes are:

$\mathbf{b_1, b_2}$	$\mathbf{b_2, b_3}$	b_3, g_1	g_1, g_2
$\mathbf{b_1, b_3}$	b_2, g_1	b_3, g_2	g_2, b_1
b_1, g_1	b_2, g_2	g_1, b_1	g_2, b_2
b_1, g_2	$\mathbf{b_3, b_1}$	g_1, b_2	g_2, b_3
$\mathbf{b_2, b_1}$	$\mathbf{b_3, b_2}$	g_1, b_3	g_2, g_1

Out of the 20 possible outcomes, 6 are favorable. These are the ones in boldface. They represent a pair of matching black socks. Thus the probability that both are black is $\frac{6}{20}$. □

EXAMPLE 4   In the previous example, what is the probability that the second sock selected is black if we know that the first sock is black?

SOLUTION   There are two ways to do this problem. One way is to list all possible outcomes and then count the favorable ones. There are 12 possible outcomes, 6 of which are favorable. These are in boldface as shown here:

$$\mathbf{b_1, b_2} \quad\quad \mathbf{b_2, b_3} \quad\quad \mathbf{b_1, g_1} \quad\quad b_2, g_2$$
$$\mathbf{b_1, b_3} \quad\quad \mathbf{b_3, b_1} \quad\quad b_1, g_2 \quad\quad b_3, g_1$$
$$\mathbf{b_2, b_1} \quad\quad \mathbf{b_3, b_2} \quad\quad b_2, g_1 \quad\quad b_3, g_2$$

Thus the probability that the second sock is black, if we know that the first sock is black, is $\frac{6}{12}$, or $\frac{1}{2}$. □

Comment   Compare this answer with the answer to Example 3. It is not the same because in Example 3 we were considering only the probability of selecting a pair of black socks out of *all* possible ways of selecting a pair of socks. In Example 4, we considered the probability of selecting a second black sock *after* we know that the first sock is already a black one.

The situation of Example 4 is called a **conditional probability** because we are interested in the probability of getting a black sock, given that (or conditional on the fact that) the first sock was black. We use a special symbol for this. We write

prob (second sock is black | first sock is black).

The vertical line "|" stands for the words "given that" or "if we know that." Using this notation, we have

$$\text{prob (second sock is black | first sock is black)} = \frac{1}{2}.$$

A second way of doing this problem involves a formula called the **conditional probability formula.**

Formula 6   *If A and B are any events, then*

$$\text{prob } (A \mid B) = \frac{\text{prob } (A \text{ and } B)}{\text{prob } (B)}$$

Applying this formula to our problem, we have

prob (second sock is black | first sock is black)

$$= \frac{\text{prob (second sock is black and first sock is black)}}{\text{prob (first sock is black)}}$$

This simplifies to

$$\frac{\text{prob (both socks are black)}}{\text{prob (first sock is black)}}.$$

Both of these numbers are easily calculated. As a matter of fact, the top part of the fraction was calculated in Example 3. It is $\frac{6}{20}$. The bottom part of the fraction is found by calculating the number of outcomes in which the first sock is black. This gives 12 out of a possible 20 outcomes. (Verify that there are actually 12 by counting them.) Thus Formula 6 gives

$$\text{prob (second sock is black | first sock is black)} = \frac{6/20}{12/20}$$

$$= \frac{6}{20} \div \frac{12}{20}$$

$$= \frac{6}{20} \cdot \frac{20}{12}$$

$$= \frac{6}{12}$$

$$= \frac{1}{2}.$$

EXAMPLE 5   We will redo Example 2 on p. 555 using Formula 6.

SOLUTION   We want prob (jack | picture card). Using Formula 6, we get

$$\text{prob (jack | picture card)} = \frac{\text{prob (jack and picture card)}}{\text{prob (picture card)}}.$$

For a card to be both a jack and a picture card it must be a jack. We therefore have that the probability that a card is a jack *and* a picture card is $\frac{4}{52}$. Since there are 12 picture cards, the probability of a picture card is $\frac{12}{52}$.

Putting these into the formula, we get

$$\text{prob (jack | picture card)} = \frac{4/52}{12/52}$$

$$= \frac{4}{52} \div \frac{12}{52}$$

$$= \frac{4}{52} \cdot \frac{52}{12}$$

$$= \frac{4}{12} = \frac{1}{3}.$$

Of course this is the same answer as the one we got before. ☐

EXAMPLE 6 An absentminded professor often forgets to put money in the meter when he parks. The probability that he will forget to put money in the meter is $\frac{7}{10}$. If the probability that he gets a ticket when he forgets to put money in the meter is $\frac{3}{7}$, what is the probability that he will forget to put money in the meter and that he will get a ticket?

SOLUTION We will use Formula 6:

prob (gets ticket | forgets money)

$$= \frac{\text{prob (forgets money and gets ticket)}}{\text{prob (forgets money)}}.$$

Using the given information, we have

$$\frac{3}{7} = \frac{\text{prob (forgets money and gets ticket)}}{7/10}.$$

We multiply both sides by $\frac{7}{10}$, getting

$$\frac{7}{10} \cdot \frac{3}{7} = \frac{7}{10} \cdot \frac{\text{prob (forgets money and gets ticket)}}{7/10}$$

$$\frac{3}{10} = \text{prob (forgets money and gets ticket)}.$$

Therefore the probability that he will forget to put money in the meter *and* that he will get a ticket is $\frac{3}{10}$. ☐

Independent Events In many cases it turns out that whether or not one event happens does not affect whether or not another will happen. For example, if a coin is tossed and if a die is rolled, the outcome of the coin toss has nothing to do with the outcome of rolling the die. Also, if two coins are tossed, then the outcome for the first coin has nothing to do with the outcome for the second coin. Such events are called **independent events.**

For independent events, Formula 6 can be simplified, as the following example shows.

EXAMPLE 7 If a die is rolled once and if a coin is tossed once, what is the probability that the die will show a 3 and that the coin will come up heads?

SOLUTION We will solve the problem in two ways. One is by counting. The other way is by the formula for independent events.

To do it by counting, we will list all the possible outcomes. There are 12 of them. These are:

1 H	**3 H**	5 H
1 T	3 T	5 T
2 H	4 H	6 H
2 T	4 T	6 T

Among the 12 possible outcomes, only one of them is favorable. This is the boldface one, **3 H.** Therefore the probability of getting a 3 and a head is $\frac{1}{12}$. Now we will do the problem by using the following formula. ▭

Formula 7   *If A and B are independent events, then*

$$prob\ (A\ and\ B) = prob\ (A) \cdot prob\ (B).$$

If we let *A* represent the event of getting a 3 when we throw a die and let *B* stand for the event of getting a head, then we are looking for

$$prob\ (A\ and\ B).$$

By the formula this is prob $(A) \cdot$ prob (B). We know that prob $(A) = \frac{1}{6}$ and that prob $(B) = \frac{1}{2}$. Thus

$$prob\ (A\ and\ B) = prob\ (A) \cdot prob\ (B)$$
$$= \frac{1}{6} \cdot \frac{1}{2} = \frac{1}{12}.$$

Comment   Formula 7 may look like a new formula to be learned. This is not so. It is really a simplified case of Formula 6. Let us see why. Formula 6 says that for *any* events *A* and *B*

$$prob\ (A\ |\ B) = \frac{prob\ (A\ and\ B)}{prob\ B}.$$

If *A* is independent of *B*, then prob $(A\ |\ B)$ is the same as prob (A). Why? Therefore for independent events, Formula 6 becomes

$$prob\ (A) = \frac{prob\ (A\ and\ B)}{prob\ (B)}.$$

If we now multiply both sides by prob (B), we get

$$prob\ (A) \cdot prob\ (B) = prob\ (A\ and\ B).$$

This is exactly the same as Formula 7.

EXERCISES

1. Thirty-eight percent of the employees of Apex Consulting Company are male systems analysts. Fifty-four percent of the Apex Consulting Company employees are male. If a male employee of the Apex Consulting Company is randomly selected, what is the probability that he is a systems analyst?

2. Jack McAllister owns and operates a huge farm in California. The probability that an employee is a migrant worker is 0.95, and the probability that the migrant worker is also an illegal alien is 0.24. If a migrant worker is randomly selected, what is the probability that the worker is an illegal alien?

3. A recent study found that the probability that a person in Bakersville has a checking account is 0.82 and the probability that the person has a checking account as well as overdraft privileges is 0.31. (Overdraft privileges allow customers to write checks for amounts that exceed their current balance.) If a customer who has a checking account is randomly selected, what is the probability that the customer has overdraft privileges?

4. *Legalizing Drugs.* A newspaper reporter conducted a nationwide survey of 1638 people to find out what they thought about legalizing drug use. The results of the survey are shown in the accompanying table:

Region of country in which respondent lives	In favor of proposal	Against proposal
East	164	204
Midwest	110	358
South	128	276
Far West	146	252

Find the probability that a randomly selected individual in the group

a) lives in the East given that he or she is against the proposal.

b) is against the proposal given that he or she lives in the East.

c) is against the proposal.

5. Consider the newspaper article below. The probability of a motorist having a mechanical breakdown on the city's highways is 0.03. What is the probability that a motorist's car will break down and that the motorist will *not* be able to summon help because of a call box that is not functioning properly?

Majority of Police Call Boxes Not Functioning

NEW YORK: A random survey by reporters for the local auto club found that 57% of the emergency call boxes on the city's highways were not functioning properly because of vandalism or were missing telephones completely. The phones in these strategically placed call boxes allow a motorist to summon help in the event of a mechanical breakdown.

DAILY PRESS May 17, 1985

6. Roberto is having trouble with the disc drive of his computer. He is also using cheap discs, which often are defective. The probability that the disc drive is not operating properly is $\frac{3}{11}$. The probability that the disc drive is not operating properly and that the cheap discs are defective is $\frac{2}{33}$. If it is known that the disc drive is not operating properly, find the probability that the cheap discs are defective.

7. Government records indicate that 17% of all senior citizens in a certain city have had some sort of trouble in the past with the Social Security system. Moreover, in this city, 51% of the population are senior citizens. If a senior citizen is selected at random, what is the probability that the individual has never had any sort of trouble in the past with the Social Security system?

8. The Bruce Mechanical Corporation owns two photocopying machines, which often do not operate properly. The probability that the first machine will produce unacceptable copies is 0.23, and the probability that the second machine will produce unacceptable copies is 0.16. What is the probability that on a given day both machines will produce copies that are acceptable?

9. Beth McGuire, Nancy Peters, and Joy Richards have applied to a bank for auto loans. The probability that Beth McGuire's application will be approved is 0.82. The probability that Nancy Peters's application will be approved is 0.73 and the probability that Joy Richards's application will be approved is 0.67. Assuming independence, what is the probability that all three applications will be approved?

10. Refer back to the previous question. What is the probability that only Nancy Peters's application will be approved?

11. In a certain mining town the following statistics have been accumulated. The probability that a miner has black lung disease is 0.53, and the probability that a miner has arthritis is 0.21. Assuming independence, what is the probability that a randomly selected miner does not have black lung disease but has arthritis?

12. Heather has just parked her car by a parking meter. She notices that someone has placed a paper bag over the meter and scribbled the words "meter out of order" on it. Is this an accurate description of the meter? Should she deposit her quarter? The probability that the parking meter is out of order is $\frac{5}{12}$. The probability that the parking meter is out of order *and* that she loses her quarter is $\frac{2}{9}$. If the parking meter is actually out of order, what is the probability that she does *not* lose her quarter?

13. *Law.* In June 1964 an elderly woman was mugged in San Pedro, California. In the vicinity of the crime a bearded, black man sat waiting in a yellow car. Shortly after the crime was committed, a young white woman, wearing her blonde hair in a ponytail, was seen running from the scene of the crime and getting into the car, which sped off. The police broadcast a description of the suspected muggers. Soon afterward a couple fitting the description was arrested and convicted of the crime. Although the evidence in the case was largely circumstantial, the prosecutor based his case on probability and the unlikelihood of another couple having such characteristics. He assumed the probabilities shown in the accompanying table.

Characteristic	Assumed probability
Drives yellow car	$\frac{1}{10}$
Black–white couple	$\frac{1}{1000}$
Black man	$\frac{1}{3}$
Man with beard	$\frac{1}{10}$
Blonde woman	$\frac{1}{4}$
Woman with ponytail	$\frac{1}{10}$

The prosecutor then multiplied the individual probabilities:

$$\left(\frac{1}{10}\right)\left(\frac{1}{1000}\right)\left(\frac{1}{3}\right)\left(\frac{1}{10}\right)\left(\frac{1}{4}\right)\left(\frac{1}{10}\right) = \frac{1}{12,000,000}.$$

He claimed that the probability is 1/12,000,000 that another couple has such characteristics. The jury agreed and convicted the couple. The conviction was overturned by the California Supreme Court in 1968. The defense attorneys got some professional advice on probability. Serious errors were found in the prosecutor's probability calculations. Some of these involved assumptions about independent events. As a matter of fact, it was demonstrated that the probability is 0.41 that another couple with the same characteristics existed in the area once it was known that there was at least one such couple. For a complete discussion of this probability case, read "Trial by Mathematics" in *Time* (January 8, 1965, p. 42; and April 26, 1968, p. 41).

10.7

**ODDS AND
MATHEMATICAL
EXPECTATION**

Gamblers are always interested in the odds of a game or a race. They are also interested in the amount of money to be won. In this section we will investigate the meaning of these ideas and learn how to calculate odds.

To best understand these ideas, let us consider a man at Aqueduct Raceway. Nine horses have been entered in the big race. Our man places $10 on the horse Liverwurst. We will assume that each horse has an equal chance of winning. Therefore the probability that the man will win is $\frac{1}{9}$. Gamblers prefer to speak in terms of **odds.** They would say that the odds in favor of winning are 1 to 8 and the odds against his winning are 8 to 1. The 8 represents the eight chances of losing. Thus we have the following definitions.

> **Definition 10.6**   *The **odds in favor** of an event occurring are p to q, where p is "the number of favorable outcomes" and q is "the number of unfavorable outcomes."*

> **Definition 10.7**   *The **odds against** an event occurring are q to p, where q and p are the same as in Definition 10.6.*

We illustrate these definitions with several examples.

EXAMPLE 1   What are the odds in favor of drawing an ace from a full deck of 52 cards on one draw?

SOLUTION   Since there are 4 aces and 48 non-aces, Definition 10.6 tells us that the odds in favor of drawing an ace are 4 to 48.   □

EXAMPLE 2   What are the odds in favor of winning at the Hopeless Wheel of Fortune, which is divided into 10 equal parts, each with a different color, if someone bets the colors red and blue and only one color wins?

SOLUTION   Since there are 2 favorable and 8 unfavorable outcomes, Definition 10.6 tells us that the odds in favor of winning are 2 to 8.   □

EXAMPLE 3   What are the odds *against* throwing a 2 or a 12 in throwing a pair of dice?

SOLUTION   When a pair of dice is thrown, there are 36 possible outcomes (p. 522). Two are favorable, and 34 are unfavorable. Thus Definition 10.7 tells us that the odds against getting a 2 or a 12 are 34 to 2.   □

When gambling (as well as in business situations), the amount of money to be won in the long run is called the **mathematical expectation.** It is defined as follows.

> **Definition 10.8** *Suppose an event has several possible outcomes with probabilities p_1, p_2, p_3, and so on. Suppose on the first event the payoff is m_1, on the second event the payoff is m_2, on the third event the payoff is m_3, etc. Then the **mathematical expectation** of the event is*
>
> $$m_1p_1 + m_2p_2 + m_3p_3 + \cdots$$

The following examples show how this definition is applied.

EXAMPLE 4   A die is tossed once. If a 1 comes up, then Joe will win $10; and if a 4 comes up, he will win $7. What is his mathematical expectation?

SOLUTION   When a die is tossed once, then the probability of getting a 1 is $\frac{1}{6}$.

Similarly, the probability of getting a 4 is $\frac{1}{6}$. Using Definition 10.8, we find that the mathematical expectation is

$$10 \cdot \frac{1}{6} + 7 \cdot \frac{1}{6} = \frac{17}{6} = \$2.83 \quad \text{(when rounded)}. \quad \square$$

EXAMPLE 5   The local chapter of the American Cancer Society is planning to hold a bazaar to raise funds. If the bazaar is held outdoors, $100,000 is expected to be raised. If the bazaar is held indoors, then $75,000 is expected to be raised. The probability that it rains, forcing the bazaar to be held indoors, is $\frac{3}{5}$ and the probability that the weather is suitable for an outdoor bazaar is $\frac{2}{5}$. How much money can they expect to raise?

SOLUTION   We will use Definition 10.8. We have

$$100{,}000 \cdot \frac{2}{5} + 75{,}000 \cdot \frac{3}{5} = 40{,}000 + 45{,}000$$
$$= 85{,}000.$$

Thus they can expect to raise $85,000. $\quad \square$

EXAMPLE 6   A wheel of fortune at an amusement park is divided into 4 colors: red, blue, yellow, and green. The probabilities of the spinner's landing in any of these colors are $\frac{3}{10}, \frac{4}{10}, \frac{2}{10},$ and $\frac{1}{10}$, respectively. A player can win $4 if the spinner stops on red and $2 if it stops on green, and lose $2 if it stops on blue and $3 if it stops on yellow. Trudy has decided to try her luck at the wheel. What is her mathematical expectation?

SOLUTION   We indicate the possible outcomes and the corresponding probabilities by the chart below.

Outcome	Probability	Amount of money won or lost
Red	$\dfrac{3}{10}$	+4
Blue	$\dfrac{4}{10}$	−2
Yellow	$\dfrac{2}{10}$	−3
Green	$\dfrac{1}{10}$	+2

Thus her mathematical expectation is

$$(4)\frac{3}{10} + (-2)\frac{4}{10} + (-3)\frac{2}{10} + (2)\frac{1}{10} = \frac{12}{10} - \frac{8}{10} - \frac{6}{10} + \frac{2}{10}$$
$$= 0.$$

Her mathematical expectation is 0. What does this 0 mean? We interpret this to mean that in the long run she will win 0 dollars or break even. □

Comment   Some gamblers base their decision whether or not to play a particular game solely on the game's mathematical expectation. Obviously, if the game has a negative mathematical expectation, a gambler should not play, since he or she will lose money in the long run. There would be little point in playing a game whose mathematical expectation is 0, since in the long run the amount of money that can be won is 0.

Mathematical expectation can also be applied to nonmoney situations, as the following example will illustrate.

EXAMPLE 7   An observer for an energy conservation group has collected the following statistics on the number of occupants per car that pass through a certain tollgate.

Number of passengers in car (including driver)	1	2	3	4	5	6
Probability	0.37	0.29	0.18	0.09	0.05	0.02

What is the expected number of occupants per car?

SOLUTION   We apply Definition 10.8 and multiply each of the possible outcomes by its probability. We get

$$1(0.37) + 2(0.29) + 3(0.18) + 4(0.09) + 5(0.05) + 6(0.02),$$

which equals 2.22. Thus the expected number of occupants per car is 2.22. ☐

Pascal used mathematical expectation to make a "wager with God." As we have said, Pascal was extremely religious. He reasoned that leading a religious life will result in eternal happiness. The value of eternal happiness is infinite. Therefore the expectation is

$$m \cdot p = \begin{pmatrix} \text{value of eternal} \\ \text{happiness} \end{pmatrix} \begin{pmatrix} \text{probability of obtaining} \\ \text{eternal happiness} \end{pmatrix}.$$

Since the value of eternal happiness is infinite, the product $m \cdot p$ is also infinite, even if the probability of obtaining eternal happiness is small. Thus it pays to lead a religious life, since the expectation is infinite.

EXERCISES

1. A card is drawn from an ordinary deck of cards. What are the odds in favor of getting a picture card?

2. A clothing store receives 100 pairs of "designer" jeans. Ninety-five of these are authentic, and 5 are cheap imitations. Maureen buys a pair of these jeans. What are the odds against her getting the designer jeans?

3. Jerry parks his car by a meter. He needs to put a quarter into the meter, but he has no change. There are 15 people standing nearby. Jerry decides to ask *only* one person to give him change for a dollar. If 6 of the people actually have change, what are the odds against his getting the change?

4. Consider the newspaper article at the right. Bill Sadowski invested a considerable amount of money in an abusive tax shelter. What are the odds in favor of his tax return being audited?

IRS To Investigate Abusive Tax Shelters

WASHINGTON: Officials of the Internal Revenue Service announced yesterday that they would carefully scrutinize all abusive tax shelters. Furthermore, an official said that an individual who invests in such shelters has a 95% chance of having his or her tax return audited.

THE TIMES, May 10, 1985

5. Barbara purchased 100 shares of stock of a company. The probability that there will be a stock split is $\frac{4}{9}$. What are the odds in favor of there being a stock split?

6. The town of Rockville is planning a flea market sale with profit going to the volunteer ambulance corps. If the weather is nice, then the flea market will be held outdoors in the baseball field. It is expected that $10,000 can then be raised. If it rains, then the flea market will be held indoors in the high school gym. If this occurs, then only $5500 can be raised, since many people do not like the tight spaces of the gym. If it snows, then the flea market sale will be canceled completely, and no money will be raised. The weather forecast is as follows:

Weather forecast	Probability
Nice weather	$\frac{5}{9}$
Rain	$\frac{3}{9}$
Snow storm	$\frac{1}{9}$

How much money can the town expect to raise?

7. A manufacturer has just introduced a new laundry detergent and would like to promote it by advertizing on television, radio, in magazines, and by distributing free samples. Market research indicates the following sales to be generated from each advertisement medium and the associated probability:

Advertisement medium	Potential sales	Probability
Television	$70,000	0.37
Radio	$37,000	0.56
Magazine	$45,000	0.49
Distributing free samples	$50,000	0.42

Find the expected sales from each medium.

8. Refer back to the previous exercise. If the manufacturer wishes to use only one of these advertisement media, which one should be selected? Explain your answer.

9. A man has just taken out a life insurance policy that will pay his beneficiaries $10,000 in the event of his death. The premium for this insurance coverage is $300. If the probability of the man dying is 0.07, what is the insurance company's mathematical expectation?

10. A clerk in the police department has compiled the following list on the number of requests for gun permits by individual citizens per day.

Number of requests	Probability
0	0.07
1	0.13
2	0.25
3	0.14
4	0.13
5	0.10
6	0.15
7	0.03

What is the expected number of requests for gun permits per day?

11. Jake purchased 100 tickets for this week's state lottery. The cost per ticket was $1. There will be one grand prize and ten second prizes as follows:

Prize amount	Probability of any one ticket winning
Grand prize: $100,000 (only 1 winner)	0.0003
Second prize: $1000 (for each winning ticket)	0.0009

What is Jake's mathematical expectation?

12. Casey is considering opening up a fast-food restaurant at one of two amusement parks. In park I he must invest $20,000, and he can expect an annual income of $100,000 with a probability of $\frac{5}{9}$ if successful. In park II he must invest $5000, and he can expect an annual income $37,000 with a probability of $\frac{7}{9}$ if successful.

a) What is his mathematical expectation from each park?

b) On the basis of mathematical expectation only, in which park should Casey open his fast-food restaurant? Explain your answer.

13. Jim is told to roll a pair of dice. He can either win or lose money, depending upon the sum of the number of dots shown on both dice together. The amount of money won or lost is as follows:

Sum of the number of dots shown	Amount of money won or lost
2 or 4	win $12
5 or 8 or 9	lose $7
3 or 6	win $9
10 or 12	lose $10
7 or 11	win $15

What is the expected amount of money that Jim could win (or lose)?

10.8

APPLICATION TO GENETICS

One very interesting application of probability is in the science of **genetics.** This science is concerned with which traits can be inherited. The pioneer in this field was Gregor Mendel. As we mentioned in Chapter 2, Mendel performed many experiments with garden peas. As a result of his experiments, Mendel was able to state the basic laws of heredity.

Specifically, Mendel crossbred plants from wrinkled seeds with plants from smooth seeds. The resulting plants all had smooth seeds. However, when he crossbred these new plants with one another, a strange thing happened. Three-fourths of the plants had smooth seeds, and one-fourth had wrinkled seeds (See Fig. 10.8).

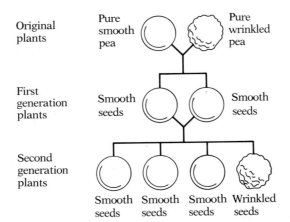

Figure 10.8

He concluded that certain *genes* are *dominant*. When a gene is dominant, then the trait that it represents will always appear, no matter which gene it is paired with. Such was the case for smooth seeds. On the other hand, there are other genes that are *recessive*. If a trait is recessive, it must be matched with the same type of gene for that trait to appear. Wrinkled seeds are recessive. This explains why when wrinkled seeds are matched with smooth seeds, the resulting plants all had smooth seeds. The smooth-seed gene "dominated" the wrinkled-seed one. On the other hand, in the second generation, "mixed" plants that now had both wrinkled and smooth genes were matched with each other. Those that contained any smooth-seed genes produced smooth-seed plants, and those that contained *only* wrinkled-seed genes produced wrinkled seeds as shown below.

	Smooth Seeds	*Wrinkled Seeds*
Smooth Seeds	Offspring have smooth seeds	Offspring have smooth seeds
Wrinkled Seeds	Offspring have smooth seeds	Offspring have wrinkled seeds

The same analysis was used by Mendel to explain why different flowers of the same kind of plant have different colors. For example, when a red-flowered four o'clock plant carrying a gene R for red was crossbred with another four o'clock plant carrying R, the offspring was RR and always had red flowers. Similarly, when a white-flowered four o'clock plant carrying a gene W for white was crossbred with another four o'clock plant carrying W, the offspring was WW and always had white flowers. On the other hand, when a plant carrying R was crossbred with a plant carrying W, a strange thing happened. The offspring contained the genes RW and was therefore pink, P. Mendel continued with these experiments and recorded the numerical percentages with which characteristics were inherited as shown in the following table.

		Genes of one plant		
		Red (RR) *Offspring will be*	*White (WW)* *Offspring will be*	*Pink (RW)* *Offspring will be*
Genes of other plant	*Red (RR)*	RR offspring— all red	RW offspring— all pink	RR—50% red RW—50% pink
	White (WW)	RW offspring— all pink	WW offspring— all white	RW—50% pink WW—50% white
	Pink (RW)	RR—50% red RW—50% pink	RW—50% pink WW—50% white	RR—25% red RW—50% pink WW—25% white

From his observations on the peas and flowers, Mendel concluded that certain characteristics (or traits) were determined by a single pair of genes (one for each parent). The same is true of human beings. For example, it is known that a baby's eye color is determined by certain genes, one obtained from each parent. There are other inherited traits that are determined by several genes from each parent. In this section we concern ourselves only with those characteristics that are determined by a single pair of genes.

Let us analyze several such traits, such as eye color and the blood factor Rh.

EXAMPLE 1 It is known that in human eyes, the color brown is dominant and the color blue is recessive. Let B represent a gene for brown eyes and let b represent a gene for blue eyes. Thus a person who has BB genes (that is, brown from each parent) or Bb genes (brown from one parent and blue from the other) will have brown eyes, since brown is dominant. On the other hand, a person who has bb genes (blue from both parents) will have blue eyes. Arlene, who is known to have Bb genes, marries Tom, who is also known to have Bb genes. What is the probability that their child will have blue eyes?

SOLUTION We set up a chart indicating how the genes can be matched and the resulting child's eye color (Fig. 10.9). In this chart, every time B is paired with any other gene, the child will have brown eyes. Only the pairing bb will result in blue eyes for the baby. Out of the 4 possible outcomes BB, Bb, bB, and bb, only 1 is favorable. Thus the probability that the child will have blue eyes is $\frac{1}{4}$. □

		Arlene Gene *B*	Arlene Gene *b*
Tom	Gene *B*	*BB*	*Bb*
Tom	Gene *b*	*bB*	*bb*

Figure 10.9

EXAMPLE 2 Another inherited trait that is known to be determined by a single gene from each parent is the Rh blood factor. When a person has this factor in his or her blood, we say that he or she is Rh positive. Otherwise he or she is Rh negative. Eighty-five percent of American Caucasians and 93 percent of blacks are born with this inherited substance. Thus if the parents of the husband and of the wife are pure Rh positive, then all the children will be Rh positive. On the other hand, if both husband and wife had one Rh-positive and one Rh-negative parent, then such a couple would produce, in every 4 children, one child who is pure Rh positive, 2 children who are partial Rh positive, and one child who is Rh negative.

In a recent court suit, Jane, who was known to be Rh negative, claimed that Bill, who was also Rh negative, was the father of her Rh-positive baby. Find the probability that Bill actually was the father.

SOLUTION   Since Jane is known to be Rh negative, it is impossible for Bill to be the father, since he is also Rh negative. Two Rh-negative parents can produce only an Rh-negative child. Thus the probability that Bill is the father of Jane's child is 0.   □

Comment   Certain traits in human beings are known to be determined by the genes received from each parent. For example, albinism (no skin color, eye color, or hair color), muscular atrophy, Tay-Sach's disease, hemophilia, and sickle-cell anemia are known to be determined by genes. With appropriate medical care and advice the probability of these occurring can be reduced.

EXERCISES

For Exercises 1–3, use the following information: An important human characteristic that is known to be transmitted by genes is albinism, in which a child is born with no skin color, no eye color, and no hair color. It is known that albinism is a recessive trait. Thus if N is the normal gene and a is a gene for albinism, then a person with Na genes will have normal color, and a child with aa genes will be an albino.

1. Mr. Gelespie has Na genes, and Mrs. Gelespie has Na genes. Find the probability that their child will have normal color.

2. Mr. Jones has NN genes, and Mrs. Jones has Na genes. Find the probability that their child will

   a) be an albino.   b) have normal color.

3. Mrs. Smith has just given birth to an albino child. Furthermore, it is known that Mrs. Smith

has aa genes. What type of genes can Mr. Smith have for him to be the father?

For Exercises 4–6, use the following information: A botanist is experimenting with flowers that have a certain trait T or t. She crossbreeds a plant that has Tt genes with a plant that has TT genes. Find the probability that the offspring will be:

4. TT 5. Tt 6. tt

For Exercises 7–9, use the following information: A researcher in a laboratory is splicing genes. She crossbreeds an organism that has a certain characteristic T (it has Tt genes) with another organism that lacks the trait (it has tt genes). Find the probability that the offspring will be:

7. TT 8. tt 9. Tt

10.9
RANDOM NUMBERS

In a large city the social services department has decided to make a detailed study of 100 welfare cases. These 100 cases are to be chosen completely *at random* from among the 20,000 families currently receiving aid. It is important to the department that each family have an equal chance of being selected for the study. How can the department select the families for the study? There are several ways in which this can be done.

1. Write each family name on a piece of paper and put the names in a box. After mixing the contents thoroughly, select 100 names from the box. While this approach might seem sensible, it is unlikely that we would obtain a truly random mix. For one thing, most people would not stick their hand in to the bottom of the box to make a selection. Thus the names on the bottom are less likely to be selected than the names on top. Moreover, if you do not replace each name after a selection, then the 100th name drawn has a greater likelihood of being selected than the first name drawn. (Can you see why this is so?) Even if we correct these difficulties, which clearly can be done, this method is obviously inefficient.

2. Arrange all the names alphabetically and number them from 1 to 20,000. Then select every 200th name. Again this does not give a truly random selection. By selecting every 200th name we are actually making sure that those families with numbers 1–199, 201–399, etc., never have a chance of being selected.

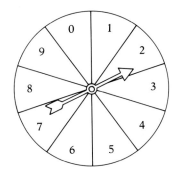

3. Assign a 5-digit case number to each family, starting with the number 00001. Then use a spinner like one shown here, to generate *random numbers* as follows. Use the spinner 5 times to obtain 5 digits in order. The 5-digit number obtained by this process is the first case number to be selected. We continue in this way until we get the 100 case numbers that we want.

It is unlikely that this method will generate completely random numbers either. Can you find some reasons why it will not? Basically, this last method is good. That is, generating random numbers and using the families whose numbers have been selected is reasonable. The difficulties arise because we generate the random numbers using the spinner. We can overcome this difficulty by using a different technique to generate the random numbers. There are various ways of doing this, most of which use computers. Random numbers generated in this manner are often listed in *Tables of Random Digits*.

Therefore the best way for the social services department to select the families for the study is to use such a table to generate random

TABLE 10.1
Table of random digits

Line	(1)	(2)	(3)	(4)	(5)	(6)	(7)	(8)	(9)	(10)	(11)	(12)	(13)	(14)
1	10480	15011	01536	02011	81647	91646	69179	14194	62590	36207	20969	99570	91291	90700
2	22368	46573	25595	85393	30995	89198	27982	53402	93965	34095	52666	19174	39615	99505
3	24130	48360	22527	97265	76393	64809	15179	24830	49340	32081	30680	19655	63348	58629
4	42167	93093	06243	61680	07856	16376	39440	53537	71341	57004	00849	74917	97758	16379
5	37570	39975	81837	16656	06121	91782	60468	81305	49684	60672	14110	06927	01263	54613
6	77921	06907	11008	42751	27756	53498	18602	70659	90655	15053	21916	81825	44394	42880
7	99562	72905	56420	69994	98872	31016	71194	18738	44013	48840	63213	21069	10634	12952
8	96301	91977	05463	07972	18876	20922	94595	56869	69014	60045	18425	84903	42508	32307
9	89579	14342	63661	10281	17453	18103	57740	84378	25331	12566	58678	44947	05585	56941
10	85475	36857	53342	53988	53060	59533	38867	62300	08158	17983	16439	11458	18593	64952
11	28918	69578	88231	33276	70997	79936	56865	05859	90106	31595	01547	85590	91610	78188
12	63553	40961	48235	03427	49626	69445	18663	72695	52180	20847	12234	90511	33703	90322
13	09429	93969	52636	92737	88974	33488	36320	17617	30015	08272	84115	27156	30613	74952
14	10365	61129	87529	85869	48237	52267	67689	93394	01511	26358	85104	20285	29975	89868
15	07119	97336	71048	08178	77233	13916	47564	81056	97735	85977	29372	74461	28551	90707
16	51085	12765	51821	51259	77452	16308	60756	92144	49442	53900	70960	63990	75601	40719
17	02368	21382	52404	60268	89368	19885	55322	44819	01188	65255	64835	44919	05944	55157
18	01011	54092	33362	94904	31273	04146	18594	29852	71585	85030	51132	01915	92947	64951
19	52162	53916	46369	58586	23216	14513	83149	98736	23495	64350	94738	17752	35156	35749
20	07056	97628	33787	09998	42698	06691	76988	13602	51851	46104	88916	19509	25625	58104
21	48663	91245	85828	14346	09172	30168	90229	04734	59193	22178	30421	61666	99904	32812
22	54164	58492	22421	74103	47070	25306	76468	26384	58151	06646	21524	15227	96909	44592
23	32639	32363	05597	24200	13363	38005	94342	28728	35806	06912	17012	64161	18296	22851
24	29334	27001	87637	87308	58731	00256	45834	15398	46557	41135	10367	07684	36188	18510
25	02488	33062	28834	07351	19731	92420	60952	61280	50001	67658	32586	86679	50720	94953

26	81525	72295	04839	96423	24878	82651	66566	14778	76797	14780	13300	87074	79666	95725
27	29676	20591	68086	26432	46901	20849	89768	81536	86645	12659	92259	57102	80428	25280
28	00742	57392	39064	66432	84673	40027	32832	61362	98947	96067	64760	64584	96096	98253
29	05366	04213	25669	26422	44407	44048	37937	63904	45766	66134	75470	66520	34693	90449
30	91921	26418	64117	94305	26766	25940	39972	22209	71500	64568	91402	42416	07844	69618
31	00582	04711	87917	77341	42206	35126	74087	99547	81817	42607	43808	76655	62028	76630
32	00725	69884	62797	56170	86324	88072	76222	36086	84637	93161	76038	65855	77919	88006
33	69011	65795	95876	55293	18988	27354	26575	08625	40801	59920	29841	80150	12777	48501
34	25976	57948	29888	88604	67917	48708	18912	82271	65424	69774	33611	54262	85963	03547
35	09763	83473	73577	12908	30883	18317	28290	35797	05998	41688	34952	37888	38917	88050
36	91567	42595	27958	30134	04024	86385	29880	99730	55536	84855	29080	09250	79656	73211
37	17955	56439	90999	49127	20044	59931	06115	20542	18059	02008	73708	83517	36103	42791
38	46503	18584	18845	49618	02304	51038	20655	58727	28168	15475	56942	53389	20562	87338
39	92157	89634	94824	78171	84610	82834	09922	25417	44137	48413	25555	21246	35509	20468
40	14577	62765	35605	81263	39667	47358	56873	56307	61607	49518	89686	20103	77490	18062
41	98427	07523	33362	64270	06138	92477	66969	98420	04880	45585	46565	04102	46880	45709
42	34914	63976	88720	82765	34476	17032	87589	40836	32427	70002	70663	88863	77775	69348
43	70060	28277	39475	46373	23219	53416	94970	25832	69975	94884	19661	72828	00102	66794
44	53976	54914	06990	67245	68360	82948	11398	42878	80287	88267	47363	46634	06541	97809
45	76072	29515	40980	07391	58745	25774	22987	80059	39911	96189	41151	14222	60697	59583
46	90725	52210	83974	29992	65831	38857	50490	83765	55657	14361	31720	57375	56228	41546
47	64364	67412	33339	31926	14883	24413	59744	92351	97473	89286	35931	04110	23726	51900
48	08962	00358	31662	25388	61642	34072	81249	35648	56891	69352	48373	45578	78547	81788
49	95012	68379	93526	70765	10592	04542	76463	54328	02349	17247	28865	14777	62730	92277
50	15664	10493	20492	38391	91132	21999	59516	81652	27195	48223	46751	22923	32261	85653

case numbers, and then use the families whose numbers come up in this way.

On pp. 572–573 we give a table of random digits. To use the table, we can start at any column and on any line. Thus if we use column 1 and read off the numbers, we get 10480, 22368, 24130, and so forth. Since the families have numbers only up to 20,000, we skip those numbers that are over 20,000. Therefore we skip the number 22368. The same is true of the number 24130. The next number that we accept is on line 13; it is 09429, then 10365, and so forth. Proceeding in this manner, we can obtain a random sample by selecting those families whose numbers are 10480, 09429, 10365, 07119, 02368, 01011, 07056. . . . When we get to the bottom of column 1, we go to column 2 and follow the same procedure. We stop when we get the 100 numbers that we need.

EXAMPLE 1   During 1985 a large auto manufacturer received 350 complaints from customers about the quality of the service performed by one particular dealer. The company decides to investigate some of these complaints by selecting a random sample of 18 of these complaints and thoroughly investigating them. By using column 4 of Table 10.1, which customers will we select?

SOLUTION   We first number the customers' complaint letters from 1 to 350. Then we use column 4 of the table of random digits. Although the table gives 5-digit numbers, we simply use the first 3 digits of the column. Thus we select the numbers 20, 166, 79, 102, 332, 34, 81, 99, 143, 242, 73, 264, 129, 301, 73, 299, 319, and 253. Therefore the customers with these numbers will have their complaints investigated.   ☐

EXAMPLE 2   Sixty students have registered for a statistics course. The chairperson of the department wishes to start a new section. She asks for 15 volunteers, but no one volunteers to transfer to the new section. She decides to randomly select 15 students. By using column 5 of Table 10.1, how can this be done?

SOLUTION   She should first assign each student a number from 1 to 60. Then she should use the first 2 digits of the numbers in column 5. Thus she should select those students whose numbers are 30, 7, 6, 27, 18, 17, 53, 49, 48, 31, 23, 42, 9, 47, and 13.   ☐

EXERCISES

1. In a certain state, license plates have 5 digits. The state motor vehicle department is interested in knowing how many of these cars are equipped with radial tires. It decides to send a letter to 100 randomly selected car owners. If columns 1 and 2 of Table 10.1 are used, which car owners will be selected to receive this letter?

2. A new drug is being tested for its ability to overcome drowsiness. Two hundred people have volunteered to take this drug to test its effectiveness. Each person is given a number from 1 to 200. For experimental purposes, only 15 people will receive the new drug. The remaining 185 will be given a sugar pill. If columns 2 and 3 of Table 10.1 are used, which of the volunteers will be selected to receive the new drug?

3. During 1985 a manufacturer produced 9000 stereo sets, each with a serial number from 1 to 9000. The company decides to send a questionnaire to 30 owners of these sets. Using the seventh column of Table 10.1, to which owners should the company send the questionnaire?

4. A new movie has just been seen by 4000 people. The producer wants to know how the people reacted to a particular horror scene. Since each person already has a ticket stub with a number from 1 to 4000 on it, the producer decides to select a random sample of 25 people. If columns 9 and 10 of Table 10.1 are used, which ticket holders will be selected?

5. There are 328 licensed restaurants in Newville. The health department has decided to investigate the sanitary conditions of these restaurants by randomly selecting 20 of these restaurants. The licenses are numbered consecutively from 1 to 328. If columns 8 and 9 of Table 10.1 are used, which restaurants should be investigated?

10.10

SUMMARY In this chapter we discussed the important idea of probability and how it is used. There are many possible ways of defining probability. Thus we defined the probability of an event as

$$\frac{\text{number of outcomes favoring the event}}{\text{number of possible outcomes}}$$

When two or more experiments are performed together, the counting principle or tree diagrams can be used to determine the total number of possible outcomes. In addition, we discussed permutations and combinations. Permutations are arrangements of objects where order is important, and combinations are distinct groups of things without regard to their order. We discussed factorial notation and formulas for calculating $_nP_r$ and $_nC_r$. Of course, $_nC_r$ can be evaluated by using Pascal's triangle.

We then went on to discuss independent events, which are events in which the occurrence of one event does not affect the occurrence of a second event, and mutually exclusive events, which are events that cannot happen at the same time. We gave several formulas for calculating probability under the conditions mentioned.

We then applied probability to determine the amount of money to be won or lost and the odds of an event happening or not happening. This is called mathematical expectation and odds. Finally, we indicated how probability can be used when studying genetics.

We also indicated how we can obtain random numbers by referring to a table of random digits.

STUDY GUIDE   You should now be able to demonstrate your knowledge of the following ideas by giving definitions, descriptions, or specific examples. Page references are given in parentheses.

Counting (p. 521)   Mutually exclusive events (p. 547)
Fundamental Principle of Counting   Conditional probability (p. 556)
  (p. 523)   prob $(A \mid B)$ (p. 556)
Tree diagram (p. 524)   Independent events (p. 558)
Permutation (p. 528)   Odds (p. 562)
Factorial notation (p. 528)   Odds in favor (p. 562)
$_nP_r$ (p. 528)   Odds against (p. 562)
Combination (p. 532)   Mathematical expectation (p. 563)
$_nC_r$ (p. 532)   Genetics (p. 567)
Pascal's triangle (p. 534)   Genes (p. 567)
Sample space (p. 539)   Dominant trait (p. 567)
Event (p. 539)   Recessive trait (p. 568)
Probability (p. 540)   Random number (p. 571)
Null event (p. 543)   Table of random digits (p. 571)
Definite event (p. 544)

**FORMULAS
TO REMEMBER**   The following list summarizes all the formulas given in this chapter.

1. Probability of an event $= \dfrac{\text{number of favorable outcomes}}{\text{number of possible outcomes}}$.

2. The number of permutations of n things taken r at a time is

$$_nP_r = \frac{n!}{(n-r)!}$$

3. The number of permutations of n things when p are alike, q are alike, r are alike, etc., is

$$\frac{n!}{p!q!r!} \qquad (\text{where } p + q + r + \ldots = n)$$

4. The number of combinations of n things taken r at a time is

$$_nC_r = \frac{n!}{(n-r)!\, r!}$$

5. prob $(A$ or $B) =$ prob $(A) +$ prob (B) if A and B are mutually exclusive.

6. prob $(A$ or $B) =$ prob $(A) +$ prob $(B) -$ prob $(A$ and $B)$ for any events.

7. prob $(A$ or B or $C) =$ prob $(A) +$ prob $(B) +$ prob $(C) -$ prob $(A$ and $B)$ $-$ prob $(A$ and $C) -$ prob $(B$ and $C) +$ prob $(A$ and B and $C)$.

8. prob $(A \mid B) = \dfrac{\text{prob } (A \text{ and } B)}{\text{prob } (B)}$.

9. prob $(A$ and $B) =$ prob $(A \mid B) \cdot$ prob (B).

10. prob $(A$ and $B)$ = prob $(A) \cdot$ prob (B) for independent events.

11. The mathematical expectation of an event is

$$m_1p_1 + m_2p_2 + m_3p_3 + \cdots .$$

12. Odds in favor of an event are p to q where p = the number of favorable outcomes and q = the number of unfavorable outcomes.

13. Odds against an event are q to p where p and q are defined as in preceding Formula 12.

MASTERY TESTS

Form A

1. Which of the following *cannot* be the probability of some event?

a) 0.003　　**b)** $\dfrac{7}{9}$　　**c)** $\dfrac{99}{100}$　　**d)** -0.001　　**e)** 0.27

2. One hundred people are being held by rioting prisoners as hostages. It is decided that 10 of these hostages will be selected to negotiate with authorities. In how many different ways can the negotiators be selected?

a) $\dfrac{100!}{90!}$　　**b)** $_{100}P_{10}$　　**c)** $\dfrac{100!}{10!}$　　**d)** $\dfrac{100!}{90!10!}$　　**e)** none of these

3. Two people who do not know each other, Cory and Steve, board a train at the same station. The train will make 5 stops before going out of service. What is the probability that they both get off at the same stop?

a) $\dfrac{1}{5}$　　**b)** $\dfrac{1}{2}$　　**c)** $\dfrac{2}{5}$　　**d)** $\dfrac{1}{25}$　　**e)** none of these

4. Given the numbers 5, 6, 7, 8, and 9. How many 3 digit numbers larger than 700 can be formed from these digits if repetition is *not* allowed?

a) 18　　**b)** 36　　**c)** 125　　**d)** 60　　**e)** none of these

5. A tax agent finds that 1 out of every 10 income tax filers in a certain city has claimed more than 7 dependents. If an income tax return is randomly selected from this group, what are the odds against the filer claiming more than 7 dependents?

a) 1 to 10　　**b)** 1 to 9　　**c)** 10 to 1　　**d)** 9 to 1　　**e)** none of these

6. There are 7 smokers and 8 nonsmokers at a city council meeting. A committee of 3 smokers and 3 nonsmokers is to be chosen from this group to study new ordinances banning smoking in restaurants. In how many different ways can they be chosen?

7. In how many different ways can the letters of the word "CALCULUS" be arranged?

a) 8!　　**b)** $\dfrac{8!}{2!}$　　**c)** $\dfrac{8!}{2!2!}$　　**d)** $\dfrac{8!}{2!2!2!}$　　**e)** none of these

For questions 8–10, use the following information. A taste test was given to 900 people at a shopping mall to determine which brand of vanilla ice cream they preferred. The results were:

	Brand A	Brand B	Brand C	Brand D
Male	128	131	118	101
Female	123	116	109	74

8. What is the probability that a person in the survey preferred brand A?

9. What is the probability that a person preferred brand A given that the person is a female?

10. What is the probability that the person is a female given that the person preferred brand A?

11. Thirty-one percent of all students who are dormitory students at Whipple University have tried drugs at one time or another. Twenty-six percent of the students are dormitory students. If a student is selected at random, what is the probability that the student lives in the dormitory and has tried drugs?

12. A biologist is experimenting with plants that have a particular trait A or a. If a plant that has Aa genes is crossbred with another plant that has Aa genes, find the probability that the offspring will be

   a) AA b) aa c) Aa

13. *Fertilizer.* The Balken Chemical Company manufactures nine different kinds of fertilizer, each containing different concentrations of nitrogen as follows:

$$4–6–3$$
$$4–6–2$$
$$4–4–1$$
$$4–4–2$$
$$4–2–3$$
$$4–8–5$$
$$12–6–1$$
$$9–3–2$$

   Earl and Helen Weaver are farmers who use the fertilizers produced by this company. They discover a bag of fertilizer in their barn. Unfortunately, only the first digit that indicates the nitrogen content is legible. If the first digit is a 4, find the probability that the bag contains a 4–8–5 mixture.

14. Mathew Priofsky is in a gambling casino and observes a card dealer selecting 3 cards (without replacement) from a deck of 52 cards. In how many different ways can the 3 cards be selected?

15. Each calculator manufactured by the Texas Calculator Co. is inscribed with a six-digit serial number preceded *and* followed by a letter. Using this serial number scheme, how many different codes are possible?

Form B

1. A chemist has mixed five different solutions together and created a new plastic compound. Unfortunately, she does not remember the order in which the chemicals were introduced into the solutions. She decides to repeat the experiment. How many possibilities are there?

2. Charlie drives up to a parking meter and notices a piece of paper on the ground that says "Meter out of order—No parking." Is this an old sign? Should he deposit money in the meter and park there? The probability that the meter is out of order is $\frac{4}{9}$. The probability that the meter is out of order *and* that he will get a ticket for parking there is $\frac{2}{13}$. If the meter is actually out of order, what is the probability that he will get a ticket?

3. To reverse its past discriminatory practices, the Hasting Corp. plans to hire 5 blacks and 6 Puerto Ricans. If 12 blacks and 18 Puerto Ricans qualify for the job, find the number of ways in which the vacancies can be filled.

4. The Geary Supermarket chain employs 16 part-time and 10 full-time clerks at one of its stores. As an economy move, the company plans to lay off 2 full-time and 2 part-time workers. In how many different ways can this be done?

5. Each year, readers of a certain magazine are asked to rank the top three best-dressed men from among a list of 12 candidates. In how many different ways can this be done?

6. There are 6 vacant conference rooms in an office building. The Bordeaux Corp. is planning 6 meetings for which it will need the 6 conference rooms. In how many different ways can these meetings be assigned to the conference rooms?

7. The Argavon Corp. believes that it is in the company's best interest to maintain the physical fitness of its 40 employees. It recently purchased 6 exercise machines to be used by the employees during their lunch break or after work. In how many different ways can 6 of the 40 employees be assigned to the different machines?

8. Matthew Valentine is the keynote speaker at tomorrow's board of directors' meeting. He can select any one of 5 pairs of trousers, 8 shirts, 8 ties, and 6 pairs of shoes to wear for the meeting. How many different outfits are possible?

9. A farmer has 5 different kinds of vegetables, 6 different kinds of fruits, and 3 different dairy products, which he is arranging on a roadside fruit stand. A customer comes along and selects one item from each of the categories mentioned. How many selections are possible?

10. In how many different ways can a circus operator arrange 10 different performances, that is, in how many different ways may the 10 performances be ordered?

11. A contestant on a network variety show must select one of three doors behind which there are different items, as shown below:

Door	Value of item(s)
1	$3 magazine
2	$7000 car
3	$2700 home entertainment center

If the contestant is likely to select any door with equal probability, find the contestant's expected amount of money to be won.

12. A visitor traveling in Europe stops at an airport in Italy. The visitor stops a guard and asks for directions. The probability that the guard speaks French is 0.22, and the probability that the guard speaks German is 0.33. What is the probability that the guard speaks either language if the probability that the guard speaks both is 0.17?

13. A Japanese camera manufacturer ships 125 cameras to an American store. Owing to a packer's error, 20 of these cameras were packed with operating instructions in Japanese only. A customer (who does not read or understand Japanese) buys one of these cameras. What are the odds against getting one of the cameras with operating instructions in Japanese only?

14. Caroline has been having trouble with her car. The probability that she turns on the air conditioning system and that the car overheats is 0.32. Furthermore, the probability that she turns on the air conditioning system is 0.68. If Caroline is observed driving the car with the air conditioning system on, what is the probability that the car will overheat?

15. Twenty-eight female and 25 male college graduates have applied for a computer training program. For various reasons, the program can accept only 22 female and 16 male applicants. In how many different ways can these applicants be selected?

11

STATISTICS

CHAPTER OBJECTIVES

To analyze what is meant by a sample, and to see how statistics are used and possibly misused. (*Sections 11.2 and 11.9*)

To study measures of central tendency. These are the mean, median, or mode, and each indicates some different way of analyzing numbers. (*Section 11.3*)

To calculate the relative standing or performance of one score with respect to others. This is called its percentile or percentile rank. (*Section 11.4*)

To learn several ways of measuring how spread out or dispersed numbers are. These measures include the range, variance, standard deviation, and average deviation. (*Section 11.5*)

To discuss how a set of numbers can be analyzed by graphical techniques. We first set up frequency distributions and then draw bar graphs, line graphs and circle graphs. (*Section 11.6*)

To introduce and apply a special type of distribution called the normal distribution. (*Section 11.7*)

To indicate how to determine whether a relationship exists between several variables and then to find the equation expressing this relationship. This is linear correlation and regression. (*Section 11.8*)

Statistics are used in making all kinds of calculations. In the graph below we have figures that compare the "average" salaries of men and women in different categories. How do we determine such averages?

Now consider the second article. Note the use of the word median, which is also some sort of indicator of a general trend.

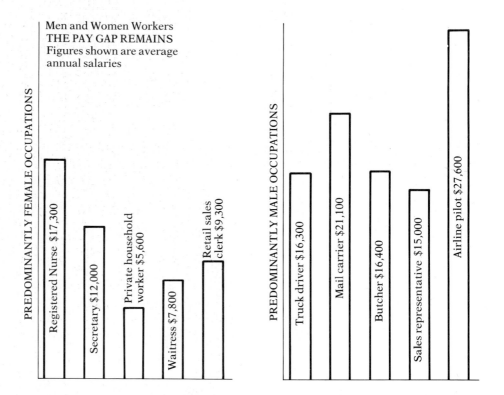

Men and Women Workers
THE PAY GAP REMAINS
Figures shown are average annual salaries

PREDOMINANTLY FEMALE OCCUPATIONS
Registered Nurse $17,300
Secretary $12,000
Private household worker $5,600
Waitress $7,800
Retail sales clerk $9,300

PREDOMINANTLY MALE OCCUPATIONS
Truck driver $16,300
Mail carrier $21,100
Butcher $16,400
Sales representative $15,000
Airline pilot $27,600

1982 Salary Settlements; First Quarter Increases

NEW YORK (Oct. 17)—Figures released by the regional office of the U.S. Bureau of Labor Statistics show that the median first-year wage increases in the first quarter of 1982 for all industries were 8.0 percent or about 48.9 cents. For the same period in 1981, the comparable figures were 9.6 percent or about 62.0 cents per hour. Figures currently available indicate that about 10 percent of the contract settlements provide for no first-year wage hike.

During 1981 the figure was one percent. A breakdown of the figures shows that the median private sector increases were 8 percent or 50.0 cents per hour, whereas the public sector median were 4.8 percent or 34.4 cents per hour. More than 75 percent of the settlements made health and welfare provision changes with about 50 percent also providing for changes in pension plans. Fringe benefit change appeared in 92 percent of the agreements.

DETROIT PLANET, October 17, 1982

Statistics can be used in making predictions and projections, as shown in the articles on pages 583 and 584. How are such projections made? Are they accurate?

Job Outlook For New Ph.D.s Very Bleak

WASHINGTON (Jan. 12)—Half of the people who obtained their doctorates through 1986 will be unable to get jobs suitable to their training, according to the latest projections released by the U.S. Bureau of Labor Statistics. This continues a trend that began with the end of the boom in the 1960s and reflects the continuing inability of our present economy to fully absorb these very highly educated people.

NATIONAL TIMES, January 12, 1986

11.1

INTRODUCTION

Most of us have heard of, and have some idea of what is meant by, the word **statistics.** We usually think of statistics as having something to do with tables or charts of numbers. Mathematicians, however, usually use this word in a more general sense. To the mathematician the subject of statistics is concerned with how to collect numerical facts, called **data,** how to organize and analyze them, and finally how to interpret the data.

The subject of statistics has been applied to many different areas, including medicine, insurance, electronics, advertising, television audiences, population growth, and student enrollment in schools. Some knowledge of statistics is fast becoming an important tool for everyone. The following examples of the use of statistics should be quite familiar.

1. Statistics show that male drivers under the age of 25 have more accidents than other drivers. On the basis of this information, insurance companies charge higher premiums for these drivers.

2. On election-day television newscasts, computers are used to "project" the winners on the basis of only very early returns. Samples from representative districts are collected, and the predictions are based on these statistics.

3. The latest statistics released by the FBI indicate that serious crime in a large northeastern city increased by 38% during the year 1985.

Enrollment in U.S. Colleges Expected to Be up in 1980s

WASHINGTON—Total enrollment in institutions of higher education in the United States increased by about 39 percent between 1970 and 1980. The National Center for Education Statistics indicates that this trend is expected to continue in the 1980s with large increases in the enrollment of part-time students in publicly controlled institutions. The Bureau of the Census predicts that enrollment increases in the 1980s are expected to be substantially higher for blacks and Hispanics than for white students. Furthermore, the areas of concentration of these students is changing as shown.

SPRINGFIELD GAZETTE, October 20, 1985

PROJECTED WINNERS AND LOSERS IN THE ACADEMIC PAPER CHASE
Change in Area of Study of Earned Degrees

FIELDS WITH GREATEST PERCENT DECREASE

Bachelor's degrees	Master's degrees	Doctor's degrees
Mathematics	Foreign languages	Mathematics
Library science	Mathematics	Engineering
Letters	Letters	Physical sciences
Foreign languages	Area studies	Letters
Social sciences	Social sciences	Foreign languages

FIELDS WITH GREATEST PERCENT INCREASE

Bachelor's degrees	Master's degrees	Doctor's degrees
Public affairs	Interdisciplinary studies	Interdisciplinary studies
Computer information sciences	Health professions	Theology
Health professions	Public affairs	Architecture
Interdisciplinary studies	Computer information sciences	Law
Communications	Business and management	Public affairs

Source: 1981 edition of National Center for Education Statistics,
THE CONDITION OF EDUCATION.

4. Statistics indicate that because of the extremely high tax rate in New York City, many citizens and businesses are moving out of the city into neighboring states. Consequently, the statistics show that an increase in taxes will result in an exodus of people and businesses from the city.

5. When the Social Security system was first put into effect, the rate of contribution was determined by statistics that predicted how long a person would live. Statistics show that people have been living longer in recent years. As a result, the rate of contribution has been steadily increasing.

6. The Nielsen television ratings show that one network has 20% more viewers than another between the hours of 7:00 P.M. and 8:00 P.M. on Monday night.

What mathematical knowledge is needed to understand and use statistics? The collection and organization of the data require little or no mathematical background. On the other hand, the interpretation of the data is another story. The statistician should have some mathematical knowledge if he or she is to interpret the data in a meaningful way.

If the data are not interpreted properly, very wrong and sometimes ridiculous conclusions can be drawn from them. As an example

HISTORICAL NOTE

The study of statistics was really begun by an Englishman, John Graunt (1620–1674). Graunt studied death records in various cities and noticed that the percentages of deaths from different causes were about the same and did not change much from year to year. Graunt was also the first to discover (using statistics) that there were more boys born than girls. Because, at the time, men were more subject to death from occupational accidents and diseases and from war, it turned out that at the age suitable for marriage the number of men and women was about equal. Graunt believed that this was a natural way of guaranteeing monogamy.

In 1662, Graunt published *Natural and Political Observations . . . upon the Bills of Mortality*, which has been said to have founded the science of statistics.

The work begun by Graunt was continued by others, who wanted to make the social sciences more "quantitative," that is, based more on mathematics. In the late seventeenth century, life insurance companies were formed, and they, of course, were also interested in the information to be obtained from statistics, such as death rates and life expectancy. The Industrial Revolution increased interest in statistics even further. Government agencies and social reformers wanted statistics on births and deaths, national and individual incomes, unemployment, occurrence of disease, etc. By the nineteenth century, statistics was also accepted as an important tool in the physical sciences.

of how data can be misinterpreted, consider the following statistics. During 1985 in a southern state there were 1246 accidents involving drunken pedestrians. There were 723 accidents involving drunken drivers. You could conclude that it is more dangerous to be a drunken pedestrian than a drunken driver. Do you agree that this is a reasonable way of interpreting the data; that is, do you agree with the conclusion?

In this chapter we will discuss how statistical data can be organized and tabulated so that meaningful results can be drawn from them.

11.2

SAMPLING

Suppose a producer is interested in knowing what percentage of the television audience enjoys watching a new show. He or she obviously will not (or cannot) ask every individual who watches television for a reaction to the new show. What he or she will do is take a **sample.** A relatively small group of TV viewers will be selected and asked for their reactions. From their comments, a generalization will be made for *all* television viewers.

Before indicating some of the difficulties involved in taking a sample, we first state formally what we mean by a sample.

> **Definition 11.1**   A **sample** is a small group of individuals (or objects) selected to stand for a larger group, usually called the **population.**

In taking a sample, a number of problems can arise. The first problem is that of sample size. If a sample is too small, then the individuals used may not be truly typical (or representative) of the population. A sample that is too large is usually very costly. A good sample should be large enough to be typical of the population it represents, but not so large that its cost is ridiculously expensive. How large a sample to select varies from situation to situation. Only by applying statistical procedures can one be reasonably confident of the correct sample size.

The most important thing in taking a sample is the requirement that it be a **random** one. This means that each individual in the population should have an equally likely chance (or probability) of being selected. Unless the sample is a random one, it may not give a true picture of the population it represents.

In 1948 the pollsters predicted that Harry Truman would lose the presidential election. This prediction was based on a sample of the population. It turned out that the pollsters were wrong. This was because their sample was not random and was not truly representative of the population.

Gary Settle/NYT Pictures

Similarly, on April 6, 1976, both ABC and NBC television networks projected that Morris Udall would win the Democratic primary in Wisconsin. Their projections were based on samples from selected precincts and did not take certain districts into consideration. When all the rural votes were counted, Jimmy Carter came out on top. Many newspapers were so confident of their predictions that they printed, erroneously, the morning editions of their newspapers with the headline "CARTER UPSET BY UDALL."

Sometimes in a statistics class a student will suggest that one could get a random sample by opening a telephone book and selecting every hundredth name appearing. But this technique does not get a random sample because not every name has an equal chance of being selected. (Why not?) Statisticians have devised techniques for conducting a random sample. Although the *table of random digits* technique discussed in the last chapter is the simplest way of obtaining a random sample, there are other techniques that can be used. A discussion of these is beyond the scope of this text.

In this chapter we will assume that all data have been collected from random samples.

11.3

MEASURES OF CENTRAL TENDENCY

In this section we shall be concerned with several methods of interpreting data. To understand what we mean by this, consider a used-car dealer, who, because of economic conditions, has decided to fire one of two salesmen, Crazy Eddie or Mad Mike. Obviously, the dealer wants to keep the better salesman. To help him decide which employee to fire, he has made a chart as shown below, indicating the number of cars sold by each man over the last seven weeks.

	Crazy Eddie	Mad Mike
Week 1	8	10
Week 2	6	12
Week 3	12	12
Week 4	8	11
Week 5	6	12
Week 6	38	12
Week 7	6	8
Total	84	77

At first glance, one would claim that Crazy Eddie is a better salesman, since he sold a total of 84 cars, whereas Mad Mike sold only 77 cars. However, let us analyze the situation a bit more care-fully. Notice that Mad Mike sold more cars than, or the same number of cars as, Crazy Eddie during every week but the sixth.

Let us compute the average number of cars sold by both salesmen by dividing each total by the number of weeks (which is 7). We get

$$\text{Crazy Eddie's average:} \quad \frac{84}{7} = 12;$$

$$\text{Mad Mike's average:} \quad \frac{77}{7} = 11.$$

Thus Crazy Eddie sold 12 cars on the average, whereas Mad Mike sold only 11 on the average. It would again appear that Crazy Eddie is a better salesman.

Another look at the data, however, shows that Mad Mike sold 12 cars most often (on 4 weeks). Crazy Eddie sold 12 cars only once. Crazy Eddie sold 6 cars most often. One might now say that Mad Mike is a better salesman in terms of consistent performance.

Suppose we were to arrange the number of cars sold by each (per week) in order from smallest to largest. We get the following table.

Crazy Eddie	Mad Mike
6	8
6	10
6	11
⑧	⑫
8	12
12	12
38	12

Note that two numbers have been circled. These are the numbers that are in the middle. For Mad Mike this number is 12, and for Crazy Eddie this number is 8. This example leads us to the following definitions.

Definition 11.2   *The **mean** or **average** of a set of numbers is found by adding them together and dividing the total by the number of numbers added.*

Definition 11.3   *The **mode** of a set of numbers is the number that occurs most often. If every number occurs only once, then we say that there is no mode. A set of numbers may have more than one mode.*

Definition 11.4   *If a set of numbers is arranged in order (from smallest to largest), then the number that is in the middle is called the **median**. This is only when there is an odd number of numbers. If there is an even number of them, then the **median** is the average of the middle two numbers (when arranged in order).*

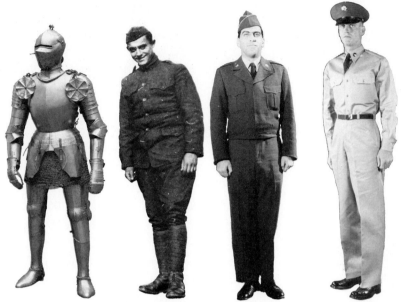

The median height of a medieval knight in armor was 5 ft 6 in. In World War I the median height of a soldier was 5 ft $7\frac{3}{4}$ in.; in World War II, 5 ft $8\frac{1}{2}$ in.; and after 1958, 5 ft $8\frac{9}{10}$ in. What is the value of knowing these statistics?

Medieval armor: courtesy of Museum of Fine Arts, Boston; soldiers of World War I, World War II, and 1958: E. P. Jones

Let us now apply these definitions to our two salesmen. We have the following:

	Crazy Eddie	*Mad Mike*
Mean	12	11
Median	8	12
Mode	6	12

Who is a better salesman, Crazy Eddie or Mad Mike? One would probably say Mad Mike, even though his mean (or average) is less than Crazy Eddie's. Eddie's average is 12 only because of the sixth-week sales, when he sold 38 cars. Mike, on the other hand, consistently sold 11 or 12 cars. In this case the mean does not tell us as much about the salesmen as does the median or the mode.

The above ideas will be illustrated further by several examples.

EXAMPLE 1   A professor recently gave a test to her statistics class of eleven students. The following results (grades) were obtained: 78, 53, 100, 27, 94, 88, 98, 93, 98, 91, and 89. Find the mean, median, and mode for this class.

SOLUTION   We first arrange the grades in order from lowest to highest. We get 27, 53, 78, 88, 89, 91, 93, 94, 98, 98, and 100.

The grade that occurred most often is 98. Thus the mode is 98.

The grade that is in the middle (now that we have arranged them in order) is 91, so the median is 91.

To find the mean, we first add all the numbers. The sum is 909. We then divide 909 by the total number of grades (which is 11), getting 909/11 = 82.64 (rounded off to two decimal places). The mean is then 82.64.

Which is a better indication of class performance in this particular example, the mean, median, or mode? ☐

EXAMPLE 2   The mathematics department at a state university consists of eight members whose salaries are given in the following chart.

Rank	Salary
Dave, chairman	$21,000
Arthur, professor	19,000
Roger, assoc. professor	18,500
Betsy, assoc. professor	17,000
Nancy, assist. professor	16,000
Alice, assist. professor	15,810
Bob, instructor	13,120
Jim, lecturer	12,810

Find the mean, median, and modal salary for the members of the mathematics department at this particular university.

SOLUTION   We notice that the salaries are already arranged in order, from highest to lowest. We can then read up the list.

To find the median, we look for the number that is in the middle. In this case there is an even number of salaries, so that no number is in the middle. The median is somewhere between $17,000 and 16,000. Definition 11.4 tells us that the median is the average of these two numbers, or that the median is

$$\frac{17,000 + 16,000}{2} = \frac{33,000}{2} = \$16,500.$$

What about the mode? Notice that no salary occurred more than once. Definition 11.3 tells us that there is no mode.

To find the mean, we first add all the salaries. The total is $133,240. We now divide the total by 8, getting

$$\frac{133,240}{8} = \$16,655.$$

For the salaries in the mathematics department of this university we have

Mean $16,655,   Median $16,500,   Mode none.

Which is a better indication of the teachers' salary, the mean or the median? ☐

Comment The mean, median, and mode are called **measures of central tendency.** The reason for this name should be obvious. Each of these (mean, median, or mode) measures some central or general trend of the data. Depending on the situation, one will usually prove to be more meaningful than the others.

Comment The word "average" is often used in newspaper or magazine articles. Note the use of the word *average* in the following article. Has the "average" SAT score increased or decreased over the years?

SAT Score Averages 1963–1983

Score averages from 1963 to 1966 represent all SAT candidates.

From 1967 to 1983 the figures represent the average scores of college-bound seniors only, which are released each year by the College Board in its national report on College-Bound Seniors.

	Verbal	Mathematics		Verbal	Mathematics
1963	478	502	1974	444	480
1964	475	498	1975	434	472
1965	473	496	1976	431	472
1966	471	496	1977	429	470
1967	466	492	1978	429	468
1968	466	492	1979	427	467
1969	463	493	1980	424	466
1970	460	488	1981	424	466
1971	455	488	1982	426	467
1972	453	484	1983	427	468
1973	445	481			

We can rewrite Definition 11.2 by using *summation notation*. In this notation the Greek letter Σ, read as sigma, stands for the operation of adding a sequence of numbers. Thus suppose that we have a group of numbers $x_1, x_2, x_3, \ldots, x_n$, where n is the total number in the group. Then we have Formula 1.

Formula 1

$$\mu = \frac{\Sigma x}{n} = \frac{x_1 + x_2 + x_3 + \cdots + x_n}{n}$$

EXAMPLE 3   There are 19 employees in the math department of Bologa University. Their salaries are determined by their rank as shown in the table. What are the mean, median, and modal annual salaries of these employees?

Rank	Number of people with this rank	Annual salary
Professor	2	$35,000
Associate professor	7	28,600
Assistant professor	4	23,500
Instructor	2	19,400
Lecturer	3	16,100
Secretary	1	10,305

SOLUTION   To find the mean, we must find the sum of the salaries of all 19 employees. We have

Number of employees	Annual Salary	Total payment
2	$35,000	$ 70,000
7	28,600	200,200
4	23,500	94,000
2	19,400	38,800
3	16,100	48,300
1	10,305	10,305
		$461,605

Thus, the mean annual salary is

$$\mu = \frac{\Sigma x}{n} = \frac{\$461,605}{19} = \$24,295.$$

The modal annual salary is $28,600, and the median annual salary is $23,500. Can you see why? ☐

Comment In Example 3 the mean annual salary is $24,295 even though nobody actually earns that amount.

EXERCISES

1. *Funeral Costs.* Consider the newspaper article on the right. During the week of November 4–8 the Jefferson Funeral Chapels performed six funerals. The charges for these funerals were $1480, $1700, $1370, $1510, $1625, and $1430. Find the mean, median, and modal prices for the charge of a funeral performed by the Jefferson Funeral Chapels.

State to Investigate Funeral Charges

MARLINGTON—Commissioner White of the State Investigatory Commission announced yesterday that he would launch an immediate investigation into the funeral costs charged by the Jefferson Funeral Chapels, Inc. Commissioner White revealed that his agency had received numerous complaints about overcharges and unexplained charges that bereaved families have been forced to pay. Most of the complaints centered around the Jefferson Funeral Chapels, Inc.

MARLINGTON NEWS, November 5, 1985.

2. An insurance company executive is analyzing the sales records of 12 brokers who work for the company. The executive has the following information for one business day.

Salesperson	Number of new policies sold
Philip	4
Cheryl	2
Ellen	8
Frank	3
Martin	4
Peter	16
Evens	6
Kim	1
Anthony	7
Evelyn	2
Patrick	3
Daniel	4

Find the mean, median, and mode for the number of new policies sold.

3. The Porterville Savings Bank employs 75 people at its downtown branch office. The salary of these employees is dependent upon the title of the worker as shown in the accompanying chart.

Category	Number of workers in this category	Annual salary
Vice-President	4	$36,000
Teller (full-time)	17	16,000
Teller (part-time)	23	7,000
Supervisor	6	25,000
Secretarial	9	10,000
Security guard	4	14,000
Other	12	9,000

Find the mean, median, and modal salaries.

4. A taxi driver claims that the average tip that she receives is $1.25. To which average is she probably referring: the mean, median, or mode?

5. *Cost Averaging.* For the past seven weeks, Jackie Zweigh has purchased shares of ABC stock at the prevailing price each Monday morning as shown below:

Number of shares purchased	Price per share
120	$33
150	29
80	31
90	30
110	28
60	35
170	25

Jackie wishes to cost average and has directed her broker to purchase 100 shares of the stock the following Monday. At what price should these be purchased so that the price for all the shares purchased will be $29 a share?

6. There are 19 mechanics on the payroll of Mel's Automatic Transmission Repair Shop. The mean weekly salary is $382, the median weekly salary is $326, and the modal weekly salary is $345. Find the total weekly payroll.

7. Refer back to Exercise 1. Officials of the Jefferson Funeral Chapels have announced that as of January 1, the cost for each of the funerals mentioned will be increased by 10%. How are the mean, median, and mode affected by the proposed price changes?

8. The union representing 812 textile workers is negotiating a new labor contract with the company. The union claims that the *average* hourly rate of pay is $5.11, which is much lower than the pay of workers in competing companies. Management claims that the *average* hourly rate of pay is $8.69, which is much higher than the pay of workers in competing companies. A labor negotiator believes that both the union claim and the management claim are accurate. How can this be? Explain your answer.

9. Seven people were hospitalized at Brookhaven Hospital last year because of a particular disease. The length of stay in the hospital is shown below. You will notice that the average length of stay at the hospital for these patients is 25 days, which is longer than the length of stay of all but one patient. What can you conclude?

Patient	Length of stay in hospital
John Bender	19 days
Grace Ahl	20 days
David Browne	4 days
Andrew Digiacomo	2 days
Pat Elsayed	13 days
George Grigoli	16 days
Chris Hardy	101 days

10. Owing to numerous boiler breakdowns, a new heating system was recently installed at Skytown Towers. Several tenants are still complaining that it is now too warm, whereas others are complaining that it is now too cold. In an attempt to satisfy everyone, management has decided to install an energy-saving thermostat and to poll each of the residents as to what temperature it should be set at. When this is done, the following data are obtained.

Mean   71°

Median   70°

Mode   69°

At what temperature should the thermostat be set so as to satisfy as many residents as possible?

11. A certain set of numbers has mean 25. What happens to the mean if each number in the set is

a) increased by 3.   **b)** decreased by 4.

c) multiplied by 7.   **d)** divided by 5.

** 12. Consider the set of numbers 1, 6, 8, 4, 9, 13, 14, 11, 12, and 10. Find

a) Σx **b)** μ

c) $\Sigma (x - \mu)$ **d)** $\Sigma (x - \mu)^2$

** **13.** For the numbers given in Exercise 12, compute

 a) Σx^2 **b)** $(\Sigma x)^2$

 c) Are the answers the same? Explain why or why not.

14. A company pays its workers the following salaries depending on title.

Title	Number of employees	Weekly salary
Manager	1	$500
Supervisor	3	300
Machinists	8	275
Other workers	14	200
Secretaries	4	150

Two students were asked to calculate the average. One student did as follows:

$$\mu = \frac{500 + 300 + 275 + 200 + 150}{5}$$

$$= \frac{1425}{5} = \$285.$$

This student then claimed that the average salary was $285 per week. A second student computed a **weighted arithmetic average** as follows:

$$\mu = \frac{1 \cdot 500 + 3 \cdot 300 + 8 \cdot 275 + 14 \cdot 200 + 4 \cdot 150}{30}$$

$$= \frac{7000}{30} = \$233.33.$$

He claimed that the average salary was $233.33. Which student is correct and why?

11.4

PERCENTILES

In the preceding section we discussed different ways of analyzing data. Quite often, one is interested in knowing the position of a score in a list of numbers. Thus if Mary Ruth got a score of 83 on a civil service exam that she has just taken, she undoubtedly would be interested in knowing how this score compares with the scores of others who have taken the exam. In such a situation she would probably be more interested in her **percentile rank** than in the mean, median, or mode.

Let us analyze the results of the civil service exam that Mary Ruth took. The exam was given to 200 people, including Mary Ruth. She finds that 70% of the people who took the exam got below 83, 10% of the people got 83, and the remaining 20% scored above 83. Since 70% of the people scored below 83 and 20% scored above 83, her percentile rank should be between 70 and 80. (Why?) We use 75, which is halfway between 70 and 80. What we have done is find the percentage of scores that are below her score and add one-half of the percentage of scores that are the same as her score. The result is her percentile rank. In our case, Mary Ruth's percentile rank is 75. This means that *approximately* 25% of the people who took the exam scored higher than she did and 75% of the people scored lower than she did. (Why is it only approximate?) Most civil service tests are graded by using such a procedure.

This leads us to the following.

> **Definition 11.5**   *The **percentile rank** of a score is found by adding the percentage of scores below it to one-half of the percentage of scores equal to it.*

Although this definition enables us to find the percentile rank of an individual score, in practice we can use a convenient formula. Let X be a given score, let B represent the *number* of scores below the given score X, and let E represent the *number* of scores equal to the given score X. If the total number of scores is n, then the percentile rank of the given score can be found by the following formula.

Formula 2

$$\text{Percentile rank of } X = \frac{B + \frac{1}{2}E}{n} \cdot 100$$

Let us illustrate the use of Formula 2 with an example.

EXAMPLE 1   There are 23 students in a statistics class. On the midterm exam the grades of the students were 79, 63, 94, 100, 83, 92, 78, 62, 53, 84, 76, 22, 17, 52, 57, 66, 83, 72, 81, 70, 69, 46, and 97. Douglas got an 83 on the exam. Find his percentile rank.

SOLUTION   Since 23 people took the exam, $n = 23$. Analyzing the individual scores, we find that 16 scores are below 83 and exactly 2 scores (including Douglas's) are equal to 83. Thus $B = 16$ and $E = 2$. We now apply Formula 2 to find Douglas's percentile rank. We have

$$
\begin{aligned}
\text{Douglas's percentile rank} &= \frac{B + \frac{1}{2}E}{n} \cdot 100 \\
&= \frac{16 + \frac{1}{2}(2)}{23} \cdot 100 \\
&= \frac{16 + 1}{23} \cdot 100 \\
&= \frac{17}{23} \cdot 100 \\
&= 73.91.
\end{aligned}
$$

Therefore Douglas's percentile rank is 73.91. This means that Douglas did better than approximately 73.91% of the students and that only about 26.09% of the students did better than Douglas.   ☐

Comment   In the example above, Douglas's percentile rank was 73.91. We sometimes say that Douglas was in the 73.91st percentile.

Statisticians have special names for certain percentiles. The 25th percentile is called the *lower or first quartile*. The 50th percentile is called the *median or middle quartile*. The 75th percentile is called the *upper or third quartile*.

EXERCISES

1. The weights of the 20 members of a soccer team are

Tim: 180	Stu: 220	Stan: 206
John: 169	Jose: 210	Maurice: 180
Joe: 198	George: 229	Cornell: 184
Bill: 176	Bob: 204	Catfish: 192
Hal: 199	Jack: 213	Sigmund: 208
Jasper: 210	Allen: 216	Roger: 186
Mike: 237	Al: 239	

   Find the percentile rank of Maurice.

2. The following are the test results for an advanced calculus final:

72	77	84	71	67	61	67
53	81	55	87	68	60	78
45	68	79	92	100	68	77
72	97	80	77	90	53	53
82	71	47	58	99	49	
67	63	93	79	88	61	

   What is the percentile rank of the student(s) who scored 79 on the exam?

3. In Maureen's psychology class there are 40 students. On a midterm exam, 24 students got lower grades than she did, and 10 students got higher grades. What is Maureen's percentile rank?

4. Refer back to Exercise 3. If the teacher decides to curve the exam by adding six points to everyone's grade, how is Maureen's percentile rank affected?

11.5

MEASURES OF VARIATION

It is very hard to find two things of any type that are identical. Two people of the same age and sex may differ a great deal in height, weight, etc. Every cook knows that the same recipe may not always result in the same quality of cake. Even things that are mass-produced are not really exactly the same. There is some variation or difference among them. In this section we will discuss some of the methods used to measure variation.

To get us started, let us consider the following. A state consumers' group is investigating the milk prices charged by two large supermarket chains in various parts of the city. Both chains claim that their average milk price is 50¢ a quart. Investigators find that the prices charged by these chains in five different neighborhoods of the city are as follows.

Chain A	Chain B
53¢	52¢
48	50
61	48
47	49
41	51

You will notice that the average (mean) price for a quart of milk at each chain is 50¢. Yet for chain A the price of a quart of milk varies from 41 to 61 cents. This gives a *range* of 61−41, or 20 cents. For chain B the prices vary from 48 to 52 cents. Their range is 52−48, or 4 cents.
This leads us to the following definition.

> **Definition 11.6**   *The **range** of a set of numbers is found by subtracting the smallest number from the largest.*

Comment   No matter how many numbers are in the original data, only two of them (the smallest and the largest) are needed to compute the range.

Unfortunately, the range does not tell us anything about how the other numbers vary. For this reason we need another measure of variation called the **standard deviation.** This tells us how "spread out" the numbers are. To find the standard deviation for each of the supermarket chains above is a relatively simple procedure. We first find the mean. (In our case we already know that it is 50¢.) We subtract the mean from each price and square the result. We then find the average of these squares. Finally, we take the square root of this average. The result is called the standard deviation.
We will now calculate the standard deviation for each supermarket chain.

Supermarket chain A

Price (in cents)	Difference from mean	Square of difference
53	53 − 50 = 3	(3)(3) = +9
48	48 − 50 = −2	(−2)(−2) = +4
61	61 − 50 = 11	(11)(11) = +121
47	47 − 50 = −3	(−3)(−3) = +9
41	41 − 50 = −9	(−9)(−9) = +81
Total 250		Total 224

Mean $= \dfrac{250}{5} = 50$

Average $= \dfrac{224}{5} = 44.8.$

Thus the standard deviation for supermarket chain A is $\sqrt{44.8}$ or approximately 6.69.[1] The standard deviation is a measure of how spread out the data are.

What about supermarket chain B? Let us compute its standard deviation.

Supermarket chain B

Price (in cents)	Difference from mean	Square of difference
52	52 − 50 = 2	(2)(2) = +4
50	50 − 50 = 0	(0)(0) = 0
48	48 − 50 = −2	(−2)(−2) = +4
49	49 − 50 = −1	(−1)(−1) = +1
51	51 − 50 = +1	(1)(1) = 1
Total 250		Total 10
Mean = $\frac{250}{5}$ = 50		Average = $\frac{10}{5}$ = 2

Therefore for supermarket chain B the standard deviation is $\sqrt{2}$, or approximately 1.41.

We can summarize the procedure to be used in finding the standard deviation by the following rule.

> **RULE**   *The (population) standard deviation of a set of numbers is the result obtained by finding (in order)*
>
> *a) the mean (or average) of the numbers,*
> *b) the difference between each number and the mean,*
> *c) the squares of each of these differences,*
> *d) the average of these squares,*
> *e) the square root of the average of these squares.*

We will illustrate this rule further with another example.

EXAMPLE 1   A large city in the south requires that its police officers be at least 5 ft 7 in. tall (67 in.). Seven police officers are selected and their heights recorded. What is the standard deviation if their heights are 69, 72, 74, 67, 68, 70, and 70 in.?

SOLUTION   We arrange their heights in order as shown in the table on the next page and then calculate the standard deviation. The standard deviation of their heights is thus $\sqrt{4.857}$, or approximately 2.2.

---

1. A knowledge of how to compute square roots is not assumed. These values can be obtained by using a calculator.

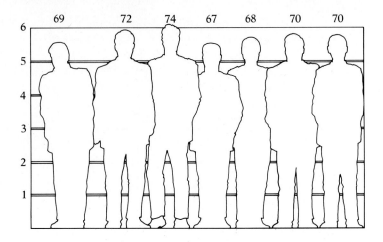

Height	Difference from mean	Square of difference
67	$67 - 70 = -3$	$(-3)(-3) = +9$
68	$68 - 70 = -2$	$(-2)(-2) = 4$
69	$69 - 70 = -1$	$(-1)(-1) = 1$
70	$70 - 70 = 0$	$(0)(0) = 0$
70	$70 - 70 = 0$	$(0)(0) = 0$
72	$72 - 70 = 2$	$(2)(2) = 4$
$\underline{74}$	$74 - 70 = 4$	$(4)(4) = \underline{16}$
490		34

$$\text{Mean} = \frac{490}{7} = 70 \qquad\qquad \text{Mean} = \frac{34}{7}$$

$$= 4.857 \text{ (rounded off)}$$

Comment   You may feel that the standard deviation is a rather complicated number to calculate, so why bother. However, it is an extremely important and useful number to the mathematician. A detailed discussion of how it is used is beyond the scope of this text.

Another measure of variation that is often used is the **variance.** It is computed in a manner similar to that for the standard deviation, with one difference. We omit step (e) of the rule on p. 599. Thus for Example 1 the standard deviation is $\sqrt{4.857}$, and the variance is 4.857.

EXAMPLE 2   A researcher is interested in determining the number of full-time women professors in the mathematics departments of several Western colleges. The researcher has obtained the following information on 9 such colleges. For every 100 faculty members, these schools employed 21, 7, 10, 2, 43, 15, 22, 14, and 1 women. Find the standard deviation and the variance.

SOLUTION   We arrange the numbers as shown in the following table and then calculate the variance and standard deviation.

Number of female employees	Difference from mean	Square of difference
21	21 − 15 = 6	(6)(6) = 36
7	7 − 15 = −8	(−8)(−8) = 64
10	10 − 15 = −5	(−5)(−5) = 25
2	2 − 15 = −13	(−13)(−13) = 169
43	43 − 15 = 28	(28)(28) = 784
15	15 − 15 = 0	(0)(0) = 0
22	22 − 15 = 7	(7)(7) = 49
14	14 − 15 = −1	(−1)(−1) = 1
1	1 − 15 = −14	(−14)(−14) = 196
Total 135		Total 1324

$$\text{Mean} = \frac{135}{9} = 15$$

$$\text{Mean} = \frac{1324}{9}$$
$$= 147.11$$

Thus the variance is 147.11, and the standard deviation is $\sqrt{147.11}$, or approximately 12.13. ▢

Another measure of variation that is often used is the **average deviation.** It is computed in a manner similar to that for the standard deviation, with one major exception. Instead of squaring the differences from the mean, we simply take the absolute value (neglect any negative signs) of the differences and find the average of these absolute values. No square roots are involved.

We illustrate the procedure with the following example.

EXAMPLE 3   Find the average deviation of the numbers 7, 11, 20, 5, 3, 4, 6, 12, 19, and 3.

SOLUTION   We arrange the numbers in a chart form and proceed as indicated by the following chart. Thus the average deviation is 5.2.

Number	Difference from mean	Absolute value of difference from mean
3	3 − 9 = −6	6
3	3 − 9 = −6	6
4	4 − 9 = −5	5
5	5 − 9 = −4	4
6	6 − 9 = −3	3
7	7 − 9 = −2	2
11	11 − 9 = +2	2
12	12 − 9 = 3	3
19	19 − 9 = 10	10
20	20 − 9 = 11	11
90		52

Mean $\dfrac{90}{10} = 9$ Average deviation $\dfrac{52}{10} = 5.2$

Notation   We denote the population standard deviation of set of numbers by the Greek symbol σ (read as sigma).

EXERCISES

1. The number of arrests for drunken driving on the Washington Turnpike during the first 10 days of 1986 was as follows:

Day	Number of arrests
Jan. 1	32
Jan. 2	26
Jan. 3	17
Jan. 4	19
Jan. 5	22
Jan. 6	12
Jan. 7	18
Jan. 8	16
Jan. 9	14
Jan. 10	20

Find the range and standard deviation for the number of arrests.

2. The number of stock transactions executed by

Jan Segal over a two-week period is as follows:

Day	No. of stock transactions executed
Mon.	123
Tues.	112
Wed.	96
Thurs.	142
Fri.	53
Mon.	88
Tues.	76
Wed.	89
Thurs.	91
Fri.	108

Find the range and standard deviation for the number of stock transactions executed.

3. The price of a share of stock of the Bil Corpo-

ration on a daily basis was as follows: $23, $26, $19, $24, $23, $27, $20, and $25. Find the range and standard deviation for the price of a share of stock of this company over this period.

4. *Home Heating Needs.* The number of degree days used in determining home heating oil needs on a daily basis over a two-week period for a northeastern city was as follows: 23, 58, 62, 29, 32, 41, 84, 99, 121, and 109. Find the range and standard deviation for the number of degree days.

5. *Car Insurance.* On June 10, 1985, Joe Frapacci called 6 different insurance companies to obtain price quotes for car insurance for his 1984 Honda. The prices quoted for the same coverages were $812, $690, $750, $902, $784, and $848. Find the mean and standard deviation of the price for car insurance.

6. *Inflation.* Refer back to Exercise 5. A year later, Joe Frapacci calls the same 6 insurance companies to obtain price quotes for car insurance. Each company informs Joe that because of inflation the rates have been increased by 10%.

   a) Calculate the mean and standard deviation for the new rates for car insurance.

   b) How do the answers compare with the answers obtained in Exercise 5?

7. *Arson.* The average cost of fire insurance for a 200-square-ft two-story loft in a certain city is $3000 annually with a standard deviation of $310. Due to the high incidence of arson in the city, the rates will be doubled next year. What will the new mean and standard deviation be?

8. *Money Market Fund.* Stacy is interested in investing some money in a money market fund. After analyzing the investment objectives of numerous companies, Stacy has narrowed down her choice to either Fund A or Fund B. For the past eight months the price of a unit share in either fund has been as follows:

Date	Unit value of a share of Fund A	Unit value of a share of Fund B
Jan. 1	16.12	32.17
Feb. 1	17.03	33.19
Mar. 1	16.82	31.78
Apr. 1	17.08	32.14
May 1	16.91	33.03
June 1	16.72	31.64
July 1	16.41	32.01
Aug. 1	15.93	31.45

   a) Find the range and standard deviation of the unit value of a share of each fund.

   b) Which fund performed more consistently?

9. Two identical stereo components manufactured by two different companies were thoroughly tested to determine the useful life of each component. The following results were obtained.

	Average life	Standard deviation
Component manufactured by Company A	140 hours	7 hours
Component manufactured by Company B	135 hours	2 hours

Which stereo component would you buy? Explain your answer.

11.6

FREQUENCY DISTRIBUTIONS AND GRAPHS There are many situations in which the data may be so numerous that it would be difficult (if not impossible) to come up with any meaningful interpretation of them. To see how this can happen, consider Michael, who is late for work many times. His boss tells him that for the past six weeks he has been late the number of minutes per day shown below.

$$
\begin{array}{cccccc}
7 & 6 & 1 & 4 & 0 & 13 \\
15 & 10 & 12 & 3 & 3 & 12 \\
10 & 12 & 11 & 13 & 2 & 11 \\
2 & 3 & 8 & 5 & 3 & 14 \\
5 & 14 & 6 & 7 & 7 & 9
\end{array}
$$

All one can say definitely at first glance is that he was on time only once and one day he was as much as 15 minutes late. Since the numbers are not arranged in order, it is somewhat difficult to conclude anything else from them. For this reason we use a frequency distribution to organize the data. First we have the following definition.

> **Definition 11.7**   *A **frequency distribution** is a convenient way of organizing data so that we may see what patterns they have. The word **frequency** will be interpreted to mean how often a number occurred.*

A frequency distribution is made very easily. We first make a list of numbers from 0 to 15 in a column to show how many minutes Michael was late. Then we make a second column for tally marks. We go through the original numbers, and each time he was late we put a tally mark in the appropriate space. Finally, we add the tally marks per line and indicate this sum in the frequency column. When we apply this to our problem, we get the table shown below.

Minutes late to work	Tally	Frequency	Minutes late to work	Tally	Frequency
0	|	1	8	|	1
1	|	1	9	|	1
2	||	2	10	||	2
3	||||	4	11	||	2
4	|	1	12	|||	3
5	||	2	13	||	2
6	||	2	14	||	2
7	|||	3	15	|	1

Once we have done this, we can come up with meaningful interpretations. We see that most latenesses were 3-minute ones (there were four of them). Michael was also late more than 5 minutes 19 times. This would represent 19 out of 30 times, or approximately 60% of the time. Still other interpretations can be given to the data shown.

One may want to draw a **bar graph** for these numbers.

To construct the bar graph, we first draw two lines, one horizontal (across) and one vertical (up–down). The horizontal line we will label "Minutes late" and the vertical line we will label "Frequency."

Once we have the frequency distribution, we can draw the bar graph very easily. The height of each bar will represent the frequency. The bar graph will also tell us at a glance that the most latenesses were the 3-minute ones. See Figure 11.1.

Figure 11.1

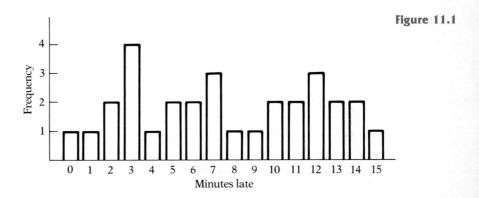

We could shorten the above frequency distribution as shown in the following chart.

Minutes late	Tally	Frequency	Minutes late	Tally	Frequency
0–1	||	2	8–9	||	2
2–3	‖‖		6	10–11	||||	4
4–5	|||	3	12–13	‖‖	5
6–7	‖‖	5	14–15	|||	3

This chart is more compact, *but* some of the information is lost in this version. For example, we can see that Michael is late 2–3 minutes 6 times. But we cannot tell exactly how often he is 2 minutes late or 3 minutes late.

When there are many numbers, listing them separately as we did in the first chart may make it difficult to look at the data and draw meaningful conclusions. If the data are grouped, as in the second chart, they may be easier to interpret.

To further illustrate the idea of a frequency distribution and a bar graph, consider the following example.

EXAMPLE 1 A large midwestern university is reviewing the performance of its star basketball player. During the past season he scored the following number of points per game.

27	16	19	24	18	23	24	18	24	25
23	16	23	24	19	22	17	25	19	27
19	29	24	25	24	18	32	23	21	30

Find the frequency distribution and draw the bar graph for the above numbers.

SOLUTION   We make three columns. The first column will contain the number of points scored, the second will have the tally, and the third will give the frequency.

Number of points	Tally	Frequency		Number of points	Tally	Frequency
16	||	2		25	|||	3
17	|	1		26		0
18	|||	3		27	||	2
19	||||	4		28		0
20		0		29	|	1
21	|	1		30	|	1
22	|	1		31		0
23	||||	4		32	|	1
24	ⅣⅠ|	6				

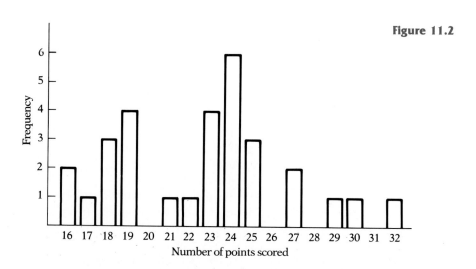

Figure 11.2

The bar graph for this distribution is shown in Figure 11.2. One thing should be immediately obvious. The player scored 24 points most often. ▯

EXAMPLE 2 George is a maintenance man in a large office building. He has compiled the following list of numbers that indicate the life length (in hours) of 25 special lightbulbs.

50	40	40	55	50
45	55	50	60	45
55	50	45	50	60
60	45	55	45	55
50	35	50	65	40

Find the frequency distribution and draw the bar graph for these numbers.

SOLUTION Again we make three columns. The first column is for the life length, the second for the tally, and the third for the frequency. We have the chart shown below.

Life length	Tally	Frequency
35	|	1
40	|||	3
45	|||||	5
50	||||| ||	7
55	|||||	5
60	|||	3
65	|	1

The bar graph for this distribution is shown in Figure 11.3.

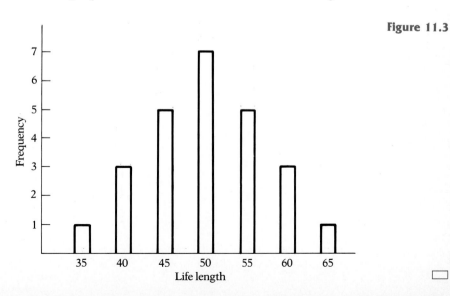

Figure 11.3

Until now we have seen how a bar graph can be used to picture information. In statistics we use not only bar graphs, but also other kinds of graphs to show all the information given in a situation. This is often of great help in arriving at meaningful conclusions.

The following examples illustrate how bar graphs and other kinds of graphs can be used.

EXAMPLE 3   The number of vehicles passing through a toll booth on a turnpike during a 24-hour period is given in the table below:

Time	Number of cars arriving
12 midnight–2 A.M.	120
2 A.M.–4 A.M.	640
4 A.M.–6 A.M.	1790
6 A.M.–8 A.M.	5780
8 A.M.–10 A.M.	3460
10 A.M.–12 noon	2010
12 noon–2 P.M.	1860
2 P.M.–4 P.M.	2000
4 P.M.–6 P.M.	4030
6 P.M.–8 P.M.	5640
8 P.M.–10 P.M.	2440
10 P.M.–12 midnight	560

To be able to draw any meaningful conclusion from these data, we construct a bar graph that contains all the given information (Figure 11.4). From this graph we can see that the greatest number

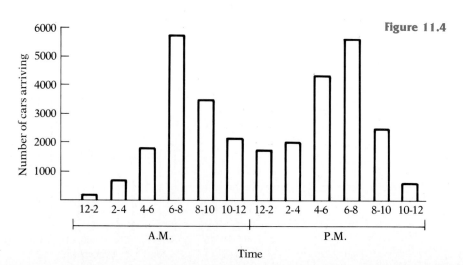

Figure 11.4

of cars passed through the toll booth during the hours of 6 to 8 A.M. Traffic then began dropping off until after 2 P.M., at which time the number of cars increased until about 8 P.M. and then dropped off again. ☐

Information of this type is needed by the authorities to determine the number of toll collectors to hire and the hours to hire them for, so that motorists will not have to wait in long lines. It is easier to determine this information from a graph than from the table of numbers.

EXAMPLE 4 The latest statistics showing the incidence of heart attacks among various age groups in a certain community in the southwest are pictured in the bar graph in Figure 11.5.

a) For this particular community, in what age group are there more than 60 heart attacks per thousand?

b) In what age group are there more than 100 heart attacks per thousand?

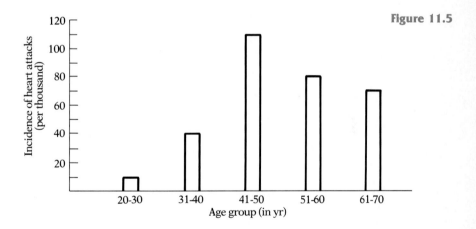

Figure 11.5

SOLUTION **a)** We see from the graph that in the age groups between 41 and 70 yr there were more than 60 heart attacks per thousand people.

b) From the graph we see that only the age group of 41–50 yr had more than 100 heart attacks per thousand people. ☐

EXAMPLE 5 A large supermarket chain is interested in knowing which flavor of soda is in greatest demand and during which months. It needs this information so that it can adequately stock its warehouses in advance. For its three most popular flavors it has available last year's statistics, which have been recorded in the form of a bar graph (Figure 11.6).

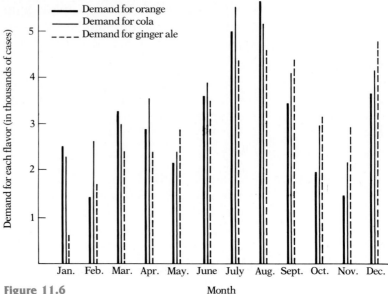

Figure 11.6

a) In which months is orange most popular?

b) Which flavor is most popular in December?

c) Which flavor is least popular in February? ☐

EXAMPLE 6   The budget of the student government at a particular university has just been approved. The **circle graph** in Figure 11.7 indicates how the students plan to spend their money. If the college administration has

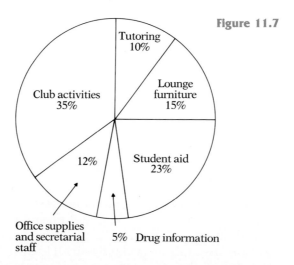

Figure 11.7

granted them $100,000, how much money will be spent for:

a) club activities?   **b)** drug information?

c) lounge furniture?   **d)** student aid?

SOLUTION   **a)** Since 35% of $100,000 will be spent on club activities, we have

$$0.35 \times 100{,}000 = 35{,}000.$$

Thus they will spend $35,000 on club activities. (We write 35% in decimal form as 0.35.)

b) They will spend 5% of $100,000 for drug information. (We write 5% as 0.05.) Thus they will spend

$$0.05 \times 100{,}000 = \$5000,$$

or $5000 for drug information.

c) They will spend

$$0.15 \times 100{,}000 = 15{,}000,$$

or $15,000 for lounge furniture.

d) They will spend

$$0.23 \times 100{,}000 = 23{,}000,$$

or $23,000 for student aid.   ☐

Comment   Circle graphs are also known as **pie charts.**

EXAMPLE 7   An electric company wants to know during which hours electrical supply is in greatest demand on a typical summer day. It needs this

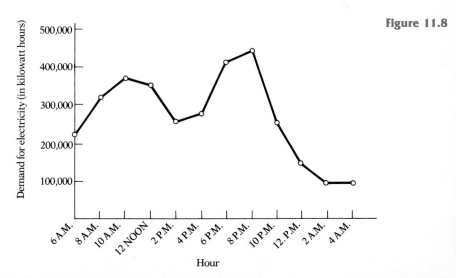

Figure 11.8

information so that it can prepare itself to satisfy consumer demands adequately. Information for one typical day has been gathered and is shown in the **line graph** in Figure 11.8.

a) During which hour(s) is electricity in greatest demand?

b) During which hours is the demand for electricity decreasing?

c) During which hour(s) is electricity in least demand? ☐

In recent years the bars of a bar graph have often been replaced by a series of equally spaced identical pictures, where each picture or symbol represents a specified quantity. Such graphs, which are called **pictographs** or **pictograms,** are used to catch the reader's eye and to make the graph more appealing.

EXAMPLE 8   *The price of gas.* The pictograph in Figure 11.9 shows how the price of gas has changed over the past few years in one particular city. ☐

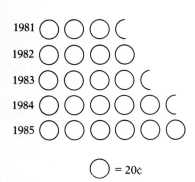

Figure 11.9
Average selling price of a gallon of gasoline.

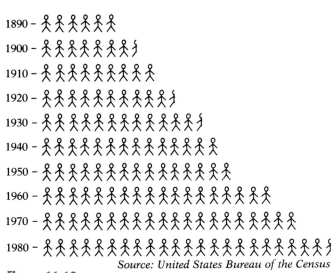

Source: United States Bureau of the Census

Figure 11.10

EXAMPLE 9   *Population changes.* The pictograph in Figure 11.10 indicates the population change of the United States from 1890 to 1980. (Each symbol represents ten million people.)

a) How many people were there in 1930?

b) By approximately how many people did the population of the United States change from 1920 to 1970?

SOLUTION   a) In 1930 there were 125 million people in the United States.

b) The population of the United States increased from 105 million in 1920 to 200 million in 1970, or by 95 million people. ☐

EXERCISES

1. *Jogging.* The ages of the 40 runners who have registered to run in this year's 26-mile marathon are as follows:

31	36	34	29	18	21	18
32	24	32	34	25	19	30
19	34	32	22	17	31	33
20	21	28	29	27	17	28
24	36	31	16	34	29	
18	26	18	18	25	22	

Construct the frequency distribution for these data and then draw its bar graph.

2. The number of flights scheduled to depart daily from Republic Airport for the month of January is as follows:

27	17	20	17	16	19	21
26	33	35	23	26	20	
25	27	25	19	27	40	
27	15	20	26	18	24	
23	32	22	21	18	35	

Construct the frequency distribution for these data and then draw its bar graph.

3. Each of the 38 students in a statistics class were asked to indicate the number of children in their families. The results are as follows:

7	8	2	6	6	1	4	3
3	5	4	2	1	9	1	2
7	9	4	5	3	2	3	2
8	1	7	4	1	2	2	
3	5	5	4	2	1	5	

Construct the frequency distribution for the data and then draw its bar graph.

4. The number of new housing permits issued during a particular 6 week period on a daily basis in Sequa Valley is as follows:

21	5	38	14	21	8
5	8	16	40	5	17
12	16	12	28	12	9
15	13	15	31	15	
10	12	10	16	21	

Construct the frequency distribution for the data and then draw its bar graph.

5. The Printex Corporation employs 823 people. Recently, management decided to change the life insurance coverage that it provides for its workers. The company analyzed the ages of its workers so that it could apply to another insurance company. The ages of the workers are as follows:

Age	Male	Female
Under 30 yrs	103	203
Between 30 and 40 yrs	91	127
Between 40 and 55 yrs	106	82
Over 55 yrs	62	49

Draw a circle graph (pie chart) to picture the above information.

6. The Hakewa Clothing Corporation sells its products throughout the world as shown in the accompanying circle graph.

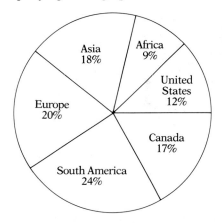

If the Hakewa Company had sales of $7,823,100 during 1985, find the sales figures for each region mentioned.

7. The graph on the following page shows the number of murders in a large southeastern city for the years 1980–1985.

   a) In 1982, how many murders were there?

   b) In 1984, how many murders were there?

   c) What was the increase between 1981 and 1985?

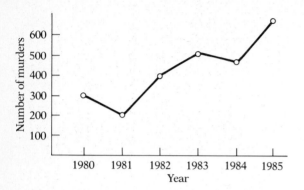

8. The graph below shows the number of businesses filing for Chapter 11 bankruptcy during 1985 in District 6.

  a) During which month(s) were there the greatest number of bankruptcies?

  b) How many bankruptcies were there during the entire year of 1985?

Unemployment Seen as Top Problem

BOSTON (Dec. 9)—The results of a national Gallup Poll of nearly half of the U.S. voters—both Democratic and Republican—disclosed unemployment as the most important problem facing their congressional districts. The survey was taken during the third week of September, much before the staggering 10.1 percent unemployment figures were released.

Forty-seven percent of the people polled indicated that unemployment was their main concern. Furthermore, when asked which party they thought would be better able to deal with this problem, the Democratic Party won handily over the GOP, 45 percent when compared with 20 percent.

Other problems named by the respondents ranked far behind unemployment. Some of these other problems of concern to voters include the following: inflation (11 percent), the economy (9 percent), local problems (8 percent), taxes (7 percent), crime (7 percent), and high interest rates (3 percent).

SAN FRANCISCO GLOBE, December 9, 1982

9. Refer to the accompanying newspaper article. The following data are available from the survey. Draw a bar graph to picture the information.

Problem of concern to you

	Democrat	Republican
Unemployment	352	118
Inflation	82	28
The economy	68	22
Local problems	60	20

10. The Jones' spend their $26,000 family income as follows:

Food	$4500
Clothing	$5000
Rent	$6000
Education	$2700
Entertainment	$2400
Travel	$3200
Miscellaneous	$2200
	$26,000

Draw a circle graph to picture this information.

11. *Real Estate Taxes.* Real estate taxes per $1000 of assessed evaluation on residential property in Charleston have changed over the past 8 years as shown in the accompanying chart. Draw a line graph to picture this information.

Year	1978	1979	1980	1981	1982	1983	1984	1985
Real estate taxes	$84	$88	$97	$105	$116	$132	$157	$194

11.7

THE NORMAL DISTRIBUTION

Recently, a survey was taken of the weights of the 800 students at Trixy College. After the data were arranged into a frequency distribution containing 14 intervals, the bar graph shown in Figure 11.11 was obtained. When the data were grouped into a frequency distribution containing 28 intervals, the bar graph shown in Figure 11.12 was obtained. If we were to continue this process of adding additional

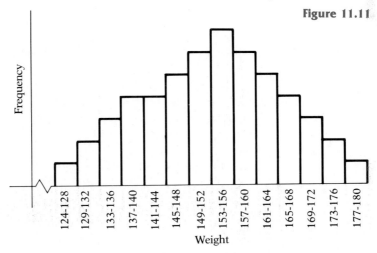

Figure 11.11

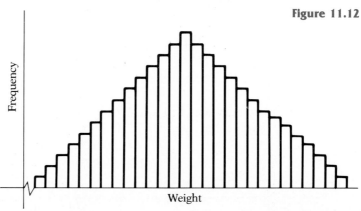

Figure 11.12

Figure 11.13
The normal curve

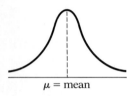

μ = mean

Figure 11.14

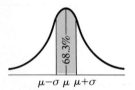

68.3%

$\mu-\sigma$ μ $\mu+\sigma$

Figure 11.15

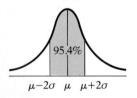

95.4%

$\mu-2\sigma$ μ $\mu+2\sigma$

Figure 11.16

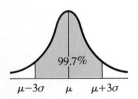

99.7%

$\mu-3\sigma$ μ $\mu+3\sigma$

Figure 11.17

intervals, the graph would tend to smooth out into a continuous curve similar to the one shown in Figure 11.13. Such a curve is called a *normal curve*, and the distribution that gives rise to it is called a *normal distribution*.

Since the normal distribution has wide-ranging applications, we need a careful description of a normal curve and some of its properties. The graph of a normal distribution is a bell-shaped curve that extends in both directions. Although the curve gets closer and closer to the horizontal axis, it never really touches it, no matter how far it is extended.

The mean of the normal distribution is at the center of the curve, and the curve is symmetric about the mean. This tells that we can fold the curve along the dotted line in Figure 11.14 and either portion of the curve will correspond with the other portion.

For a normal distribution the mean, the median, and the mode are all equal. Remember that μ represents the mean and σ represents the population standard deviation.

The data that make up the normal distribution tend to cluster around the middle with very few values more than three standard deviations from the mean on either side. As a matter of fact, about 68.3% of the data will fall within one standard deviation of the mean on either side (see Figure 11.15). Also, approximately 95.4% of the data will fall within two standard deviations of the mean on either side (see Figure 11.16), and approximately 99.7% of the data will fall within three standard deviations of the mean on either side (see Figure 11.17).

Thus in our case, if the weights of the college students are normally distributed with a mean of 150 pounds and a standard deviation of 10 pounds, then we would expect approximately 68.3% of the students to weigh between

$$\mu - \sigma \quad \text{and} \quad \mu + \sigma$$
$$150 - 10 \quad \text{and} \quad 150 + 10$$

or between 140 and 160 pounds. Similarly, we would expect approximately 95.4% of the students to weigh between

$$\mu - 2\sigma \quad \text{and} \quad \mu + 2\sigma$$
$$150 - 2(10) \quad \text{and} \quad 150 + 2(10)$$

or between 130 and 170 pounds. Also 99.7% of the students should weigh between

$$\mu - 3\sigma \quad \text{and} \quad \mu + 3\sigma$$
$$150 - 3(10) \quad \text{and} \quad 150 + 3(10)$$

or between 120 and 180 pounds.

A normal distribution is completely specified by its mean and standard deviation. Thus although all normal distributions are basically bell-shaped, different means and different standard deviations will describe different bell-shaped curves (see Figure 11.18). It is possible, however, to convert each of these different normal distributions into one standard form. You may be wondering, why bother? The answer is rather simple as we shall see.

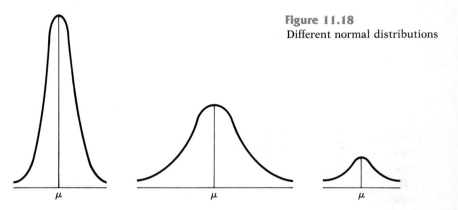

Figure 11.18
Different normal distributions

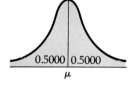

Figure 11.19

The total area under the normal curve is exactly one (one square unit). Since the normal curve is symmetric about the mean, we can immediately conclude that $\frac{1}{2}$ (or 0.5000) of *all* the area is to the right of the mean and $\frac{1}{2}$ (or 0.5000) of the area is to the left of the mean (see Figure 11.19).

*Area under the normal curve is associated with probability. Thus if a measurement x, the weights of college students in our case, is normally distributed, then the probability that x will fall between the values of a and b is equal to the **area** under the normal curve between a and b.*

Since areas under a normal distribution are related to probability, we can use special normal distribution tables for calculating probabilities. Since the mean and the standard deviation can be any values, however, it would seem that we need an endless number of tables. Fortunately, this is not the case; we need only one standardized table. (See Table 11.1 page 618.) Thus the area under the curve between 40 and 60 of a normal distribution with a mean of 50 and a standard deviation of 10 will be the same as the area between 70 and 80 of another normally distributed variable with mean 75 and standard deviation 5. They are both within one standard deviation unit from the mean. It is for this reason that statisticians use a standard normal distribution.

Definition 11.8 *A **standard normal distribution** is a normal distribution with a mean of 0 and a standard deviation of 1.*

TABLE 11.1
Areas of a Standard Normal Distribution

An entry in the table is the proportion under the entire curve that is between $z = 0$ and a positive value of z. Areas for negative values of z are obtained by symmetry.

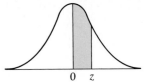

z	0.00	0.01	0.02	0.03	0.04	0.05	0.06	0.07	0.08	0.09
0.0	0.0000	0.0040	0.0080	0.0120	0.0160	0.0199	0.0239	0.0279	0.0319	0.0359
0.1	0.0398	0.0438	0.0478	0.0517	0.0557	0.0596	0.0636	0.0675	0.0714	0.0753
0.2	0.0793	0.0832	0.0871	0.0910	0.0948	0.0987	0.1026	0.1064	0.1103	0.1141
0.3	0.1179	0.1217	0.1255	0.1293	0.1331	0.1368	0.1406	0.1443	0.1480	0.1517
0.4	0.1554	0.1591	0.1628	0.1664	0.1700	0.1736	0.1772	0.1808	0.1844	0.1879
0.5	0.1915	0.1950	0.1985	0.2019	0.2054	0.2088	0.2123	0.2157	0.2190	0.2224
0.6	0.2257	0.2291	0.2324	0.2357	0.2389	0.2422	0.2454	0.2486	0.2517	0.2549
0.7	0.2580	0.2611	0.2642	0.2673	0.2704	0.2734	0.2764	0.2794	0.2823	0.2852
0.8	0.2881	0.2910	0.2939	0.2967	0.2995	0.3023	0.3051	0.3078	0.3106	0.3133
0.9	0.3159	0.3186	0.3212	0.3238	0.3264	0.3289	0.3315	0.3340	0.3365	0.3389
1.0	0.3413	0.3438	0.3461	0.3485	0.3508	0.3531	0.3554	0.3577	0.3599	0.3621
1.1	0.3643	0.3665	0.3686	0.3708	0.3729	0.3749	0.3770	0.3790	0.3810	0.3830
1.2	0.3849	0.3869	0.3888	0.3907	0.3925	0.3944	0.3962	0.3980	0.3997	0.4015
1.3	0.4032	0.4049	0.4066	0.4082	0.4099	0.4115	0.4131	0.4147	0.4162	0.4177
1.4	0.4192	0.4207	0.4222	0.4236	0.4251	0.4265	0.4279	0.4292	0.4306	0.4319
1.5	0.4332	0.4345	0.4357	0.4370	0.4382	0.4394	0.4406	0.4418	0.4429	0.4441
1.6	0.4452	0.4463	0.4474	0.4484	0.4495	0.4505	0.4515	0.4525	0.4535	0.4545
1.7	0.4554	0.4564	0.4573	0.4582	0.4591	0.4599	0.4608	0.4616	0.4625	0.4633
1.8	0.4641	0.4649	0.4656	0.4664	0.4671	0.4678	0.4686	0.4693	0.4699	0.4706
1.9	0.4713	0.4719	0.4726	0.4732	0.4738	0.4744	0.4750	0.4756	0.4761	0.4767
2.0	0.4772	0.4778	0.4783	0.4788	0.4793	0.4798	0.4803	0.4808	0.4812	0.4817
2.1	0.4821	0.4826	0.4830	0.4834	0.4838	0.4842	0.4846	0.4850	0.4854	0.4857
2.2	0.4861	0.4864	0.4868	0.4871	0.4875	0.4878	0.4881	0.4884	0.4887	0.4890
2.3	0.4893	0.4896	0.4898	0.4901	0.4904	0.4906	0.4909	0.4911	0.4913	0.4916
2.4	0.4918	0.4920	0.4922	0.4925	0.4927	0.4929	0.4931	0.4932	0.4934	0.4936
2.5	0.4938	0.4940	0.4941	0.4943	0.4945	0.4946	0.4948	0.4949	0.4951	0.4952
2.6	0.4953	0.4955	0.4956	0.4957	0.4959	0.4960	0.4961	0.4962	0.4963	0.4946
2.7	0.4965	0.4966	0.4967	0.4968	0.4969	0.4970	0.4971	0.4972	0.4973	0.4974
2.8	0.4974	0.4975	0.4976	0.4977	0.4977	0.4978	0.4979	0.4979	0.4980	0.4981
2.9	0.4981	0.4982	0.4982	0.4983	0.4984	0.4984	0.4985	0.4985	0.4986	0.4986
3.0	0.4987	0.4987	0.4987	0.4988	0.4988	0.4989	0.4989	0.4989	0.4990	0.4990

Thus, what we attempt to do is to convert any normal curve with mean μ and standard deviation σ into a normal curve with mean 0 and standard deviation 1. For example, suppose we consider the weights of college students (discussed earlier) whose mean $\mu = 150$ and whose standard deviation $\sigma = 10$. From the original data, one standard deviation to the right of the mean corresponds to 150 + 10, or 160. On the standard normal curve this simply corresponds to +1. Also, from the original data, two standard deviations to the right of the mean corresponds to 150 + 2(10), or 170. On the standard normal curve this simply corresponds to +2. The corresponding value on the standard normal curve is usually expressed in units of z (see Figure 11.20). Similarly, one standard deviation to the left of the mean corresponds to a weight of 150 − 10, or 140 pounds. On the standard normal curve this corresponds to a negative z-value of −1. In a similar manner a weight of 151 pounds corresponds to $z = 0.1$, a weight of 152 corresponds to $z = 0.2$, and so on.

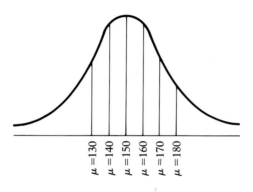

Figure 11.20
The correspondence between the original data (top) and the standard normal curve (bottom)

More generally, we can convert any raw score x into a z-score by using Formula 3.

Formula 3   *The **z-score** or **z-value** of any number x in a normal distribution is given by*

$$z = \frac{x - \mu}{\sigma},$$

where x = original score, μ = mean, and σ = standard deviation.

Comment   A z-score tells us how many standard deviation units a particular number is away from the mean as well as whether this number is above or below the mean.

EXAMPLE 1   A certain brand of flashlight battery has a mean life μ of 40 hours and a standard deviation σ of 5 hours. Find the z-score of a battery that lasts for

    a) 50 hours   **b)** 35 hours   **c)** 40 hours

SOLUTION   Since $\mu = 40$ and $\sigma = 5$, we use Formula 3.

    a) The z-score of 50 is

$$z = \frac{x - \mu}{\sigma} = \frac{50 - 40}{5} = 2.$$

Since the z-score is a positive number, we know that $x = 50$ lies to the right of the mean.

    b) The z-score of 35 is

$$z = \frac{x - \mu}{\sigma} = \frac{35 - 40}{5} = -1.$$

Since the z-score is a negative number, we know that $x = 35$ lies to the left of the mean.

    c) The z-score of 40 is

$$z = \frac{40 - 40}{5} = \frac{0}{5} = 0.$$

Since the z-score is 0, we know that $x = 40$ coincides with the mean. $\square$

Comment   Because σ is always a positive number, z will be a negative number whenever x is less than μ, as $x - \mu$ is then a negative number. A z-score of 0 implies that the term has the same value as the mean.

Let us now apply the idea of z-score to find the areas under a normal curve and thus probability calculations.

Table 11.1 on page 618 gives the areas of a standard normal distribution between $z = 0$ and $z = 3.09$. We read the table as follows: The first two digits of the z-score are under the column headed by z; the third digit heads the other columns. To find the area from $z = 0$ to $z = 2.59$, we first look under z to 2.5 and then across from $z = 2.5$ to the column headed by 0.09. The area is 0.4952 or 49.52%.

Similarly, to find the area from $z = 0$ to $z = 1.94$, we first look under $z = 1.9$ and then move across to the column headed by 0.04. The area is 0.4738.

EXAMPLE 2   Find the area between $z = 0$ and $z = 1.13$ in a standard normal curve.

SOLUTION   We first draw a sketch as shown in Figure 11.21. Then using Table 11.1 for $z = 1.13$, we find that the area between $z = 0$ and $z = 1.13$ is 0.3708. This means that the probability of a score with this normal distribution falling between $z = 0$ and $z = 1.13$ is 0.3708.   ▫

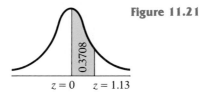

Figure 11.21

0.3708

$z = 0$ $z = 1.13$

EXAMPLE 3   Find the area between $z = -1.29$ and $z = 0$ in a standard normal distribution.

SOLUTION   We first draw a sketch as shown in Figure 11.22. Then using Table 11.1, we look up the area between $z = 0$ and $z = 1.29$. The area is 0.4015, not -0.4015. A negative value of z tells us that the value of z is to the left of the mean. The area under the curve (and the resulting probability) is *always* a positive number. Thus the probability of getting a z–score between 0 and -1.29 is 0.4015.   ▫

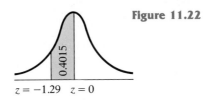

Figure 11.22

0.4015

$z = -1.29$ $z = 0$

EXAMPLE 4   Find the area between $z = -1.37$ and $z = 1.78$ in a standard normal distribution.

SOLUTION   We draw the sketch as shown in Figure 11.23. Since Table 11.1 gives the area only from $z = 0$ on, we first look under the normal curve from $z = 0$ to $z = 1.78$. We get 0.4625. Then we look up the area between $z = 0$ and $z = -1.37$. We get 0.4147. Finally, we add these two together and get $0.4625 + 0.4147 = 0.8772$. Thus the probability that a z–score is between $z = -1.37$ and $z = 1.78$ is 0.8772.   ▫

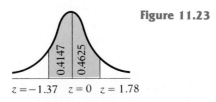

Figure 11.23

0.4147   0.4625

$z = -1.37$ $z = 0$ $z = 1.78$

By following a procedure similar to that used in Example 4, you should verify the following:

1. The probability that a z-score falls within one standard deviation of the mean on either side, that is, between $z = -1$ and $z = 1$, is approximately 68%.

2. The probability that a z-score falls within two standard deviations of the mean, that is, between $z = -2$ and $z = 2$, is approximately 95%.

3. The probability that a z-score falls within three standard deviations of the mean is 99.7%. Thus approximately 99.7% of the z-scores fall between $z = -3$ and $z = 3$.

In many cases we have to find areas between two given values of z or areas to the right or left of some value of z. Finding these areas is an easy task provided that we remember that the area under the entire normal distribution is 1. Thus since the normal distribution is symmetrical about $z = 0$, we conclude that the area to the right of $z = 0$ and the area to the left of $z = 0$ are both equal to 0.5000.

EXAMPLE 5   Find the area between $z = 0.96$ and $z = 2.83$ in a standard normal distribution.

SOLUTION   We cannot look this up directly, since the chart starts at 0, not at $z = 0.96$. However, we can look up the area between $z = 0$ and $z = 2.83$ and get 0.4977 and then the area between $z = 0$ and $z = 0.96$ and get 0.3315 (see Figure 11.24). We then take the difference between the two and get $0.4977 - 0.3315 = 0.1662$. ☐

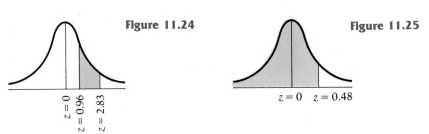

Figure 11.24   Figure 11.25

$z = 0$ $z = 0.96$ $z = 2.83$ $z = 0$ $z = 0.48$

EXAMPLE 6   Find the probability of getting a z-value that is less than 0.48 in a standard normal distribution.

SOLUTION   The probability of getting a z-value that is less than 0.48 really refers to the area under the curve to the left of $z = 0.48$. This represents the shaded portion of Figure 11.25. We look up the area from $z = 0$ to $z = 0.48$ and get 0.1844. We add this to 0.5000, getting $0.1844 + 0.5000 = 0.6844$. Thus the probability of getting a z-value less than 0.48 is 0.6844. ☐

If we are given a normal distribution whose mean is different from 0 and whose standard deviation is different from 1, we can convert this normal distribution into a standard normal distribution by converting each of its scores into a standard score by using Formula 3.

Expressing the scores of a normal distribution as standard scores allows us to calculate different probabilities, as Example 7 shows.

EXAMPLE 7 In a normal distribution with $\mu = 40$ and $\sigma = 5$, find the probability of obtaining the following.

a) A value greater than 50

b) A value less than 25

SOLUTION

a) We use Formula 3. We have $\mu = 40$, $x = 50$, and $\sigma = 5$, so that

$$z = \frac{x - \mu}{\sigma} = \frac{50 - 40}{5} = \frac{10}{5} = 2.$$

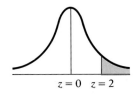

$z = 0 \quad z = 2$

Figure 11.26

See Figure 11.26. Thus we are really interested in the area to the right of $z = 2$ of a standard normal curve. The area from $z = 0$ to $z = 2$ is 0.4772. The area to the right of $z = 2$ is then 0.5000 − 0.4772 = 0.0228. Therefore the probability of obtaining a value greater than 50 is 0.0228.

b) We use Formula 3. We have $\mu = 40$, $x = 25$, and $\sigma = 5$, so that

$$z = \frac{x - \mu}{\sigma} = \frac{25 - 40}{5} = -\frac{15}{5} = -3.$$

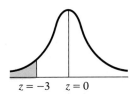

$z = -3 \quad z = 0$

Figure 11.27

See Figure 11.27. Thus we are interested in the area to the left of $z = -3$. The area from $z = 0$ to $z = -3$ is 0.4987. Thus the area to the left of $z = -3$ is 0.5000 − 0.4987 = 0.0013. The probability of obtaining a value less than 25 is therefore 0.0013. ▫

Let us now apply the normal distribution to some examples.

EXAMPLE 8 From past experience it has been found that the weight of a newborn at a maternity hospital is normally distributed with a mean of $6\frac{1}{2}$ pounds (104 ounces) and a standard deviation of 21 ounces. If a newborn baby is selected at random, find the probability that the baby weighs more than 8 pounds (128 ounces).

SOLUTION   We use Formula 3. Here $\mu = 104$, $\sigma = 21$, and $x = 128$, so that

$$z = \frac{x - \mu}{\sigma} = \frac{128 - 104}{21} = \frac{24}{21} = 1.14.$$

Figure 11.28

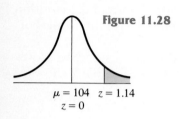

$\mu = 104 \quad z = 1.14$
$z = 0$

Thus, we are interested in the area to the right of $z = 1.14$. The area from $z = 0$ to $z = 1.14$ is 0.3729, so the area to the right of $z = 1.14$ is $0.5000 - 0.3729 = 0.1271$. See Figure 11.28. Therefore the probability that a randomly selected baby weighs more than 8 pounds is 0.1271. ▢

EXAMPLE 9   In a recent study in a certain town it was found that the number of hours that a typical ten-year-old child watches television per week is normally distributed with a mean of 7 hours and a standard deviation of 0.83 hours. Lester is a ten-year-old child in this town. What is the probability that he watches between 5 and 8 hours of television per week?

SOLUTION   We first find the probability that Lester will watch television between 7 and 8 hours and add to this the probability that he will watch television between 5 and 7 hours per week. Using Formula 3, we get

$$z = \frac{x - \mu}{\sigma} = \frac{8 - 7}{0.83} = 1.20.$$

The area between $z = 0$ and $z = 1.20$ is 0.3849. See Figure 11.29.

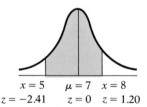

$x = 5 \qquad \mu = 7 \quad x = 8$
$z = -2.41 \qquad z = 0 \quad z = 1.20$ **Figure 11.29**

Also,

$$z = \frac{x - \mu}{\sigma} = \frac{5 - 7}{0.83} = -2.41.$$

The area between $z = 0$ and $z = -2.41$ is 0.4920. Adding these two probabilities, we get $0.4920 + 0.3849 = 0.8769$. Thus the probability[2] that Lester watches between 5 and 8 hours of television per week is 0.8769. ▢

---

2. For a more detailed discussion of the applications and uses of the normal distribution, see *Statistics and Probability in Modern Life* by Joseph Newmark, 4th ed., chap. 8 (Philadelphia: Saunders College Publishing, 1987).

EXERCISES

1. Which of the following would be likely to be normally distributed? (In each case, assume a random sample of size 5000.)

   a) height of statisticians

   b) useful life of a car

   c) age at which American men marry for the first time

   d) age at which Americans have their first heart attack

2. The following distribution represents the number of accidents (per year) occurring in a factory which employs 10,000 workers. (The mean is 10 and the standard deviation is 2.)

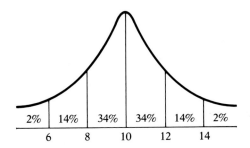

   a) What percent of the employees had 8–12 accidents per year?

   b) What percent of the employees had more than 8 accidents per year?

   c) What percent of the employees had fewer than 8 accidents per year?

   d) Approximately how many employees had between 6 and 14 accidents per year?

3. In a standard normal distribution, find the area

   a) between $z = 0$ and $z = 2.78$.

   b) between $z = -0.91$ and $z = 0$.

   c) to the right of $z = 1.37$.

   d) to the left of $z = -1.98$.

   e) to the right of $z = 2.76$.

   f) between $z = -1.63$ and $z = 1.94$.

   g) between $z = -1.93$ and $z = 2.79$

   h) between $z = -2.46$ and $z = 2.24$.

4. Find the percentage of z-scores in a standard normal distribution that are

   a) above $z = -1.69$.

   b) below $z = 2.83$.

   c) between $z = 0.89$ and $z = 3.01$.

   d) above $z = 2.01$.

   e) below $z = -2.86$.

   f) between $z = -2.27$ and $z = -1.49$.

   g) between $z = -2.08$ and $z = 2.81$.

5. In a recent weight-lifting contest the average weight lifted was 215 lb with a standard deviation of 12 lb. Find the z-score of an individual who

   a) lifted 227 lb.   **b)** lifted 251 lb.

   c) lifted 191 lb.   **d)** lifted 215 lb.

6. Find the percentage of z-scores in a normal distribution with $\mu = 79$ and $\sigma = 4$ that are

   a) between 75 and 83.

   b) between 81 and 84.

   c) less than 85.

   d) more than 76.

7. In a recent experiment, numerous volunteers were given 5 different brands of beer and were asked to determine which beer maintained its head the longest. The volunteers rated the beers in terms of z-scores.

Brand	Rating
A	−1.12
B	3.72
C	1.59
D	0.18
E	−0.93

   a) Rank these beers from highest to lowest.

   b) Which brands were above average?

   c) Which brands were below average?

8. A manufacturer claims that the outdoor paint being marketed by the company will require an average of 6 hours drying time with a standard deviation of 1.2 hours. Furthermore, the drying time is normally distributed. Find the probability that a house that is painted with this brand of paint will require more than 8 hours to dry.

9. According to medical officials, the number of hours elapsed before any reaction to a particular vaccination takes place is normally distributed with a mean of 36 hours and a standard deviation of 2.3 hours. Find the probability that a person who was given this vaccination will develop a reaction before 30 hours.

10. One brand of video game is manufactured by an automated process that gives the game an average useful life of 3000 hours with a standard deviation of 350 hours. The average useful life is known to be normally distributed.

George buys one such video game. Find the probability that it will last between 2700 hours and 3500 hours.

11. Marjorie Johnson is the personnel director for the General's Department Store Chain. Each day she interviews prospective workers. It has been found that the number of minutes that she needs to interview an applicant is normally distributed with a mean of 20 minutes and a standard deviation of 4 minutes. What percentage of the time will it take Marjorie more than 25 minutes to interview an applicant?

12. Bank officials claim that the number of days required to process a mortgage application is normally distributed with a mean of 12 days and a standard deviation of 2.4 days. Martha has applied for a mortgage. Find the probability that it will be processed in at most 14 days.

11.8

LINEAR CORRELATION AND REGRESSION[3]

Many colleges require students to take a mathematics placement test before allowing them to enroll in any calculus courses. Presumably, a student who scores well on such a placement test is more likely to do better in a calculus course than a student who scores poorly on such an exam. College officials may be interested in determining the reliability of such tests. Furthermore, they may be interested in being able to predict a student's performance in the calculus course when his or her performance on the placement exam is known.

Similarly, many colleges administer vocational aptitude tests. The officials may be interested in knowing whether there is any relationship between the math aptitude score and the business aptitude score. Do students who score well on the math part of the aptitude exam also do well on the business part? If we know a student's math score, can we predict the student's business score? Questions of this nature frequently arise when we have many variables and are interested in determining relationships between them.

Sir Francis Galton, a cousin of Charles Darwin, undertook a detailed study of human characteristics. He was interested in determining whether a relationship exists between the heights of fathers

---

3. Examples and exercises in this section should be solved with the aid of a hand-held calculator.

and the heights of their sons. Do tall parents have tall children? Do intelligent parents or successful parents have intelligent or successful children? In *Natural Inheritance*, Galton introduced the idea that we today refer to as **correlation**. This mathematical idea allows us to measure the closeness of the relationship between two variables. Galton found that there exists a very close relationship between the heights of fathers and the heights of their sons. On the question of whether intelligent parents have intelligent children, it has been found that the correlation is 0.55. As we shall see, this means that it is not necessarily true that intelligent parents have intelligent children.

To understand what is meant by correlation, let us consider the different aptitude scores that were obtained by ten students at State Tech College as shown in the accompanying table. Is there any relationship between math aptitude scores and business aptitude scores? If a student scores well in math, will the student also score well in business or in language?

Student	Math score	Language score	Music score	Business score
A	44	75	20	40
B	70	31	9	72
C	32	22	50	31
D	49	11	17	50
E	51	19	24	49
F	63	67	13	59
G	28	31	54	24
H	26	48	57	27
I	49	53	23	49
J	52	26	22	48

We can analyze the situation pictorially by means of a *scatter diagram*. We simply draw two lines (called *axes*), one vertical and one horizontal. On one of these axes we indicate the math scores, and on the other we indicate the business scores. After both axes are labeled, we use a point to represent each student's score. The point is placed directly above the student's math score and directly to the right of the business score. Thus the point for student A's scores is placed directly above the 44 score on the math axis and to the right of 40 on the business axis. This is pictured in Figure 11.30. In a similar manner we plot the other scores. You will notice that these points form an approximate straight line. When this happens, we say that

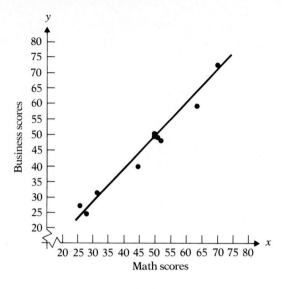

Figure 11.30
Scatter diagram for the math and business scores

there is a *linear correlation* between the two variables. Notice that the higher the math score, the higher is the business score. The line moves in a direction from lower left to upper right. When this happens, we say that there is a *positive correlation* between the math scores and the business scores. This means that a student with a higher math score will also have a higher business score.

Now let us draw the scatter diagram for the business aptitude scores and the music aptitude scores. It is given in Figure 11.31. In this case you will notice that the higher the business score, the lower the music score. Again the points arrange themselves in the form of a line, but this time the line moves in a direction from upper left to lower right. When this happens, we say that there is a *negative cor-*

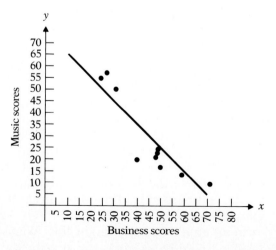

Figure 11.31
Scatter diagram for the business and music scores

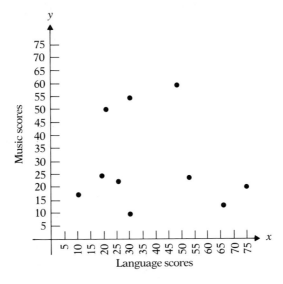

Figure 11.32
Scatter diagram for the language and music scores

relation between the business scores and the music scores. This means that a student with a high business score will have a low music score.

Now let us draw the scatter diagram for the language scores and the music scores. It is given in Figure 11.32. In this case the points do not form a straight line. When this happens, we say that there is little or no correlation between the language scores and the music scores.

Once we have determined that there is a linear correlation between two variables, we may be interested in determining the strength of the linear relationship. Karl Pearson developed a *coefficient of linear correlation*, which measures the strength of the relationship between two variables. The value of the coefficient of linear correlation is calculated by means of a formula.

Formula 4   *The **coefficient of linear correlation** is given by*

$$r = \frac{n(\Sigma\, xy) - (\Sigma\, x)(\Sigma\, y)}{\sqrt{n(\Sigma\, x^2) - (\Sigma\, x)^2}\ \sqrt{n(\Sigma\, y^2) - (\Sigma\, y)^2}},$$

where

x = *label for one of the variables,*

y = *label for the other variable,*

n = *number of pairs of scores.*

When using Formula 4 the coefficient of correlation will always have a value between $+1$ and -1. A value of $+1$ means perfect positive correlation and corresponds to the situation in which all the dots lie exactly on a straight line. A value of -1 means perfect negative correlation and again corresponds to the situation in which all the points lie exactly on a straight line. Correlation is considered high when it is close to +1 or −1 and low when it is close to 0. If the coefficient of linear correlation is 0, we say that there is no linear correlation. These possibilities are indicated in Figures 11.33 and 11.34.

Although Formula 4 looks complicated, it is rather easy to use. The only new symbol that appears is $\Sigma\,xy$. This value is found by multiplying the corresponding values of x and y and then adding all the products. The following examples illustrate the procedure.

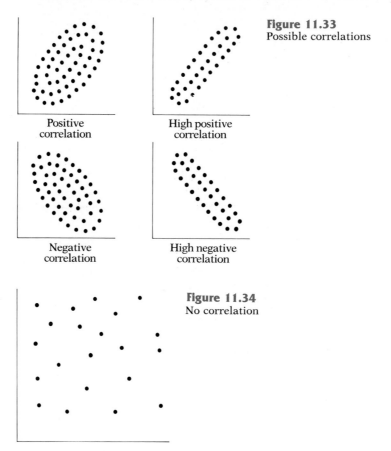

Figure 11.33
Possible correlations

Positive correlation

High positive correlation

Negative correlation

High negative correlation

Figure 11.34
No correlation

EXAMPLE 1   Find the coefficient of linear correlation between the business and music scores given here.

Business score	40	72	31	50	49	59	24	27	49	48
Music score	20	9	50	17	24	13	54	57	23	22

SOLUTION We let x represent the business score and y the music score. Then we arrange the data in tabular form.

x	y	x^2	y^2	xy
40	20	1600	400	800
72	9	5184	81	648
31	50	961	2500	1550
50	17	2500	289	850
49	24	2401	576	1176
59	13	3481	169	767
24	54	576	2916	1296
27	57	729	3249	1539
49	23	2401	529	1127
48	22	2304	484	1056
$\Sigma x = 449$	$\Sigma y = 289$	$\Sigma x^2 = 22{,}137$	$\Sigma y^2 = 11{,}193$	$\Sigma xy = 10{,}809$

Now we use Formula 4. We have

$$r = \frac{n(\Sigma\, xy) - (\Sigma\, x)(\Sigma\, y)}{\sqrt{n(\Sigma\, x^2) - (\Sigma\, x)^2}\,\sqrt{n(\Sigma\, y^2) - (\Sigma\, y)^2}}$$

$$= \frac{10(10{,}809) - (449)(289)}{\sqrt{10(22{,}137) - 449^2}\,\sqrt{10(11{,}193) - 289^2}}$$

$$= \frac{-21{,}671}{\sqrt{19{,}769}\sqrt{28{,}409}} = \frac{-21{,}671}{(140.60)(168.55)} = \frac{-21{,}671}{23{,}698.13}$$

$$= -0.9145.$$

Thus the cofficient of correlation is -0.9145. Since this value is close to -1, we say that there is a high degree of negative correlation. Figure 11.31 also indicated this. ▢

EXAMPLE 2 A fire department official believes that as the temperature decreases, the number of fires increases. To support this claim, she has gathered the statistics shown in the accompanying table. Compute the coefficient of linear correlation and comment.

Temperature, x	40	35	30	25	20	15	10	5
Number of fires, y	33	37	40	44	56	60	61	71

SOLUTION   We arrange the data in tabular form.

x	y	x^2	y^2	xy
40	33	1600	1089	1320
35	37	1225	1369	1295
30	40	900	1600	1200
25	44	625	1936	1100
20	56	400	3136	1120
15	60	225	3600	900
10	61	100	3721	610
5	71	25	5041	355
$\Sigma x = 180$	$\Sigma y = 402$	$\Sigma x^2 = 5100$	$\Sigma y^2 = 21{,}492$	$\Sigma xy = 7900$

Now we use Formula 4. We have, with $n = 8$

$$r = \frac{n(\Sigma\, xy) - (\Sigma\, x)(\Sigma\, y)}{\sqrt{n(\Sigma\, x^2) - (\Sigma\, x)^2}\,\sqrt{n(\Sigma\, y^2) - (\Sigma\, y)^2}}$$

$$= \frac{8(7900) - (180)(402)}{\sqrt{8(5100) - 180^2}\,\sqrt{8(21{,}492) - 402^2}}$$

$$= \frac{-9160}{\sqrt{8400}\sqrt{10{,}332}} = \frac{-9160}{(91.65)(101.65)}$$

$$= -0.9832.$$

Thus the coefficient of correlation is -0.9832, a rather high negative correlation. Therefore the data would seem to support the fire official's claim. ☐

Although the coefficient of correlation is usually the first number that is calculated when we are given several sets of scores, great care must be used in interpreting the results. It can undoubtedly be said that among all the statistical measures discussed in this chapter the correlation coefficient is the one that is most misused. One reason for this is the assumption that because the two variables are related, a change in one will result in a change in the other. Frequently, two variables may appear to have a high correlation, even though they are not directly associated with each other. There may be a third variable that is highly correlated to these two variables.

Regression Lines   Once we determine that there is a correlation between two variables, we might be interested in finding the equation of the line that best fits the set of points. If a relationship in the form of an equation can be found between the two variables, we can use this equation to *predict* the value of one of the variables if the value of the other is known.

Fitting a line to a set of numbers is by no means an easy task. Nevertheless, methods have been designed to handle such prediction problems. The **least squares method** determines the line in such a way that the sum of the squares of the y distances between the given points and the line is a minimum. Such a line is called a **regression line** of y on x.

> **Formula 5**   The **equation of the regression line** is given by
>
> $$y = m_y + b(x - m_x),$$
>
> where m_x and m_y are the means of x and y, respectively, and
>
> $$b = \frac{n(\Sigma\ xy) - (\Sigma\ x)(\Sigma\ y)}{n(\Sigma\ x^2) - (\Sigma\ x)^2}$$
>
> and n is the number of pairs of scores.

Let us use Formula 5 in the next examples to find the regression equation connecting two variables.

EXAMPLE 3   Fifteen students have been receiving special instruction prior to taking a state civil service exam. In an effort to determine the effectiveness of the program, a comparison is made between the grade on the exam and the number of weeks each student received the special instruction. The results are given in the table on the top of page 634.

Number of weeks, x	0.50	0.75	1.00	1.25	1.50	1.75	2.00	2.25	2.50	2.75	3.00	3.25	3.50	3.75	4.00
Grade, y	57	64	59	68	74	76	79	83	85	86	88	89	90	94	96

a) Compute the correlation coefficient between x and y.

b) Find the regression equation that will predict a student's score if we know how many weeks of special instruction he or she received.

c) If a student receives 0.65 weeks of special instruction, what is the student's predicted score on the exam?

SOLUTION We arrange the data in tabular form.

x	y	x^2	y^2	xy
0.50	57	0.2500	3249	28.50
0.75	64	0.5625	4096	48.00
1.00	59	1.0000	3481	59.00
1.25	68	1.5625	4624	85.00
1.50	74	2.2500	5476	111.00
1.75	76	3.0625	5776	133.00
2.00	79	4.0000	6241	158.00
2.25	83	5.0625	6889	186.75
2.50	85	6.2500	7225	212.50
2.75	86	7.5625	7396	236.50
3.00	88	9.0000	7744	264.00
3.25	89	10.5625	7921	289.25
3.50	90	12.2500	8100	315.00
3.75	94	14.0625	8836	352.50
4.00	96	16.0000	9216	384.00
$\Sigma x = 33.75$	$\Sigma y = 1188$	$\Sigma x^2 = 93.4375$	$\Sigma y^2 = 96,270$	$\Sigma xy = 2863.00$

a) To compute the coefficient of correlation, we use Formula 4:

$$r = \frac{n(\Sigma\ xy) - (\Sigma\ x)(\Sigma\ y)}{\sqrt{n(\Sigma\ x^2) - (\Sigma\ x)^2}\ \sqrt{n(\Sigma\ y^2) - (\Sigma\ y)^2}}$$

$$= \frac{15(2863) - (33.75)(1188)}{\sqrt{15(93.4375) - 33.75^2}\ \sqrt{15(96,270) - 1188^2}}$$

$$= \frac{2850}{\sqrt{262.5}\sqrt{32,706}} = \frac{2850}{(16.20)(180.85)} = 0.9728$$

Thus the coefficient of correlation is 0.9728.

b) To find the regression equation, we must first calculate the values of m_y, m_x, and b. We have

$$m_y = \frac{\Sigma y}{n} = \frac{1188}{15} = 79.2$$

$$m_x = \frac{\Sigma x}{n} = \frac{33.75}{15} = 2.25$$

$$b = \frac{n(\Sigma xy) - (\Sigma x)(\Sigma y)}{n(\Sigma x^2) - (\Sigma x)^2} = \frac{15(2863) - (33.75)(1188)}{15(93.4375) - 33.75^2}$$

$$= \frac{2850}{262.5} = 10.86$$

Thus the regression equation predicting a student's score when we know how many weeks of special instruction he or she received is, using Formula 5,

$$y = m_y + b(x - m_x)$$
$$= 79.2 + 10.86(x - 2.25).$$

Therefore the regression equation is $y = 79.2 + 10.86(x - 2.25)$.

c) If the student receives 0.65 weeks of instruction, then $x = 0.65$. Substituting this value of x into the regression equation gives

$$y = 79.2 + 10.86(0.65 - 2.25)$$
$$= 79.2 + 10.86(-1.6)$$
$$= 79.2 - 17.38$$
$$= 61.82.$$

Thus the student's predicted score on the exam is 61.82. ☐

EXAMPLE 4 As we indicated earlier, Galton believed that there exists a very close relationship between the heights of fathers and the heights of their sons. To test this claim, a scientist selects eight men at random and records their heights and the heights of their sons (in inches) as shown in the table.

Father, x	66	68	71	72	69	69	73	70
Son, y	63	66	70	74	70	68	73	70

a) Find the regression equation.

b) If a father is 74 inches tall, what is the predicted height of his son?

SOLUTION We arrange the data in tabular form.

x	y	x^2	xy
66	63	4356	4158
68	66	4624	4488
71	70	5041	4970
72	74	5184	5328
69	70	4761	4830
69	68	4761	4692
73	73	5329	5329
70	70	4900	4900
$\Sigma x = 558$	$\Sigma y = 554$	$\Sigma x^2 = 38{,}956$	$\Sigma xy = 38{,}695$

a) To find the regression equation, we calculate the values of m_y, m_x, and b. We have

$$m_y = \frac{\Sigma y}{n} = \frac{554}{8} = 69.25$$

$$m_x = \frac{\Sigma x}{n} = \frac{558}{8} = 69.75$$

$$b = \frac{n(\Sigma\ xy) - (\Sigma\ x)(\Sigma\ y)}{n(\Sigma\ x^2) - (\Sigma\ x)^2} = \frac{8(38{,}695) - (558)(554)}{8(38{,}956) - 558^2}$$

$$= \frac{428}{284} = 1.51$$

Thus the regression equation is

$$y = m_y + b(x - m_x)$$
$$= 69.25 + 1.51(x - 69.75).$$

b) If a father is 74 inches tall, then $x = 74$. Substituting this value of x into the regression equation gives

$$y = 69.25 + 1.51(74 - 69.75)$$
$$= 69.25 + 1.51(4.25)$$
$$= 69.25 + 6.42$$
$$= 75.67.$$

Thus the predicted height of the son is 75.67 inches.   ▢

    After determining the least squares regression equation that predicts a value of y when x has a particular value, we may be interested in determining how the predicted value of y and the observed value of y compare. Quite often there may be large differences between the two.

Fortunately, statisticians have devised a method for measuring the differences between the predicted and the observed values. This is the *standard error of the estimate* and is given by the formula

$$\sqrt{\frac{\Sigma\,(Y - Y_p)^2}{n - 2}}$$

where Y_p is the predicted value, Y is the observed value, and n is the number of pairs of scores.[4]

---

4. A complete discussion of this formula is beyond the scope of this book. The interested reader can find a detailed discussion in *Statistics and Probability in Modern Life* by Joseph Newmark, 4th ed., chap. 9 (Philadelphia: Saunders College Publishing, 1987).

EXERCISES

1. *Advertising.* A cereal manufacturer has determined that sales of his cereals are influenced by the number of times per week that commercials advertising his product are seen on television as shown in the following table:

Number of times that commercial is seen, x	Sales of product, y (in millions of dollars)
7	85
10	87
11	92
12	93
14	95
17	98
19	100
25	110

a) Draw the scatter diagram for the data.

b) Compute the coefficient of correlation.

c) Find the regression equation.

d) If the commercial is televised 22 times per week, what is the predicted sales?

2. The City of Cornwall wishes to sell $175 million of 20-year bonds. The comptroller has approached several banks to act as underwriting syndicates. The comptroller has found that the number of banks expressing an interest in the bonds is dependent upon the annual rate of interest to be paid as shown in the accompanying table:

Interest rate, x	Number of banks, y
7%	8
8%	10
9%	12
10%	13
$10\frac{1}{2}$%	15
11%	16
$11\frac{1}{2}$%	17
12%	19

a) Draw the scatter diagram for the data.

b) Compute the coefficient of correlation.

c) Find the regression equation.

d) If the annual interest rate is $9\frac{1}{2}$%, what is the predicted number of banks that will participate in the underwriting?

3. *Height and Starting Salaries.* Numerous studies have shown that taller people land better jobs and make more money than shorter people. To investigate this claim further, Halsey and Cobb obtained the following data for numerous people who obtained identical jobs.

Height, x (in inches)	Averaging starting salary, y
65	$21,000
66	24,000
68	25,000
70	27,000
72	28,000
73	29,000
74	30,000

a) Draw the scatter diagram.

b) Compute the coefficient of correlation.

c) Find the regression equation.

d) What is the predicted starting salary of a person who is 69 inches tall?

4. Several scientists are experimenting with chemicals that generate heat as they are blended together. The temperature of the solution, y, measured in degrees Celsius is dependent upon the number of minutes elapsed, x, after the chemicals are mixed together as shown below:

Time, x	Temperature, y
1	80°
2	84°
4	85°
6	87°
9	90°
11	94°
12	95°
15	98°
17	100°

a) Compute the coefficient of correlation.

b) Find the regression equation.

c) What is the predicted temperature 10 minutes after the chemicals are mixed together?

5. *Age and Weight.* The following table shows the age and weights of boys in different age groups at the Carvano Elementary School.

Age, x (in years)	Weight, y (in lb)
5	55
6	58
7	63
8	67
9	69
10	75
11	79
12	85
13	90

a) Compute the coefficient of correlation.

b) Find the regression equation.

c) What is the predicted weight of a $10\frac{1}{2}$-year-old boy at this school?

6. Medical researchers at a hospital have found that the number of days required for post-operative convalescence in the hospital after undergoing a particular operation is directly related to the age of the patient as shown below:

Age of patient, x (in years)	Number of days, y
25	2
28	3
31	4
35	5
40	6
45	8
53	9
60	11
65	12

a) Compute the coefficient of correlation.

b) Find the regression equation.

c) What is the predicted number of days of convalescence for a patient who is 50 years old?

7. As a result of a new state bottling law, which requires that all soda bottles sold in the state be recycled, industry officials are uncertain as to how much deposit to charge for each bottle. Past experience in other states indicates that the percentage of bottles returned depends upon the amount of money required as a deposit as shown below:

Per bottle deposit charge, x (in cents)	Percentage of bottles returned, y
1	75
2	77
3	80
4	82
5	90
8	94
10	96

a) Compute the coefficient of correlation.

b) Find the regression equation.

c) What is the predicted percentage of bottles that will be returned if the per bottle deposit charge is 7¢?

8. According to city government officials, the number of new housing starts is dependent upon the prevailing mortgage rate of interest as shown below:

Prevailing mortgage rate of interest, x (in percent)	Monthly number of new housing starts in the city, y
8	112
9	109
10	103
11	94
12	79
13	60
14	48

a) Compute the coefficient of correlation.

b) Find the regression equation.

c) What is the predicted number of new hous-ing starts if the mortgage rate of interest is $13\frac{1}{2}\%$?

9. Local AAA records indicate that the number of calls for assistance depends upon the outside temperature and is greatest when the temperature is lowest as shown below:

Outside temperature (in °F), x	Average number of calls for assistance, y
40°	227
35°	240
30°	260
25°	290
20°	340
15°	400
10°	470
5°	550
0°	700

a) Compute the coefficient of correlation.

b) Find the regression equation.

c) If the outside temperature drops to 3°F, what is the predicted number of calls for assistance?

10. Health officials claim that as the quality of the air we breathe increases (on a scale from 1 to 10), then the reported number of cases of upper respiratory infections decreases as shown below:

Air quality, x	Number of reported cases, y
1	80
2	65
3	48
4	37
5	25
6	19
8	15
9	12
10	7

a) Find the regression equation.

b) If the air quality is 7, what is the predicted

number of cases of upper respiratory infections that will be reported?

11.9
MISUSES OF STATISTICS

Statistics can be very meaningful and useful when applied properly, but great care must be taken to make sure that we do not read too much into them. This is the job of the statistician. It is important to know the size and extent of the sample, whether it was selected randomly, the kinds of analysis used, and so on.

An interesting example of how statistics can be misused is the following: A university in Texas has three female faculty members in the mathematics department. Recently, one of them married one of her students. The student newspaper then printed an article under the following headline.

$33\frac{1}{3}$ % OF OUR FEMALE MATHEMATICS FACULTY

MEMBERS MARRY THEIR STUDENTS

What, if anything, is wrong with this headline?

In April 1969 the *New York Times* stated that the Mets, who had finished tenth in the National League five times and ninth twice, were trying for third or fourth place that year but were not considered strong contenders. They won first place in 1969. Are statistics reliable information in making predictions?

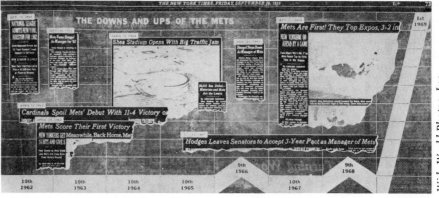

Wide World Photos, Inc.

There are many other examples of how statistics can be misused. For example, consider the following argument. There were at least two automobile accidents for every driver per 10,000 miles driven during 1985. Also, there were two electrical storms during the year, but there were no airplane accidents during these storms. Can we conclude that it is safer to be in an airplane during an electrical storm than it is to drive a car at any time?

Sometimes statistics can lead to contradictory conclusions. For example, in 1970 the Nobel Prize winner Dr. Linus Pauling claimed, on the basis of statistical data that he had collected, that large doses of vitamin C are quite effective in preventing the occurrence of the common cold and also in reducing its severity. To test this claim, many other studies have been made. Specifically, several doctors at the University of Toronto's School of Hygiene conducted such a study during the winter of 1971–1972. They found that vitamin C had no significant effect in preventing colds but did seem to reduce the severity of colds. So in one respect the findings of the Toronto group contradicted those of Dr. Pauling, whereas in another respect they confirmed them.

Other studies have both supported and contradicted Dr. Pauling's results. Thus whether or not you believe that vitamin C is a "cure" for the common cold may depend on how you collect and interpret your statistics.

Different Horizontal and/or Vertical Scales

Consider the two graphs shown in Figure 11.35. Both indicate how the crime rate has changed in a particular city over the past six years. One paints a more alarming picture than the other. Can you see why?

Figure 11.35

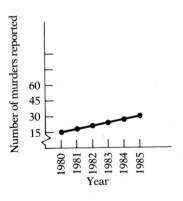
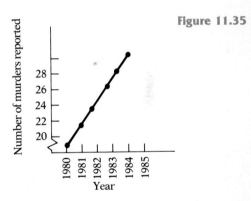

Correlation

Many people have applied a positive correlation to prove a cause-and-effect relationship that may not even exist. To illustrate the point, it has been shown that there is a high positive correlation between teachers' salaries and the use of drugs on campus. Does this mean that the more money a teacher earns, the more prevalent the use of drugs will be on campus?

Disregarding Unfavorable Data

Often statistics are used in the area of consumer products. For example, a drug manufacturer may be interested in claiming that 90% of the doctors surveyed recommend her particular product. She then has her advertising agent poll groups of 100 doctors until she finds a group in which the overwhelming majority use the product. The unfavorable results are then discarded.

Units Not Indicated Consider the following headline that appeared in a local newspaper: "In 1984 Murder in Our City Increased by 100%." Upon careful analysis it was found that in 1983 there were four murders committed in a city of 5 million and in 1984 there were eight murders committed. Thus although the headline may be mathematically accurate, it definitely does give one the impression that crime is more widespread in the city than it actually is.

For further examples of how statistics can be misused or misinterpreted, see *How to Lie with Statistics* by Darrell Huff (New York: W. W. Norton, 1954).

11.10

SUMMARY In this chapter we discussed different ways of analyzing a mass of data so that meaningful interpretations can be made. One method we mentioned is to use one of the measures of central tendency along with some measure of the dispersion involved. We indicated that the most commonly used measure of dispersion is the standard deviation. In discussing measures of central tendency we introduced summation notation, a notation used to indicate the operation of a sum of numbers.

A large quantity of numbers can also be analyzed by graphical techniques. We simply set up a frequency distribution and then draw the bar graph to picture the information contained in the frequency distribution. We also indicated how pictographs, circle graphs, and line graphs can be used, each in its own right, to picture information visually.

One particular distribution that we discussed in detail is the normal distribution. We indicated how it can be applied to a variety of problems.

Finally, we discussed a method for determining whether a relationship exists among several variables and, if so, how strong this relationship is. We indicated how to find the equation of the regression line so that we can make predictions.

It should be apparent that our discussion was by no means a thorough analysis of all the ideas in the field of statistics. Our discussions were merely intended to introduce the basic ideas of statistics and how they can be misused.

A more detailed discussion of the topics covered in this chapter can be found in any elementary statistics book.

STUDY GUIDE You should now be able to demonstrate your knowledge of the following ideas by giving definitions, descriptions, or specific examples. Page numbers are given in parentheses.

Statistics (p. 583)
Data (p. 583)
Sample (p. 586)
Population (p. 586)
Random sample (p. 586)
Mean (p. 588)
Mode (p. 588)
Median (p. 588)
Measures of central tendency (p. 591)
Summation notation (p. 592)
Weighted arithmetic average (p. 595)
Percentile (p. 595)
Percentile rank (p. 596)
Lower or first quartile (p. 597)
Middle quartile (p. 597)
Upper or third quartile (p. 597)
Measures of variation (p. 597)
Range (p. 598)
Standard deviation (p. 599)
Variance (p. 600)
Average deviation (p. 601)
Frequency distribution (p. 604)

Frequency (p. 604)
Bar graph (p. 605)
Graph (p. 608)
Circle graph (p. 610)
Line graph (p. 612)
Pictograph (p. 612)
Normal curve (p. 616)
Normal distribution (p. 616)
Standard normal distribution (p. 617)
z-score or z-value (p. 619)
Correlation (p. 627)
Axes (p. 627)
Scatter diagram (p. 627)
Positive correlation (p. 628)
Negative correlation (p. 628)
Linear correlation (p. 629)
Coefficient of linear correlation (p. 629)
Least squares method (p. 633)
Regression line (p. 633)
Standard error of the estimate (p. 637)
Misuses of statistics (p. 640)

FORMULAS TO REMEMBER The following list summarizes the formulas given in this chapter.

$\left.\begin{array}{l}\text{Mean} \\ \text{Median} \\ \text{Mode}\end{array}\right\}$ See pp. 587–592 Also $\mu = \dfrac{\Sigma x}{n}$

Percentile rank of $X = \dfrac{B + \frac{1}{2}E}{n} \cdot 100$

$\left.\begin{array}{l}\text{Standard deviation} \\ \text{Variance} \\ \text{Average deviation}\end{array}\right\}$ See pp. 597–601

Range = largest number − smallest number.

z-score or z-value: $z = \dfrac{x - \mu}{\sigma}$

Coefficient of linear correlation:

$$r = \frac{n(\Sigma\ xy) - (\Sigma\ x)(\Sigma\ y)}{\sqrt{n(\Sigma\ x^2) - (\Sigma\ x)^2}\ \sqrt{n(\Sigma\ y^2) - (\Sigma\ y)^2}}$$

Regression equation: $y = m_y + b(x - m_x)$ where

$$m_y = \frac{\Sigma\ y}{n}$$

$$m_x = \frac{\Sigma\ x}{n}$$

$$b = \frac{n(\Sigma\ xy) - (\Sigma\ x)(\Sigma\ y)}{n(\Sigma\ x^2) - (\Sigma\ x)^2}$$

Standard error of the estimate: $\sqrt{\dfrac{\Sigma\ (Y - Y_p)^2}{n - 2}}$.

MASTERY TESTS

Form A

1. The proportion of area under the normal curve from $z = 0.4$ to $z = 0.7$ is

   a) 0.1554   **b)** 0.2580   **c)** 0.4134   **d)** 0.1026   **e)** none of these

   For questions 2–4, use the following information: Dr. James Conway is a cardiologist. The following chart shows the number of patients that Dr. Conway visited in the hospital and also the number of hours that Dr. Conway spent at the hospital.

Number of hours, x	Number of patients, y
4	15
5	20
8	25
10	30
11	31

2. Find the coefficient of correlation for this data.

3. Find the least squares prediction equation.

4. What is the predicted number of patients that Dr. Conway will see if he spends 9 hours at the hospital?

5. The area under a standardized normal curve below $z = 2.45$ is

   a) 0.0071   **b)** 0.4929   **c)** 0.9929   **d)** 0.4918   **e)** none of these

6. If $x_1 = 12$, $x_2 = 13$, $x_3 = 17$, $x_4 = 21$, and $x_5 = 25$, find

   a) $\Sigma\ x^2$ **b)** $(\Sigma\ x)^2$

7. In a standard normal distribution, which z-score cuts off the top 13%?

8. A correlation coefficient of -0.15

    a) indicates a strong negative correlation

    b) indicates a weak negative correlation   **c)** is impossible

    d) is insignificant   **e)** none of these

9. What type of correlation exists between lung cancer and smoking?

    a) strong negative correlation   **b)** zero correlation

    c) positive correlation   **d)** depends upon the individual

    e) none of these

10. On a recent dancing skills test, the mean was 250 and the standard deviation was 25. What percent of the contestants scored *below* 225?

    a) 68.16%   **b)** 15.87%   **c)** 34.13%   **d)** 84.13%

    e) none of these

11. Marianne is one of 50 players on a volleyball team. In one particular tournament she performs better than exactly 36 players on the team, and 12 players perform better than her. What is her percentile rank?

12. Consider the distribution 16, 12, 8, 4. The standard deviation of this distribution is

    a) 5   **b)** $\sqrt{5}$ **c)** 20   **d)** $\sqrt{20}$ **e)** none of these

13. Consider a distribution where $\mu = 10$ and $\sigma = 3$. The z-score of 10 is

    a) 0   **b)** 10   **c)** 3   **d)** 9   **e)** 1

14. In a normal distribution, the median is at

    a) $z = -1$ **b)** $z = 0$ **c)** $z = +1$ **d)** $z = 0.5$ **e)** none of these

15. In a certain health club the weights of the members are normally distributed with a mean of 160 lb and a standard deviation of 10 lb. The percentage of members whose weights are above 170 lb is approximately

    a) 16   **b)** 14   **c)** 34   **d)** 84   **e)** none of these

Form B   **1.** The number of births and the number of deaths for a small midwestern city during the first five days of February were as follows:

Number of births, x	Number of deaths, y
4	7
7	9
5	6
9	11
6	8

Find the coefficient of correlation for this data.

2. The circle graph on the top of the next page gives the breakdown of the 500 vehicles parked in a city-owned garage.

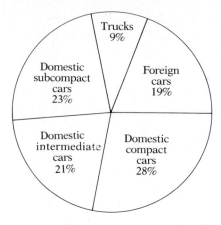

a) How many vehicles are trucks?

b) How many vehicles are subcompact cars?

3. A new computer terminal is to be installed on the tenth floor of a large office building to serve the needs of the numerous offices located on the floor. In an effort to determine where the machine should be placed, a list is made of the distance from each office to one particular spot. The mean, median, and modal distances are then computed. Which of these *average* distances would be more helpful in deciding where the terminal should be placed? Explain your answer.

4. Consider the accompanying newspaper clipping. Explain why the median is used when discussing age, salary and number of years teaching experience.

New York State Public School Teachers

	1976–77	1981–82
Median Age	37	39
Median Salary	$17,150	$23,437
Median Number of Years		
Total Education Experience	11	14
Local District Experience	9	13
Percent with:		
Master's Degree	58%	65%
Permanent Certification	85%	87%
Percent:		
Female	59%	61%
Male	41%	39%
Number of teachers	173,975	168,516

Source: New York State Education Department, Bureau of Basic Educational Data System (BEDS)

** **5.** Consider a set of numbers $x_1, x_2, x_3, \ldots, x_n$. Add a constant amount c to each number, thereby forming a new set of numbers $x_1 + c$, $x_2 + c$, $x_3 + c, \ldots, x_n + c$. What effect does this operation have on the standard deviation? Explain your answer.

6. *Birth and Death Rates.* The birth and death rates for several countries are given in the accompanying table. Draw the bar graph to picture this information.

Country	Birth rate	Death rate
Egypt	34.8	13.1
Israel	27.2	7.1
Japan	19.4	6.6
France	16.4	10.7
Philippines	24.8	7.3
Mexico	44.7	9.1
United States	15.0	9.4
USSR	17.7	8.7
Puerto Rico	24.1	6.7

7. *Natural Resources.* Although the United States has 6% of the world population, it consumes very high percentages of the world production of certain materials as shown in the accompanying table. Draw a bar graph to picture this information.

Material	Percentage
Natural gas	57
Silver	42
Aluminum	36
Petroleum	32
Tin	32
Nickel	30
Copper	27
Steel	19

8. *Interest Rates.* The following line graph shows the interest rates paid on short-term loans over the years. Is it true that interest rates are highest during a recession? Explain your answer.

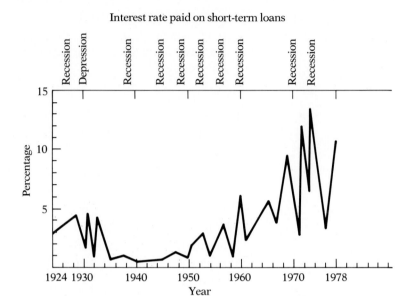

Interest rate paid on short-term loans

Source: *Federal Reserve Board, U.S. Dept. of Commerce*

9. *Stolen Cars.* The following pictograph indicates the number of cars that were reported stolen in a city over the past six years. By how many cars did the number of reported stolen cars in 1984 exceed the number of reported stolen cars in 1979?

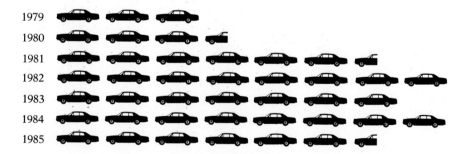

Note: Each symbol represents 10,000 cars.

10. *Stay in School.* According to one study, the average salary of an individual in a certain state is influenced by the number of years of schooling beyond high school that the individual has as shown below:

Number of years of schooling beyond high school, x	Average salary, y
0	$11,150
1	14,000
2	16,500
3	17,880
4	19,150
5	24,000
6	29,000

Find the regression equation.

11. Refer back to question 10. Find the predicted salary of an individual who has completed $4\frac{1}{2}$ years of schooling beyond high school.

12

MATRICES

CHAPTER OBJECTIVES

To define a matrix. (*Section 12.2*)

To learn how to organize data into matrix format. (*Sections 12.2, 12.3, and 12.5*)

To discuss the notation used for matrices and to specify what is meant by the dimension of a matrix. (*Section 12.2*)

To indicate how we add, subtract, and multiply matrices. (*Sections 12.3 and 12.4*)

To apply matrices to business problems. (*Section 12.5*)

To find the inverse of a matrix and then to apply this idea to solve a system of equations. (*Section 12.6*)

To understand what an augmented matrix is and to use it to solve a system of equations. This is called the Gauss–Jordan method. (*Section 12.6*)

To see how matrices can be used in understanding communications networks. (*Section 12.7*)

Project NYS School Enrollment Increases

Total enrollments in public elementary and secondary schools are expected to begin a gradual upward shift in 1989–90, according to the State Education Department.

Public school elementary enrollment (K–6) started to increase in New York State in 1984–85; in New York City, the increases began in 1983–84.

Public school secondary enrollment (7–12) began declining in New York State in 1976–77; in New York City, the first dip was in 1971–72, followed by an upturn in 1974–76, with another decline beginning in 1977–78.

Projections of public school enrollment by region, New York State

Region	1985–86	1988–89	1994–95
New York City	935,786	967,343	1,075,033
Nassau-Suffolk	421,589	396,840	398,835
Mid-Hudson	287,937	276,199	289,233
Upper Hudson	147,829	142,207	142,573
Lake George-Lake Champlain	40,465	38,667	38,183
Black River-St. Lawrence	50,241	48,661	48,646
Upper Mohawk Valley	51,931	49,145	48,118
Central	131,284	127,403	131,776
Southern Tier East	83,435	81,168	83,893
Southern Tier Central	36,117	34,001	32,888
Southern Tier West	51,631	50,266	50,234
Genesee-Finger Lakes	173,416	166,927	174,455
Western	175,870	164,473	162,258
Total State	**2,587,531**	**2,543,300**	**2,676,125**

Source: "Projections of Public and Nonpublic School Enrollment and HS Graduates to 1994–95, New York State," State Education Department.

Often we must analyze lists of numbers in order to properly understand newspaper articles. To facilitate such comparative studies, in this chapter we introduce matrices (which are nothing more than rectangular arrays or tables of numbers) and discuss methods of working with them.

12.1
INTRODUCTION

The **matrix** (plural, **matrices**) is a powerful tool of mathematics and the sciences. It has its beginnings in the work of the remarkable Irish genius William Rowan Hamilton (1804–1865). By the age of thirteen he knew thirteen different languages, among them Greek, Latin, Hebrew, Sanskrit, Arabic, and Persian. When he was seventeen, he had already begun his first great work in mathematical physics.

Hamilton's theory of **quaternions** led to the theory of matrices, which was developed by Arthur Cayley (1821–1895) and James Joseph Sylvester (1814–1897). Cayley and Sylvester were lifelong friends, although their personalities were quite different. Sylvester was hot-tempered, whereas Cayley was calm and rarely lost his temper. Each demonstrated his exceptional mathematical talent as a child, and each continued his mathematical activities until practically the day of his death.

Cayley and Sylvester continued the work begun by Hamilton. Although Hamilton's work was concerned with geometric situations, it has been found that the theory of matrices can be applied to almost all areas of mathematics and physics. As we shall see, it has many interesting applications in business and economics. Of course, matrices are also used extensively in pure mathematics.

You are probably wondering what a matrix is. It is just any list (or array) of numbers, such as the following:

$$\begin{pmatrix} 1 & 7 \\ 6 & 3 \end{pmatrix} \quad \begin{pmatrix} 1 & 7 & 6 \\ 0 & -2 & 4 \\ 4 & 7 & 9 \\ 1 & 3 & 3 \end{pmatrix} \quad \begin{pmatrix} 1 \\ 4 \\ 3 \\ -7 \end{pmatrix} \quad (3 \quad 2 \quad 9 \quad 10).$$

We enclose each matrix within parentheses, as shown.

In the next section we will restate this definition and discuss it in detail. We will also investigate some of the properties of matrices and their applications.

The concept of a matrix can arise in familiar nonmathematical situations. Suppose the student council of Zeesa University is sponsoring a music concert, the proceeds of which are to go to the local drug rehabilitation clinic. There are two performances, one on Friday and one on Saturday. Seat prices are as follows:

$3.00   Front orchestra
 2.50   Back orchestra
 2.00   Mezzanine
 1.50   Lower balcony
 1.00   Upper balcony

The table below shows the number of tickets sold and the amount of money collected on Friday.

Type of seat	Number of seats		Price per seat	Total collected
Front orchestra	250	×	$3.00	$ 750.00
Back orchestra	120	×	2.50	300.00
Mezzanine	73	×	2.00	146.00
Lower balcony	208	×	1.50	312.00
Upper balcony	124	×	1.00	124.00
				$1632.00

On Saturday night the numbers of seats sold were: 300 front orchestra, 110 back orchestra, 78 mezzanine, 220 lower balcony, and 113 upper balcony.

The treasurer wants to figure out how much money has been collected altogether for the two concerts. He can do this in two ways.

1. He can calculate the amount collected on Saturday by setting up a table similar to the one above. He would get $1774.00. He would now add the totals together, getting $1632.00 plus $1774.00, or $3406.00.

2. He can also calculate the total number of each kind of ticket sold for both nights, multiply each total by the price of that ticket, and then add. Such a calculation would look like this:

Type of seat	Seats sold Friday	Seats sold Saturday	Total seats sold	Price per seat	Total collected
Front orchestra	250	300	550 ×	3.00	$1650.00
Back orchestra	120	110	230 ×	2.50	575.00
Mezzanine	73	78	151 ×	2.00	302.00
Lower balcony	208	220	428 ×	1.50	642.00
Upper balcony	124	113	237 ×	1.00	237.00
					$3406.00

In performing these calculations, we have taken lists of objects (seats) and prices, and added and multiplied them in various ways. When we work with matrices, we do the same thing in similar ways, as we shall see. The difference is that, with matrices, we use only the *numbers* involved. We do not consider the *type* of objects involved in the lists.

In this chapter we will investigate some of the properties of matrices and their applications.

12.2
DEFINITION AND NOTATION

We begin our discussion with the definition of a matrix.

Definition 12.1 *A **matrix** is a rectangular array (table) of numbers. We enclose the matrix within parentheses.*

Some examples of matrices are as follows:

a) $\begin{pmatrix} 1 & 2 \\ 4 & -7 \end{pmatrix}$ b) $\begin{pmatrix} -1 & 0 & 3 \\ 5 & 0 & -9 \\ 1 & 3 & 8 \end{pmatrix}$ c) $\begin{pmatrix} 1 & 0 \\ 0 & 1 \end{pmatrix}$

d) $(0 \quad 0 \quad 0)$ e) $\begin{pmatrix} 1 \\ 2 \\ -5 \\ 7 \end{pmatrix}$ f) $\begin{pmatrix} 1 & 7 & 8 \\ -2 & -1 & -3 \\ 1 & -4 & 9 \\ -8 & 0 & 1 \end{pmatrix}$

g) (4).

```
'MAXIMIZE: Z = 60X1 + 35X2 + 50X3 + 45X4'
SUBJECT TO:  '  5X1 + 4X2 + 1X3 + 5X4 <=  600'
             '  4X1 + 2X2 + 3X3 + 2X4 <=  800'
             '  3X1 + 5X2 + 5X3 + 2X4 <= 1200'
```

CB	CJ → BASIS	-60.00 X 1	-35.00 X 2	-50.00 X 3	-45.00 X 4	0.00 X 5	0.00 X 6	0.00 X 7	CONSTANTS
0.00	X 5	5.00	4.00	1.00	5.00	1.00	0.00	0.00	600.00
0.00	X 6	4.00	2.00	3.00	2.00	0.00	1.00	0.00	800.00
0.00	X 7	3.00	5.00	5.00	2.00	0.00	0.00	1.00	1200.00
	C̄ ROW	-60.00	-35.00	-50.00	-45.00	0.00	0.00	0.00	Z = 0.00
-60.00	X 1	1.00	0.80	0.20	1.00	0.20	0.00	0.00	120.00
0.00	X 6	0.00	-1.20	2.20	-2.00	-0.80	1.00	0.00	320.00
0.00	X 7	0.00	2.60	4.40	-1.00	-0.60	0.00	1.00	840.00
	C̄ ROW	0.00	13.00	-38.00	15.00	12.00	0.00	0.00	Z = 7200.00
-60.00	X 1	1.00	0.91	0.00	1.18	0.27	-0.09	0.00	90.91
-50.00	X 3	0.00	-0.55	1.00	-0.91	-0.36	0.45	0.00	145.45
0.00	X 7	0.00	5.00	0.00	3.00	1.00	-2.00	1.00	200.00
	C̄ ROW	0.00	-7.73	0.00	-19.55	-1.82	17.27	0.00	Z = 2727.27
-60.00	X 1	1.00	-1.06	0.00	0.00	-0.12	0.70	-0.39	12.12
-50.00	X 3	0.00	0.97	1.00	0.00	-0.06	-0.15	0.30	206.06
-45.00	X 4	0.00	1.67	0.00	1.00	0.33	-0.67	0.33	66.67
	C̄ ROW	0.00	24.85	0.00	0.00	4.70	4.24	6.52	Z = 4030.30

```
THE OPTIMUM SOLUTION IS:4030.30
THE VALUE OF THE VARIABLES ARE: X 1 = 12.12, X 3 = 206.06, X 4 = 66.67, ALL OTHER X'S = 0.00
THE OPTIMUM SOLUTION IS UNIQUE
```

Pictured above is the computer printout of a linear programming problem. Note how the data are arranged in matrix format.

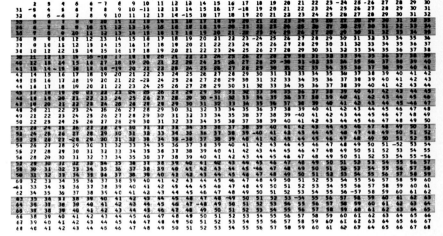

Much information that is put into a computer can be fed in and processed in matrix form. Computers do addition and subtraction, as well as multiplication, of matrices. The computer can also provide results in matrix form, as shown here in a section of a printout.

The matrix of (a) has 2 rows and 2 columns. We call it a 2×2 matrix. The matrix of (b) has 3 rows and 3 columns. We call it a 3×3 matrix. The matrix of (f) has 4 rows and 3 columns. We call it a 4×3 matrix. The matrix of (c) is a 2×2 matrix; (d) is a 1×3 matrix; (e) is a 4×1 matrix; and (g) is a 1×1 matrix.

In general, *a matrix that has m rows and n columns is called an* $m \times n$ *matrix*, read as "m by n matrix." We say that the **dimension** of the matrix is $m \times n$.

If a matrix has the same number of rows and columns, we call it a **square matrix.** Both (a) and (c) above are 2×2 square matrices, and (b) is a 3×3 square matrix. On the other hand, (d), (e), and (f) are not square matrices. Is (g) a square matrix?

In our discussion, we will denote matrices by capital letters, such as $A, B, C, \ldots$.

12.3

ADDITION AND SUBTRACTION OF MATRICES

Adding two matrices is a simple procedure. We just add corresponding elements. Let $A = \begin{pmatrix} 1 & 3 \\ -2 & 0 \end{pmatrix}$, and let $B = \begin{pmatrix} 5 & 4 \\ 1 & -7 \end{pmatrix}$. Then $A + B$ is

$$\begin{pmatrix} 1 & 3 \\ -2 & 0 \end{pmatrix} + \begin{pmatrix} 5 & 4 \\ 1 & -7 \end{pmatrix} = \begin{pmatrix} 1 + 5 & 3 + 4 \\ -2 + 1 & 0 + (-7) \end{pmatrix} = \begin{pmatrix} 6 & 7 \\ -1 & -7 \end{pmatrix}.$$

EXAMPLE 1   Add the matrices $\begin{pmatrix} 2 & 4 \\ -1 & 0 \\ 6 & 6 \end{pmatrix}$ and $\begin{pmatrix} 1 & 1 \\ 10 & 8 \\ 12 & -3 \end{pmatrix}$.

$$SOLUTION \quad \begin{pmatrix} 2 & 4 \\ -1 & 0 \\ 6 & 6 \end{pmatrix} + \begin{pmatrix} 1 & 1 \\ 10 & 8 \\ 12 & -3 \end{pmatrix} = \begin{pmatrix} 2+1 & 4+1 \\ -1+10 & 0+8 \\ 6+12 & 6+(-3) \end{pmatrix} = \begin{pmatrix} 3 & 5 \\ 9 & 8 \\ 18 & 3 \end{pmatrix} \quad \square$$

EXAMPLE 2   Add the matrices $(1 \quad 7 \quad 9)$ and $(2 \quad -3 \quad 0)$.

$$SOLUTION \quad (1 \quad 7 \quad 9) + (2 \quad -3 \quad 0) = (1+2 \quad 7+(-3) \quad 9+0)$$
$$= (3 \quad 4 \quad 9). \quad \square$$

EXAMPLE 3   Add the matrices $\begin{pmatrix} 1 & -2 & 4 \\ 8 & 0 & -1 \\ 0 & 3 & 4 \end{pmatrix}$ and $\begin{pmatrix} 7 & 7 & -7 \\ -5 & 2 & 1 \\ 4 & 3 & 4 \end{pmatrix}$.

$$SOLUTION \quad \begin{pmatrix} 1 & -2 & 4 \\ 8 & 0 & -1 \\ 0 & 3 & 4 \end{pmatrix} + \begin{pmatrix} 7 & 7 & -7 \\ -5 & 2 & 1 \\ 4 & 3 & 4 \end{pmatrix} = \begin{pmatrix} 1+7 & -2+7 & 4+(-7) \\ 8+(-5) & 0+2 & -1+1 \\ 0+4 & 3+3 & 4+4 \end{pmatrix}$$

$$= \begin{pmatrix} 8 & 5 & -3 \\ 3 & 2 & 0 \\ 4 & 6 & 8 \end{pmatrix}. \quad \square$$

Notice that you can add together two matrices *only* if they are of the same *dimension*. This means that the number of rows in each is the same, and the number of columns in each is the same.

EXAMPLE 4   The matrices $\begin{pmatrix} 1 & 4 \\ -3 & 9 \end{pmatrix}$ and $\begin{pmatrix} 4 & 9 & -3 \\ -2 & 0 & 1 \end{pmatrix}$ *cannot* be added because the first matrix has 2 columns and the second matrix has 3 columns. $\square$

Subtraction of matrices is similar to addition. If the matrices are of the same dimension, we subtract the corresponding elements, as the following examples illustrate.

EXAMPLE 5   Subtract the matrix $\begin{pmatrix} 2 & 0 \\ -3 & -5 \end{pmatrix}$ from $\begin{pmatrix} 4 & 8 \\ -3 & 5 \end{pmatrix}$.

$$SOLUTION \quad \begin{pmatrix} 4 & 8 \\ -3 & 5 \end{pmatrix} - \begin{pmatrix} 2 & 0 \\ -3 & -5 \end{pmatrix} = \begin{pmatrix} 4-2 & 8-0 \\ -3-(-3) & 5-(-5) \end{pmatrix} = \begin{pmatrix} 2 & 8 \\ 0 & 10 \end{pmatrix}. \quad \square$$

EXAMPLE 6   Perform the subtraction $\begin{pmatrix} 4 & 1 & 1 \\ -2 & 0 & -8 \end{pmatrix} - \begin{pmatrix} 1 & 5 & 9 \\ 8 & 3 & -5 \end{pmatrix}$.

$$SOLUTION \quad \begin{pmatrix} 4 & 1 & 1 \\ -2 & 0 & -8 \end{pmatrix} - \begin{pmatrix} 1 & 5 & 9 \\ 8 & 3 & -5 \end{pmatrix} = \begin{pmatrix} 4-1 & 1-5 & 1-9 \\ -2-8 & 0-3 & -8-(-5) \end{pmatrix}$$

$$= \begin{pmatrix} 3 & -4 & -8 \\ -10 & -3 & -3 \end{pmatrix}. \quad \square$$

Comment   Wherever addition is possible, the commutative and associative laws hold. This means that if A, B, and C are matrices of the same dimension, then

$$A + B = B + A, \qquad \text{(Commutative law)}$$
$$A + (B + C) = (A + B) + C. \qquad \text{(Associative law)}$$

Definition 12.2   *Two matrices are said to be **equal** if they are of the same dimension and if their corresponding entries are equal. When matrices A and B are equal, we write this as A = B.*

EXAMPLE 7   If $A = \begin{pmatrix} 9 & 3 \\ 2 & 7 \end{pmatrix}$ and $B = \begin{pmatrix} 9 & 3 \\ 2 & 7 \end{pmatrix}$, then $A = B$. $\square$

EXAMPLE 8   If $A = \begin{pmatrix} 8 & 2 \\ 16 & 4 \end{pmatrix}$ and $B = \begin{pmatrix} x & y \\ 16 & 4 \end{pmatrix}$, then $A = B$ only if $x = 8$ and $y = 2$. Otherwise, matrix $A \neq$ matrix B. $\square$

EXAMPLE 9   If $A = \begin{pmatrix} 5x + 7y \\ 3x - 2y \end{pmatrix}$ and $B = \begin{pmatrix} 22 \\ 5 \end{pmatrix}$, then $A = B$ only if $5x + 7y = 22$ and $3x - 2y = 5$. This represents the pair of simultaneous equations

$$\begin{cases} 5x + 7y = 22 \\ 3x - 2y = 5 \end{cases}$$

that must be solved for values of x and y. This will be done in Section 12.6. $\square$

EXAMPLE 10   If $A = \begin{pmatrix} -3 & 2 \\ 9 & 4 \end{pmatrix}$ and $B = \begin{pmatrix} -3 & 2 & 1 \\ 9 & 4 & 5 \end{pmatrix}$, then $A \neq B$ because A and B are not of the same dimension. $\square$

EXERCISES

Find the dimension of each of the matrices given in Exercises 1–8.

1. $\begin{pmatrix} 1 & 7 & 6 \\ 3 & -2 & 1 \\ -9 & 7 & 8 \\ 4 & 0 & 2 \end{pmatrix}$
2. $\begin{pmatrix} 3 & 9 & 2 \\ 1 & 0 & -7 \end{pmatrix}$
3. $\begin{pmatrix} 9 & 2 \\ 5 & -3 \\ -6 & 1 \\ 8 & 2 \end{pmatrix}$
4. $\begin{pmatrix} 7 & 9 & 3 & 8 \end{pmatrix}$

5. (9)

6. $\begin{pmatrix} 1 & 0 & 0 & 0 \\ 0 & 1 & 0 & 0 \\ 0 & 0 & 1 & 0 \\ 0 & 0 & 0 & 1 \end{pmatrix}$

7. $\begin{pmatrix} 3 & 7 & 6 \\ -4 & -2 & 9 \\ 8 & 3 & 1 \end{pmatrix}$

8. $\begin{pmatrix} 5 \\ 0 \\ -2 \\ -3 \\ 1 \end{pmatrix}$

9. Which (if any) of the matrices given in Exercises 1–8 are square matrices?

10. Are the following matrices equal? Explain your answer.

$$\begin{pmatrix} 5 & 3 & 2 & 7 \\ 4 & 9 & 0 & -1 \end{pmatrix} \qquad \begin{pmatrix} 5 & 4 \\ 3 & 9 \\ 2 & 0 \\ 7 & -1 \end{pmatrix}$$

In Exercises 11–16, find a value for the variable(s) so that matrices A and B will be equal. If this is not possible, explain why.

11. $A = \begin{pmatrix} 7 & 3 \\ 4 & x \end{pmatrix}$　　　$B = \begin{pmatrix} 7 & 3 \\ 4 & 9 \end{pmatrix}$

12. $A = \begin{pmatrix} -3 & 9 \\ -14 & 2y \end{pmatrix}$　　$B = \begin{pmatrix} -3 & 9 \\ -14 & 20 \end{pmatrix}$

13. $A = \begin{pmatrix} 10 & 17 \\ 16 & -5y \\ 4 & -9 \end{pmatrix}$　$B = \begin{pmatrix} 10 & 17 \\ 2x & 25 \\ 4 & -9 \end{pmatrix}$

14. $A = \begin{pmatrix} 11 & 7 & 2x \\ 6 & 4 & 5y \\ 2 & 9 & 3z \end{pmatrix}$　$B = \begin{pmatrix} 11 & 7 & 10 \\ 6 & 4 & 0 \\ 2 & 9 & 10 \end{pmatrix}$

15. $A = \begin{pmatrix} 1 & 7 & x \\ 6 & 2 & 3 \end{pmatrix}$　$B = \begin{pmatrix} 1 & 7 & 8 & 3 \\ 6 & 2 & 3 & 9 \end{pmatrix}$

16. $A = \begin{pmatrix} 4 & 2x+7 & 9 \\ 3 & 5z-3 & 7 \end{pmatrix}$　$B = \begin{pmatrix} 4 & 4x-1 & 9 \\ 3 & 3z+7 & 2y \end{pmatrix}$

In each of the Exercises 17–27, perform the indicated operations or indicate why the operation cannot be performed.

17. $\begin{pmatrix} 1 & 7 \\ -3 & 4 \end{pmatrix} + \begin{pmatrix} 2 & 0 \\ -8 & 9 \end{pmatrix}$

18. $\begin{pmatrix} 1 \\ 7 \\ 9 \end{pmatrix} + \begin{pmatrix} 3 \\ 0 \\ -2 \end{pmatrix}$

19. $\begin{pmatrix} 4 & 1 & 9 \\ 10 & -3 & 7 \end{pmatrix} + \begin{pmatrix} 0 & -2 & -5 \\ 3 & 8 & 4 \end{pmatrix}$

20. $\begin{pmatrix} 7 & 8 & -3 \end{pmatrix} - \begin{pmatrix} -4 & -2 & 9 \end{pmatrix}$

21. $\begin{pmatrix} -4 & -2 \\ 8 & 3 \end{pmatrix} + \begin{pmatrix} 2 & 7 & 8 \\ 1 & 6 & 0 \end{pmatrix}$

22. $\begin{pmatrix} 1 & 8 & 3 \\ 9 & 10 & -12 \end{pmatrix} - \begin{pmatrix} 8 & 0 & 11 \\ 6 & -2 & 7 \\ 3 & 5 & 4 \end{pmatrix}$

23. $\begin{pmatrix} 8 & 1 & 10 \\ 11 & -2 & -\frac{1}{2} \\ 3 & 7 & -4 \end{pmatrix} - \begin{pmatrix} 5 & -3 & -4 \\ 2 & 0 & \frac{3}{2} \\ 2 & -8 & 9 \end{pmatrix}$

24. $\begin{pmatrix} 7 & -1 & 8 \\ 10 & 12 & -7 \end{pmatrix} - \begin{pmatrix} 0 & -6 & -5 \\ -2 & 3 & 8 \end{pmatrix}$

25. $\begin{pmatrix} 5 & 3 & 2 \\ 1 & 4 & 7 \\ 0 & 2 & 9 \end{pmatrix} + \begin{pmatrix} 0 & 0 & 0 \\ 0 & 0 & 0 \\ 0 & 0 & 0 \end{pmatrix}$

26. $\begin{pmatrix} 4 & 7 \\ x & y \end{pmatrix} + \begin{pmatrix} 0 & 0 \\ 0 & 0 \end{pmatrix}$

27. $\begin{pmatrix} 0 & 0 \\ 0 & 0 \end{pmatrix} + \begin{pmatrix} 4 & 7 \\ x & y \end{pmatrix}$

28. Find x and y such that

$$\begin{pmatrix} 2 & 7 \\ 9 & x \end{pmatrix} + \begin{pmatrix} y & -3 \\ 6 & -5 \end{pmatrix} = \begin{pmatrix} 8 & 4 \\ 15 & 10 \end{pmatrix}$$

29. If

$$A = \begin{pmatrix} 4 & 3 & 7 \\ 6 & -2 & 0 \\ 1 & 9 & 8 \end{pmatrix}, B = \begin{pmatrix} 1 & -8 & 2 \\ -3 & 9 & 10 \\ 5 & 1 & 8 \end{pmatrix},$$

$$C = \begin{pmatrix} -1 & 7 & 9 \\ 2 & 8 & 4 \\ 6 & 0 & 5 \end{pmatrix},$$

verify that the associative law for addition, $A + (B + C) = (A + B) + C$, holds for these matrices.

30. Refer back to the previous exercise. Verify that the commutative law for addition, $A + B = B + A$, holds for these matrices.

31. *Inflation.*   The accompanying chart indicates how the average prices for several auto repair items have changed from the previous years in each of two years for three cities. In these charts a plus sign indicates an increase in price, whereas a minus sign indicates a decrease in price. (*Note:* All price changes are in dollars.)

a) Find the net effect of inflation over the two-year period by adding the matrices.

b) Which item's price increased the most and in which city?

c) Which item's price increased the least or decreased the most and in which city?

1985

	Engine tune-up	Transmission tune-up	Exhaust system repair	Charging air conditioning system	New tires
City A	+6	+7	+13	+1	+8
City B	+8	+3	+8	−2	0
City C	+5	0	−1	−3	0

1986

	Engine tune-up	Transmission tune-up	Exhaust system repair	Charging air conditioning system	New tires
City A	+8	+2	0	+1	−1
City B	+5	+8	+4	+3	0
City C	−1	−3	+6	0	+7

32. *Inventory Control.* A large supermarket chain has four warehouses from which it ships various items to its stores located within the city. It keeps accurate records on the inventories at each of these warehouses as shown in the accompanying charts:

a) How many items were removed from the inventory during June?

b) Which warehouse shipped the most merchandise in June?

c) Which warehouse shipped the least merchandise in June?

June 1 Inventory
Warehouse location

Item		Main St.	West St.	Beck St.	Ave. L
	Coffee	1258	373	984	262
	Detergent	1728	841	636	954
	Sugar	894	1381	1292	651
	Oil	1523	1281	990	657

July 1 Inventory
Warehouse location

Main St.	West St.	Beck St.	Ave. L
375	250	225	130
870	726	432	298
695	977	973	179
603	527	408	199

33. The president of the Back Corporation is analyzing the number of hours that each of the company's 5 computer programmers spends on the computer terminals. On Monday the programmers Bill, Cindy, Gail, Jake, and Jason spent 6, 4, 5, 3, and 9 hours, respectively, on the terminals. On Tuesday they spent 8, 6, 5, 7, and 2 hours, respectively, and on Wednesday they spent 9, 6, 8, 3, and 7 hours, respectively.

a) Rewrite this information in matrix form.

b) Find the total amount of time spent by each of these programmers on the terminals over the three days.

c) What is the total number of hours spent by these employees on the terminals over the three day period?

12.4

MULTIPLICATION OF MATRICES

It may seem that multiplication of matrices should be done in a manner similar to that of addition of matrices. However, if multiplication is done this way, then the range of applications is extremely limited. Since our main interest in matrices is in their applications, we will multiply matrices in a way that turns out to be useful. The procedure for multiplication will at first appear strange and unnecessarily complicated. However, you will find after working with it that it is very useful.

To best understand the procedure, let us start with a row matrix A and a column matrix B, that is, let A be the row matrix $A = (a_1 \ a_2 \ a_3 \ \ldots \ a_n)$ and let B be the column matrix

$$B = \begin{pmatrix} b_1 \\ b_2 \\ \cdot \\ \cdot \\ \cdot \\ b_n \end{pmatrix}$$

Then the product $A \cdot B$ is the 1×1 matrix (written in parentheses) whose entries are calculated by multiplying corresponding entries of A and B and forming the sum; that is,

$$A \cdot B = (a_1 \ a_2 \ a_3 \ \ldots \ a_n) \cdot \begin{pmatrix} b_1 \\ b_2 \\ b_3 \\ \cdot \\ \cdot \\ \cdot \\ b_n \end{pmatrix}$$

$$= (a_1b_1 + a_2b_2 + a_3b_3 + \cdots + a_nb_n).$$

Thus if $A = (8 \quad 5 \quad 6)$ and $B = \begin{pmatrix} 4 \\ 1 \\ 2 \end{pmatrix}$, then

$$A \cdot B = \begin{pmatrix} 8 \cdot 4 + 5 \cdot 1 + 6 \cdot 2 \end{pmatrix} = \begin{pmatrix} 49 \end{pmatrix}.$$

The above procedure works nicely since matrix A is 1×3 and matrix B is 3×1. Notice that the number of columns of the left matrix equals the number of rows of the right matrix. How would we multiply the following matrices?

$$\begin{pmatrix} 1 & 2 \\ 3 & 5 \end{pmatrix} \cdot \begin{pmatrix} 2 & 7 \\ 3 & 4 \end{pmatrix}$$

In this case we first multiply the rows of the left matrix by the columns of the right matrix. We must be careful to write the rows or columns as they appear in the original matrices. Thus we initially multiply the first row of the left matrix $(1 \quad 2)$ with the first column of the right matrix $\begin{pmatrix} 2 \\ 3 \end{pmatrix}$. We get

$$\begin{pmatrix} 1 & 2 \end{pmatrix} \cdot \begin{pmatrix} 2 \\ 3 \end{pmatrix} = \begin{pmatrix} 1 \cdot 2 + 2 \cdot 3 \end{pmatrix} = \begin{pmatrix} 8 \end{pmatrix}.$$

This will be the entry in the first row, first column of the product matrix. Thus we have

$$\begin{pmatrix} 1 & 2 \\ 3 & 5 \end{pmatrix} \cdot \begin{pmatrix} 2 & 7 \\ 3 & 4 \end{pmatrix} = \begin{pmatrix} 8 & \\ & \end{pmatrix}.$$

Then we multiply the first row of the left matrix by the second column of the right matrix. We get

$$\begin{pmatrix} 1 & 2 \end{pmatrix} \cdot \begin{pmatrix} 7 \\ 4 \end{pmatrix} = \begin{pmatrix} 1 \cdot 7 + 2 \cdot 4 \end{pmatrix} = \begin{pmatrix} 15 \end{pmatrix}.$$

This will be the entry in the first row, second column of the product matrix. So we now have

$$\begin{pmatrix} 1 & 2 \\ 3 & 5 \end{pmatrix} \cdot \begin{pmatrix} 2 & 7 \\ 3 & 4 \end{pmatrix} = \begin{pmatrix} 8 & 15 \\ & \end{pmatrix}.$$

Now we multiply the second row of the left matrix by the first column of the right matrix. We get

$$\begin{pmatrix} 3 & 5 \end{pmatrix} \begin{pmatrix} 2 \\ 3 \end{pmatrix} = \begin{pmatrix} 3 \cdot 2 + 5 \cdot 3 \end{pmatrix} = \begin{pmatrix} 21 \end{pmatrix}.$$

Therefore

$$\begin{pmatrix} 1 & 2 \\ 3 & 5 \end{pmatrix} \cdot \begin{pmatrix} 2 & 7 \\ 3 & 4 \end{pmatrix} = \begin{pmatrix} 8 & 15 \\ 21 & \end{pmatrix}.$$

Finally, we multiply the second row of the left matrix by the second column of the right matrix. This gives

$$\begin{pmatrix} 3 & 5 \end{pmatrix} \cdot \begin{pmatrix} 7 \\ 4 \end{pmatrix} = \begin{pmatrix} 3 \cdot 7 + 5 \cdot 4 \end{pmatrix} = \begin{pmatrix} 41 \end{pmatrix}.$$

Our final answer is

$$\begin{pmatrix} 1 & 2 \\ 3 & 5 \end{pmatrix} \cdot \begin{pmatrix} 2 & 7 \\ 3 & 4 \end{pmatrix} = \begin{pmatrix} 8 & 15 \\ 21 & 41 \end{pmatrix}.$$

Since every row of the left matrix has been multiplied by every column of the right matrix we are finished.

More generally, let us multiply the two matrices

$$\begin{pmatrix} 2 & 4 \\ 3 & -2 \\ 9 & 0 \end{pmatrix} \cdot \begin{pmatrix} 3 & 0 & 2 & 7 \\ 2 & 4 & -1 & 2 \end{pmatrix}.$$

The first matrix has 3 rows and 2 columns. It is a 3×2 matrix. The second matrix is 2×4. Notice that to multiply matrices, it is *not* necessary that they have the same dimension. *What is necessary is that the number of columns of the first matrix be exactly the same as the number of rows of the second matrix.* In our case this number is 2. Our answer will be a 3×4 matrix.

The following diagram illustrates the relationship between the number of rows and columns of the matrices to be multiplied and the answer.

$$3 \times ② \text{ and } ② \times 4$$

Do they match?

If yes, our answer is

$$3 \times 4.$$

Now how do we obtain the numbers for the 3×4 matrix that is our answer? Let us rewrite the problem with blanks in the answer.

$$\begin{pmatrix} 2 & 4 \\ 3 & -2 \\ 9 & 0 \end{pmatrix} \cdot \begin{pmatrix} 3 & 0 & 2 & 7 \\ 2 & 4 & -1 & 2 \end{pmatrix} = \begin{pmatrix} ? & ? & ? & ? \\ ? & ? & ? & ? \\ ? & ? & ? & ? \end{pmatrix}.$$

To find out what belongs in the first row, first column (the circled one), we multiply each element of row 1 of the first matrix by the *corresponding* elements of column 1 of the second matrix, and then add. This gives

$$2 \cdot 3 + 4 \cdot 2 = 6 + 8 = 14.$$

To find what belongs in the third row, fourth column, we multiply row 3 of the first matrix by corresponding elements of column 4 of the second matrix, and then add. This gives

$$9 \cdot 7 + 0 \cdot 2 = 63 + 0 = 63.$$

In a similar manner we find the number that belongs in the second row, third column. We multiply row 2 of the first matrix by column 3 of the second matrix. We get

$$3 \cdot 2 + (-2)(-1) = 6 + 2 = 8.$$

Proceeding in the same way for all the other blanks, we get

$$\begin{pmatrix} 2 & 4 \\ 3 & -2 \\ 9 & 0 \end{pmatrix} \cdot \begin{pmatrix} 3 & 0 & 2 & 7 \\ 2 & 4 & -1 & 2 \end{pmatrix}$$

$$= \begin{pmatrix} 2 \cdot 3 + 4 \cdot 2 & 2 \cdot 0 + 4 \cdot 4 & 2 \cdot 2 + 4(-1) & 2 \cdot 7 + 4 \cdot 2 \\ 3 \cdot 3 + (-2)2 & 3 \cdot 0 + (-2)4 & 3 \cdot 2 + (-2)(-1) & 3 \cdot 7 + (-2)(2) \\ 9 \cdot 3 + 0 \cdot 2 & 9 \cdot 0 + 0 \cdot 4 & 9 \cdot 2 + 0(-1) & 9 \cdot 7 + 0 \cdot 2 \end{pmatrix}$$

$$= \begin{pmatrix} 14 & 16 & 0 & 22 \\ 5 & -8 & 8 & 17 \\ 27 & 0 & 18 & 63 \end{pmatrix}.$$

Notice that if we try to multiply

$$\begin{pmatrix} 3 & 0 & 2 & 7 \\ 2 & 4 & -1 & 2 \end{pmatrix} \cdot \begin{pmatrix} 2 & 4 \\ 3 & -2 \\ 9 & 0 \end{pmatrix},$$

it cannot be done, since the first matrix is 2×4 and the second matrix is 3×2. The number of columns of the first matrix is *not* the same as the number of rows of the second.

EXAMPLE 1   Multiply $\begin{pmatrix} 1 & 4 \\ 3 & -2 \end{pmatrix}$ by $\begin{pmatrix} 1 & 0 & 2 \\ -1 & 3 & 5 \end{pmatrix}.$

SOLUTION   The matrices are 2 × 2 and 2 × 3, respectively. We draw a diagram.

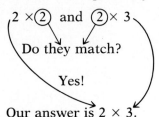

2 ×② and ②× 3
Do they match?
Yes!
Our answer is 2 × 3.

Computing the entries for the 2 × 3 matrix, we get

$$\begin{pmatrix} 1 & 4 \\ 3 & -2 \end{pmatrix} \cdot \begin{pmatrix} 1 & 0 & 2 \\ -1 & 3 & 5 \end{pmatrix}$$

$$= \begin{pmatrix} 1 \cdot 1 + 4(-1) & 1 \cdot 0 + 4 \cdot 3 & 1 \cdot 2 + 4 \cdot 5 \\ 3 \cdot 1 + (-2)(-1) & 3 \cdot 0 + (-2)3 & 3 \cdot 2 + (-2)5 \end{pmatrix}$$

$$= \begin{pmatrix} -3 & 12 & 22 \\ 5 & -6 & -4 \end{pmatrix} \quad \square$$

EXAMPLE 2   Multiply $(1 \quad 3 \quad -1)$ by $\begin{pmatrix} 7 \\ -8 \\ 0 \end{pmatrix}$.

SOLUTION   The matrices are

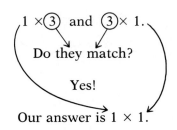

1 ×③ and ③× 1.
Do they match?
Yes!
Our answer is 1 × 1.

Computing the answer, we get

$$\left(1 \cdot 7 + 3 \cdot (-8) + (-1) \cdot 0 \right) = (-17).$$

It would be wrong to write the answer as -17. Since the answer is a matrix, the parentheses are necessary. $\square$

EXAMPLE 3   Multiply $\begin{pmatrix} 7 \\ -8 \\ 0 \end{pmatrix}$ by $\begin{pmatrix} 1 & 3 & -1 \end{pmatrix}$.

SOLUTION   The matrices are

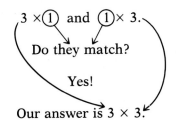

$$3 \times \textcircled{1} \text{ and } \textcircled{1} \times 3.$$

Do they match?

Yes!

Our answer is 3×3.

Computing the answer, we have

$$\begin{pmatrix} 7 \\ -8 \\ 0 \end{pmatrix} \cdot \begin{pmatrix} 1 & 3 & -1 \end{pmatrix} = \begin{pmatrix} 7 \cdot 1 & 7 \cdot 3 & 7(-1) \\ (-8)1 & (-8)3 & (-8) \cdot (-1) \\ 0 \cdot 1 & 0 \cdot 3 & 0 \cdot (-1) \end{pmatrix}$$

$$= \begin{pmatrix} 7 & 21 & -7 \\ -8 & -24 & 8 \\ 0 & 0 & 0 \end{pmatrix}. \square$$

William Rowan Hamilton

Compare this problem with Example 2. What does it tell us about the commutative law for multiplication of matrices? We can obviously conclude that the commutative law does not hold. When Hamilton discovered this, he was so excited about it that he scratched the result on a bridge in Dublin, where he happened to be walking at the time.

EXAMPLE 4   Multiply $\begin{pmatrix} 5 & 7 \\ -3 & 2 \end{pmatrix}$ by $\begin{pmatrix} 4 & 3 & 8 \\ -2 & 5 & 9 \\ 1 & 0 & 1 \end{pmatrix}$.

SOLUTION   The matrices are

$$2 \times \textcircled{2} \text{ and } \textcircled{3} \times 3.$$

Do they match?

No!

Since they do not match, this multiplication is impossible. $\square$

EXAMPLE 5   Multiply $\begin{pmatrix} 5 & 7 \\ -3 & 2 \end{pmatrix}$ by $\begin{pmatrix} 1 & 0 \\ 0 & 1 \end{pmatrix}$.

SOLUTION   The matrices are

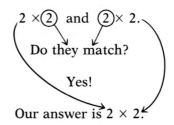

2 ×② and ②× 2.

Do they match?

Yes!

Our answer is 2 × 2.

The product is

$$\begin{pmatrix} 5 & 7 \\ -3 & 2 \end{pmatrix} \cdot \begin{pmatrix} 1 & 0 \\ 0 & 1 \end{pmatrix} = \begin{pmatrix} 5 \cdot 1 + 7 \cdot 0 & 5 \cdot 0 + 7 \cdot 1 \\ (-3)1 + 2 \cdot 0 & (-3)0 + 2 \cdot 1 \end{pmatrix}$$

$$= \begin{pmatrix} 5 & 7 \\ -3 & 2 \end{pmatrix}. \quad \square$$

In the last example the answer was the same matrix as the one we started with. Multiplying by $\begin{pmatrix} 1 & 0 \\ 0 & 1 \end{pmatrix}$ did nothing. If you multiply $\begin{pmatrix} 1 & 0 \\ 0 & 1 \end{pmatrix}$ by $\begin{pmatrix} 5 & 7 \\ -3 & 2 \end{pmatrix}$, you will find that the answer is also $\begin{pmatrix} 5 & 7 \\ -3 & 2 \end{pmatrix}$. Verify this. We say that $\begin{pmatrix} 1 & 0 \\ 0 & 1 \end{pmatrix}$ is the **identity for multiplication** of 2 × 2 matrices. Multiplying any 2 × 2 matrix by the matrix $\begin{pmatrix} 1 & 0 \\ 0 & 1 \end{pmatrix}$ does not change anything.

EXAMPLE 6   **a)** Multiply $\begin{pmatrix} 4 & 3 \\ 5 & 7 \\ -1 & 0 \end{pmatrix} \cdot \begin{pmatrix} 1 & 0 \\ 0 & 1 \end{pmatrix}$.

b) Multiply $\begin{pmatrix} 1 & 0 \\ 0 & 1 \end{pmatrix} \cdot \begin{pmatrix} 4 & 3 \\ 5 & 7 \\ -1 & 0 \end{pmatrix}$.

SOLUTION   **a)** $\begin{pmatrix} 4 & 3 \\ 5 & 7 \\ -1 & 0 \end{pmatrix} \cdot \begin{pmatrix} 1 & 0 \\ 0 & 1 \end{pmatrix} = \begin{pmatrix} 4 & 3 \\ 5 & 7 \\ -1 & 0 \end{pmatrix}$

b) This multiplication cannot be done. Why not? $\square$

In Example 5, $\begin{pmatrix} 1 & 0 \\ 0 & 1 \end{pmatrix}$ was the identity matrix because no matter which way we performed the multiplication, our answer was always $\begin{pmatrix} 5 & 7 \\ -3 & 2 \end{pmatrix}$. In Example 6, $\begin{pmatrix} 1 & 0 \\ 0 & 1 \end{pmatrix}$ is *not* the identity matrix, since we can multiply only from one side. Thus we define an identity matrix in the following way.

> **Definition 12.3**   An **identity** matrix for multiplication is a **square** matrix that, when multiplied by another square matrix A on the left side or on the right side, leaves the matrix A unchanged. Thus $\begin{pmatrix} 1 & 0 \\ 0 & 1 \end{pmatrix}$ is the 2 × 2 identity matrix, and it can be shown that
>
> $$\begin{pmatrix} 1 & 0 & 0 \\ 0 & 1 & 0 \\ 0 & 0 & 1 \end{pmatrix}$$

is the 3 × 3 identity matrix.

Sometimes we want to multiply a matrix by an ordinary number. For example, suppose we want to multiply the matrix $\begin{pmatrix} 3 & 5 \\ -1 & 7 \end{pmatrix}$ by the number 4. This is done by multiplying each entry of the matrix by 4. Thus we have the following.

$$4 \begin{pmatrix} 3 & 5 \\ -1 & 7 \end{pmatrix} = \begin{pmatrix} 4 \cdot 3 & 4 \cdot 5 \\ 4(-1) & 4(7) \end{pmatrix}$$

$$= \begin{pmatrix} 12 & 20 \\ -4 & 28 \end{pmatrix}.$$

In general terms, if k is any number and A is any matrix, then kA means another matrix whose elements are obtained by multiplying each element of matrix A by the number k.

EXAMPLE 7   **a)** $3 \begin{pmatrix} 4 & 7 \\ 2 & -2 \\ 7 & 4 \end{pmatrix} = \begin{pmatrix} 3 \cdot 4 & 3 \cdot 7 \\ 3 \cdot 2 & 3(-2) \\ 3 \cdot 7 & 3 \cdot 4 \end{pmatrix} = \begin{pmatrix} 12 & 21 \\ 6 & -6 \\ 21 & 12 \end{pmatrix}$

b) $-2 \begin{pmatrix} 5 \\ 0 \\ -4 \\ 9 \end{pmatrix} = \begin{pmatrix} (-2)5 \\ (-2)0 \\ (-2)(-4) \\ (-2)9 \end{pmatrix} = \begin{pmatrix} -10 \\ 0 \\ +8 \\ -18 \end{pmatrix}$ □

Perhaps you are wondering why matrix multiplication is defined in such a complicated fashion, especially since multiplication is not commutative. As we indicated earlier, it is the applications for matrices that lead us to define multiplication in such an unusual way. The following problem illustrates one simple application. In Section 12.5 we will give many other applications.

Suppose Mary is in a supermarket comparing the prices of several cans of vegetables. She obtains the following information on the cost per can of several nationally advertised brand name items: beans 49¢, corn 59¢, beets 55¢. This information can be written in a 1×3 matrix as

$$\begin{pmatrix} 49 & 59 & 55 \end{pmatrix}$$

If Mary decides to purchase 7 cans of beans, 5 cans of corn, and 6 cans of beets, then we can arrange this information as a 3×1 column matrix

$$\begin{pmatrix} 7 \\ 5 \\ 6 \end{pmatrix}$$

The product of these two matrices represents the total cost. Thus,

$$\begin{pmatrix} 49 & 59 & 55 \end{pmatrix} \cdot \begin{pmatrix} 7 \\ 5 \\ 6 \end{pmatrix} = \begin{pmatrix} 49 \cdot 7 + 59 \cdot 5 + 55 \cdot 6 \end{pmatrix}$$

$$= \begin{pmatrix} 968 \end{pmatrix}.$$

Mary would spend a total of 968¢ or $9.68 for these items.

The supermarket also carries a no-name label (generic label) for these items. The prices for these items along with the brand name ones can be arranged in matrix form as shown.

	Beans	Corn	Beets	
Nationally advertised brand name	49	59	55	$= A$
No-name generic label	41	53	47	

Since Mary is not sure about the quality of the no-name generic-labeled cans, she decides to purchase several cans of each as indicated in matrix B.

	Nationally advertised brand name	No-name generic label	
Beans	4	3	
Corn	3	2	$= B$
Beets	4	2	

Let us now perform the multiplication $A \cdot B$.

$$A \cdot B = \begin{pmatrix} 49 & 59 & 55 \\ 41 & 53 & 47 \end{pmatrix} \cdot \begin{pmatrix} 4 & 3 \\ 3 & 2 \\ 4 & 2 \end{pmatrix}$$

$$= \begin{pmatrix} 49(4) + 59(3) + 55(4) & 49(3) + 59(2) + 55(2) \\ 41(4) + 53(3) + 47(4) & 41(3) + 53(2) + 47(2) \end{pmatrix}$$

$$= \begin{pmatrix} 593 & 375 \\ 511 & 323 \end{pmatrix}.$$

Can you interpret these results? The 593, for example, means that she spent \$5.93 on the nationally advertised brand name items, and the 323 means that she spent \$3.23 on the no-name generic-labeled items. What interpretation can you find for the 375 and the 511?

We can also perform the multiplication $B \cdot A$.

$$B \cdot A = \begin{pmatrix} 4 & 3 \\ 3 & 2 \\ 4 & 2 \end{pmatrix} \cdot \begin{pmatrix} 49 & 59 & 55 \\ 41 & 53 & 47 \end{pmatrix}$$

3 × ② and ② × 3

Do they match?
Yes. The answer is
of dimension

3 × 3

We have

$$B \cdot A = \begin{pmatrix} 4(49) + 3(41) & 4(59) + 3(53) & 4(55) + 3(47) \\ 3(49) + 2(41) & 3(59) + 2(53) & 3(55) + 2(47) \\ 4(49) + 2(41) & 4(59) + 2(53) & 4(55) + 2(47) \end{pmatrix}$$

$$= \begin{pmatrix} 319 & 395 & 361 \\ 229 & 283 & 259 \\ 278 & 342 & 314 \end{pmatrix}.$$

In this matrix the 319 means that Mary spent \$3.19 for beans, the 283 means that she spent \$2.83 for corn, and the 314 means that she spent \$3.14 for beets. Can you interpret the other numbers?

Comment The preceding problem illustrates that even when two matrices are compatible for multiplication on both sides, the results and the corresponding interpretations are completely different.

In Section 12.5 we will discuss further applications of matrices.

EXERCISES

In each of Exercises 1–15, perform the indicated operations or explain why it is not possible.

1. $\begin{pmatrix} 2 & 7 \\ -3 & 6 \end{pmatrix} \cdot \begin{pmatrix} 4 & -1 \\ -2 & 9 \end{pmatrix}$

2. $\begin{pmatrix} 1 & 7 \\ 0 & 2 \end{pmatrix} \cdot \begin{pmatrix} 8 & -3 \\ -9 & 6 \end{pmatrix}$

3. $\begin{pmatrix} 1 & 7 & 6 \\ 2 & -3 & 1 \\ 4 & 8 & 3 \end{pmatrix} \cdot \begin{pmatrix} 1 & 0 \\ -3 & -7 \\ 2 & 9 \end{pmatrix}$

4. $\begin{pmatrix} 3 & -8 & 2 \\ 6 & -9 & 1 \end{pmatrix} \cdot \begin{pmatrix} -2 & 7 \\ 6 & -1 \\ 0 & 9 \end{pmatrix}$

5. $\begin{pmatrix} 1 & -8 \\ -3 & 7 \\ 6 & 5 \end{pmatrix} \cdot \begin{pmatrix} 2 & -3 & 5 \\ -4 & 8 & 1 \end{pmatrix}$

6. $(1 \quad 3 \quad 7 \quad 10) \cdot \begin{pmatrix} 2 \\ -1 \\ -6 \\ -9 \end{pmatrix}$

7. $\begin{pmatrix} 2 \\ -1 \\ 6 \\ 9 \end{pmatrix} \cdot (1 \quad 3 \quad 7 \quad 10)$

8. $\begin{pmatrix} 4 & -5 & 3 \\ 8 & 1 & 0 \end{pmatrix} \cdot \begin{pmatrix} 21 & 6 & 8 \\ 10 & -2 & 5 \\ -11 & 12 & 0 \end{pmatrix}$

9. $-3 \begin{pmatrix} 5 & -1 & 6 \\ 2 & -1 & 8 \\ 7 & 0 & 9 \end{pmatrix}$

10. $-6 \begin{pmatrix} 0 & 2 & 5 \\ 1 & 7 & 2 \\ 6 & 6 & 3 \\ 5 & 8 & -7 \end{pmatrix}$

11. $4 \begin{pmatrix} 7 & -1 & 10 \\ 3 & -6 & 2 \\ 5 & 1 & 3 \end{pmatrix} \cdot \begin{pmatrix} 8 & 2 & 6 \\ -1 & -4 & -2 \\ 0 & 3 & 5 \end{pmatrix}$

12. $5 \begin{pmatrix} 3 & 7 \\ 1 & 6 \end{pmatrix} + 3 \begin{pmatrix} 8 & 2 \\ 0 & 3 \end{pmatrix}$

13. $8 \begin{pmatrix} 2 & 1 \\ 6 & 9 \end{pmatrix} - 4 \begin{pmatrix} 3 & 8 \\ 4 & 2 \end{pmatrix}$

14. $3 \begin{pmatrix} -1 & 4 & 7 \\ 6 & -9 & 3 \\ 1 & -4 & 5 \end{pmatrix} - 4 \begin{pmatrix} -3 & 4 & -7 \\ 6 & 1 & 0 \\ 2 & 5 & -3 \end{pmatrix}$

15. $-3 \begin{pmatrix} 9 & 6 \\ 4 & 2 \\ 0 & 1 \end{pmatrix} \cdot \begin{pmatrix} -2 & -9 \\ 3 & 4 \end{pmatrix}$

****16.** Find a 2×2 matrix A such that

$$A^2 = A \cdot A = \begin{pmatrix} 1 & 0 \\ 0 & 1 \end{pmatrix}$$

(Assume that A is not the identity matrix.)

****17.** Consider any matrix

$$A = \begin{pmatrix} x & y \\ z & w \end{pmatrix}$$

Under what conditions will it be true that

$$A^2 = A \cdot A = \begin{pmatrix} 1 & 0 \\ 0 & 1 \end{pmatrix}$$

****18.** Let

$$A = \begin{pmatrix} x & y \\ z & w \end{pmatrix} \quad \text{and} \quad B = \begin{pmatrix} 1 & 1 \\ -1 & 1 \end{pmatrix}$$

where A is not the 2×2 identity matrix. Under what circumstances will it be true that $A \cdot B = B \cdot A$?

****19.** Can you find two 2×2 matrices A and B such that

$$A \neq \begin{pmatrix} 0 & 0 \\ 0 & 0 \end{pmatrix} \quad B \neq \begin{pmatrix} 0 & 0 \\ 0 & 0 \end{pmatrix} \quad \text{and}$$

$$A \cdot B = \begin{pmatrix} 0 & 0 \\ 0 & 0 \end{pmatrix}$$

****20.** Can you find a 2×2 matrix A such that $A \cdot A = A$? (Assume that

$$A \neq \begin{pmatrix} 0 & 0 \\ 0 & 0 \end{pmatrix}$$

12.5

APPLICATIONS TO BUSINESS PROBLEMS

Matrices can be very useful in analyzing such business problems as determining cost, profit, and amounts of materials needed. The following examples illustrate how this can be done.

A drug and cosmetics company markets the following products: Eagle hair tonic, Victor toothpaste, and Laury deodorant. The following table indicates the number of cases of each product ordered by four different supermarkets.

	Hair tonic	*Toothpaste*	*Deodorant*
Market I	5	4	7
Market II	8	0	1
Market III	13	3	9
Market IV	2	6	3

The hair tonic costs $12 per case, the toothpaste costs $9 a case, and the deodorant costs $15 a case.

There are many things that may be of interest to both the manufacturer and store managers. Each of these can be expressed by some combination of matrices. Suppose the manufacturer wants to know the total income from each supermarket and also the total income from all supermarkets together.

From market I the income is $5(12) + 4(9) + 7(15) = \$201$.

From market II the income is $8(12) + 0(9) + 1(15) = \$111$.

From market III the income is $13(12) + 3(9) + 9(15) = \$318$.

From market IV the income is $2(12) + 6(9) + 3(15) = \$123$.

We can put the final result in matrix form as

$$\begin{pmatrix} 201 \\ 111 \\ 318 \\ 123 \end{pmatrix}.$$

The entire computation can be done in a much more efficient way using matrices as follows.

Let A represent the amounts ordered of each product. We then have

$$A = \begin{pmatrix} 5 & 4 & 7 \\ 8 & 0 & 1 \\ 13 & 3 & 9 \\ 2 & 6 & 3 \end{pmatrix}.$$

Let C represent the cost of each item. Then

$$C = \begin{pmatrix} 12 \\ 9 \\ 15 \end{pmatrix},$$

and

$$A \cdot C = \begin{pmatrix} 5 & 4 & 7 \\ 8 & 0 & 1 \\ 13 & 3 & 9 \\ 2 & 6 & 3 \end{pmatrix} \cdot \begin{pmatrix} 12 \\ 9 \\ 15 \end{pmatrix} = \begin{pmatrix} 5(12) + 4(9) + 7(15) \\ 8(12) + 0(9) + 1(15) \\ 13(12) + 3(9) + 9(15) \\ 2(12) + 6(9) + 3(15) \end{pmatrix} = \begin{pmatrix} 201 \\ 111 \\ 318 \\ 123 \end{pmatrix}.$$

The matrix AC represents the total income from each supermarket. Market I paid \$201 for all three products; \$111 is the total income from market II. Similarly, \$318 and \$123 are the incomes from markets III and IV, respectively.

The total income from all the supermarkets is obtained by adding the numbers down. This gives \$753.

Some further examples illustrate how matrices can be used in business problems.

EXAMPLE 1 Let us refer back to the concerts discussed on p. 653. The computations performed there can be done in matrix form as follows: Let the matrix A represent the number of tickets sold on Friday night. Let the matrix B stand for the number of tickets sold on Saturday. Let C be the cost matrix. Then

$$A = \begin{pmatrix} 250 \\ 120 \\ 73 \\ 208 \\ 124 \end{pmatrix}, B = \begin{pmatrix} 300 \\ 110 \\ 78 \\ 220 \\ 113 \end{pmatrix}, C = (3.00 \quad 2.50 \quad 2.00 \quad 1.50 \quad 1.00),$$

$$A + B = \begin{pmatrix} 250 + 300 \\ 120 + 110 \\ 73 + 78 \\ 208 + 220 \\ 124 + 113 \end{pmatrix} = \begin{pmatrix} 550 \\ 230 \\ 151 \\ 428 \\ 237 \end{pmatrix} = \begin{matrix} \text{Total seats sold of each type} \\ \text{Friday and Saturday,} \end{matrix}$$

$$C \cdot (A + B) = \begin{pmatrix} 3.00 & 2.50 & 2.00 & 1.50 & 1.00 \end{pmatrix} \cdot \begin{pmatrix} 550 \\ 230 \\ 151 \\ 428 \\ 237 \end{pmatrix} = (3406).$$

This represents the total income from both performances. ☐

EXAMPLE 2 The following table lists the number of summonses issued for various traffic violations for the first three months of 1985 in one midwestern town.

	Speeding	Illegal U-turn	Double parking	Drunken driving
January	2064	210	5314	206
February	3018	342	3709	421
March	1997	78	4112	308

Furthermore, the fines for these offenses are $50, $10, $20, and $30, respectively.

Mayor Peters wants to know how much money the town can expect from these summonses. She may also be interested in determining how much money will be collected each month. Using matrices she can proceed in the following way.

Let the matrix A represent the number of summonses, and let the matrix C be the fine for each offense. We have

$$A = \begin{pmatrix} 2064 & 210 & 5314 & 206 \\ 3018 & 342 & 3709 & 421 \\ 1997 & 78 & 4112 & 308 \end{pmatrix}, \quad C = \begin{pmatrix} 50 \\ 10 \\ 20 \\ 30 \end{pmatrix},$$

$$A \cdot C = \begin{pmatrix} 217,760 \\ 241,130 \\ 192,110 \end{pmatrix} \begin{matrix} \leftarrow \text{amount collected in January.} \\ \leftarrow \text{amount collected in February.} \\ \leftarrow \text{amount collected in March.} \end{matrix}$$

The total amount collected for all three months is found by adding the monthly totals together. It is $651,000. ☐

EXAMPLE 3 Three salesmen for a dress manufacturer submit the following orders for the month of January. The results are coded in the form of matrices where the rows stand for the dress sizes: 8, 10, 12, and 14. The columns represent the colors: black, red, green, and beige.

Salesman 1

$$\begin{pmatrix} 12 & 10 & 5 & 1 \\ 8 & 2 & 8 & 5 \\ 7 & 0 & 17 & 9 \\ 5 & 4 & 19 & 0 \end{pmatrix}$$

Salesman 2

$$\begin{pmatrix} 2 & 3 & 4 & 1 \\ 5 & 0 & 9 & 6 \\ 3 & 14 & 17 & 8 \\ 10 & 7 & 3 & 2 \end{pmatrix}$$

Salesman 3

$$\begin{pmatrix} 0 & 4 & 8 & 12 \\ 5 & 4 & 3 & 9 \\ 8 & 7 & 12 & 6 \\ 0 & 1 & 5 & 0 \end{pmatrix}$$

From these reports the manufacturer can obtain much information. Some of these results are summarized below. In each case, you should verify the results given.

108 size 12 dresses were sold,

59 beige dresses were sold,

65 black dresses were sold,

84 dresses were sold by salesman 3,

290 dresses were sold altogether. ☐

Comment In doing these examples it may be possible to write a matrix in two different ways. For instance, in Example 2 the matrix *C* was written as a 4 × 1 matrix,

$$C = \begin{pmatrix} 50 \\ 10 \\ 20 \\ 30 \end{pmatrix}.$$

We could have written it as a 1 × 4 matrix: (50   10   20   30). However, note that if we had written it as a 1 × 4, it would not have been possible to multiply it by *A*. (Why not?) We would then have been unable to solve this problem using matrices. Care is needed in deciding which matrices are to be used and how they are to be written. You must consider what is to be done with the matrix when making these decisions.

Perhaps you are wondering why we use matrices to solve these problems when general arithmetic would work just as well. One reason is that the same type of problem may have to be solved many times with different numbers. For example, a salesman may send in monthly sales reports, all of which have different figures and all of which have to be analyzed by the same procedure. Matrices provide a *mechanical* procedure for doing this. All we have to do is "plug in" the new numbers each time and work out the answer mechanically. We do not have to rethink the method each time.

Another reason for using matrices is that the lists may be very long. There may be many different things that we want to analyze. In such situations we could use a computer to help us with the calculations. Matrix operations can be very easily performed by a computer, and there are many software programs prepared to do such operations.

EXERCISES

1. *Long Distance.* Marjorie has made several long-distance calls for her boss as shown in the accompanying table.

Country to which call was made	Number of minutes	Cost per minute
France	7	$3.21
Brazil	8	2.69
Egypt	4	4.92
Sweden	6	6.52
Holland	9	5.02

   a) Write the number of minutes as a row matrix for these items.

   b) Write the cost as a column matrix for these items.

   c) Using matrix multiplication, find the total cost for these long distance calls.

2. *Watching Your Calories.* Glen is celebrating his twentieth wedding anniversary and is in a restaurant where he has just ordered a 14-ounce steak, 3 pieces of garlic bread, $\frac{2}{3}$ cup of vegetables, two scoops of mashed potatoes, three slices of apple pie, and two drinks of liquor. The number of calories in each of these items is as follows:

Item	Calories
Steak	105 per ounce
Garlic bread	115 per piece
Vegetables	72 per cup
Mashed potatoes	110 per scoop
Apple pie	350 per slice
Liquor	100 per drink

   a) Write the quantity ordered of each item as a row matrix.

   b) Write the calorie content of each item as a column matrix.

   c) Using matrix multiplication, find the total number of calories in the dinner.

3. The AZE Publishing Company publishes books and magazines that must be processed by each of its three factories. The following matrix indicates the hourly production by each factory for a shipment of magazines and a shipment of books:

$$\begin{array}{c} \quad\quad \text{Factory I } \text{Factory II } \text{Factory III} \\ \begin{array}{c} \text{Books} \\ \text{Magazines} \end{array} \begin{pmatrix} 8 & 6 & 5 \\ 10 & 3 & 4 \end{pmatrix} = A \end{array}$$

   Let
$$B = \begin{pmatrix} 8 \\ 9 \\ 7 \end{pmatrix}$$

   be the number of hours per day that each factory operates. Compute $A \cdot B$ and interpret the results.

Gov't to Prosecute Bilo for Illegal Dumping

MARLBERG—Officials of the state's Environmental Protection Agency announced yesterday that they would begin prosecuting the Bilo Chemical Corp. for illegally dumping chemical wastes from its Patchaw plant into neighboring lakes and streams. Said a spokesperson for the agency, Bilo has callously disregarded the environmental impact of such dumping. The adverse effects are beginning to show up. Last week, many dead fish were washed up on the shoreline.

   Bilo claims that it already has installed antipolluting devices.

THE GLOBE, August 26, 1982

4. Consider the newspaper article above. It is known that the Bilo Chemical Corp. manufactures five different items. Furthermore, the amount and kind of pollution generated in the

production of each of these items are shown in the following matrix:

	Pollutant I	Pollutant II	Pollutant III
Item I	9	5	3
Item II	6	11	4
Item III	10	8	2
Item IV	6	9	8
Item V	7	3	11

$= A$

The company claims that its costs per unit associated with removing these pollutants after installing the antipolluting devices are $12,000, $36,000, and $20,000, respectively.

a) Write the cost as a column matrix for these antipolluting devices. Call it B.

b) Compute the product of the two matrices $A \cdot B$ and interpret the results.

5. The Mark School operates two day-care centers, one in Newton and one in Baylif. Each of the centers operates independently of the other. On February 1 the food manager of the Newton center purchased 112 cans of peaches, 48 pounds of beef, and 23 pounds of vegetables, while the food manager of the Baylif center purchased 84 cans of peaches, 53 pounds of beef, and 51 pounds of vegetables. A can of peaches costs 59¢, a pound of beef costs $1.32, and a pound of vegetables costs 32¢. Using matrix multiplication, find the amount of money spent by the Mark School in purchasing the food for these two day-care centers.

6. *Cost Averaging.* Refer back to Exercise 5. In order to cost average, the food managers of the two day-care centers order the same quantities of food on February 25. However, the prices are now different. A can of peaches costs 62¢, a pound of beef costs $1.29, and a pound of vegetables costs 49¢.

a) Using matrix multiplication, find the amount of money spent by both centers on these food items.

b) How much money was spent by the Mark School on food for these two orders? (Use matrix techniques.)

12.6

APPLICATIONS TO SYSTEMS OF EQUATIONS

If we are asked to solve the system of equations

$$\begin{cases} 2x + 3y = 11 \\ 7x + 5y = 33 \end{cases}$$

for x and y, then we are looking for one value of x and one value of y that satisfy both equations. The solution for these equations is $x = 4$ and $y = 1$. This can be verified by substituting $x = 4$ and $y = 1$ into each of the equations. This answer can be obtained using matrices in the following way.

We first write the numbers on the left-hand side, in matrix form. Call this matrix A. It consists of the numbers in front of the letters. If no number is indicated, it is understood to be 1. In our case we have

$$A = \begin{pmatrix} 2 & 3 \\ 7 & 5 \end{pmatrix}.$$

The numbers on the right-hand side of the equation can be written as a matrix C:

$$C = \begin{pmatrix} 11 \\ 33 \end{pmatrix}.$$

The unknowns x and y can be written as a matrix M:

$$M = \begin{pmatrix} x \\ y \end{pmatrix}.$$

If we multiply matrices A and M (by matrix multiplication), we get

$$AM = \begin{pmatrix} 2 & 3 \\ 7 & 5 \end{pmatrix} \cdot \begin{pmatrix} x \\ y \end{pmatrix} = \begin{pmatrix} 2x + 3y \\ 7x + 5y \end{pmatrix}.$$

Notice that the result is the same as C, since $2x + 3y = 11$ and $7x + 5y = 33$. Therefore

$$AM = C.$$

It can be shown that a 2×2 matrix, such as A, may have an **inverse.** This means that there may exist another matrix which when multiplied by A will give the identity matrix (see Definition 12.3). Such an inverse will be denoted by A^{-1}. We can verify (see Example 3, p. 682) by direct matrix multiplication that if matrix $T = \begin{pmatrix} a & b \\ c & d \end{pmatrix}$, then the inverse, T^{-1}, is $\begin{pmatrix} d & -b \\ -c & a \end{pmatrix}$ multiplied by the number $\dfrac{1}{ad - bc}$, provided that $ad - bc$ is not 0.[1] Using this result, we find that the inverse of $A = \begin{pmatrix} 2 & 3 \\ 7 & 5 \end{pmatrix}$ is $A^{-1} = \begin{pmatrix} 5 & -3 \\ -7 & 2 \end{pmatrix}$ multiplied by $\dfrac{1}{2(5) - 3(7)}$, or $A^{-1} = \begin{pmatrix} 5 & -3 \\ -7 & 2 \end{pmatrix}$ multiplied by $-\dfrac{1}{11}$. This gives

$$A^{-1} = \begin{pmatrix} \dfrac{-5}{11} & \dfrac{+3}{11} \\ \dfrac{+7}{11} & \dfrac{-2}{11} \end{pmatrix}.$$

(If you have forgotten how this type of multiplication is performed, see p. 669). You should verify that this is the inverse by multiplying it with A. You should get

$$\begin{pmatrix} 2 & 3 \\ 7 & 5 \end{pmatrix} \cdot \begin{pmatrix} \dfrac{-5}{11} & \dfrac{+3}{11} \\ \dfrac{+7}{11} & \dfrac{-2}{11} \end{pmatrix} = \begin{pmatrix} 1 & 0 \\ 0 & 1 \end{pmatrix}.$$

---

1. If $ad - bc = 0$, then the number $1/(ad - bc)$ would be $1/0$. This, of course, is undefined.

Let us go back to our equation $AM = C$. Multiplying both sides by A^{-1} gives us

$$A^{-1}(AM) = A^{-1}C$$
$$(A^{-1}A)M = A^{-1}C \quad \text{(by the associative law).}$$

Since $A^{-1}A$ is the 2×2 identity matrix, and the identity matrix (Definition 12.3) does nothing when you multiply with it, we get

$$M = A^{-1}C.$$

(Be careful! $A^{-1}C$ may not equal CA^{-1}, since matrix multiplication is not always commutative.)

This means that

$$\begin{pmatrix} x \\ y \end{pmatrix} = \begin{pmatrix} \dfrac{-5}{11} & \dfrac{+3}{11} \\ \dfrac{+7}{11} & \dfrac{-2}{11} \end{pmatrix} \cdot \begin{pmatrix} 11 \\ 33 \end{pmatrix}$$

$$= \begin{pmatrix} \dfrac{-5}{11} \cdot (11) + \dfrac{3}{11} \cdot (33) \\ \dfrac{+7}{11} \cdot (11) + \dfrac{-2}{11} \cdot (33) \end{pmatrix}$$

$$= \begin{pmatrix} -5 + 9 \\ 7 - 6 \end{pmatrix} = \begin{pmatrix} 4 \\ 1 \end{pmatrix}.$$

Thus we have $x = 4$ and $y = 1$.

EXAMPLE 1   Solve $\begin{cases} 3x - 2y = 0 \\ 2x + y = 7 \end{cases}$ for x and y.

SOLUTION   We have

$$A = \begin{pmatrix} 3 & -2 \\ 2 & 1 \end{pmatrix}, \quad M = \begin{pmatrix} x \\ y \end{pmatrix}, \quad C = \begin{pmatrix} 0 \\ 7 \end{pmatrix},$$

$$A^{-1} = \begin{pmatrix} 1 & 2 \\ -2 & 3 \end{pmatrix} \quad \text{multiplied by } \frac{1}{3 \cdot (1) - (-2)2},$$

$$A^{-1} = \begin{pmatrix} 1 & 2 \\ -2 & 3 \end{pmatrix} \quad \text{multiplied by } \frac{1}{+7}.$$

Therefore

$$A^{-1} = \begin{pmatrix} \dfrac{1}{7} & \dfrac{2}{7} \\ \dfrac{-2}{7} & \dfrac{3}{7} \end{pmatrix}.$$

Our answer is then

$$M = A^{-1}C,$$

$$\begin{pmatrix} x \\ y \end{pmatrix} = \begin{pmatrix} \dfrac{1}{7} & \dfrac{2}{7} \\ \dfrac{-2}{7} & \dfrac{3}{7} \end{pmatrix} \cdot \begin{pmatrix} 0 \\ 7 \end{pmatrix}$$

$$= \begin{pmatrix} \dfrac{1}{7} \cdot (0) + \dfrac{2}{7} \cdot (7) \\ \dfrac{-2}{7} \cdot (0) + \dfrac{3}{7} \cdot (7) \end{pmatrix}$$

$$= \begin{pmatrix} 0 + 2 \\ 0 + 3 \end{pmatrix}$$

$$= \begin{pmatrix} 2 \\ 3 \end{pmatrix}.$$

Finally, we have $x = 2$ and $y = 3$. (You should check these answers by substituting $x = 2$ and $y = 3$ into the original equations.) ☐

This technique can be extended to solve a system with any number of equations and the same number of unknowns.

EXAMPLE 2 Solve $\begin{Bmatrix} 4x - 3y = 5 \\ x + 2y = 4 \end{Bmatrix}$ for x and y.

SOLUTION We have

$$A = \begin{pmatrix} 4 & -3 \\ 1 & 2 \end{pmatrix}, \qquad M = \begin{pmatrix} x \\ y \end{pmatrix}, \qquad C = \begin{pmatrix} 5 \\ 4 \end{pmatrix}$$

$$A^{-1} = \begin{pmatrix} 2 & 3 \\ -1 & 4 \end{pmatrix} \qquad \text{multiplied by } \frac{1}{4(2) - (-3)(1)},$$

$$A^{-1} = \begin{pmatrix} 2 & 3 \\ -1 & 4 \end{pmatrix} \qquad \text{multiplied by } \frac{1}{11}.$$

Therefore

$$A^{-1} = \begin{pmatrix} \dfrac{2}{11} & \dfrac{3}{11} \\ \dfrac{-1}{11} & \dfrac{4}{11} \end{pmatrix}.$$

Our answer is then

$$M = A^{-1}C,$$

$$\begin{pmatrix} x \\ y \end{pmatrix} = \begin{pmatrix} \dfrac{2}{11} & \dfrac{3}{11} \\ \dfrac{-1}{11} & \dfrac{4}{11} \end{pmatrix} \cdot \begin{pmatrix} 5 \\ 4 \end{pmatrix}$$

$$= \begin{pmatrix} \dfrac{2}{11} \cdot (5) + \dfrac{3}{11} \cdot (4) \\ \dfrac{-1}{11} \cdot (5) + \dfrac{4}{11} \cdot (4) \end{pmatrix}$$

$$= \begin{pmatrix} \dfrac{10}{11} + \dfrac{12}{11} \\ \dfrac{-5}{11} + \dfrac{16}{11} \end{pmatrix}$$

$$= \begin{pmatrix} \dfrac{22}{11} \\ \dfrac{11}{11} \end{pmatrix}$$

$$= \begin{pmatrix} 2 \\ 1 \end{pmatrix}.$$

Thus we have $x = 2$ and $y = 1$. (Again you should check these answers by substituting $x = 2$ and $y = 1$ into the original equations.) □

EXAMPLE 3 Verify, by matrix multiplication, that if matrix

$$T = \begin{pmatrix} a & b \\ c & d \end{pmatrix},$$

then its inverse is

$$\begin{pmatrix} d & -b \\ -c & a \end{pmatrix} \quad \text{multiplied by} \quad \frac{1}{ad - bc}.$$

(We will assume that $ad - bc$ is not equal to 0. Otherwise the inverse does not exist.)

SOLUTION We first multiply $\begin{pmatrix} a & b \\ c & d \end{pmatrix}$ by $\begin{pmatrix} d & -b \\ -c & a \end{pmatrix}$, getting

$$\begin{pmatrix} a & b \\ c & d \end{pmatrix} \cdot \begin{pmatrix} d & -b \\ -c & a \end{pmatrix} = \begin{pmatrix} ad + b(-c) & a(-b) + ba \\ cd + d(-c) & c(-b) + da \end{pmatrix}$$

$$= \begin{pmatrix} ad - bc & -ab + ba \\ cd - dc & -bc + da \end{pmatrix} = \begin{pmatrix} ad - bc & 0 \\ 0 & ad - bc \end{pmatrix}.$$

Finally, multiplying this matrix by $\dfrac{1}{ad - bc}$ gives

$$\begin{pmatrix} \dfrac{ad - bc}{ad - bc} & \dfrac{0}{ad - bc} \\[2mm] \dfrac{0}{ad - bc} & \dfrac{ad - bc}{ad - bc} \end{pmatrix} = \begin{pmatrix} 1 & 0 \\ 0 & 1 \end{pmatrix},$$

which is the 2×2 identity matrix. Also

$$\begin{pmatrix} d & -b \\ -c & a \end{pmatrix} \cdot \begin{pmatrix} a & b \\ c & d \end{pmatrix} = \begin{pmatrix} da - bc & 0 \\ 0 & -cb + ad \end{pmatrix}.$$

Multiplying this matrix by $\dfrac{1}{ad - bc}$ gives $\begin{pmatrix} 1 & 0 \\ 0 & 1 \end{pmatrix}$. Thus the inverse of $\begin{pmatrix} a & b \\ c & d \end{pmatrix}$ is $\begin{pmatrix} d & -b \\ -c & a \end{pmatrix}$ multiplied by $\dfrac{1}{ad - bc}$. □

Elementary Row Operations

We can also solve a system of equations by performing what are known as **elementary row operations.** These are defined as follows.

Definition 12.4 *An **elementary row operation** on any matrix consists of any one of the following three operations:*

1. *Interchanging any two rows.*

2. *Multiplying all elements of any row by a given nonzero number.*

3. *Multiplying each element of a row by a number and then adding the results to the corresponding elements of another row.*

EXAMPLE 4   The matrix $\begin{pmatrix} 3 & 7 \\ 9 & 4 \end{pmatrix}$ can be transformed into the matrix $\begin{pmatrix} 9 & 4 \\ 3 & 7 \end{pmatrix}$ by interchanging the first and the second rows. □

EXAMPLE 5   The matrix $\begin{pmatrix} 4 & 8 \\ 6 & 9 \end{pmatrix}$ can be transformed into the matrix $\begin{pmatrix} 2 & 4 \\ 2 & 3 \end{pmatrix}$ by multiplying each element of row 1 by $\frac{1}{2}$ and each element of row 2 by $\frac{1}{3}$. □

Comment   The elementary row operations correspond to operations that we can perform with a system of equations without changing the solution to the equations. For example, we may interchange any two equations, or we may multiply both sides of an equation by a number and then add it to the other equation.

Let us solve the system of equations

$$\begin{cases} 4x - 3y = 5 \\ x + 2y = 4 \end{cases}$$

for x and y by using elementary row operations. (These equations were already solved in Example 2.) We already know that

$$A = \begin{pmatrix} 4 & -3 \\ 1 & 2 \end{pmatrix} \quad \text{and} \quad C = \begin{pmatrix} 5 \\ 4 \end{pmatrix}.$$

We combine these two matrices into an *augmented matrix* using a dashed line to separate them:

$$\left(\begin{array}{cc|c} 4 & -3 & 5 \\ 1 & 2 & 4 \end{array} \right).$$

Now we perform a series of elementary row operations on the augmented matrix to arrive at the answer. We indicate the operations applied to the equations and to the augmented matrix so that you can see the correspondence.

Equation	*Matrix*
$4x - 3y = 5,$ $x + 2y = 4$	Our objective is to apply elementary row operations to the augmented matrix so that we get the identity matrix to the left of the dashed line. Thus we try to get a 1 in the row 1, column 1 position, and then a 0 in the row 2, column 1 position. To accomplish this, we interchange row 2 and row 1. This results in
$x + 2y = 4,$ $4x - 3y = 5$	$\left(\begin{array}{cc|c} 1 & 2 & 4 \\ 4 & -3 & 5 \end{array} \right).$
$x + 2y = 4,$ $4x - 4x - 3y - 8y = 5 - 16$ or	Now we multiply each element of row 1 by -4 and add the results to row 2. We get
$x + 2y = 4,$ $0x - 11y = -11$	$\left(\begin{array}{cc|c} 1 & 2 & 4 \\ 0 & -11 & -11 \end{array} \right)$
	(Note that row 1 remains unchanged) Next we want to get a 1 in the row 2, column 2 position. To accomplish this, we multiply each element of row 2 by $\dfrac{-1}{11}$.

This gives us

$$x + 2y = 4$$
$$0x + y = 1$$

$$\begin{pmatrix} 1 & 2 & | & 4 \\ 0 & 1 & | & 1 \end{pmatrix}.$$

Finally, we want to have a 0 in the row 1, column 2 position. To accomplish this, we multiply each element of row 2 by -2 and add the results to row 1. We get

$$x + 0y = 2$$
$$0x + y = 1$$

$$\begin{pmatrix} 1 & 0 & | & 2 \\ 0 & 1 & | & 1 \end{pmatrix}.$$

The final augmented matrix gives us the solution to the problem. We have

Thus $x = 2,$
$\quad\quad y = 1.$

$$1x + 0y = 2,$$
$$0x + 1y = 1$$

or simply

$$x = 2, \quad y = 1.$$

This is exactly the same answer that we obtained using matrix inverses.

This new technique, which is called the *Gauss–Jordan method* in honor of the two famous mathematicians Carl F. Gauss (1777–1855) and Camille Jordan (1838–1922), is illustrated in Example 6.

EXAMPLE 6 Using matrix methods, solve the following system of equations for x and y:

$$\begin{Bmatrix} 5x + 3y = -1 \\ 3x - 4y = 11 \end{Bmatrix}.$$

SOLUTION Again we will indicate the operations applied to the equations and to the augmented matrix so that you can see the correspondence.

| *Equation* | *Matrix* |

We first form the augmented matrix

$$5x + 3y = -1,$$
$$3x - 4y = 11$$

$$\begin{pmatrix} 5 & 3 & | & -1 \\ 3 & -4 & | & 11 \end{pmatrix}.$$

Then we perform several operations so that one of the variables drops out. We divide the first equation by 5 to get

$$x + \frac{3}{5}y = -\frac{1}{5},$$
$$3x - 4y = 11$$

We multiply the first equation by -3 and add the result to the second equation, getting

$$x + \frac{3}{5}y = -\frac{1}{5},$$
$$0x - \frac{29}{5}y = \frac{58}{5}.$$

We multiply the second equation by $\frac{-5}{29}$. We get

$$x + \frac{3}{5}y = -\frac{1}{5},$$
$$0x + y = -2.$$

We multiply the second equation by $\frac{-3}{5}$ and add the result to the first equation, getting

$$x + 0y = 1,$$
$$0x + y = -2.$$

Then we perform several elementary row operations so that the numbers to the left of the dashed line will make up an identity matrix. To get started, we try to get a 1 in the row 1, column 1 position. This can be accomplished by multiplying each element of row 1 by $\frac{1}{5}$. This gives us

$$\begin{pmatrix} 1 & \frac{3}{5} & \vdots & \frac{-1}{5} \\ 3 & -4 & \vdots & 11 \end{pmatrix}.$$

Now we try to get a 0 in the row 2, column 1 position. To accomplish this, we multiply each element of row 1 by -3 and add the result to row 2, getting

$$\begin{pmatrix} 1 & \frac{3}{5} & \vdots & \frac{-1}{5} \\ 0 & \frac{-29}{5} & \vdots & \frac{58}{5} \end{pmatrix}.$$

Now we try to get a 1 in the row 2, column 2 position. We multiply each element of row 2 by $\frac{-5}{29}$. Our result is

$$\begin{pmatrix} 1 & \frac{3}{5} & \vdots & \frac{-1}{5} \\ 0 & 1 & \vdots & -2 \end{pmatrix}.$$

Finally we try to get a 0 in the row 1, column 2 position. To accomplish this, we multiply each element of row 2 by $\frac{-3}{5}$ and add the results to row 1. This gives us

$$\begin{pmatrix} 1 & 0 & \vdots & 1 \\ 0 & 1 & \vdots & -2 \end{pmatrix}.$$

We notice that we have an identity matrix to the left of the dashed line. We can now read our answer from the extreme right column:

$$1x + 0y = 1,$$
$$0x + 1y = -2$$

Our answer is

$$x = 1, \quad y = -2.$$

or

$$x = 1, \quad y = -2. \quad \square$$

Comment As we indicated earlier, after we obtain the augmented matrix, our main objective is to change the numbers to the left of the dashed line into the identity matrix so that the numbers to the right of this line will become solutions for x and y.

We summarize the above procedure as follows:

To solve a system of equations using row reduction techniques, you must:

1. Form the augmented matrix.
2. Perform the row operations necessary to get a 1 in the upper left-hand corner of the matrix.
3. Perform row operations to get an identity matrix to the left of the dashed line if this matrix is a square matrix. (If the matrix is not square, then 1's must be gotten along the main diagonal, and all other entries must be zero.)
4. Read the solutions on the right of the dashed line.

EXERCISES

In each of Exercises 1–8, find the appropriate elementary row operation(s) that can be used to transform matrix A into matrix B.

1. $A = \begin{pmatrix} 4 & 1 \\ 3 & 8 \end{pmatrix}$ $\qquad B = \begin{pmatrix} -12 & -3 \\ -9 & -24 \end{pmatrix}$

2. $A = \begin{pmatrix} 5 & 7 & 8 & 14 \\ 3 & 0 & 2 & -6 \end{pmatrix}$

$\qquad B = \begin{pmatrix} -10 & -14 & -16 & -28 \\ -6 & 0 & -4 & 12 \end{pmatrix}$

3. $A = \begin{pmatrix} 15 & 18 \\ 12 & 9 \\ 27 & 21 \end{pmatrix}$ $\qquad B = \begin{pmatrix} 10 & 12 \\ 8 & 6 \\ 18 & 14 \end{pmatrix}$

4. $A = \begin{pmatrix} 1 & 4 & 7 \\ 3 & 9 & 2 \\ -2 & 1 & 7 \end{pmatrix}$ $\qquad B = \begin{pmatrix} 1 & 4 & 7 \\ 0 & -3 & -19 \\ 0 & 9 & 21 \end{pmatrix}$

5. $A = \begin{pmatrix} 2 & 3 & 9 \\ 1 & 0 & 1 \\ -3 & 0 & 4 \end{pmatrix}$ $\qquad B = \begin{pmatrix} 0 & 3 & 7 \\ 1 & 0 & 1 \\ 0 & 0 & 7 \end{pmatrix}$

6. $A = \begin{pmatrix} 7 & 3 & 2 \\ 1 & 2 & 1 \\ 5 & 2 & 0 \end{pmatrix}$ $\qquad B = \begin{pmatrix} 1 & 2 & 1 \\ 0 & -11 & -5 \\ 0 & -8 & -5 \end{pmatrix}$

**7. $A = \begin{pmatrix} 5 & 3 & 8 \\ -2 & 7 & 5 \\ 1 & 0 & 2 \end{pmatrix}$ $\qquad B = \begin{pmatrix} 1 & 0 & 0 \\ 0 & 1 & 0 \\ 0 & 0 & 1 \end{pmatrix}$

**8. $A = \begin{pmatrix} 4 & 12 & 70 \\ 6 & -9 & 1 \\ 3 & 2 & 4 \end{pmatrix}$ $B = \begin{pmatrix} 1 & 0 & 0 \\ 0 & 1 & 0 \\ 0 & 0 & 1 \end{pmatrix}$

Write the augmented matrix for each set of equations in Exercises 9–18. Do not attempt to solve them.

9. $7x + 3y = 11$
$5x - 2y = 19$

10. $12x - 2y = 13$
$5x + 9y = 23$

11. $2x - 3y = 5$
$7x + 12y = 19$

12. $5x - 3y = 15$
$4x + 3y = -12$

13. $2x + 3y - 5z = 10$
$3x - 7y + 8z = -15$

14. $2x + 3y - 4z = 24$
$8x - 4y + 7z = 16$
$3x - 5y + 8z = 12$

15. $2x + 3y = 12$
$7x + 8y = 27$
$2x - 9y = 18$

16. $= 7$
$y - 3z = 10$
$x \quad - 8z = 23$

17. $4x + y \quad = 10$
$3y - z = 17$
$2x \quad + 7z = 15$

18. $2x + 3y + 5z - 8w = 7$
$3x - 2y - 4z + 7w = 0$
$4x + 7y - 3z - 4w = -3$
$2x \quad + 3z - w = 1$

19. Write a linear system to correspond to each of the following augmented matrices.

a) $\begin{pmatrix} 5 & 7 & | & 10 \\ 3 & 9 & | & 14 \end{pmatrix}$

b) $\begin{pmatrix} 8 & 7 & | & 24 \\ 9 & 5 & | & 45 \end{pmatrix}$

c) $\begin{pmatrix} 2 & 3 & -7 & 6 & | & 10 \\ 5 & 4 & 8 & 1 & | & 0 \end{pmatrix}$

d) $\begin{pmatrix} 2 & 9 & 6 & | & 1 \\ 5 & -4 & -7 & | & 0 \\ 3 & -2 & 9 & | & 10 \end{pmatrix}$

e) $\begin{pmatrix} 2 & 5 & 1 & 2 & | & 8 \\ 1 & 3 & 0 & 3 & | & 10 \\ 9 & 2 & 7 & 4 & | & 14 \end{pmatrix}$

In Exercises 20–31, use matrix operations to find solutions (if they exist) for the set of equations.

20. $\begin{cases} 5x + 3y = 13 \\ 2x - 5y = -1 \end{cases}$ **21.** $\begin{cases} 4x - y = 13 \\ 2x + 3y = 3 \end{cases}$

22. $\begin{cases} 8x + 5y = -28 \\ 4x - 6y = 20 \end{cases}$ **23.** $\begin{cases} 5x - 3y = 26 \\ x + 8y = 31 \end{cases}$

24. $\begin{cases} 6x + 8y = 48 \\ 3x - 5y = 24 \end{cases}$ **25.** $\begin{cases} 4x - 16y = 14 \\ 10x + 4y = 2 \end{cases}$

26. $\begin{cases} 7x - 3y = 16 \\ x + 2y = -5 \end{cases}$ **27.** $\begin{cases} 2x + 3y = 800 \\ 0.06x + 0.04y = 140 \end{cases}$

28. $\begin{cases} 3x - y = 7 \\ 6x + 2y = 22 \end{cases}$ **29.** $\begin{cases} 2x + 3y = -4 \\ x - 2y = 19 \end{cases}$

30. $\begin{cases} x - 7y = 27 \\ 3x + 4y = -19 \end{cases}$ **31.** $\begin{cases} 7x + 9y = 35 \\ 5x - 3y = 25 \end{cases}$

32. Verify that the inverse of matrix $\begin{pmatrix} a & b \\ c & d \end{pmatrix}$ is $\begin{pmatrix} d & -b \\ -c & a \end{pmatrix}$ multiplied by $\dfrac{1}{ad - bc}$ by using the following procedure. Consider

$$\begin{pmatrix} a & b \\ c & d \end{pmatrix} \cdot \begin{pmatrix} w & x \\ y & z \end{pmatrix} = \begin{pmatrix} 1 & 0 \\ 0 & 1 \end{pmatrix}.$$

Using multiplication on the left-hand side, we get

$$\begin{pmatrix} aw + by & ax + bz \\ cw + dy & cx + dz \end{pmatrix} = \begin{pmatrix} 1 & 0 \\ 0 & 1 \end{pmatrix}.$$

This means that

$$aw + by = 1,$$
$$ax + bz = 0,$$
$$cw + dy = 0,$$
$$cx + dz = 1.$$

Solve the first two equations and the last two equations in pairs for w, x, y, and z. You should get

$$w = \frac{d}{ad - bc}, \qquad x = \frac{-b}{ad - bc},$$

$$y = \frac{-c}{ad - bc}, \qquad z = \frac{a}{ad - bc}.$$

Compare this with the answer we got. They are the same.

12.7

APPLICATIONS OF MATRICES TO COMMUNICATIONS NETWORKS

We spend a good part of our lives communicating with other people in many different ways. When you call a friend, write a letter, or watch television, you are communicating or being communicated with. In writing this book we hope we are communicating some ideas to you.

Some communication links are quite complicated. For example, suppose Steve wants to contact Florence, whom he met at a party last year. He doesn't have her phone number, so he calls his friend John, who gave the party. John does not have her number either, but he calls Ellen and Marie, who went with Florence to the party. Neither has her number, but Ellen calls Stan, who does. He calls Steve and gives him the number. Through this *network*, Steve succeeds in getting Florence's number. We can illustrate this situation with the diagram shown below.

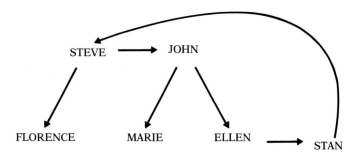

Matrices can be helpful in analyzing communication networks. The following examples will illustrate how this is done. In our discussion we will assume that *one cannot communicate with oneself.*

EXAMPLE 1   On a battlefield, four soldiers—Bob, Ed, Mike, and Al—have walkie-talkies with which they communicate with each other. Some of the walkie-talkies are broken. Bob's is completely broken. He can neither receive nor transmit any information. Ed's walkie-talkie can only receive but not transmit. Mike's is working properly. Al's can only transmit but not receive any information.

We form a matrix with rows and columns for each soldier. All entries are either 1 or 0. An entry of 1 means that the soldier in that row can transmit to the soldier in that column. Otherwise we write a 0. The matrix would then be the following:

	Bob	Ed	Mike	Al
Bob	0	0	0	0
Ed	0	0	0	0
Mike	0	1	0	0
Al	0	1	1	0

□

EXAMPLE 2   Suppose in the situation of Example 1 the walkie-talkies are not completely broken but are not working properly. There is also some interference. The following matrix represents the communications that are possible:

	Bob	Ed	Mike	Al
Bob	0	1	0	0
Ed	1	0	1	0
Mike	0	0	0	1
Al	0	1	0	0

The matrix gives us the following information:

Bob can transmit only to Ed.

Ed can transmit to both Bob and Mike.

Mike can transmit only to Al.

Al can transmit only to Ed.

If Al wants to send a message to Bob, then he could transmit it by way of Ed. Is there any way for Mike to send a message to Bob?

□

You may be wondering why one would use matrices at all for such a situation. The matrix provides a quick and easy-to-read diagram of the entire network. By looking at the appropriate row and column we can tell immediately who can communicate with whom.

Furthermore, by studying these matrices in greater detail it is possible to learn a lot about a network. For example, in a large corporation with many levels of supervision (some of them overlapping), one can find who has the most power, who has the least power, who is the second most powerful, and so on.

EXAMPLE 3 In Chad there are four neighbors, Mrs. Young, Mr. Babson, Miss Hall, and Mrs. Gordon. Mrs. Young passes on all the gossip she hears to Mr. Babson and Mrs. Gordon. Mr. Babson repeats gossip only to Miss Hall. Miss Hall tells all to Mrs. Young. Mrs. Gordon never repeats any gossip. The matrix below represents the neighborhood gossip network.

	Mrs. Young	Mr. Babson	Miss Hall	Mrs. Gordon
Mrs. Young	0	1	0	1
Mr. Babson	0	0	1	0
Miss Hall	1	0	0	0
Mrs. Gordon	0	0	0	0

An important application of matrices and their inverses is in *cryptography,* or *code theory.* Secret codes are used by governments when sending messages. There are many specialists who are experts at breaking codes. Yet secret codes are quite difficult to break when they are written in matrix form. They can be decoded by using matrix inverses.

One way of coding messages is to associate each letter of the alphabet with some other letter or with some number. Such codes are rather easy to break since certain letters of the alphabet occur more frequently than others. Nevertheless, this is done quite often. When this system is used, we associate a different coding symbol with each letter to avoid confusion.

EXAMPLE 4 *Coding.* Use the following coding scheme to code the message DESTROY.

A B C D E F G H I J K L M N O P Q R S T U V W X Y Z
↓ ↓
F H Q S A T J B N W C U I O V D L X E Y R G Z M P K

SOLUTION Using the coding scheme, the message DESTROY would be transmitted as SAEYXVP.

EXAMPLE 5 *Coding—Alternate Scheme.* Use the following coding scheme to code the message DESTROY.

A B C D E F G H I J K L M N O P Q R S T U V W X Y Z
↓ ↓
8 2 9 13 20 1 10 17 5 18 22 3 15 19 11 26 21 25 4 24 12 23 7 14 16 6

SOLUTION   The message DESTROY would be transmitted as

$$13 \quad 20 \quad 4 \quad 24 \quad 25 \quad 11 \quad 16.$$ ☐

EXAMPLE 6   Use the coding scheme given in Example 4 to decode the following message: SV-OVY-JNGA-RD.

SOLUTION   The message SV-OVY-JNGA-RD is DO NOT GIVE UP.   ☐

Messages can also be coded by using matrices. We must decide in advance whether the message is to be grouped in pairs of letters, triplets of letters, or some other pattern. This determines the dimension of the matrix that has to be used. For example, suppose we wish to send the message

THE DEAL IS OFF

We would rewrite this message in groups of two letters as

TH   ED   EA   LI   SO   FF

With the code given in Example 5 this message can be written in matrix form as

$$\begin{pmatrix} 24 \\ 17 \end{pmatrix} \begin{pmatrix} 20 \\ 13 \end{pmatrix} \begin{pmatrix} 20 \\ 8 \end{pmatrix} \begin{pmatrix} 3 \\ 5 \end{pmatrix} \begin{pmatrix} 4 \\ 11 \end{pmatrix} \begin{pmatrix} 1 \\ 1 \end{pmatrix}$$

Now we select any 2×2 matrix that has an inverse and contains no fractions. One such matrix is

$$A = \begin{pmatrix} 2 & 3 \\ 1 & 2 \end{pmatrix}.$$

Its inverse is

$$A^{-1} = \begin{pmatrix} 2 & -3 \\ -1 & 2 \end{pmatrix}.$$

The product of A and each of the code matrices is

$$\begin{pmatrix} 2 & 3 \\ 1 & 2 \end{pmatrix} \cdot \begin{pmatrix} 24 \\ 17 \end{pmatrix} \quad \begin{pmatrix} 2 & 3 \\ 1 & 2 \end{pmatrix} \cdot \begin{pmatrix} 20 \\ 13 \end{pmatrix} \quad \begin{pmatrix} 2 & 3 \\ 1 & 2 \end{pmatrix} \cdot \begin{pmatrix} 20 \\ 8 \end{pmatrix}$$

$$\begin{pmatrix} 2 & 3 \\ 1 & 2 \end{pmatrix} \cdot \begin{pmatrix} 3 \\ 5 \end{pmatrix} \quad \begin{pmatrix} 2 & 3 \\ 1 & 2 \end{pmatrix} \cdot \begin{pmatrix} 4 \\ 11 \end{pmatrix} \quad \begin{pmatrix} 2 & 3 \\ 1 & 2 \end{pmatrix} \cdot \begin{pmatrix} 1 \\ 1 \end{pmatrix}$$

The coded message then is transmitted as

$$99, 58, 79, 46, 64, 36. \ldots$$

The process of translating a message into a matrix is called *encoding*. To *decode*, we simply regroup the numbers into a matrix and then multiply by the matrix A^{-1}. For the above example we have

$$A^{-1} = \begin{pmatrix} 2 & -3 \\ -1 & 2 \end{pmatrix}$$

so that

$$\begin{pmatrix} 2 & -3 \\ -1 & 2 \end{pmatrix} \begin{pmatrix} 99 \\ 58 \end{pmatrix} = \begin{pmatrix} 24 \\ 17 \end{pmatrix}$$

$$\begin{pmatrix} 2 & -3 \\ -1 & 2 \end{pmatrix} \begin{pmatrix} 79 \\ 46 \end{pmatrix} = \begin{pmatrix} 20 \\ 13 \end{pmatrix}$$

$$\begin{pmatrix} 2 & -3 \\ -1 & 2 \end{pmatrix} \begin{pmatrix} 64 \\ 36 \end{pmatrix} = \begin{pmatrix} 20 \\ 8 \end{pmatrix}$$

$$\vdots$$

etc.

The message in coded matrix form then is

$$\begin{pmatrix} 24 \\ 17 \end{pmatrix} \begin{pmatrix} 20 \\ 13 \end{pmatrix} \begin{pmatrix} 20 \\ 8 \end{pmatrix} \cdots$$

To finish the decoding process, we use the code given in Example 5.

EXERCISES

1. *Planning College Courses.* There are three colleges in the State University system, each offering degrees in different programs as indicated in matrix A.

	Liberal arts majors	Engineering majors	Business majors	
College I	1	1	0	
College II	1	0	1	$= A$
College III	1	1	1	

(A 1 represents a yes, and 0 represents a no.) Furthermore, each different program requires that certain math courses be completed as a prerequisite for graduation as indicated in matrix B.

	Only math proficiency courses	College algebra course	Calculus course	
Liberal arts	1	0	0	
Engineering	0	1	1	$= B$
Business	0	1	0	

Find the product $A \cdot B$ and explain how this helps the college decide which courses to offer.

2. *Radio Communications.* Five soldiers—Jed, Max, Catfish, Augustus, and Frenchie—have parachuted into different parts of an island. Their radio receivers and transmitters are not functioning properly. Suppose that the radios have been adjusted so that the accompanying matrix represents the communications that are now possible.

	Jed	Max	Catfish	Augustus	Frenchie
Jed	0	0	1	1	1
Max	1	0	1	0	1
Catfish	1	0	0	1	1
Augustus	0	0	1	0	1
Frenchie	1	1	1	1	0

Can a message be transmitted so that everyone can receive it? By whom must it be transmitted?

3. *Faculty Evaluations.* Several students in the Beta Fraternity were asked to indicate which of their professors they would recommend to their friends. The results were then tabulated and put into matrix form. An entry of 1 corresponds to a favorable recommendation, and an entry of 0 corresponds to an unfavorable recommendation.

	Prof. Hoffman	Prof. Lichtenfeld	Prof. Blois	Prof. Rogers	Prof. Kennedy
José	1	1	1	0	1
Richard	0	0	1	1	0
Ralph	1	0	1	1	1
John	1	1	0	0	0
Phil	0	0	0	0	1
Jamie	0	1	1	0	0

Based on this matrix, answer the following questions:

   a) Which teacher is most recommended?
   b) Which teacher is least recommended?
   c) Which teacher(s) does John recommend?
   d) Which teacher(s) does Richard not recommend?

4. *Camp Records.* A camp director is reviewing the preference records of the campers and the corresponding data file. The accompanying coded information has been obtained from the camp office for six campers. An

entry of 1 corresponds to a yes.

	Swimming enthusiast	Boating enthusiast	Baseball enthusiast	Hiking enthusiast	
Mark	1	1	1	1	
Steve	0	0	0	0	
Mary	0	1	1	0	= A
Sue	1	1	1	1	
Louis	0	1	0	1	
Gail	1	0	0	1	

	Submitted complete medical history records	Has Red Cross swimmers certificate	Has backpack equipment	Has baseball equipment	
Swimming enthusiast	1	0	0	0	
Boating enthusiast	0	1	0	0	= B
Baseball enthusiast	0	0	0	1	
Hiking enthusiast	0	0	1	0	

The product of these matrices, $A \cdot B$, is

$$\begin{pmatrix} 1 & 1 & 1 & 1 \\ 0 & 0 & 0 & 0 \\ 0 & 1 & 1 & 0 \\ 1 & 1 & 1 & 1 \\ 0 & 1 & 0 & 1 \\ 1 & 0 & 0 & 1 \end{pmatrix} \begin{pmatrix} 1 & 0 & 0 & 0 \\ 0 & 1 & 0 & 0 \\ 0 & 0 & 0 & 1 \\ 0 & 0 & 1 & 0 \end{pmatrix} = \begin{pmatrix} 1 & 1 & 1 & 1 \\ 0 & 0 & 0 & 0 \\ 0 & 1 & 0 & 0 \\ 1 & 1 & 1 & 1 \\ 0 & 1 & 1 & 0 \\ 1 & 0 & 1 & 0 \end{pmatrix}$$

Can you interpret the entries in the product matrix $A \cdot B$?

5. *Tracing Mononucleosis.* Bill has been admitted to the hospital with symptoms of mononucleosis. Doctors are anxious to find out the names of all the people who have had contact with Bill within the past few days. Questioning him, the doctors discover the following information: Bill has seen his wife Marion, his brother Jeff, and his daughter Pam with her boyfriend. Marion has seen her parents. Pam has seen only her boyfriend. Jeff has seen his wife Molly. Bill, Marion, Pam, Molly, Jeff, the boyfriend, and Marion's parents have seen no one else. On the basis of this information, set up the communications matrix to indicate who can transmit the disease to whom.

6. Use the code given in Example 4 to encode the message

JONES-IS-AN-ENEMY-AGENT-KILL-HIM.

7. Use the coding scheme given in Example 5 to encode the message given in the previous exercise.

8. Use the coding scheme given in Example 5 to decode the following message

4, 5, 19, 22, 24, 17, 20, 4, 17, 5, 26

9. Arrange the following message into groups of two letters:

BILLY-IS-YOUR-CONTACT

Find an appropriate 2×2 matrix and indicate how the coded message can be transmitted.

**10. *Counterespionage.* An enemy agent transmits the following message,

187 65 409 143 361 126 77 26 66 23 113 39,

which has been coded with $\begin{pmatrix} 3 & 17 \\ 1 & 6 \end{pmatrix}$ as the coding matrix. Use matrix inverses to decode the message.

12.8

CONCLUDING REMARKS

There are many other interesting and useful applications of matrix theory. Some of the ways in which matrices can be used are in

1. predicting population growth,
2. analyzing marriage rules of various societies (in an anthropology discussion),
3. various problems in economics,
4. studying heredity (genetics),
5. code theory.

The interested reader can consult the suggested further readings.

12.9

SUMMARY

In this chapter we discussed the concept of a matrix. After defining what a matrix is, we learned how to add and subtract them, provided that they are of the same dimension. Our rule for multiplying matrices may have seemed a bit unusual in that we do not multiply corresponding elements. Furthermore, we can multiply matrices even when they are not of the same dimension, provided that the number of columns of the first matrix is the same as the number of rows of the second matrix.

We applied matrices to business situations as well as to communications networks. Although we used matrices to help us solve only a pair of simultaneous equations in two unknowns, they can be similarly applied to solving three equations in three unknowns, etc.

We also indicated how matrices can be used in cryptography (code theory).

STUDY GUIDE   You should now be able to demonstrate your knowledge of the following ideas by giving definitions, descriptions, or specific examples. Page references are given in parentheses.

Basic Ideas

Matrix (p. 655)
Square matrix (p. 657)
Addition and subtraction of matrices (p. 657)
Dimension (p. 658)
Equal matrix (p. 659)
Multiplication of matrices (p. 662)
Identity matrix (p. 669)

Multiplication of a matrix by a number (p. 669)
Inverse of a matrix (p. 679)
Elementary row operations (p. 683)
Augmented matrix (p. 684)
Gauss–Jordan method (p. 685)
Cryptography (code theory) (p. 691)
Encoding (p. 692)

Applications Were Given For:

a) business problems (p. 673)
c) communications networks (p. 689)

b) systems of equations (p. 678)
d) cryptography (p. 691)

FORMULAS TO REMEMBER   The following list summarizes the formulas given in this chapter.

1. When adding matrices, they must be of the same dimension.

2. When multiplying a matrix that is $m \times n$ with a matrix that is $n \times p$, the resulting matrix will be $m \times p$.

MASTERY TESTS

Form A

1. If $A = \begin{pmatrix} 1 & 7 \\ 6 & 9 \\ 3 & -2 \end{pmatrix}$ and $B = \begin{pmatrix} 1 & 0 \\ 0 & 1 \end{pmatrix}$, find $A \cdot B$.

2. If $\begin{pmatrix} 25 & 16 \\ 18 & 40 \end{pmatrix} = \begin{pmatrix} 5x + 10 & 16 \\ 2y + 5 & 40 \end{pmatrix}$, find x and y.

3. If $A = \begin{pmatrix} 4 & 7 \\ 3 & -9 \end{pmatrix}$ find A^{-1}.

4. For the system of equations $\begin{cases} 4x + 3y = 17 \\ 2x - 9y = 12 \end{cases}$ find the augmented matrix.

5. If $A = \begin{pmatrix} 3 & 9 & -2 \\ 7 & -6 & 1 \\ 4 & 0 & 5 \end{pmatrix}$ and $B = \begin{pmatrix} 9 & -3 & 5 \\ 0 & 2 & 6 \\ 7 & 1 & -4 \end{pmatrix}$ find $3A - 4B$.

6. Write a linear system to correspond to the following matrix:

$$\begin{pmatrix} 7 & 12 & | & 5 \\ 3 & -4 & | & 11 \end{pmatrix}$$

7. If $A = \begin{pmatrix} 1 & 7 \\ 1 & 7 \end{pmatrix}$ find A^{-1}.

8. If $A = \begin{pmatrix} 5 & 3 \\ -9 & 7 \end{pmatrix}$ find A^2.

9. *True or False?* In any communications matrix, the diagonal from upper left to lower right has only l's.

10. If $A = \begin{pmatrix} 4 & 6 \\ -3 & 5 \end{pmatrix}$, find

    a) A^{-1}

    b) $(A^{-1})^{-1}$. What can you conclude?

Form B   **1.** If $A = \begin{pmatrix} 5 & -9 \\ 6 & 1 \end{pmatrix}$, $B = \begin{pmatrix} -1 & 6 \\ 4 & 3 \end{pmatrix}$, and $C = \begin{pmatrix} 0 & 8 \\ 9 & -7 \end{pmatrix}$, verify that the distributive law, $A(B + C) = A \cdot B + A \cdot C$, holds for these matrices.

2. *Construction Costs.* A construction firm hired 7 plumbers and 8 helpers for a day to do a certain job. The total cost was $1190. At the same rate of pay, the construction firm hired 5 plumbers and 9 helpers for a day and paid $1080. How much does a plumber and how much does a helper earn each day? (Use only matrix techniques to solve the problem.)

3. *Phone Booth.* A telephone company employee has just removed the 235 coins from the coin box of a public pay phone. The value of the coins (which consisted of only nickels and dimes) amounted to $20.25. How many nickels and how many dimes were in the coin box? (Use only matrix techniques to solve the problem.)

4. An insurance salesperson is analyzing the marital status of the 450 females of the George Corporation. Some information about them is summarized in the accompanying matrix.

Marital status

		Single	Divorced	Widowed
	19–25	27	36	7
Age	26–34	31	38	27
(in years)	35–44	16	81	43
	45–60	21	65	58

If the salesperson wishes to sell life insurance to only single women, how many women does the salesperson have to interview?

5. Dr. Jones and Dr. Smith are both surgeons. During 1985, they performed the following operations:

$$\begin{array}{c} & \text{Appendectomy} & \text{Hernia} \\ \text{Dr. Jones} \\ \text{Dr. Smith} & \begin{pmatrix} 84 & 63 \\ 58 & 73 \end{pmatrix} = A \end{array}$$

Furthermore, the charge for each operation was

$$\begin{array}{c} & \text{Cost for operation} \\ \text{Appendectomy} \\ \text{Hernia} & \begin{pmatrix} \$1600 \\ 1000 \end{pmatrix} = B \end{array}$$

Find $A \cdot B$ and interpret the results.

6. Using the coding scheme given in Example 4 of Section 12.7, code the following message:

THE-MICROFILM-IS-IN-THE-CAMERA.

7. *Women's Gains.* Consider the following table[2] listing the number of elective jobs held by women.

Job	1974	1978
Members of Congress	16	20
Governors	0	2
Lieutenant governors	0	3
Secretaries of state	7	11
State treasurers	8	8
State legislators	466	702
Mayors	566	735
County commissioners	456	660
Municipal council members	5365	6961

a) Rewrite the information in matrix form.

b) Compute the total number of elective jobs held by women in 1974 and in 1978. Compare the figures and comment.

8. On January 10, Molly bought 120 shares of BIM stock, 85 shares of TAT stock and 175 shares of ACME stock. Mike bought 95, 150, and 290 shares of these stocks, respectively. BIM sold at $60 a share, TAT at $78 a share, and ACME at $52 a share. Use matrix multiplication to find the amount of money spent by Molly and Mike on these stocks.

9. *Cost Averaging.* Refer back to Exercise 8. To average their costs, Molly and Mike buy the same amounts of these stocks on February 10. However,

---

2. *Source:* Center for the American Woman and Politics.

the prices are now $55 a share for BIM, $79 a share for TAT, and $50 a share for ACME.

a) Using matrix multiplication, find the amount of money spent by Molly and Mike on these stocks.

b) How much was spent by Molly and Mike for these stocks over the two-month period? (Use matrix techniques.)

10. *Legal Costs.* Valerie is a lawyer. Over a four-month period she reported handling the number of cases shown in the accompanying matrix. Furthermore, Valerie charged $175 for each uncontested divorce, $85 for writing a will, $475 for a simple house closing, and $125 for each personal bankruptcy.

	Uncontested divorces	Writing wills	House closings	Personal bankruptcy
Month 1	3	6	5	1
Month 2	2	7	3	0
Month 3	7	4	2	2
Month 4	4	6	4	1

Using matrix multiplication, find Valerie's income from each category for each month. What is Valerie's total income for this period?

13

THE NATURE OF COMPUTERS

CHAPTER OBJECTIVES

To learn how the computer has evolved to its present state. (*Section 13.2*)

To discuss what a computer is, what it can do, and what its limitations are. (*Section 13.3*)

To indicate several different programming languages in which computer programs can be written. (*Section 13.4*)

To describe how a computer actually works by analyzing its various parts. (*Section 13.5*)

To point out what personal computers are and why they are gaining in popularity. (*Section 13.6*)

Suspect Arrested

NEW YORK: The Federal Bureau of Investigation announced today the arrest of a suspect in the recent burglary of a Brooklyn bank. Extensive computer traces of fingerprints and criminal records by the National Crime and Information Center of the FBI in Washington led to the arrest. The name of the suspect was not disclosed pending further investigation.

THE BULLETIN, March 11, 1985

Computer Goof

BERGENVILLE—Bank officials were at a loss to explain how a $100,000 mistake went unnoticed for several days. According to Al Carlton, President of County Savings Bank, an unidentified depositor closed his savings account last week. The computer inadvertently made out a check in the amount of $100,000 instead of $1000. The mistake went unnoticed until a routine audit disclosed the mistake. In the meantime, the check had been cashed. The unidentified depositor has since moved out of town without leaving a forwarding address.

THE DAILY, August 20, 1984

Computers are becoming more a part of our lives with each passing day. When used properly, they can be extremely helpful in such areas as criminal investigations. This is shown by the first newspaper article. When used improperly, they can produce great monetary losses as illustrated by the second newspaper article. Computers can also be used in a variety of other ways both in the scientific field and in the general business field.

Since a computer is only a huge piece of metal, how does the computer do it? Can a computer think? Will computers replace people?

13.1

INTRODUCTION

We live in a computer age. The computer is a very important work machine. At home, as well as at the office or in school, personal computers are appearing in greater numbers and are being used for a variety of purposes such as word processing, record keeping, accounting, and other workaday tasks. Additionally, computer games have become very popular. Although a good deal of the initial home computer use was generated by games, the widespread availability of personal computers today and the supporting software allow the user to complete office or school work at home.

Computers have been on the market for only a little more than 35 years; nevertheless, in this short time they have become an integral part of our society and our lives. Computers are not electronic beasts as some people claim but are rather very efficient servants of humans (when used properly). As is true of learning to drive a car, operating a computer (mainframe or personal) is an easy task, once you learn how to navigate the keyboard, how to negotiate the disc operating system (DOS), what each applications program can do, and which key gets it to do those things. Computer manufacturers have tried to make computers more "user friendly." However, the newcomer to computers can look forward to a long period of study and trial-and-error.

Our objective in this chapter is to help you become computer literate and to give you an overview of what a computer is and what it is not. After presenting some historical facts about computers, we will discuss how a computer works.

In the next chapter we will actually learn how to do some programming in the BASIC programming language.

13.2

THE HISTORY OF COMPUTERS

Ever since people learned how to count, they have been looking for ways to make calculations easier. One of the earliest devices invented for this purpose was the **abacus,** or **counting board** (see Figure 13.1).

Computations using the abacus are performed by moving the beads back and forth along the rods. Although the abacus can be used to perform many calculations quickly, it requires a great deal of skill

Figure 13.1 An abacus

Courtesy of IBM.

to operate properly. For example, if we add 5 and 8, we get 3 in the "ones" column and must "carry" a 1 to the "tens" column. The abacus cannot do this mechanically. The user must move a bead on the next (tens) rod by hand. This may be one reason why the abacus was not accepted by Western merchants who often had to add large columns of numbers.

The problems encountered in using the abacus led to the invention of other calculating devices in the seventeenth century. One such invention was Napier's bones, which were used to perform multiplication. See the discussion on page 128. The first mechanical adding machine was invented by Blaise Pascal. His adding machine was similar to many modern-day inexpensive desk calculators. The user had to enter each number on a dial by hand and then pull a handle to register the number. This had to be repeated for each number entered. This was a slow and time-consuming process, and various improvements were made on it in the years that followed.

About 150 years later, the Englishman Charles Babbage (1791–1871) invented his "difference engine," which was an automatic calculator capable of doing calculations with numbers of up to 20 decimal places. Although Babbage obtained the financial backing of the British government, his project was not completely successful because the technology of the time had not advanced sufficiently to produce the parts needed for his machine.

In 1833, Babbage began to work on the "analytical engine," which would have the capacity to read data from punched cards, and was to be powered by steam. It also was to have a "memory unit" and a unit for doing arithmetic operations. The results were to be printed out. These ideas are all part of the structure of our modern-day computers. Unfortunately, Babbage did not succeed because neither the funds nor the technology to complete the project was available.

The first major advance in computers came about as a direct result of the 1890 United States Census. It had taken the U.S. Census Bureau almost nine years to tabulate the results of the 1880 census. The population of the United States had grown considerably in the years from 1880 to 1890, and it was rather obvious that considerably more time would be needed to process the 1890 results. At the time, Herman Hollerith (1860–1927) was working in the U.S. Patent Office. Hollerith knew that Joseph Marie Jacquard had invented a weaving loom that used punched cards. In 1804, Jacquard had built a loom capable of producing the most complicated designs and controlled by means of punched cards (see the photograph on p. 146). "Instructions" to the machine were punched on a card, which then wove the appropriate design. Hollerith was convinced that punched cards could be used to enter numbers into an adding machine. This would make the tallying of the census figures much easier.

Pascal's adding machine. In 1642 the French philosopher Blaise Pascal, 19 years old and tired of totaling up figures for his tax-collecting father, invented this fancy machine for adding and subtracting. Its cylinders and gears were housed in a small box. The wheels on top of this box corresponded to units, 10s, 100s, and so on. Each wheel could register the digits 0 through 9. *Courtesy of IBM.*

Charles Babbage, a nineteenth-century English inventor, designed the difference engine to calculate and print mathematical tables. *The Bettmann Archive*

Babbage's multiplier. The computing element of Babbage's complex machine for multiplying was a series of toothed wheels on shafts. They worked like the wheels of a modern mileage indicator. *Courtesy of IBM.*

An early keypunch machine.
Courtesy of IBM.

Several years later, Hollerith did indeed invent such a machine. He invented the first **automatic data-processing machine.** In 1896, Hollerith started the Tabulating Machine Corporation. This company was later to become the International Business Machines Corporation, commonly known as IBM. With the help of Hollerith's machine the 1890 census was completed in three years.

In 1944, Howard Aiken and IBM completed the MARK I computer. It was the first completely automatic electromechanical computer. With it, computations were carried out by means of electrical impulses instead of by mechanical devices such as rotating wheels. In 1946 the ENIAC (Electronic Numerical Integrator and Calculator) computer was completed by the Remington Rand Corporation. Unlike the MARK I computer, the ENIAC used vacuum tubes to replace electric relays. This was the first purely electronic computer. It contained 18,000 vacuum tubes and was capable of adding 5000 numbers in one second. Unfortunately, neither the MARK I nor the ENIAC computer was capable of storing instructions internally. The instructions had to be read into the computer one at a time by means of punched cards or paper tapes.

Historically, most of the rapid changes that took place in computer technology occurred around the time of World War II. The military demand of the Army and Navy required accurate and often time-consuming computations.

In 1946, John Von Neumann (1903–1957), who had worked on the atomic bomb, began work on the EDVAC computer. When completed in 1952, it became the first computer capable of operating at electronic speeds and of storing its programs internally. This important advance was a milestone in the development of the computer. It meant that a problem could be placed into the computer and left there. Because the computer can store the information and instructons necessary to process the program, the programmer can do something else while the program is being run. The operation of the computer need not be supervised.

An example of a computer with less ability to store information internally is the hand-held calculator, which was discussed in an earlier chapter. Suppose you want to perform the calculation $[(2 + 3) - 1] \times 7$ using a calculator. You cannot just enter these numbers with one push of a button and expect the calculator to solve the problem. You must enter each number and operation by hand at the appropriate time.

Since World War II, technological advances have come so fast that computer professionals often classify computers as belonging to a particular "generation." All of the early computers were called **first-generation computers.** They all used vacuum tubes and performed calculations in a few milliseconds (thousandths of a second). Examples of first-generation computers are the IBM 650 and the business-oriented version of the ENIAC, developed by Remington Rand and called the UNIVAC.

Although transistors were invented in 1948, the first transistorized computer did not appear on the market until 1954. Computers built with transistors instead of vacuum tubes are called **second-generation computers.** These computers perform calculations in a few microseconds (millionths of a second). The IBM 1410 is an example of a second-generation computer.

Beginning in 1965, **third-generation computers** appeared on the market. These use groups of transistors and integrated circuits and operate at such a speed (billionths of a second) that they are capable of working on several programs at the same time. The IBM 360 is an example of a third-generation computer. Third-generation computers that use integrated circuits are capable of performing more than 500,000 calculations in a second. This immense speed enables them to work on several programs at the same time through time-sharing plans.

Computer progress was further advanced with the development of **Monolithic Systems Technology,** which made it possible for computers to be built with microscopic logic circuits that are capable of performing over a billion computations per second. Indeed, in 1971 a miniature computer (as opposed to the original 30-ton ENIAC computer of 1946) utilizing a microprocessor was invented. This invention has opened the way for computers that are much smaller but just as efficient.

Many people believe that these new computers, which use microprocessors and very large scale miniaturized circuits, actually represent the beginnings of **fourth-generation computers.** These computers are capable of processing 100 million to 1 billion instructions per second.

The United States and Japan are currently in a technological race to be the first country to develop an **advanced mainframe supercomputer** that would run 1000 times as fast as our present-day supercomputer, which runs about 500 million operations per second in a short burst. Today, supercomputers are used in designing aircraft, in geological expeditions for oil, and even in designing circuits for other computers. Currently, the CRAY X-MP computer built by the Cray Research Company is one of the fastest and most powerful supercomputers in the world.

Today, a large high-speed computer is capable of doing far more work than most users actually need. Moreover, such computers are costly, and often a company cannot afford to buy or rent such a machine. It is particularly costly and wasteful if the machine is left unused for large periods of time. Therefore many companies do not

At the computer exhibition at the Sheraton Centre in New York City, an inventor helps his computerized mouse navigate a maze. *Edward Hausner/NYT Pictures*

buy or rent their own computers. Instead they buy time on another computer. This is called **time-sharing.** In time-sharing, a company has a *terminal* that looks like a large typewriter. The terminal is connected (by telephone or other means) to a computer that may be many miles away from the terminal. The user types programs on the terminal and receives results back the same way. The user pays only for the time that the computer is used or the time that he or she wants it available for use. Because of the computer's tremendous capacity, many different terminals can be hooked up to the same computer. With such an arrangement the computer can work on several programs at the same time.

In the future, we can expect increased use of the computer. It is even possible that every home will actually have its own terminal hooked up to a large computer.

EXERCISES

1. What were some of the disadvantages of the early first-generation computers compared to our modern fourth-generation computers?

2. Who invented the first mechanical adding machine?

3. What is the significance of the EDVAC computer?

4. What are some of the advantages in using a computer built with transistors instead of vacuum tubes?

5. Why would a company consider time-sharing?

6. Who invented the first automatic data-processing machine?

7. Refer back to question 6. Why was this invention so significant?

8. In what ways have computing devices changed over the years?

9. What was the earliest device people used to help them calculate? What were some of the advantages and disadvantages of this device?

10. Why have some people spent time and money building new and more complex devices for calculations?

13.3
COMPUTER USES

Computers have been on the market for about 35 years. In this short period of time they have already become a necessity for many businesses, schools, hospitals, and government agencies. This is because of the computer's wide-ranging capabilities, as we shall see shortly. Even small companies today are finding it harder and harder to compete in business unless they have access to a computer. They do not necessarily have to own a computer outright. They can have a terminal installed in their offices on some time-sharing plan. Present predictions are that more than three and a half billion time-sharing terminals will be in use by 1990 alone.

In the next few paragraphs we will discuss some specific situations in which the computer has been applied.

Identifying the Hit-and-Run Driver

An old man was injured at the intersection of Broadway and Main Street by a car that went through a red light. The driver did not stop. Witnesses at the scene told the police that the car was either a 1980 or 1981 green Oldsmobile Cutlass with a vinyl top. The last three letters of the license plate were KJP.

The police department relays this information to the State Motor Vehicle Bureau where all car registration records are stored. Through the computer, these records are searched to find all the registrations whose last license plate letters are KJP. Then this list is further narrowed down by car make, model, year, and color. Finally, the computer prints out a list of 45 addresses of owners whose cars fit the description given. The speed with which the Motor Vehicle Bureau computers can make such information available often allows the police department to track down a hit-and-run driver within hours of an accident. Without the aid of a computer it would take days, perhaps even weeks, to search all the registrations in the state. By that time the driver could easily have escaped or disposed of the car. Police departments, the FBI, and various government agencies use computers to search through their records to obtain essential information quickly.

Election Predictions

In November 1952, people watching the election-night returns were informed shortly after the polls closed and long before all the ballots were counted that the next president would be Eisenhower. The television newscasters had used an IBM computer to project winners. This was the first time that a computer had been used to do this.

On the basis of careful statistical analysis, samples of returns from key districts were fed into the computer. The computer had been programmed to analyze these returns and to make projections based on them. This technique was so successful that it was not long before this became a standard procedure on election night. In 1960 an IBM RAMAL computer predicted victory for Kennedy at 8:12 P.M. on election night. Since then, predictions of victory before the closing of all the polls nationwide have become commonplace, leading many to lobby for change in the electoral process.

Computer-Assisted Instruction

Today there is considerable interest in computer-assisted instruction. In this method of teaching, the student works at a terminal and responds to questions asked by the machine. If the student answers a question correctly, the machine will tell the student so and give a slightly more difficult problem. Otherwise, the machine will give another problem of the same type to try to solve again.

The computer gives the student the kind of individual attention that would be difficult or impossible for a teacher working with several students at the same time. Because the machine keeps an up-to-date and accurate record of the student's success or failure in handling each concept, it can regulate the rate at which to proceed. Different students, all learning the same material, can proceed at their own pace. Such individualized instruction would not otherwise be available unless each student had a private tutor.

The major disadvantage in using computers for this purpose is that personal communication between the teacher and student is lacking. The student cannot ask the machine any questions.

Simulation and Air Traffic Control

The most important part of any airport is the air traffic control center. From this center the controllers direct air traffic into and out of the airport by means of radar, which makes use of a computer. Traffic controllers are trained for this demanding job by a process called **simulation.** Different traffic patterns and flight situations are randomly generated by computers to which the controller is trained to react.

The flow of automobile traffic on major roadways can also be controlled by means of computers that instantly analyze the traffic patterns and change the traffic signals accordingly.

The building of highways themselves (in fact, the entire analysis of urban transportation) often requires extensive computer analysis. Some of the important questions that can be analyzed by the computer are:

1. How many cars can be allowed on a particular street or bridge at any one time?

2. How many entrance and exit ramps to a highway are required at specific points?

3. What are the maximum and minimum speeds to be allowed on a highway?

Medical Care   In the intensive care units of hospitals, patients are "hooked up" to various devices that monitor their vital signs. This information is displayed visually on a screen either at the patient's bedside or in the nurse's station or both. This enables the nursing staff to see immediately if there has been any change in a patient's condition requiring immediate attention. The doctor can also recall this information from computer memory when it is needed.

   Computers are also used to determine the amounts of medication to be administered in cases where extreme accuracy is necessary. For example, in radiation treatment of cancer patients, too much radiation could be dangerous, and too little could be ineffective. Computers are used to determine the exact amount needed.

Analysis of   Many children are born with birth defects and disorders that can be
Genetic Defects   traced back to defective chromosomes. The chromosomes carry the genes that determine which characteristics are inherited from parents. Thus it is extremely important for researchers to analyze chromosomes and be able to recognize any that are abnormal. This kind of analysis takes time and requires a great deal of training. Because of the speed and accuracy with which computers operate, they are now being used to carry out this research in laboratories throughout the country.

The Internal   When an individual files his or her tax return with the Internal Rev-
Revenue Service   enue Service, the computer takes over. Some of the items that the computer will check are (1) the arithmetic calculations, (2) the use of the correct tax tables, and (3) the itemized deductions. If the computer finds irregularities in any of these, it will automatically flag the return for further review. In addition, the computer has been programmed to randomly select a certain percentage of returns for automatic review.

The Computer   To apply the computer to the arts may seem strange. After all, the
in the Arts   arts express human individuality and creativity. The computer cannot create and has no "mind of its own" at all; it does only what it is told to do. Yet the computer is being used with interesting results in several types of artistic activity. Possibly the most familiar is computer art. Everyone has seen "computer graphics" like the one shown in Figure 13.2.

   A picture like this is created by programming the computer to print a character in certain places and leave blanks in others. Computers can also draw pictures on paper using automatic plotters or on TV screens using cathode-ray tubes to draw lines by means of an electron beam. The program controls the direction of the beam. Computers have also been programmed to design sculptures such as the one shown in Figure 13.3.

Figure 13.2

Figure 13.3
A computer graphic
AT&T Bell Laboratories

Artificial Intelligence   Literature does not seem to be a promising field for the computer, since of all the arts it depends the most on the logical order of the words. To write a properly constructed and meaningful sentence is difficult enough for a human being. How can a computer, which cannot think, do it? For this reason there has been little success up to now in getting computers to write intelligent prose. There has been more success with certain types of poetry where the meaning can be more unstructured. For example, Japanese haiku poems can be written by programming a basic form into the computer and letting it choose the words to fill in at random. For example, we might program in the following structure:

The _____ _____ _____ is not _____
_____ is _____ _____ of the _____
All _____ is _____

and let the computer fill in the blanks at random. The results of this type of "poem" can be very pleasing to both the ear and eye. There are, of course, other methods of "writing" poetry by computer.

In music, computers have been programmed both to write music and to play it. In dance, they have even been programmed to do choreography.

We have only scratched the surface of the vast range of computer applications. You can see from these that computers have become an essential part of our lives. Many people regard a computer as the enemy—it is seen as an impersonal monster that sends incorrect bills and junk mail. It never answers when we write to it. However, computers are not monsters that are going to take over the world and reduce us all to numbers. The are actually very efficient servants of humanity and will perform only functions that human beings program them to do. When properly used, computers can make all our lives easier and better.

EXERCISES

1. Explain how computers are used by many hospitals in CAT scans. (Consult the library if necessary.)

2. Both on-board and land-based computers play an important role in space exploration. Explain how. (Consult the library if necessary.)

3. What is the difference between a supercomputer and a fourth-generation computer?

4. Give three applications of the computer to each of the following areas. (Do not mention any of the applications given in the text.) Write a sentence or two about each application.

    a) Technology and science

    b) Government

    c) National defense

    d) The arts

    e) Business

    f) Medicine

    g) Ecology

    h) Crime control and prevention

    i) Education

13.4
PROGRAMMING LANGUAGES

Before a computer can begin performing functions, it must be told exactly what has to be done. This is accomplished by writing a **program.** A program is a set of instructions telling the computer exactly what has to be done and in what order. This is the job of the **computer programmer** who actually writes the program.

Depending upon the task that is to be done, programs can be written in many different languages, each suited to a particular need. Great care must be exercised when writing computer programs. With the slightest mistake, even a minor spelling mistake, the machine will not be able to proceed. The machine will then send you a **syntax error** message, meaning that you have made a programming error. This will occur no matter what programming language is used. The program then has to be **debugged** to find the error. Often it is a difficult task to find the error.

Computing languages can generally be divided into two main categories: **high-level languages** and **low-level languages.** High-level languages, such as BASIC, FORTRAN, PILOT, Logo, and Pascal, use commands and concepts in a manner that is comparable to everyday human language and thought. No extensive background knowledge of the computer or of programming is needed to use these languages.

On the other hand, the low-level languages, that is assembly language and machine language, require that the user have some background knowledge of the computer. Generally speaking, they take longer to learn. However, the reward is a detailed knowledge of what the computer is actually doing.

Some of the more commonly used programming languages are the following:

1. **BASIC,** which represents the words Beginner's All-purpose Symbolic Instruction Code, was invented by John Kemeny at Dartmouth College. BASIC is used widely by business and personal computer systems because of its simplicity. No prior programming ability is needed to quickly learn and easily use this language. In the next chapter we will discuss some details of BASIC.

2. **PL/1,** which represents the words Programming Language One, is a general-purpose language that can be used by both the experienced and beginning programmer. It is used primarily when working with the large mainframe computer rather than with the personal microcomputer.

3. **COBOL,** which represents the words Common Business Oriented Language, is a business-oriented programming language. As with PL/1, it is used primarily when working with the large mainframe computers.

4. **FORTRAN,** which represents the words Formula Translator, is used primarily in scientific-oriented or mathematical research. Its combination of mathematical simplicity and power accounts for its popularity in solving many scientific and business problems. FORTRAN's weakness lies in its ability to handle input and output, for example, in generating reports or manipulating text.

5. **Pascal** was developed to teach programming as a clear, systematic approach to problem solving. Statements in Pascal are put together much like statements in English.

6. **Logo,** a recently created language, introduces programming to children in an intuitive way. A turtle (a small triangle that moves around the screen) is used to draw different geometric figures. The language is designed to introduce programming concepts in a problem-solving manner.

7. **PILOT** was designed for teachers. It enables the teacher to generate computer-aided instruction lessons, including graphics, for classroom use.

Today there are well over 175 different programming languages in use, and new languages are constantly being developed to suit some particular need.

EXERCISES

1. Why is machine language referred to as a low-level language?

2. When using programming languages, when is debugging necessary?

3. What is the difference between a high-level and a low-level language?

Each of Exercises 4–8 gives an alternative programming language. In each case, state what the abbreviation stands for, and then discuss when the language might be used. (Consult the library if necessary.)

4. RPG
5. ALGOL
6. APL
7. LISP
8. SNOBOL

9. When would a syntax error message appear on a computer monitor?

13.5
HOW A COMPUTER WORKS

Although a computer can perform difficult and time-consuming calculations at lightning speeds, we must realize that it is not capable of doing its own thinking. In this chapter we will discuss only the type of computer called **digital.** Most of the computers produced in the United States today are of this type. There is another type of computer called an **analog computer,** which we will not discuss at all. Such computers are widely used in science.

How does a computer work? All computers must have an **input unit.** The input mechanism enters the data and instructions into the machine. "Input" into the machine can be accomplished by using punched cards, paper tapes, console typewriters, discs, magnetic tapes, or optical scanners such as those used in many supermarkets. Input might even be at a terminal in an office miles away from the computer. If a punched card is used, it is placed into a card reader, and the machine will "read" the information from the card by noting the presence or absence of holes in certain positions. This information is punched on the card using one of the many possible programming languages. The information on the card is then transferred by means of electrical impulses into a **compiler.**

Although we perform all our arithmetic calculations in the decimal system, computers perform these calculations in the binary system (see Chapter 3). With the modern-day computer, it is not necessary for the programmer to convert decimal numbers into binary system numbers. All of this is done by the compiler or coder. The compiler converts numbers and instructions into binary numbers so that the machine can understand them. Before the compiler became part of the computer, programmers had to do all of the translation themselves; that is, they had to type in instructions and data that were already written in the binary system.

After the input information has been translated into a language that the machine can understand, the information is passed on (by way of electrical signals) to the **control unit.** The control unit receives and interprets the instructions that it receives from the input unit. Then the control unit sends these signals to the various other units, called the **arithmetic unit,** the **memory unit,** and the **output unit.** The control unit determines which calculations are necessary for a given problem and in which order these calculations must be performed in order to solve the problem. The control unit also keeps track of the proper sequencing of the mathematical computations performed in the arithmetic unit. The control unit with the processing unit make up the **central processing unit (CPU).**

The **arithmetic unit** performs all the calculations that are needed to solve a given problem. All computations that are performed here are done in the binary system.

The **memory unit** stores all the data, whether a number, an instruction for performing a calculation, or the results of the calculation itself. Information can be stored in several ways. Many computers store information by using magnetic cores. These tiny doughnut-shaped pieces of metal, which are called *bits*, are no larger than a pinhead. The memory unit of a typical machine has thousands of such cores. These cores can be magnetized in one of two directions. When magnetized in one direction, the number 1 is represented, and when magnetized in the other direction the number 0 is represented. With improvements in technology the size of these magnetic cores is constantly decreasing.

As we mentioned earlier, computers use the binary system, in which only the digits 0 and 1 are used. Each 0 or 1 in the binary system is called a **bit.** By themselves, single bits cannot store all of the numbers, letters, and special characters that a computer must process. Instead, the bits are put together in groups usually containing six or eight bits. Such groups are called **bytes.**

The storage capacity of a computer is often expressed in terms of the letter K. A **kilobyte** is K bytes, that is, 1024 bytes. Most microcomputers have between 4K and 64K bytes. Minicomputers have anywhere from 64K bytes to 1 megabyte (millions of bytes), and the large mainframe computers have from 512K bytes up to 16 megabytes and even more.

Many third-generation computers store information using cores grouped into small **chips** that are no larger than one-eighth of an inch on each side. There are 256 cores on a typical chip. Such storage ability greatly increases the capabilities of the machine. The more storage places a memory unit has, the more information the computer can store. The size of these magnetic cores, which are usually grouped into small *chips,* has been constantly decreasing, thereby increasing the memory capabilities of the machine.

Some computers store information by using **magnetic drums** or **discs.** Such drums are usually about one foot in diameter. Information is stored in these drums by magnetizing certain spots on the drums or discs.

After the computer solves the problem, the control unit sends the solution to the **output unit.** Before reaching the output unit, the solution is decoded from machine language back into the programming language being used. The output unit may give the answer by printing the result on paper, by punching holes in a card, by displaying it visually on a video screen, or by plotting a graph. These all depend on how the machine has been programmed.

In Figure 13.4 we show the parts of a computer and how they interact with each other. The physical parts of a computer are referred to as the **hardware.** This includes the machine itself, the terminals, the typewriters, card readers and printers, magnetic tapes and discs, etc. On the other hand, the programming techniques and the actual programs are referred to as the **software.** In recent years there has been remarkable progress in both the hardware and software areas of the computer industry.

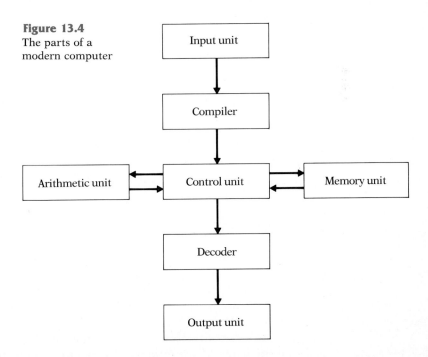

Figure 13.4
The parts of a modern computer

EXERCISES

1. How does the control unit or central processing unit (CPU) get its input?

2. Which part of the computer is in charge of arithmetic, logic, decision-making, and the like?

3. Name several ways in which a computer can receive information or instructions.

4. What is the significance of having a computer with internal storage capabilities?

5. Why might the CPU be called a *micro-processor*?

6. Which part of the computer can be called the "brain" of the computer? Why?

7. How does the computer let a human see the results of its calculations?

8. What do card readers, disc drives, even telephone hook-ups (modems) all have in common?

9. What is the difference between a digital and an analog computer? (Consult the library if necessary.)

13.6

USING PERSONAL COMPUTERS

More and more people are purchasing personal computers these days than ever before. Such **microcomputers,** as they are called, have microprocessors for their CPU (central processing unit) and usually have memory of 1024K or less. People use these computers for a multitude of functions. Some of these are the following:

a) Business accounting

b) Home accounting

c) Writing programs to sell or use

d) Playing games

e) Education (e.g., spelling, math, vocabulary)

f) Helping learn more about computers

g) Learn programming

h) Word processing

i) Filing/record keeping

j) Technical calculations

k) Telecommunication linking to other computer information systems.

Recent surveys show that the most frequent application for computers at home is word-processing, and not game-playing. A personal computer combined with a word processor allows the user to compose documents on a screen. A document can be reorganized, revised, and corrected as the user moves along. More sophisticated word processing programs designed for business and professional use even allow the user to combine several documents into one or to produce fancy printing and layouts. Some word-processing programs even automatically check for spelling or typing errors against a dictionary of words stored by the program.

All personal computer systems (no matter which brand) consist of certain basic components called **hardware.** Some computer systems may have more components than others, but the most important parts are the following:

1. **Microprocessor.** The microprocessor carries out sequences of instructions (called programs) stored in an area of the computer's main memory.

2. **Main memory.** This is where a computer stores programs while it is running them. Whenever only a portion of the main memory (that is running a program) is being used, then the remaining portion is available for storing information entered at the keyboard or other information that the computer is working on.

3. **The keyboard.** The keyboard allows the user to send information to the computer. The user can type text or push certain control characters.

Additionally, to make the computer useful, various input/output devices must be connected to the computer to complete the system. These are called the **peripherals** of the system and include the following:

4. **A computer monitor or display device.** A monitor enables the computer to convey information to the user either in words or numbers (text) or in pictures (graphics) on a screen.

5. **A disc drive.** A disc drive reads and writes information on a magnetic disc in much the same way that a tape recorder can record and play back music. Information saved on a disc can be recalled and loaded back into main memory. Otherwise, when a computer is turned off, everything in main memory is lost.

6. **Flexible (floppy) or hard discs.** A disc is similar to tape used with a tape recorder. Some discs are completely blank, and the user stores any information created on them. Others have prerecorded programs on them. These are usually purchased in a computer supply store.

Depending upon the computer's microprocessor, a computer needs a master control program that synchronizes the execution of computer programs. This is known as the computer's **operating system.** Different computers use different operating systems. Often a computer that has been designed for use with one operating system can be made to run on a different operating system. This can be accomplished by purchasing a special circuit board. It is important to know in advance the type of operating system that a computer has, since not all of the thousands of commercially available programs run on all operating systems.

Regardless of the brand of computer used, one usually needs a **printer** to transfer information from the computer to paper. Often, in addition to hardware incompatability, there is at least partial incompatability between a printer and the programs being used. The computer program simply does not understand the printer's codes.

Printers can be used to generate graphics. **Computer graphics** can include drawings, maps, photo-images, graphs, and shapes with varying details. To accomplish this, a printer must be able to produce tiny dots on the paper within close tolerances. These tiny dots are created when the print head fires one pin at a time.

A computer printer can be of the **dot-matrix** type. A printer of this type has a print head consisting of one or more vertical rows of tiny pins that strike an inked ribbon against the paper as the print head moves back and forth. Such printers can create letters of any shape or graphics such as charts and maps of any kind.

Computer printers can also be of the **daisy-wheel** type. In this type of printer a wheel spins until the desired letter is in position. A hammer then strikes it against the ribbon, which in turn strikes the paper. Generally speaking, the characters produced by a daisy-wheel printer are of much better quality than those of the dot-matrix printer, though the daisy-wheel printer cannot produce graphics. Today there are even laser printers.

If you determine that you need a personal computer, then you must decide in advance what you will use it for. In turn, this will help you decide what type of computer to buy, what type of printer and/or peripherals to look for, and what software or commercially available programs are needed.

EXERCISES

1. What is the function of a microprocessor?

2. What is the difference between a dot-matrix computer printer and a daisy-wheel computer printer?

3. Why is it necessary to have an operating system for a personal computer?

4. Personal computers are often used for data base management and for electronic spreadsheets. Explain how personal computers accomplish this. (Use a library to obtain additional information if necessary.)

5. A **modem** is a device that enables a user to "hook up" with another computer through a telephone connection. Explain the advantages of using a modem.

6. Computer "hackers" are individuals who use their personal computers to illegally obtain information from another computer's memory.

Can you explain how this is accomplished and what can be done to prevent it?

13.7

SUMMARY

In this chapter we traced the development of the computer from its infancy to its present stage. We discussed numerous ways in which the computer is used today in a variety of fields. We also discussed how a computer works from both a hardware and a software point of view.

Since computers are fast becoming part of our normal life, it is important to know how a computer works, its capabilities, and its limitations. Our discussion was not intended to make expert programmers out of anybody. Rather, it was intended to give you an appreciation of what a computer is.

Finally, we indicated what a personal computer is and the parts that are needed to use it efficiently.

STUDY GUIDE

You should now be able to demonstrate your knowledge of the following ideas presented in the chapter by giving definitions, descriptions, or specific examples. Page references (in parentheses) are given for each term so that you can check your answer.

Abacus (p. 703)
Counting board (p. 703)
Automatic data processing machines (p. 706)
First-generation computers (p. 706)
Second-generation computers (p. 707)
Third-generation computers (p. 707)
Monolithic Systems Technology (p. 707)
Fourth-generation computers (p. 707)
Advance mainframe supercomputers (p. 707)
Time-sharing (p. 708)
Simulation (p. 710)
Program (p. 713)
Computer programmer (p. 713)
Syntax error (p. 713)
Debugged (p. 713)
High-level computer languages (p. 713)

Low-level computer languages (p. 713)
BASIC (p. 714)
PL/1 (p. 714)
COBOL (p. 714)
FORTRAN (p. 714)
Pascal (p. 714)
Logo (p. 714)
PILOT (p. 714)
Digital computer (p. 715)
Analog computer (p. 715)
Input unit (p. 715)
Compiler (p. 715)
Control unit or central processing unit (CPU) (p. 716)
Arithmetic unit (p. 716)
Memory unit (p. 716)
Output unit (p. 716)
Bits (p. 716)
Bytes (p. 716)
Kilobytes (p. 716)
Chips (p. 716)

Magnetic drums (p. 717)
Discs (p. 717)
Hardware (p. 717)
Software (p. 717)
Personal computers (p. 718)
Microcomputers (p. 718)
Microprocessor (p. 719)
Main memory (p. 719)
Keyboard (p. 719)
Peripherals (p. 719)

Computer monitor or display device
  (p. 719)
Disc drive (p. 719)
Flexible (floppy) or hard discs
  (p. 719)
Operating systems (p. 719)
Computer printers (p. 720)
Computer graphics (p. 720)
Dot-matrix printer (p. 720)
Daisy-wheel printer (p. 720)

MASTERY TESTS

Form A

1. Computers that use(d) vacuum tubes are called
   a) first-generation computers   **b)** second-generation computers
   c) third-generation computers   **d)** fourth-generation computers
   e) none of these

2. What is time-sharing?

3. What do the letters of the word BASIC represent?

4. If a computer sends you a message that you have made a programming error, then this is called a
   a) syntax error   **b)** debugging error   **c)** simulation error
   d) data base error   **e)** none of these

5. A set of instructions telling the computer exactly what has to be done and in what order is known as a
   a) compiler   **b)** decoder   **c)** program   **d)** translator
   e) none of these

6. A computer language that uses commands and concepts in a manner that is comparable to everyday language and thought is called
   a) assembly language   **b)** machine language
   c) low-level language   **d)** high-level language
   e) none of these

7. Which of the following computer languages is used primarily in scientific-oriented or mathematical research?
   a) Pascal   **b)** BASIC   **c)** COBOL   **d)** FORTRAN
   e) none of these

8. A computer language designed for teachers and used to generate computer-aided instruction lessons is
   a) Logo   **b)** PILOT   **c)** PL/1   **d)** FORTRAN
   e) none of these

9. The part of the computer that receives and interprets the instructions that it receives from the input unit is called the

   a) control unit   **b)** memory unit   **c)** arithmetic unit

   d) peripheral unit   **e)** none of these

10. The physical parts of a computer are referred to as the

    a) hardware   **b)** software   **c)** peripherals

    d) hard discs   **e)** none of these

Form B

1. A personal computer is an example of a

   a) mainframe computer   **b)** supercomputer

   c) first-generation computer   **d)** microcomputer

   e) none of these

2. Which part of a personal computer carries out sequences of instructions stored in an area of the computer's memory?

   a) main memory   **b)** computer monitor

   c) microprocessor   **d)** keyboard   **e)** none of these

3. A computer programmer can send information to the computer with a

   a) monitor   **b)** keyboard   **c)** microprocessor

   d) decoder   **e)** none of these

4. A computer's master control program that synchronizes the execution of the computer program is called the

   a) operating system   **b)** disc drive   **c)** main memory

   d) peripheral   **e)** none of these

5. A computer printer that has a print head consisting of one or more vertical rows of tiny pins is called

   a) thermal   **b)** dot-matrix   **c)** daisy-wheel

   d) continuous roller-feed   **e)** none of these

6. When a computer is turned off, everything in main memory is

   a) temporarily lost but will reappear when the computer is turned on again   **b)** partially lost   **c)** completely lost

   d) automatically transferred to a blank disc in the disc drive

   e) none of these

7. A person who actually writes computer programs is called a

   a) systems analyst   **b)** keypunch operator

   c) computer programmer   **d)** compositor   **e)** none of these

8. The first computer built that was capable of operating at electronic speeds and of storing its programs internally was the

   a) EDVAC   **b)** ENIAC   **c)** MARK I   **d)** UNIVAC

   e) none of these

9. Which generation of computers use microprocessors and very large scale miniaturized circuits?

    a) first   **b)** second   **c)** third   **d)** fourth   **e)** none of these

10. Which of the following computer parts is in charge of performing a process?

    a) keyboard   **b)** screen or monitor   **c)** printer   **d)** CPU

    e) none of these

14

INTRODUCTION TO BASIC

CHAPTER OBJECTIVES

To indicate what computer programs are and why they are needed. (*Section 14.2*)

To analyze what flowcharts are and how they are used. (*Section 14.3*)

To study programming and the computer language of BASIC, which was developed by John Kemeny. (*Section 14.4*)

To learn how to actually write computer programs using BASIC. (*Section 14.4*)

To apply the ideas of computer programs to solve several simple problems. (*Section 14.4*)

725

Computer Crime up Again

WASHINGTON—Government officials announced yesterday the establishment of a special task force to combat the increase in computer crime. Under the scheme, some shady employees were bilking banks out of millions of dollars by programming the computer to register loans to various customers when such transactions never actually occurred. The employees would then pocket the money.

When the customers complained, the computers were then programmed to transfer the loan to a second customer, and so on.

The task force will thoroughly investigate the security system that is being used by the banks and will recommend some restrictions on who has access to the programming of such transactions on a computer.

COMPUTER NEWS, February 13, 1985

When used effectively, computers can be great servants as they perform tasks at lightning speeds. Nevertheless, as the newspaper article indicates, some people program the computer to do the wrong thing. Thus they can make the computer an accomplice to their embezzlement schemes.

14.1
INTRODUCTION

To communicate with a computer, we must use one of the many computer languages available. In this chapter we concentrate on a very popular conversational language called BASIC. This language consists of short, easy-to-learn commands that are quite similar to everyday English. Thus many people can learn to use this language quickly, since no prior computing background knowledge is needed. One can then write computer programs for almost any personal computer. This explains the popularity of the BASIC language.

In this chapter we discuss in detail the BASIC computer language.

14.2
PROGRAMMING A COMPUTER: BASIC

As we mentioned in the preceding chapter, if you want a computer to do a job for you, it must be given detailed, step-by-step instructions. If one step is left out, then the machine cannot do the job properly. These instructions are contained in the **computer program**, which is fed into the machine.

A computer program can be written in many different languages. Some of the more common programming languages are FORTRAN, BASIC, PL/1, COBOL, ALGOL, and APL. Each of these languages may be used for different applications. For example, COBOL is commonly used in business. In scientific applications, programs are often written in ALGOL. BASIC is a popular language frequently used with personal computers or terminals. It is a very simple language to learn and is convenient for people who do not intend to use the computer extensively. It gives them quick access to the personal computer or terminal without their having to spend a considerable amount of time learning one of the more complicated languages. FORTRAN (an abbreviation of FORmula TRANslator) is a general-purpose language and can be used in many different areas. BASIC is perhaps the most widely used of computer languages; for this reason it is the language that we will discuss here.

The BASIC (Beginner's All-purpose Symbolic Instruction Code) language was developed by John Kemeny at Dartmouth College during the 1960s.[1] BASIC is easy to learn because it is similar in many respects to ordinary English and mathematics. It should be pointed out that our discussion is not intended to be a complete and detailed analysis of the BASIC language but is merely intended to give you a feeling for the kind of detail that is necessary to get a program to run on a machine. A more complete discussion can be found in any manual on BASIC.

14.3
FLOWCHARTS

Often the number of steps involved in writing a program is lengthy and quite complicated. In such cases it is usually advisable to first draw a *flowchart*. This is a diagram indicating the logical sequence of steps to be followed by the computer; that is, flowcharts show us what is to be done and in what order. They also indicate when decisions have to be made. Flowcharts can help us analyze programs, since they provide a visual insight into what is happening.

When drawing a flowchart, it is standard to use circles (or ovals) to indicate the beginning or end of a program and rectangles to indicate the processing of a program. Input/output is usually indicated by using the parallelogram symbol. We also use the diamond symbol to indicate that a decision must be made. These symbols are illustrated in Table 14.1.

---

1. Recently published are new "structured" forms of **BASIC** such as *TRUE BASIC* by Kemeny and Kurtz.

TABLE 14.1

Symbols commonly used when drawing flowcharts

Symbol	When used
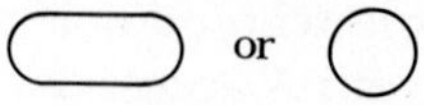 ⬭ or ◯	to indicate the beginning or end of a program
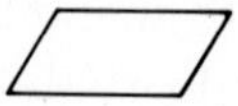 ▱	to indicate input or output of data
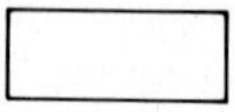 ▭	to indicate the processing of the program
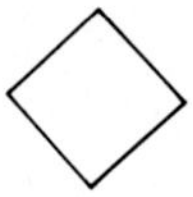 ◇	to indicate that a decision must be made and to show the branching to different alternatives depending on the decision made
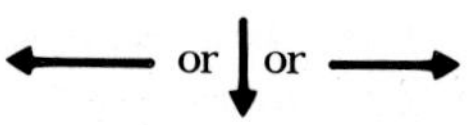 ← or ↓ or →	to indicate the direction to go in the flowchart to find the next step

Let us construct flowcharts for several different situations.

EXAMPLE 1   Construct a flowchart to compute the area A of a triangle whose base B is 8 cm and whose height H is 12 cm.

SOLUTION   This is a rather simple program since no decisions must be made. The flowchart is shown in Figure 14.1.

A program written from this flowchart will tell the computer to find the area by multiplying one-half of the base, 8, by the height,

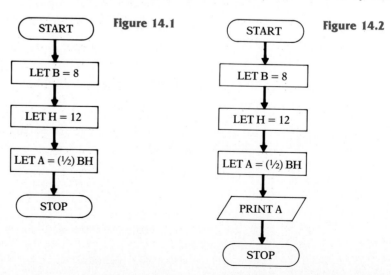

Figure 14.1

START → LET B = 8 → LET H = 12 → LET A = ($\frac{1}{2}$) BH → STOP

Figure 14.2

START → LET B = 8 → LET H = 12 → LET A = ($\frac{1}{2}$) BH → PRINT A → STOP

12, obtaining 48. However, since the computer cannot talk to us, how are we to be told the answer? Obviously, we must tell the computer to print out the answer that it obtains. The flowchart for this problem would now be as shown in Figure 14.2. ☐

Comment When using flowcharts, we follow the arrows from step to step.

EXAMPLE 2 *Consumer Price Index.* The Consumer Affairs Department of New York City keeps accurate records on the prices of different items. It then publishes a weekly cost-of-living index. It usually has its shoppers purchase the same items on a weekly basis in local supermarkets. It then computes the average cost of these items. Assume that the shoppers have purchased 25 different items. Write a flowchart to help the department find the average cost of these items.

SOLUTION To calculate the mean (average), we must first tell the computer what the prices are. This is accomplished by a READ statement, such as READ $A_1, A_2, A_3, \ldots, A_{25}$, where the A_i values are the prices of the items. This READ instruction gets the numbers into the computer. Once the prices have been entered, we have the computer add them and then divide the sum by 25. Finally, we want the computer to print out the result. The flowchart that accomplishes these things is shown in Figure 14.3. ☐

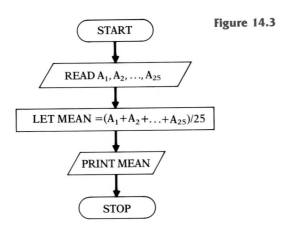

Figure 14.3

EXAMPLE 3 In the preceding example, suppose the department would like to reuse the program and compare the results for several weeks. Write a flowchart to help the department find the average price for several weeks.

SOLUTION In this case the same program can be used over and over again for each week. The only thing that will be different is the data for the different weeks. We indicate that the computer is to do the same

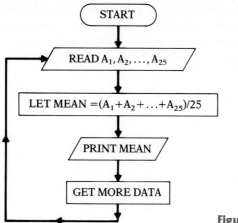

Figure 14.4

thing by *looping* it back to the READ statement to start again. The flowchart that accomplishes this is shown in Figure 14.4. When the computer runs out of data it will stop automatically. ☐

Let us now draw a flowchart for a situation in which a decision is involved.

EXAMPLE 4   Write a flowchart to print the square roots of the first 25 (natural) numbers.

SOLUTION   In this case the flowchart will involve a loop and a decision symbol. It is shown in Figure 14.5. Let us analyze this flowchart carefully.

Figure 14.5

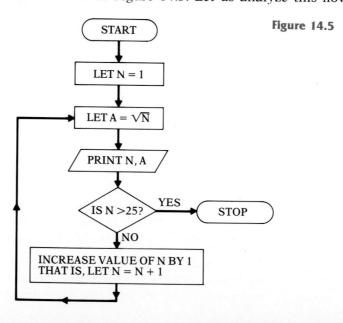

The first step tells the computer that $N = 1$ and then that $A = \sqrt{N}$ $= \sqrt{1} = 1$. The computer prints this information. The computer must now decide if $N > 25$ or not. If the answer is yes, then the program is done, and the computer stops. If the answer is no, then the machine increases the value of N by 1. In BASIC computer language we write $N = N + 1$. This is not to be interpreted as an algebraic equation. Instead, it is an instruction to the computer that the new value of N, wherever it appears from now on, is to be the old value with one added to it. With this change in mind the computer now loops back to find the value of A. Since N is now 2, the value of A is $\sqrt{2}$, or 1.414. The same procedure is followed until the value of N reaches 25 when the program ends and the computer stops. ☐

Comment When using the diamond symbol to indicate that a decision is to be made, we must tell the computer what to do with each alternative, that is, how to proceed if the answer to the question is yes or if the answer is no.

In Section 14.4, when we actually write computer programs in BASIC, we suggest that you always draw a flowchart appropriate for the problem so that you can actually see how a logical thought or process becomes an instruction to be executed by the computer.

EXERCISES

1. Write a flowchart to find the area of a trapezoid whose bases are 10 cm and 16 cm and whose height is 8 cm. [*Hint:* $A = \frac{1}{2}$ height $\times$ (base 1 + base 2).]

2. Write a flowchart to find the average of the numbers 23, 79, 68, 41, 83, and 72.

3. Write a flowchart to find the median of the numbers 23, 79, 68, 41, 83, and 72.

4. Write a flowchart to find the mode of the numbers 23, 79, 68, 41, 83, and 72.

5. Write a flowchart to find the average deviation of the numbers 23, 79, 68, 41, 83, and 72.

6. Write a flowchart to find the standard deviation of the numbers 23, 79, 68, 41, 83, and 72.

7. Write a flowchart to convert an angle measured in radians to an angle measured in degrees.

8. Write a flowchart to convert the measure of a liquid from gallons to liters.

9. Write a flowchart to convert the weight of a person from pounds to grams.

**10. Write a flowchart to convert the number $1101101_{(2)}$ written in base 2 to a number in base 10.

11. Write a flowchart to determine the cost of sending a telegram of 33 words if the cost for the first 8 words is 86¢ and the cost for each additional word is 7¢.

12. Write a flowchart to find the square root of the integers from 10 to 25.

13. Write a flowchart to find the cube root of the integers from 15 to 25.

14. The State Lottery Commission pays one particular vendor 8¢ for each lottery ticket that she sells. Additionally, if she sells more than 1000 lottery tickets for any one drawing, she is given an extra 3¢ for every lottery ticket over 1000 that she sells. Assuming that she sells more than 1000 tickets in a week, write a flowchart to determine the number of dollars D in her weekly income in terms of the number N of tickets sold.

15. Write a flowchart to determine the number of passengers N on an airplane if there are D double seats and T triple seats, each of which is occupied.

16. A cashier earns $8 an hour for the first 30 hours per week worked and $13 for each hour over 30 hours worked per week. Write a flowchart that the company can use to compute the weekly earnings E of its cashiers if a cashier works H hours per week.

** 17. Write a flowchart to determine the maximum finance charge that can be charged by banks and oil companies on any outstanding indebtedness as shown to the right:

State	Average daily balance	Monthly rate of interest
Ohio	Less than $400	$1\frac{1}{2}\%$
	$400 and over	1%
Minnesota	All	1%
New York	Less than $500	$1\frac{1}{2}\%$
	$500 and over	1%
Arizona	All	$\frac{5}{6}\%$

In each of the Exercises 18–21, tell what the flowchart does.

18.

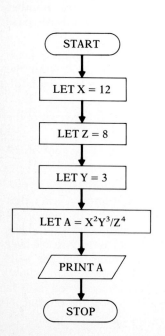

19.

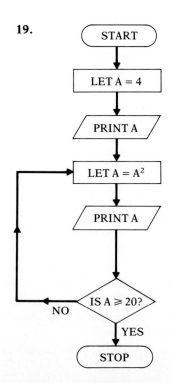

20.

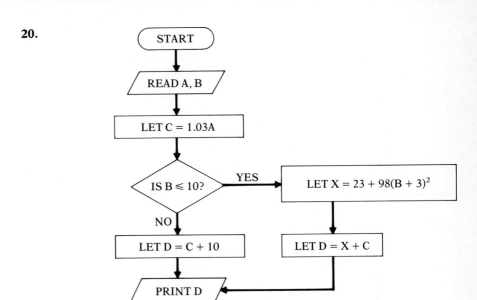

21.

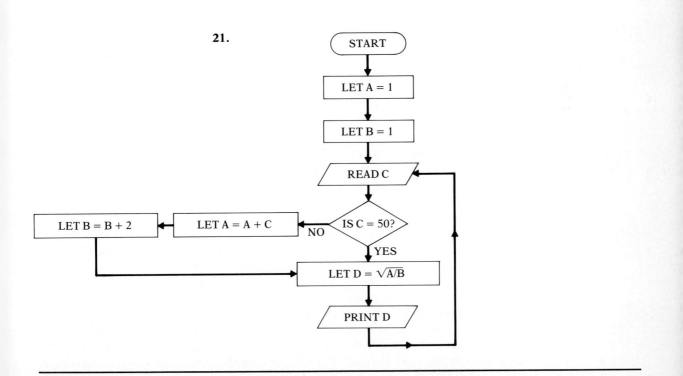

14.4

SOME SAMPLE BASIC PROGRAMS

In this section we will actually write some sample BASIC programs. It should be pointed out that, depending on the capabilities and design of the computer machines at different schools, the actual programs may have to be changed slightly to conform to a machine's specifications before they will actually run. However, the programs we give should work on the vast majority of the computers in use today.

Let us consider the Abel Office Equipment Company, which rents photocopying machines for a flat rental fee of $100 a month plus a charge of 4¢ for each copy made. Since each customer pays the same rental fee of $100 and the same 4¢ for each copy made, the company is interested in writing a BASIC program that would determine the amount of money to bill each customer depending on the number of copies made. How can such a program be written?

As we indicated in the last section, it is usually advisable to first draw a flowchart that indicates the logical sequence to be followed by the computer. It is important to remember that a program will run *only* if the commands of the programming language are used.

Since personal computers are gaining in popularity, let us pause for a moment to discuss how we input information into the computer. The first thing we do is turn on the computer and the monitor. When this is done, a flashing square, flashing rectangle, or flashing line will appear on the monitor. This flashing square is called a **cursor.** At this point you should begin typing your program much as you would do on any typewriter. The cursor will move and will appear where the next typed item will appear on the monitor. Try this with the programs given in this section. As a statement is typed, it will appear on the monitor. However, it is not entered into the computer's memory until the return key or enter key is pressed. Thus if a typing mistake is discovered before the return key is pressed, just move the cursor to the appropriate position and make the necessary corrections. If a mistake is discovered after the return key has been pressed, then we proceed as follows: As we shall see shortly, each step of a program has a line number. We can correct an error on a line that has already been entered into the computer's memory by retyping the same line, making any necessary corrections and using the same line number as the original line that had an error. The computer then replaces the line that has an error with the corrected version.

If you attempt to run a program and you have made an error in programming, then the computer will inform you that you have made a **syntax error.** Usually, this means that you made a mistake in typing or that you left something out. Be careful. Make any necessary corrections as indicated above, or else the program will not run or you will not obtain the desired results.

In what follows, we give some sample BASIC programs. It is suggested that if you have access to a computer, you actually use it. Type and then run the programs given. Such hands-on experience is very valuable.

EXAMPLE 1　Suppose one customer of the Abel Office Equipment Company made 1505 copies on the rented photocopier during a particular month. Write a BASIC program to determine the amount of money that the company should bill the customer.

SOLUTION　Let N = the number of copies made during the month and let C = the charge for these N copies. A flowchart and a BASIC program that can be used to determine the amount of money A that the customer should be billed are shown below.

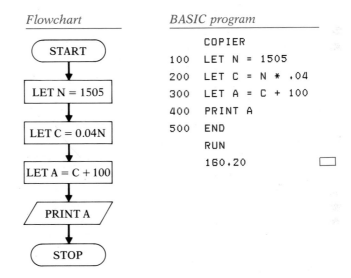

Although the above BASIC program will accomplish the job, there are many steps that should be added so that the program is really useful. As we proceed with our discussion of this program, we will add many extra steps so that by the time we are done we will have a "useful" program. In each case we will indicate why the extra step (instruction to the computer) was added.

Let us now analyze our simple program. We notice that this program has been named COPIER. Each program that is run on the computer should have a name, which can be no longer than six characters. Although we are free to choose any name we want for any program, it is best to select a name that will identify the program being run. In our case we have selected the six-letter word COPIER as the name of the program for obvious reasons. A program *must* have a name if one wishes to save it and reuse it at a later date.

Notice that each line in the program has a number. We can use any numbers from 1 to 99999 as the line numbers, provided that they are chosen in numerical order with respect to the order of the steps and that each line number is different. Most programmers skip numbers between lines to allow for the insertion of other lines between two existing lines without having to change all the line numbers in the program. Skipping numbers also makes corrections a rather simple process. For example, in our case, if the company decides to charge 6¢ a copy instead of 4¢, we can simply insert the following new line 250:

```
        COPIER
100     LET N = 1505
200     LET C = N * .04
250     LET C = N * .06
300     LET A = C + 100
400     PRINT A
500     END
        RUN
        190.30
```

The computer will now use this new line 250 to replace the previous value of C that was obtained when line 200 was used.

Line 100 contains a LET statement. This is a very important instruction, since it tells the computer to replace N by 1505. Similarly, line 200 tells the computer to replace C by the value obtained when N is multiplied by .04. Line 300 tells the computer that A is to be replaced by the value obtained for C + 100. In BASIC an instruction like N = N + 1 is also perfectly meaningful, since it tells the computer that N now has the value N + 1.

Line 300, 300 LET A = C + 100, contains two types of characters: a variable and a constant. A *constant* is a number that never changes, whereas a *variable* may change in value as the problem proceeds. In BASIC a variable name can consist of either a single letter or a single letter followed by a single digit. In our example, N, C, and A are one-character variable names. Examples of two-character BASIC names are X6, A7, B3, etc.

In BASIC, all mathematical calculations are performed much like ordinary algebraic or arithmetic calculations. The following symbols are used in BASIC.

BASIC symbol	Meaning	Example of BASIC expression	Similar algebraic expression
+	Addition	A + B	$a + b$
−	Subtraction	A − B	$a - b$
=	Equal	A = B	$a = b$
*	Multiplication	A * B	ab
/	Division	A / B	$\dfrac{a}{b}$ or $a \div b$
↑	Powers	A ↑ 3	a^3
( )	Grouping symbol	(A + B)/3	$\dfrac{a + b}{3}$

In BASIC, each symbol must be indicated. Nothing is taken for granted. Thus although we may write *ab* in ordinary algebra to stand for *a* times *b*, in BASIC we *must* write A*B. The symbol AB does not mean A times B.

Generally speaking, when the computer must perform several calculations, powers are evaluated first, then multiplications and divisions (from left to right), and finally addition and subtraction (again from left to right). Parentheses are used to group or to change the order of operations in a problem. Thus the statement A + B/3 tells the computer to first divide B by 3 and then to add the results to A, as opposed to the statement (A + B)/3, which tells the computer first to add A and B together and then to divide the sum by 3.

EXAMPLE 2 Which of the following are correct BASIC statements? If incorrect, explain why.

a) 100   LET A = B + 400

b) 300   LET X34 = B + 75

c) LET T = 5

d) 400   LET A + B = 31

e) 500   LET N = N + 1

SOLUTION   **a)** Correct

b) Incorrect; the BASIC variable has three characters. In BASIC, at most two character variables are permitted.

c) Incorrect; there is no line number.

d) Incorrect; we cannot have two variables to the left of the equal sign.

e) Correct. ☐

EXAMPLE 3   Translate each of the following mathematical equations or formulas into BASIC statements.

a) $y = 3x + 2$

b) $a = \dfrac{2b + c}{3}$

c) $a = \dfrac{b^2}{3}$

d) $u = \dfrac{x_1 + x_2 + x_3 + x_4}{4}$

e) $y = \dfrac{(x + 2)^3}{7x}$

f) $y = \sqrt{x + 1}$ or $y = (x + 1)^{1/2}$

g) $x = \dfrac{-b + \sqrt{b^2 - 4ac}}{2a}$

h) $v = \dfrac{4}{3}\pi r^3$ (Use $\pi = 3.14159$.)

SOLUTION   The BASIC statements corresponding to each of the mathematical equations or formulas are as follows.

a) 100   LET Y = 3 * X + 2

b) 120   LET A = (2 * B + C)/3

c) 130   LET A = (B ↑ 2)/3

d) 140   LET U = (X1 + X2 + X3 + X4)/4

e) 150   LET Y = ((X + 2) ↑ 3)/(7 * X)

f) 160   LET Y = (X + 1) ↑(1/2)   or
   161   LET Y = (X + 1) ↑ .5

g) 170   LET X = (−B + (B ↑ 2 − 4 * A * C) ↑ .5)/(2 * A)

h) 180   LET V = (4 * 3.14159 * R ↑ 3)/3 ▢

EXAMPLE 4   What is wrong with each of the following BASIC statements? Can you suggest a correction?

a) 100   LET A = 2(B + 7)

b) 110   A = 2

c) 400   LET A = (X + 5) ↑ 1/2

SOLUTION   a) There is no multiplication symbol. A possible correction is as follows.

101   LET A = 2*(B + 7)

b) The LET statement has been left out. The correction is as follows.

111   LET A = 2

c) If the square root is intended, then parentheses symbols must be used. A possible correction is as follows.

401   LET A = (X + 5) ↑(1/2)

As it stands, statement 400 tells the computer to raise X + 5 to the first power and divide the result by 2. In other words, evaluate $(x + 5)/2$.   □

Let us now return to line 400 of our original problem (Example 1). It is

```
400   PRINT A
```

When the computer determines that the cost is $160.20, it will simply print 160.20. We could change line 400 to read as follows:

```
401   PRINT N, C, A
```

This would cause the computer to print the following:

```
1505              60.20              160.20
```

The commas in the PRINT statement tell the computer to skip 15 spaces before printing the next numbers or words. By using this type of PRINT statement we are now able to have the original data printed out along with the result.

It is also possible to use a PRINT statement to have the computer print out words. We simply enclose the message the be printed within quotation marks followed by a semicolon. The semicolon tells the computer what to print next and to disregard the predetermined *fields*. Thus in our case we could replace line 400 by

```
402   PRINT "TOTAL COST TO CUSTOMER IS"; A
```

The computer output for this would then be

```
TOTAL COST TO CUSTOMER IS 160.20
```

Suppose the Abel Office Equipment Corporation wants to use this program to calculate the total cost for several of its customers. With the program currently available it would have to reenter a new program that contains different data for each customer. How can we use this same program and change only the data? This can be accomplished by using READ and DATA statements as shown in the following example.

EXAMPLE 5   Write a BASIC program that would determine the total cost to a customer of the Abel Office Equipment Corporation who made 1505 photocopies during the month and that can be used to compute the costs to be charged to other customers.

SOLUTION   The BASIC program accomplishing this is as follows.

```
      COPIER
100   PRINT "CUSTOMER", "COPIES USED", "TOTAL COST"
200   PRINT
300   READ M, N
400   LET C = N * .04
500   LET A = C + 100
600   PRINT M, N, A
700   DATA 1, 1505
800   END
      RUN
```

The READ *statement* tells the computer to find the values of M and N in the DATA *statement*. We always list the variables to be read after the READ statement and the values of these variables in the DATA statement. The numbers in the DATA statement must be separated by commas, and the number of data terms must be a multiple of the number of variables used. DATA statements are generally placed right before the END statement.

The BASIC program above will result in the following printout:

```
CUSTOMER                 COPIES USED              TOTAL COST

1                        1505                     160.20
```

Notice that line 100 tells the computer to print headings, whereas line 200, which contains only the word PRINT, causes the computer to skip a line. Line 800 is the END statement. Every program written in BASIC must have an END statement that tells the computer that the program has been completed.

You will also notice the use of the RUN statement. We never assign a line number to a RUN statement. A RUN statement is always placed after an END statement and tells the computer to begin executing the program. ☐

All the programs that we have discussed so far were relatively simple. None of them required the computer to make any decisions or to repeat the same type of calculation several times. The computer can be programmed to perform the same type of operation many times by using a GO TO statement, as shown in the following example.

EXAMPLE 6   Write a BASIC program that would allow the Abel Office Equipment Corporation to determine the total cost to customer 1 who made 1505 photocopies during the month, customer 2 who made 1227 copies, and customer 3 who made 1807 copies.

SOLUTION   The BASIC program accomplishing this is as follows.

```
        COPIER
100     READ M, N
200     LET C = N * .04
300     LET A = C + 100
350     PRINT "CUSTOMER NUMBER"; M
375     PRINT "NUMBER OF COPIES USED"; N
400     PRINT "TOTAL COST $"; A
425     PRINT
450     PRINT
475     PRINT
600     PRINT
700     GO TO 100
800     DATA 1, 1505, 2, 1227, 3, 1807
900     END
        RUN
```

Notice that in line 700 we have added a GO TO statement. This causes the computer to go to the line number indicated. After the computer determines the total charge for customer 1, we want the computer to go to the READ statement, read new values for M and N, and determine the total cost for these values. Thus we insert the GO TO 100 statement right after the PRINT statement and before the DATA statement. The computer goes to line 100, computes the total cost for customer 2, and then goes through the same procedure for customer 3. The computer is now in a *loop.*

When all the data in line 800 are used up, the computer will stop and tell us "out of data in line 100". This means that there are no more numbers in the data list to read.

The previous program will cause the computer to print out the following results:

```
CUSTOMER NUMBER 1
NUMBER OF COPIES USED 1505
TOTAL COST $160.20



CUSTOMER NUMBER 2
NUMBER OF COPIES USED 1227
TOTAL COST $149.08
```

```
CUSTOMER NUMBER 3
NUMBER OF COPIES USED 1807
TOTAL COST $172.28


OUT OF DATA IN LINE 100  ☐
```

EXAMPLE 7   Write a BASIC program to compute the area of three different circles, one with a radius of 7, one with a radius of 9, and one with a radius of 10.6. (Remember that the area of a circle is given by $A = \pi r^2$ or $A = 3.14159 r^2$, where r is the radius of the circle and where we have used 3.14159 as the value of π.)

SOLUTION   The BASIC program that will accomplish this is as follows.

```
      AREA
100   READ R
125   LET A = 3.14159 * R ↑ 2
150   PRINT "IF RADIUS OF CIRCLE IS"; R
175   PRINT "AREA OF CIRCLE IS "; A
200   PRINT
300   PRINT
400   PRINT
500   PRINT
600   GO TO 100
700   DATA 7, 9, 10.6
800   END
      RUN
```

When the computer executes this program, it will print out the following.

```
IF RADIUS OF CIRCLE IS 7
AREA OF CIRCLE IS 153.93791



IF RADIUS OF CIRCLE IS 9
AREA OF CIRCLE IS 254.46879



IF RADIUS OF CIRCLE IS 10.6
AREA OF CIRCLE IS 352.9890524



OUT OF DATA IN LINE 100  ☐
```

EXAMPLE 8   Write a BASIC program to compute the squares and cubes of the first five numbers.

SOLUTION   As indicated earlier, it is advisable to first draw a flowchart to picture the logical steps that are necessary to execute the program. Using the symbols given in Section 14.3, we have the flowchart shown in Figure 14.6. A BASIC program based on this flowchart is as follows.

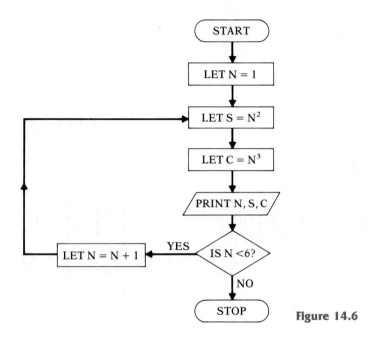

Figure 14.6

```
      SQUARE
100   PRINT
200   PRINT
300   LET N = 1
400   LET S = N ↑ 2
500   LET C = N ↑ 3
525   PRINT "IF NUMBER IS"; N
550   PRINT "SQUARE IS"; S
575   PRINT "CUBE IS"; C
600   PRINT
625   PRINT
650   PRINT
700   LET N = N + 1
800   IF N < 6 THEN 400
900   END
      RUN
```

Notice that in line 800 we have used an IF THEN *statement*. Such a statement tells the computer to decide whether the relationship given is true or false. If the relation given is true, the computer will jump to the line number given in the statement. If the relation is false, the computer proceeds to the next BASIC statement in sequence.

When the computer executes this program, the following will be printed out:

```
IF NUMBER IS 1
SQUARE IS 1
CUBE IS 1


IF NUMBER IS 2
SQUARE IS 4
CUBE IS 8


IF NUMBER IS 3
SQUARE IS 9
CUBE IS 27


IF NUMBER IS 4
SQUARE IS 16
CUBE IS 64


IF NUMBER IS 5
SQUARE IS 25
CUBE IS 125
```

For Example 8 a BASIC program can also be written that will have the computer print out the results in the following format:

NUMBER	SQUARE OF NUMBER	CUBE OF NUMBER
1	1	1
2	4	8
3	9	27
4	16	64
5	25	125

The following list summarizes the various relational symbols that can be used in BASIC IF THEN statements.

Table 14.2
Some BASIC symbols and their meanings

BASIC symbol	Meaning
=	is equal to
<	is less than
>	is greater than
< =	is less than or equal to
> =	is greater than or equal to
< >	is not equal to

EXAMPLE 9 The Cal Finance Company charges different interest rates on outstanding monthly balances depending on the amount of money involved as shown below:

Outstanding monthly balance	Annual interest rate
Up to $200.00	18%
From $200.01 to $500.00	15%
Above $500.00	12%

Write a BASIC program to determine the finance charge for five accounts whose outstanding monthly balances are $58.17, $423.17, $312.98, $1400.16, and $817.64.

SOLUTION It is best to first draw a flowchart for the problem. Let B = the outstanding balance and let I = the finance charge. Then a flowchart for the problem is shown in Figure 14.7 on the following page. Using this flowchart, we set up the following BASIC program:

```
        CHARGE
 100    PRINT
 200    PRINT
 300    READ B
 400    IF B > 200.00 THEN 700
 500    LET I = B * 0.18
 600    GO TO 1100
 700    IF B > 500.000 THEN 1000
 800    LET I = B * 0.15
 900    GO TO 1100
1000    LET I = B * 0.12
```

```
1100    PRINT = "IF OUTSTANDING BALANCE IS $"; B
1125    PRINT "FINANCE CHARGE IS $"; I
1130    PRINT
1150    PRINT
1175    PRINT
1180    PRINT
1200    GO TO 300
1300    DATA 58.17, 423.17, 312.98, 1400.16, 817.64
1400    END
        RUN
```

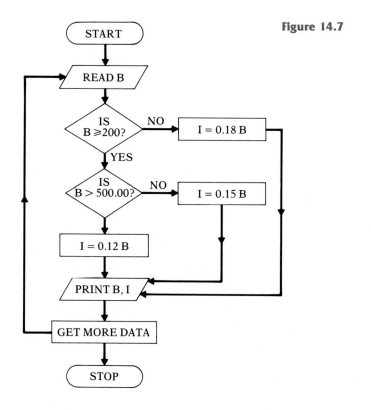

Figure 14.7

When this program is executed, the computer printout will be as follows:

```
IF OUTSTANDING BALANCE IS $58.17
FINANCE CHARGE IS $10.47



IF OUTSTANDING BALANCE IS $423.17
FINANCE CHARGE IS $63.48
```

```
IF OUTSTANDING BALANCE IS $312.98
FINANCE CHARGE IS $46.95



IF OUTSTANDING BALANCE IS $1400.16
FINANCE CHARGE IS $168.02



IF OUTSTANDING BALANCE IS $817.64
FINANCE CHARGE IS $98.12



OUT OF DATA IN LINE 300  ▭
```

EXERCISES

Which of the variable names given in Exercises 1–6 are not acceptable in BASIC?

1. A23
2. KKK
3. B12
4. 10X
5. B4
6. JFK

In Exercises 7–11, write each of the given algebraic statements as equivalent BASIC statements.

7. $D = RT$

8. $y = x^2 - 5x + 3$

9. $y = z(7x - 3)$

10. $F = \frac{9}{5}C + 32$

11. $A = \frac{B - C}{D + E}$

In Exercises 12–16, what is the value of A in each BASIC statement? (Use X = 10, Y = 12, and Z = 2.)

12. `20  LET A = 3 * (X + Y)`
13. `30  LET A = Z ↑ 3`
14. `40  LET A = (X + Y + 1)/Z`
15. `50  LET A = Z ↑ 2 + X ↑ 2`
16. `60  LET A = (Z + X) ↑ 2`

17. Is the following a correct BASIC statement? Explain your answer.

```
10  LET  A12 = X - 3
```

18. Is the following a correct BASIC statement? Explain your answer.

```
100  PRINT "THE ANSWER IS A
```

19. Is the following a correct BASIC statement? Explain your answer.

```
200  READ A, B, C, D
250  DATA 8, 9, 20, 35, 76
```

In Exercises 20–22, find at least one error in each BASIC program.

20.
```
     EXMPLE
10   LET A = 10
20   LET B = 20
30   LET N = (A * (B - 5)
40   PRINT A, B, N
50   END
     RUN
```

21.
```
     MEAN
10   LET A = 5
20   LET B = 15
30   LET AV = A + B/2
40   PRINT "THE MEAN OF A AND B IS"
50   PRINT AV
60   END
     RUN
```

22.
```
        EXMPLE
  10    READ B2
  20    READ MN
  30    LET A = MN + B2
  40    PRINT A
  50    DATA 20, 30
  60    END
        RUN
```

In Exercises 23–25, explain what the computer will print out when it executes the indicated BASIC programs.

23.
```
        EXMPLE
  10    LET A = 5
  20    LET B = 75
  30    LET C = 8
  40    LET M = (A * B) ↑ 2 * C
  50    PRINT M
  60    END
        RUN
```

24.
```
        EXMPLE
  10    LET A = 5
  20    LET B = 3 ↑ A
  30    LET M = B - A
  40    PRINT "THE VALUE OF M IS "; M
  50    END
        RUN
```

25.
```
        EXMPLE
 100    READ A, B
 120    IF A > B THEN 220
 150    PRINT B "IS BIGGER THAN"; A
 200    GO TO 100
 220    PRINT A, "IS BIGGER THAN"; B
 250    GO TO 100
 280    DATA 220, −75, 570, 320, −5, 50
 300    END
        RUN
```

26. Write a BASIC program for computing the area of several rectangles whose bases and heights

are as follows.

Base, in cm	Height, in cm
53	23
17	55
61.3	18.2
847.7	62.73
7298.3	509.67

27. Write a BASIC program to compute the area of a square.

28. Write a BASIC program to find the sum of the first 25 integers.

29. Write a BASIC program to find the product of the integers 20 through 30.

30. Write a BASIC program to compute the compound interest on a bank deposit of $2000 that earns annual interest of 8%, compounded annually for eight years.

31. In a math course, three tests are given, plus a final exam. The final is counted as two tests. The grade in the course is the average of all the tests (with the final counted twice). For example, if a student's three test scores are 85, 61, and 92 and his or her final exam is 74, then the grade is

$$\frac{85 + 61 + 92 + 2(74)}{5},$$

which equals 77.2. Write a BASIC program to compute a student's grade in this course.

32. Write a BASIC program to calculate the value of 9!.

33. Write a BASIC program to find the sum of the cubes of the integers from 10 to 20 inclusive.

34. Write a BASIC program to compute the value of the hypotenuse c shown in Figure 14.8. (*Hint:*

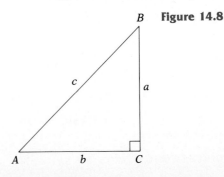

Figure 14.8

Remember that for the right triangle shown, $c = \sqrt{a^2 + b^2}$.)

A first-degree equation of the form $ax + b = c$ has as its solution

$$x = \frac{c - b}{a}.$$

Write a BASIC program that can be used to solve each of the equations given in Exercises 35–39.

***35.** $7x + 5 = 26$ ***36.** $0.4x + 9 = 17$

***37.** $5x - 4 = 19$ ***38.** $\frac{2}{3}x + 8 = 22$

***39.** $\frac{5}{7}x - 9 = 50$

A quadratic equation of the form $ax^2 + bx + c = 0$ has as one of its solutions

$$x = \frac{-b + \sqrt{b^2 - 4ac}}{2a}.$$

Write a BASIC program to solve each of the equations given in Exercises 40–43.

***40.** $x^2 - 9x + 18 = 0$

***41.** $x^2 + 5x + 6 = 0$

***42.** $x^2 - x - 6 = 0$

***43.** $x^2 - 10x + 24 = 0$

14.5

SUMMARY Among the many different computing languages available, we chose to discuss BASIC in detail in this chapter. BASIC is the simplest language to learn. We indicated how a computer program is written, first making a flowchart and then translating it into the appropriate BASIC form. We actually analyzed several programs that you might find helpful in your studies. You should now be able to write flowcharts and some simple programs in BASIC and be able to recognize what a program that has already been written in BASIC actually accomplishes.

STUDY GUIDE You should now be able to demonstrate your knowledge of the following ideas presented in this chapter by giving definitions, descriptions, or specific examples. Page references (in parentheses) are given for each term so that you can check your answer.

Computer program (p. 726)	END statement (p. 740)
BASIC (p. 727)	READ statement (p. 740)
Flowchart (p. 727)	DATA statement (p. 740)
Looping (p. 730)	GO TO statement (p. 741)
LET statement (p. 736)	Field (p. 739)
BASIC constant (p. 736)	Loop (p. 741)
BASIC variable (p. 736)	RUN statement (p. 740)
PRINT statement (p. 739)	IF THEN statement (p. 744)

MANUAL FOR QUICK REFERENCE The following list of symbols or key words summarizes the BASIC statements and symbols or flowchart symbols discussed in this chapter.

Flowchart symbols		*BASIC statements*

Flowchart symbols

⬭ or ◯	start or end of program
▱	input or output of data
▭	processing of program
◇	decision symbol
⬅ ↓ ➡	direction to go in flowchart

BASIC statements

LET   tells the computer that a variable has a particular value.

PRINT   tells the computer to print out the results that it obtained.

READ   tells the computer to obtain data from the DATA statement in sequential order.

DATA   stores the data to be used as input information.

GO TO   tells the computer to go out of sequence and go to the line number indicated.

IF THEN   tells the computer to make a decision and to proceed according to the answer to the decision made.

END   tells the computer that the program is done. It is always the last statement in the program.

RUN   tells the computer to start executing the program.

BASIC operations		*BASIC decision symbols*	
+	Addition	=	is equal to
−	Subtraction	<	is less than
*	Multiplication	>	is greater than
/	Division	< =	is less than or equal to
↑	Exponents (powers)	> =	is greater than or equal to
( )	Grouping symbol	< >	is not equal to
=	Equality		

MASTERY TESTS

Form A   1. When using flowcharts, the ▭ symbol is used to indicate
   a) the beginning or end of a program
   b) input or output of data
   c) the processing of the program

    d) that a decision must be made

    e) none of these

2. What is the BASIC symbol for less than or equal to?

3. Rewrite the following BASIC statement into an ordinary algebraic expression:

```
20   LET A = B * C / D ↑ (1/5)
```

4. Translate the following into a BASIC statement:

$$y = 2x^3z^4$$

5. What will be printed when the following BASIC program is run?

```
     EXMPLE
10   LET A = 7.6
15   LET B = 11.6
18   LET C = A/(B - A)
25   PRINT A, B, C
30   END
     RUN
```

6. Find at least one error in the following BASIC program.

```
     EXMPLE
10   LET A1 = 10
20   LET A2 = 6
30   PRINT A1 ↑ A2
40   END
     RUN
```

7. What does the following computer instruction mean?

```
20   LET A = A + 5
```

8. A computer program to be saved for later use should have a name that is no longer than _____ characters.

9. When writing computer programs, any numbers from 1 to _____ can be used as line numbers.

10. In BASIC we write a^2b^5 as _____.

Form B   **1.** What will be printed when the following program is run?

```
     EXMPLE
10   READ M, N, P, Q
20   LET A = (M - Q)/(N + P) ↑ 2
30   PRINT A
40   DATA 7, 8, 12, 6
50   END
     RUN
```

2. Is the following a correct BASIC statement? If incorrect, explain why.

```
20   LET X15 = Y - 3
```

3. Translate the following into a BASIC statement:

$$y = -a + \sqrt[3]{7m^2y^5}$$

4. *True or false?* In BASIC, AB stands for A times B.

5. What does the statement $A - B/2$ tell the computer to do?

6. In BASIC a variable name can consist of either a single letter or a single letter followed by a _____.

7. Find at least one error in the following BASIC program.

```
     EXMPLE
10   LET X = 7
20   LET Y = 10
30   LET A = (X + 6)/(Y - 8) ↑ 2
40   PRINT "A EQUALS; A
50   END
     RUN
```

8. What does the following BASIC program accomplish?

```
     EXMPLE
10   LET A = 1
20   PRINT A
30   LET A = A + 1
40   IF A > 100 THEN 60
50   GO TO 20
60   END
     RUN
```

9. What does the following BASIC program accomplish?

```
     EXMPLE
10   LET A = 100
20   PRINT A
30   LET A = A - 1
40   IF A < 1 THEN 60
50   GO TO 20
60   END
     RUN
```

10. What does the following BASIC program accomplish?

```
       EXMPLE
10     LET A = 1
20     PRINT "I AM A SMART COMPUTER"
30     LET A = A + 1
40     IF A > 10 THEN 60
50     GO TO 20
60     END
       RUN
```

11. What does the following BASIC program accomplish?

```
       EXMPLE
10     PRINT "WHO IS PRESIDENT OF ꞏTHE USA?"
20     PRINT "1. NIXON  2. FORD  3. CARTER  4. REAGAN"
30     PRINT "THE NUMBER OF THE CORRECT CHOICE IS"; N
40     IF N = 4 THEN 70
50     PRINT "INCORRECT, TRY AGAIN."
60     GO TO 10
70     PRINT "CORRECT."
80     END
       RUN
```

12. What does the following flowchart accomplish?

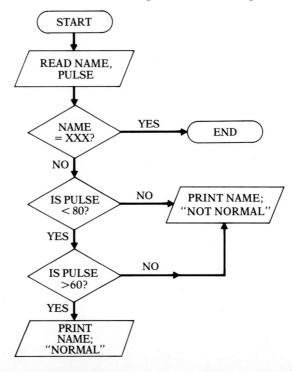

SUGGESTED FURTHER READING

CHAPTER 1   Bell, E. T., *Men of Mathematics*. New York: Simon & Schuster, 1961. Chapter 29 contains a biography of G. Cantor.

Dinkines, F., *Elementary Theory of Sets*. New York: Appleton-Century-Crofts, 1964. Contains a basic but thorough discussion of sets with applications.

Mathematics in the Modern World (Readings from *Scientific American*). San Francisco: W. H. Freeman, 1968. Article 28 discusses paradoxes, and Article 30 discusses a set theory different from Cantor's.

Stoll, R., *Sets, Logic, and the Axiomatic Method*. San Francisco: W. H. Freeman, 1974. A good discussion of set theory.

CHAPTER 2   Copi, I. M., *Introduction to Logic*. New York: Macmillan, 1961. This text contains a clear discussion of inductive and deductive logic and the use of inductive logic in science.

Dantzig, T., *Number, the Language of Science*. New York: Macmillan, 1954. Chapter 4 discusses inductive reasoning.

May, K. O., "The Origin of the Four-Color Problem" in *Isis* **56**:346–348 (1965).

Nagel, E., "Symbolic Notation, Haddocks' Eyes and the Dog-Walking Ordinance," in James R. Newman, ed., *The World of Mathematics*, Vol. 1, pp. 1878–1900. New York: Simon & Schuster, 1956. This article shows how logic can be applied to language.

Sawada, Diayo, "Magic Squares: Extensions into Mathematics," in *The Arithmetic Teacher*, March 1974, pp. 183–188.

S-1

Schaaf, William, *A Bibliography of Recreational Mathematics*. Washington, D.C.: National Council of Teachers of Mathematics, 1970.

Williams, Horace, "A Note on Magic Squares," in *The Mathematics Teacher*, October 1974, pp. 511–513.

CHAPTER 3   Aaboe, A. S. *Episodes From the Early History of Mathematics*. New York: Random House, 1964. Chapter 1 discusses the Babylonian number system.

Bergamini, D. *et al.*, *Life Mathematics* (*Life* Science Library). New York: Time-Life Books, 1970.

Computers and Computation (Readings from *Scientific American*). San Francisco: W. H. Freeman, 1971.

Conant, L. L., "Counting," in James R Newman, ed., *The World of Mathematics*, vol. I, Part III, Chap. 2. New York: Simon & Schuster, 1956. This article discusses the human ability to count.

Gillings, R. J., *Mathematics in the Time of the Pharaohs*. Cambridge, Mass.: MIT Press, 1962.

Ore, O., *Graphs and Their Uses*. New York: Random House, 1963. Pages 73–75 discuss *Nim*.

CHAPTER 4   Boyer, Carl, *A History of Mathematics*. New York: Wiley, 1968. This book contains a complete history of mathematics from ancient to modern times.

Cooley, Hollis R., and H. E. Wahlert, *Introduction to Mathematics*. Boston: Houghton-Mifflin, 1968. Chapters 2, 3, and 4 discuss the real number system.

Gardner, M. See the following articles in *Scientific American*: "Calculating Prodigies" (April 1967); "Divisibility Rules" (September 1962); "Fibonacci Numbers" (August 1959, December 1966, March 1969); "Perfect numbers" (March 1968); "Prime Numbers" (March 1964, March 1968, August 1970); "Unsolved Problems in Elementary Number Theory" (December 1973).

CHAPTER 5   *Best's Insurance Reports—Life/Health*, published annually by the A. M. Best Company. A good source of information about life insurance policies and companies. This book is available in the reference room of most public libraries.

CHAPTER 6   Bell, E. T., *Men of Mathematics*. New York: Simon & Schuster, 1961. Chapters 17 and 20 contain biographies of Abel and Galois.

Courant, R., and H. Robbins, *What Is Mathematics?* New York: Oxford University Press, 1960. Pages 31–40 discuss congruences.

Dinkines, F., *Abstract Mathematical Systems*. New York: Appleton-Century-Crofts, 1964. A general basic discussion of groups.

Lindsey, G. R., "An Investigation of Strategies in Baseball," in *Operations Research* **2**:477–501 (1963).

MacDonald, J., *Strategy in Poker, Business, and War*. New York: W. W. Norton, 1950.

Mathematics in the Modern World (Readings from *Scientific American*). San Francisco: W. H. Freeman, 1968. Articles 40 and 41 discuss the theory of games and the misuses of game theory.

Newmark, J., *Using Finite Mathematics*. New York: Harper and Row, 1982.

Rapoport, A., *Flights, Games, and Debates*. Ann Arbor: University of Michigan Press, 1960. Discusses debates and strategies to outwit opponents.

CHAPTER 7   Ablon, L., *et al.*, *Series in Mathematical Modules*, 2nd edition. Menlo Park, Calif.: Benjamin/Cummings, 1981. Modules I & II contain a clear basic treatment of algebraic operations.

CHAPTER 8   Dressler, I., and Rich, B., *Algebra*. New York: Amsco School Publications, 1976.

Kemeny J., L. Snell, and J. Thompson, *Introduction to Finite Mathematics*, 3rd edition, Chapter 7. Englewood Cliffs, N.J.: Prentice-Hall, 1974.

Kemeny, J. *et al.*, *Finite Mathematics with Business Applications*, Chapter 7, Englewood Cliffs, N.J.: Prentice-Hall, 1968.

Newmark, J., *Using Finite Mathematics*. New York: Harper and Row, 1982.

Wheeler, R., *Modern Mathematics for Business Students*, Chapter 8. Monterey, Calif.: Brooks/Cole, 1968.

CHAPTER 9   Choquet, G., *Geometry in a Modern Setting*. Paris: Hermann, 1961. Section 57 deals with the definition of angles.

Jacobs, H. R., *Geometry*. San Francisco: W. H. Freeman, 1974.

Kline, M., *Mathematics: A Cultural Approach*. Reading, Mass.: Addison-Wesley, 1962. Chapter 6 discusses the nature and uses of Euclidean geometry, and Chapter 26 discusses non-Euclidean geometries and their significance.

Synge, J. L., *Science: Sense and Non-sense*. New York: W. W. Norton, 1950. Pages 26–30 contain an imaginary discussion between Euclid and a 12-year-old boy.

Toth, L., "Non-Euclidean Geometry before Euclid," in *Scientific American* (November 1969). This article suggests that the ancient Greeks knew about non-Euclidean geometry.

CHAPTER 10   Bailey, William, "Friday-the-Thirteenth" in *Mathematics Teacher*, vol. LXII, no. 5 (1969), pp. 363–364.

Caesar, Allen J., "The Saint Petersburg Paradox and Some Related Series," in *Two-Year College Mathematics Journal*, Nov. 1981. This article discusses the mathematical expectation of flipping a coin.

David, F. N., *Games, Gods, and Gambling*. New York: Hafner, 1962.

Epstein, R. A., *Theory of Gambling and Statistical Logic*. New York: Academic Press, 1967. Contains an interesting discussion on the fairness of coins.

Gallup, G., *The Sophisticated Poll Watcher's Guide*. Princeton, N. J.: Princeton Opinion Press, 1972. Contains an interesting discussion on taking and interpreting polls.

Havermann, E., "Wonderful Wizard of Odds," in *Life* **51**(14):30*ff.* (Oct. 6, 1961).

Huff, Darrell, *How to Take a Chance*. New York: Norton, 1959.

Ritter, G. L., *et al.*, "An Aid to the Superstitious," in *Mathematics Teacher*, **70**:5(456–457).

CHAPTER 11 Cooke, W. P., "Beginning Statistics at the Track," in *Mathematics Magazine* **46**:250–255 (November–December 1973).

Huff, Darrell, *How to Lie with Statistics*. New York: W. W. Norton, 1954. Discusses how statistics can be misused.

Los Angeles Times, "Nielsen Raters' Views Decide Your TV Fare," June 9, 1974.

New York Times, "On the Nielsen Families," September 15, 1974.

Newmark, J., *Statistics and Probability in Modern Life*, 4th edition. Philadelphia: Saunders, 1987. Chapters 2–3 contain a thorough discussion on how to analyze data by graphical and arithmetic techniques.

Tanur, J. M., ed., *Statistics: A Guide to the Unknown*. San Francisco: Holden Day, 1972. Contains a discussion on the uses of statistics.

CHAPTER 12 Campbell H., *An Introduction to Matrices, Vectors, and Linear Programming*, 2nd edition. New York: Appleton-Century-Crofts, 1971. An excellent introduction to matrix theory.

Kemeny, J., L. Snell, and J. Thompson, *Introduction to Finite Mathematics*, 3rd edition. Englewood Cliffs, N.J.: Prentice-Hall, 1974. Chapter 4 discusses matrices and their applications.

Mathematics in the Modern World (Readings from *Scientific American*), San Francisco: W. H. Freeman, 1968. Article 7 contains a biographical sketch of Hamilton.

Newmark, J. *Using Finite Mathematics*, Chapter 11, Harper and Row, 1982.

Sawyer, W. W., *Prelude to Mathematics*. Baltimore: Penguin Books, 1959. Chapter 8 discusses matrix algebra.

CHAPTERS 13 AND 14 *Computers and Computation* (Readings from *Scientific American*). San Francisco: W. H. Freeman, 1971. Contains many interesting articles on the history, uses, and future of the computer. Articles 23–26 discuss the applications of the computer to technology, organization, education, and science.

Eames, C., and R. Eames, *A Computer Perspective*. Cambridge, Mass.: Harvard University Press, 1973. Gives a pictorial history of the development of computers.

Hawkes, N., *The Computer Revolution*. New York: World of Science Library, Dutton, 1972. Contains detailed discussions on how the computer is applied. Specifically, Chapter 2 discusses computers in business, Chapter 3 computers in science, Chapter 4 the computer and the arts, and Chapter 8 the future.

Kemeny, J. G., *Man and the Computer*. New York: Scribner, 1972.

Reichardt, J., *The Computer in Art*. New York: Van Nostrand, Reinhold, 1971.

Rosen, S., "Electronic Computers: A Historical Survey," in *Computing Surveys* **1**(1): March 1969.

Springer, C. H., *et al.*, *The Mathematics for Management Series*. Homewood, Ill.: Richard D. Irwin, 1966. Volume 3 discusses several computer applications.

ANSWERS TO SELECTED EXERCISES

CHAPTER 1

Section 1.2 (pages 9–10)

1. The set of the months of the year that begin with the letter J.
2. The set of all multiples of 4 between 4 and 28 inclusive.
3. The set of all prime numbers between 2 and 19 inclusive.
4. The set of the cubes of the first five counting numbers.
5. The set of the days of the week beginning with the letter T.
6. The set of all multiples of 3 between 3 and 999 inclusive.
7. The set of some computers on the market today.   8. The set of all even numbers larger than 1.
9. {M, I, S, P}   10. {February, April, June, September, November}
11. {Truman, Eisenhower, Kennedy, Johnson, Nixon, Ford, Carter, Reagan}
12. {1, 4, 9, 16, 25, 36, 49, 64}   13. {14, 16, 18}   14. Ø
15. {Florida, Indiana, Virginia, Texas, Alabama, Mississippi, Louisiana, North Carolina, Georgia, South Carolina, Nevada}
16. We suggest that you check with *Information Please Almanac* or with a current encyclopedia as the answer is constantly changing.   17. Ø   18. Ø   19. Not well defined.   20. Not well defined
21. Not well defined   22. Not well defined   23. Well defined   24. Not well defined
25. Finite nonempty set   26. Null set   27. Finite nonempty set   28. Finite nonempty set
29. Null set   30. Infinite set   31. Finite nonempty set   32. Finite nonempty set   33. Null set
34. Infinite set   35. False   36. True   37. False
38. True, although some people consider tomato to be a vegetable.   39. False   40. False   41. False
42. False   43. Yes   44. Not necessarily   45. Equivalent and equal   46. Neither   47. Neither

48. Equal and equivalent　**49.** Neither　**50.** Equal and equivalent. Both are null sets.
51. Equivalent but not equal　**52.** Equal and equivalent　**53.** Neither
54. One possibility is shown below. Others are possible.

　　a) {x,　q,　John,　b} **b)** {Heather,　Jason,　Marlene}
　　　↕　↕　↕　↕　　　↕　　　↕　　　↕
　　{△, □,　α,　c}　　{Marlene,　Joe,　Bill}

55. 2　**56.** 0
57. 33 women have been executed since 1930. 307 women have been executed in the U.S. from 1608 through 1985.
58. 0　**59.** 10　**60.** Infinite　**61.** No　**62.** Yes　**63.** 2　**64.** $n(n-1)(n-2)\cdots 1$　or　$n!$

Section 1.3 (pages 17–20)

1. {Reagan} and Ø　**2.** {True} and {False}　**3.** {Kennedy} and {Moynihan}　**4.** {a, e, i} and {o, u}
5. {1, 2, α} and {β}　**6.** {Water pollution} and {air pollution}　**7.** {Hamburgers} and {beefburgers}
8. {Honda} and Ø　**9.** {The regional telephone companies created by the divestiture of AT&T}
10. {famous female tennis players}　**11.** {popular computer software packages}　**12.** {imported cars}
13. {next-day delivery service companies}　**14.** {fish}　**15.** {space programs}
16. {popular soft drinks sold in the U.S.}　**17.** Neither　**18.** Proper　**19.** Improper　**20.** Proper
21. Neither　**22.** Proper　**23.** Improper　**24.** Proper　**25.** {9}, {11}, {12}
26. {9, 11}, {9, 12}, {11, 12}　**27.** {9, 11, 12}　**28.** Ø　**29.** { }　**30.** {a} and { }
31. {a}, {b}, {a, b} and Ø　**32.** {a, b, c}, {a}, {b}, {c}, {a, b}, {a, c}, {b, c}, Ø
33. {a, b, c, d}, {a}, {b}, {c}, {d}, {a, b}, {a, c}, {a, d}, {b, c}, {b, d}, {c, d}, {a, c, d}, {a, b, d}, {b, c, d}, {a, b, c} and Ø
34. {a, b, c, d, e}, {a}, {b}, {c}, {d}, {e}, {a, b}, {a, c}, {a, d}, {a, e}, {b, c}, {b, d}, {b, e}, {c, d}, {c, e}, {d, e}, {a, b, c}, {a, b, d}, {a, b, e}, {a, c, d}, {a, c, e}, {a, d, e}, {b, c, d}, {b, c, e}, {b, d, e}, {c, d, e}, {a, b, c, d}, {a, b, c, e}, {a, c, d, e}, {b, c, d, e}, {a, b, d, e} and Ø.
35. 1　**36.** 3　**37.** 7　**38.** 15

39.

Number of elements in set	6	10	100	n
Number of possible subsets	64	1024	2^{100}	2^n

40. True　**41.** True　**42.** True　**43.** True　**44.** False　**45.** False　**46.** True　**47.** False
48. True　**49.** {a, b, c, d, e, f, g, i}　**50.** {f, g}　**51.** {b, c, e, h, i}　**52.** {b, c, e, f, g, h, i}　**53.** {a, d}
54. {b, c, e, i}　**55.** Ø　**56.** {a, d, g}
57. {(a, b), (a, f), (a, i), (d, b), (d, f), (d, i), (f, b), (f, f), (f, i), (g, b), (g, f), (g,i)}
58. {(b, a), (b, d), (b, f), (b, g), (f, a), (f, d), (f, f), (f, g), (i, a), (i, d), (i, f), (i, g)}　**59.** {b, c, e, f, i}　**60.** Ø
61. {h}　**62.** {a, c, d, e, g, h}　**63.** {c, d, f, h}　**64.** A = {a, b, e, g}　**65.** U = {a, b, c, d, e, f, g, h}
66. A = {a, b, e, g}　**67.** Ø　**68.** A = {a, b, e, g}　**69.** A = {a, b, e, g}　**70.** Ø
71. A = {a, b, e, g}　**72.** Ø　**73.** A' = {c, d, f, h}　**74.** Ø　**75.** A is a subset of B.
76. B is a subset of A.　**77.** B is a subset of A.　**78.** A is a subset of B.
79. {x | x is a number between 1 and 20 inclusive or between 60 and 80 inclusive}　**80.** {20}　**81.** Ø
82. {x | x is a number between 1 and 80 inclusive}　**83.** {x | x is a number between 20 and 80 inclusive}
84. {x | x is a number between 60 and 80 inclusive}　**85.** A = Ø　**86.** Yes, if A = B　**87.** B
88. Ø　**89.** B　**90.** C　**91.** C　**92.** D　**93.** B　**94.** B　**95.** U
96. **c)** All men in the town who shave themselves or who do not shave themselves.　**d)** Yes
97. **a)** 2　**b)** 3　**c)** 6　**98.** pq
99. {(finished basement, solar heat), (finished basement, oil heat), (finished basement, gas heat), (unfinished basement, solar heat), (unfinished basement, oil heat), (unfinished basement, gas heat)}
100. 427 + 309 + 296 + 203 = 1235　**101.** 522　**102.** 2895　**103.** 309　**104.** 3879　**105.** 784

106. 1211   **107.** 805   **108.** 1490   **109.** 301   **110.** 234   **111.** 48   **112.** 125   **113.** 140
114. 48   **115.** 128   **116.** 27   **117.** 172   **118.** 97   **119.** 46   **120.** 154   **121.** 119
122. 154   **123.** 106   **124.** 126   **125.** 316   **126.** 0

Section 1.4 (pages 25–26)

1. 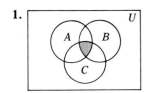   2. 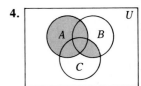   3. 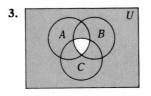   4.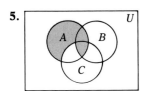

5. 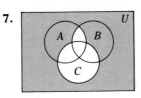   6. 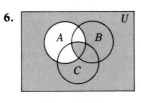   7.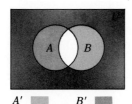

8. The set of all tall intelligent people who are not pleasant
9. The set of all pleasant people who are also either tall or intelligent
10. The set of all tall people who are also either not intelligent or pleasant
11. The set of all people who are not tall or who are intelligent and pleasant
12. The set of all pleasant people who are also either not tall or intelligent
13. The set of all tall intelligent people excluding those who are pleasant
14. $27 + 36 + 22 + 53 + 17 + 41 + 39 = 235$ **15.** 22   **16.** $27 + 53 + 39 + 58 = 177$
17. $27 + 53 + 17 + 41 + 39 + 58 = 235$ **18.** 121   **19.** $27 + 53 + 36 + 41 + 58 = 215$ **20.** 39
21. 12   **22.** 28   **23.** 58   **24.** 61   **25.** 2   **26.** 28
27. region $2 = A' \cap B' \cap C$
region $3 = A' \cap B \cap C$
region $4 = A' \cap B \cap C'$
region $5 = A \cap B \cap C'$
region $6 = A \cap B \cap C$
region $7 = A \cap B' \cap C$
region $8 = A \cap B' \cap C'$

28.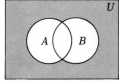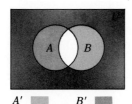

$(A \cup B)'$ is the shaded
portion

A' B'
$A' \cup B'$ is the region
that has both kinds of
shading

29. $A' \cap B \cap C'$ **30.** $A' \cap B \cap C$ **31.** $A \cap B' \cap C$ or $A \cap B'$ **32.** $A \cap C$

33. Not true   **34.** Not true   **35.** Not true   **36.** Not true   **37.** Not true   **38.** Not true
39. True   **40.** Not true

41. 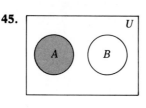   **42.** 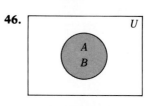   **43.**

44. 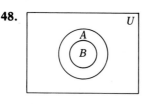   **45.** 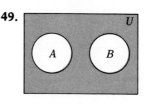   **46.**

47.   **48.**   **49.**

50. $A' \cup B'$ **51.** $\emptyset$ **52.** $A \cap B'$ **53.** $A \cup B$ **54.** U **55.** $A \cup B'$

Section 1.5 (pages 31–34)

1. 102   **2.** 189   **3.** 79   **4.** 156   **5.** 203   **6.** 62   **7.** 21   **8.** 19   **9.** 88   **10.** 26   **11.** 59
12. 29   **13.** 39   **14.** 27   **15.** 18   **16.** 0   **17.** 93   **18.** 57   **19.** 17   **20.** 30   **21.** 49
22. 30   **23.** 100   **24.** 10   **25.** 50   **26.** 30   **27.** 70   **28.** 60   **29.** 30   **30.** 50   **31.** 70
32. 159   **33.** 107   **34.** 177   **35.** 86   **36.** 58   **37.** 27
38. The numbers are inconsistent. They do not add up to 100.   **39.** Yes. The numbers are inconsistent.
40. Phillies *vs.* Cubs; Pirates *vs.* Cardinals; Mets *vs.* Astros; Dodgers *vs.* Giants   **41.** 1420   **42.** 620
43. Any committee that consists of at least four of these people
44. Any committee that consists of at most two of these people
45. Any committee that consists of exactly four of these people
46. Yes. Any committee that consists of exactly three of these people
47. Any committee that consists of at least four of these people or three of these people if Ann is one of the members also
48. Any committee that consists of at most two of these people or any committee of three people if Ann is *not* one of the members
49. Any three person committee of which Ann is a member   **50.** No   **51.** No   **52.** Yes
53. Any coalition of members that has 12 votes among them   **54.** There are none
55. Minimal winning: any coalition consisting of the "Big Five" and four others
  Losing: any coalition consisting of six or fewer members that does not include any of the "Big Five" countries
  Blocking: any coalition consisting of at least one "Big Five" country

Section 1.6 (pages 40–41)

1. $\{3, 6, 9, 12, \ldots, n, \ldots\}$ ⇕ ⇕ ↖ ↖ ↖
$\{6, 12, 18, 24, \ldots, 2n, \ldots\}$

2. $\{7, 14, 21, 28, \ldots, 7n\}$ ⇕ ⇕ ⇕ ⇕ ⇕
$\{8, 16, 24, 32, \ldots, 8n\}$

3. $\{1, 8, 27, 64, \ldots, n\}$ ⇕ ⇕ ⇕ ⇕ ↗
$\{1, \frac{1}{8}, \frac{1}{27}, \frac{1}{64}, \ldots, \frac{1}{n}\}$

4. $\{1, 8, 27, 64, \ldots, n^3\}$ ⇕ ↖ ↖ ↖ ↖
$\{1, 16, 81, 256, \ldots, n^4\}$

5. $\{4, 8, 12, 16, 20, \ldots, 4n\}$ ⇕ ↖ ↖ ↖ ↖
$\{4, 16, 64, 256, 1024, \ldots, 4^n\}$

6. Draw lines from the center of the circle to the square.

7. Draw lines from the center of the inner figure to the outside figure.

9. No elements can be the same for the result to be unique. **10.** 8 **11.** 15 **12.** 6 **13.** $\aleph_0$

14. $\aleph_0$ **15.** $\aleph_0$ **16.** $\aleph_1$

17. An infinite number, as the guest that is in room n is now moved to room $n + 1$. We then have the following 1:1 correspondence between the sets.

$$\{1, 2, 3, \ldots, n\}$$
↑ ↑ ↑ ↖
$$\{2, 3, 4, \ldots, n + 1\}$$

Mastery Tests
Form A (pages 42–44)

1. (d) **2.** (a) **3.** (b) **4.** (b) **5.** (a) **6.** (b) **7.** (d) **8.** (d) **9.** (e) **10.** (d) **11.** (d)

12. (b) **13.** (c) **14.** (b) **15.** (b)

Form B (pages 44–45)

1. (b) **2.** (c) **3.** (a) **4.** (a) **5.** (c) **6.** 530 **7.** (c) **8.** (c) **9.** (a) **10.** (c) **11.** (b)

12. 7 **13.** $A' \cap B$ **14.** 15 **15.** 45

CHAPTER 2

Section 2.2 (pages 53–54)

1. Inductive **2.** Deductive **3.** Inductive **4.** Inductive

5. The one that offers the semiannual raise

6.

7. 25 **8.** 8 **9.** 21 **10.** 1 **11.** 17 **12.** T **13.** 50 **14.** 3 **15.** $\frac{6}{9}$

Section 2.3 (pages 58–59)

1.

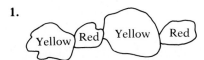

2. 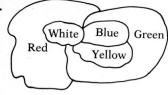 **3.**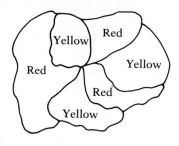

4.

Number of points on the circle	Number of regions
5	16
6	31
7	57
8	99

5. 8   **6.** 14   **7.** 22

8. If there are n circles, the plane is divided into a maximum of $n^2 - n + 2$ regions.
9. No natural numbers are known to exist.
10. $5^3 = 15^2 - 10^2$ **11.** $1^3 + 2^3 + 3^3 + 4^3 + \cdots + n^3 = (1 + 2 + 3 + 4 + \cdots + n)^2$
$\quad 6^3 = 21^2 - 15^2$
15. Row
$\quad\quad$ 6 $\quad$ 1 $\;$ 6 $\quad$ 15 $\quad$ 20 $\quad$ 15 $\quad$ 6 $\;$ 1
$\quad\quad$ 7 $\;$ 1 $\;$ 7 $\;$ 21 $\quad$ 35 $\quad$ 35 $\quad$ 21 $\;$ 7 $\;$ 1
16. 2 $\quad$ **17.** 4 $\quad$ **18.** 8 $\quad$ **19.** 16 $\quad$ **20.** 2^n

Section 2.4 (pages 64–65)

1. 8

2.

4	3	8
9	5	1
2	7	6

3.

8	1	6
3	5	7
4	9	2

4.

16	3	2	13
5	10	11	8
9	6	7	12
4	15	14	1

5.

17	24	1	8	15
23	5	7	14	16
4	6	13	20	22
10	12	19	21	3
11	18	25	2	9

6.

39	46	23	30	37
45	27	29	36	38
26	28	35	42	44
32	34	41	43	25
33	40	47	24	31

7.

6	2	34	33	35	1
25	11	27	10	8	30
19	23	16	15	20	18
13	14	22	21	17	24
12	29	9	28	26	7
36	32	3	4	5	31

8.

1	14	15	4
12	7	6	9
8	11	10	5
13	2	3	16

4	5	9	16
15	6	10	3
14	7	11	2
1	16	4	13

9.

32	41	50	3	12	21	30
40	49	9	11	20	29	31
48	8	10	19	28	37	39
7	16	18	27	36	38	47
15	17	26	35	44	46	6
23	25	34	43	45	5	14
24	33	42	51	4	13	22

10. $\dfrac{n(n^2 + 1)}{2}$ if $n \leq 6$

Section 2.5 (pages 70–72)

1. Invalid   **2.** Invalid   **3.** Valid   **4.** Valid   **5. a)** Valid **b)** Valid **c)** Valid **d)** Invalid
6. a) Invalid **b)** Invalid **c)** Valid **d)** Invalid   **7. a)** Invalid **b)** Invalid **c)** Invalid
8. a) Valid **b)** Invalid **c)** Invalid **d)** Invalid **e)** Valid
9. a) Invalid **b)** Invalid **c)** Invalid **d)** Invalid **e)** Invalid
10. a) Valid **b)** Valid **c)** Valid **d)** Invalid **e)** Invalid **f)** Invalid
11. a) Invalid **b)** Invalid **c)** Valid **d)** Invalid **e)** Invalid
12. a) Invalid **b)** Invalid **c)** Invalid **d)** Invalid   **13.** Valid
14. (1) Invalid (2) Invalid (3) Invalid (4) Invalid (5) Invalid
15. (1) Valid (2) Valid (3) Valid (4) Invalid (5) Invalid

Section 2.6 (pages 75–76)

1. Invalid   **2.** Invalid   **3.** Invalid   **4.** Invalid   **5.** Invalid   **6.** Invalid   **7.** Invalid
8. Invalid   **9.** Invalid   **10.** Invalid   **11.** Invalid   **12.** Valid

Section 2.7 (pages 87–88)

1. Not a proposition   **2.** Proposition   **3.** Not a proposition   **4.** Proposition
5. Not a proposition   **6.** Proposition   **7.** Not a proposition   **8.** Proposition
9. I love music and I attend rock concerts.   **10.** I do not attend rock concerts.
11. Either I love music or I attend rock concerts.
12. I do not love music but (and) I attend rock concerts.
13. I do not love music and I do not attend rock concerts.
14. I love music but (and) I do not attend rock concerts.
15. It is not true that I love music or attend rock concerts.
16. Either I do not love music or I attend rock concerts.
17. Either I do not love music or I do not attend rock concerts.
18. It is not true that I love music and attend rock concerts.
19. $p \rightarrow q$ **20.** $p \wedge (\sim q)$ **21.** $(\sim p) \wedge (\sim q)$ **22.** $\sim (p \wedge q)$ **23.** $q \wedge (\sim p)$ **24.** $(\sim q) \rightarrow (\sim p)$
25. No airlines have a better safety record than others.
26. Some instant coffees do not raise the blood cholesterol level of humans.
27. All cardiologists carry extensive medical malpractice insurance.
28. Some presidents of the U.S. have died of cancer while in office.
29. If I am afraid of heights then I have flown in an airplane.
30. If I have flown in an airplane then I am afraid of heights.
31. If I am afraid of heights then I have not flown in an airplane.

32. If I am not afraid of heights then I have flown in an airplane.

33. If I am not afraid of heights then I have not flown in an airplane.

34. If I have not flown in an airplane then I am not afraid of heights.

35. *Converse:* If the seats are floating, then an airplane has crashed into the ocean.
Inverse: If an airplane does not crash into the ocean, then the seats will not float.
Contrapositive: If the seats are not floating, then an airplane has not crashed into the ocean.

36. *Converse:* If you use a lot of aspirins, then you suffer from migraine headaches.
Inverse: If you don't suffer from migraine headaches, then you don't use a lot of aspirins.
Contrapositive: If you don't use a lot of aspirins, then you don't suffer from migraine headaches.

37. *Converse:* If college tuition costs will increase moderately, then the consumer price index has risen by more than 4%.
Inverse: If the consumer price index does not rise by more than 4%, then college tuition costs will not increase moderately.
Contrapositive: If the college tuition costs will not increase moderately, then the consumer price index will not rise by more than 4%.

38. (b) and (d)   **39.** No   **40.** No

41. a) If your luggage does not usually arrive safely at its destination, then it was not fastened securely on all overseas flights.   **b)** If your luggage is fastened securely on all overseas flights, then it will usually not arrive safely at its destination.

42. Joan has good recommendations, and if Joan has a high college grade point average then she was accepted to medical school.

43. Either Joan has good recommendations or Joan has a high college grade point average, and she was accepted to medical school.

44. Either Joan does not have good recommendations or she does not have a high college grade point average nor was she accepted to medical school.

45. Either Joan has good recommendations or if she does have a high college grade point average then she was not accepted to medical school.

46. If Joan does not have good recommendations, then she was accepted to medical school, and she has a high college grade point average.

47. Joan does not have a high college grade point average or was not accepted to medical school and has good recommendations.

Section 2.8 (pages 94–95)

1.

p	q	p → q	~(p → q)	
T	T	T	F	
T	F	F	T	Neither
F	T	T	F	
F	F	T	F	

3.

p	q	p ∧ q	p ∨ q	(p ∧ q) → (p ∨ q)	
T	T	T	T	T	
T	F	F	T	T	Tautology
F	T	F	T	T	
F	F	F	F	T	

5.

p	q	p ∨ q	q ∨ p	(p ∨ q) → (q ∨ p)	~[(p ∨ q) → (q ∨ p)]	
T	T	T	T	T	F	
T	F	T	T	T	F	Self-contradiction
F	T	T	T	T	F	
F	F	F	F	T	F	

7.

~p	q	r	q ∨ r	(~p) → (q ∨ r)	
T	T	T	T	T	
T	T	F	T	T	
T	F	T	T	T	
T	F	F	F	F	Neither
F	T	T	T	T	
F	T	F	T	T	
F	F	T	T	T	
F	F	F	F	T	

9.

p	q	~s	p → q	(p → q) ∧ (~s)	
T	T	T	T	T	
T	T	F	T	F	
T	F	T	F	F	
T	F	F	F	F	Neither
F	T	T	T	T	
F	T	F	T	F	
F	F	T	T	T	
F	F	F	T	F	

11.

p	q	r	~r	p → q	(p → q) ∧ (~r)	[(p → q) ∧ (~r)] → r	
T	T	T	F	T	F	T	
T	T	F	T	T	T	F	
T	F	T	F	F	F	T	
T	F	F	T	F	F	T	Neither
F	T	T	F	T	F	T	
F	T	F	T	T	T	F	
F	F	T	F	T	F	T	
F	F	F	T	T	T	F	

13.

~p	~q	r	(~q) ∨ r	(~p) → ((~q) ∨ r)	~[(~p) → ((~q) ∨ r)]	
T	T	T	T	T	F	
T	T	F	T	T	F	
T	F	T	T	T	F	
T	F	F	F	F	T	Neither
F	T	T	T	T	F	
F	T	F	T	T	F	
F	F	T	T	T	F	
F	F	F	F	T	F	

15.

p	~q	r	~r	p ∧ (~q)	(p ∧ (~q)) → r	[(p ∧ (~q)) → r] → (~r)	
T	T	T	F	T	T	F	
T	T	F	T	T	F	T	
T	F	T	F	F	T	F	
T	F	F	T	F	T	T	
F	T	T	F	F	T	F	Neither
F	T	F	T	F	T	T	
F	F	T	F	F	T	F	
F	F	F	T	F	T	T	

17.

p	q	~r	s	p → q	s → (~r)	(p → q) ∨ [s → (~r)]
T	T	T	T	T	T	T
T	T	T	F	T	T	T
T	T	F	T	T	F	T
T	T	F	F	T	T	T
T	F	T	T	F	T	T
T	F	T	F	F	T	T
T	F	F	T	F	F	F
T	F	F	F	F	T	T
F	T	T	T	T	T	T
F	T	T	F	T	T	T
F	T	F	T	T	F	T
F	T	F	F	T	T	T
F	F	T	T	T	T	T
F	F	T	F	T	T	T
F	F	F	T	T	F	T
F	F	F	F	T	T	T

Neither

19.

p	~p	q	~r	p ∧ q	(p ∧ q) → (~r)	[(p ∧ q) → (~r)] → (~p)
T	F	T	T	T	T	F
T	F	T	F	T	F	T
T	F	F	T	F	T	F
T	F	F	F	F	T	F
F	T	T	T	F	T	T
F	T	T	F	F	T	T
F	T	F	T	F	T	T
F	T	F	F	F	T	T

Neither

21.

p	q	~p	~q	p → q	(~p) → (~q)
T	T	F	F	T	T
T	F	F	T	F	T
F	T	T	F	T	F
F	F	T	T	T	T

These don't match.

22. 16   **23.** 32   **24.** 2^n **25.** True   **26.** False   **27.** False   **28.** True   **29.** True   **30.** True
31. True   **32.** False   **33.** False

34.

p	q	~ (p ∨ q)	(~p) ∧ q	~p
T	T	F	F	F
T	F	F	F	F
F	T	F	T	T
F	F	T	F	T

Other answers are possible.

35.

p	q	$p \to q$	$[p \land (\sim q)] \lor [(\sim p) \land q]$	$\sim(p \land q)$
T	T	T	F	F
T	F	F	T	T
F	T	T	T	T
F	F	T	F	T

36. False **37.** True **38.** True **39.** False
40. Not equivalent since the truth table for $\sim(p \lor q)$ and for $(\sim p) \lor (\sim q)$ are different.
41. Not equivalent **42.** Not equivalent **43.** Equivalent **44.** Not equivalent
45. Not equivalent

46.

p	q	$p \veebar q$
T	T	F
T	F	T
F	T	T
F	F	F

47.

p	q	$p \downarrow q$
T	T	F
T	F	F
F	T	F
F	F	T

48.

p	q	$\sim p$	$p \downarrow q$	$\sim(p \downarrow q)$	$(\sim p) \veebar q$	$\sim(p \downarrow q) \land [(\sim p) \veebar q]$
T	T	F	F	T	T	T
T	F	F	F	T	F	F
F	T	T	F	T	F	F
F	F	T	T	F	T	F

Section 2.9 (pages 98–99)

1. Valid **2.** Invalid **3.** Invalid **4.** Valid **5.** Invalid **6.** Valid

7. *Hypotheses:* $p \to q$

$\dfrac{\qquad q \qquad}{}$ Invalid

Conclusion: p

8. *Hypotheses:* $p \lor q$

$\dfrac{\qquad \sim q \qquad}{}$ Valid

Conclusion: p

9. *Hypotheses:* $p \to q$

$\dfrac{\qquad \sim q \qquad}{}$ Valid

Conclusion: $\sim p$

10. *Hypotheses:* $p \to q$

$\dfrac{\qquad q \to r \qquad}{}$ Valid

Conclusion: $p \to r$

11. *Hypotheses:* $p \to q$

$q \to r$ Valid

$\dfrac{\qquad \sim q \qquad}{}$

Conclusion: $p \to (\sim r)$

12. Hypotheses: $p \to (q \lor r)$

$\dfrac{\qquad q \to (\sim r) \qquad}{}$ Invalid

Conclusion: $\sim p$

13. *Hypotheses:* $p \to q$

$q \to (\sim r)$ Invalid

$\dfrac{\qquad q \lor r \qquad}{}$

Conclusion: $p \to r$

14. *Hypotheses:* $p \to (q \lor r)$

$r \to (\sim p)$ Invalid

$\dfrac{\qquad p \qquad}{}$

Conclusion: q

15. *Hypotheses:* $p \lor q$

$$\frac{\quad\quad \sim p \quad\quad}{}$$ Valid

Conclusion: q

16. *Hypotheses:* $(\sim p) \to q$

$$\frac{\quad\quad p \to r \quad\quad}{\sim q}$$ Invalid

Conclusion: $\sim r$

17. *Hypotheses:* $p \to (q \lor r)$

$$\frac{\quad\quad (\sim r) \to (\sim q) \quad\quad}{}$$ Invalid

Conclusion: $\sim p$

18. *Hypotheses:* $p \to q$

$$\frac{\quad\quad q \to [r \land (\sim s)] \quad\quad}{}$$ Valid

Conclusion: $s \to (\sim q)$

19. *Hypotheses:* $(\sim p) \to q$

$$\frac{\quad\quad q \to (r \land s) \quad\quad}{}$$ Valid

Conclusion: $(\sim s) \to p$

20. *Hypotheses:* $p \to (q \lor r)$ Invalid

Conclusion: $(\sim q) \to (\sim p)$

Section 2.10 (pages 104–105)

1. $[P \land (\sim Q)] \lor Q$

P	Q	$\sim Q$	$P \land (\sim Q)$	$[P \land (\sim Q)] \lor Q$
T	T	F	F	T
T	F	T	T	T
F	T	F	F	T
F	F	T	F	F

2. $[P \land (\sim Q)] \lor [(\sim P) \land Q]$

P	Q	$\sim P$	$\sim Q$	$P \land (\sim Q)$	$(\sim P) \land Q$	$[P \land (\sim Q)] \lor [(\sim P) \land Q]$
T	T	F	F	F	F	F
T	F	F	T	T	F	T
F	T	T	F	F	T	T
F	F	T	T	F	F	F

3. $P \land Q \land [(P \land (\sim Q)) \lor Q]$

P	Q	$\sim Q$	$P \land (\sim Q)$	$(P \land (\sim Q)) \lor Q$	$P \land Q \land [(P \land (\sim Q)) \lor Q]$
T	T	F	F	T	T
T	F	T	T	T	F
F	T	F	F	T	F
F	F	T	F	F	F

4. $P \lor (\sim Q) \lor (P \land Q)$

P	Q	$\sim Q$	$P \land Q$	$P \lor (\sim Q) \lor (P \land Q)$
T	T	F	T	T
T	F	T	F	T
F	T	F	F	F
F	F	T	F	T

5. $[(P \land (\sim Q)) \lor ((\sim P) \land Q)] \lor (P \land Q)$

P	Q	$\sim P$	$\sim Q$	$P \land (\sim Q)$	$(\sim P) \land Q$	$(P \land (\sim Q)) \lor ((\sim P) \land Q)$	$P \land Q$	$[(P \land (\sim Q)) \lor ((\sim P) \land Q)] \lor (P \land Q)$
T	T	F	F	F	F	F	T	T
T	F	F	T	T	F	T	F	T
F	T	T	F	F	T	T	F	T
F	F	T	T	F	F	F	F	F

6. $(P \wedge Q) \vee (R \vee (\sim P))$

P	Q	R	$\sim P$	$P \wedge Q$	$R \vee (\sim P)$	$(P \wedge Q) \vee (R \vee (\sim P))$
T	T	T	F	T	T	T
T	T	F	F	T	F	T
T	F	T	F	F	T	T
T	F	F	F	F	F	F
F	T	T	T	F	T	T
F	T	F	T	F	T	T
F	F	T	T	F	T	T
F	F	F	T	F	T	T

7. $[(P \wedge (\sim Q)) \vee (R \vee Q)] \wedge (\sim P)$

P	Q	R	$\sim P$	$\sim Q$	$P \wedge (\sim Q)$	$R \vee Q$	$[(P \wedge (\sim Q)) \vee (R \vee Q)]$	$[(P \wedge (\sim Q)) \vee (R \vee Q)] \wedge (\sim P)$
T	T	T	F	F	F	T	T	F
T	T	F	F	F	F	T	T	F
T	F	T	F	T	T	T	T	F
T	F	F	F	T	T	F	T	F
F	T	T	T	F	F	T	T	T
F	T	F	T	F	F	T	T	T
F	F	T	T	T	F	T	T	T
F	F	F	T	T	F	F	F	F

8. Not electrically equivalent   **9.** Electrically equivalent

10.

11.
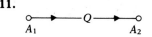

12.
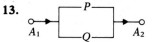

13.
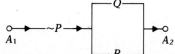

14. Exercise 1 can be simplified to $P \vee Q$
Exercise 3 can be simplified to $P \wedge Q$
Exercise 5 can be simplified to $P \vee Q$
Exercise 7 can be simplified to $P \wedge (\sim P)$

15.
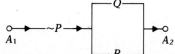

16.

17. 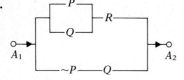 **18.**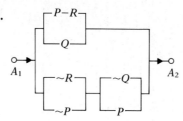

Section 2.11 (pages 107–110)

1.

p	q	$p \wedge q$
T	T	T
T	F	F
T	M	M
M	T	M
M	F	F
M	M	M
F	T	F
F	F	F
F	M	F

2.

p	q	$p \vee q$
T	T	T
T	F	T
T	M	T
M	T	T
M	F	M
M	M	M
F	T	T
F	F	F
F	M	M

3.

p	$\sim p$	q	$(\sim p) \wedge q$	$[(\sim p) \wedge q] \vee p$
T	F	T	F	T
T	F	F	F	T
T	F	M	F	T
M	M	T	M	M
M	M	F	F	M
M	M	M	M	M
F	T	T	T	T
F	T	F	F	F
F	T	M	M	M

4.

Logic expression	Set expression
$p \wedge q$	$P \cap Q$
$p \vee q$	$P \cup Q$
$\sim p$	P'
$p \rightarrow q$	$P' \cup Q$

6. Bill Holland is going to England; Pat Canada is going to Holland; Debbie English is going to Canada.
7. The blind person wore a white hat.   **9.** No   **14.** Mr. Gusher did it.
15. Pointing to one particular road, she would ask, "Would a member of the other tribe tell me to take this road?"

Mastery Tests
Form A (pages 112–113)

1. (c)   **2.** (c)   **3.** (c)   **4.** (a)   **5.** (b)   **6.** (a)   **7.** (b)   **8.** (b)   **9.** (a)   **10.** (a)   **11.** (b)
12. (b)   **13.** (d)   **14.** (a)   **15.** (a)

Form B (pages 113–115)

1. (c)   **2.** (a)

3.

4. (b)   **5.** (b)   **6.** (c)   **7.** (b)   **8.** (a)   **9.** (b)   **10.** (a)
11. $[(Q \lor (\sim P)) \land R] \lor [(\sim P) \land Q] \lor [(P \land (\sim Q)) \lor (\sim R)]$ **12.** (a)
13. If you are audited by the IRS, then you did not donate money to the United Way Fund.
14. 4   **15.** 111

CHAPTER 3

Section 3.2 (pages 121–124)

1. 336   **2.** 22,135   **3.** 1,247   **4.** 12,434   **5.** 1119   **6.** 11,229

7. 999 ∩∩∩∩∩∩ ||| 　**8.** ⌇⌇⌇⌇ 99 ∩∩∩ |||| ||| 　**9.** ∩∩∩∩∩∩∩∩ ||||

10. ⫽⫽ ⌇⌇⌇⌇⌇⌇⌇ 999999 ∩∩∩∩∩∩∩∩ ||| || 　**11.** ⌇⌇⌇⌇⌇ 9999 ∩∩||

12. ⌇⌇⌇⌇⌇⌇|| 　**13.** ⟋ ∩ ||| ||| 　**14.** ⟋ ⌇9∩|

15. ⌇⌇⌇⌇ ⌇9999999 ∩∩∩∩∩∩ || ||| ;5775

16. ⫽⫽⫽ ⌇⌇⌇⌇ 999999999 ∩∩∩∩∩∩ |||| ;34,978   **17.** 99 ∩ ;210

18. ⌇⌇⌇⌇⌇⌇⌇ 999999999 ∩∩∩∩∩∩ |||| ||| ;7988

19. Since position of symbols was not significant

20. (symbols)   **21.** (symbols)   **22.** (symbols)

23. (symbols)   **24.** (symbols)   **25.** (symbols)

26. (symbols)   **27.** (symbols)

28. (symbols)   **29.** (symbols)   **30.** (symbols)

31. (symbols)   **32.** (symbols)

33. 〒 ⋓ 冖 㫖 冖 ⋔ 冖 ⋓ 〒 **34.** 〒 㫖 冖 ⋔ ㊀ ⋓ 冖

35. 〒 ⋓ 〒 㫖 2 ⋔ 2 ⋓ 〒 **36.** (dots) **37.** (dots) **38.** (bars) **39.** (dots)

40. (dots) **41.** (dots) **42.** • moved up one space **43.** • moved up two spaces **44.** 28

Section 3.3 (pages 127–132)

1. LIX **2.** XCIV **3.** LXIX **4.** MCCXXXIV **5.** 84 **6.** 209 **7.** 1149 **8.** 1822
9. MDCLX **10.** MCCLXVI **12.** 1,932 **13.** 19,872 **14.** 39,474 **15.** 1,052,296 **16.** 3,056,318
17. 4,884,420 **19.** 424 **20.** 585 **21.** 23,936 **22.** 44,536 **23.** 36,423 **24.** 41,472
25. 222,385 **26.** 187,813 **27.** 3,114,501 **28.** 44,061,424 **29. a)** 3,166,000 **30.** 2394
31. 1326 **32.** 1485

33. 999999 ∩∩∩ ‖‖' **34.** ⸿⸿⸿ 99999999 ∩∩∩∩∩∩ ‖‖‖‖

35. 1,716 **36.** 16,497 **37.** 13,166 **38.** 37,196 **39.** 11,466 **40.** 52,353 **41.** 53,244
42. 12,264 **43.** 30,772 **44.** 36,103 **45.** 20,332 **47.** 63 **48.** 45 **49.** 72 **50.** 54
52. 3603 **53.** 2248 **54.** 4678 **55.** 2624 **56.** 24

Section 3.4 (pages 137–138)

1. $111100_{(2)}$ **2.** $22201_{(3)}$ **3.** $2258_{(9)}$ **4.** $23t0_{(12)}$ **5.** $2130_{(4)}$ **6.** $443_{(7)}$ **7.** $2567_{(8)}$ **8.** $20303_{(4)}$
9. $105_{(6)}$ **10.** $24222_{(6)}$ **11.** $52551_{(7)}$ **12.** $2t6487_{(12)}$ **13.** $221020202_{(3)}$ **14.** $56252_{(9)}$
15. $2261_{(9)}$ **16.** $314_{(5)}$ **17.** $451_{(6)}$ **18.** $11111_{(2)}$ **19.** $14t_{(12)}$ **20.** $322_{(4)}$ **21.** 9 **22.** 9
23. 9 **24.** 6 **25.** Otherwise there would be only one digit.
26. The answer should be $5247_{(9)}$, since in base 9 we borrow nines, not tens. **28.** Base 7 **29.** Base 8
30. Base 8 **31.** Base 12 **32.** $A = 0, M = 1, D = 2$
33. $A = 0, M = 1, D = 2, P = 3$ or $A = 0, M = 2, D = 1, P = 3$

Section 3.5 (pages 141–142)

1. $12213_{(4)}$ **2.** $1203_{(5)}$ **3.** $11210_{(3)}$ **4.** $652_{(8)}$ **5.** $666_{(9)}$ **6.** $1252_{(7)}$ **7.** $1442_{(6)}$ **8.** $235_{(12)}$
9. $116_{(16)}$ **10.** $2123_{(5)}$ **11.** $1010100_{(2)}$ **12.** $364_{(7)}$ **13.** $10000001_{(2)}$ **14.** $1e2_{(12)}$ **15.** $1163_{(8)}$
16. $27e4_{(12)}$

18. $217_{(10)} = 11011001_{(2)} = 331_{(8)}$. When regrouped 11011001 becomes $\underbrace{11}\ \underbrace{011}\ \underbrace{001}$ as
$$\quad 3\quad 3\quad 1$$
powers of 2, or 331 as powers of 8. Also $301_{(10)} = 100101101_{(2)} = 455_{(8)}$.
19. Advantages: fewer digits
Disadvantages: the numerals become very long because there are fewer digits available.

Section 3.6 (page 144)

1. 414 **2.** 123 **3.** 7465 **4.** 1583 **5.** 1149 **6.** 549 **7.** 14,810 **8.** 4047 **9.** 791
10. 409 **11.** 5500 **12.** 146 **13.** 580 **14.** 109 **15.** 961 **16.** 44,773 **17.** $230_{(5)}$
18. No, as $11_{(64)} = 65$ years old **19.** Bill Sommers

Section 3.8 (page 152)

1. Good combination   **2.** Good combination   **3.** Bad combination   **4.** Good combination
5. Good combination

Section 3.9 (pages 157–158)

1. 377, 610, 987, 1597, 2584, 4181, 6765, 10946   **2. a)** 7, one less   **b)** 12, one less   **c)** 20
3. 1, 1, 2, 3, 1, 0, 1, 1, 2, 3, 1, 0, 1, 1, 2, 3, 1, 0, 1, 1, 2, 3, 1, 0, 1, 1, 2, 3, 1, 0
5. To cell 5—13 possible paths
   To cell 6—21 possible paths
   To cell 7—34 possible paths
6. 29, 47, 76, 123, 199, 322   **8.** The divisors of 10 are 1, 2, and 5 (excluding 10 itself) and $1 + 2 + 5 \neq 10$
9. $6 = 110_{(2)}$ **10.** 120   **11.** 129   **12.** 246   **13.** 8   **14.** 55   **15.** 390,625
    $28 = 11100_{(2)}$
   $496 = 111110000_{(2)}$
  $8128 = 1111111000000_{(2)}$
16. Arithmetic   **17.** Geometric   **18.** Arithmetic   **19.** Arithmetic   **20.** Geometric
21. Not a wise decision as 2^{30} cents $\approx \$10,737,418 > \$25,000$

Section 3.10 (pages 163–165)

1. 2, 3, 5, 7, 11, 13, 17, 19, 23, 29, 31, 37, 41, 43, 47, 53, 59, 61, and 67   **2.** 2 and 3   **3.** 8, 9, and 10
4. 24, 25, 26, and 27
5. 2, 3, 5, 7, 11, 13, 17, 19, 23, 29, 31, 37, 41, 43, 47, 53, 59, 61, 67, 71, 73, 79, 83, 89 and 97
6. $2 \times 2 \times 2 \times 3 \times 3$ **7.** 7×7 **8.** $2 \times 2 \times 2 \times 5 \times 5 \times 5$ **9.** $3 \times 5 \times 5 \times 7$ **10.** $2 \times 7 \times 43$
11. 3×109 **12.** $2 \times 2 \times 2 \times 2 \times 5 \times 5$ **13.** $2 \times 2 \times 3 \times 5 \times 13$ **14.** $5 \times 5 \times 37$
15. $2 \times 2 \times 5 \times 5 \times 11$ **16.** Never in a 6, but it may end in a 5
17. $17 = 4 \cdot 4 + 1$; $29 = 4 \cdot 7 + 1$; $37 = 4 \cdot 9 + 1$
18. 11 can be written as $4 \cdot 2 + 3$; 19 can be written as $4 \cdot 4 + 3$; and 23 can be written as $4 \cdot 5 + 3$.
19. $3 = 2^2 - 1$; $7 = 2^3 - 1$; $31 = 2^5 - 1$ and $127 = 2^7 - 1$. Other answers are possible.
20. 2 and 3 are two possible answers.   **21.** 4 and 6 are two possible answers.
22. 11 and 13; 17 and 19; 29 and 31 are three possible answers.   **23.** 3, 5, and 7   **24.** No
25. a) 2, 3, 5, 7, 11, 13, 17, 19, 23, 29, 31, 37, 41, 43, and 47. There are 15 of them.
   b) Those indicated in part (a) and also, 53, 59, 61, 67, 71, 73, 79, 83, 89 and 97. There are 25 of them.
27. a) 19   **b)** 257   **c)** 2999

28. a)   **b)**

Prime number	Prime number +1	Prime number −1		Prime number +1	Prime number −1
5	6	4		0	4
7	8	6		2	0
11	12	10	Remainders	0	4
13	14	12		2	0
17	18	16		0	4
19	20	18		2	0

   c) When any prime number +1 is divided by 6, the remainders are always 0 or 2 if the prime
   number is larger than 3. When any prime number −1 is divided by 6, the remainders are always 0
   or 4 when the prime number is larger than 3.
29. 1979, 1981, and 1999   **30.** $18 = 5 + 13$; $18 = 7 + 11$
31. $10 = 5 + 5$; $12 = 5 + 7$; $14 = 7 + 7$; $16 = 5 + 11$; $18 = 7 + 11$; $20 = 7 + 13$; $22 = 11 + 11$;
   $24 = 11 + 13$; $26 = 7 + 19$; $28 = 11 + 17$; $30 = 11 + 19$; $32 = 13 + 19$; $34 = 17 + 17$; $36 = 17 + 19$

32. $21 = 3 + 7 + 11$; $23 = 5 + 7 + 11$; $25 = 5 + 7 + 13$; $27 = 3 + 7 + 17$; $29 = 5 + 7 + 17$; $31 = 3 + 11 + 17$; $33 = 5 + 11 + 17$; $35 = 7 + 11 + 17$; $37 = 7 + 13 + 17$
33. $2 = 1^2 + 1$; $17 = 4^2 + 1$ and $37 = 6^2 + 1$ **34.** $143 = 12^2 - 1$; $323 = 18^2 - 1$ **35.** Yes   **36.** Yes
37. Yes   **38.** Yes   **39.** Yes

Section 3.11 (pages 168–169)

1. $(7 \times 10^1) + (8 \times 10^0) + (1 \times 10^{-1})$ **2.** $(3 \times 10^2) + (1 \times 10^1) + (6 \times 10^0) + (0 \times 10^{-1}) + (1 \times 10^{-2})$
3. $(4 \times 10^2) + (1 \times 10^1) + (2 \times 10^0) + (3 \times 10^{-1}) + (7 \times 10^{-2})$
4. $(5 \times 10^1) + (1 \times 10^0) + (0 \times 10^{-1}) + (0 + 10^{-2}) + (1 \times 10^{-3})$
5. $(6 \times 10^2) + (1 \times 10^1) + (9 \times 10^0) + (8 \times 10^{-1}) + (2 \times 10^{-2}) + (3 \times 10^{-3})$
6. $(3 \times 10^3) + (4 \times 10^2) + (7 + 10^1) + (2 \times 10^0) + (1 \times 10^{-1}) + (3 \times 10^{-2})$
7. $(4 \times 10^3) + (6 \times 10^2) + (0 \times 10^1) + (9 \times 10^0) + (1 \times 10^{-1}) + (8 \times 10^{-2}) + (2 \times 10^{-3})$
8. $(3 \times 10^3) + (2 \times 10^2) + (4 \times 10^1) + (7 \times 10^0) + (1 \times 10^{-1}) + (9 \times 10^{-2}) + (3 \times 10^{-3})$
9. $(4 \times 10^2) + (5 \times 10^1) + (3 \times 10^0) + (0 \times 10^{-1}) + (2 \times 10^{-2}) + (5 \times 10^{-3}) + (8 \times 10^{-4})$ **10.** 4×10^2
11. 7×10^3 **12.** 3×10^{-3} **13.** 5×10^{-4} **14.** 1.7×10^7 **15.** 8×10^6 **16.** 7.8×10^{-7}
17. 8.79×10^{-8} **18.** 7.65432×10^3 **19.** 4.4×10^{-1} **20.** 6.123×10^{-4} **21.** -2 **22.** 2   **23.** 5
24. 6   **25.** 3   **26.** -11 **27.** -5 **28.** 2   **29.** -10 **30.** 10   **31.** 907
32. 1,800,000,000,000,000,000,000,000,000   **33.** 30,000,000,000   **34.** 26,000,000,000,000
35. 5,000,000,000
36. 13,200,000,000,000,000,000,000,000 pounds or 5,990,000,000,000,000,000,000,000 kilograms
37. 0.0000054   **38.** 93,000,000   **39.** 3,022,387   **40.** 0.00000000437   **41.** 9.5×10^{12} **42.** 2×10^9
43. 7.5×10^{-4} **44.** 1.016×10^{-2} **45.** 6×10^{23} **46.** 3.6×10^9 **47.** 3.15576×10^7 **48.** 1.8×10^4
49. 2×10^9 **50.** 9×10^6

Mastery Tests
Form A (pages 171–172)

1. (b)   **2.** (b)   **3.** (a)   **4.** (b)   **5.** (a)   **6.** (a)   **7.** (d)   **8.** (a)   **9.** (b)   **10.** (b)   **11.** (c)
12. (a)   **13.** $t5756_{(12)}$ **14.** $67_{(9)}$ **15.** $31_{(4)}$

Form B (pages 172–173)

1. $470_{(8)}$ **2.** (d)   **3.** (a)   **4.** 4,230,000   **5.** (a)   **6.** $2 \times 2 \times 2 \times 2 \times 3$ **7.** There are 16 of them.
8. $(4 \times 10^2) + (5 \times 10^1) + (1 \times 10^0) + (6 \times 10^{-1}) + (2 \times 10^{-2}) + (3 \times 10^{-3})$ **9.** 1.43×10^{11} **10.** $726_{(12)}$
11. 16   **12.** 31   **13.** (c)   **14.** $155_{(6)}$ **15.** $101110_{(2)}$

CHAPTER 4

Section 4.2 (pages 189–191)

1. Commutative law for addition   **2.** Associative law of multiplication
3. Distributive law of multiplication   **4.** Law of closure for addition   **5.** Associative law for addition
6. Commutative law for addition   **7.** Commutative law for multiplication
8. Associative law of multiplication   **9.** Distributive law of multiplication
10. Law of closure for multiplication   **11.** 1734   **12.** 1600   **13.** 3914   **14.** 2100   **15.** No
16. Yes   **17.** Yes   **18.** Yes   **19.** Yes   **20.** No   **21.** Yes   **22.** Not commutative
23. Commutative (in most cases)   **24.** Commutative   **25.** Commutative
26. Not commutative (legally)   **27.** Commutative   **28.** Not commutative
29. a) Yes b) Yes c) Yes   **30.** a) Yes b) No c) Yes
31. The cost of the gloves is the same on each day so that the total cost is $19 \times 6 + 19 \times 17$ or $19(6 + 17)$.

32. a) Yes **b)** Yes **33.** 1900 **34.** 9300 **35.** 1134

36.

```
        9
       / \
      6   4
     /     \
    8——5——10
```

(Other answers may be possible.)

37. 21, 28, and 36

38. $1 \cdot 9 + 2 = 11$
$12 \cdot 9 + 3 = 111$
$123 \cdot 9 + 4 = 1111$
.
.
.

39. $9 \cdot 9 + 7 = 88$
$98 \cdot 9 + 6 = 888$
$987 \cdot 9 + 5 = 8888$
.
.
.

40. $1 \cdot 8 + 1 = 9$
$12 \cdot 8 + 2 = 98$
$123 \cdot 8 + 3 = 987$
.
.
.

41. $3 \cdot 37 = 111$ and $1 + 1 + 1 = 3$
$6 \cdot 37 = 222$ and $2 + 2 + 2 = 6$
$9 \cdot 37 = 333$ and $3 + 3 + 3 = 9$
.
.
.

42. $1 \cdot 1 = 1$
$11 \cdot 11 = 121$
$111 \cdot 111 = 12321$
.
.
.

43. $7 \cdot 7 = 49$
$67 \cdot 67 = 4489$
$667 \cdot 667 = 444889$
.
.
.

44. $7 \cdot 15873 = 111111$
$14 \cdot 15873 = 222222$
$21 \cdot 15873 = 333333$

45. a) When the digits of *any* three digit number are rewritten in the same order to form a six digit number and the result divided by 13, there will never be any remainder.

b) Let the number be represented by *abcabc*. Then the number

$$abcabc = a \times 10^5 + b \times 10^4 + c \times 10^3 + a \times 10^2 + b \times 10 + c$$
$$= a(10^5 + 10^2) + b(10^4 + 10) + c(10^3 + 1)$$
$$= 10^2(1001a) + 10(1001b) + 10^0(1001c)$$
$$= 1001(100a + 10b + c)$$

The coefficient 1001 is divisible by both 13 and 7 as

$$\frac{1001}{13} = 77 \quad \text{and} \quad \frac{1001}{7} = 143.$$

46. There will never be a remainder. See the answer to Exercise 45.

47. 225 **48.** 1849 **49.** 1656 **50.** No. This is coincidental.

51. a) $\cup$ and $\cap$, that is, union and intersection **b)** $\cup$, that is, union only
c) None are, although there are special types of associative rules for $\cap$ and $\cup$.

Section 4.3 (pages 195–196)

1. Yes; 1 **2.** Yes **3.** Yes **4.** Commutative law of multiplication **5.** Place holder
6. a) Yes **b)** No **7.** When we divide both sides by $(b-a)$, we are dividing by 0.
8. When we divide both sides by $(a-b)$, we are dividing by 0.

Section 4.4 (pages 202–204)

1. +24   **2.** −11   **3.** +5   **4.** +1   **5.** −5   **6.** −1   **7.** −13   **8.** +8   **9.** −1   **10.** −24
11. +24   **12.** −24   **13.** −7   **14.** 0   **15.** +14   **16.** 0   **17.** No; we cannot divide by 0.
18. If x and y are any integers, then $x + y$ is an integer. If x, y, and z are any integers,
     then $x + (y + z) = (x + y) + z$.
19. Yes   **20.** Yes   **22.** Commutative law for addition   **23.** Distributive law of multiplication
24. Associative law for addition   **25.** Law of closure for multiplication
26. Associative law of multiplication   **27.** Distributive law of multiplication
28. Multiplication property of zero or law of closure for multiplication
29. Additive inverse property or law of closure for addition   **30.** Commutative law of multiplication
31. −1600   **32.** −900   **33.** +1900   **35. a)** Positive   **b)** Negative   **36.** 2411   **37.** Aristotle
38. a) $W \subset I$ **b)** $N \subset W$ **c)** $N \subset I$ **39. a)** W **b)** N **c)** N **d)** W **e)** I **f)** I **40.** 11   **41.** 8   **42.** 7
43. 0   **44.** 0   **45.** 4   **46.** 3   **47.** 5   **48.** 0
49. a) Let $a = 3$ and $b = -2$ so that $|3| > |-2|$; then $3 + (-2) = 1 = |3| - |-2| = 3 - 2 = +1$
     b) Let $a = 2$ and $b = -3$ so that $|-3| > |2|$. Then $2 + (-3) = -1 = -(|-3| - |2|) = -(3 - 2) = -1$

Section 4.5 (page 208)

1. 1, 2, 4, 7, 8, 14, 28, 56   **2.** 1, 2, 43, 86   **3.** 1, 2, 7, 14, 49, 98   **4.** 1, 2, 31, 62
5. 1, 2, 3, 6, 11, 12, 22, 44, 66, 132   **6.** 1, 2, 4, 5, 8, 10, 16, 20, 32, 40, 80, 160   **7.** 1, 149
8. 1, 2, 241, 482   **9.** 1, 2, 4, 8, 47, 94, 188, 376   **10.** 1, 2, 3, 6, 9, 18, 31, 62, 93, 186, 279, 558
11. Every nonzero number is a divisor of 0.   **12.** Divisible by 2, 3, 4, and 6   **13.** Divisible by 2 and 4
14. Divisible by 2, 3, 4, and 6   **15.** Divisible by 5   **16.** Not divisible by 2, 3, 4, 5, or 6
17. Divisible by 2, 3, 4, and 6   **18.** Divisible by 2 and 5   **19.** Divisible by 2, 3, and 6
20. Divisible by 2   **21.** Divisible by 2 and 4
22. b) Only 324, 864, 456, and 3402 are each divisible by 6.
23. a) A number is divisible by 8 if and only if the integer formed by its last 3 digits is divisible by 8.
     b) Only 864 and 456 are divisible by 8.
24. 72000 is one such number.   **25.** 181,440 is one such number.   **26.** 350,035 is one such number.
27. A number is divisible by 50 if its last 2 digits are both zeros or if its last 2 digits form the integer 50.
28. 144 is one such number. 128 is another such number.   **29.** Leap year   **30.** Not a leap year
31. Leap year   **32.** Not a leap year   **33.** Not a leap year   **34.** Leap year   **35.** Leap year
36. Not a leap year
37. The divisors of 220 are 1, 2, 4, 5, 55, 110, 44, 10, 22, 11, and 20. The sum of these divisors is 284. The
     divisors of 284 are 1, 142, 2, 71, and 4. The sum of these divisors is 220.
38. 1210
39. a) 35   **b)** 455   **c)** 2925   **d)** Always divisible by 6 when they are consecutive
     e) Also divisible by 6 when they are consecutive
40. False

Section 4.6 (pages 222–224)

1. $\dfrac{35}{36}$ **2.** $\dfrac{-13}{18}$ **3.** $\dfrac{1}{6}$ **4.** $\dfrac{1}{9}$ **5.** 2   **6.** −1   **7.** 5   **8.** $\dfrac{-6}{5}$ **9.** Equal   **10.** Not equal
11. Equal   **12.** Equal   **13.** Not equal   **14.** Not equal   **15.** Not equal   **16.** Equal
17. $\dfrac{16}{18}$ and $\dfrac{24}{27}$ **18.** $\dfrac{8}{26}$ and $\dfrac{12}{39}$ **19.** $\dfrac{10}{14}$ and $\dfrac{15}{21}$ **20.** $\dfrac{8}{22}$ and $\dfrac{12}{33}$ **21.** 3   **22.** $\dfrac{5}{9}$ **23.** $\dfrac{14}{9}$
24. $\dfrac{13}{12}$ **25.** $\dfrac{11}{63}$ **26.** $\dfrac{25}{12}$ **27.** $\dfrac{12}{17}$ **28.** $\dfrac{-104}{187}$ **29.** +1   **30.** $\dfrac{28}{15}$ **31.** $\dfrac{-64}{121}$ **32.** 128   **33.** $\dfrac{7}{54}$

34. $\dfrac{-25}{12}$ **36.** Division by 0 is not permissible.   **40.** Terminating

41. Nonterminating but repeating   **42.** Terminating   **43.** Terminating
44. Nonterminating but repeating   **45.** Terminating   **46.** Nonterminating and nonrepeating
47. Nonterminating and nonrepeating   **48.** Terminating   **49.** Not necessarily   **50.** Not necessarily
51. 0.625   **52.** 0.222 . . .   **53.** 0.6428571 . . .   **54.** 1.25   **55.** 0.8461538 . . .   **56.** 1.714285 . . .
57. 0.818181 . . .   **58.** 0.333 . . .   **59.** $\dfrac{83}{100}$ **60.** $\dfrac{1}{5}$ **61.** $\dfrac{58321}{1000}$ **62.** $\dfrac{29}{100,000,000}$ **63.** 3.03
64. 2.57   **65.** 166.36   **66.** 0.03   **67.** 1.63   **68.** 29.61   **69.** $\dfrac{28}{99}$ **70.** $\dfrac{89}{99}$ **71.** $\dfrac{19}{99}$ **72.** $\dfrac{356}{999}$
73. $\dfrac{832}{999}$ **74.** $\dfrac{722}{99}$ **75.** $\dfrac{681}{99}$ **76.** $\dfrac{57,729}{9900}$ **77.** $\dfrac{2}{3}$ **78.** 20.48 miles per gallon   **79.** \$31.79
80. $\dfrac{5}{3}$ and $\dfrac{3}{5}$ **81.** $\dfrac{1}{3}$ and $\dfrac{2}{3}$ **82.** $\dfrac{5}{3} - \dfrac{2}{3}$ **83.** $\dfrac{10}{3} \div \dfrac{5}{3}$

84. a) Definition of addition of rational numbers   **b)** Commutative law of multiplication
  c) Commutative law of addition   **d)** Definition of addition of rational numbers.
85. a) $a = 0, b = 1, c = 0,$ and $d = 1$ **b)** $a = 1, b = 2, c = 3,$ and $d = 4$ **86.** No

Section 4.7 (pages 230–231)

1. a) No   **b)** No   **c)** Yes
2. The product of $5\sqrt{3}$ and $4\sqrt{3}$ is rational, and the product of $5\sqrt{3}$ and $3\sqrt{7}$ is irrational.
12. 8 feet   **13.** 10 feet

Section 4.8 (pages 237–238)

1. a) I **b)** I **c)** $\emptyset$ **d)** R **e)** R **f)** W **g)** R **h)** Q **i)** I **2.** True   **3.** False   **4.** True   **5.** False
6. True   **7.** True

8–13.

14. $4.5 < 4.6$ **15.** $+5 > -6$ **16.** $-7 > -8$ **17.** $\dfrac{1}{4} > \dfrac{1}{5}$ **18.** $4 = 4$ **19.** $\dfrac{1}{8} = 0.125$

20. $\sqrt{3} > 1.7$ **21.** $\dfrac{1}{6} = 0.1666 \ldots$

Number	Natural number	Whole number	Integer	Rational number	Irrational number	Real number
22. 5	Yes	Yes	Yes	Yes	No	Yes
23. $\dfrac{-1}{8}$	No	No	No	Yes	No	Yes
24. 3.021021 . . .	No	No	No	Yes	No	Yes
25. 7.1	No	No	No	Yes	No	Yes
26. 0	No	Yes	Yes	Yes	No	Yes
27. $6\sqrt{3}$	No	No	No	No	Yes	Yes
28. 2.13587 . . .	No	No	No	No	Yes	Yes
29. $\dfrac{2}{\sqrt{3}}$	No	No	No	No	Yes	Yes

30. a)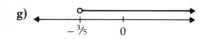

b) (number line: open circle at -6, arrow to right; -6 0)

c) (number line: solid from left to open circle at $\frac{3}{4}$; 0 $\frac{3}{4}$)

d) (number line: solid dot at -8, arrow to right; -8 0)

e) (number line: solid from left to open circle at -3; -3 0)

f) (number line: solid dot at 5, line to right; 0 5)

g) (number line: open circle at $-\frac{3}{5}$, line to right; $-\frac{3}{5}$ 0)

h) (number line: solid from left to solid dot at 0.8; 0 0.8)

32. b) Division by 0 is not permissible.   **33. a)** $\dfrac{7}{10}$ **34. a)** 2 **b)** 2.01001 . . .

35. In order to add or multiply, each must be converted to a fraction.

Thus $2.131313\ldots = \dfrac{211}{99}$, and $0.76767676 = \dfrac{76767676}{100{,}000{,}000}$.

a) $\dfrac{211}{99} + \dfrac{76767676}{100{,}000{,}000}$ **b)** $\left(\dfrac{211}{99}\right)\left(\dfrac{76767676}{100{,}000{,}000}\right)$

36. 50   **37.** 8   **38.** 21   **39.** 51   **40.** -20 **41.** 34   **42.** 89   **43.** 131   **44.** $\dfrac{1}{2}$ **45.** 85

46. 144   **47.** 324   **48.** -10 **49.** 63   **50.** 2142

Mastery Tests
Form A (pages 240–241)

1. $\dfrac{-4}{33}$ **2.** $\dfrac{18}{7}$ **3.** -5 **4.** Choice (c)   **5.** False   **6.** Yes   **7.** 3   **8.** $\dfrac{18}{99}$ **9.** False   **10.** $\dfrac{19}{56}$

11. $\dfrac{-35}{3}$ **12.** $\dfrac{11}{6}$ **13.** 0.41666 . . .   **14.** 48   **15.** $+3$

Form B (pages 241–242)

1. 2.5   **2.** Answer is 999,999
3. The divisors are 1, 2, 4, 8, 16, 23, 32, 46, 64, 92, 184, 368, 736, and 1472
4. $\dfrac{13}{64}$ **5.** 12 feet   **6.** $68(100 + 2) = 6800 + 136 = 6936$ **7.** Yes   **8.** $\dfrac{104}{15}$ **9.** (b)   **10.** (c)

11. (a)   **12.** $\dfrac{1}{2}$ **13.** 2713 years old   **14.** \$34.22   **15.** $\dfrac{8}{5}$

CHAPTER 5

Section 5.2 (pages 250–251)

1. 1680   **2.** 1.4   **3.** 855   **4.** 0.8032786 or 0.8033   **5.** 829   **6.** 9.45   **7.** 0.02001
8. 28,629,151   **9.** Most calculators will display an error message. Others will blink or display 0.
10. Most calculators will display an error message. Others will blink or display 0.   **11.** 16.37
12. 410,478.52   **13.** 0.00   **14.** 0.37   **15.** 5804.87   **16.** 173.94   **17.** 707.32   **18.** 83.31
19. 2945.50   **20.** 43.92   **21.** 5103   **22.** 81; No   **23.** 1   **24.** hELL   **25.** BELLS
26. 732.7391304   **27.** The same three-digit number that you started with

Section 5.3 (pages 255–256)

1. $\frac{16}{9}$ **2.** $\frac{1}{4}$ **3.** $\frac{3}{4}$ **4.** $\frac{12}{1}$ **5.** $\frac{7}{16}$ **6.** $\frac{20}{1}$ **7.** 16   **8.** Both are equal   **9.** $\frac{3}{5}$ **10.** 2

11. The one containing 10 bars for $2.25
12. Rent = $300; Food = $250; Education = $150; Entertainment = $100; and Miscellaneous items = $200.
13. 15 gallons of white paint and 10 gallons of blue paint
14. Advertisements = 156 pages; articles = 36 pages
15. Dried apricots = 160 pieces; dried apples = 140 pieces   **16.** Valid   **17.** Not valid
18. Not valid   **19.** Valid   **20.** Valid   **21.** Valid   **22.** 9   **23.** 40   **24.** 1   **25.** 13.5   **26.** 2.5
27. 4   **28.** $1075   **29.** 65 km/hr   **30.** $96   **31.** 20.54   **32.** 221.67 calories   **33.** $8.25
34. $11.10   **35.** $2460

Section 5.4 (pages 260–261)

1. 0.80 or 80%   **2.** 0.25 or 25%   **3.** 0.6667 or 66.67%   **4.** 0.7143 or 71.43%   **5.** 2.00 or 200%
6. 3.00 or 300%   **7.** 1.40 or 140%   **8.** 2.3333 or 233.33%   **9.** 1.6667 or 166.67%
10. 0.265 or 26.5%   **11.** 12%   **12.** 56.7%   **13.** 1%   **14.** 0.2%   **15.** 510%   **16.** 623%
17. 101%   **18.** 100%   **19.** 0.055   **20.** 0.0225   **21.** 0.06667   **22.** 0.125   **23.** 0.18   **24.** 0.015
25. 0.101   **26.** 0.1623   **27.** 6.65   **28.** 51.12   **29.** 47.396   **30.** 33.3   **31.** 80   **32.** 200
33. $4.12   **34.** $212.85   **35.** $526,000   **36.** 5.81%   **37.** 5.36%   **38. a)** 38.35%   **b)** 15.67%
39. $169   **40.** 20%   **41.** 8.33%   **42.** 34.62%   **43.** 1.26%
44. 403.76 or approximately 404 people   **45.** $161.40   **46.** $583.33

Section 5.5 (pages 269–270)

1. $4320   **2.** $2640   **3.** 32.42%   **4.** 27%   **5.** 0.3889 years   **6.** $4825.31   **7.** 7.25%
8. 6.45%   **9.** $14,265.58   **10.** $9843.09   **11.** $11,860.29
12. a) $14,707.67   **b)** $14,939.26   **c)** $15,019.98   **d)** $15,020.88   **13.** $3272.73   **14.** $2,000
15. Julia, as she will have $2180.92 and Miguel will have $2180.83
16. In a bank that pays 7% interest compounded semiannually

Section 5.6 (pages 286–287)

1. $2150   **2.** $3640   **3.** 10 months   **4.** $415.18   **5.** The second method where interest is 2%
6. $456.51   **7.** $568.86   **8.** $1397.83   **9.** $351.81   **10.** $879.75   **11.** $731.39   **12.** $124.92
13. $356.59   **14.** $33.56   **15.** $1062.36   **16.** 13.25%   **17. a)** $89.70   **b)** 14.50%   **18.** 35%

Section 5.7 (pages 294–295)

1. 0.9452   **2.** 0.2298   **3.** 0.9436   **4.** 0.3071   **5.** 0.7930   **6.** 0.0878   **7. a)** 2.38 years   **b)** 0.4857
8. a) 12.31 years   **b)** 0.0715   **9.** $1859   **10.** $1668   **11.** $5314.01
12. Year 1 = $636; Year 2 = $695; Year 3 = $760.01; Year 4 = $832; Year 5 = $910.99
13. Year 1 = $255.85; Year 2 = $276.25; Year 3 = $300.05; Year 4 = $326.40; Year 5 = $354.45;
    Year 6 = $385.05
14. $5,439.51

Mastery Tests
Form A (page 297)

1. Yes   **2.** 20   **3.** 88.89%   **4.** 53.2%   **5.** 0.0275   **6.** 16.2   **7.** 120   **8.** 3 : 10
9. Augustine got $65,000 and Lena got $35,000.   **10.** Hardware = $55,000 and software = $15,000.
11. $1687.50   **12.** $1750   **13.** 57.14%   **14.** $1000   **15.** 9.417%

Form B (pages 297–298)

1. $5545.24   **2.** $7,971.98   **3.** $3800   **4.** $1524.72   **5.** $10,336.08   **6.** $54.37   **7.** $2679.95
8. 0.7125   **9.** $801.68   **10.** 0.8969   **11.** $4714.80   **12.** 22.88%   **13.** 25%
14. The one that costs $10.23   **15.** 176.43 calories

CHAPTER 6

Section 6.2 (pages 307–308)

1. a)

+	0	1	2	3	4	5	6
0	0	1	2	3	4	5	6
1	1	2	3	4	5	6	0
2	2	3	4	5	6	0	1
3	3	4	5	6	0	1	2
4	4	5	6	0	1	2	3
5	5	6	0	1	2	3	4
6	6	0	1	2	3	4	5

b)

+	0	1	2	3	4	5	6	7
0	0	1	2	3	4	5	6	7
1	1	2	3	4	5	6	7	0
2	2	3	4	5	6	7	0	1
3	3	4	5	6	7	0	1	2
4	4	5	6	7	0	1	2	3
5	5	6	7	0	1	2	3	4
6	6	7	0	1	2	3	4	5
7	7	0	1	2	3	4	5	6

c)

+	0	1
0	0	1
1	1	0

d)

+	0	1	2	3	4
0	0	1	2	3	4
1	1	2	3	4	0
2	2	3	4	0	1
3	3	4	0	1	2
4	4	0	1	2	3

2. a)

−	0	1	2	3	4
0	0	4	3	2	1
1	1	0	4	3	2
2	2	1	0	4	3
3	3	2	1	0	4
4	4	3	2	1	0

b)

−	0	1	2	3	4	5	6	7	8
0	0	8	7	6	5	4	3	2	1
1	1	0	8	7	6	5	4	3	2
2	2	1	0	8	7	6	5	4	3
3	3	2	1	0	8	7	6	5	4
4	4	3	2	1	0	8	7	6	5
5	5	4	3	2	1	0	8	7	6
6	6	5	4	3	2	1	0	8	7
7	7	6	5	4	3	2	1	0	8
8	8	7	6	5	4	3	2	1	0

3. 1   **4.** 0   **5.** 6   **6.** 8   **7.** 2   **8.** 1   **9.** 0   **10.** 9   **11.** 2   **12.** 2   **13.** 0   **14.** 7
15. 5   **16.** 3   **17.** 2   **18.** 0   **19.** 5   **20.** 10   **21.** Mod 6   **22.** Mod 7   **23.** Mod 5
24. Mod 7   **25.** Mod 12   **26.** Mod 8
27. The inverse of 0 is 0; the inverse of 1 is 5; the inverse of 2 is 4; the inverse of 3 is 3; the inverse of 4 is 2; the inverse of 5 is 1.
28. The inverse of 0 is 0; the inverse of 1 is 8; the inverse of 2 is 7; the inverse of 3 is 6; the inverse of 4 is 5; the inverse of 5 is 4; the inverse of 6 is 3; the inverse of 7 is 2; the inverse of 8 is 1.
29. Valid   **30.** 9   **31.** 7   **32.** 4   **33.** 15   **34.** 2   **35.** 3   **36.** 5   **37.** 5

Section 6.3 (pages 311–312)

1. 0   **2.** 6   **3.** 8   **4.** 3   **5.** 0   **6.** 3   **7.** 0   **8.** 2   **9.** 2   **10.** 4   **11.** True   **12.** True
13. True   **14.** True   **15.** False   **16.** False   **17.** True   **18.** True   **19.** True   **20.** True
21. 3   **22.** 4   **23.** 3 ·   **24.** 3   **25.** 8   **26.** 11   **27.** 9   **28.** 2   **29.** 12   **30.** 3   **31.** 5
32. 11   **33.** 10   **34.** 5   **35.** 2   **36.** 10   **37.** 14   **38.** 4   **39.** 20   **40.** 25   **41.** Sunday
42. Thursday   **43.** Monday, assuming this year is not a leap year   **44.** Wednesday   **45.** No
46. 89   **47.** 78

Section 6.4 (page 315)

1. 1191   **2.** 19,680   **3.** 289   **4.** 18,889   **5.** 895   **6.** 210,915   **7.** 15,708   **8.** 1,485,348
9. 13,289   **10.** 2625   **11.** 12,695,822   **12.** 1,522,756
13. She merely interchanged two digits. The sum of the digits will be the same.
14. The remainders are the same.   **15.** Yes, the sum of the digits must be divisible by 9.   **16.** No

Section 6.5 (page 317)

1. 865   **2.** 1,079   **3.** 13,617   **4.** 22,775   **5.** 58   **6.** 3556   **7.** 6472   **8.** 1587   **9.** 33,995
10. 613,958   **11.** 5,810,454   **12.** 48,887,696
14. Checks by casting out 11's and not by casting out 9's.   **15. a)** $1041_{(8)}$ **b)** $1523_{(8)}$ **c)** $53122_{(8)}$

Section 6.6 (pages 321–322)

1. No   **2.** No   **3.** No   **4.** No   **5.** No   **6.** Yes   **7.** No   **8.** Yes   **9.** No   **10.** Yes
11. No   **12.** Yes   **13.** No   **14.** No   **15.** Yes
17. a) a **b)** The inverse of a is a; the inverse of b is b; and the inverse of c is c.   **c)** No   **18.** No
19. Yes   **20. a)** Yes   **b)** 1   **c)** 1 and 4   **d)** 1   **e)** Yes
21. a) Yes   **b)** Yes   **c)** α
   d) The inverse of α is α; the inverse of β is β; the inverse of γ is γ; the inverse of Δ is Δ.
22. a) z **b)** y **c)** Yes   **d)** Yes   **23.** No   **24.** No

25.

	(i)	(ii)	(iii)	(iv)
a)	No	No	No	No
b)	Yes	No	Yes	No
c)	No	No	No	No
d)	Yes	Yes	Yes	Yes

26. a) Yes   **b)** Yes   **27.** No   **28.** No   **29.** No

Section 6.7 (pages 328–329)

2.

$r \circ (s \circ t) = (r \circ s) \circ t$	$q \circ (u \circ v) = (q \circ u) \circ v$	$v \circ (w \circ r) = (v \circ w) \circ r$
$r \circ v = q \circ t$	$q \circ s = v \circ v$	$v \circ v = r \circ r$
$w = w$	$p = p$	$p = p$

3. The inverse of X is X; the inverse of Y is Z; the inverse of Z is Y; the inverse of P is P; the inverse of Q is Q; the inverse of R is R.

4. Let X, Y, Z, and W be the following symmetries.

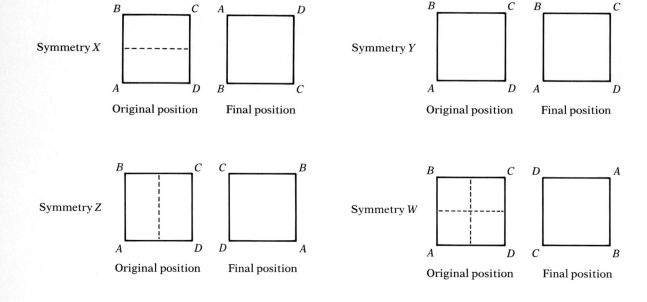

∘	X	Y	Z	W
X	Y	X	W	Z
Y	X	Y	Z	W
Z	W	Z	Y	X
W	Z	W	X	Y

This forms an Abelian group.

5. Yes

6. Yes

7. a)

∘	r	l	a	s
r	a	s	l	r
l	s	a	r	l
a	l	r	s	a
s	r	l	a	s

b) Yes

c) Yes

d) Yes, s

e) The inverse of r is l; the inverse of l is r. The inverse of a is a; the inverse of s is s.

f) r for both

g) Yes

8. $Z \circ P$, which is Q, is not equal to $P \circ Z$, which is R.

Section 6.8 (pages 335–336)

1. Each player should show card 2 part of the time, card 4 (a different) part of the time, and card 5 the remaining part of the time. This strategy should be followed not according to any pattern.

2. a) Show 1 finger part of the time, and show 2 fingers the rest of the time.
   b) Show 1 finger part of the time, and show 2 fingers the rest of the time.

3. Since the value is +1   **4. a)** No

5. Since the minimum of the first row, which is 0, is also at the same time the maximum of the column that it is in.

Strictly determined	Value	Fair	Optimal Strategy for R	Optimal Strategy for C
6. Yes	4	No	Play row 1	Play column 1
7. Yes	0	Yes	Play row 1	Play column 2
8. No	—	—	—	—
9. Yes	0	Yes	Play row 1	Play column 1
10. Yes	0	Yes	Play row 1	Play column 1
11. Yes	−2	No	Play row 1	Play column 2
12. No	—	—	—	—
13. Yes	2	No	Play row 1	Play column 3

14. a)

		Felix	
	Black 3	Red 7	Black 10
Black 4	7	−11	14
Lisa Red 8	−11	15	−18
Red 9	−12	16	19

b) No

15. a)

		Frank hides	
		one $1 bill	two $1 bills
Karen hides	one $1 bill	2	−2
	two $1 bills	−2	4

b) No

16. a)

		Expresso opens in	
		Blakeville	Bushtown
Yellow opens in	Blakeville	0	+10
	Bushtown	−8	0

b) Yes **c)** 0 **d)** Yes **e)** Yellow Cab Co. opens in Blakeville and Expresso opens in Blakeville.

17. a)

		Gail	
		Promoted	Not promoted
Bill	Promoted	−100	40
	Not promoted	40	40

b) 40

Section 6.9 (pages 341–343)

	R_1	R_2	C_1	C_2	V
1.	$\frac{1}{4}$	$\frac{3}{4}$	$\frac{1}{2}$	$\frac{1}{2}$	5
2.	$\frac{1}{5}$	$\frac{4}{5}$	$\frac{3}{5}$	$\frac{2}{5}$	$\frac{4}{5}$
3.	$\frac{3}{14}$	$\frac{11}{14}$	$\frac{1}{14}$	$\frac{13}{14}$	$\frac{59}{7}$
4.	$\frac{7}{8}$	$\frac{1}{8}$	$\frac{15}{16}$	$\frac{1}{16}$	$\frac{65}{4}$
5.	$\frac{10}{21}$	$\frac{11}{21}$	$\frac{1}{3}$	$\frac{2}{3}$	$\frac{-4}{3}$
6.	$\frac{8}{13}$	$\frac{5}{13}$	$\frac{8}{13}$	$\frac{5}{13}$	$\frac{-80}{13}$
7.	$\frac{2}{9}$	$\frac{7}{9}$	$\frac{4}{9}$	$\frac{5}{9}$	$\frac{-1}{9}$
8.	$\frac{5}{9}$	$\frac{4}{9}$	$\frac{11}{18}$	$\frac{7}{18}$	$\frac{23}{9}$

9. a)

	Jane guesses	
	$5 bill	$10 bill
Diana hides — $5 bill	-5	$+10$
$10 bill	$+5$	-10

c) 0   **d)** Yes

e) Diana: Hide $5 bill $\frac{1}{2}$ of time and $10 bill $\frac{1}{2}$ of time. Jane: Guess $5 bill $\frac{2}{3}$ of time and $10 bill $\frac{1}{3}$ of time.

10. a)

	Jim shows	
	1 finger	2 fingers
Sam shows — 1 finger	1	-1
2 fingers	-1	4

b) $\frac{3}{7}$ **c)** No, Sam

d) Sam: Show 1 finger $\frac{5}{7}$ of time and 2 fingers $\frac{2}{7}$ of time. Jim: Show 1 finger $\frac{5}{7}$ of time and 2 fingers $\frac{2}{7}$ of time.

11. a) *Jim shows* **b) No c) No**

		1 finger	2 fingers	3 fingers
	1 finger	+2	−3	+4
Sam shows	2 fingers	−3	+4	−5
	3 fingers	+4	−5	+6

12. a)

		Enemy gunners	
		Short range	Long range
Navy ship	Comes in close	1	29
	Stays out far	20	−8

b) Navy ship: Come in $\frac{1}{2}$ of time and stay out $\frac{1}{2}$ of time. Enemy gunners: Set guns at short range $\frac{37}{56}$ of time and long range $\frac{19}{56}$ of time.

c) The navy ship can use a coin and the enemy gunners can use a spinner.

13. a) John: Open on north side $\frac{4}{5}$ of time and on south side $\frac{1}{5}$ of time. Peter: Open on north side $\frac{3}{5}$ of time and on south side $\frac{2}{5}$ of time.

b) Each should use a spinner.

14. A player's strategies are not likely to be random and are bound to be detected by the opponent.

15. a)

		Moriarty gets off at	
		Canterbury	Dover
Holmes gets off at	Canterbury	−300	+40
	Dover	+150	−300

b) $-\frac{8400}{79}$; No **c)** Get off at Canterbury $\frac{45}{79}$ of the time and at Dover $\frac{34}{79}$ of the time.

16. a)

		Nina	
		Confess	Don't confess
Al	Confess	2	−30
	Don't confess	−30	26

b) Al: Confess $\frac{7}{11}$ of the time and don't confess $\frac{4}{11}$ of the time. Nina: Confess $\frac{7}{11}$ of the time and don't confess $\frac{4}{11}$ of the time.

17. a)

		Police car turns	
		Right	Left
Robber turns	Right	0.80	0.20
	Left	0.50	0.50

b) Robber: Turn left always
Police car: Turn left always

Mastery Tests
Form A (pages 347–348)

1. Choice (b) **2.** Choice (a) **3.** Choice (a) **4.** Choice (a) **5.** Choice (d) **6.** Choice (e)
7. Choice (c) **8.** Choice (a) **9.** Choice (a) **10.** Choice (b) **11.** Choice (b) **12.** Choice (b)
13. Choice (e) **14.** Choice (c) **15.** 76

Form B (pages 349–350)

1. Choice (b) **2.** $\dfrac{1}{2}$ **3.** $\dfrac{-9}{2}$ **4. a)** 895 **b)** 289 **c)** 252,540 **5.** $\dfrac{136}{15}$ **6.** $\dfrac{2}{15}$ **7.** $\dfrac{8}{15}$
8. $\dfrac{13}{15}$ **9.** Choice (b) **10.** Choice (d) **11.** Monday **12.** 77
13. $\begin{cases} \text{Player R: Play row 1} \\ \text{Player C: Play column 1} \end{cases}$ **14.** False **15.** Choice (d)

CHAPTER 7

Section 7.2 (page 356–357)

1. 64 **2.** $\dfrac{1}{512}$ **3.** -32 **4.** $\dfrac{1}{2}$ **5.** 8 **6.** $5^6 = 15,625$ **7.** 1 **8.** $4^5 = 1024$ **9.** 16
10. $2^{12} = 4096$ **11.** $\dfrac{1}{343}$ **12.** 18 **13.** 10,077,696 **14.** 5 **15.** 320 **16.** $\dfrac{1}{8}$

Section 7.3 (page 364)

1. $16x^3 - 3x^2 + 7x - 1$ **2.** $-8x^2 + 3x + 4y$ **3.** $-x^3 - 6x^2y + 5xy^2 - 10y^3$ **4.** 10 **5.** $-x^2 + 4x - 15$
6. $2x^3 - 3x^2 + 6x - 6$ **7.** $15 - 9x$ **8.** $3x^2 + 10x + 6$ **9.** $4x^2 - 6x - 8$ **10.** $10x^2 - 17x + 5$
11. $-3x - 4$ **12.** $-9x^2 + 10x - 11$ **13.** $-40x^3 + 24x^2 - 56x$ **14.** $-35x^4y^5z^3 + 15x^3y^3z^3 - 40x^2y^2z^3$
15. $-25x^2 + 35xy^3 - 60y^3$ **16.** $-50x^9y^9 - 30x^5y^2 + 170x^4y$ **17.** $-2x^2y^5 + 6x^3y^6 - 2x^2y^{10}$
18. $15x^2 + 14x - 16$ **19.** $x^2 - 2xy - 15y^2$ **20.** $10x^2 + xy - 21y^2$
21. $24x^3 - 15x^2y + 9xy^2 - 40x^2 + 25xy - 15y^2$ **22.** $2x^3 - 3x^2 + 25x + 51$ **23.** $2x^3 + 11x^2y - xy^2 - 3y^3$
24. $10x^3 - 31x^2y + 29xy^2 - 35y^3$ **25.** $8x^2 + 20x + 3$ **26.** $-14x - 1$ **27.** $-21x^2 - 22x + 8$
28. $-2a + 6a^2b$ **29.** $5x^2y^4 - 9x^2$ **30.** $-2x^4y^2 + 1$ **31.** $-x^3y^3 + xy + 1$ **32.** $8y^3 + 6x - 4y$
33. $x - 7$ **34.** $2x + 1$ **35.** $x^2 - 3x + 2$ **36.** $x^2 - 2x + 10$ Remainder -16
37. $8x^2 + 16x + 96$ Remainder 487 **38.** $x + 9y$ Remainder $26y^2$ **39.** $x^2 - 4x + 16$
40. $2x^3 - 3x + 6$ Remainder $-14x + 4$ **41.** $15x^3 + 15x^2 + 15x + 15$ Remainder 14
42. $2x^3 + 3x^2 + 4x + 4$ Remainder $-4x + 2$ **43.** $1000x^2 + 215x + 3$ **44.** $315x^2 + 192x + 21$
45. $300x + 2y + \dfrac{9}{x}$

Section 7.4 (pages 367–368)

1. Yes. **2.** No

	$f(1)$	$f(0)$	$f(-3)$	$f\left(\frac{1}{2}\right)$	$f(a)$	$f(x + h)$
3.	1	-3	-15	-1	$4a - 3$	$4x + 4h - 3$
4.	5	2	-7	3.5	$3a + 2$	$3x + 3h + 2$
5.	4	3	12	3.25	$a^2 + 3$	$x^2 + 2xh + h^2 + 3$
6.	$\frac{4}{7}$	$\frac{1}{2}$	$\frac{8}{13}$	$\frac{5}{9}$	$\frac{3a + 1}{5a + 2}$	$\frac{3x + 3h + 1}{5x + 5h + 2}$
7.	3	0	27	$\frac{3}{4}$	$3a^2$	$3x^2 + 6xh + 3h^2$
8.	8	14	20	$\frac{45}{4}$	$(2 - a)(7 + a)$	$(2 - x - h)(7 + x + h)$
9.	14	6	-6	$\frac{39}{4}$	$a^2 + 7a + 6$	$x^2 + 2xh + h^2 + 7x + 7h + 6$
10.	17	5	-43	9.5	$2a^2 + 3a^2 + 7a + 5$	$2(x + h)^3 + 3(x + h)^2 + 7(x + h) + 5$
11.	9	3	-3	Can't be done	$\frac{5a^2 + 7a - 3}{2a - 1}$	$\frac{5x^2 + 10xh + 5h^2 + 7x + 7h - 3}{2x + 2h - 1}$

12. a) Yes **b)** $y = 15 + 6(x - 1)$ **c)** \$33

13. a) Yes **b)** $y = 0.06x$ when $x \le 25,000$
$$\left\{\begin{array}{l} y = 0.06(25,000) + 0.08(x - 25,000) \\ \text{or } y = 0.08x - 500 \end{array}\right\} \text{ when } x > 25,000\}$$
c) January \$1453.68; February \$1042.08; March \$1581.12; April \$1750.56; May \$1858.80; June \$2595.36

14. a) Yes **b)** $y = 428(1.01)^a$ where $a = \frac{x}{6}$, a is a nonnegative integer, and $x = 0, 6, 12, 18, \ldots$
c) $428(1.01)^{10} = \$472.78$

	Domain	Range
15.	All real numbers	All real numbers
16.	All real numbers larger than or equal to $\frac{-2}{7}$	All nonnegative numbers
17.	All real numbers except $\frac{-1}{2}$	All real numbers
18.	All real numbers	All nonnegative numbers

Section 7.5 (pages 373–375)

1. 3   **2.** -4 **3.** 2   **4.** 2   **5.** 10   **6.** -6 **7.** 10   **8.** -2 **9.** 21   **10.** $\frac{4}{7}$ **11.** 6

12. $\frac{-65}{4}$ **13.** 1   **14.** 2   **15.** 6   **16.** 6   **17.** $\frac{c + 5}{b}$ **18.** 15   **19.** $\frac{a - 6}{b}$ **20.** $\frac{9}{2}b$ **21.** $2a$

22. 4   **23.** 2   **24.** $\frac{-51}{7}$ **25.** 5   **26.** -3

27. $a \to$ viii; $b \to$ ix; $c \to$ vi; $d \to$ iii; $e \to$ v; $f \to$ ii; $g \to$ iv; $h \to$ i   **28.** \$360   **29.** $33\frac{1}{3}\%$

30. 150 children tickets and 300 adult tickets   **31.** Seven $10 bills, ten $5 bills, and forty $1 bills

32. Madeline 40, Phyllis 122   **33.** Monday = 2; Tuesday = 4; Wednesday = 4   **34.** $8\frac{1}{4}\%$

35. $45,733.94   **36.** $1000 @ 6% and $1500 @ 10%   **37.** Secretary = 22 years
                                                                              President = 66 years

38. Width = 4 feet and length = 12 feet   **39.** Bill = 19; Mag = 18; Chris = 31

Section 7.6 (pages 378–379)

1. $x = 5, y = 3$ **2.** $x = 2, y = -3$ **3.** $x = -3, y = 1$ **4.** $x = 16, y = 28$ **5.** $x = 4, y = 3$
6. $x = 2, y = 3$ **7.** $x = 3, y = 2$ **8.** $x = 11, y = 4$ **9.** $x = 3, y = -2$ **10.** $x = 4, y = -3$
11. $x = \frac{1}{3}, y = \frac{1}{2}$ **12.** $x = 5, y = -6$ **13.** No solution   **14.** Adult = $7, child = $4
15. Fifteen $10 bills and ten $5 bills   **16.** Fifty 22¢ stamps and 25 20¢ stamps
17. Bat costs $12; ball costs $4   **18.** Man = $75, Woman = $50
19. Prelaw = 1500; Premedical = 4500   **20.** Rowboat = $3, motorboat = $7
21. Male = 1200, female = 1000   **22.** 20 gallons of the 4% solution and 16 gallons of the 40% solution
23. Milk = 2.78 cups, orange juice = 0.48 cups.

Section 7.7 (pages 382–383)

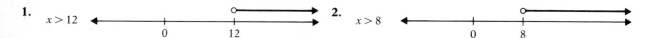

1. $x > 12$ **2.** $x > 8$

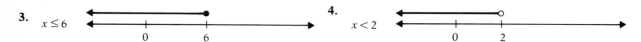

3. $x \leq 6$ **4.** $x < 2$

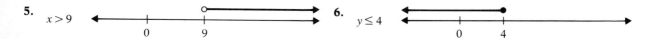

5. $x > 9$ **6.** $y \leq 4$

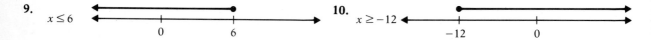

7. $x \geq -4$ **8.** $x \leq 3$

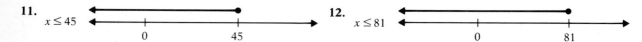

9. $x \leq 6$ **10.** $x \geq -12$

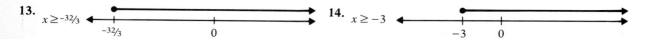

11. $x \leq 45$ **12.** $x \leq 81$

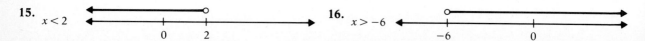

13. $x \geq -32/3$ **14.** $x \geq -3$

15. $x < 2$ **16.** $x > -6$

17. $x \geq \frac{3}{2}$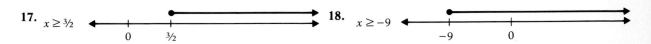

18. $x \geq -9$

19. $x \geq -2$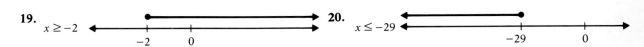

20. $x \leq -29$

21. 8   **22.** Mary: at least \$84; Stephanie at least 14   **23.** 6 pounds   **24.** 23000 gallons   **25.** \$3
26. 24,000 miles   **27.** 12

Section 7.8 (pages 388–389)

1. $10(x + 2)$ **2.** $3(x - 2y)$ **3.** $6(2 - 3y)$ **4.** $6x(x - 2)$ **5.** $11x^2y(3xy - 4)$ **6.** $16x^2y^3z(5x^2y^2z^2 + 4)$
7. $3xy(8y^2 - 6xy + 3)$ **8.** $6xy(8x^2y - 10xy^4 + 9)$ **9.** $(12 + x)(12 - x)$ **10.** $(4 + x^2)(2 - x)(2 + x)$
11. $(x^2 + 8)(x^2 - 8)$ **12.** $(12x + 13y)(12x - 13y)$ **13.** $(xy + 7)(xy - 7)$ **14.** $\left(\frac{1}{8} + x\right)\left(\frac{1}{8} - x\right)$
15. $\left(\frac{4}{5}xy^4 - \frac{2}{3}\right)\left(\frac{4}{5}xy^4 + \frac{2}{3}\right)$ **16.** $16(x + 1)(x - 1)$ **17.** $3(x + 3y)(x - 3y)$ **18.** $x(x + 1)(x - 1)$
19. $(x^2 + 9)(x + 3)(x - 3)$ **20.** $x^2(x^2 + 1)(x + 1)(x - 1)$ **21.** $4(x + 3)(x - 3)$ **22.** $(7x + 8y)(7x - 8y)$
23. $(x + 6)(x + 1)$ **24.** $(x + 6)(x - 4)$ **25.** $(x - 7)(x - 3)$ **26.** $(x - 3)(x + 2)$ **27.** $(x + 3)(x - 2)$
28. $(x + 12)(x - 3)$ **29.** $(2x + 3)(x + 2)$ **30.** $(3x + 4)(x + 2)$ **31.** $(3x + 4)(2x - 1)$
32. $(2x - 5)(2x - 1)$ **33.** $(5x - 2)(2x - 1)$ **34.** $(4x + 3y)(x - 2y)$ **35.** $4(x - 4)(x + 2)$
36. $(x + 4)(x - 4)(x + 2)(x - 2)$ **37.** $3(x + 3)(x + 2)$ **38.** $4(x - 6)(x - 2)$ **39.** $x(x + 6)(x + 4)$
40. $-3(x - 9)(x + 2)$

Section 7.9 (pages 394–395)

1. $8, -9$ **2.** $2, \frac{-5}{2}$ **3.** $-2, \frac{-3}{2}$ **4.** $\frac{1}{3}, 3$ **5.** $\frac{2}{3}, 2$ **6.** $\frac{-5}{2}, 2$ **7.** $3, 9$ **8.** $-1, 4$
9. $6.1, -4.1$ **10.** $-5.9, 2.9$ **11.** $0, -7$ **12.** $0, -12$ **13.** $+3, -3$ **14.** $5, -5$ **15.** 8
16. $1.1, -6.1$ **17.** $11.8, 0.2$ **18.** $2, 6$ **19.** $5.8, -0.3$ **20.** $4.6, -0.6$ **21.** $1.4, -1.9$ **22.** $3.6, 0.4$
23. $5.4, 0.1$ **24.** $1, \frac{2}{3}$ **25.** $1.8, -0.3$ **26.** $0.7, -0.5$ **27.** $\dfrac{5 \pm \sqrt{25 - 8a}}{4}$ **28.** $\dfrac{-1 \pm \sqrt{1 + 20c}}{10}$
29. $2.6, -0.6$ **30.** $-0.8, 0.4$ **31.** 4 hours   **32.** 27 shares @ \$50 each   **33.** 13 and 15
34. 25.4 feet   **35.** Length = 30 meters, width = 40 meters   **36.** Length = 22 feet, width = 9 feet
37. 13 feet   **38.** 4.4 seconds

Mastery Tests
Form A (pages 396–397)

1. $(x + 0.8y)(x - 0.8y)$ **2.** -3 **3.** 12   **4.** Choice (d)   **5.** Choice (b)   **6.** -2 **7.** $x > \dfrac{-1}{11}$

8. Choice (d)   **9.** $8, -3$ **10.** $(2x - 9)(x - 2)$ **11.** Choice (b)   **12.** $\dfrac{11}{2}$ **13.** $\dfrac{t - a}{n - 1}$

14. Choice (c)   **15.** $x \leq -8$

Form B (pages 397–398)

1. Choice (c)   **2.** 2.8, 0.7   **3.** $7x^2 + 2x + 5 + \dfrac{9}{x}$ **4.** $-2.4, 0.4$ **5.** $-0.2, -7.3$

6. Mechanic \$12/hr; helper \$3/hr   **7.** 3 nickels and 17 quarters   **8.** 5.45 hrs   **9.** $\dfrac{2x-5}{24}$

10. \$4000 @ 8%, \$1000 @ 6%   **11.** 9 and 18 hrs   **12.** 6,250,000 square yards   **13.** \$1.92
14. 12.96 hours   **15.** 118.51 units

CHAPTER 8

Section 8.2 (page 405)

1. $(2, 1)$ **2.** $(-3, 4)$ **3.** $(5, -3)$ **4.** $(1, 0)$ **5.** $(-2, 0)$ **6.** $(0, 5)$ **7.** $(0, -2)$ **8.** $(-3, -3)$
21. IV   **22.** III   **23.** II   **24.** I   **25.** $(0, -7)$ **26.** $(5, 0)$ **27.** 0   **28.** 0   **29.** $(0, 0)$ **30.** 8

Section 8.3 (page 411)

1.

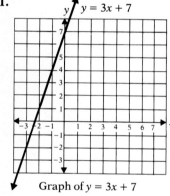

Graph of $y = 3x + 7$

2.

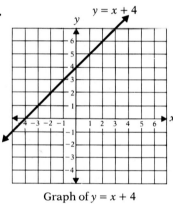

Graph of $y = x + 4$

3.

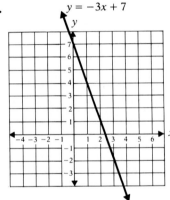

Graph of $y = -3x + 7$

4.

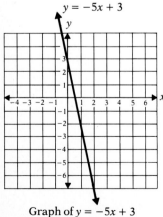

Graph of $y = -5x + 3$

5.

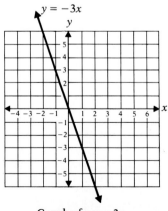

Graph of $y = -3x$

6.

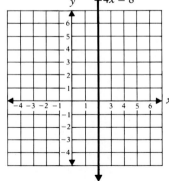

Graph of $4x = 8$

7.

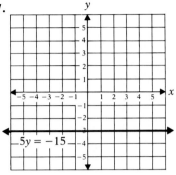

Graph of $5y = -15$

8.

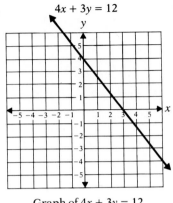

Graph of $4x + 3y = 12$

9.

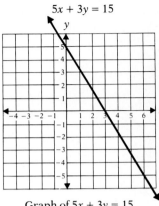

Graph of $5x + 3y = 15$

10.

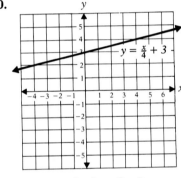

Graph of $y = \frac{x}{4} + 3$

11.

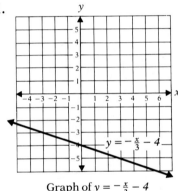

Graph of $y = -\frac{x}{3} - 4$

12.

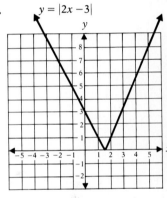

Graph of $y = |2x - 3|$

13.

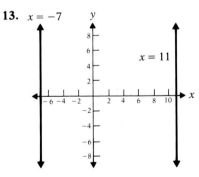

Graph of $|x-2| = 9$

14.

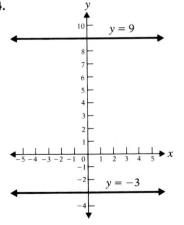

Graph of $|y-3| = 6$

15.

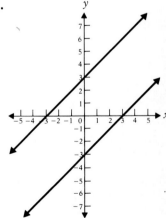

Graph of $|x-y| = 3$

16.

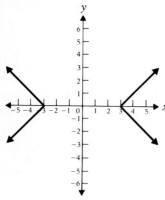

Graph of $|x| - |y| = 3$

17.

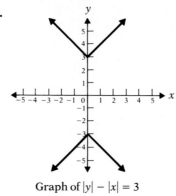

Graph of $|y| - |x| = 3$

Section 8.4 (pages 419–420)

1. a) $\frac{1}{5}$ **b)** No slope **c)** 0   **2. a)** 2 **b)** $\frac{1}{3}$ **c)** $\frac{-7}{9}$ **d)** $\frac{7}{5}$ **e)** $\frac{7}{3}$ **f)** 3 **g)** $\frac{-5}{4}$ **h)** -1 **i)** $\frac{5}{4}$ **j)** 0

3. a) 3: -1 **b)** $\frac{1}{3}$; 4 **c)** $\frac{3}{5}$; 3 **d)** $\frac{8}{5}$; -8 **e)** 1; 0 **f)** 0: 7 **g)** none; none   **4.** No

5. a) 3; -4 **b)** -3; $\frac{3}{2}$ **c)** 0; 0 **d)** $\frac{5}{2}$; -5 **e)** 3; none **f)** none; $\frac{-14}{3}$

6. a) $y = -3x - 8$ **b)** $4y - 5x = 0$ **c)** $3y + 2x = 34$ **d)** $y + 8x + 17 = 0$ **e)** $7y + 5x = 0$

7. $y = 4x - 14$ **8.** $y = 2x - 5$ **9.** $y = 3x$ **10. a)** Yes **b)** Yes **c)** Yes

11. a) $\frac{1}{2}$ **b)** $\frac{-2}{x-7}$ **c)** $x = 3$ **d)** $y = -6x + 18$

Section 8.5 (page 423)

1.

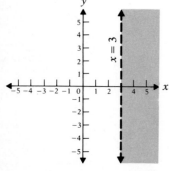

Graph of $x > 3$

2.

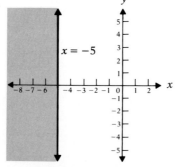

Graph of $x \leq -5$

3.

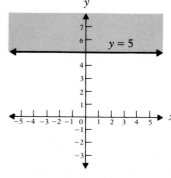

Graph of $y \geq 5$

4.

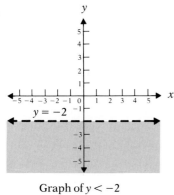

Graph of $y < -2$

5.

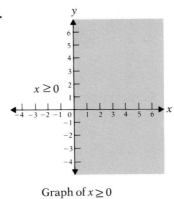

Graph of $x \geq 0$

6.

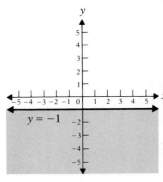

Graph of $y \leq -1$

7.

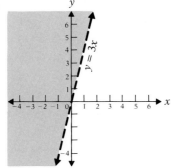

Graph of $y > 3x$

8.

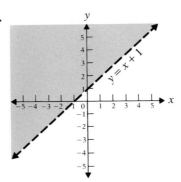

Graph of $y > x + 1$

9.

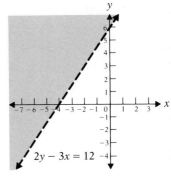

Graph of $2y - 3x > 12$

10.

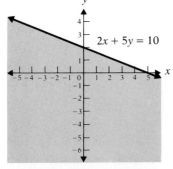

Graph of $2x + 5y \leq 10$

11.

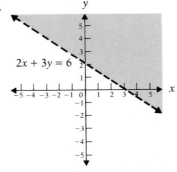

Graph of $2x + 3y > 6$

12.

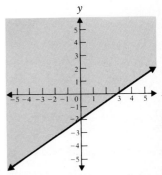

Graph of $2x - 3y \leq 6$

13.

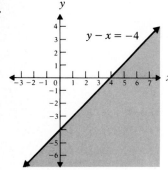

Graph of $y - x \le -4$

14.

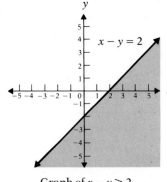

Graph of $x - y \ge 2$

15.

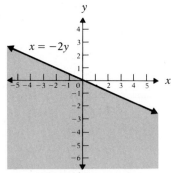

Graph of $x \le -2y$

16.

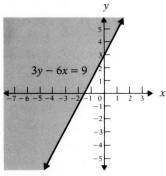

Graph of $3y - 6x \ge 9$

17.

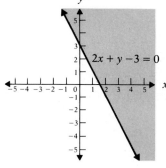

Graph of $2x + y - 3 \ge 0$

18.

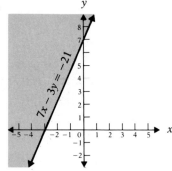

Graph of $7x - 3y \le -21$

19.

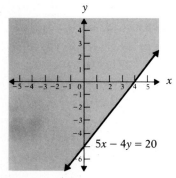

Graph of $5x - 4y \le 20$

20.

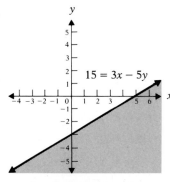

Graph of $15 \le 3x - 5y$

21. $6y \ge 5x - 12$ **22.** $3y \ge 7x - 8$ **23.** $y \ge x$ **24.** $y \le 2x$ **25.** $2y \le 3x + 4$
26. $18y \ge 15x + 10$

Section 8.6 (page 428)

1. $(3, 12)$ **2.** $(-2, 1)$ **3.** $(0, 1)$ **4.** $(3, 6)$ **5.** $(2, -1)$ **6.** $(4, 3)$ **7.** $(1, 2)$ **8.** $(-3, -1)$
9. $(5, 5)$ **10.** $(4, 1)$ **11.** $(-6, 3)$ **12.** $(5, 5)$ **13.** $(-3, -2)$ **14.** $(1, -1)$ **15.** Inconsistent
16. Consistent   **17.** Dependent   **18.** Consistent   **19.** Dependent   **20.** Consistent

Section 8.7 (page 431)

1.

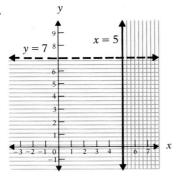

2.

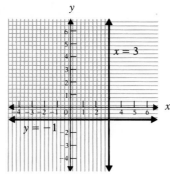

3.

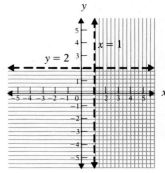

4.

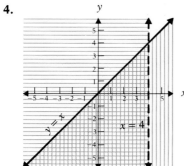

5.

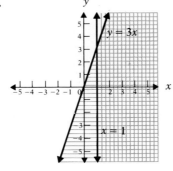

6.

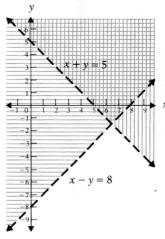

7.

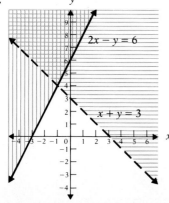

8.

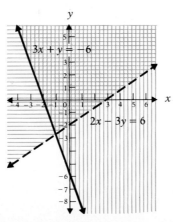

9.

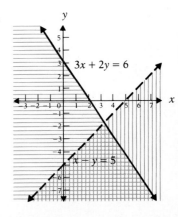

10.

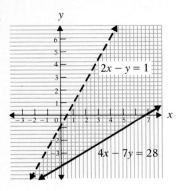

$2x - y = 1$

$4x - 7y = 28$

11.

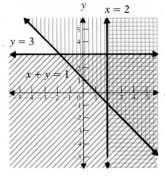

$x = 2$

$y = 3$

$x + y = 1$

12.

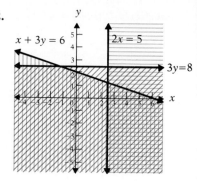

$x + 3y = 6$

$2x = 5$

$3y = 8$

13.

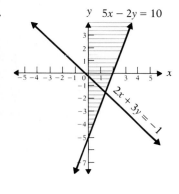

$5x - 2y = 10$

$2x + 3y = -1$

14.

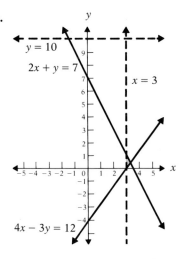

$y = 10$

$2x + y = 7$

$x = 3$

$4x - 3y = 12$

15.

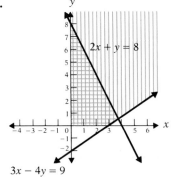

$2x + y = 8$

$3x - 4y = 9$

16.

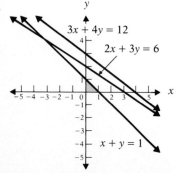

$3x + 4y = 12$

$2x + 3y = 6$

$x + y = 1$

Section 8.8 (pages 438–440)

1. Let x = number of 12″ sets to be produced.
   y = number of 19″ sets to be produced

   max $P = 38x + 48y$, subject to
   $$\begin{cases} x + 3y \le 12 \\ 2x + y \le 18 \\ x \ge 0 \\ y \ge 0 \end{cases}$$

2. Let x = number of jeeps to be manufactured
   y = number of tanks to be manufactured

   max $P = 360x + 480y$, subject to
   $$\begin{cases} 3x + 4y \le 120 \\ 4x + 3y \le 90 \\ x \ge 0 \\ y \ge 0 \end{cases}$$

3. Let x = number of suits to be manufactured
   y = number of sport coats to be manufactured

   max $P = 40x + 24y$, subject to
   $$\begin{cases} 4x + 2y \le 12 \\ 2x + 4y \le 12 \\ x \ge 0 \\ y \ge 0 \end{cases}$$

4. Let x = number of buses to be serviced
   y = number of vans to be serviced

   max $P = 60x + 40y$, subject to
   $$\begin{cases} 4x + 3y \le 24 \\ 2x + y \le 16 \\ x \ge 0 \\ y \ge 0 \end{cases}$$

5. Let x = number of pairs of shoes to be produced
   y = number of racquets to be produced
   z = number of pairs of shorts to be produced

   max $P = 4x + 3y + 2z$, subject to
   $$\begin{cases} 2x + 3y + 6z \le 30 \\ 5x + 4y + 6z \le 48 \\ 3x + y + 2z \le 48 \\ x \ge 0 \\ y \ge 0 \\ z \ge 0 \end{cases}$$

6. Let x = number of one-family houses to be completed
   y = number of two-family houses to be completed
   z = number of small businesses to be completed

   max $P = 200x + 300y + 500z$, subject to
   $$\begin{cases} 12x + 10y + 28z \le 40 \\ 10x + 14y + 26z \le 40 \\ 8x + 12y + 20z \le 40 \\ x \ge 0, y \ge 0, z \ge 0 \end{cases}$$

7. Let x = number of packages of type A to be produced
   y = number of packages of type B to be produced
   z = number of packages of type C to be produced

   max $P = 24x + 30y + 36z$, subject to
   $$\begin{cases} 4x + 5y + 3z \le 4000 \\ 5x + 3y + 4z \le 5000 \\ 4x + 3y + 5z \le 6000 \\ x \ge 0, y \ge 0, z \ge 0 \end{cases}$$

8. Let x = number of days for mill A to be open
y = number of days for mill B to be open

$$\max P = 4000x + 6000y, \text{ subject to } \begin{cases} 20x + 14y \le 300 \\ 10x + 12y \le 400 \\ 14x + 18y \le 360 \\ x \ge 0, y \ge 0 \end{cases}$$

9. Let x = number of acres of wheat to be planted
y = number of acres of corn to be planted

$$\max P = 400x + 300y, \text{ subject to } \begin{cases} x + y \le 300 \\ 3x + 4y \le 200 \\ 60x + 80y \le 4000 \\ x \ge 0, y \ge 0 \end{cases}$$

10. Let x = number of full teams to be used
y = number of half teams to be used

$$\max P = 300x + 200y, \text{ subject to } \begin{cases} x + y \le 50 \\ 3x + 2y \le 120 \\ x \ge 0, y \ge 0 \end{cases}$$

Section 8.9 (pages 444–446)

1. $x = 6, y = 0$; max $z = 30$ **2.** $x = \dfrac{6}{17}, y = \dfrac{60}{17}$; max $z = \dfrac{204}{17} = 12$ **3.** $x = 0, y = 5$; max $z = 35$

4. $x = 3, y = 0$; max $z = 24$ **5.** $x = \dfrac{7}{2}, y = 0$; min $z = \dfrac{21}{2}$ **6.** $x = 4, y = 0$, max $z = 32$

7. $x = 0, y = 3$; min $z = 24$ **8.** $x = \dfrac{36}{17}, y = \dfrac{40}{17}$; min $z = 36$ **9.** $x = 6, y = 2$; min $z = 74$

10. $x = 12, y = 0$; min $z = 24$ **11.** $x = 5, y = 0$; max $z = 15$ **12.** $x = 2, y = \dfrac{3}{2}$; min $z = 45$

13. One type A meal, 0 type B meal, minimum cost = 73¢ **14.** 12 ounces of cheese and 0 slices of bread
15. 40 cheap mopeds, 4 expensive mopeds. Cost $3656
16. 3095.238 gallons of yogurt and 0 gallons of ice cream. Profit = $2785.71
17. 60 part-time workers and 0 full-time workers. Cost = $240
18. 12 dry-cleaning machines and 0 dryers. Profit = $9.00
19. $50,000 in stocks and $50,000 in bonds. Return = $12,500

Mastery Tests
Form A (pages 448–449)

1. Choice (c)   **2.**

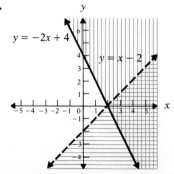

3. $(0, 0)$ **4.** Choice (a)   **5.** Choice (a)   **6.** $(-2, 1)$ **7.** Choice (a)

8.

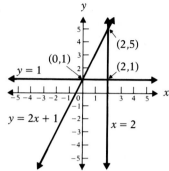

9. $(0, 1)$, $(2, 1)$, and $(2, 5)$ **10.** -1 **11.** $\dfrac{6}{7}$ **12.** $2y = 3x - 2$ **13.** $y = 8x - 6$ **14.** $\left(0, \dfrac{9}{4}\right)$

15. $y = 3x + 7$

Form B (pages 449–451)

1. Choice (c)

2.

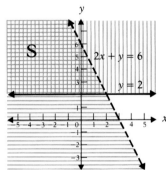

3. 0   **4.** Undefined   **5.** Choice (d)   **6.** $x = 0, y = 4$; Max value is 12

7. $x = 0, y = 2$; Min value is 4   **8.** Produce 0 ski poles and 3 skate boards; Max profit $= \$24$

9. Produce $\dfrac{40}{3}$ jeeps and $\dfrac{95}{3}$ tanks; Profit $= \$36{,}500$

10. Manufacture 0 normal traps and 3 big traps; Profit $= \$12$

11. Produce 150 type I discs and 200 type II discs daily; Profit $= \$2200$

12. Advertise 18.2 minutes on radio and 9.1 minutes of TV.

13. $x = \dfrac{12}{5}, y = \dfrac{6}{5}$; Max $P = \dfrac{48}{5}$ **14.** $x = 0, y = 4$; Max $P = 32$ **15.** $(0, 8)$

CHAPTER 9

Section 9.2 (pages 459–460)

1. $\overline{BC}$ **2.** $\overline{BC}$ **3.** $\overline{AC}$ **4.** $\overline{AD}$ **5.** $\overline{BC}$ **6.** $\overleftrightarrow{AB}$ **7.** $\overline{AD}$ **8.** $\overrightarrow{BC}$ **9.** Point B **10.** $\overline{AC}$

11. $\overline{CD}$ **12.** $\overleftrightarrow{CE}$ **13.** point D **14.** $\overline{CD}$ **15.** $\overrightarrow{EA}$ **16.** $\emptyset$
17. All points on line segment $\overline{BC}$ or line segment $\overline{DE}$ **18.** $\overleftrightarrow{BC}$
19. All points on line segment $\overline{BC}$ or line segment $\overline{CD}$ **20.** All points on the sides of $\triangle BCD$
21. Point C **22.** Point E **23.** All points on the sides of quadrilateral $ABCD$ **24.** Point C
25. Point B **26.** All points on the line segment $\overline{AD}$ or line segment $\overline{DC}$ **27.** 3   **28.** 6   **29.** 6
30. No; Yes   **31.** Not necessarily   **32.** 4

Section 9.3 (pages 469–470)

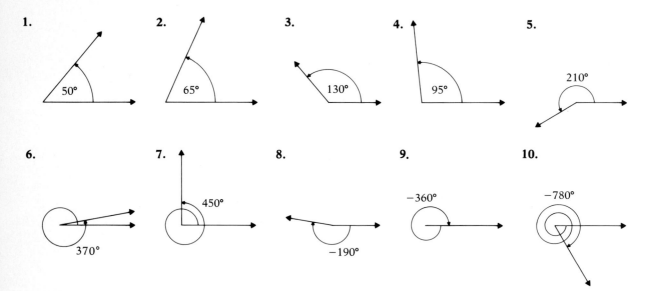

11. a and d, b and c, e and h, and f and g are vertical angles; a and b, b and d, d and c, e and f, f and h, h and g, e and g, and a and c are adjacent angles.
12. i and l, j and k, m and p, n and o are vertical angles; i and j, j and l, l and k, i and k, m and n, n and p, p and o, and m and o are adjacent angles.
13. No vertical angles. s and r, r and q are adjacent angles.
14. No vertical angles. u and v are adjacent angles. u and t are adjacent angles.

15. No vertical angles. d and e, d and c, c and b, b and a are adjacent angles.   **16.** $\frac{\pi}{18}$ **17.** $\frac{7\pi}{6}$

18. $\frac{7\pi}{4}$ **19.** $\frac{-\pi}{4}$ **20.** $\frac{19\pi}{18}$ **21.** -2π **22.** $\frac{52\pi}{9}$ **23.** $\frac{-19\pi}{6}$ **24.** 5π **25.** $\frac{-\pi}{3}$ **26.** $\frac{-11\pi}{6}$

27. $\frac{80\pi}{9}$ **28.** 135°   **29.** 90°   **30.** 210°   **31.** 165°   **32.** −160°   **33.** −432°   **34.** $\frac{\pi}{18}°$ **35.** 900°
36. −84°   **37.** $\overrightarrow{AB}$ **38.** $\angle DAE$ **39.** Point A **40.** Point A **41.** False   **42.** False   **43.** True
44. a) 30° **b)** 140° **c)** 60° **d)** 50°
45. $m(\angle 1) = 70°$, $m(\angle 2) = 50°$, $m(\angle 3) = 60°$, $m(\angle 4) = 120°$, $m(\angle 5) = 60°$, $m(\angle 6) = 120°$, $m(\angle 9) = 110°$, $m(\angle 10) = 110°$, $m(\angle 11) = 70°$
46. 63°   **47.** 60°   **48.** 20° and 70°   **49.** 30°   **50.** 110°

Section 9.4 (pages 474–476)

1. Closed   **2.** Not simple   **3.** Closed   **4.** Closed   **5.** Not closed   **6.** Not simple   **7.** Polygon

8. Not a polygon   **9.** Polygon   **10.** Not a polygon   **11.** Polygon   **12.** Polygon
13. It can be done. It is an acute triangle.   **14.** It can't be done.   **15.** 34 cm   **16.** 16 cm
17. Can't be done

18.    **19.**    **20.**

21. No   **22. a)** 360°   **b)** 360°

23. a) 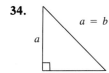   **b)**   **c)**

25. a) False   **b)** False   **c)** False   **d)** True   **e)** False   **f)** True   **g)** False   **26.** 15   **27.** 17
28. $\sqrt{369}$ **29.** $\sqrt{52} = 2\sqrt{13}$ **30.** $\sqrt{91}$ **31.** $\sqrt{72} = 6\sqrt{2}$ **32.** 22   **33.** 93

34.   **35.**

36. No   **37.** No

Section 9.5 (pages 480–482)

1. 24   **2.** $m(\angle C) = 60°, m(\angle D) = 68°$ **3.** $\overline{BC} = 6, \overline{DE} = 10$
4. $m(\angle C) = 55°, m(\angle D) = 55°, m(\angle E) = 70°$ **5.** 12   **6.** 2.25   **7.** Yes   **8.** Yes   **9.** Yes   **10.** No
11. Yes   **12.** Yes   **13.** Yes   **14.** Yes   **15.** No   **16.** No   **17.** Yes   **18.** No   **19.** Yes
20. No   **21.** No   **22.** No   **23.** No   **24.** No   **25.** $y = 18, x = 4, m = 14$ **26.** 22.5   **27.** 25
28. a) Yes **b)** 40   **29.** $2\frac{5}{8}$ meters   **30.** 10   **31.** $x = 4; \overline{AB} = \overline{DE} = 29$
32. They have angle A in common, and they both have right angles.

Section 9.6 (pages 488–489)

1. 8.226
2. 248,850 if you convert from pounds to kg and then to grams. 250,398.4 if you convert from pounds to ounces and then to grams.
3. 7.35   **4.** 5.06   **5.** 1728   **6.** 0.037   **7.** 365.76   **8.** 15.84
9. 0.4675 if you convert from mL to ounces and then to gallons. 0.4576 if you convert from mL to liters and then to gallons.
10. 7,700,000   **11.** 275,000   **12.** 5.25   **13.** 400.4   **14.** 300.3   **15.** 20,421.6   **16.** 0.083
17. 118.8   **18.** 68.88   **19.** 14.56   **20.** 620   **21.** 125.952   **22.** 18,150   **23.** 389.4   **24.** 18.755
25. 70,000   **26.** 0.0860932   **27.** 56 km   **28.** 3375 kg   **29.** 195.58 cm, 87.75 kg, 96.52 cm.
30. 32,800   **31.** 0.042   **32.** 1,047.75   **33.** 36¢   **34.** 20°C   **35.** Choice (a)   **36.** −9.4°F
37. −88.28°C

Section 9.7 (pages 492–493)

1. 30   **2.** 29   **3.** 24π **4.** 39   **5.** 37   **6.** 56   **7.** 60   **8.** 81   **9.** 56   **10.** 9π **11.** 192
12. 91   **13.** 196   **14.** The first room   **15.** bh **16.** $\frac{1}{2}h(2a + b + d)$

Section 9.10 (pages 505–507)

1. a)   **b)**   **c)**   **d)**   **e)**   **f)**

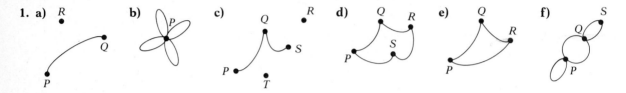

2. a) P and Q are odd, R is even.
  b) P even, no odd
  c) Vertices Q, R, and T are even, and the others are odd.   **d)** All are even.   **e)** All are even.
  f) All are even.
3. Can be drawn   **4.** Cannot be drawn   **5.** Cannot be drawn   **6.** Can be drawn   **7.** Can be drawn
8. Can be drawn   **9.** Cannot be drawn   **10.** Can be drawn   **11.** Cannot be drawn
12. Can be drawn   **13.** Yes   **14. b)** Yes   **15.** Yes   **16.** No

Section 9.11 (pages 511–512)

1. Connected, $V = 3, E = 3$, and $F = 2$ **2.** Connected, $V = 2, E = 1$, and $F = 1$
3. Connected, $V = 1, E = 2$, and $F = 3$ **4.** Not connected   **5.** Connected, $V = 5, E = 6$, and $F = 3$
6. Not connected   **7.** Connected, $V = 2, E = 2$, and $F = 2$ **8.** Connected, $V = 10, E = 15$, and $F = 7$
9. Connected, $V = 8, E = 7$, and $F = 1$ **10.** There are several possibilities. One possibility is as follows:

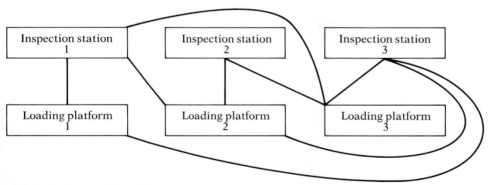

Section 9.12 (pages 513–514)

1. You get a two-sided strip as shown below:

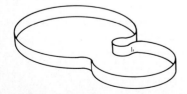

2. You get two strips intertwined. One is a Möbius strip, the other is a two-sided strip as shown below.

Mastery Tests
Form A (pages 516–517)

1. Choice (c)   **2.** 5   **3.** 15   **4.** 15°   **5.** $3x - 4$ **6.** 24   **7.** Choice (b)   **8.** 24

9. $x^2 - 5x - 24$ **10.** $\frac{5\pi}{12}$ **11.** 105°   **12.** 6   **13.** $\frac{p}{4}$ **14.** 2   **15.** 40°

Form B (pages 517–518)

1. $\sqrt{34}$ **2.** $\sqrt{32}$ or $4\sqrt{2}$ **3.** 8   **4.** 24   **5.** 16   **6.** More
7.   **8.** Yes   **9.** 429   **10.** 48 km per hour   **11.** 23.89°C   **12.** $8.37   **13.** 7000

14. 109,200 square cm   **15.** $\frac{440}{12}$ or 36.67

CHAPTER 10

Section 10.2 (pages 526–527)

1. 72   **2.** 72   **3.** 720   **4.** 67,600   **5.** 5040   **6.** 40,320   **7.** 1,000,000   **9.** 72   **10.** 18
11. 720   **12.** 100   **13. a)** 256; 24   **b)** 3125; 120   **c)** 823,543; 5040   **14.** 4032
15.

	Child 1	Child 2	Child 3	Child 4	Family consists of:
				Boy	Boy, boy, boy, boy
			Boy	Girl	Boy, boy, boy, girl
		Boy		Boy	Boy, boy, girl, boy
			Girl	Girl	Boy, boy, girl, girl
	Boy			Boy	Boy, girl, boy, boy
			Boy	Girl	Boy, girl, boy, girl
		Girl		Boy	Boy, girl, girl, boy
			Girl	Girl	Boy, girl, girl, girl
Start				Boy	Girl, boy, boy, boy
			Boy	Girl	Girl, boy, boy, girl
		Boy		Boy	Girl, boy, girl, boy
			Girl	Girl	Girl, boy, girl, girl
	Girl			Boy	Girl, girl, boy, boy
			Boy	Girl	Girl, girl, boy, girl
		Girl		Boy	Girl, girl, girl, boy
			Girl	Girl	Girl, girl, girl, girl

Section 10.3 (pages 537–538)

1. 20   **2.** 42   **3.** 1   **4.** $\frac{1}{7}$ **5.** 20, 160   **6.** 5040   **7.** 360   **8.** 6,652,800   **9.** 720   **10.** 1

11. 3024   **12.** 1   **13.** 35   **14.** 56   **15.** 6   **16.** 720   **17.** 1   **18.** 8   **19.** 8   **20.** 10

21. 10   **22.** 1   **23.** 1   **24.** 126   **25.** Impossible   **26.** 1   **27.** 3,326,400   **28.** 45,360

29. 302,400   **30.** 34,650   **31.** 5040   **32.** 5040   **33.** $_{30}P_{10}$ **34.** 18,564   **35.** 11,880

36. a) 362,880   **b)** 1728   **37. a)** 5880   **b)** 8156   **38.** $_{52}C_7 = 133{,}784{,}560$ **39. a)** 5040   **b)** 40,320

40. a) 5040   **b)** 720   **c)** 120   **41.** 184,800   **42.** $_{17}C_5 = 6188$ **43.** 20,160

44. a) $_{12}C_5 = 792$ **b)** $_{12}C_7 = 792$ **c)** They are the same.

45. a) 1   **b)** 2   **c)** 3   **d)** 5   **e)** They form a Fibonacci sequence.

Section 10.4 (pages 544–546)

1. a) 0   **b)** $\frac{1}{6}$ **c)** $\frac{6}{36} = \frac{1}{6}$ **d)** $\frac{1}{2}$ **e)** $\frac{1}{36}$ **2. a)** $\frac{13}{52} = \frac{1}{4}$ **b)** $\frac{4}{52} = \frac{1}{13}$ **c)** $\frac{12}{52} = \frac{3}{13}$ **d)** $\frac{24}{52} = \frac{6}{13}$ **3.** $\frac{1}{4}$

4. a) $\frac{8}{27}$ **b)** $\frac{9}{27} = \frac{1}{3}$ **c)** $\frac{6}{27} = \frac{2}{9}$ **d)** $\frac{4}{27}$ **5. a)** $\frac{109}{418}$ **b)** $\frac{203}{418}$ **c)** $\frac{235}{418}$ **6.** $\frac{1}{6}$ **7.** Choices (a) and (b)

8. a) $\frac{177}{500}$ **b)** $\frac{245}{500}$ **c)** $\frac{78}{500}$ **9. a)** $\frac{37}{1000}$ **b)** $\frac{411}{1000}$ **c)** $\frac{425}{1000}$ **d)** $\frac{207}{1000}$ **10.** $\frac{216}{223}$ **11.** $\frac{100}{1000} = \frac{1}{10}$

12. a) Let L – lemon, C = cherry, and A = apple. The outcomes are *LCA, LAC, LLL, LCC, LAA, LAL, LLA, LLC, LCL, ACA, AAC, ALL, ACC, AAA, AAL, ALA, ALC, ACL, CCA, CAC, CLL, CCC, CAA, CAL, CLA, CLC,* and *CCL.*

   b) $\frac{1}{27}$

13. a) $\frac{688}{4800}$ **b)** Greater

Section 10.5 (pages 553–554)

1. Not mutually exclusive   **2.** Mutually exclusive   **3.** Not mutually exclusive

4. Mutually exclusive   **5.** Not mutually exclusive   **6.** Not mutually exclusive   **7.** $\frac{19}{32}$ **8.** $\frac{79}{143}$

9. 0.92   **10.** 0.46   **11.** 63%   **12.** $\frac{1}{68}$ **13.** $\frac{18}{21} = \frac{6}{7}$ **14.** $\frac{2}{9}$ **15.** 0.35   **16.** 0.95   **17.** $\frac{2381}{2771}$

18. $\frac{1}{168}$ **19.** 0.70

Section 10.6 (pages 560–561)

1. $\frac{19}{27}$ **2.** $\frac{24}{95}$ **3.** $\frac{31}{82}$ **4. a)** $\frac{204}{1090} = \frac{102}{545}$ **b)** $\frac{204}{368} = \frac{51}{92}$ **c)** $\frac{1090}{1638} = \frac{545}{819}$ **5.** 0.0171

6. $\frac{2}{9}$ **7.** $\frac{2}{3}$ **8.** 0.6468   **9.** 0.4011   **10.** 0.0434   **11.** 0.0987   **12.** $\frac{7}{15}$

Section 10.7 (pages 565–567)

1. 12:40 or 3:10   **2.** 5:95 or 1:19   **3.** 9:6 or 3:2   **4.** 95:5 or 19:1   **5.** 4:5   **6.** $7388.89

7. Television = $25,900; Radio = $20,720; Magazine = $22,050; Distributing free samples = $21,000

8. Television   **9.** $-421   **10.** 3.18   **11.** $-61

12. a) Park I = $35,555.56; Park II = $23,777.78   **b)** Park II   **13.** $\frac{+100}{36} = \$+2.78$

Section 10.8 (page 570)

1. $\frac{3}{4}$ **2. a)** 0 **b)** 1 **3.** Either aa or Na **4.** $\frac{1}{2}$ **5.** $\frac{1}{2}$ **6.** 0 **7.** 0 **8.** $\frac{1}{2}$ **9.** $\frac{1}{2}$

Section 10.9 (pages 574–575)

1. There are 100 numbers listed in columns 1 and 2. Send letters to those 100 owners whose license plate numbers appears on this list.
2. Those volunteers whose numbers are 150, 69, 143, 127, 42, 47, 185, 75, 3, 104, 15, 62, 110, 54, and 55.
3. Those owners whose serial numbers are 6917, 2798, 1517, 3944, 6046, 1860, 7119, 5774, 3886, 5686, 1866, 3632, 6768, 4756, 6075, 5532, 1859, 8314, 7698, 7646, 4583, 6095, 6656, 8976, 3283, 3793, 3997, 7408, 7622, and 2657.
4. Those whose numbers are 2533, 0815, 3001, 0151, 0118, 2349, 3580, 0599, 1805, 2816, 0488, 3242, 3991, 0234, 2719, 3620, 3409, 3208, 1505, 1256, 1798, 3159, 2084, 0827, and 2635.
5. Those restaurants whose license numbers are 141, 248, 187, 058, 176, 298, 136, 047, 263, 287, 153, 147, 222, 086, 205, 254, 258, 253, 081, and 300.

Mastery Tests
Form A (pages 577–579)

1. Choice (d) **2.** Choice (d) **3.** Choice (a) **4.** Choice (b) **5.** Choice (d) **6.** $_7C_3 \cdot {_8}C_3 = 1960$
7. Choice (d) **8.** $\frac{251}{900}$ **9.** $\frac{123}{422}$ **10.** $\frac{123}{251}$ **11.** 0.0806 **12. a)** $\frac{1}{4}$ **b)** $\frac{1}{4}$ **c)** $\frac{1}{2}$ **13.** $\frac{1}{6}$
14. 132,600 **15.** 67,600,000

Form B (pages 579–580)

1. $5! = 120$ **2.** $\frac{9}{26}$ **3.** $_{12}C_5 \cdot {_{18}}C_6 = 14{,}702{,}688$ **4.** $_{16}C_2 \cdot {_{10}}C_2 = 5400$ **5.** $_{12}P_3 = 1320$
6. $_6P_6 = 720$ assuming order counts, or $_6C_6 = 1$, if order does not count **7.** $_{40}C_6 = 3{,}838{,}380$ **8.** 1920
9. 90 **10.** $10! = 3{,}628{,}800$ **11.** \$3234.33 **12.** 0.38 **13.** 21 : 4 **14.** $\frac{8}{17}$ **15.** $_{28}C_{22} \cdot {_{25}}C_{16}$

CHAPTER 11

Section 11.3 (pages 593–595)

1. Mean = \$1,519.17; Median = \$1495; Mode = None **2.** Mean = 5, Median = 4, Mode = 4
3. Mean = \$13,080; Median = \$10,000; Mode = \$7,000 **4.** Probably the mode **5.** \$26 **6.** \$7,258
7. Each will be increased by 10% **8.** One is using the mode and the other is using the mean.
9. The mean is affected by extreme scores. **10.** 69°
11. a) It is increased by 3. **b)** It is decreased by 4. **c)** It is multiplied by 7. **d)** It is divided by 5.
12. a) 88 **b)** 8.8 **c)** 0 **d)** 153.6 **13. a)** 928 **b)** 7744 **c)** No
14. The second student, because that student computed a weighted arithmetic average

Section 11.4 (page 597)

1. 15th percentile **2.** 67.5th percentile **3.** 67.5th percentile **4.** Stay the same

Section 11.5 (pages 602–603)

1. Range = 20; standard deviation = 5.589 **2.** Range = 89; standard deviation = 23.663
3. Range = \$8; standard deviation = \$2.595 **4.** Range = 98; standard deviation = 33.713
5. Mean = \$797.667; standard deviation = \$67.916

6. a) Mean = $877.43; standard deviation = $74.707   **b)** Increased by 10%   **7.** New mean = $6000
8. a) Fund A: Range = $1.15; standard deviation = 0.401
  Fund B: Range = $1.74; standard deviation = 0.587
  b) Fund A
9. The one from company B

Section 11.6 (pages 613–615)

1.

Ages	Tally	Frequency
16–17	|||	3
18–19	ⅧII	7
20–21	|||	3
22–23	||	2
24–25	||||	4
26–27	||	2
28–29	Ⅷ	5
30–31	||||	4
32–33	||||	4
34–36	ⅧI	6
		40

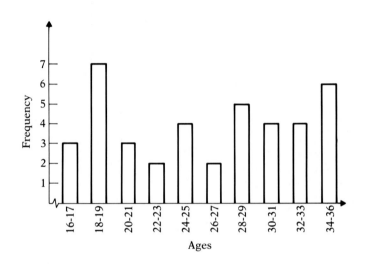

2.

Number of flights	Tally	Frequency
15–16	||	2
17–18	||||	4
19–20	Ⅷ	5
21–22	|||	3
23–24	|||	3
25–26	Ⅷ	5
27–28	||||	4
29–30		0
31–32			1
33–34			1
35–36	||	2
37–38		0
39–40			1
		31

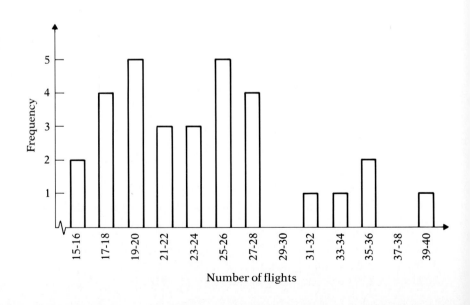

3.

Number of children	Tally	Frequency
1	卌 I	6
2	卌 III	8
3	卌	5
4	卌	5
5	卌	5
6	II	2
7	III	3
8	II	2
9	II	2
		38

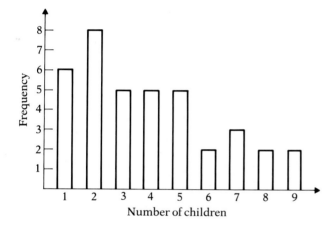

4.

Number of new housing permits	Tally	Frequency
5–8	卌	5
9–12	卌 II	7
13–16	卌 III	8
17–20	I	1
21–24	III	3
25–28	I	1
29–32	I	1
33–36		0
37–40	II	2
		28

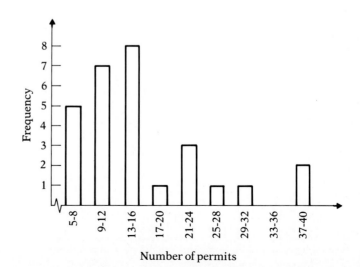

5.

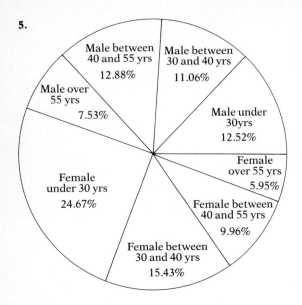

Male between 40 and 55 yrs 12.88%

Male between 30 and 40 yrs 11.06%

Male over 55 yrs 7.53%

Male under 30yrs 12.52%

Female over 55 yrs 5.95%

Female under 30 yrs 24.67%

Female between 40 and 55 yrs 9.96%

Female between 30 and 40 yrs 15.43%

6.

Region	Sales
Africa	$704,079
United States	938,772
Canada	1,329,927
South America	1,877,544
Europe	1,564,620
Asia	1,408,158

7. a) 400 **b)** 450 **c)** from 200 to 700 or by 500
8. a) June **b)** 20 + 30 + 10 + 60 + 50 + 95 + 75 + 65 + 45 + 35 + 40 + 45 = 570

9.

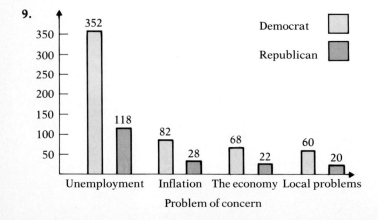

Democrat ☐

Republican ◼

Unemployment 352 / 118
Inflation 82 / 28
The economy 68 / 22
Local problems 60 / 20

Problem of concern

10.

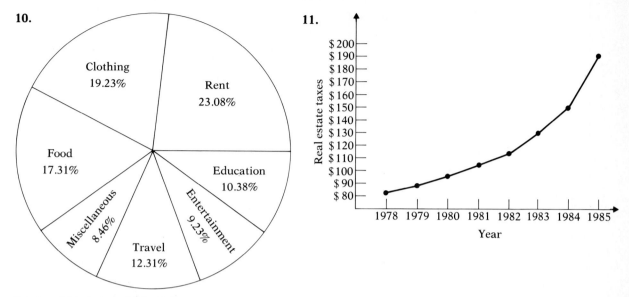

11.

Section 11.7 (pages 625–626)

1. Only (a)   **2. a)** 68% **b)** 84% **c)** 16% **d)** 9600
3. a) 0.4973 **b)** 0.3186 **c)** 0.0853 **d)** 0.0239 **e)** 0.0029 **f)** 0.9222 **g)** 0.9706 **h)** 0.9806
4. a) 0.9545 **b)** 0.9977 **c)** 0.1854 **d)** 0.0222 **e)** 0.0021 **f)** 0.0565 **g)** 0.9787
5. a) 1 **b)** 3 **c)** −2 **d)** 0   **6. a)** 0.6826 **b)** 0.2029 **c)** 0.9332 **d)** 0.7734
7. a) B, C, D, E, A **b)** $B, C,$ and D **e)** A and E **8.** 0.0475   **9.** 0.0045   **10.** 0.7287   **11.** 0.1056
12. 0.7967

Section 11.8 (pages 637–640)

1. a)

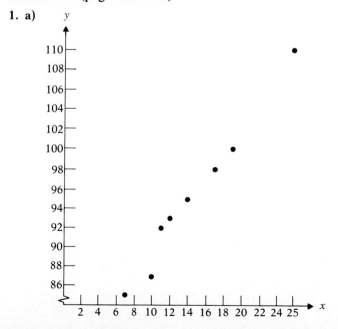

b) 0.988 **c)** $y = 75.53 + 1.354x$ **d)** 105.318 **2. b)** 0.99 **c)** $y = -6.92 + 209.357x$ **d)** 12.97

3. a)

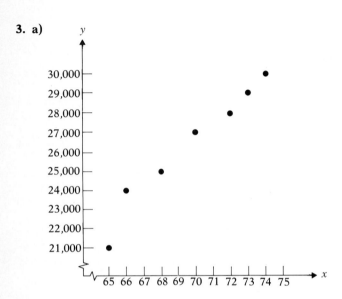

b) 0.977 **c)** $y = -35,019.45 + 879.377x$ **d)** \$25,657.56
4. a) 0.992 **b)** $y = 80.12 + 1.194x$ **c)** 92.06
5. a) 0.996 **b)** $y = 32.07 + 4.35x$ **c)** 77.745 **6. a)** 0.996 **b)** $y = -3.77 + 0.246x$ **c)** 8.53
7. a) 0.967 **b)** $y = 73.06 + 2.502x$ **c)** 90.574 **8. a)** −0.972 **b)** $y = 209.79 - 11.214x$ **c)** 58.401
9. a) −0.952 **b)** $y = 609.8 - 11.173x$ **c)** 576.281 **10. a)** −0.939 **b)** $y = 74.47 - 7.546x$ **c)** 21.648

Mastery Tests
Form A (pages 644–645)

1. Choice (d) **2.** 0.987 **3.** $y = 7.57 + 2.188x$ **4.** 27.262 **5.** Choice (c) **6. a)** 1668 **b)** 7744
7. 1.13 **8.** Choice (b) **9.** Choice (c) **10.** Choice (b) **11.** 74th percentile **12.** Choice (d)
13. Choice (a) **14.** Choice (b) **15.** Choice (a)

Form B (pages 645–649)

1. 0.932 **2. a)** 45 **b)** 115 **3.** Probably the mode
4. since it is truly in the middle, with 50% above the median number and 50% below
5. The standard deviation remains unchanged.

6.

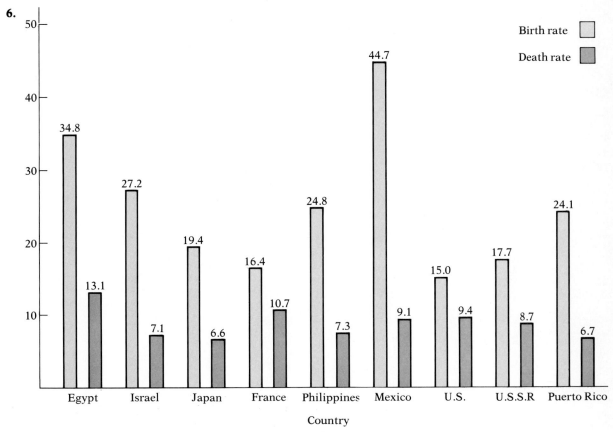

7.

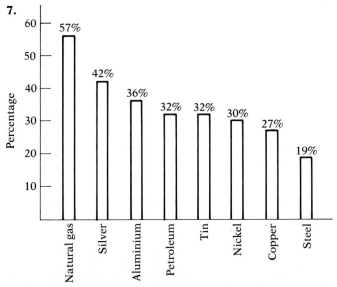

8. Yes, based on this graph   **9.** 50,000   **10.** $y = 10,647.14 + 2721.429x$ **11.** \$22,893.57

CHAPTER 12

Section 12.3 (pages 659–662)

1. 4×3 **2.** 2×3 **3.** 4×2 **4.** 1×4 **5.** 1×1 **6.** 4×4 **7.** 3×3 **8.** 5×1
9. Those matrices given in Exercises 5, 6, and 7 are square matrices.
10. No, one is 2×4 and the other is 4×2.   **11.** $x = 9$ **12.** $y = 10$ **13.** $x = 8, y = -5$

14. $x = 5, y = 0, z = \dfrac{10}{3}$ **15.** Not possible   **16.** $x = 4, y = \dfrac{7}{2}, z = 5$ **17.** $\begin{pmatrix} 3 & 7 \\ -11 & 13 \end{pmatrix}$ **18.** $\begin{pmatrix} 4 \\ 7 \\ 7 \end{pmatrix}$

19. $\begin{pmatrix} 4 & -1 & 4 \\ 13 & 5 & 11 \end{pmatrix}$ **20.** $(11 \quad 10 \quad -12)$ **21.** Cannot be done   **22.** Cannot be done

23. $\begin{pmatrix} 3 & 4 & 14 \\ 9 & -2 & -2 \\ 1 & 15 & -13 \end{pmatrix}$ **24.** $\begin{pmatrix} 7 & 5 & 13 \\ 12 & 9 & -15 \end{pmatrix}$ **25.** $\begin{pmatrix} 5 & 3 & 2 \\ 1 & 4 & 7 \\ 0 & 2 & 9 \end{pmatrix}$ **26.** $\begin{pmatrix} 4 & 7 \\ x & y \end{pmatrix}$ **27.** $\begin{pmatrix} 4 & 7 \\ x & y \end{pmatrix}$

28. $x = 15, y = 6$ **29.** In either case our answer is $\begin{pmatrix} 4 & 2 & 18 \\ 5 & 15 & 14 \\ 12 & 10 & 21 \end{pmatrix}$

30. In either case our answer is $\begin{pmatrix} 5 & -5 & 9 \\ 3 & 7 & 10 \\ 6 & 10 & 16 \end{pmatrix}$

31. a)

	Engine tune-up	Transmission tune-up	Exhaust system repair	Charging air conditioning system	New tires
City A	+14	9	+13	+2	+7
City B	+13	+11	+12	+1	0
City C	+4	−3	+5	−3	+7

b) Engine tune-up in city A **c)** transmission tune-up and charging air conditioning system in city C
32. a) 7838   **b)** Main St.   **c)** West St.
33. a)

	Mon.	Tues.	Wed.
Bill	6	8	9
Cindy	4	6	6
Gail	5	5	8
Jake	3	7	3
Jason	9	2	7

b) Bill = 23; Cindy = 16; Gail = 18; Jake = 13; Jason = 18   **c)** 88

Section 12.4 (page 672)

1. $\begin{pmatrix} -6 & 61 \\ -24 & 57 \end{pmatrix}$ **2.** $\begin{pmatrix} -55 & 39 \\ -18 & 12 \end{pmatrix}$ **3.** $\begin{pmatrix} -8 & 5 \\ 13 & 30 \\ -14 & -29 \end{pmatrix}$ **4.** $\begin{pmatrix} -54 & 47 \\ -66 & 60 \end{pmatrix}$ **5.** $\begin{pmatrix} 34 & -67 & -3 \\ -34 & 65 & -8 \\ -8 & 22 & 35 \end{pmatrix}$

6. (-133) **7.** $\begin{pmatrix} 2 & 6 & 14 & 20 \\ -1 & -3 & -7 & -10 \\ 6 & 18 & 42 & 60 \\ 9 & 27 & 63 & 90 \end{pmatrix}$ **8.** $\begin{pmatrix} 1 & 70 & 7 \\ 178 & 46 & 69 \end{pmatrix}$ **9.** $\begin{pmatrix} -15 & 3 & -18 \\ -6 & 3 & -24 \\ -21 & 0 & -27 \end{pmatrix}$

10. $\begin{pmatrix} 0 & -12 & -30 \\ -1 & -7 & -4 \\ -30 & -48 & 42 \end{pmatrix}$ **11.** $\begin{pmatrix} 228 & 192 & 376 \\ 120 & 144 & 160 \\ 156 & 60 & 172 \end{pmatrix}$ **12.** $\begin{pmatrix} 39 & 41 \\ 5 & 39 \end{pmatrix}$ **13.** $\begin{pmatrix} 4 & -24 \\ 32 & 64 \end{pmatrix}$

14. $\begin{pmatrix} 9 & -4 & 49 \\ -6 & -31 & 9 \\ -5 & -32 & 27 \end{pmatrix}$ **15.** $\begin{pmatrix} 0 & 171 \\ 6 & 84 \\ -9 & -12 \end{pmatrix}$ **16.** $\begin{pmatrix} 1 & 1 \\ 0 & -1 \end{pmatrix}$ **17.** $x = y = 1, z = 0, w = -1$

18. $x = w$, and $z = -y$ **19.** One possible answer is $\begin{pmatrix} 0 & 0 \\ 2 & -2 \end{pmatrix}$ and $\begin{pmatrix} 2 & 0 \\ 2 & 0 \end{pmatrix}$. Other answers are possible.

20. $\begin{pmatrix} 1 & 0 \\ 1 & 0 \end{pmatrix}$

Section 12.5 (pages 677–678)

1. a) $(7 \quad 8 \quad 4 \quad 6 \quad 9)$ **b)** $\begin{pmatrix} 3.21 \\ 2.69 \\ 4.92 \\ 6.52 \\ 5.02 \end{pmatrix}$ **c)** $147.97

2. a) $(14 \quad 3 \quad \frac{2}{3} \quad 2 \quad 3 \quad 2)$ **b)** $\begin{pmatrix} 105 \\ 115 \\ 72 \\ 110 \\ 350 \\ 100 \end{pmatrix}$ **c)** 3333

3. $A \cdot B = \begin{pmatrix} 153 \\ 135 \end{pmatrix}$ **4. a)** $\begin{pmatrix} 12,000 \\ 36,000 \\ 20,000 \end{pmatrix}$ **b)** $\begin{pmatrix} 348,000 \\ 548,000 \\ 448,000 \\ 556,000 \\ 412,000 \end{pmatrix}$ **5.** $272.64   **6. a)** $288.07   **b)** $560.71

Section 12.6 (pages 687–689)

1. Multiply each row by -3 **2.** Multiply each row by -2 **3.** Multiply each row by $\frac{2}{3}$

4. New row 2 equals original row 2 added to -3 times row 1; new row 3 equals original row 3 added to 2 times row 1. Row 1 remains unchanged.

5. New row 3 equals 3 times row 2 added to original row 3; new row 1 equals original row 1 added to -2 times row 2. Row 2 remains unchanged.

6. Interchange rows 1 and 2. New row 2 equals interchanged row 2 added to -7 times interchanged row 1; new row 3 equals original row 3 added to -5 times interchanged row 1. Row 1 now remains unchanged.

7. Interchange rows 1 and 3. Row 2 equals original row 2 added to 2 times row 1; row 3 equals original row 3 added to -5 times row 1. Interchange rows 2 and 3. Multiply row 2 by $\frac{1}{3}$. Row 3 equals original row 3 added to -7 times row 2. Multiply row 3 by $\frac{3}{41}$. Row 2 equals original row 2 added to $\frac{2}{3}$ times row 3; row 1 equals original row 1 added to -2 times row 3.

9. $\begin{pmatrix} 7 & 3 & \vdots & 11 \\ 5 & -2 & \vdots & 19 \end{pmatrix}$ **10.** $\begin{pmatrix} 12 & -2 & \vdots & 13 \\ 5 & 9 & \vdots & 23 \end{pmatrix}$ **11.** $\begin{pmatrix} 2 & -3 & \vdots & 5 \\ 7 & 12 & \vdots & 19 \end{pmatrix}$ **12.** $\begin{pmatrix} 5 & -3 & \vdots & 15 \\ 4 & 3 & \vdots & -12 \end{pmatrix}$

13. $\begin{pmatrix} 2 & 3 & -5 & \vdots & 10 \\ 3 & -7 & 8 & \vdots & -15 \end{pmatrix}$ **14.** $\begin{pmatrix} 2 & 3 & -4 & \vdots & 24 \\ 8 & -4 & 7 & \vdots & 16 \\ 3 & -5 & 8 & \vdots & 12 \end{pmatrix}$ **15.** $\begin{pmatrix} 2 & 3 & \vdots & 12 \\ 7 & 8 & \vdots & 27 \\ 2 & -9 & \vdots & 18 \end{pmatrix}$ **16.** $\begin{pmatrix} 1 & 0 & 0 & \vdots & 7 \\ 0 & 1 & -3 & \vdots & 10 \\ 1 & 0 & -8 & \vdots & 23 \end{pmatrix}$

17. $\begin{pmatrix} 4 & 1 & 0 & \vdots & 10 \\ 0 & 3 & -1 & \vdots & 17 \\ 2 & 0 & 7 & \vdots & 15 \end{pmatrix}$ **18.** $\begin{pmatrix} 2 & 3 & 5 & -8 & \vdots & 7 \\ 3 & -2 & -4 & 7 & \vdots & 0 \\ 4 & 7 & -3 & -4 & \vdots & -3 \\ 2 & 0 & 3 & -1 & \vdots & 1 \end{pmatrix}$

19. a) $5x + 7y = 10$ **b)** $8x + 7y = 24$ **c)** $2x + 3y - 7z + 6w = 10$ **d)** $2x + 9y + 6z = 1$
      $3x + 9y = 14$ $9x + 5y = 45$ $5x + 4y + 8z + w = 0$ $5x - 4y - 7z = 0$
                                                                                        $3x - 2y + 9z = 10$

   e) $2x + 5y + z + 2w = 8$
        $x + 3y + 0z + 3w = 10$
      $9x + 2y + 7z + 4w = 14$

20. $x = 2, y = 1$ **21.** $x = 3, y = -1$ **22.** $x = -1, y = -4$ **23.** $x = 7, y = 3$ **24.** $x = 8, y = 0$

25. $x = \frac{1}{2}, y = -\frac{3}{4}$ **26.** $x = 1, y = -3$ **27.** $x = 3880, y = -2320$ **28.** $x = 3, y = 2$

29. $x = 7, y = -6$ **30.** $x = -1, y = -4$ **31.** $x = 5, y = 0$

Section 12.7 (pages 693–696)

1. $\begin{pmatrix} 1 & 1 & 1 \\ 1 & 1 & 0 \\ 1 & 2 & 1 \end{pmatrix}$ **2.** Yes

3. a) Prof. Blois   **b)** Prof. Rogers   **c)** Professors Hoffman and Lichtenfeld
   d) Professors Hoffman, Lichtenfeld, and Kennedy

5.

	Bill	Marion	Jeff	Molly	Pam	Pam's boyfriend	Marion's parents
Bill	0	1	1	0	1	1	0
Marion	1	0	0	0	0	0	1
Jeff	1	0	0	1	0	0	0
Molly	0	0	1	0	0	0	0
Pam	1	0	0	0	0	1	0
Pam's boyfriend	1	0	0	0	1	0	0
Marion's parents	0	1	0	0	0	0	0

6. WVOAE—NE—FO—AOAIP—FJAOY—CNUU—BNI **7.** 18, 11, 19, 20, 4—5, 4—8, 19—20, 19, 20, 15, 16—8, 10, 20, 19, 24—22, 5, 3, 3—17, 5, 15 **8.** SINK—THE—SHIP

9. BI—LL—YI—SY—OU—RC—ON—TA—CT; $\begin{pmatrix} 4 & 1 \\ 7 & 2 \end{pmatrix}$ Message transmitted as (using coding scheme given in Example 5) 13, 24, 15, 27, 69, 122, 32, 60, 56, 101, 109, 193, 63, 115, 104, 184, 60, 111.

10. Assuming the coding scheme is the one given in Example 5, the message is HAVE—THE—FILMS.

Mastery Tests
Form A (pages 697–698)

1. $\begin{pmatrix} 1 & 7 \\ 6 & 9 \\ 3 & -2 \end{pmatrix}$ **2.** $x = 3, y = \dfrac{13}{2}$ **3.** $\begin{pmatrix} \dfrac{9}{57} & \dfrac{7}{57} \\ \dfrac{3}{37} & \dfrac{-4}{57} \end{pmatrix}$ **4.** $\left(\begin{array}{cc|c} 4 & 3 & 17 \\ 2 & -9 & 12 \end{array} \right)$ **5.** $\begin{pmatrix} -27 & 39 & -26 \\ 21 & -26 & -21 \\ -16 & -4 & 31 \end{pmatrix}$

6. $7x + 12y = 5$
   $3x - 4y = 11$ **7.** There is none. **8.** $\begin{pmatrix} -2 & 36 \\ -108 & 22 \end{pmatrix}$ **9.** False

10. a) $A^{-1} = \begin{pmatrix} \dfrac{5}{38} & \dfrac{-6}{38} \\ \dfrac{3}{38} & \dfrac{4}{38} \end{pmatrix}$ **b)** $(A^{-1})^{-1} = A$

Form B (pages 698–700)

1. In both cases our answer is $\begin{pmatrix} -122 & 106 \\ 7 & 80 \end{pmatrix}$. **2.** Plumber $90, helper $70

3. 65 nickels and 170 dimes **4.** 95 **5.** $\begin{pmatrix} \$197,\!400 \\ 165,\!800 \end{pmatrix}$

6. YBA—INQXVTNUI—NE—NO—YBA—QFIAXF

7. a) $\begin{pmatrix} 16 & 20 \\ 0 & 2 \\ 0 & 3 \\ 7 & 11 \\ 8 & 8 \\ 466 & 702 \\ 566 & 735 \\ 456 & 660 \\ 5365 & 6961 \end{pmatrix}$ **b)** 9102 in 1978 as opposed to 6884 in 1974 **8.** Molly $22,930; Mike $32,480

9. a) Molly $22,065; Mike $31,575 **b)** Molly $44,995; Mike $64,055

10.

	Uncontested divorces	Writing wills	House closings	Personal bankruptcy
Month 1	$525	$510	$2375	$125
Month 2	$350	595	1425	0
Month 3	$1225	340	950	250
Month 4	$700	510	1900	125

Total income $11,905

CHAPTER 13

Section 13.2 (page 708)

1. Early first generation computers were not capable of storing instructions internally.   **2.** Blaise Pascal
3. It was the first computer capable of operating at electronic speeds and of storing its programs internally.
4. Computers built with transistors perform calculations in a few microseconds.
5. Since large high-speed computers are costly, companies cannot afford to buy or rent such machines, especially if the machines are left unused for a period of time.   **6.** Herman Hollerith
7. It enabled punched cards to be used to enter numbers into an adding machine, thereby making the tallying of the census easier.   **8.** Size of device and its huge capabilities
9. An abacus or counting board can be used to perform many calculations quickly. However, great skill is needed to operate it; for example, in "carrying," which must be done manually.
10. To enable us to perform immense and complicated calculations in a short period of time and to be able to do many such calculations simultaneously

Section 13.3 (page 713)

3. A supercomputer can run about 500 million operations per second in a short burst. It is hoped that an advanced mainframe supercomputer would run 1000 times as fast as our present day supercomputer. Fourth generation computers are capable of processing 100 million to 1 billion instructions per second.

Section 13.4 (page 715)

1. Because machine languages require that the user have some background knowledge of how the computer works, its capabilities, and its limitations   **2.** When a programming error occurs
3. High-level languages use commands and concepts in a manner comparable to everyday human language and thought as opposed to low-level languages, which require that the user have some background knowledge of the computer. Prior knowledge of programming is needed.
4. RPG stands for <u>R</u>eport <u>P</u>rogram <u>G</u>enerator and is used mostly to generate business reports.
5. ALGOL stands for <u>Al</u>gorithmic <u>L</u>anguage and is used primarily in Europe for scientific programming.
6. APL stands for <u>A</u> <u>P</u>rogramming <u>L</u>anguage and is mostly used for handling large groups of tables of related numbers.
7. LISP stands for <u>Lis</u>t <u>P</u>rocessor and was designed for the processing of non-numeric data (for example, symbols, words, characters). It is often used with programs dealing with artificial intelligence.
8. SNOBOL is often used for manipulating alphanumeric characters.
9. When a programming error has been made

Section 13.5 (page 718)

1. The control unit receives its input from the input unit after passing through the compiler, which translates any input information into a language that the machine can understand.   **2.** The control unit
3. Input into a computer can be accomplished by using punched cards, paper tapes, console typewriters, magnetic tapes, and optical scanners.
4. It means that the computer can store the information and instructions necessary to process the program so that the programmer can do something else while the program is being run. The operation of the computer need not be supervised.
5. Because of size and capabilities; it carries out the sequences of instructions stored in the computer's main memory.   **6.** The control unit
7. The control unit sends the results of its calculations to the output unit.
8. They are all mechanisms for entering data and instructions into the computer.

Section 13.6 (pages 720–721)

1. It carries out sequences of instructions stored in the computer's main memory.
2. A dot-matrix printer has a print head consisting of one or more vertical rows of tiny pins that strike an inked ribbon against the paper as the head moves back and forth. Such printers can create letters of any shape or graphics. In a daisy wheel printer, a wheel spins until the desired letter is in position, at which point a hammer strikes it against the ribbon, which in turn strikes the paper.
3. A personal computer must have an operating system as a master control program to synchronize the execution of computer programs.
5. The user can be many miles away from the computer.

Mastery Tests
Form A (pages 722–723)

1. Choice (a)
2. In time sharing, a person or company buys time on another company's computer. The user pays only for the time the computer is used or the time that he or she wants it available for use.
3. Beginner's All-purpose Symbolic Instruction Code   **4.** Choice (a)   **5.** Choice (c)   **6.** Choice (d)
7. Choice (d)   **8.** Choice (b)   **9.** Choice (a)   **10.** Choice (a)

Form B (pages 723–724)

1. Choice (d)   **2.** Choice (c)   **3.** Choice (b)   **4.** Choice (a)   **5.** Choice (b)   **6.** Choice (c)
7. Choice (c)   **8.** Choice (a)   **9.** Choice (d)   **10.** Choice (d)

CHAPTER 14

Section 14.3 (pages 731–733)

1.

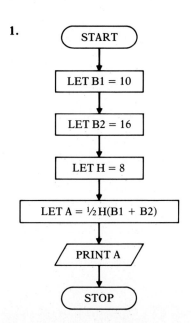

2.

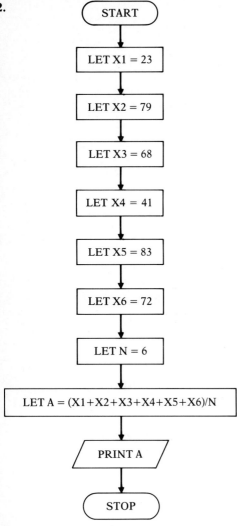

START

LET X1 = 23

LET X2 = 79

LET X3 = 68

LET X4 = 41

LET X5 = 83

LET X6 = 72

LET N = 6

LET A = (X1+X2+X3+X4+X5+X6)/N

PRINT A

STOP

3–17. Answers similar to those given for Exercises 1 and 2.

18. It computes the value of $A = \dfrac{x^2 y^3}{z^4}$ when $x = 12$, $y = 3$ and $z = 8$, and prints out $A = 0.94921875$.

19. It computes the squares of each of the integers from 4 to 20 inclusive. It then stops.

20. It reads values for A and B and then computes a value for D, depending upon the value of B. Finally it prints out the value of D.

21. Starting with values of $A = B = 1$, it reads a value for C. It then computes a value for D and prints the result. It then reads a new value for C, after which it computes a new value for D.

Section 14.4 (pages 747–749)

1. Not acceptable　　**2.** Not acceptable　　**3.** Not acceptable　　**4.** Not acceptable　　**5.** Acceptable
6. Not acceptable　　**7.** 100　LET $D = R * T$　　**8.** 105　LET $Y = (X \uparrow 2) - 5 * X + 3$

9. 120   LET $Y = Z * (7 * X - 3)$ **10.** 130   LET $F = (9/5) * C + 32$

11. 140   LET $A = (B - C)/(D + E)$ **12.** 66   **13.** 8   **14.** $\frac{23}{2}$ or 11.5   **15.** 104   **16.** 144

17. No, BASIC variable name wrong   **18.** No, missing end quotation marks

19. No, there are 4 variables in the READ statement and 5 values in the DATA statement.

20. Mistake in line 30.   **21.** Mistake in line 30.   **22.** Mistake in line 20.   **23.** 1,125,000

24. THE VALUE OF M is 238

25. 220 IS BIGGER THAN -75 570 IS BIGGER THAN 320   50 IS BIGGER THAN -5

30. \$3701.86   **32.** 362,880   **33.** 42,075   **35.** 3   **36.** 20   **37.** 4.6   **38.** 21   **39.** 82.6

40. 6   **41.** -2 **42.** 3   **43.** 6

Mastery Tests
Form A (pages 750–751)

1. Choice (c)   **2.** $< =$ **3.** $a = \dfrac{bc}{\sqrt[5]{d}}$ **4.** 100   LET $Y = 2 * (X \uparrow 3) * (Z \uparrow 4)$ **5.** 7.6, 11.6, 1.9

6. Error in line 30.   **7.** Replace a by $a + 5$, or add 5 to the previous value of a.   **8.** Six   **9.** 99999

10. $(A \uparrow 2) * (B \uparrow 5)$

Form B (pages 751–753)

1. 0.0025   **2.** No, variable name 3 characters long

3. 100   LET $Y = (-A) + (7 * (M \uparrow 2) * (Y \uparrow 5)) \uparrow (1/3)$ **4.** False

5. First to divide the value of B by 2 and then to subtract the result from A **6.** A single digit

7. Mistake in line 40, missing end of quotation marks

8. Prints the integers in order from 1 to 100 inclusive

9. Prints the integers from 100 to 1 inclusive, starting from 100

10. Prints I AM A SMART COMPUTER 10 times

11. Asks the user WHO IS THE PRESIDENT OF THE USA? If the user inputs the correct choice (from among those given) it prints CORRECT; otherwise, it prints INCORRECT, TRY AGAIN. The user must input that $N = 4$ is the correct choice.

12. Prints the name and pulse of a person and whether the pulse is NORMAL or NOT NORMAL, depending on whether pulse is less than 80 and greater than 60.

INDEX

A